COMPLEX FUNCTION THEORY

COMPLEX FUNCTION THEORY

Namita Das
Reader in Mathematics
P.G. Dept. of Mathematics
Utkal University
Bhubaneswar, Orissa

Allied Publishers Pvt. Ltd.
NEW DELHI • MUMBAI • KOLKATA
CHENNAI • BANGALORE • HYDERABAD

ALLIED PUBLISHERS PRIVATE LIMITED

D-5, Sector-2, **Noida**–201 301
Ph. Nos.: 0120-4320295/2542557/4352866
• E-mail: delhi.books@alliedpublishers.com

17 Chittaranjan Avenue, **Kolkata**–700072
Ph.: 033-22129618 • E-mail: cal.books@alliedpublishers.com

15 J.N. Heredia Marg, Ballard Estate, **Mumbai**–400001
Ph.: 022-42126969 • E-mail: mumbai.books@alliedpublishers.com

No. 25/10 Commander-in-Chief Road, Ethiraj Lane (Next to Post Office) Egmore, **Chennai**–600008
Ph.: 044-28223938 • E-mail: chennai.books@alliedpublishers.com

P.B No. 9932, No. 15, 3rd Floor (Next to Vijaya Bank), 5th Cross, Gandhinagar, Karnataka, **Bangalore**–560009
Ph.: 080-41530285 / 22386239
• E-mail: bngl.journals@alliedpublishers.com / apsabng@airtelmail.in

Sri Jayalakshmi Nilayam, No. 3-4-510, 3rd Floor (Above More Super Market) Barkatpura, **Hyderabad**–500027
Ph.: 040-27551811, 040-27551812
• E-mail: hyd.books@alliedpublishers.com

Website: www.alliedpublishers.com

ISBN: 978-81-8424-580-6

Published by Sunil Sachdev and printed by Ravi Sachdev at Allied Publishers Pvt. Ltd. D-5, Sector-2, Noida 201 301

PREFACE

Complex analysis is a subject which has wide range of applications in other parts of analysis. The aim of this book is to help students of mathematics and related sciences acquire a basic understanding of the concepts in complex function theory and to know how vastly different the subjects of complex analysis and real analysis are.

This book is intended for students pursuing master degree in Mathematics as well as those from technical branches who would like to refine their knowledge of Complex analysis. The book can be used as a textbook for Post Graduate course in Indian Universities. The prerequisites for the book are kept to a minimum. An undergraduate course in real analysis should sufficiently prepare the reader for the book. Basic facts on metric spaces is collected in a chapter of the book. Some exercises are provided at the end of each chapter. Most of these problems are workable. These exercises should be suitable for homework assignments if the book is used as a textbook.

Chapter 1 begins with Complex numbers, its geometrical interpretations. Many familiar inequalities between Complex numbers are also presented.

Chapter 2 deals with elementary of metric spaces and the topology of $\mathbb{C}$. In this chapter topology of $\mathbb{C}$ is thoroughly analyzed. Chapter 3 is devoted to analysis on $\mathbb{C}$.

Chapter 4 studies properties of the most important function called exponential function. Exponentiation of complex numbers is introduced using the properties of the convergence of infinite series of complex numbers. It is relevant to point out that many textbooks straight away write $z = re^{i\theta}$ where r is the modulus of z and θ its argument. Exponentiation of a real number by a complex number is something that has to be defined. The function e^z has a meaning only in the sense of its power series expansion and the algebraic properties like $e^{z_1}e^{z_2} = e^{z_1+z_2}$ can be justified by a fact that the Cauchy product of two absolutely convergent series converges to the product of their sums. This has to be pointed out before we can manipulate terms of the product. Even argument has to be properly defined.

Chapter 5 begins with differentiation of a complex valued function on an open set. In this chapter the properties of complex valued functions are contrasted with analogous properties of real valued functions. Here we pointed out that for the differentiation to be meaningful the domain of the function has to be open. Through this we define holomorphic function. Before jumping to any new ideas we look at such holomorphic functions like polynomials and rational functions. Then we pose the question if there are other examples. Here we show that a power series represents a holomorphic function. Then we define an analytic function, which immediately implies that every analytic function is holomorphic. Here we ask the question if every holomorphic function is analytic? Now we have arrived at the heart of Cauchy theory: any holomorphic function f on an open set is locally a power series, i.e., an analytic function. Not much attempt has been made in books written in India to distinguish between analyticity and holomorphy. No doubt in $\mathbb{C}$ a function is holomorphic if and only if it is analytic. But when a student comes across a term like real analytic he is confused. Even the Cauchy theorem which is key to establishing the fact that every holomorphic function is analytic is seldom dealt with properly. That the Cauchy theorem has a topological context is hardly mentioned. That this topological aspect of the Cauchy's theorem is related to the question as to when a closed differential form is also an exact differential form is not mentioned at all. This could be a natural path to de Rham cohomology for those who want to pursue this.

Chapter 6 deals with Complex integration. Chapter 7 studies the simple version of Cauchy's theorem and its consequences. Here we are in a position to prove Liouville's theorem, Morera's theorem, Identity theorem or principle of analytic continuation, maximium modulus principle etc.

Chapter 8 begins with studies on isolated singularities, Meromorphic functions and Casorati-Weierstrass theorem. A section is also devoted to Laurent series expansion and characterization of singularities.

In Chapter 9 we introduce the concept of winding numbers of closed curves and present many important results involving the winding numbers. We prove the homology version of Cauchy's theorem and homotopy version of Cauchy's theorem in Chapter 10.

In Chapter 11 we prove the residue theorem and demonstrate some of its applications. Some of the mapping properties of holomorphic functions is also presented in this Chapter and we prove Hurwitz's theorem, local mapping theorem, open mapping theorem and inverse function theorem etc.

Chapter 12 is devoted to computation of definite real integrals and presented the techniques for finding the summation of infinite series using residue theorem.

This is followed by Chapter 13 where we present a discussion on extended complex plane, fractional linear transformations, conformal maps and prove Schwarz's lemma. In Chapter 14 we define Harmonic, subharmonic functions and derive some of their properties and discuss about the Dirichlet problem in the unit disk.

Chapter 15 begins with a discussion on normal families. In this Chapter we prove the Riemann mapping theorem and present many applications of Schwartz-Christoffel formula. Chapter 16 begins with Schwarz reflection principle and is followed by a discussion on meromorphic and analytic continuation and we prove the monodromy theorem.

Chapter 17 studies representation of meromorphic functions. Here we prove Mittag Leffler theorem, Weierstrass factorization theorem, introduce the concepts order, genus, type of entire functions and prove Hadamard theorem, Lucas theorem etc. We also discuss in detail about gamma function and

Riemann ζ function. Chapter 18 deals with elliptic functions and their properties.

Most of the analysts with whom I have discussed, or whose books I have read, have contributed in one way or another to this book. The book owe a great deal to the lecture notes of Prof. S. Kumaresan and Prof. J.R. Partington and the teachings of Prof. S. Pattanayak. I express my deep sense of gratitude to Dr. C.K. Mohapatra for his encouragement and indispensable help throughout the preparation of the book. I also deeply appreciate the assistance of Sri Rajendra Prasad Lal (CSIR scholar) and Miss Debasmita Dash in computer graphics. This book is dedicated to my parents. Their encouragement is most precious for me.

Namita Das

Dedicated to my parents

Contents

Chapter 1

Complex Numbers

We take for granted the acquaintance with the set of positive integers $\mathbb{N} = \{1, 2, 3, \cdots\}$, the set of integers $\mathbb{Z} = \{0, \pm 1, \pm 2, \cdots\}$, the set of rational numbers $\mathbb{Q} = \{\frac{p}{q} : p \in \mathbb{Z}, q \in \mathbb{N}\}$ together with the operations of addition, multiplication and ordering on them. It is presumed that the readers, at this stage, have some working knowledge of the algebraic and topological properties of real numbers. We shall construct the set $\mathbb{C}$ of complex numbers in this chapter.

It is important to realize that the simple equation $a + x = b$, $a, b \in \mathbb{N}$ is solvable in $\mathbb{N}$ if $a < b$ and not solvable in $\mathbb{N}$ if $a = b$ or $a > b$; but it is solvable in $\mathbb{Z}$. Similarly, the equation $bx = a$, $a, b \in \mathbb{Z}$ is not in general solvable in $\mathbb{Z}$ but solvable in $\mathbb{Q}$ provided $b \neq 0$. The above observation highlights the inadequacies of the system $\mathbb{N}$ and $\mathbb{Z}$ and justifies the extension from $\mathbb{N}$ to $\mathbb{Z}$ and $\mathbb{Z}$ to $\mathbb{Q}$. Though the rational system has rich algebraic structures, it is also inadequate for many purposes. One such inadequacy is exhibited by the fact that the equation $x^2 = 2$ has no rational solution. So our realm of numbers should be larger than $\mathbb{Q}$. Yet every such numbers should be approximable as closely as possible to rational numbers. Thus we tried as far as possible to build our number system to include rationals and preserve all the essential properties of rationals (its algebraic and order relation). This extended system of numbers shall be called the real numbers and the collection of all real numbers shall be denoted as $\mathbb{R}$. We shall call an element of $\mathbb{R} - \mathbb{Q}$ an irrational number.

Consider the quadratic equation

$$ax^2 + bx + c = 0 \tag{1.1}$$

where $a, b, c \in \mathbb{R}$. We know if $b^2 - 4ac \geq 0$ then this has real solutions. But when $b^2 < 4ac$ there is no real number which satisfies the above equation. The problem came from the fact that the square of every real number is nonnegative. So the simple equation $x^2 + 1 = 0$ does not have any solution in $\mathbb{R}$. If we agree to extend our scheme of numbers to include such numbers whose square can be a negative real number then how will be the algebra of numbers be modified.

It is obvious the formula $\frac{-b\pm\sqrt{b^2-4ac}}{2a}$ breaks down when $b^2 < 4ac$ if we have only real numbers in our scheme. But if we agree that negative real numbers also have square roots (the numbers so obtained are called "imaginary numbers") then we can use the above formula. To systematize this let us write $\sqrt{-1} = i$ and agree that $\sqrt{ab} = \sqrt{a}\sqrt{b}$ for a, b real. Hence square root of a negative number will be a real number multiplied by i.

Thus even if the real number system is adequate and comprehensive for many purposes of mathematical analysis, its inadequacy is exhibited by the fact that the simple algebraic equation

$$x^2 + 1 = 0 \tag{1.2}$$

has no solution in $\mathbb{R}$. Admitting the mysterious symbol i, called imaginary unit [due to Euler (1707-1783)] such that

$$i^2 = -1 \tag{1.3}$$

we now define the set $\mathbb{C}$ of complex numbers by

$$\mathbb{C} = \{z = x + iy : x \in \mathbb{R}, y \in \mathbb{R}\}.$$

The number x is called the real part and y is called the imaginary part of z, and written as $x = \text{Re}z$, $y = \text{Im}z$. The complex number z is real if $\text{Im}z = 0$ and z is purely imaginary if $\text{Re}z = 0$. With suitable definition of addition and multiplication in $\mathbb{C}$, it is possible to make $\mathbb{C}$ a field which contains $\mathbb{R}$ as a subfield and in which equation (1.2) is solvable.

1.1 The algebra of complex numbers

We now proceed to develop the algebra of $\mathbb{C}$ starting with the following definitions of equality, addition and multiplication. Let $z_1, z_2 \in \mathbb{C}$ and let $z_1 = x_1 + iy_1$, $z_2 = x_2 + iy_2$, $x_1, x_2, y_1, y_2 \in \mathbb{R}$.

Equality

Two complex numbers z_1 and z_2 are called equal, written as $z_1 = z_2$ if and only if $x_1 = x_2$ and $y_1 = y_2$.

Addition

The sum or addition of two complex numbers z_1 and z_2 is defined by

$$z_1 + z_2 = (x_1 + x_2) + i(y_1 + y_2). \tag{1.4}$$

Multiplication

The product or multiplication of two complex numbers z_1 and z_2 is defined by

$$z_1 z_2 = (x_1 x_2 - y_1 y_2) + i(y_1 x_2 + x_1 y_2). \tag{1.5}$$

In particular, we obtain

$$\begin{aligned}(x_1 + i0) + (x_2 + i0) &= (x_1 + x_2) + i0;\\ (x_1 + i0).(x_2 + i0) &= x_1x_2 + i0.\end{aligned}$$

In view of this we identify $\mathbb{R}$ with $\{x + i0 : x \in \mathbb{R}\}$ via the field isomorphism h from $\mathbb{R}$ into $\mathbb{C}$ given by $h(x) = x + i0$.

An alternative but equivalent way to define a complex number $z = x + iy$ is to identify it with the ordered pair $(x, y) \in \mathbb{R}^2$ with the following addition and multiplication and with $(0, 1) = i$.

$$\begin{aligned}(A) \quad (x, y) + (u, v) &= (x + u, y + v);\\ (M) \quad (x, y)(u, v) &= (xu - yv, xv + yu).\end{aligned}$$

The set $\mathbb{C}$ is a field with the addition and multiplication defined in (A) and (M) above and it contains $\mathbb{R}$ as a subfield, with $0 + i0 = 0$ serving as additive identity, $1 + i0 = 1$ as multiplicative identity. Moreover, the distributive law holds:

$$(a, b)[(x, y) + (u, v)] = (a, b)(x, y) + (a, b)(u, v)$$

for all $(a, b), (x, y), (u, v) \in \mathbb{C}$. The inverse of the element $x + iy$ where both x and y are not zero is given by $(x+iy)^{-1} = \frac{x}{x^2+y^2} - i\frac{y}{x^2+y^2}$. At this stage, we may remark that the introduction of complex number system is quite satisfying, marked by its elegance and generality, and its accomplishments are manifold and spectacular. For example, the following two remarkable results are available only within the fold of $\mathbb{C}$ and not within $\mathbb{R}$; (a) The fundamental theorem of algebra (C.F.Gauss, 1777-1855) which states that every nonconstant polynomial with coefficients in $\mathbb{C}$ can be factored over $\mathbb{C}$ into linear factors. That is $\mathbb{C}$ is a algebraically closed field. (b) Euler's formula: $e^{ix} = \cos x + i \sin x$.

But in the process of extension of field $\mathbb{R}$ to $\mathbb{C}$, though certain advantages have been gained, we have sacrificed the property of order. In fact $\mathbb{C}$ is not an ordered field.

Order in the complex field

An order in a field F is a subset $\mathcal{P}$ such that (1) F is the disjoint union of $\mathcal{P}$, $-\mathcal{P} = \{-x : x \in \mathcal{P}\}$ and $\{0\}$ and (2) whenever $x, y \in \mathcal{P}$, then $x + y$ and xy are in $\mathcal{P}$. The set $\mathcal{P}$ is called the positive elements of F with respect to the order $\mathcal{P}$. For $x, y \in F$ and using this $\mathcal{P}$ we say $x < y$ if $y - x \in \mathcal{P}$.

One can show that there is no order on $\mathbb{C}$ which is compatible with the one on $\mathbb{R}$, that is, if $\mathcal{P}$ is the set of positive elements in $\mathbb{C}$, then $\mathcal{P} \cap \mathbb{R}$ is the set of positive reals in $\mathbb{R}$. Let, if possible, such an order exist with $\mathcal{P}$ as the set of positive elements in $\mathbb{C}$. Now either $i \in \mathcal{P}$ or $-i \in \mathcal{P}$. In either case, $-1 = i^2 \in \mathcal{P}$. This means that -1 is a positive real !

In view of this we should never write inequalities between two complex numbers unless both of them are real.

Remark 1.1. Notice that we can define an order relation on $\mathbb{C}$ as follows. We shall say $x_1 + iy_1 \prec x_2 + iy_2$ if $x_1 < x_2$, or if $x_1 = x_2$ and $y_1 < y_2$. Thus with respect to this order relation $0 = 0 + i \cdot 0 < 0 + i \cdot 1 = i$. That is, i is a positive element in $\mathbb{C}$ and therefore $-1 = i^2 > 0$ (if $\mathbb{C}$ is an ordered field with respect to this relation). Hence the order relation '$\prec$' is not compatible with the one that is defined on $\mathbb{R}$.

1.2 Geometry of complex numbers

Geometrical representation of complex numbers is made by identifying the point $z = x+iy$ with the point $(x, y) \in \mathbb{R}^2$ of two-dimensional co-ordinate plane and with this identification $\mathbb{C}$ is called a **complex or Argand plane**. The set

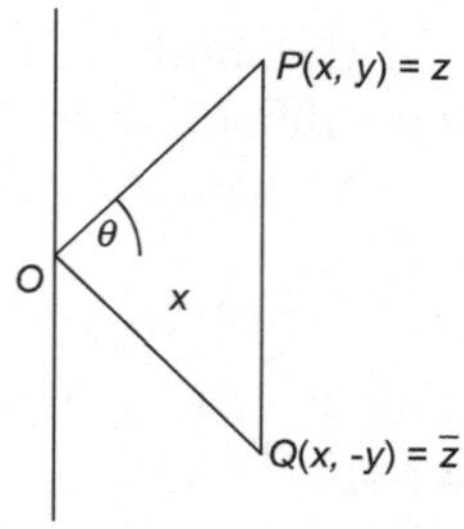

Fig. 1.1: Argand plane

$\{(x, 0) : x \in \mathbb{R}\}$ is called the x-axis or real axis or real number line; similarly the set $\{(0, y) : y \in \mathbb{R}\}$ is called the y-axis or imaginary axis. The point $(-1, 1)$ represents point $-1 + i$ in the complex plane. If $z = x + iy, x, y \in \mathbb{R}$, then **conjugate** of z, denoted as $\overline{z}$ is the complex number $\overline{z} = x - iy$. If P is the point (x, y) that is identified with z in the complex plane, then its conjugate $\overline{z}$ represents a point $Q(x, -y)$ which is the reflection of P about x-axis. Thus $\text{Re}z = \frac{z+\overline{z}}{2}$ and $\text{Im}z = \frac{z-\overline{z}}{2i}$. Given $z = x + iy$, we define the **modulus** or **absolute value** of z by

$$|z| = \sqrt{x^2 + y^2}.$$

Thus $|z|$ represents the distance of the point $z \in \mathbb{C}$ from the origin O. We can easily compute the following:

Theorem 1.1. *For all complex numbers $z, w \in \mathbb{C}$,*

(i)	$z\overline{z} = \|z\|^2$,	*(ii)*	$\|z\| = \|\overline{z}\| = \|-z\|$,
(iii)	$\frac{1}{z} = \frac{\overline{z}}{\|z\|^2} (z \neq 0)$,	*(iv)*	$\|zw\| = \|z\|\|w\|$,
(v)	$\|z^n\| = \|z\|^n$,	*(vi)*	$\overline{z + w} = \overline{z} + \overline{w}$,
(vii)	$\overline{zw} = \overline{z}\ \overline{w}$,	*(viii)*	$\overline{\overline{z}} = z$,
(ix)	*z is real iff $\overline{z} = z$,*	*(x)*	*z is purely imaginary iff $z = -\overline{z}$,*
(xi)	$-\|z\| \leq Re z \leq \|z\|$.		

In (xi) the first equality holds if and only if $z \leq 0$ and the second if and only if $z \geq 0$.

Notice that $|z|^2 \neq z^2$ unless z is real. For example, $|i|^2 = 1$, $i^2 = -1$. If we think of $z = x + iy$ as a vector in $\mathbb{R}^2$, then $|z| = \|(x, y)\|$, the norm of the vector with respect to the dot product on $\mathbb{R}^2$. (Given two vectors (x, y) and (u, v) in $\mathbb{R}^2$, $(x, y) \cdot (u, v) = xu + yv$ and $\|(x, y)\|^2 = (x, y) \cdot (x, y) = x^2 + y^2$.)

Theorem 1.2. *Every complex number $a = c + id$ has a square root in $\mathbb{C}$.*

Proof. If $z = x+iy$ and $z^2 = a$, then we get $x^2 - y^2 = c$ and $2xy = d$. Therefore $x = \frac{\sqrt{c+\sqrt{c^2+d^2}}}{\sqrt{2}}$ and $y = \frac{d}{\sqrt{2}\sqrt{c+\sqrt{c^2+d^2}}}$. □

Thus it follows that every quadratic polynomial with complex coefficients has roots in $\mathbb{C}$. Let $P(z) = az^2 + bz + c$ be a polynomial with $a \neq 0$. Then $P(z) = a(z + \frac{b}{2a})^2 - \frac{b^2}{4a} + c = 0$ implies $z = \frac{-b\pm\sqrt{b^2-4ac}}{2a}$. Further note that the nonreal roots of a polynomial with real coefficients occur in conjugate pairs. That is, if $a_0 z^n + a_1 z^{n-1} + \cdots + a_n = 0, a_0, a_1, \cdots, a_n \in \mathbb{R}$ then $0 = \bar{a}_0 \bar{z}^n + \bar{a}_1 \bar{z}^{n-1} + \cdots + \bar{a}_n = a_0 \bar{z}^n + a_1 \bar{z}^{n-1} + \cdots + a_n$. Thus $\bar{z}$ is a root of the polynomial.

Theorem 1.3. *For any $z, w \in \mathbb{C}$, we have*

(a) ***Parallelogram Law***

$$|z + w|^2 + |z - w|^2 = 2(|z|^2 + |w|^2).$$

(b) ***Triangle Inequality***

$$|z + w| \leq |z| + |w|.$$

Equality holds if and only if $z = 0$ or $\frac{w}{z} \geq 0$(i.e. w is a nonnegative multiple of z).

(c) ***Continuity of modulus***

$$|\,|z| - |w|\,| \leq |z - w|.$$

The equality holds if and only if $z = 0$ or $\frac{w}{z} \geq 0$.

Proof.

$$\begin{aligned} |z + w|^2 &= (z + w)\overline{(z + w)} \\ &= (z + w)(\overline{z} + \overline{w}) \\ &= z\overline{z} + w\overline{w} + z\overline{w} + w\overline{z} \\ &= |z|^2 + |w|^2 + 2\mathrm{Re}(z\overline{w}). \end{aligned}$$

Similarly, we can compute that

$$|z - w|^2 = |z|^2 + |w|^2 - 2\mathrm{Re}(z\overline{w}).$$

The result (a) follows by adding the two expressions above. Since $|x| \leq \sqrt{x^2 + y^2} = |z|$, it follows that $|\mathrm{Re} z| \leq |z|$. Hence $|\mathrm{Re}(z\overline{w})| \leq |z\overline{w}| = |z||w|$. Now it follows from (a) that $|z + w|^2 \leq |z|^2 + |w|^2 + 2|z||w| = (|z| + |w|)^2$. The result (b) follows from the above inequality. Equality in (b) holds if and only if in the above chain of inequalities, we have $\text{Re} z\overline{w} = |z||w|$. By the equality part of (xi) of the Theorem 1.1, this happens if $z\overline{w} \geq 0$. That is if either $z = 0$ or $z\overline{w} = |z|^2 \frac{\overline{w}}{\overline{z}} \geq 0$, the assertion in (b) follows.

Since $|z| = |z - w + w| \leq |z - w| + |w|$ by triangle inequality, it follows that

$$|z| - |w| \leq |z - w|. \tag{1.6}$$

Interchanging the role of w and z, we obtain

$$|w| - |z| \leq |w - z| = | - (z - w)| = |z - w|. \tag{1.7}$$

The result (c) follows by combining (1.6) and (1.7). The equality case is left to the reader. □

The reader is invited to appreciate why the results (a) and (b) are respectively called Parallelogram and Triangle laws. The main advantage of

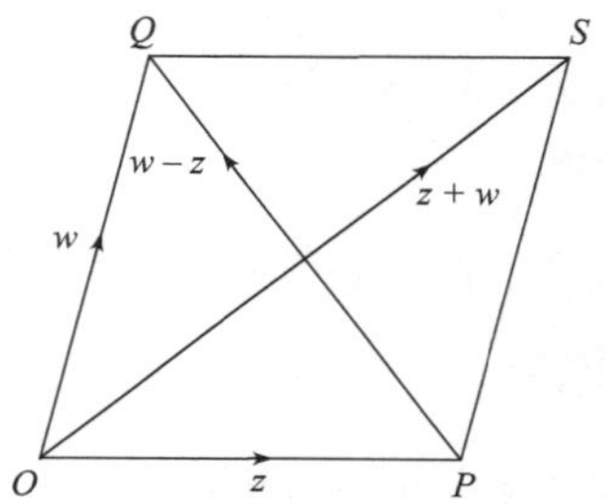

Fig. 1.2: Geometric representation of Parallelogram law

the geometrical representation of complex numbers is that the addition of two complex numbers can be visualized as vector addition, i.e. the sum of two complex numbers z and w represented by points P and Q corresponds to a complex number represented by a point S so that $OPSQ$ is a parallelogram. If $\overrightarrow{OP} = z$ and $\overrightarrow{OQ} = w$ then $\overrightarrow{OS} = z + w$. The magnitude of the vector $\overrightarrow{OS}$ is $OS = |z + w|$ and that of $\overrightarrow{PQ}$ is $|w - z| = PQ$.

Hence the result (a) tells that in a parallelogram, the sum of the squares of diagonals is equal to the sum of the squares of sides which is a well-known geometrical fact. Similarly the triangle law (b) tells that in a triangle, the sum of any two sides is greater than the third.

We know that the inequality between two complex numbers does not exist unless the numbers are real. But since the modulus of a complex number $z = x + iy$ is a nonnegative real number $\sqrt{x^2 + y^2}$, it follows that complex numbers may appear in inequalities involving modulus, for example, the triangle inequality. Thus we can now define the least upper bound, greatest lower bound and the maximum of the modulus of a subset of $\mathbb{C}$ and therefore we can define bounded subset of $\mathbb{C}$.

1.3 Polar form, De Moivre's formula and roots of unity

In order to give a geometrical interpretation of the product of two complex numbers, we first introduce the polar co-ordinates.

If $z = x + iy \in \mathbb{C}$ is identified with $(x, y) \in \mathbb{R}^2$, then we have the polar coordinates r and θ of (x, y). They are called the polar co-ordinates of z. If θ is the angle between the positive real axis and the line segment joining origin with P representing $z = x + iy$, with the usual sign convention (positive in the counter-clockwise sense and negative in the clockwise sense) then $x = \text{Re}z = r\cos\theta$, $y = \text{Im}z = r\sin\theta$, where we write $r = |z| = \sqrt{x^2 + y^2}$. Hence

$$z = r(\cos\theta + i\sin\theta). \tag{1.8}$$

This is called the polar form (representation) of the complex number z. Thus the point $z = x + iy$ can now be represented by the polar co-ordinates (r, θ) where the polar angle θ is called the argument or amplitude of z and denoted as $\theta = \arg z$. Observe that the polar angle θ is indeterminate since $\theta + 2n\pi$ can be substituted for θ in relation (1.8) above; that is, for $n \in \mathbb{N}$, $z = r(\cos\theta + i\sin\theta) = r(\cos(2n\pi + \theta) + i\sin(2n\pi + \theta))$. Therefore, it is convenient to have the principal value of the argument and this is defined by θ where $-\pi < \theta \leq \pi$.

Definition 1.1. *We define the argument argz of a nonzero complex number z to be any $\theta \in \mathbb{R}$ such that $z = |z|(\cos\theta + i\sin\theta)$. The principal argument Argz of z is defined to be that argument of z which lies in the interval $(-\pi, \pi]$. The relation between argz and Argz is given by*

$$\arg z = \text{Arg} z + 2n\pi$$

for some $n \in \mathbb{Z}$.

For example, $\text{Arg}1 = 0$, $\text{Arg}(-1) = \pi$, $\text{Arg}i = \frac{\pi}{2}$, $\text{Arg}(-i) = -\frac{\pi}{2}$. We now give a significant result concerning argument:

Theorem 1.4. *(a) Let $z_1, z_2 \in \mathbb{C}$. Then (i) $\arg(z_1z_2) = \arg z_1 + \arg z_2$, (ii) $\arg\left(\frac{z_1}{z_2}\right) = \arg z_1 - \arg z_2$ if $z_2 \neq 0$. (b) If (r, θ) is the polar co-ordinates of $z \in \mathbb{C}$, then for $n \in \mathbb{Z}$, (i) $z^n = r^n(\cos n\theta + i\sin n\theta)$, (ii) $(\cos\theta + i\sin\theta)^n = \cos n\theta + i\sin n\theta, n \in \mathbb{Z}$.*

The formula b(ii) is known as De Moivre's formula.

Proof. Let $z_1 = r_1(\cos\theta_1 + i\sin\theta_1)$, $z_2 = r_2(\cos\theta_2 + i\sin\theta_2)$. Then the product

$$\begin{aligned} z_1z_2 &= r_1r_2(\cos\theta_1 + i\sin\theta_1)(\cos\theta_2 + i\sin\theta_2) \\ &= r_1r_2(\cos(\theta_1 + \theta_2) + i\sin(\theta_1 + \theta_2)). \end{aligned} \tag{1.9}$$

Hence $\arg(z_1z_2) = \theta_1 + \theta_2 = \arg z_1 + \arg z_2$.

Similarly if $z_2 \neq 0$, then

$$\begin{aligned}\frac{z_1}{z_2} &= \frac{z_1\overline{z_2}}{|z_2|^2}\\ &= \frac{r_1}{r_2}(\cos\,\theta_1 + i\sin\,\theta_1)(\cos\,\theta_2 - i\sin\,\theta_2)\\ &= \frac{r_1}{r_2}(\cos\,(\theta_1-\theta_2) + i\sin\,(\theta_1-\theta_2)).\end{aligned}$$

Hence

$$\arg\left(\frac{z_1}{z_2}\right) = \theta_1 - \theta_2 = \arg z_1 - \arg z_2. \tag{1.10}$$

By repeated arguments as in (1.9), we get

$$z_1 z_2 \cdots z_k = r_1 r_2 \cdots r_k(\cos(\theta_1+\theta_2+\cdots+\theta_k) + i\sin(\theta_1+\theta_2+\cdots+\theta_k)) \tag{1.11}$$

where $z_j = r_j(\cos\theta_j + i\sin\theta_j), j = 1, 2, \cdots, k$. Putting $z_1 = z_2 = \cdots = z_k = z$ in (1.11) we obtain

$$z^n = r^n(\cos n\theta + i\ \sin n\theta), n \in \mathbb{N}. \tag{1.12}$$

Since $\frac{1}{z} = \frac{\overline{z}}{|z|^2} = \frac{\cos\theta - i\ \sin\theta}{r} = \frac{\cos(-\theta)+i\ \sin(-\theta)}{r}$, it follows from (1.11) that

$$\frac{1}{z^n} = \frac{1}{r^n}(\cos(-n\theta) + i\ \sin(-n\theta)), n \in \mathbb{N}. \tag{1.13}$$

Now combining (1.12) and (1.13) we obtain b(i). Putting $r = 1$ in b(i) we obtain b(ii). □

The geometric interpretation of the product of two numbers is evident from equation (1.9) above. It tells that the number z_1z_2 has the modulus r_1r_2 and also it has the argument $\theta_1 + \theta_2$. Thus the argument of the product of two complex numbers is equal to the sum of the arguments of the factors.

De Moivre's formula enables us to find the nth roots of any complex number. Suppose we seek to find the cube roots of 1. Any such cube root has absolute value 1, so it has the form $\cos\alpha + i\sin\alpha$ for some angle α. By De Moivre's formula we must have $\cos 3\alpha + i\sin 3\alpha = 1$, which is true if and only if $3\alpha = 2k\pi$ for some $k \in \mathbb{Z}$. The cube roots of 1 are thus the numbers $\cos\frac{2k\pi}{3} + i\sin\frac{2k\pi}{3}$ with $k \in \mathbb{Z}$. Two different choices of k yield the same root if and only if the choices are congruent modulo 3. There are thus three distinct cube roots of 1, which we obtain, for example, from the choices $k = 0, 1, 2$; besides 1 itself, they are

$$\begin{aligned}\cos\frac{2\pi}{3} + i\sin\frac{2\pi}{3} &= \frac{-1+i\sqrt{3}}{2},\\ \cos\frac{4\pi}{3} + i\sin\frac{4\pi}{3} &= \frac{-1-i\sqrt{3}}{2}.\end{aligned}$$

By similar reasoning we are in a position to prove the following theorem. We shall show that the field $\mathbb{C}$ contains not only square roots of each of its elements (Theorem 1.2), but even nth roots for every positive integer n.

Theorem 1.5. *Let $a \in \mathbb{C} - \{0\}$, $n \in \mathbb{N}$. Then the equation*

$$z^n = a \tag{1.14}$$

has n distinct nth roots.

Proof. Let (ρ, ϕ) be the polar co-ordinates of a. Then $a = \rho(\cos\phi + i\sin\phi)$. Let $z = r(\cos\theta + i\sin\theta)$. Then the equation (1.13) reduces to, by theorem 1.4,

$$\begin{aligned}
&r^n(\cos n\theta + i\sin n\theta)\\
&= \rho(\cos\phi + i\sin\phi)\\
&= \rho(\cos(2k\pi + \phi) + i\sin(2k\pi + \phi)), k = 0, 1, 2, \cdots\\
&= \rho\left[\cos\frac{2k\pi + \phi}{n} + i\sin\frac{2k\pi + \phi}{n}\right]^n.
\end{aligned}$$

Hence

$$z = \rho^{\frac{1}{n}}\left[\cos\frac{2k\pi + \phi}{n} + i\sin\frac{2k\pi + \phi}{n}\right]. \tag{1.15}$$

But for $k = 0, 1, 2, \cdots, n-1$ we get distinct values and for $k = n, n+1, \cdots$ the values are repeated. So the n roots of the equation (1.14) are given by (1.15) for $k = 0, 1, 2, \cdots, n-1$. □

We give below two examples which are interesting consequences of the above.

Example 1.1. *Consider the equation $x^5 = 1$. Besides the solution $x = 1$ the other solutions are the solutions of the equation $x^4 + x^3 + x^2 + x + 1 = 0$. So the other four solutions of the equation $x^5 = 1$ are*

$$cos\frac{2\pi}{5} + isin\frac{2\pi}{5}, \quad cos\frac{4\pi}{5} + isin\frac{4\pi}{5},$$

$$cos\frac{6\pi}{5} + isin\frac{6\pi}{5}, \quad cos\frac{8\pi}{5} + isin\frac{8\pi}{5}.$$

Observe that

$$\begin{aligned}
cos\frac{2\pi}{5} + isin\frac{2\pi}{5} &= \frac{1}{cos\frac{8\pi}{5} + isin\frac{8\pi}{5}},\\
cos\frac{4\pi}{5} + isin\frac{4\pi}{5} &= \frac{1}{cos\frac{6\pi}{5} + isin\frac{6\pi}{5}}.
\end{aligned}$$

Now $x^4+x^3+x^2+x+1 = x^4+1+x^3+x+x^2 = x^2(x^2+\frac{1}{x^2}+x+\frac{1}{x}+1)$. *Thus solution of* $x^4+x^3+x^2+x+1=0$ *are also the solutions of the equation* $x^2+\frac{1}{x^2}+x+\frac{1}{x}+1=0$. *To solve this write* $x+\frac{1}{x}=2y$, *so* $x^2+\frac{1}{x^2}=4y^2-2$. *Hence the equation reduces to* $4y^2+2y-1=0$. *This is a quadratic equation having two real solutions* $\frac{-1\pm\sqrt{5}}{4}$. *In fact they correspond to the two values of* y, *e.g.,* $cos\frac{2\pi}{5}$ *and* $cos\frac{4\pi}{5}$. *Thus we obtain*

$$cos72^\circ = cos\frac{2\pi}{5} = \frac{-1+\sqrt{5}}{4};$$

$$cos36^\circ = -cos144^\circ = -cos\tfrac{4\pi}{5} = \frac{1+\sqrt{5}}{4}.$$

Example 1.2. *Consider a circle of radius* 1; *and let* $Q_1, Q_2, \cdots, Q_n$ *be the vertices of a regular* n*-gon inscribed in the circle (see Fig.1.3). Join* Q_1 *to* Q_2, $Q_3, \cdots, Q_n$ *by segments of a straight line. We obtain* $(n-1)$ *segments of lengths* $\lambda_2, \lambda_3, \cdots, \lambda_n$. *We shall show that* $\prod_{i=2}^{n}\lambda_i = n$.
Solution: Let $P_{n-1}(x) = \frac{x^n-1}{x-1} = x^{n-1}+\cdots+1$ *for* $x \neq 1$, *so* $P_{n-1}(1)=n$. *Let* $p_k = e^{\frac{2\pi i(k-1)}{n}}$ *for* $k=1,\cdots,n$ *be the nth roots of* 1. *As* $p_1=1$, *we have*

$$\prod_{i=2}^{n}(z-p_k) = P_{n-1}(z).$$

Letting $z=1$ *we get the desired result as* $P_{n-1}(1)=n$.

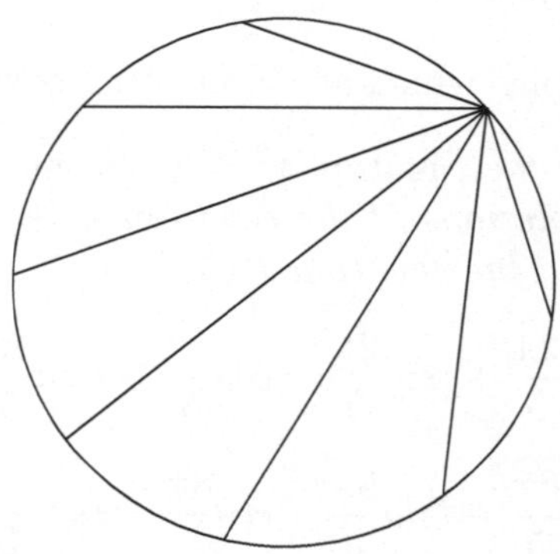

Fig. 1.3: The regular n-gon

1.4 Inequalities

In this section we shall prove certain important inequalities.

Theorem 1.6. *(Cauchy-Schwarz inequality): For complex numbers $z_1, z_2, \cdots, z_n$ and $w_1, w_2, \cdots, w_n$ we have*

$$|z_1\bar{w}_1 + z_2\bar{w}_2 + \cdots + z_n\bar{w}_n|^2 \leq (|z_1|^2 + \cdots + |z_n|^2)(|w_1|^2 + \cdots + |w_n|^2).$$

Proof. To prove this we appeal to the fact that the quadratic expression $ax^2 + bx + c$ for $a, b, c \in \mathbb{R}$ does not change its sign for any real value of x if and only if $b^2 < 4ac$. Notice that

$$\begin{aligned} 0 &\leq |z_1 + xw_1|^2 + |z_2 + xw_2|^2 + \cdots + |z_n + xw_n|^2 \\ &= (|z_1|^2 + |z_2|^2 + \cdots + |z_n|^2) + x^2(|w_1|^2 + |w_2|^2 + \cdots + |w_n|^2) \\ &\quad + 2xRe(z_1\bar{w}_1 + z_2\bar{w}_2 + \cdots + z_n\bar{w}_n). \end{aligned}$$

Thus $0 \leq (|z_1|^2 + |z_2|^2 + \cdots + |z_n|^2) + x^2(|w_1|^2 + |w_2|^2 + \cdots + |w_n|^2) + 2x|z_1\bar{w}_1 + z_2\bar{w}_2 + \cdots + z_n\bar{w}_n|$. So we have by the above fact $|z_1\bar{w}_1 + z_2\bar{w}_2 + \cdots + z_n\bar{w}_n|^2 \leq (|z_1|^2 + \cdots + |z_n|^2)(|w_1|^2 + \cdots + |w_n|^2)$.

An alternative way to prove this fact is as follows: Let λ denote an arbitrary complex number. Notice that

$$0 \leq \sum_{i=1}^{n} |z_i - \lambda w_i|^2 = \sum_{i=1}^{n} |z_i|^2 + |\lambda|^2 \sum_{i=1}^{n} |w_i|^2 - 2Re\bar{\lambda} \sum_{i=1}^{n} z_i\bar{w}_i. \tag{1.16}$$

We can choose

$$\lambda = \frac{\sum_{i=1}^{n} z_i\bar{w}_i}{\sum_{i=1}^{n} |w_i|^2},$$

for if the denominator should vanish there is nothing to prove. Substituting this value of λ in 1.16 we find, after simplifications,

$$\sum_{i=1}^{n} |z_i|^2 - \frac{|\sum_{i=1}^{n} z_i\bar{w}_i|^2}{\sum_{i=1}^{n} |w_i|^2} \geq 0$$

which proves our claim. Notice further that the sign of equality holds if and only if the z_i are proportional to the w_i. □

Lemma 1.1. *If $|z| < 1$ and $|a| < 1$, then*

$$\left|\frac{z - a}{1 - \bar{a}z}\right| \leq \frac{|z| + |a|}{1 + |a||z|} < 1.$$

Proof. If $|z| < 1$ and $|a| < 1$ then $(1 - |z|)(1 - |a|) > 0$. Hence $1 + |a||z| > |z| + |a|$. Therefore

$$1 > \frac{|z| + |a|}{1 + |a||z|}. \tag{1.17}$$

Let

$$r_1 = 1 - \left|\frac{z - a}{1 - \bar{a}z}\right|^2 = \frac{(1 - |a|^2)(1 - |z|^2)}{|1 - \bar{a}z|^2}$$

and

$$r_2 = 1 - \left(\frac{|z| + |a|}{1 + |a||z|}\right)^2 = \frac{(1 - |a|^2)(1 - |z|^2)}{(1 + |a||z|)^2}.$$

By the triangle inequality $|1 - \overline{a}z| \leq 1 + |a||z|$, and so $r_2 \leq r_1$. Therefore, $1 - r_1 \leq 1 - r_2$ or

$$\left|\frac{z - a}{1 - \overline{a}z}\right|^2 \leq \frac{(|a| + |z|)^2}{(1 + |a||z|)^2}$$

which gives us the first inequality. Combining this with (1.17) we proved the claim. The inequality (1.17) can also be proved in the following way.

Fixing $|a|$, the function

$$\Psi(r) = \frac{r + |a|}{1 + |a|r}$$

has derivative

$$\Psi'(r) = \frac{1 - |a|^2}{(1 + |a|r)^2} > 0.$$

Thus $\Psi(r)$ increases with r. Since $\Psi(1) = 1$, the inequality (1.17) follows. □

Lemma 1.2.

$$\left|\frac{\alpha - \beta}{1 - \overline{\beta}\alpha}\right| \geq \frac{|\alpha| - |\beta|}{1 - |\alpha||\beta|}$$

if $|\alpha| < 1$ *and* $|\beta| < 1$.

Proof. Let

$$r_1 = 1 - \left|\frac{\alpha - \beta}{1 - \overline{\beta}\alpha}\right|^2$$

and

$$r_2 = 1 - \left(\frac{|\alpha| - |\beta|}{1 - |\alpha\beta|}\right)^2.$$

Then simplifying we obtain

$$r_1 = \frac{(1 - |\alpha|^2)(1 - |\beta|^2)}{|1 - \overline{\beta}\alpha|^2}$$

and

$$r_2 = \frac{(1 - |\alpha|^2)(1 - |\beta|^2)}{(1 - |\alpha\beta|)^2}.$$

Now $0 < 1 - |\alpha\beta| \leq |1 - \overline{\beta}\alpha|$. Thus $r_1 \leq r_2$. Hence

$$1 - \left|\frac{\alpha - \beta}{1 - \overline{\beta}\alpha}\right|^2 \leq 1 - \left(\frac{|\alpha| - |\beta|}{1 - |\alpha\beta|}\right)^2.$$

Thus

$$\left|\frac{\alpha - \beta}{1 - \overline{\beta}\alpha}\right|^2 \geq \left(\frac{|\alpha| - |\beta|}{1 - |\alpha||\beta|}\right)^2.$$

Taking positive square roots of the terms in both sides of the inequality we obtain the result. □

The above result can also be proved in the following way:

It is enough to assume $\alpha = a > 0$. Then

$$\left|\frac{a - re^{i\theta}}{1 - are^{i\theta}}\right|^2 = \frac{a^2 - 2ar\cos\theta + r^2}{1 - 2ar\cos\theta + a^2r^2} = \frac{A + x}{B + x}$$

where $A = \frac{a^2+r^2}{2ar}$; $B = \frac{1+a^2r^2}{2ar}$; $x = -\cos\theta$. The problem is to find the minimum of

$$\Phi(x) = \frac{A + x}{B + x}$$

for $-1 \le x \le 1$. But in this interval

$$\Phi'(x) = \frac{B - A}{(B + x)^2} > 0,$$

since $1 \le A < B$. Hence

$$\Phi(x) \ge \Phi(-1) = \frac{(a - r)^2}{(1 - ar)^2},$$

which proves the result.

For any $x = (x_1, x_2, \cdots, x_n)$ and $y = (y_1, y_2, \cdots, y_n)$ belong to $\mathbb{R}^n$, define $\|x\| = (x_1^2 + x_2^2 + \cdots + x_n^2)^{\frac{1}{2}}$ and $\langle x, y\rangle = \sum_{i=1}^n x_iy_i$. Thus $\|x\|^2 = \langle x, x\rangle$. The following holds.

Example 1.3. *For any nonzero x and y in $\mathbb{R}^n$,*

$$\left\| \frac{y}{\|y\|} - \|y\|x \right\| = \left\| \frac{x}{\|x\|} - \|x\|y \right\|.$$

*This result we call the **symmetry lemma** in the Euclidean plane.*

Solution: Let $x = (x_1, x_2, \cdots, x_n)$ and $y = (y_1, y_2, \cdots, y_n)$. Then $\|x\| = (x_1^2 + x_2^2 + \cdots + x_n^2)^{\frac{1}{2}}$ and $\|y\| = (y_1^2 + y_2^2 + \cdots + y_n^2)^{\frac{1}{2}}$. Now

$$\left\| \tfrac{y}{\|y\|} - \|y\|x \right\|^2 = \langle \tfrac{y}{\|y\|} - \|y\|x, \tfrac{y}{\|y\|} - \|y\|x\rangle$$

$$= \langle \tfrac{y}{\|y\|}, \tfrac{y}{\|y\|}\rangle + \langle \tfrac{y}{\|y\|}, -\|y\|x\rangle - \langle \|y\|x, \tfrac{y}{\|y\|}\rangle + \langle \|y\|x, \|y\|x\rangle$$

$$= 1 - \langle y, x\rangle - \langle x, y\rangle + \|y\|^2\|x\|^2$$

$$= 1 - \langle x, y\rangle - \langle x, y\rangle + \|x\|^2\|y\|^2 \qquad (\text{as } \langle x, y\rangle = \langle y, x\rangle)$$

$$= \langle \tfrac{x}{\|x\|} - \|x\|y, \tfrac{x}{\|x\|} - \|x\|y\rangle$$

$$= \left\| \tfrac{x}{\|x\|} - \|x\|y \right\|^2.$$

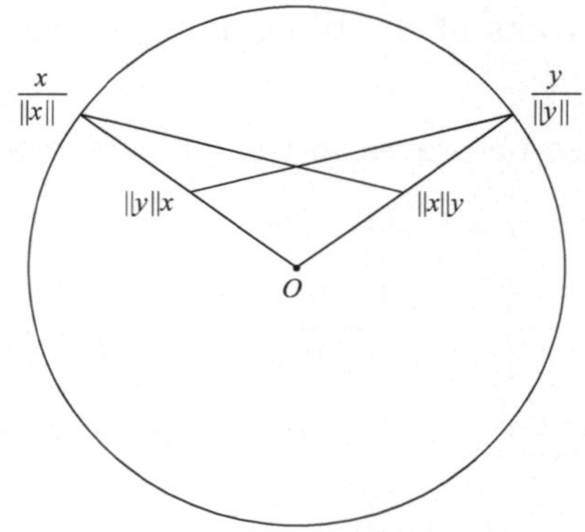

Fig. 1.4: Illustration of the symmetry lemma

In particular if z and w are two complex numbers then we have

$$\left| \frac{z}{|z|} - |z|w \right| = \left| \frac{w}{|w|} - |w|z \right|.$$

Theorem 1.7. *Let $z_1, z_2, \cdots$ be distinct points in the open unit disk $\mathbb{D}$. If there is a constant $c < 1$ such that*

$$1 - |z_{n+1}| \leq c(1 - |z_n|), n = 1, 2, \cdots$$

then

$$\left|\frac{z_k - z_j}{1 - \overline{z}_j z_k}\right| \geq \frac{1 - c^{j-k}}{1 + c^{j-k}}, j > k$$

and

$$\left|\frac{z_k - z_j}{1 - \overline{z}_j z_k}\right| \geq \frac{1 - c^{k-j}}{1 + c^{k-j}}, j < k.$$

Proof. The condition $1 - |z_{n+1}| \leq c(1 - |z_n|), n = 1, 2, \cdots$ implies that $1 - |z_j| \leq c^{j-k}(1 - |z_k|), j > k; k = 1, 2, \cdots$. Therefore it follows that for $j > k$

$$|z_j| - |z_k| \geq (1 - c^{j-k})(1 - |z_k|)$$

and

$$1 - |z_j z_k| = 1 - |z_j| + |z_j|(1 - |z_k|) \leq (1 + c^{j-k})(1 - |z_k|).$$

Hence

$$\begin{aligned} \left|\frac{z_k - z_j}{1 - \overline{z}_j z_k}\right| &\geq \frac{|z_j| - |z_k|}{1 - |z_j z_k|} \qquad \text{(by Lemma 1.2)} \\ &\geq \frac{1 - c^{j-k}}{1 + c^{j-k}}, j > k. \end{aligned}$$

For $j < k$, this inequality takes the form

$$\left|\frac{z_k - z_j}{1 - \overline{z}_j z_k}\right| \geq \frac{1 - c^{k-j}}{1 + c^{k-j}}, j < k.$$

□

Example 1.4. *For $a, b \in \mathbb{C}$,*

$$|a+b|^p \le \begin{cases} |a|^p + |b|^p & \text{if } 0 < p < 1; \\ 2^{p-1}(|a|^p + |b|^p) & \text{if } p > 1. \end{cases}$$

Solution: In these cases if either a or b is zero then there is nothing to prove. So without loss of generality we shall assume that neither a nor b is zero. When a, b are positive integers without loss of generality we shall assume $a \le b$ (so that $x = \frac{b}{a} \ge 1$ and $x < \infty$ as $a, b \ne 0$). Similarly when $a, b \in \mathbb{C}$, we assume without loss of generality that $|a| \le |b|$. It is enough to find the supremum of $g(x) = \frac{(1+x)^p}{1+x^p}$ for $x \ge 1$. Differentiation gives $g'(x) = p(1+x)^{p-1}(1+x^p)^{-2}(1-x^{p-1})$. Thus $g(x)$ is decreasing in the interval $1 \le x < \infty$ if $p > 1$ and $g(1) = 2^{p-1}$. Hence $g(x) \le 2^{p-1}$. Suppose a, b are arbitrary positive numbers and $p > 1$. Then $(a+b)^p = a^p(1+\frac{b}{a})^p \le a^p 2^{p-1}(1+(\frac{b}{a})^p) = 2^{p-1}(a^p + b^p)$. Now if $a, b \in \mathbb{C}$, then

$$\begin{aligned} |a+b|^p &\le (|a|+|b|)^p \text{ (by triangle inequality)} \\ &\le 2^{p-1}(|a|^p + |b|^p) \text{ if } p > 1. \end{aligned}$$

So for $p > 1$, the result follows. Again $g(x)$ is increasing in $1 \le x < \infty$ if $0 < p < 1$ and $g(x) \to 1$ as $x \to \infty$. Thus for $0 < p < 1$ and $1 \le x < \infty$, $g(x) = \frac{(1+x)^p}{1+x^p} \le 1$. So if a and b are arbitrary positive numbers then $(a+b)^p = a^p(1+\frac{b}{a})^p \le a^p(1+\frac{b^p}{a^p}) \le a^p + b^p$ if $0 < p < 1$. If $a, b \in \mathbb{C}$ and $0 < p < 1$ we obtain $|a+b|^p \le (|a|+|b|)^p \le |a|^p + |b|^p$.

Notice that it is not difficult to prove $|a+b|^p \le 2^p(|a|^p + |b|^p)$ if $p > 1$: Let $a, b \in \mathbb{C}$. Then either $|a| \le |b|$ or $|b| \le |a|$. Suppose $|a| \le |b|$. Then $|a+b| \le |a| + |b| \le 2|b|$. So $|a+b|^p \le 2^p(|b|^p) \le 2^p(|a|^p + |b|^p)$.

Example 1.5. *Let a_{jk} $(j, k = 1, 2, \cdots, n)$ be complex numbers such that $a_{kj} = \overline{a_{jk}}$ and*

$$\sum_{j=1}^{n} |a_{jk}| \le M, k = 1, 2, \cdots, n.$$

Then for any numbers $x_1, x_2, \cdots, x_n$ belonging to $\mathbb{C}$,

$$\left| \sum_{j,k=1}^{n} a_{jk} x_j \overline{x_k} \right| \le M \sum_{j=1}^{n} |x_j|^2.$$

Solution: Write $|a_{jk} x_j \overline{x_k}| = (|a_{jk}|^{\frac{1}{2}} |x_j|)(|a_{jk}|^{\frac{1}{2}} |x_k|)$ and apply the Cauchy-Schwarz inequality twice. Since the calculation is straightforward the details will be omitted.

Example 1.6. *For $a, b \in \mathbb{C}$ and for $0 < p \le 1$,*

$$\left| log^+|a| - log^+|b| \right| \le \frac{1}{p} \big| \, |a| - |b| \, \big|^p.$$

Solution:

$$\log^+ x = \begin{cases} \log x & \text{if } x \geq 1 \\ 0 & \text{if } 0 \leq x < 1. \end{cases}$$

It is enough to assume $1 \leq |b| < |a|$. The result then follows from the inequality $\log x \leq \frac{1}{p}(x-1)^p, x \geq 1$, by setting $x = \frac{|a|}{|b|}$. The latter inequality can be proved by noting that the difference of the two expressions has a positive derivative and vanishes at $x = 1$.

We shall now give the geometric description of the set of points z satisfying the following equations or inequalities.

(1) $|z - a| = r$.

Let $z = x + iy$ and $a = c + id$ where x, y, c, d are real. Since $|z - a| = r$, we obtain $(x - c)^2 + (y - d)^2 = r^2$ which represents a circle with center at $a = (c, d)$ and radius r.

(2) (i) $|z - a| < r$ and (ii) $|z - a| > r$.

The first inequality gives the interior of a circle of radius r with center at the point a and the second gives the exterior of the same circle.

(3) (i) $|z - 2| + |z + 2| = 5$ (ii) $|z - 2| - |z + 2| > 3$.

The equation in (i) represents an ellipse with foci at the points $z = \pm 2$ and semi major axis $\frac{5}{2}$. The inequality (ii) gives the interior of the left hand branch of a hyperbola with foci at points $z = \pm 2$ and real semi axis $\frac{3}{2}$. In general one can show that the set $E = \{z \in \mathbb{C} : |z - a| + |z - a| = 2c\}$ describes an ellipse. If $z \in E$, then we see that the sum of distances from the point z to the given points a and $-a$, is equal to constant $2c$. This means that E represents an ellipse with focii at $\pm a$. For instance if $a \in \mathbb{R}$, then the equation of the ellipse using rectangular coordinates is given by $\frac{x^2}{c^2} + \frac{y^2}{c^2 - a^2} = 1$. Its center is at the origin and its semi-major and semi-minor axes are equal to c and $\sqrt{c^2 - a^2}$ respectively.

If $a \in \mathbb{R}, z = x + iy$ and $|z - a| + |z - a| = 2c$ then $|x + iy - a| + |x + iy + a| = 2c$. That is, $|x + iy - a|^2 = (2c - |x + iy + a|)^2$. This implies, $(x - a)^2 + y^2 = 4c^2 + (x + a)^2 + y^2 - 4c|x + a + iy|$. Hence $(4c|x + a + iy|)^2 = (4c^2 + 4ax)^2$ and therefore $(c^2 - a^2)x^2 + c^2y^2 + c^2(c^2 - a^2)$. Thus $\frac{x^2}{c^2} + \frac{y^2}{c^2 - a^2} = 1$. Further, by triangle inequality, it follows that $|2a| = |z + a - (z - a)| \leq |z + a| + |z - a| = 2c$; hence $|a| \leq |c|$ and if $z \in E$, then $2|z| = |z + a + z - a| \leq |z + a| + |z - a| = 2c$. That is, $\max_{z \in E} |z| = c$. Here the maximum is attained at $z = ce^{-i\theta}$ where $\theta = \arg \frac{c}{a}$. Similarly it can be shown that the set $F = \{z \in \mathbb{C} : |z - a| - |z - a| = 2c\}$ describes a hyperbola with focii at $z = a, z = -a$.

(4) $|z - a| = |z - b|$.

This describes a straight line perpendicular to the segment joining points a and b and passing through the mid point of this segment.

(5) (i) $\text{Re}z \geq C$ and (ii) $\text{Im}z < C$.

The inequality (i) represents the set formed by the straight line $x = C$ and the half plane located to the right of it. The inequality (ii) represents the half plane located below the straight line $y = C$.

(6) $0 < \text{Re}(iz) < 1$.

This gives the strip $-1 < y < 0$.

(7) (i) $\theta < \arg z < \theta_0 \quad (-\pi < \theta < \theta_0 \leq \pi)$
(ii) $\theta < \arg(z - a) < \theta_0 \quad (-\pi < \theta < \theta_0 \leq \pi)$.

The inequality (i) gives us the interior of an angle (containing the positive real axis) with vertex at the origin and with sides forming angles θ and θ_0 with the real axis. Further (ii) represents the interior of a similar angle with vertex at point a.

(8) $|z| = \text{Re}z + 1$.

This represents the parabola $y^2 = 2x + 1$.

(9) $\text{Re}z + \text{Im}z < 1$.

This gives the half plane bounded by the straight line $x + y = 1$ and containing the origin.

(10) (i) $\text{Im}\frac{z-a}{z-b} = 0$ and (ii) $\text{Re}\frac{z-a}{z-b} = 0$.

The equation (i) describes the straight line passing through points a and b (with point b excluded) and the equation (ii) describes a circle whose diameter is a segment joining points a and b (with point b removed).

(11) $|2z| > |1 + z^2|$.

The inequality represents the set which is formed by the interior of the circles $|z - i| = \sqrt{2}$ and $|z + i| = \sqrt{2}$ with the exception of their common part.

1.5 Exercises

1. Verify the following: (i) $(1 + i)^8 = 16$; (ii) $\frac{(1+i)}{(2-i)(3+i)} = \frac{3}{25} + i\frac{4}{25}$; (iii) $\frac{52i}{(1-i)(3-2i)} = 2i - 10$; (iv) $\frac{i^4+i^9+i^{16}}{2-i^5+i^{10}-i^{15}} = 2 + i$; (v) $3(\frac{1+i}{1-i})^2 - 2(\frac{1-i}{1+i})^3 = -3 - 2i$.

2. Compute: (i) i^n for $n \in \mathbb{N}$; (ii) $\left(\frac{1-i}{1+i}\right)^n, n \in \mathbb{Z}$; (iii) $(1 + i)^n + (1 - i)^n$, $n \in \mathbb{Z}$.

3. (a) Find the real and imaginary part of (i) $\frac{1+z}{1-z}$, $z \neq 1$; (ii) $z + \frac{1}{z}$ $(z \neq 0)$;

   (b) Find real numbers x and y such that $2x - 3iy + 4ix - 2y - 5 - 10i = (x + y + z) - (y - x + 3)i$;

(c) If $z_1 = 1-i$, $z_2 = -2+4i$, $z_3 = \sqrt{3}-2i$, evaluate each of the following: (i) $|z_1\bar{z}_2 + z_2\bar{z}_1|$; (ii) Im $(\frac{z_1 z_2}{z_3})$; (iii) $\overline{(z_2 + z_3)(z_1 - z_3)}$.

4. (a) Find all the roots of $z^4 + z^2 + 1 = 0$. (b)Prove that the equation $z^4 + z + 4$ has no solution in $\{z| \, |z| < 1\}$.

5. Prove $|z+1| > |z-1|$ if and only if Re$z > 0$.

6. (a) Prove that $|\text{Re}\, z| + |\text{Im}\, z| \le \sqrt{2}|z|$; (b) Show that if Im$a > 0$ and Im$b > 0$, then $\left|\frac{a-b}{a-\bar{b}}\right| < 1$.

7. If $a \in \mathbb{C}$ with $|a| < 1$, prove that $\left|\frac{z-a}{1-\bar{a}z}\right| \le 1$ if and only if $|z| \le 1$. Equality holds if and only if $|z| = 1$.

8. If $|z| = 1$, prove that $\left|\frac{az+b}{\bar{b}z+\bar{a}}\right| = 1$ for all $a, b \in \mathbb{C}$.

9. If $|a| = 1$ and $b \ne a$ then show that $\left|\frac{a-b}{1-\bar{b}a}\right| = 1$.

10. (a) Prove that the equation of a circle in the complex plane is given by

$$z\bar{z} + a\bar{z} + \bar{a}z + b = 0.$$

(b) Show that the equation of an ellipse with major axis of length 10 and foci at $(-4, 0)$ and $(4, 0)$ is given by $|z+4| + |z-4| = 10$.

11. (a) Let p and q be distinct points of $\mathbb{C}$. Let $\lambda \ne 1$ be a positive real number. Then show that the set $\{z \in \mathbb{C} : |z-p| = \lambda|z-q|\}$ is a circle.

(b) Describe the locus represented by each of the following: (i) $|z-3i| = 5$; (ii) $|z+3i|+|z-3i| = 7$; (iii) $|z-4|-|z+4| = 5$; (iv) $z(\bar{z}+2) = 5$; (v) Im $z^2 = 4$.

12. Given complex numbers $a_1, \cdots, a_n$, show that

$$|1-(1+a_1)(1+a_2)\cdots(1+a_n)| \le e^{|a_1|+\cdots+|a_n|} - 1.$$

13. Let $\alpha > 0$ and Γ be the set of points z satisfying $|z-\alpha| = c\text{Re}z$. Show that Γ is

(i) an ellipse if $0 < c < 1$;

(ii) a parabola if $c = 1$;

(iii) a hyperbola if $1 < c < \infty$.

14. Give a geometric description of the set of points z satisfying the following equations or inequalities:

(i)	$\mathrm{Im}z > 0$	(ii)	$\lvert z-i\rvert < \lvert z+i\rvert$
(iii)	$1 < \lvert z\rvert < 2$	(iv)	$\lvert z\rvert < 1$ and $\mathrm{Im}z > 0$
(v)	$\lvert z-2+3i\rvert < 5$	(vi)	$\lvert z\rvert^2 = \mathrm{Im}z$
(vii)	$\frac{1}{z} = \overline{z}$	(viii)	$\left\lvert\frac{z-1}{z+1}\right\rvert = 1$
(ix)	$\lvert z-2\rvert > \lvert z-3\rvert$	(x)	$\mathrm{Re}[(2+3i)z] > 0$
(xi)	$\mathrm{Im}z \geq \mathrm{Re}z$	(xii)	$\mathrm{Re}z = \lvert z-2\rvert$
(xiii)	$\lvert z-i\rvert = \lvert z+1\rvert$	(xiv)	$\lvert z-1\rvert + \lvert z+1\rvert \leq 4$
(xv)	$\lvert z-3\rvert - \lvert z+3\rvert > 4$	(xvi)	$\lvert z\rvert > \lvert z-1\rvert$
(xvii)	$\lvert z\rvert^2 > z+\overline{z}$	(xviii)	$\lvert z+2\rvert > 1+\lvert z-2\rvert$

(xix) $|z| < \arg z$ if $0 < \arg z \leq 2\pi$

(xx) $|z+\frac{2}{3}| < \frac{1}{3}$ and $|z+\frac{1}{3}| < |z+1|$

(xxi) $|z| < \arg z$ if $0 \leq \arg z < 2\pi$

(xxii) $|z-a| + |z-b| < 1, a, b \in \mathbb{C}, |a-b| < 1$

(xxiii) $|\,|z-a| - |z-b|\,| > 1.$

15. Prove that if a polynomial with real coefficients has the complex root z, then it also has $\overline{z}$ as a root.

16. Let z_1, z_2, z_3 be points in the complex plane, with $z_1 \neq z_2$. Prove that the distance from z_3 to the line determined by z_1 and z_2 equals

$$\frac{1}{2|z_2 - z_1|}|z_1(\overline{z_2} - \overline{z_3}) + z_2(\overline{z_3} - \overline{z_1}) + z_3(\overline{z_1} - \overline{z_2})|;$$

in particular, the points z_1, z_2, z_3 are collinear if and only if $z_1(\overline{z_2} - \overline{z_3}) + z_2(\overline{z_3} - \overline{z_1}) + z_3(\overline{z_1} - \overline{z_2}) = 0$.

17. Prove that if z_1, z_2, z_3 are noncollinear points in the complex plane then the medians of the triangle with vertices z_1, z_2, z_3 intersect at the point $\frac{1}{3}(z_1 + z_2 + z_3)$.

18. Prove that the distinct complex numbers z_1, z_2, z_3 are the vertices of an equilateral triangle if and only if $z_1^2 + z_2^2 + z_3^2 = z_1z_2 + z_2z_3 + z_3z_1$.

19. For any nonzero $z \in \mathbb{C}$, prove that

$$\arg z = \begin{cases} \sin^{-1}(\frac{y}{|z|}) & \text{if } x \geq 0; \\ \pi - \sin^{-1}(\frac{y}{|z|}) & \text{if } x < 0 \text{ and } y \geq 0; \\ -\pi - \sin^{-1}(\frac{y}{|z|}) & \text{if } x < 0 \text{ and } y < 0. \end{cases}$$

20. Find the polar form of $-1+i, 1+i, \sqrt{3}+i, -1+\sqrt{3}i, -2-2i, \frac{1+i}{1-i}, 3i, 2, -3-4i, 1-2i, 1-\sqrt{3}i, -2i, -1, 2+i$.

21. Let $z = r(\cos\theta + i\sin\theta)$ and $w = R(\cos\phi + i\sin\phi), 0 < r < R$. Prove that

$$\mathrm{Re}\left(\frac{z+w}{z-w}\right) = \frac{|w|^2 - |z|^2}{|w-z|^2} = \frac{R^2 - r^2}{R^2 - 2Rr\cos(\theta - \phi) + r^2}.$$

22. Show that (i) $(1+i\sqrt{3})^{99} = -2^{99}$; (ii) $(\frac{1+i}{\sqrt{2}})^{10} = i$; (iii) $\frac{3i^{30}-i^{19}}{2i-1} = 1+i$; (iv) $(\frac{1+\sqrt{3}i}{1-\sqrt{3}i})^{10} = -\frac{1}{2} + \frac{\sqrt{3}}{2}i$.

23. Let $\theta = \arg z$ and $\phi = \arg w$.

    (a) Suppose $zw \neq 0$. Prove $z\bar{w} = |z|\,|w|$ if and only if $\theta - \phi = 2k\pi$ for some $k \in \mathbb{Z}$;

    (b) Suppose $zw \neq 0$. Prove $|z+w| = |z| + |w|$ if and only if $\theta - \phi = 2k\pi$ for some $k \in \mathbb{Z}$.

24. Find the 4th roots of unity and find all cube roots of i.

25. (a) Compute: (i) $i^{\frac{3}{2}}$; (ii) $(-1+i)^{\frac{1}{3}}$; (iii) $(-2\sqrt{3}-2i)^{\frac{1}{4}}$.

    (b) Find each of the indicated roots: (i) $i^{\frac{2}{3}}$; (ii) $(-16i)^{\frac{1}{4}}$; (iii) $(-4+4i)^{\frac{1}{5}}$.

26. Show that

$$\sum_{k=0}^{n} \cos k\theta = \frac{1}{2} + \frac{\sin(n+\frac{1}{2})\theta}{2\sin\frac{\theta}{2}};$$

$$\sum_{k=1}^{n} \sin k\theta = \frac{\cos\frac{\theta}{2} - \cos(n+\frac{1}{2})\theta}{2\sin\frac{\theta}{2}}.$$

(Try by making the substitution $z = \cos\theta + i\sin\theta$ in the identity $1 + z + z^2 + \cdots + z^n = \frac{z^{n+1}-1}{z-1}$, $z \neq 1$.)

27. Let $w = \cos\frac{2k\pi}{n} + i\sin\frac{2k\pi}{n}$ for an integer $n > 1$. Show that $\sum_{k=1}^{n} w^k = 0$.

28. Describe the set $\{z \in \mathbb{C} : |z^2 - 1| < 1\}$.

29. Use De Moivre's formula to find expressions for $\cos 3\theta$, $\sin 3\theta$, $\cos 5\theta$, $\sin 5\theta$ as polynomials in $\cos\theta$ and $\sin\theta$.

30. Let w be an n-th root of 1 different from 1 itself. Establish the formulas

    (i) $1 + 2w + 3w^2 + \cdots + nw^{n-1} = \frac{n}{w-1}$,

    (ii) $1 + 4w + 9w^2 + \cdots + n^2 w^{n-1} = \frac{n^2}{w-1} - \frac{2n}{(w-1)^2}$.

31. Find the value of $\cos 144^{\circ}$ algebraically.

32. If $w_1, w_2, \cdots, w_{n-1}$ are the roots of $z^n = 1$ besides $z = 1$, evaluate $(1-w_1)(1-w_2)\cdots(1-w_{n-1})$.

33. Show that the sets $\mathbb{R} \times \mathbb{R}$ and $\mathbb{C}$ are uncountable and $\text{card}(\mathbb{R} \times \mathbb{R}) = \text{card}(\mathbb{C}) = \text{card}(\mathbb{R})$ where $\text{card}(A)$ means the cardinality of the set A.

34. Determine families of lines in the complex plane specified by each of the following equations:

    (i) $\left|\frac{z-a}{z-b}\right| = \lambda$, $(\lambda > 0)$ (ii) $\arg\frac{z-a}{z-b} = \alpha$ $(-\pi < \alpha \leq \pi)$

    (iii) $\text{Re}z^2 = C$, $-\infty < C < \infty$ (iv) $\text{Im}z^2 = C$, $-\infty < C < \infty$

    (v) $\text{Re}\frac{1}{z} = C$, $-\infty < C < \infty$ (vi) $\text{Im}\frac{1}{z} = C$, $-\infty < C < \infty$.

35. Prove that

$$\left|\sum_{k=0}^{n} \cos k\theta\right| \leq \frac{1}{|\sin(\frac{\theta}{2})|}$$

and

$$\left|\sum_{k=0}^{n} \sin k\theta\right| \leq \frac{1}{|\sin(\frac{\theta}{2})|}$$

for all $n \in \mathbb{N}$ and $\theta \neq 2k\pi, k \in \mathbb{Z}$. (Hint: Use triangle inequality in the identity of exercise 26.)

36. Describe graphically the region bounded by each of the following:

    (i) $1 < |z+i| \leq 2$;

    (ii) Re $z^2 > 1$;

    (iii) $|z+3i| > 4$;

    (iv) $|z+2-3i| + |z-2+3i| < 15$.

37. Find an equation of the circle passing through the points $1-i, 2i, 1+i$.

38. Show that the locus of z such that $|z-b||z+b| = b^2, b > 0$ is a lemniscate.

39. Let $p, q \in \mathbb{C}$ such that $p \neq q$ and let L be the perpendicular bisector of the line joining p and q. Prove that an equation of L is $\bar{\beta}z + \beta\bar{z} + \gamma = 0$ where $\beta = q-p$ and $\gamma = |p|^2 - |q|^2$. (Hint: L is given by $|z-p|^2 = |z-q|^2$; that is, by $(z-p)(\bar{z}-\bar{p}) = (z-q)(\bar{z}-\bar{q})$.)

40. Let $az^2 + bz + c$ be a polynomial with nonzero complex coefficients. Show that the zeros of the polynomial lie in the closed disk $|z| \leq \left|\frac{b}{a}\right| + \left|\frac{c}{b}\right|$.

41. If a and b are two complex numbers with $|\arg a - \arg b| \leq \theta \leq \pi$, then prove that for every positive integer n,

$$|a-b|^n \leq (|a|^n + |b|^n)\max(1, 2^{n-1}\sin^n\frac{\theta}{2}).$$

42. If z_1 and z_2 are complex numbers, and if c is a positive number, then prove that $|z_1+z_2|^2 \leq (1+c)|z_1|^2 + (1+\frac{1}{c})|z_2|^2$, with equality if and only if $z_2 = cz_1$.

43. If $a_1, \cdots, a_n$ are positive numbers such that $\sum_{k=1}^{n}\frac{1}{a_k} = 1$, then prove that $|z_1+\cdots+z_n|^2 \leq a_1|z_1|^2 + \cdots + a_n|z_n|^2$ for complex numbers $z_1, z_2, \cdots, z_n$.

44. If a and b are nonzero complex numbers, then prove that

$$|a-b| \geq \frac{1}{2}(|a|+|b|)\left|\frac{a}{|a|} - \frac{b}{|b|}\right|$$

and equality holds if and only if $|a| = |b|$.

45. If Re$z \geq 1$, then prove that for any positive integer n,

$$\left|z^{n+1} - 1\right| > |z|^n|z - 1|.$$

46. If z and w are complex numbers and $\lambda \geq 2$, then prove that

$$2(|z|^\lambda + |w|^\lambda) \leq |z + w|^\lambda + |z - w|^\lambda.$$

47. Let $z_1, \cdots, z_n$ be arbitrary complex numbers. Prove that there exists a subset of $\{1, \cdots, n\}$ such that

$$\left|\sum_{k \in M} z_k\right| \geq \frac{1}{4\sqrt{2}} \sum_{k=1}^{n} |z_k|.$$

48. If $z_1, \cdots, z_n$ are complex numbers then prove that there exists a subset I of $\{1, \cdots, m\}$ such that

$$\left|\sum_{k \in I} z_k\right| \geq \frac{1}{\pi} \sum_{k=1}^{m} |z_k|.$$

Chapter 2

Metric Spaces and the Topology of $\mathbb{C}$

In this chapter we develop the fundamentals of point set topology and metric spaces. Our main concern is to derive all topological properties of $\mathbb{C}$ that are essential for the study of complex-valued functions.

"Nearness" can be defined if we have a concept of "distance". Let us try to analyze in what context we use the term distance. We say distance between Sambalpur and Bhubaneswar is 320 kilometers. We don't say distance of Bhubaneswar is 320 kilometers. In other words distance is a number associated with a pair of objects. Thus if X is a set, a distance on X ought to be a rule which associates to a pair $(x, y), x, y \in X$, a number. How we associate the numbers depends on the context.

In real life we, sometimes, do come across cases when travelling between two points is dependent on the order. Take a part of a road where there is a restriction of travelling between two places x and y. One can travel from x to y but if one has to travel from y to x then one has to take a detour. Hence distance from x to y may not be the same as that between y and x.

Further when we say the sequence $\{x_n\}$ of real numbers converges to x as $n \to \infty$ we mean that x_n can be brought as near to x as we like by choosing n sufficiently large. Thus when we defined the concept of convergence on $\mathbb{R}$ we tacitly assumed that there was a concept of nearness on the scheme of real numbers.

Topological considerations are extremely important for the foundation of the study of complex valued functions, and the first systematic study of topology was motivated by this need.

2.1 Elementary of metric spaces

Definition 2.1. *A metric space is a pair (X, d) where X is a set and d is a function from $X \times X$ into $\mathbb{R}$, called a distance function or metric, which satisfies the following conditions for x, y, and z in X:*

(i) $d(x, y) \geq 0$,

(ii) $d(x, y) = 0$ *if and only if* $x = y$,

(iii) $d(x, y) = d(y, x)$,

(iv) $d(x, z) \leq d(x, y) + d(y, z)$ *(triangle inequality).*

Example 2.1. *The following are some examples of metric spaces:*

1. *Let* $X = \mathbb{R}$ *and define* $d(x, y) = |x - y|$. *Then* $(\mathbb{R}, d)$ *is a metric space.*

2. *Let* $X = \mathbb{C}$ *and define (i)* $d(z, w) = |z - w|$; *(ii)* $\rho(z, w) = |Rez - Rew| + i|Imz - Imw|$; *(iii)* $D(z, w) = max\{|Rez - Rew|, |Imz - Imw|\}$.

Then $(\mathbb{C}, d), (\mathbb{C}, d')$ *and* $(\mathbb{C}, D)$ *are metric spaces. So we can define more than one metric on a set.*

3. *Let* $X = \mathbb{R}^k$. *If* $x = (x_1, \cdots, x_k), y = (y_1, \cdots, y_k)$, *define* $d(x, y) = (\sum_{i=1}^{k} |x_i - y_i|^2)^{\frac{1}{2}}$. *Then* d *is a metric on* X *called the Euclidean metric on* X.

4. *If* (X, d) *is a metric space and* $Y \subset X$ *then* (Y, d) *is also a metric space.*

5. *Given a set* X, *define* $d(x, y) = 1$ *if* $x \neq y$; $d(x, y) = 0$; *if* $x = y$. *It is not difficult to check that d is a metric on* X.

6. *Let* X *be a metric space with metric d. Define* $\overline{d}$: $X \times X \to \mathbb{R}$ *by the equation* $\overline{d}(x, y) = min\{d(x, y), 1\}$. *Then* $\overline{d}$ *is a metric on* X. *The metric* $\overline{d}$ *is called the standard bounded metric corresponding to d.*

7. *Let* X *be a metric space with metric d. Define* $d'(x, y) = \frac{d(x,y)}{1+d(x,y)}$. *Then* d' *is a metric on* X.

Thus if a metric d is defined on a set X it gives rise to many other metrics on X. For example, see (6) and (7) in Example 2.1.

Definition 2.2. *Let* X *be a metric space with metric d. A subset* E *of* X *is said to be bounded if there is some number* M *such that* $d(x, y) \leq M$ *for all* $x, y \in E$. *If* E *is bounded, the diameter of* E *is defined to be the number* $diamE = lub\{d(x, y) : x, y \in E\}$.

Given a metric d on X, the number $d(x, y)$ is often called the distance between x and y in the metric d. Given $\epsilon > 0$, consider the set $B_d(x, \epsilon) = \{y \in X : d(x, y) < \epsilon\}$. It is called the open ball with center at x and radius ϵ or a neighborhood of x with radius ϵ. Sometimes we omit the metric d from the notation and write simply $B(x, \epsilon)$, when no confusion will arise. Further, suppose $Y \subset X$, then (Y, d) is also a metric space. If $x \in Y$, we shall write $B_X(x, \epsilon) = \{y \in X : d(x, y) < \epsilon\}$, to denote the open ball in X with center at x and radius ϵ and $B_Y(x, \epsilon) = \{y \in Y : d(x, y) < \epsilon\}$ to denote the open ball in Y with center at x and radius ϵ. When there is no confusion about the sets we simply write $B(x, \epsilon)$.

Example 2.2. *If* $X = \mathbb{C}$, *then for* $a \in \mathbb{C}$ *and* $r > 0$, $B(a,r) = \{z \in \mathbb{C} : |z-a| < r\}$.

Example 2.3. *Take* $X = \mathbb{R}$. *Then for* $a \in \mathbb{R}$ *and* $r > 0$, $B(a,r) = (a-r, a+r) = \{x \in \mathbb{R} : a-r < x < a+r\}$.

Definition 2.3. *Let* X *be a metric space with metric* d. *Let* $E \subset X$.

(a) *A point* $x \in X$ *is an interior point of* E *if there exists* $r > 0$ *such that* $B(x,r) \subset E$. *Thus if* x *is an interior point of* E *then* $x \in E$.

(b) *The set* E *is open if every point of* E *is an interior point of* E. *Let* E^0 *be the set of all interior points of* E. *Hence* $E^0 \subset E$. *Thus* E *is open in* X *if and only if* $E^0 = E$.

(c) *If* E *is open and* $x \in E$, *then* E *is called a neighborhood of* x. *In fact* E *is a neighborhood of each of its points.*

(d) *A point* $p \in X$ *is a limit point (sometimes called accumulation point or cluster point) of the set* E *if every neighborhood* p *contains a point of* E *other than* p.

(e) *If* $p \in E$ *and* p *is not a limit point of* E, *then* p *is called an isolated point of* E.

(f) *The set* E *is closed if it contains all its limit points. Let* E' *be the set of all limit points of* E *in* X. *The set* E' *is called the derived set of* E *in* X. *Thus the set* E *is closed in* X *if* $E' \subset E$.

Remark 2.1. Thus in a metric space a set is open if and only if it contains a neighborhood of each of its points. In other words, any open subset of a metric space is a union of open balls.

Theorem 2.1. *Let* (X,d) *be a metric space. Let* $x \in X$ *and* $\epsilon > 0$. *Then* $B(x,\epsilon)$ *is an open subset of* X.

Proof. Let $y \in B(x,\epsilon)$. Hence $d(x,y) < \epsilon$. We have to show y is an interior point of $B(x,r)$. Let $\delta = \epsilon - d(x,y)$. Then $\delta > 0$ and $B(y,\delta) \subset B(x,\epsilon)$. This is because if $z \in B(y,\delta)$, then $d(y,z) < \delta = \epsilon - d(x,y)$, from which we conclude that $d(x,z) \le d(x,y) + d(y,z) < \epsilon$. See Fig. 2.1. □

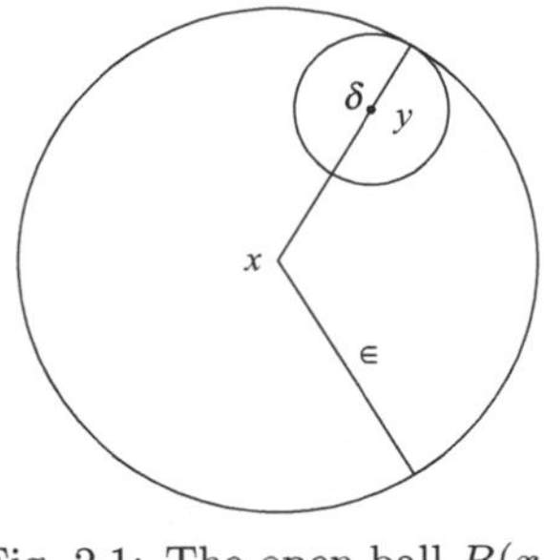

Fig. 2.1: The open ball $B(x,\epsilon)$

Example 2.4. *The set $\mathcal{H} = \{z \in \mathbb{C} : Imz > 0\}$ is open in $\mathbb{C}$. To see this, let $z \in \mathcal{H}$. Let $r = Imz > 0$. We show that $B(z,r) \subset \mathcal{H}$. Let $w \in B(z,r)$. From the inequalities $|Imw - Imz| \leq |w - z| < r$, we get $-r < Imw - Imz < r$. Hence $0 < Imw < 2r$. Thus $Imw > 0$ and $w \in \mathcal{H}$.*

Example 2.5. *It is not difficult to verify that the set $\mathbb{C}^* = \mathbb{C} - \{0\}$, $B(z,r) \setminus \{z\}$ are open in $\mathbb{C}$.*

Theorem 2.2. *Every open interval is an open subset of $\mathbb{R}$ but not open in $\mathbb{C}$.*

Proof. Let $(a,b) = \{x \in \mathbb{R} : a < x < b\}$ be an open interval in $\mathbb{R}$. Let $c = \frac{a+b}{2}$ and $r = \frac{b-a}{2}$. Then

$$B_{\mathbb{R}}(c,r) = \{x \in \mathbb{R} : |x - c| < r\} = (c - r, c + r) = (a,b).$$

Hence (a,b) is open in $\mathbb{R}$ by Theorem 2.1. To show that (a,b) is not open in $\mathbb{C}$, take $x_0 \in (a,b)$ and consider the open disk $B_{\mathbb{C}}(x_0,\epsilon)$ in $\mathbb{C}$. It is obvious that $B_{\mathbb{C}}(x_0,\epsilon) \not\subset (a,b)$; for example, $z = x_0 + i\frac{\epsilon}{2}$ belongs to $B_{\mathbb{C}}(x_0,\epsilon)$ but it does not belong to (a,b). □

Definition 2.4. *Let (X,d) be a metric space and $E \subset X$. The interior of E is defined as the union of all open sets contained in E, and the closure of E is defined as the intersection of all closed sets containing E. The interior of E is denoted as $IntE$ and the closure of E is denoted by clE or by $\overline{E}$. Thus if E is open then $IntE = E$; while if E is closed, $E = \overline{E}$.*

Theorem 2.3. *Let (X,d) be a metric space and $E \subset X$. Then $E^0 = IntE$.*

Proof. Let $x \in \text{Int}E$. By definition of Int$E$ there exists an open set U contained in E such that $x \in U$. The set U is open and so all its elements are interior points of it. Hence x is an interior point of U. Thus there exists $\epsilon > 0$ such that $B(x,\epsilon) \subset U$. But $U \subset E$. It follows therefore that $B(x,\epsilon) \subset E$. Hence $x \in E^0$. Thus we obtain

$$\text{Int}E \subset E^0. \tag{2.1}$$

Conversely, suppose $x \in E^0$. Then there exists $\epsilon > 0$ such that $B(x,\epsilon) \subset E$. By Theorem 2.1, $B(x,\epsilon)$ is an open set. Hence $B(x,\epsilon) \subset \text{Int}E$. This implies

$$x \in \text{Int}E. \tag{2.2}$$

From (2.1) and (2.2), Int$E = E^0$. □

The reader must note that E^0 is the largest open set contained in E and $\overline{E}$ is the smallest closed set containing E.

Theorem 2.4. *Let (X,d) be a metric space; let E be a subset of X. Then the point $x \in X$ is a limit point of E if and only if every neighborhood of x contains infinitely many points of E.*

Proof. Suppose there is a neighborhood $B(x,\epsilon)$ $(\epsilon > 0)$ of x which contains only a finite number of points of E. Let $B(x,\epsilon) \cap E/\{x\} = \{q_1, q_2, \cdots, q_n\}$. Let

$$r = \min\{d(x,q_1), d(x,q_2), \cdots, d(x,q_n)\}.$$

Then $r > 0$ since $q_i \neq x$ for $1 \leq i \leq n$. It follows therefore that $B(x,r)$ contains no point of E other than x. Thus x is not a limit point of E which is a contradiction. □

Corollary 2.1. *Let (X,d) be a metric space and let E be a finite subset of X. Then E has no limit points.*

Definition 2.5. *Let (X,d) be a metric space and let $E \subset X$. The boundary of E is the closure of E minus the interior of E in X. A point $x \in X$ belongs to the boundary of E if and only if all its neighborhoods intersect both E and E^C. We use the notation BdE or ∂E to denote boundary of E. The exterior of E is the interior of E^C. It is also the complement of the closure of E.*

Throughout let (X,d) be a metric space and let E be a subset of X.

Theorem 2.5. *A set E is open if and only if E^c(the complement of E in X) is closed.*

Proof. Follows from the definition.

□

Corollary 2.2. *A set F is closed if and only if its complement is open.*

Proof. Let $F^c = E$. Therefore $E^c = F$. From Theorem 2.5, E^c is closed if and only if $E = F^c$ is open. □

Remark 2.2. Note $X, \emptyset$ are open in X. Therefore both X and $\emptyset$ are closed in X.

Theorem 2.6. *Let I be an indexing set and (X,d) be a metric space. Then*

(i) For any collection $\{U_\alpha\}_{\alpha\in I}$ of open sets of X, $\cup_{\alpha\in I}U_\alpha$ is open in X.

(ii) For any collection $\{F_\alpha\}_{\alpha\in I}$ of closed sets of X, $\cap_{\alpha\in I}F_\alpha$ is closed in X.

(iii) For any finite collection $U_1, U_2, \cdots, U_n$ of open sets of X, $\cap_{i=1}^n U_i$ is open in X.

(iv) For any finite collection F_1, F_2, $\cdots$, F_n of closed sets of X, $\cup_{i=1}^n F_i$ is closed in X.

Proof. Let $U = \cup_{\alpha\in I}U_\alpha$. If $x \in U$, then $x \in U_\alpha$ for some $\alpha \in I$. Since U_α is open hence x is an interior point of U_α. Thus there exists $r > 0$ such that $B(x,r) \subset U_\alpha$. But $U_\alpha \subset U$. Hence $B(x,r) \subset U$. This implies x is an interior point of U and the set U is open. This proves (i).

Since

$$(\cap_{\alpha\in I}F_\alpha)^c = \cup_{\alpha\in I}F_\alpha^c \tag{2.3}$$

and F_α^c is open (by Theorem 2.5), hence (i) implies that (2.3) is open so that $\cap_{\alpha\in I}F_\alpha$ is closed.

To prove (iii) let $V = \cap_{i=1}^n U_i$. Let $x \in V$. This implies $x \in U_i$ for all $i = 1, 2, \cdots, n$. Hence there exists open balls $B(x, r_i), r_i > 0$ such that $x \in B(x, r_i) \subset U_i$ for all $i = 1, 2, \cdots, n$. Let $r = \min\{r_1, r_2, \cdots, r_n\}$. Then it follows that $x \in B(x, r) \subset U_i$ for all $i = 1, 2, \cdots, n$. Hence $x \in B(x, r) \subset \cap_{i=1}^n U_i = V$ and the set V is open. By taking complements, (iv) follows from (i): $(\cup_{i=1}^n F_i)^c = \cap_{i=1}^n F_i^c$. □

The property (iii) in Theorem 2.6 does not hold if finite intersection is replaced by arbitrary intersection.

Example 2.6. *(a) Let $U_m = \left(-\frac{1}{m}, \frac{1}{m}\right), m \in \mathbb{N}$ be a sequence of open intervals and so they are open in $\mathbb{R}$ by Theorem 2.1. But their intersection $\cap_{m=1}^\infty \left(-\frac{1}{m}, \frac{1}{m}\right) = \cap_{m=1}^\infty U_m = \{0\}$ is a singleton set and hence it is not open in $\mathbb{R}$. In fact $\{0\}$ is closed in $\mathbb{R}$. Similarly, let $V_m = \left\{z \in \mathbb{C} : |z| < \frac{1}{m}\right\}, m \in \mathbb{N}$. The sets V_m are open in $\mathbb{C}$ but $\cap_{m=1}^\infty V_m = \{0\}$ is closed in $\mathbb{C}$.*

In the following we demonstrate that the result of (iv) in Theorem 2.6 is false if the term "finite union" is replaced by "arbitrary union".

(b) Let $F_n = \left[\frac{1}{n}, 1 - \frac{1}{n}\right], n \geq 3$ be a sequence of closed intervals. By Theorem 2.5, F_n is a closed set for each n. But

$$\cup_{n=3}^\infty F_n = \cup_{n=3}^\infty \left[\frac{1}{n}, 1 - \frac{1}{n}\right] = (0, 1)$$

which is open.

Definition 2.6. *A topology on a set X is a collection $\mathcal{T}$ of subsets of X having the following properties:*

(1) $\emptyset$ and X are in $\mathcal{T}$.

(2) The union of the elements of any subcollection of $\mathcal{T}$ is in $\mathcal{T}$.

(3) The intersection of the elements of any finite subcollection of $\mathcal{T}$ is in $\mathcal{T}$.

A set X for which a topology $\mathcal{T}$ has been specified is called a topological space. If X is a topological space with topology $\mathcal{T}$, we say that a subset U of X is an open set of X if U belongs to the collection $\mathcal{T}$.

Definition 2.7. *Let X be a topological space with topology $\mathcal{T}$. If Y is a subset of X, the collection $\mathcal{T}_Y = \{Y \cap U : U \in \mathcal{T}\}$ is a topology on Y, called the subspace topology. With this topology, Y is called a subspace of X; its open sets consist of all intersections of open sets of X with Y.*

Definition 2.8. *If X is a set, a basis for a topology on X is a collection $\mathcal{B}$ of subsets of X(called basis elements) such that (1) For each $x \in X$, there is at least one basis element B containing x. (2) If x belongs to the intersection of two basis elements B_1 and B_2, then there is a basis element B_3 containing x such that $B_3 \subset B_1 \cap B_2$.*

Definition 2.9. *If $\mathcal{B}$ is a basis for a topology on X, the topology $\mathcal{T}$ generated by $\mathcal{B}$ is described as follows: A subset U of X is said to be open in X(that is, to be an element of $\mathcal{T}$) if for each $x \in U$, there is a basis element $B \in \mathcal{B}$ such that $x \in B$ and $B \subset U$.*

Note that each element of $\mathcal{B}$ is open in X under this definition, so that $\mathcal{B} \subset \mathcal{T}$. It is easy to check that this collection of subsets of X is a topology on X.

Definition 2.10. *If d is a metric on the set X, then the collection of all $\epsilon-$balls $B_d(x, \epsilon)$, for $x \in X$ and $\epsilon > 0$, is a basis for a topology on X, called the metric topology induced by d.*

The first condition for a basis is trivial, since $x \in B(x, \epsilon)$ for any $\epsilon > 0$. Before checking the second condition for a basis, we show that if y is a point of the basis element $B(x, \epsilon)$, then there is a basis element $B(y, \delta)$ centered at y that is contained in $B(x, \epsilon)$. Define δ to be the positive number $\epsilon - d(x, y)$. Then $B(y, \delta) \subset B(x, \epsilon)$, for if $z \in B(y, \delta)$, then $d(y, z) < \epsilon - d(x, y)$, from which we conclude that $d(x, z) \leq d(x, y) + d(y, z) < \epsilon$. Now to check the second condition for a basis, let B_1 and B_2 be two basis elements and let $y \in B_1 \cap B_2$. We have just shown that we can chose positive numbers δ_1 and δ_2 so that $B(y, \delta_1) \subset B_1$ and $B(y, \delta_2) \subset B_2$. Letting δ be the smaller of δ_1 and δ_2, we conclude that $B(y, \delta) \subset B_1 \cap B_2$.

Thus every metric space is a topological space. Using what we have just proved, we can rephrase the definition of the metric topology as follows:

Remark 2.3. A set U is open in the metric topology induced by d if and only if for each $y \in U$, there is a $\delta > 0$ such that $B_d(y, \delta) \subset U$.

Clearly this condition implies that U is open. Conversely, if U is open, it contains a basis element $B = B_d(x, \epsilon)$ containing y, and B in turn contains a basis element $B_d(y, \delta)$ centered at y. Thus the two definition of open sets coincide (see Definition 2.9, Definition 2.10, Remark 2.3) in a metric space.

Definition 2.11. *A topological space X is called a Hausdorff space if for each pair x, y of distinct points of X, there exist open sets U and V of X such that $x \in U, y \in V$ and $U \cap V = \emptyset$.*

Every metric space (X, d) is a Hausdorff space because if $x, y \in X$ and $x \neq y$, then $d(x, y) > 0$. Let $\delta = \frac{d(x,y)}{3}$. Then $\delta > 0$. Now $B(x, \delta)$ and $B(y, \delta)$ are open sets containing x and y respectively and $B(x, \delta) \cap B(y, \delta) = \emptyset$.

We define the set $\{z \in X : d(z,a) \leq r, r > 0\} = \overline{B(a,r)}$ as the closed disk with center at a and radius r. Note that the closed disk in $\mathbb{R}$ reduces to a closed interval $[a-r, a+r]$.

Theorem 2.7. *Every closed disk in X is a closed set in X.*

Proof. The proof follows from Theorem 2.5. □

Remark 2.4. If a set is not open, then it is not necessary that it should be closed. In fact there are sets which are both open and closed. Such sets are called clopens. For example, X and ϕ are clopens in X. There are also sets which are neither open nor closed. For example, consider the set $E = \{z \in \mathbb{C} : \text{Re}z > 0\} \cup \{0\}$. The set is neither open nor closed. The set is not open because $B(0,\epsilon)$ is not contained in this set, no matter how small we choose ϵ. The set is not closed because the point $z = i$ is a limit point of the set E but $i \notin E$.

Example 2.7. *The set $U = (-\infty, 1) \cup \{2\}$ is neither open nor closed in $\mathbb{R}$. The interval $(2-\epsilon, 2+\epsilon)$ however small $\epsilon > 0$ may be, contains other points of $\mathbb{R}$ which are not in U. So 2 is not an interior point of U. Hence U is not open. Next, we observe that $U^c = [1,\infty) \cap (\mathbb{R} - \{2\}) = [1,\infty)$ but the interval $[1,\infty)$ is not open in $\mathbb{R}$ as the point 1 is not an interior point. Hence U^c is not open and this implies that U is not closed in $\mathbb{R}$.*

Theorem 2.8. *Let A be a subset of the metric space (X,d). Then $x \in \overline{A}$ if and only if for every $\epsilon > 0$, $B(x,\epsilon) \cap A \neq \emptyset$.*

Proof. The proof follows from the definition of $\overline{A}$.

□

Thus we can make the following definition.

Definition 2.12. *Let (X,d) be a metric space and $A \subset X$. A point $x \in X$ is said to be a limit point of A if $B(x,r) \cap (A - \{x\}) \neq \phi$ for all $r > 0$. (That is, if every neighborhood of x intersects A at a point different from x.) A point $x \in X$ belongs to $\overline{A}$ if every neighborhood of x intersects A.*

It is not difficult to verify the following:

(i) The limit points of the set $\mathbb{D} = \{z \in \mathbb{C} : |z| < 1\}$ are the points of the set $\overline{\mathbb{D}} = \{z \in \mathbb{C} : |z| \leq 1\}$.

(ii) The limit points of the set $A = \{z \in \mathbb{C} : z = 1 + (-1)^n \frac{n}{n+1}, n \in \mathbb{N}\}$ are $z = 0$ and $z = 2$.

(iii) The limit points of the set $B = \{z \in \mathbb{C} : z = \frac{1}{m} + \frac{i}{n}, m, n \in \mathbb{Z}\}$ are $z = 0, z = \frac{1}{m}, z = \frac{i}{n}$ (m and n any integers).

(iv) The limit points of the set $C = \{z \in \mathbb{C} : z = \frac{p}{m} + i\frac{q}{n}, m, n, p, q \in \mathbb{Z}\}$ are all points of the complex plane $\mathbb{C}$.

Theorem 2.9. *Let A be a subset of the metric space X; let A' be the derived set of A. Then $\overline{A} = A \cup A'$.*

Proof. The proof follows from the definition of $\overline{A}$ and A'. □

Corollary 2.3. *A subset of a metric space is closed if and only if $\overline{A} = A$.*

Proof. The result follows from Theorem 2.9 because the set A is closed if and only if $A' \subset A$. □

Remark 2.5. The reader must note the following:

If A is a subset of a metric space X then (1) $\overline{A}$ is a closed set (follow from the definition). (2) $A = \overline{A}$ if and only if A is closed. (3) $\overline{A}$ is the smallest closed set containing A, that is if F is a closed subset of X containing A then $\overline{A} \subset F$.

Theorem 2.10. *Every finite point set in a metric space X is closed.*

Proof. Let $A = \{x_0, x_1, \cdots, x_n\}$ be a finite set in X. Then $A = \cup_{i=0}^{n}\{x_i\}$. By Theorem 2.6, it suffices to show that every singleton set $\{y\}$ in X is closed in X. If x is a point of X different from y, then x and y have disjoint neighborhoods U and V, respectively. This is possible because X is Hausdorff. Since U does not intersect $\{y\}$, the point x cannot belong to the closure of the set $\{y\}$. Thus $\overline{\{y\}} = \{y\}$. Hence $\{y\}$ is closed. □

Remark 2.6. Let (X, d) be a metric space. Suppose $Y \subset X$. Then Y is also a metric space with the metric d. Let $E \subset Y$. To say E is open in X means that given any $x \in X$, there exists $\epsilon > 0$ such that $\{y \in X : d(x, y) < \epsilon\} = B_X(x, \epsilon) \subset E$. We shall say E is open relative to Y if to each $x \in E$, there is associated an $\epsilon > 0$ such that $B_Y(x, \epsilon) = \{y \in Y : d(x, y) < \epsilon\} \subset Y$. Note here that a set may be open relative to Y without being an open subset of X. For example, we have already shown in Theorem 2.2, that the open interval (a, b) is open in $\mathbb{R}$ but not open in $\mathbb{C}$. Thus we observe that if $E \subset Y \subset X$, then E may be open relative to Y without being open relative to X. The property of being open thus depends on the space in which E is embedded.

However, there is a simple relation between these concepts which we now state.

Theorem 2.11. *Let (X, d) be a metric space. Suppose $Y \subset X$. A subset E of Y is open relative to Y if and only if $E = Y \cap U$ for some open subset U of X.*

Proof. Suppose the set E is open relative to Y. To each $p \in E$, there exists $r_p > 0$ such that $B_Y(p, r_p) \subset E$. Let $V_p = B_X(p, r_p)$ and $U = \cup_{p \in E} V_p$. Then U is open in X by Theorem 2.1 and Theorem 2.6 and $V_p \cap Y = B_Y(p, r_p) \subset E$ for every $p \in E$. Now let $p \in E$, then $p \in Y$ and $p \in V_p$. Since $V_p \subset U$, it follows that $p \in U \cap Y$. So it is clear that

$$E \subset U \cap Y. \tag{2.4}$$

Since $V_p \cap Y \subset Y$ for all $p \in E$ hence

$$U \cap Y = (\cup_{p\in E} V_p) \cap Y = \cup_{p\in E}(V_p \cap Y) \subset E. \tag{2.5}$$

Thus $E = U \cap Y$ and one half of the theorem is proved.

Conversely, let U be open in X and $E = U \cap Y$. Suppose $p \in E$. Then $p \in U$. Since U is open, there exists $r_p > 0$ such that $B_X(p, r_p) \subset U$. Then $B_X(p, r_p) \cap Y \subset U \cap Y = E$. Thus $B_Y(p, r_p) \subset E$. Hence E is open relative to Y. □

Similar result also holds for closed sets.

Theorem 2.12. *Let (X, d) be a metric space and let Y be a subset of X. A subset A of Y is closed relative to Y if and only if $A = C \cap Y$ for some closed subset C of X.*

Proof. The proof follows from Theorem 2.11.

□

Theorem 2.13. *Let (X, d) be a metric space and let Y be a subset of X. The following holds. (i) If U is open in Y and Y is open in X, then U is open in X. (ii) If A is closed in Y and Y is closed in X, then A is closed in X.*

Proof. Since U is open in Y, by Theorem 2.11, $U = V \cap Y$ for some set V open in X. Thus $V \cap Y$ is open in X by Theorem 2.6. So (i) holds. Similarly, one can prove (ii). □

When dealing with a metric space X and a subset Y of X, one needs to exercise care in taking closure of sets. If A is a subset of Y, the closure of A in Y and the closure of A in X will in general be different. In such a situation, we reserve the notation $\overline{A}$ to stand for the closure of A in X. The closure of A in Y can be expressed in terms of $\overline{A}$ as the following theorem shows.

Theorem 2.14. *Let Y be a subset of a metric space X; let A be a subset of Y; let $\overline{A}$ denote the closure of A in X. Then the closure of A in Y equals $\overline{A} \cap Y$.*

Proof. Let B denote the closure of A in Y. The set $\overline{A}$ is closed in X, so $\overline{A} \cap Y$ is closed in Y by Theorem 2.12. Since $A \subset \overline{A} \cap Y$, and since by definition B is the smallest closed subset of Y containing A hence

$$B \subset \overline{A} \cap Y. \tag{2.6}$$

On the other hand, we know that B is closed in Y. Hence by Theorem 2.12, $B = C \cap Y$ for some set C closed in X. Then C is a closed set of X containing A; because $\overline{A}$ is the smallest closed subset of X containing A, we conclude that $\overline{A} \subset C$. Then

$$\overline{A} \cap Y \subset C \cap Y = B. \tag{2.7}$$

From (2.6) and (2.7), it follows that $B = \overline{A} \cap Y$. □

Theorem 2.15. *Every uncountable subset of $\mathbb{C}$ contains a limit point of itself.*

Proof. Let S be an uncountable subset of $\mathbb{C}$ and S contains no limit point of itself. Then for each $x \in S$, there is a $\delta_x > 0$ such that $B(x, \delta_x) \cap S = \{x\}$. Let $\epsilon_x = \frac{\delta_x}{2}$. The balls $B(x, \epsilon_x)$ are disjoint, so we can chose a distinct point from each one with rational coordinates (that is, a complex number of the form $a + ib, a, b \in Q$). Since the set of points in $\mathbb{C}$ with rational coordinates is countable, the set S must be countable, a contradiction. Hence, S must contain one of its limit points. □

Corollary 2.4. *An uncountable subset of $\mathbb{C}$ contain uncountable number of its limit points.*

Notice that we only say that uncountable subset of $\mathbb{C}$ (for that matter uncountable subset of $\mathbb{R}^n, n \geq 1$) contains a limit point of itself. However, the proof carries over to any separable metric space instead of $\mathbb{C}$.

Theorem 2.16. *Let (X, d) be a metric space and $E \subset X$. Then E', the set of all limit points of E is closed.*

Proof. Let x be a limit point of E' and let N be an arbitrary neighborhood of x. Then N intersects E' at a point different from x. Let $y \in N \cap (E' - \{x\})$. Thus $y \in E'$ and $y \neq x$. Hence every neighborhood M of y intersects E at a point different from y. That is, $M \cap (E - \{y\}) \neq \emptyset$. In particular, $N \cap \{x\}^c$ is a neighborhood of y and $N \cap \{x\}^c \cap (E - \{y\}) \neq \emptyset$. Thus $N \cap (E - \{x\}) \neq \emptyset$. Hence x is a limit point of E and $x \in E'$. □

Definition 2.13. *Let (X, d) be a metric space and let E be a subset of X. Then we say E is dense in X if $\overline{E} = X$.*

In the above definition since $E \subset X$ and X is closed, it is clear that $\overline{E} \subset X$. Thus $\overline{E} = X$ if and only if $X \subset \overline{E}$. That is if $x \in X$ then x must belong to $\overline{E}$. In other words, $\overline{E} = X$ if and only if given any $x \in X, \epsilon > 0$, $B(x, \epsilon) \cap E \neq \emptyset$. Hence $\overline{E} = X$ if and only if given $x \in X$ and $\epsilon > 0$ there exists a $y \in E$ such that $d(x, y) < \epsilon$. Thus for future reference we state the following:

Theorem 2.17. *Let (X, d) be a metric space and $E \subset X$. The set E is dense in X if and only if given any $x \in X$ and $\epsilon > 0$, there exists a $y \in E$ such that $d(x, y) < \epsilon$.*

Notice also that a subset A of a metric space X is dense in X if for every nonempty open set U in X, we have $U \cap A \neq \emptyset$.

Example 2.8. *The set of rational numbers Q is dense in $\mathbb{R}$ and $Q(i) = \{a + ib : a, b \in Q\}$ is dense in $\mathbb{C}$ (verify).*

Definition 2.14. *Let (X, d) be a metric space. The metric space X is said to be separable if there exists a countable subset of X which is dense in X.*

For example $\mathbb{R}$ and $\mathbb{C}$ are separable metric spaces.

Definition 2.15. *A topological space X is said to have a countable basis at x if there is a countable collection $\mathcal{B}$ of neighborhoods of x (open sets containing the point x) such that each neighborhood of x contains at least one of the elements of $\mathcal{B}$. A topological space that has a countable basis at each of its points is said to satisfy the first countability axiom. A topological space is said to satisfy the second countability axiom if X has a countable basis for its topology.*

Note that second countability axiom implies the first countability axiom and every metric space satisfies the first countability axiom. The second axiom is, in fact, much stronger than the first; it is so strong that not even every metric space satisfies it.

Definition 2.16. *Let X and Y be topological spaces. The product topology on $X \times Y$ is the topology having as basis the collection $\mathcal{B}$ of all sets of the form $U \times V$, where U is an open subset of X and V is an open subset of Y.*

It is not difficult to verify that if $\mathcal{B}$ is a basis for the topology of X, and $\mathcal{C}$ is a basis for the topology of Y, then the collection $\mathcal{D} = \{B \times C : B \in \mathcal{B} \text{ and } C \in \mathcal{C}\}$ is a basis for the product topology on $X \times Y$.

2.2 Connectedness

Definition 2.17. *Let (X, d) be a metric space. A separation of X is a pair U, V of disjoint nonempty open subsets of X whose union is X. The space X is said to be connected if there does not exist a separation of X.*

Another way of formulating the definition of connectedness is the following:

Theorem 2.18. *A metric space X is connected if and only if the only subsets of X that are both open and closed in X are the empty set and X itself.*

Proof. The proof follows from the definition of connected spaces. □

If (X, d) is a metric space and $Y \subset X$ then Y is a connected subset of X if the metric space (Y, d) is connected. For a subset Y of a metric space X, there is another useful way of formulating the definition of connectedness:

Theorem 2.19. *Two nonempty subsets A and B of the metric space X form a separation of Y if $Y = A \cup B$ and if both $A \cap \overline{B}$ and $\overline{A} \cap B$ are empty (i.e., if no point of A lies in the closure of B and no point of B lies in the closure of A.) The metric space Y is connected if there exists no separation of Y.*

Proof. The proof follows from Corollary 2.3 and Theorem 2.14. □

Remark 2.7. Notice that separated sets are disjoint but disjoint sets need not be separated. For example, let $Y = [0, 1] \cup (1, 2)$. The two intervals $[0, 1]$ and $(1, 2)$ do not form a separation of Y as 1 is a limit point of $(1, 2)$ and it belongs to the interval $[0, 1]$. But the two intervals are disjoint.

Now we show that the nonempty connected subsets of the real line $\mathbb{R}$ are the intervals. But first we shall show that $\mathbb{R}$ is connected.

Suppose that the real line $\mathbb{R}$ is not connected. Then $\mathbb{R}$ has a separation. That is, $\mathbb{R} = A \cup B$ where A and B are two nonempty, disjoint,closed sets. Suppose $a_1 \in A$ and $b_1 \in B$ and $a_1 < b_1$. We bisect the interval (a_1, b_1) and note that one of the two halves has its left end points in A and its right end point in B. We denote this interval by (a_2, b_2) and continue the process indefinitely. In this way obtain a sequence of nested intervals (a_n, b_n) with $a_n \in A$ and $b_n \in B$. The sequences (a_n) and (b_n) have a common limit c. Since A and B are closed, c would have to be a common point of A and B. But $A \cap B = \emptyset$. This contradiction shows that $\mathbb{R}$ is connected.

Lemma 2.1. *Let E be a nonempty set of real numbers which is bounded above. Let $y = sup_{x \in E} x$. Then $y \in \overline{E}$.*

Proof. If $y \in E$ then $y \in \overline{E}$. Assume $y \notin E$. for every $\epsilon > 0$, there exists then a point $x \in E$ such that $y - \epsilon < x < y$, for otherwise $y - \epsilon$ would be an upper bound of E. Thus y is a limit point of E and $y \in \overline{E}$. □

Theorem 2.20. *A subset $E \subset \mathbb{R}$ is connected if and only if it has the following property: If $x \in E, y \in E$, and $x < z < y$, then $z \in E$.*

Proof. If there exist $x \in E, y \in E$, and some $p \in (x, y)$ such that $p \notin E$, then $E = A_p \cup B_p$ where $A_p = E \cap (-\infty, p), B_p = E \cap (p, \infty)$. Since $x \in A_p$ and $y \in B_p$, A and B are nonempty. Since $A_p \subset (-\infty, p)$ and $B_p \subset (p, \infty)$, they are separated. Hence E is not connected. To prove the converse, suppose E is not connected. Then there are nonempty separated sets A and B such that $A \cup B = E$. Pick $x \in A, y \in B$, and assume (without loss of generality) that $x < y$. Define $p = \sup(A \cap [x, y])$. By Lemma 2.1, $p \in \overline{A}$; hence $p \notin B$. In particular, $x \leq p < y$. If $p \notin A$, it follows that $x < p < y$ and $p \notin E$. If $p \in A$, then $p \notin \overline{B}$, hence there exists p_1 such that $p < p_1 < y$ and $p_1 \notin B$. Then $x < p_1 < y$ and $p_1 \notin E$. □

Thus any interval any ray in $\mathbb{R}$ is connected.

Definition 2.18. *If z and w are two complex numbers then we denote the straight line segment from z to w by*

$$[z, w] = \{tw + (1 - t)z : 0 \leq t \leq 1\}.$$

A polygon from a to b is a set

$$P = [a, z_1] \cup [z_1, z_2] \cup \cdots \cup [z_n, b].$$

We shall sometime write $[a, z_1, \cdots, z_n, b]$ to denote the polygon P.

Theorem 2.21. *An open set $\Omega \subset \mathbb{C}$ is connected if and only if for any two points a, b in Ω there is a polygon from a to b lying entirely inside Ω.*

Proof. Let Ω be an open connected set in $\mathbb{C}$. Choose a point $a \in \Omega$. Let

$$A = \{b \in \Omega : \text{there is a polygon } P \subset \Omega \text{ from } a \text{ to } b\},$$

and

$$B = \{c \in \Omega : \text{there is no polygon } P \text{ in } \Omega \text{ from } a \text{ to } c\}.$$

We shall prove that A and B are both open. First, if $b \in A$ there exists $\epsilon > 0$ such that $B(b, \epsilon) \subset \Omega$ (as Ω is open). All points in this ball can be joined to b by a line segment, and from there to a by a polygon. Hence $B(b, \epsilon) \subset A$, and A is open.

Secondly, if $c \in B$, then there exists $\delta > 0$ such that $B(c, \delta) \subset \Omega$ (as Ω is open). If a point in $B(c, \delta)$ could be joined to a by a polygon, then c could be joined to this point by a line segment, and from there to a. This is contrary to the definition of B, and we conclude that B is open. Since Ω was connected either A or B must be empty. But A contains the point a, hence $B = \emptyset$, and all points can be joined to a. Finally, any two points in A can be joined by way of a and we have proved that the condition is necessary. In order to prove the sufficiency we assume that Ω is not connected. We will obtain a contradiction.

As Ω is not connected we can write $\Omega = A \cup B$, where A and B are two disjoint nonempty open sets of Ω. Choose $b \in A$ and $c \in B$ and suppose that these points can be joined by a polygon in Ω. One of the sides of the polygon must then join a point in A to a point in B, and for this reason it is sufficient to consider the case where b and c are joined by a line segment. This segment has a parametric representation $z = tc + (1 - t)b$ where $0 \leq t \leq 1$. Let $S = \{t \in [0, 1] : tc + (1 - t)b \in A\}$ and $T = \{t \in [0, 1] : tc + (1 - t)b \in B\}$. Then $S \cap T = \emptyset, S \cup T = [0, 1], 0 \in S$ and $1 \in T$. It is not difficult to show that both S and T are open. This contradicts the connectedness of $[0, 1]$. Thus Ω must be connected. $\square$

For future reference we state the following fact as a corollary.

Corollary 2.5. *If $\Omega \subset \mathbb{C}$ is open and connected and a and b are points in Ω then there is a polygon P in Ω from a to b which is made up of line segment that are parallel to the coordinate axes.*

Proof. The proof is almost similar to the proof of Theorem 2.21. Define the set A and B as in the necessary part of Theorem 2.21 but add the restriction that the polygon's segments are all parallel to one of the coordinate axes. The remainder of the proof will be valid with one exception. If $z \in B(b, \epsilon)$ then $[b, z]$ may not be parallel to an axis. But it is easy to see that if $z = x + iy$ and $b = m + in$ then the polygon $[b, m + iy] \cup [m + iy, z] \subset B(b, \epsilon)$ and has segments parallel to an axis. $\square$

Lemma 2.2. *If the sets C and D from a separation of the metric space (X, d), and if Y is a connected subset of X, then Y lies entirely within either C or D.*

Proof. Since C and D are both open in X by Theorem 2.11 , the sets $C \cap Y$ and $D \cap Y$ are open in Y. These two sets are disjoint and their union is Y; if they were both nonempty, they would constitute a separation of Y. Therefore, one of them is empty. Hence Y must lie entirely in C or in D. □

Theorem 2.22. *Let (X, d) be a metric space. Let $\{A_\alpha\}$ be a collection of connected subsets of the space X such that $\cap_\alpha A_\alpha \neq \emptyset$. Then $A = \cup_\alpha A_\alpha$ is connected.*

Proof. Let $p \in \cap_\alpha A_\alpha$. Suppose $A = C \cup D$ where C, D are two disjoint, nonempty, open subsets of A. The point p is in one of the sets C or D; suppose $p \in C$. Since the set A_α is connected, it must lie entirely in either C or D, and it cannot lie in D because it contains the point p of C. Hence $A_\alpha \subset C$ for every α. Thus $A = \cup_\alpha A_\alpha \subset C$, contradicting the fact that D is nonempty. □

Theorem 2.23. *Suppose A is a connected subset of the metric space (X, d). If $A \subset B \subset \overline{A}$, then B is also connected.*

Proof. Suppose $B = C \cup D$ is a separation of B. By Lemma 2.2, A must lie entirely in C or in D. Suppose that $A \subset C$. Then $\overline{A} \subset \overline{C}$. By Theorem 2.19, $\overline{C} \cap D = \emptyset$. Since $B \subset \overline{A}$ hence $B \cap D = \emptyset$. This contradicts the fact that D is a nonempty subset of B. □

Let (X, d) be a metric space. There is a natural way to break it up into pieces that are connected.

Definition 2.19. *Given a metric space X, define an equivalence relation on X by setting $x \sim y$ if there is a connected subset of X containing both x and y. The equivalence classes are called the components (or "connected components") of X.*

Theorem 2.24. *Let (X, d) be a metric space. The components of X are connected disjoint subsets of X whose union is X, such that each connected subset of X intersects only one of them (i.e., if C_1 and C_2 are two components of X, and A is a connected subset of X then $A \cap C_1 \neq \emptyset$ and $A \cap C_2 \neq \emptyset$ implies $C_1 = C_2$.)*

Proof. Being equivalence classes, the components of X are disjoint and their union is X. Each connected set A in X intersects only one of them. For if A intersects the components C_1 and C_2 of X, say in points x_1 and x_2, respectively, then $x_1 \sim x_2$ by definition; this cannot happen unless $C_1 = C_2$.

To show the component C is connected, choose a point x_0 of C. For each point x of C, we know (by definition of component) that $x_0 \sim x$. Hence there is a connected set A_x containing x_0 and x. Since $A_x \cap C \neq \emptyset$ it follows that $A_x \subset C$. Therefore, $C = \cup_{x \in C} A_x$. Since the sets A_x are connected and have the point x_0 in common, their union is connected. □

Corollary 2.6. *If C is a component of a metric space (X, d) then C is closed.*

Proof. Note that a component C of X is a maximal connected subset of X. But $C \subset \overline{C}$. By Theorem 2.23, $\overline{C}$ is connected. Since C is a maximal connected set therefore $C = \overline{C}$ and C is closed. □

Theorem 2.25. *Let Ω be open in $\mathbb{C}$. Then the components of Ω are open and there are only a countable number of them.*

Proof. To see that the number of components is countable, let $N = \{a + ib \in \Omega : a, b \in Q\}$. Then N is countable and each component of Ω contains a point of N. So there are only countable number of components of Ω. Now, let C be a component of Ω and let $x_0 \in C$. Then there exists an $\epsilon > 0$ such that $B(x_0, \epsilon) \subset \Omega$ (as Ω is open). By Theorem 2.22, the set $B(x_0, \epsilon) \cup C$ is connected and must be equal to C as C is maximal. Hence $B(x_0, \epsilon) \subset C$. Therefore, the component C is open. □

2.3 Sequences and completeness

A sequence in a metric space (X, d) is a function f from $\mathbb{Z}_+$ into X. We usually denote a sequence by (x_n) or $(x_1, x_2, \cdots)$.

Definition 2.20. *A sequence (x_n) in a metric space X is said to converge to the point x (in symbols $x = \lim x_n$ or $x_n \to x$) if for each $\epsilon > 0$ there is an integer N such that $d(x_n, x) < \epsilon$ whenever $n \geq N$. Alternatively, $x = \lim x_n$ if $\lim d(x_n, x) = 0$. If the sequence (x_n) does not converge to any point of X then (x_n) is said to diverge.*

Theorem 2.26. *Let (x_n) be a sequence in a metric space X. Then the following holds: (i) The sequence (x_n) converges to $x \in X$ if and only if every neighborhood of x contains all but finitely many of the terms x_n. (ii) If $x, y \in X$ and if (x_n) converges to x and to y, then $x = y$. (iii) If (x_n) converges then (x_n) is bounded.*

Proof. The proof follows from Definition 2.20. □

Lemma 2.3. *(The sequence lemma): Let (X, d) be a metric space and let A be a subset of X. Then $x \in \overline{A}$ if and only if there is a sequence of points of A converging to x.*

Proof. Suppose that $x_n \to x$, where $x_n \in A$. Then every neighborhood U of x contains a point of A, so $x \in \overline{A}$ by Theorem 2.8. Conversely, suppose that $x \in \overline{A}$. For each positive integer n, take the neighborhood $B(x, \frac{1}{n})$ of x, and choose x_n to be a point of its intersection with A (this is possible because $B(x, \frac{1}{n}) \cap A \neq \emptyset$). We assert that the sequence x_n converges to x. Given $\epsilon > 0$, choose N large enough such that $\frac{1}{N} < \epsilon$. Consider the neighborhood $B(x, \epsilon)$ of x. Then $x_n \in B(x, \epsilon)$ for all $n \geq N$. Thus if $n \geq N$ then $d(x_n, x) < \epsilon$. Hence (x_n) converges to x. □

Definition 2.21. *Given a sequence (x_n) in a metric space X, consider a sequence (n_k) of positive integers, such that $n_1 < n_2 < n_3 < \cdots$. Then the sequence (x_{n_i}) is called a subsequence of (x_n). It is clear that (x_n) converges to x if and only if every subsequence of (x_n) converges to x.*

Definition 2.22. *A sequence (x_n) in a metric space (X, d) is said to be a Cauchy sequence if for every $\epsilon > 0$ there is an integer N such that $d(x_n, x_m) < \epsilon$ for all $n, m \geq N$.*

Definition 2.23. *The metric space (X, d) is said to be complete if every Cauchy sequence in X converges.*

In any metric space every convergent sequence is necessarily a Cauchy sequence, of course; completeness requires that the converse hold. To see this suppose $x_n \to x$ in X. If $\epsilon > 0$, there is an integer N such that $d(x_n, x) < \epsilon$ for all $n \geq N$. Hence $d(x_n, x_m) \leq d(x_n, x) + d(x, x_m) < 2\epsilon$ when $n \geq N$ and $m \geq N$. Thus (x_n) is a Cauchy sequence.

Note that a closed subset of a complete metric space (X, d) is necessarily complete in the restricted metric. For a Cauchy sequence in A is also a Cauchy sequence in X, whence it converges in X. Because A is a closed subset of X, the limit must lie in A.

A useful criterion for a metric space to be complete is the following:

Lemma 2.4. *A metric space (X, d) is complete if every Cauchy sequence in X has a convergent subsequence.*

Proof. The proof follows from Definition 2.23. □

Remark 2.8. If (x_n) is a sequence in a metric space X and if $E_N = \{x_N, x_{N+1}, x_{N+2}, \cdots\}$ it is clear that (x_n) is a Cauchy sequence if and only if $\lim_{N\to\infty} \text{diam} E_N = 0$.

Theorem 2.27. *Let (X, d) be a metric space and let A be a subset of X. Then $diam A = diam \overline{A}$.*

Proof. Since $A \subset \overline{A}$, it is clear that

$$\text{diam} A \leq \text{diam} \overline{A}. \tag{2.8}$$

Now fix $\epsilon > 0$, and choose $p, q \in \overline{A}$. By Theorem 2.8, $B(p, \epsilon) \cap A \neq \emptyset$ and $B(q, \epsilon) \cap A \neq \emptyset$. Let $p' \in B(p, \epsilon) \cap A$ and $q' \in B(q, \epsilon) \cap A$. Then

$$d(p, q) \leq d(p, p') + d(p', q') + d(q', q) < 2\epsilon + d(p', q') \leq 2\epsilon + \text{diam} A.$$

It follows therefore that $\text{diam}\overline{A} \leq 2\epsilon + \text{diam} A$. Since ϵ was arbitrary, we obtain

$$\text{diam}\overline{A} \leq \text{diam} A. \tag{2.9}$$

From (2.8) and (2.9), we have $\text{diam} A = \text{diam}\overline{A}$. □

Definition 2.24. *Let $S \subset \mathbb{R} = (-\infty, \infty)$. Then the supremum of a set S, written as $\sup S$ or $\sup_{x \in S} x$ is defined by*

$$\sup S = \begin{cases} \textit{lub } S & \textit{if} \quad S \textit{ is bounded above;} \\ \infty & \textit{if} \quad S \textit{ is not bounded above.} \end{cases}$$

Similarly, the infimum of a set S, written as $\inf S$ or $\inf_{x \in S} x$ is defined by

$$\inf S = \begin{cases} \textit{glb } S & \textit{if} \quad S \textit{ is bounded below;} \\ -\infty & \textit{if} \quad S \textit{ is not bounded below.} \end{cases}$$

It is easy to check that

$$\sup_{x \in S} x = -\inf_{x \in S}(-x),$$
$$\inf_{x \in S} x = -\sup_{x \in S}(-x).$$

Definition 2.25. *Let (x_n) be any real sequence. Then the limit superior or upper limit of (x_n), written as $\limsup x_n$ or $\overline{\lim} x_n$ is defined on $\mathbb{R}^\sharp = \mathbb{R} \cup \{\infty\} \cup \{-\infty\}$ as*

$$\overline{\lim} x_n = \lim_{n \to \infty} (\sup_{k \geq n} x_k) = \inf_n (\sup_{k \geq n} x_k).$$

The limit inferior of (x_n) denoted as $\liminf x_n$ or $\underline{\lim} x_n$ is defined on $\mathbb{R}^\sharp$ as $\underline{\lim} x_n = \lim_{n \to \infty} (\inf_{k \geq n} x_k) = \sup_n (\inf_{k \geq n} x_k)$.

Theorem 2.28. *Let (x_n) and (y_n) be real sequences. The following holds.*

(1) $\inf_n x_n \leq \underline{\lim} x_n \leq \overline{\lim} x_n \leq \sup_n x_n$

(2) $\overline{\lim}(x_n + y_n) \leq \overline{\lim} x_n + \overline{\lim} y_n$ *and* $\underline{\lim}(x_n + y_n) \geq \underline{\lim} x_n + \underline{\lim} y_n$

(3) $\overline{\lim}(\alpha x_n) = \begin{cases} \alpha \overline{\lim} x_n & \textit{if} \quad \alpha \geq 0; \\ \alpha \underline{\lim} x_n & \textit{if} \quad \alpha < 0. \end{cases}$

(4) $\underline{\lim} x_n = -\overline{\lim}(-x_n)$

(5) If $x_n \leq y_n$ *then* $\overline{\lim} x_n \leq \overline{\lim} y_n, \underline{\lim} x_n \leq \underline{\lim} y_n$.

Theorem 2.29. *Let (x_n) be a real bounded sequence and let $L = \overline{\lim} x_n$, $l = \underline{\lim} x_n$. Then*

(a) for every $\epsilon > 0$, there exists $n_0 \in \mathbb{N}$ such that, for all $n > n_0, x_n < L + \epsilon$.

(b) for every $\epsilon > 0, x_n > L - \epsilon$ for infinitely many values of n.

(c) for every $\epsilon > 0$, there exists $n_1 \in \mathbb{N}$ such that for $n > n_1$, $x_n > l - \epsilon$.

(d) for every $\epsilon > 0$, there are infinitely many values of n such that $x_n < l + \epsilon$.

(e) for given $\epsilon > 0$, there exists $n_2 \in \mathbb{N}$ such that, for $n > n_2$, $l - \epsilon < x_n < L + \epsilon$.

(f) $\overline{\lim} x_n, \underline{\lim} x_n$ are cluster points of (x_n).

(g) $\overline{\lim} x_n$ is the largest and $\underline{\lim} x_n$ is the smallest cluster points of (x_n).

(h) The sequence (x_n) converges if and only if $\underline{\lim} x_n = \overline{\lim} x_n = \lim x_n$.

For proof of these see any standard book in real analysis.

Example 2.9. *$\mathbb{R}$ is complete. The proof of it is closely related with the definition of real numbers. However, we wish to use only the property that every bounded monotone sequence of real numbers has a limit. Let (x_n) be a Cauchy sequence in $\mathbb{R}$ and consider $\overline{lim}x_n$. Since (x_n) is Cauchy, it is bounded so that $\overline{lim}x_n \in \mathbb{R}$. By the definition of $\overline{lim}$, there is a subsequence of (x_n) which converges to $\overline{lim}x_n$, and (x_n) being Cauchy, (x_n) itself converges to $\overline{lim}x_n$. Thus $\mathbb{R}$ is complete.*

Example 2.10. *$\mathbb{C}$ is complete. If $(z_n) = (x_n + iy_n)$ is a Cauchy sequence in $\mathbb{C}$ then $\{x_n\}$ and $\{y_n\}$ are Cauchy sequence in $\mathbb{R}$. Since $\mathbb{R}$ is complete, $x_n \to x$ and $y_n \to y$ for points x, y in $\mathbb{R}$. It follows that $x + iy = \lim(x_n + iy_n)$, and so $\mathbb{C}$ is complete.*

Theorem 2.30. *A metric space (X, d) is complete if and only if for any sequence $\{F_n\}$ of nonempty closed sets with $F_1 \supset F_2 \supset \cdots$ and $diamF_n \to 0$, $\cap_{n=1}^{\infty} F_n$ consists of a single point.*

Proof. The proof follows from the definition of completeness and Theorem 2.27. □

Proposition 2.1. *Let (X, d) be a complete metric space and let Y be a subset of X. Then (Y, d) is complete metric space if and only if Y is closed in X.*

Proof. The proof follows from Lemma 2.3. □

2.4 Compact sets

Definition 2.26. *A collection $\mathcal{A} = \{U_\alpha\}$ of subsets of a metric space X is said to be a covering of X if $X \subset \cup_\alpha U_\alpha$. The collection $\mathcal{A} = \{U_\alpha\}$ is called an open covering of X if its elements are open subsets of X.*

Definition 2.27. *A metric space X is said to be compact if every open covering $\mathcal{A} = \{U_\alpha\}$ of X contains a finite subcollection $\{U_{\alpha_1}, U_{\alpha_2}, \cdots, U_{\alpha_n}\}$ that also covers X, i.e., $X \subset U_{\alpha_1} \cup U_{\alpha_2} \cup \cdots \cup U_{\alpha_n}$.*

Example 2.11. *(i) $\mathbb{R}$ is not compact, because the open covering $\mathcal{A} = \{(-n, n) : n \in \mathbb{N}\}$ contains no finite subcollection that covers $\mathbb{R}$. (ii) $\mathbb{C}$ is not compact because the open covering $\mathcal{A} = \{G_n : n \in \mathbb{N}\}$ where $G_n = B(0, n) = \{z \in \mathbb{C} : |z| < n\}$ contains no finite subcollection that covers the complex plane $\mathbb{C}$. (iii) Let $\mathbb{D} = \{z \in \mathbb{C} : |z| < 1\} = B(0, 1)$, the open unit disk in $\mathbb{C}$. The set $\mathbb{D}$ is not compact because the open covering $\mathcal{A} = \{G_n : n \in \mathbb{N}\}$ where $G_n = \{z \in \mathbb{C} : |z| < 1 - \frac{1}{n}\}$ has no finite subcollection that also covers $\mathbb{D}$. (iv) Every finite set in a metric space X is compact.*

By an open cover of a set K in a metric space X we mean a collection $\{U_\alpha\}$ of open subsets of X such that $K \subset \cup_\alpha U_\alpha$. A subset K of a metric space X is said to be compact (relative to X) if every open cover of K (formed by open sets of X) contains a finite subcover. That is, if $\{U_\alpha\}$ is an open cover of K, then there are finitely many indices $\alpha_1, \alpha_2, \cdots, \alpha_n$ such that $K \subset U_{\alpha_1} \cup U_{\alpha_2} \cup \cdots \cup U_{\alpha_n}$.

Theorem 2.31. *Let (X, d) be a metric space. Suppose $K \subset Y \subset X$. Then K is compact relative to X if and only if K is compact relative to Y.*

Proof. Suppose K is compact relative to X. We need to show K is compact relative Y. So let $\{V_\alpha\}$ be a collection of sets open relative to Y, such that $K \subset \cup_\alpha V_\alpha$. By Theorem 2.11, there are sets U_α, open relative to X, such that $V_\alpha = U_\alpha \cap Y$, for all α. Hence $K \subset \cup_\alpha U_\alpha$ and since K is compact relative to X, we have

$$K \subset U_{\alpha_1} \cup U_{\alpha_2} \cup \cdots \cup U_{\alpha_n} \tag{2.10}$$

for some choice of finitely many indices $\alpha_1, \alpha_2, \cdots, \alpha_n$. Now because $K \subset Y$, (2.10) implies $K \subset V_{\alpha_1} \cup V_{\alpha_2} \cup \cdots \cup V_{\alpha_n}$. Thus K is compact relative to Y. Conversely, suppose K is compact relative to Y, let $\{U_\alpha\}$ be a collection of open subsets of X which cover K, and let $V_\alpha = U_\alpha \cap Y$. Then $K \subset \cup_\alpha V_\alpha$. Since K is compact relative to Y, hence $K \subset V_{\alpha_1} \cup V_{\alpha_2} \cup \cdots \cup V_{\alpha_n}$ for some choice of $\alpha_1, \alpha_2, \cdots, \alpha_n$; and now since some $V_\alpha \subset U_\alpha$ for all α, hence $K \subset U_{\alpha_1} \cup U_{\alpha_2} \cup \cdots \cup U_{\alpha_n}$. This completes the proof. □

Theorem 2.32. *Compact subsets of metric spaces are closed.*

Proof. Let K be a compact subset of a metric space X. We shall show K^c(the complement of K) is open in X. So let $p \in K^c$. This implies $p \in X$ but $p \notin K$. If $q \in K$ then $p \neq q$ and therefore $d(p, q) > 0$. Let

$$\delta = \frac{d(p, q)}{2}. \tag{2.11}$$

Let $U_q = B(q, \delta)$ and $V_q = B(p, \delta)$. Then U_q and V_q are neighborhoods of q and p respectively and $U_q \cap V_q = \emptyset$ (for if $w \in U_q$ and $w \in V_q$ then $d(w, q) < \delta$ and $d(p, w) < \delta$. Hence by triangle inequality $d(p, q) \leq d(p, w) + d(w, q) < 2\delta$ which is false by (2.11)). Note that $K \subset \cup_{q \in K}\{q\} \subset \cup_{q \in K} U_q$. Since K is compact, there are finitely many points $q_1, q_2, \cdots, q_n$ in K such that $K \subset U_{q_1} \cup \cdots \cup U_{q_n} = U$. If $V = V_{q_1} \cap \cdots \cap V_{q_n}$, then V is a neighborhood of p. Note $U \cap V = \emptyset$, for if, $z \in U \cap V$ then $z \in U$ and therefore $z \in U_{q_i}$ for some $i = 1, 2, \cdots, n$. This implies $z \notin V_{q_i}$ and hence $z \notin V$. Hence $V \subset U^c \subset K^c$. Thus $p \in V \subset K^c$. Hence p is an interior point of K^c. The theorem follows. □

Corollary 2.7. *Suppose X is a metric space, $K \subset X$, K is compact, and $p \in K^c$. Then there are open sets U and V such that $p \in V, K \subset U$, and $U \cap V = \emptyset$.*

Theorem 2.33. *Closed subsets of compact sets are compact in a metric space X.*

Proof. Let K be a compact subset of X, $F \subset K$ and F is closed in X. We have to show F is compact (relative to X). Let $\{U_\alpha\}$ be an open cover of F. Then $F \subset \cup_\alpha U_\alpha$. Hence $K \subset (\cup_\alpha U_\alpha) \cup F^c$. Thus we obtain an open cover $\mathcal{A}$ of K which consists of the open sets U_α for all α and the open set F^c. Since K is compact, there is a finite subcollection $\mathcal{B}$ of $\mathcal{A}$ which covers K and hence F. If F^c is a member of $\mathcal{B}$, we may remove it from $\mathcal{B}$ and still retain an open cover of F. We have thus shown that a finite subcollection of $\{U_\alpha\}$ covers F. □

Corollary 2.8. *If F is closed and K is a compact subset of a metric space X then $F \cap K$ is compact.*

Proof. Theorem 2.32 and Theorem 2.11 show that $F \cap K$ is closed. Since $F \cap K \subset K$, by Theorem 2.33 $F \cap K$ is compact. □

Theorem 2.34. *If $\{K_\alpha\}$ is a collection of compact subsets of a metric space X and if $\cap_\alpha K_\alpha = \emptyset$, then some finite subcollection of $\{K_\alpha\}$ also has empty intersection.*

Proof. Put $V_\alpha = K_\alpha^c$. Fix a member K_1 of $\{K_\alpha\}$. Since no point of K_1 belongs to every K_α hence $\{V_\alpha\}$ is an open cover of K_1. Since K_1 is compact, $K_1 \subset V_{\alpha_1} \cup \cdots \cup V_{\alpha_n}$ for some finite subcollection $\{V_{\alpha_i}\}_{i=1}^n$ of $\{V_\alpha\}$. This implies that $K_1 \cap K_{\alpha_1} \cap \cdots \cap K_{\alpha_n} = \emptyset$. □

If we take the contrapositive of both the necessary and sufficient part of the above theorem we obtain the following which we also put in a remark for future reference.

Remark 2.9. If $\{K_\alpha\}$ is a collection of compact subsets of a metric space X such that the intersection of every finite subcollection of $\{K_\alpha\}$ is nonempty, then $\cap K_\alpha$ is nonempty.

Corollary 2.9. *If $\{K_n\}$ is a sequence of nonempty compact sets such that $K_n \supset K_{n+1} (n = 1, 2, 3, \cdots)$, then $\cap_{n=1}^\infty K_n$ is nonempty.*

Theorem 2.35. *If E is an infinite subset of a compact set K, then E has a limit point in K.*

Proof. Suppose no point of K is a limit point of E. Then each $q \in K$ has a neighborhood V_q which contains at most one point of E (namely q, if $q \in E$). Thus there is no finite subcollection of $\{V_q\}_{q\in K}$ which can cover E as E is an infinite set. Hence there is no finite subcollection of $\{V_q\}_{q\in K}$ which can cover K as $E \subset K$. This contradicts the compactness of K. □

Using the concepts of order in $\mathbb{R}$, we can define the following four types of sets called intervals.

Let $a, b \in \mathbb{R}$ such that $a < b$. Then the open interval (a, b), closed interval $[a, b]$, the semi-closed (or semi-open) intervals $[a, b), (a, b]$ are defined as follows:

$$(a, b) = \{x \in \mathbb{R} : a < x < b\}; \ [a, b] = \{x \in \mathbb{R} : a \leq x \leq b\}$$

$$[a, b) = \{x \in \mathbb{R} : a \leq x < b\} \text{ and } (a, b] = \{x \in \mathbb{R} : a < x \leq b\}.$$

Theorem 2.36. *If $\{I_n\}$ is a sequence of closed intervals in $\mathbb{R}$, such that $I_n \supset I_{n+1}(n = 1, 2, 3, \cdots)$, then $\cap_{n=1}^{\infty} I_n$ is not empty.*

Proof. The reader is advised to work out a proof. □

Definition 2.28. *A sequence of closed intervals $\{I_n\}$ is called a nest if*

(i) $I_1 \supset I_2 \supset I_3 \supset I_4 \cdots$,

(ii) length of $I_n \to 0$ as $n \to \infty$.

Corollary 2.10. *Let $I_n = [a_n, b_n]$ be a nest of closed intervals. Then $\cap_{n=1}^{\infty} I_n$ consists of a single point; in fact,*

$$\cap_{n=1}^{\infty} I_n = \lim_{n \to \infty} a_n = \lim_{n \to \infty} b_n.$$

Proof. The proof is immediate. □

Let k be a positive integer and $\mathbb{R}^k = \mathbb{R} \times \cdots \times \mathbb{R}$, the cartesian product of $\mathbb{R}$ with itself k times. Thus if $x \in \mathbb{R}^k$ then $x = (x_1, \cdots, x_k)$ where $x_i \in \mathbb{R}, 1 \leq i \leq k$ is called the ith coordinate of x. By a k-cell we shall mean a set of the form $[a_1, b_1] \times \cdots \times [a_k, b_k]$ which is a subset of $\mathbb{R}^k$. Thus if $x = (x_1, \cdots, x_k)$ is in the k-cell $[a_1, b_1] \times \cdots \times [a_k, b_k]$ then $x_i \in [a_i, b_i]$ for $1 \leq i \leq k$. Note that a 1-cell is a closed, bounded interval, a 2-cell is a rectangle.

Given $x = (x_1, \cdots, x_k) \in \mathbb{R}^k$, $y = (y_1, \cdots, y_k) \in \mathbb{R}^k$, define $d(x, y) = (\sum_{i=1}^k |x_i - y_i|^2)^{\frac{1}{2}}$. We have already noted that d is a metric on $\mathbb{R}^k$. The set $\mathbb{R}^k$ is a vector space over the field $\mathbb{R}$ with vector-addition and scalar multiplication defined as follows: $x+y = (x_1+y_1, \cdots, x_k+y_k)$; $\alpha x = (\alpha x_1, \cdots, \alpha x_k)$ for $\alpha \in \mathbb{R}$.

Define $\| \ \| : \mathbb{R}^k \to \mathbb{R}$ such that $\|x\| = (\sum_{i=1}^k |x_i|^2)^{\frac{1}{2}}$. Observe the following: (i) $\|x\| \geq 0$; (ii) $\|x + y\| \leq \|x\| + \|y\|$ (use Cauchy-Schwarz inequality); (iii) $\|\alpha x\| = |\alpha| \|x\|$ for $\alpha \in \mathbb{R}$.

Further note that $d(x, y) = \|x - y\|$ and for $\epsilon > 0$, $B_d(x, \epsilon) = \{y \in \mathbb{R}^k : d(x, y) < \epsilon\} = \{y \in \mathbb{R}^k : \|x - y\| < \epsilon\}$.

Theorem 2.37. *Let k be a positive integer. If (I_n) is a sequence of k-cells such that $I_n \supset I_{n+1}(n = 1, 2, 3, \cdots)$, then $\cap_{n=1}^{\infty} I_n$ is not empty.*

Proof. The proof follows from Theorem 2.36. □

Theorem 2.38. *Every k-cell in $\mathbb{R}^k$ is compact.*

Proof. For proof see [5]. □

Definition 2.29. *A metric space X is said to be limit point compact if every infinite subset of X has a limit point.*

Definition 2.30. *A metric space X is sequentially compact if every sequence in X has a convergent subsequence.*

Now we prove that compactness, limit point compactness (Fréchet compactness) and sequentially compactness are equivalent in metric spaces.

Theorem 2.39. *Let (X, d) be a metric space. Then the following are equivalent:*

(1) X is compact.
(2) X is limit point compact.
(3) X is sequentially compact.

Proof. (1) $\Rightarrow$ (2). Let X be a compact metric space. Given a subset A of X, we wish to prove that if A is infinite, then A has a limit point. We prove the contrapositive–if A has no limit point, then A must be finite. So suppose the set A has no limit points and is therefore closed. Being a closed subset of a compact metric space, by Theorem 2.33, it is compact. For each $a \in A$, we can choose a neighborhood U_a of a such that $U_a \cap (A - \{a\}) = \emptyset$, since a is not a limit point of A. Thus each set U_a contains only one point of A. The collection $\{U_a\}_{a \in A}$ is an open covering for A. Since A is compact, there exist finite number of sets, say $U_1, U_2, \cdots, U_n$ in the open covering $\{U_a\}_{a \in A}$ such that $A \subset U_1 \cup U_2 \cup \cdots \cup U_n$. Thus A contains at most n points and hence A is a finite set.

(2) $\Rightarrow$ (3). We shall prove that limit point compactness of X in turn implies that every infinite sequence (x_n) in X has a convergent subsequence. That is, there is an increasing sequence $n_1 < n_2 < \cdots < n_i < \cdots$ of positive integers such that the sequence $x_{n_1}, x_{n_2}, \cdots, x_{n_i}, \cdots$ converges.

Given the sequence (x_n), let us consider the set $A = \{x_n : n \in \mathbb{Z}_+\}$ where $\mathbb{Z}_+$ is the set of positive integers. Suppose that the set A is finite and define a function $f : \mathbb{Z}_+ \to A$ such that $f(n) = x_n$. Then $\mathbb{Z}_+ = \cup_{x \in A} f^{-1}(x)$ and since A is a finite set hence $\mathbb{Z}_+$ is the union of the finite collection of sets $f^{-1}(x)$. Thus at least one of the sets $f^{-1}(x)$ must be infinite. Therefore the sequence (x_n) has a subsequence that is constant, and therefore converges automatically.

Now suppose that A is infinite. Then A has a limit point x. We shall obtain a subsequence of (x_n) converging to x as follows: First choose n_1 so that $x_{n_1} \in B(x, 1)$. Then suppose that the positive integer n_{i-1} is given. Because the ball $B(x, \frac{1}{i})$ contains infinitely many points of A (Theorem 2.4), we can choose an index $n_i > n_{i-1}$ such that $x_{n_i} \in B(x, \frac{1}{i})$. Then the subsequence $x_{n_1}, x_{n_2}, \cdots$ converges to x.

To prove (3) $\Rightarrow$ (1) we need to prove the following lemma.

Lemma 2.5. *(Lebesgue's covering lemma): Let $\mathcal{A}$ be an open covering of the metric space (X, d). If X is compact, there is a $\delta > 0$ such that for each subset* *of X having diameter less than δ, there exists an element of $\mathcal{A}$ containing it. (The number δ is called the Lebesgue number for the covering $\mathcal{A}$.)*

Proof. For proof see [5]. □

(3) $\Rightarrow$ (1). Suppose X is sequentially compact. We wish to show X is compact.

Step 1: First we show that for every $\epsilon > 0$, there exists a finite covering of X by balls of radius ϵ. And once again, we prove the contrapositive: if for some $\epsilon > 0$, X cannot be covered by finitely many ϵ-balls, then X is not sequentially compact.

So suppose that X cannot be covered by finitely many ϵ-balls. Construct a sequence of points x_n of X as follows: First choose x_1 to be any point of X. Noting that the ball $B(x_1, \epsilon)$ is not all of X (otherwise X could be covered by a single ϵ-ball), choose x_2 to be a point of X not in $B(x_1, \epsilon)$. In general, given $x_1, \cdots, x_n$, choose x_{n+1} to be a point not in the union $B(x_1, \epsilon) \cup \cdots \cup B(x_n, \epsilon)$, using the fact that these balls do not cover X. Note that by construction $d(x_{n+1}, x_i) \geq \epsilon$ for $i = 1, \cdots, n$. Therefore, the sequence (x_n) can have no convergent subsequence; in fact, any ball of radius $\frac{\epsilon}{2}$ can contain x_n for at most one value of n.

Step 2: Now we prove X is compact. Let $\mathcal{A}$ be an open covering of X. Since X is sequentially compact, the covering $\mathcal{A}$ has a Lebesgue number δ. Using step 1, choose a finite covering of X by balls of radius $\frac{\delta}{3}$. Each of these balls has diameter at most $\frac{2\delta}{3}$, so we can choose for each of these balls an element of $\mathcal{A}$ containing it. We thus obtain a finite subcollection of $\mathcal{A}$ that covers X. □

Theorem 2.40. *Let (X, d) be a metric space. Let K be a compact subset of X. Then K is bounded.*

Proof. To see this, choose a point $x_0 \in K$ and consider all balls $B(x_0, r), r > 0$ in X. They form an open covering of K and since K is compact, it contains a finite subcovering; in other words, $K \subset B(x_0, r_1) \cup \cdots \cup B(x_0, r_n)$. Let $r = \max\{r_1, r_2, \cdots, r_n\}$. Then $K \subset B(x_0, r)$. Now for any $x, y \in K$, it follows that $d(x, y) \leq d(x, x_0) + d(x_0, y) < 2r$, and we have proved that K is bounded. □

Definition 2.31. *A metric space (X, d) is totally bounded if, for every $\epsilon > 0$, X can be covered by finitely many balls of radius ϵ.*

Theorem 2.41. *Every totally bounded set is bounded.*

Proof. Suppose the metric space (X, d) is totally bounded. Then for any $\epsilon > 0$, X can be covered by finitely many balls of radius ϵ. That is, $X \subset B(x_1, \epsilon) \cup \cdots \cup B(x_m, \epsilon)$ for some points $x_1, \cdots, x_m \in X$. Now let $x, y \in X$. Then $x \in B(x_i, \epsilon)$ and $y \in B(x_j, \epsilon)$ for some $i, j \in \{1, \cdots, m\}$. Then $d(x, y) \leq d(x, x_i) + d(x_i, x_j) + d(x_j, y) < 2\epsilon + d(x_i, x_j) \leq 2\epsilon + \max_{1 \leq i,j \leq m} d(x_i, x_j)$. Hence X is bounded. □

So we see that total boundedness of a metric space clearly implies boundedness but the converse does not hold.

Example 2.12. *Under the metric $d(z, w) = |z - w|$, the complex plane $\mathbb{C}$ is neither bounded nor totally bounded. Under the metric $\bar{d}(z, w) = \min\{|z - w|, 1\}$, the complex plane $\mathbb{C}$ is bounded but not totally bounded.*

Under the metric $d(z, w) = |z-w|$, the open unit disk $\mathbb{D} = \{z \in \mathbb{C} : |z| < 1\}$ is totally bounded. But $\mathbb{D}$ is not complete with respect to the metric d. The closed unit disk $\overline{\mathbb{D}} = \{z \in \mathbb{C} : |z| \leq 1\}$ is both complete and totally bounded.

Theorem 2.42. *A metric space (X, d) is compact if and only if it is complete and totally bounded.*

Proof. For proof see [5]. □

Theorem 2.43. *(Heine-Borel theorem) A subset E of $\mathbb{R}^k$ $(k \geq 1)$ is compact if and only if E is closed and bounded.*

Proof. If E is compact then E is closed and bounded by Proposition 2.1, Theorem 2.41 and Theorem 2.42. Now suppose E is closed and bounded. Hence there are real numbers $a_1, \cdots, a_k$ and $b_1, \cdots, b_k$ such that $E \subset F = [a_1, b_1] \times \cdots \times [a_k, b_k]$, i.e., E is contained in a k-cell F. By Theorem 2.38, F is compact. Since E is closed it follows from Theorem 2.33 that E is compact. □

Theorem 2.44. *The metric space $\mathbb{R}^k (k \geq 1)$ is complete.*

Proof. Let (x_n) be a Cauchy sequence in $\mathbb{R}^k$. Let $E_N = \{x_N, x_{N+1}, x_{N+2}, \cdots\}$. Then $\lim_{n\to\infty} \text{diam} E_n = 0$. Hence given $\epsilon = 1 > 0$, there exists N such that $\text{diam} E_n < 1$ for all $n \geq N$. In particular, $\text{diam} E_N < 1$. The range of the sequence $\{x_n\}$ is equal to $E_N \cup \{x_1, \cdots, x_{N-1}\}$. Hence $\{x_n\}$ is bounded. Let $E = \{x_1, x_2, x_3, \cdots\}$, the range of the sequence (x_n). Since E is bounded hence $\overline{E}$ is closed and bounded by Remark 2.5 and Theorem 2.27. Hence by Theorem 2.43, $\overline{E}$ is compact and hence it is complete(by Theorem 2.42). Since (x_n) is a Cauchy sequence in $\overline{E}$ and $\overline{E}$ is complete hence (x_n) converges to an element of $\overline{E}$ and therefore to an element of $\mathbb{R}^k$. □

Theorem 2.45. *(Weierstrass): Every bounded infinite subset of $\mathbb{R}^k (k \geq 1)$ has a limit point in $\mathbb{R}^k$.*

Proof. Let E be a bounded, infinite subset of $\mathbb{R}^k$. Then E is contained in a k-cell $F = [a_1, b_1] \times \cdots \times [a_k, b_k]$. By Theorem 2.38, F is compact and so by Theorem 2.35, E has a limit point in F and hence in $\mathbb{R}^k$. □

Theorem 2.46. *If K_n is a sequence of nonempty compact sets in a metric space (X, d) such that $K_n \supset K_{n+1} (n = 1, 2, 3, \cdots)$ and if $\lim\limits_{n\to\infty} diam K_n = 0$, then $\cap_{n=1}^{\infty} K_n$ consists of exactly one point.*

Proof. The proof follows from Corollary 2.9. □

Definition 2.32. *(Local compactness) A metric space X is said to be locally compact at x if there is some compact subset C of X that contains a neighborhood of x. If X is locally compact at each of its points, X is said to be locally compact.*

Note that a compact metric space is locally compact. The real line $\mathbb{R}$ is locally compact. The point x lies in some interval (a, b), which in turn is contained in the compact set $[a, b]$. The set Q of rational numbers is not locally compact.

Definition 2.33. *Let (X, d_X) and (Y, d_Y) be two metric spaces and let $f : X \to Y$. If $a \in X$ and $w \in Y$, then $\lim_{x\to a} f(x) = w$ if for every $\epsilon > 0$ there is a $\delta > 0$ such that $d_Y(f(x), w) < \epsilon$ whenever $0 < d_X(x, a) < \delta$.*

Theorem 2.47. *Let (X, d_X) and (Y, d_Y) be two metric spaces, $f : X \to Y$. Suppose $\{p_n\}$ is a sequence in X such that $p_n \neq p$, $\lim_{n\to\infty} p_n = p$. Then $\lim_{x\to p} f(x) = q$ if and only if $\lim_{n\to\infty} f(p_n) = q$.*

Proof. The proof follows from the definition of limit. □

2.5 Continuous functions

Definition 2.34. *(Continuity): Let (X, d_X) and (Y, d_Y) be two metric spaces and let $f : X \to Y$. Let $a \in X$. The function f is continuous at the point a if $\lim_{x\to a} f(x) = f(a)$. If f is continuous at each point of X then f is said to be a continuous function from X into Y.*

Observe that f is said to be continuous on X if f is continuous at every point of X. Thus it is evident from the definition that the domain of the function should be such that every point in it has a neighborhood contained in it. Note that such a set is called an open set.

Theorem 2.48. *Let $f : X \to Y$, where X and Y are metric spaces with metrics d_X and d_Y respectively. The function f is continuous at a if and only if for every $\epsilon > 0$, there exists a $\delta > 0$ such that $f(B_{d_X}(a, \delta)) \subset B_{d_Y}(f(a), \epsilon)$.*

Proof. Follows from Definition 2.34. □

Theorem 2.49. *Let (X, d_X) and (Y, d_Y) be two metric spaces and let $f : X \to Y$. The function f is continuous if and only if for every convergent sequence $x_n \to x$ in X, the sequence $f(x_n)$ converges to $f(x)$.*

Proof. Assume that f is continuous. Given $x_n \to x$. We wish to show that $f(x_n) \to f(x)$. Let $\epsilon > 0$. Since f is continuous at x by Theorem 2.48, there exists $\delta > 0$ such that

$$f(B_{d_X}(x, \delta)) \subset B_{d_Y}(f(x), \epsilon). \tag{2.12}$$

Now since $x_n \to x$ and $B_{d_X}(x, \delta)$ is a neighborhood of x, there exist N such that $x_n \in B_{d_X}(x, \delta)$ for all $n \geq N$. Hence from (2.12), $f(x_n) \in B_{d_Y}(f(x), \epsilon)$ for all $n \geq N$. Thus the sequence $f(x_n)$ converges to $f(x)$.

To prove the converse, assume that the convergent sequence condition is satisfied. We show that f is continuous. First we shall prove that $f(\overline{A}) \subset \overline{f(A)}$ if A is a subset of X.

Let $x \in \overline{A}$. By sequence lemma (Lemma 2.3), there is a sequence (x_n) of points of A converging to x. By assumption, the sequence $f(x_n)$ converges to $f(x)$. Since $f(x_n) \in f(A)$, the sequence lemma (Lemma 2.3) implies that $f(x) \in \overline{f(A)}$. Hence $f(\overline{A}) \subset \overline{f(A)}$. Let $x \in X$ and $\epsilon > 0$. Consider the ball $B_{d_Y}(f(x), \epsilon)$ which is an open set in Y. Let $B = Y - B_{d_Y}(f(x), \epsilon)$. Then by Theorem 2.1, B is closed in Y. Let $A = f^{-1}(B)$. We shall prove now that A is closed in X. Since $f(A) \subset B$ hence if x is a point of $\overline{A}$, $f(x) \in f(\overline{A}) \subset \overline{f(A)} \subset \overline{B} = B$. Hence $x \in f^{-1}(B) = A$. Thus $\overline{A} \subset A$. Since $A \subset \overline{A}$ hence $\overline{A} = A$. Thus $A = f^{-1}(B) = f^{-1}(Y - B_{d_Y}(f(x), \epsilon)) = X - f^{-1}(B_{d_Y}(f(x), \epsilon))$ is closed in X. Hence by Theorem 2.5, $f^{-1}(B_{d_Y}(f(x), \epsilon))$ is open in X and $x \in f^{-1}(B_{d_Y}(f(x), \epsilon))$. Hence there exists $\delta > 0$ such that $f(B_{d_X}(x, \delta)) \subset B_{d_Y}(f(x), \epsilon)$. Therefore by Theorem 2.48, f is continuous at $x \in X$. Since the point $x \in X$ is arbitrarily chosen, we proved that f is continuous on X. □

Remark 2.10. Let $X, Y,$ and Z be metric spaces.The following are easy to verify. (i) Any constant function $f : X \to Y$ is continuous on X. That is, if $b \in Y$, and f is a function from X into Y such that $f(x) = b$ for all $x \in X$ then f is continuous. (ii) Let f be the identity function of X into itself. Then f is continuous on X. (iii) Let d be the metric on X. Fix $a \in X$. Define $f : X \to \mathbb{R}$ by setting $f(x) = d(a, x)$. We show that f is continuous. Let $x \in X$ be arbitrary. Let (x_n)be a sequence such that $x_n \to x$. Then we need to show that $f(x_n) \to f(x)$ in $\mathbb{R}$ as $n \to \infty$. That is, $|f(x_n) - f(x)| \to 0$ as $n \to \infty$. But note that $|f(x_n) - f(x)| = |d(a, x_n) - d(a, x)| \leq d(x, x_n) \to 0$ as $n \to \infty$. (iv) If $A \subseteq X$, the inclusion function $j : A \longrightarrow X$ is continuous. (v) If $f : X \longrightarrow Y$ is continuous, and if $A \subset X$, then the restricted function $f|_A : A \longrightarrow Y$ is continuous. (vi) Let $f : X \longrightarrow Y$ be continuous. If Z is a subspace of Y containing the image set $f(X)$, then the function $g : X \longrightarrow Z$ obtained by restricting the range of f is continuous. If Z is a metric space that contains Y as a subset, then the function $h : X \longrightarrow Z$ obtained by expanding the range of f is continuous. (vii) The map $f : X \longrightarrow Y$ is continuous if $X = \cup_\alpha U_\alpha$ such that $f|_{U_\alpha}$ is continuous for each α.

Theorem 2.50. *Let $f, g : X \to Y$ be continuous. Let A be dense in X. Assume that $f(a) = g(a)$ for all $a \in A$. Then $f = g$ on X.*

Proof. Note A is dense in X if and only if $\overline{A} = X$. This implies given $x \in X$, there exists a sequence (a_n) in A such that $\lim_{n\to\infty} a_n = x$. By continuity, $\lim_{n\to\infty} f(a_n) = f(x)$ and $\lim_{n\to\infty} g(a_n) = g(x)$. But $f(a_n) = g(a_n)$ for all n. Therefore $f(x) = g(x)$. □

Proposition 2.2. *Let (X, d_X) and (Y, d_Y) be two metric spaces and let $f : X \longrightarrow Y$. The following are equivalent.*

(1) f is continuous;

(2) If U is open in Y then $f^{-1}(U)$ is open in X;

(3) If B is closed in Y then $f^{-1}(B)$ is closed in X.

Proof. The proof follows from Theorem 2.48 and Theorem 2.5. □

Example 2.13. *Let $\mathcal{H} = \{z \in \mathbb{C} : Imz > 0\}$. The function $f : \mathbb{C} \to \mathbb{R}$ given by $f(z) = Imz$ is continuous. If we take $V = (0, \infty)$, then V is open in $\mathbb{R}$ and hence $f^{-1}(V) = \mathcal{H}$ is open in $\mathbb{C}$.*

Proposition 2.3. *Let X, Y, Z be metric spaces. If $f : X \longrightarrow Y$ and $g : Y \longrightarrow Z$ are continuous then $g \circ f$ where $(g \circ f)(x) = g(f(x))$ is a continuous function from X into Z.*

Proof. If U is open in Z then $g^{-1}(U)$ is open in Y; hence, $f^{-1}(g^{-1}(U)) = (g \circ f)^{-1}(U)$ is open in X. □

Definition 2.35. *Let (X, d_X) and (Y, d_Y) be two metric spaces and let $f : X \longrightarrow Y$. The function f is uniformly continuous on X if for every $\epsilon > 0$, there is a $\delta > 0$ (depending only on ϵ) such that for any two points x, y of X, $d_X(x, y) < \delta$ implies $d_Y(f(x), f(y)) < \epsilon$.*

We say that f is a Lipschitz function if there is a constant $M > 0$ such that $d_Y(f(x), f(y)) \leq M d_X(x, y)$ for all x and y in X. Note that every Lipschitz function is uniformly continuous. In fact, given $\epsilon > 0$ choose $\delta = \frac{\epsilon}{M}$.

Remark 2.11. Every uniformly continuous function is continuous but the converse is not true. For example, let $X = Y = \mathbb{R}$ and define $f : X \longrightarrow Y$ as $f(x) = x^2$. The function f is continuous but not uniformly continuous.

Example 2.14. *Let $X = Y = [0, 1]$ and $f : X \longrightarrow Y$ is defined by $f(x) = x^{\frac{1}{2}}$. It is not so difficult to show that f is uniformly continuous but is not a Lipschitz function.*

Theorem 2.51. *(Uniform continuity theorem): Let $f : X \longrightarrow Y$ be a continuous map from the compact metric space (X, d_X) to the metric space (Y, d_Y). Then f is uniformly continuous.*

Proof. Given $\epsilon > 0$, $\mathcal{B} = \{B_{d_Y}(y, \frac{\epsilon}{2}) : y \in Y\}$. Then $\mathcal{B}$ is an open covering for Y. Let $\mathcal{A}$ be an open covering of X by the inverse images of these balls in $\mathcal{B}$ under f. That is, $\mathcal{A} = \{f^{-1}(B_{d_Y}(y, \frac{\epsilon}{2})) : y \in Y\}$. (Since f is continuous, $\mathcal{A}$ is an open covering of X.) The metric space X is compact. By Lemma 2.5, we can find a number $\delta > 0$ which is a Lebesgue number for the covering $\mathcal{A}$. Then if $x_1, x_2 \in X$ are such that $d_X(x_1, x_2) < \delta$, then the two point set $\{x_1, x_2\}$ has diameter less than δ. Hence there exist an element of $\mathcal{A} = \{f^{-1}(B_{d_Y}(y, \frac{\epsilon}{2})) : y \in Y\}$ containing the set $\{x_1, x_2\}$. So $\{f(x_1), f(x_2)\}$ lies in the ball $B_{d_Y}(y, \frac{\epsilon}{2})$ for some $y \in Y$ and hence $d_Y(f(x_1), f(x_2)) < \epsilon$, as desired. □

Definition 2.36. *Let (X, d) be a metric space. Let $A \subset X$ and $x \in X$. We define the distance from x to the set A, $d(x, A)$, by $d(x, A) = \inf\{d(x, y) : y \in A\}$.*

Proposition 2.4. *Let (X, d) be a metric space and let $A \subset X$. Then (i) $d(x, A) = d(x, \overline{A})$; (ii) $d(x, A) = 0$ if and only if $x \in \overline{A}$; (iii) The map $f : X \longrightarrow \mathbb{R}$ defined by $f(x) = d(x, A)$ is continuous on X.*

Proof. The proof is easy and is left to the reader. □

Theorem 2.52. *Let (X, d_X) and (Y, d_Y) be two metric spaces and let $f : X \longrightarrow Y$ be a continuous function. If X is compact then $f(X)$ is a compact subset of Y.*

Proof. Let $\mathcal{A}$ be a covering of the set $f(X)$ by sets open in Y. The collection $\{f^{-1}(A) : A \in \mathcal{A}\}$ is a collection of sets covering X; these sets are open in X because f is continuous. Hence finitely many of them, say $f^{-1}(A_1), \cdots, f^{-1}(A_n)$ cover X as X is compact. Then the sets $A_1, \cdots, A_n$ cover $f(X)$. □

Definition 2.37. *(Homeomorphism): Let X and Y be metric spaces; let $f : X \longrightarrow Y$ be a bijection. If both the function f and the inverse function $f^{-1} : Y \longrightarrow X$ are continuous, then f is called a homeomorphism.*

Example 2.15. *The map (i) $\Phi(z) = \frac{z}{1+|z|}$ is a homeomorphism from $\mathbb{C}$ onto $\mathbb{D} = \{z \in \mathbb{C} : |z| < 1\}$. (ii) The function $f(x) = \frac{x}{1-x^2}$ is a homeomorphism from $(-1, 1)$ onto $\mathbb{R}$. (iii) The map $g(x) = \frac{x-\frac{1}{2}}{x-x^2}$ is a homeomorphism from $(0, 1)$ onto $\mathbb{R}$. (iv) The map $h(x) = (\frac{x-a}{b-a})(d-c) + c$ is a homeomorphism from (a, b) onto (c, d).*

Theorem 2.53. *Let (X, d_X) and (Y, d_Y) be two metric spaces and let $f : X \longrightarrow Y$ be a bijective continuous function. If X is compact, then f is a homeomorphism.*

Proof. We shall prove that images of closed sets of X under f are closed in Y; this will prove continuity of the map f^{-1}. If A is closed in X, then A is compact by Theorem 2.33. Therefore, as f is continuous, by Theorem 2.52, $f(A)$ is compact. Since Y is a metric space by Theorem 2.32, $f(A)$ is closed in Y. □

Definition 2.38. *Let X and Y be metric spaces. A map $f : X \longrightarrow Y$ is said to be an open map if for each open set U of X, the set $f(U)$ is open in Y. It is said to be a closed map if for each closed set A of X, the set $f(A)$ is closed in Y.*

Remark 2.12. Note that the condition that f^{-1} be continuous is equivalent to say by Proposition 2.2 that for each open U of X, the inverse image of U under the map $f^{-1} : Y \longrightarrow X$ is open in Y. But the inverse image of U under the map f^{-1} is the same as the image of U under the map f. So we can define a homeomorphism as follows: A continuous map $f : X \longrightarrow Y$ is a homeomorphism if it is a bijective open map. Similarly by Proposition 2.2, we can define a homeomorphism as follows: A continuous map $f : X \longrightarrow Y$ is a homeomorphism if it is a bijective, closed map.

Example 2.16. *A bijective function f from a topological space X into a topological space Y can be continuous without being a homeomorphism. Let $S^1 = \{z \in \mathbb{C} : |z| = 1\} = \{x \times y : x^2 + y^2 = 1\}$ considered as a subspace of $\mathbb{R}^2$. Let $f : [0, 1) \to S^1$ be the map defined by $f(t) = (cos2\pi t, sin2\pi t)$. The fact that f is bijective and continuous follows from familiar properties of the trigonometric functions. But the function f^{-1} is not continuous. The function f^{-1} is not continuous at the point $(1, 0) = f(0)$. The image under f of the open set $U = [0, \frac{1}{4})$ of the domain for instance, is not open in S^1, for the point $p = f(0)$ lies in no open set V of $\mathbb{R}^2$ such that $V \cap S^1 \subset f(U)$. Here $X = [0, 1)$ is not compact. It may be of interest to observe that f^{-1} fails to be continuous in spite of the fact that $Y = S^1$ is compact.*

Example 2.17. *Let X be the subspace $[0, 1] \cup [2, 3]$ of $\mathbb{R}$, and let Y be the subspace $[0, 2]$ of $\mathbb{R}$. It is not difficult to show that the map $p : X \to Y$ defined by*

$$p(x) = \begin{cases} x & if \;\; x \in [0,1]; \\ x - 1 & if \;\; x \in [2,3] \end{cases}$$

is a surjective, continuous, closed map which is not an open map.

Theorem 2.54. *(The pasting lemma): Let X, Y be metric spaces and $X = A \cup B$, where A and B are closed in X. Let $f : A \longrightarrow Y$ and $g : B \longrightarrow Y$ be continuous. If $f(x) = g(x)$ for every $x \in A \cap B$, then f and g combine to give a continuous function $h : X \longrightarrow Y$, defined by setting $h(x) = f(x)$ if $x \in A$, and $h(x) = g(x)$ if $x \in B$.*

Proof. Let C be a closed subset of Y. Now $h^{-1}(C) = f^{-1}(C) \cup g^{-1}(C)$. Since f is continuous by Proposition 2.2, $f^{-1}(C)$ is closed in A and therefore closed in X by Theorem 2.13. Similarly, $g^{-1}(C)$ is closed in B and therefore closed in X. Their union is thus closed in X by Theorem 2.6. □

This theorem also holds if A and B are open sets in X.

Theorem 2.55. *Let X and Y be two metric spaces and let $f : X \longrightarrow Y$ be a continuous function. If X is connected then $f(X)$ is a connected subset of Y.*

Proof. It is given that $f : X \longrightarrow Y$ is a continuous map and X is connected. We wish to prove the image set $Z = f(X)$ is connected. Since the map obtained from f by restricting its range to the space Z is also continuous (by Remark 2.10), it suffices to consider the case of a continuous surjective map $g : X \longrightarrow Z$. Suppose that $Z = A \cup B$ is a separation of Z into two disjoint nonempty sets open in Z. Then $g^{-1}(A)$ and $g^{-1}(B)$ are disjoint sets whose union is X; they are open in X because g is continuous, and nonempty because g is surjective. Therefore, they form a separation of X, contradicting the assumption that X is connected. □

Corollary 2.11. *If $f : X \longrightarrow \mathbb{R}$ is continuous and X is connected then $f(X)$ is an interval.*

Proof. This is because connected subsets of $\mathbb{R}$ are intervals. □

Notice again that a metric space X is connected if and only if any continuous function from X to $\{0,1\} \subset \mathbb{R}$ is a constant. This is so because $\{0,1\} = \{0\} \cup \{1\}$ is not a connected set in $\mathbb{R}$.

Theorem 2.56. *(Intermediate value theorem): If $f : [a,b] \longrightarrow \mathbb{R}$ is continuous and $f(a) \leq c \leq f(b)$ then there is a point $x \in [a,b]$ such that $f(x) = c$.*

Proof. Since $[a,b]$ is connected and f is continuous by Theorem 2.55, $f([a,b])$ is a connected subset of $\mathbb{R}$. But as connected subsets of $\mathbb{R}$ are only intervals hence $f([a,b])$ is an interval I in $\mathbb{R}$. Note $f(a), f(b) \in I$ and $c \in [f(a), f(b)] \subset I$. Thus $c \in I = f([a,b])$. Hence there exists $x \in [a,b]$ such that $f(x) = c$. □

Corollary 2.12. *If $f : X \longrightarrow \mathbb{R}$ is continuous and $K \subset X$ is compact then there are points a and b in K with $f(a) = \sup\{f(x) : x \in K\}$ and $f(b) = \inf\{f(x) : x \in K\}$.*

Proof. The set K is compact and f is continuous. By Theorem 2.52, $f(K)$ is a compact subset of $\mathbb{R}$. Hence $f(K)$ is a closed and bounded subset of $\mathbb{R}$ by Theorem 2.43. If $\alpha = \sup\{f(x) : x \in K\}$ then $\alpha \in f(K) \subset \mathbb{R}$. Thus there exist $a \in K$ such that $\alpha = f(a)$. Similarly, $\beta = \inf\{f(x) : x \in K\}$ is in $f(K) \subset \mathbb{R}$. So there exists $b \in K$ such that $\beta = f(b)$. Thus a continuous function on a compact set attains its minimum and maximum. □

Corollary 2.13. *Let X be a metric space and K be a compact subset of X. Let $f : X \longrightarrow \mathbb{C}$ be continuous. Then there are points a and b in K with $|f(a)| = \sup\{|f(x)| : x \in K\}$ and $|f(b)| = \inf\{|f(x)| : x \in K\}$.*

Proof. Note $g(x) = |f(x)|$ is continuous from X into $\mathbb{R}$. The corollary follows from the Corollary 2.12. □

Corollary 2.14. *Let X be a metric space and let K be a compact subset of X. If $x \in X$ then there is a point $y \in K$ with $d(x,y) = d(x,K)$.*

Proof. Define $f : K \longrightarrow \mathbb{R}$ by $f(p) = d(x,p)$. Then f continuous because by triangle inequality, $|f(p) - f(q)| = |d(x,p) - d(x,q)| \leq d(p,q)$. By Corollary 2.12, f attains its minimum at some point y in K. Thus $f(y) \leq f(z)$ for all $z \in K$. This gives $d(x,y) \leq d(x,z)$ for all $z \in K$. Thus $d(x,y) = \inf_{z \in K} d(x,z) = d(x,K)$. □

Definition 2.39. *If A and B are subsets of a metric space X then define the distance from A to B, $d(A,B)$, by $d(A,B) = \inf\{d(a,b) : a \in A$ and $b \in B\}$.*

Notice that (1) If $A = \{x\}$, a singleton set then $d(\{x\}, B) = d(x,B)$. (2) If $A = \{x\}$ and $B = \{y\}$ then $d(\{x\},\{y\}) = d(x,y)$. (3) If $A \cap B \neq \emptyset$ then $d(A,B) = 0$. But $d(A,B) = 0$ does not imply that $A \cap B \neq \emptyset$.

Example 2.18. *Let $A = \{(x, 0) : x \in \mathbb{R}\}$ and $B = \{(x, e^x) : x \in \mathbb{R}\}$. Notice that A and B are both closed in $\mathbb{R}^2$ and $A \cap B = \emptyset$. But $d(A, B) = 0$.*

Theorem 2.57. *Let X be a metric space. Suppose A and B are two disjoint sets in X such that A is compact and B is closed in X. Then $d(A, B) > 0$.*

Proof. By Proposition 2.4, the function $f : X \longrightarrow \mathbb{R}$ defined by $f(x) = d(x, B)$ is continuous. Since $A \cap B = \emptyset$ and B is closed, $f(a) > 0$ for each a in A. But since A is compact there is a point c in A such that $0 < f(c) = \inf\{f(x) : x \in A\} = d(A, B)$. □

Definition 2.40. *(Uniform convergence): Let $f_n : X \longrightarrow Y$ be a sequence of functions from the set X to the metric space Y. Let d be the metric for Y. We say that the sequence (f_n) converges uniformly to the function $f : X \longrightarrow Y$ if given $\epsilon > 0$, there exists an integer N such that $d(f_n(x), f(x)) < \epsilon$ for all $n \geq N$ and all x in X.*

Consider the following examples: (i) Define $f_n : \mathbb{C} \to \mathbb{C}$ as $f_n(z) = \frac{1}{1+nz}, n \in \mathbb{N}$. The sequence $(f_n(z))$ is uniformly convergent to zero for all z such that $|z| \geq 2$. The reason is as follows. We have $|\frac{1}{1+nz} - 0| < \epsilon$ when $\frac{1}{|1+nz|} < \epsilon$ or $|1+nz| > \frac{1}{\epsilon}$. Now $|1+nz| \leq 1+|nz| = 1+n|z|$ and $1+n|z| \geq |1+nz| > \frac{1}{\epsilon}$ for $n > \frac{\frac{1}{\epsilon}-1}{|z|}$. Thus the sequence $(f_n(z))$ converges to 0 for $|z| > 2$. To determine whether it converges uniformly to zero, note that the largest value of $\frac{\frac{1}{\epsilon}-1}{|z|}$ in $|z| \geq 2$ occurs for $|z| = 2$ and is given by $\frac{1}{2}\{\frac{1}{\epsilon} - 1\} = N$. It follows that $|\frac{1}{1+nz} - 0| < \epsilon$ for all $n > N$ where N depends only on ϵ and not on the particular z in $|z| \geq 2$. Thus the sequence is uniformly convergent to zero in this region.

(ii) The sequence $(f_n(z)) = (1+\frac{z}{n})$ converges to 1 for all $z \in \mathbb{C}$ as $\lim_{n\to\infty}(1+\frac{z}{n}) = 1$ for all z. Hence given $\epsilon > 0$, we must find N such that $|1 + \frac{z}{n} - 1| < \epsilon$ for $n > N$. Then $|\frac{z}{n}| < \epsilon$, i.e., $\frac{|z|}{n} < \epsilon$ if $n > \frac{|z|}{\epsilon} = N$. Thus $(f_n(z))$ does not converge to 1 uniformly on $\mathbb{C}$.

We have the following theorem about uniformly convergent sequences:

Theorem 2.58. *(Uniform limit theorem): Let (X, d_X) and (Y, d_Y) be two metric spaces and let $f_n : X \longrightarrow Y$ be a sequence of continuous functions. If (f_n) converges uniformly to f, then f is continuous.*

Proof. Let V be open in Y; let x_0 be a point of $f^{-1}(V)$. We wish to find a neighborhood U of x_0 such that $f(U) \subset V$.

Let $y_0 = f(x_0)$. Then $y_0 \in V$. Since V is open there exists $\epsilon > 0$ such that $B_{d_Y}(y_0, \epsilon) \subset V$. Now since (f_n) converges uniformly to f, there exists N such that for all $n \geq N$ and all $x \in X$,

$$d_Y(f_n(x), f(x)) < \frac{\epsilon}{4}. \tag{2.13}$$

Finally, using continuity of f_N; there exists $\delta > 0$ such that

$$f_N(B_{d_X}(x_0, \delta)) \subset B_{d_Y}(f_N(x_0), \frac{\epsilon}{2}). \tag{2.14}$$

Now if $x \in B_{d_X}(x_0, \delta)$, then from (2.13) and (2.14) we obtain

$$d_Y(f_N(x), f_N(x_0)) < \frac{\epsilon}{2},$$

$$d_Y(f_N(x), f(x)) < \frac{\epsilon}{4},$$

and

$$d_Y(f_N(x_0), f(x_0)) < \frac{\epsilon}{4}.$$

Using triangle inequality, we have $d_Y(f(x), f(x_0)) < \epsilon$ whenever $d_X(x, x_0) < \delta$. Thus given $\epsilon > 0$,there exists $\delta > 0$ such that $f(B_{d_X}(x_0, \delta)) \subset B_{d_Y}(f(x_0), \epsilon)$. Hence $B_{d_X}(x_0, \delta) \subset f^{-1}(B_{d_Y}(f(x_0), \epsilon))$. It follows therefore that $f^{-1}(V)$ is open in X. By Proposition 2.2, f is continuous on X. □

Definition 2.41. *A function $f : (X, d) \to (Y, d)$ is said to be locally constant if for every $x \in X$ there exists an $r > 0$ such that $f(p) = f(x)$ for all $p \in B(x, r)$.*

The proof of the following result is a typical use of connectedness argument.

Theorem 2.59. *Let (X, d) be connected. Then any locally constant continuous function from X to any metric space Y is a constant.*

Proof. Let f be a locally constant map from X to Y. Fix an $x \in X$. Consider the set $U = \{z \in X : f(z) = f(x)\}$. Then $x \in U$, so $U \neq \emptyset$. The set U is closed. $U = f^{-1}\{f(x)\}$. This is so because $U = f^{-1}\{f(x)\}$, the singleton set $\{f(x)\}$ is closed and f is continuous. The set U is open. The reason is as follows. Let $w \in U$. Then $w \in X$. Because f is locally constant, there exists $r > 0$ such that $f(w) = f(z)$ for all $z \in B(w, r)$. But $w \in U$ implies $f(w) = f(x)$. Thus $f(x) = f(z)$ for all $z \in B(w, r)$. Hence $B(w, r) \subset U$. Therefore, U is open in X. Because X is connected, $U = X$. Thus f is a constant function. □

Definition 2.42. *A subset $U \subseteq \mathbb{C}$ is called a region if U is open and connected.*

A very useful analytical characterization of connected spaces is the following:

Proposition 2.5. *A metric space X is connected if and only if any continuous function from X to $\{0, 1\} \subset \mathbb{R}$ is a constant.*

Proof. If X is connected and if $f : X \to \{0, 1\}$ is continuous and onto, then $U = f^{-1}(-\frac{1}{2}, \frac{1}{2})$ and $V = f^{-1}(\frac{1}{2}, \frac{3}{2})$ are nonempty open subsets of X(as f is continuous and $(-\frac{1}{2}, \frac{1}{2})$ and $(\frac{1}{2}, \frac{3}{2})$ are open in $\mathbb{R}$) such that $U \cup V = X$ and $U \cap V = \emptyset$. That is X has a separation which is a contradiction. Hence f is not onto. Conversely, if X is not connected, there exist U and V which provide a separation for X. Define

$$f(x) = \begin{cases} 0 & \text{if} \quad x \in U; \\ 1 & \text{if} \quad x \in V. \end{cases}$$

Then f is continuous on X and onto $\{0, 1\}$. □

Remark 2.13. Let $X = [a, b] \subset \mathbb{R}$. Then using the above proposition we can show that X is connected. The argument goes as follows: Suppose the interval $[a, b]$ is not connected. Then (see the proof of Proposition 2.5), there exists a map $f : [a, b] \to \{0, 1\}$ which is continuous and onto. By intermediate value theorem the function f must take all value between 0 and 1, that is $f([a, b]) = [0, 1]$ which is a contradiction to our hypothesis that $f([a, b]) = \{0, 1\}$.

Definition 2.43. *Given points x and y of a metric space X, a path in X from x to y is a continuous map $f : [a, b] \to X$ of some closed interval in the real line into X, such that $f(a) = x$ and $f(b) = y$. A space X is said to be path connected if every pair of points can be joined by a path in X.*

Example 2.19. *Let $X = [a, b] \subset \mathbb{R}$. Then X is path connected. For, if $x, y \in [a, b]$, then $\gamma : [0, 1] \to [a, b]$ given by $\gamma(t) = ty + (1 - t)x$ is a path from x to y.*

Example 2.20. *Any disk $B(a, r) \subset \mathbb{C}$ is path connected. For, if $z, w \in B(a, r)$, consider $\gamma(t) = tw + (1 - t)z$ for $0 \leq t \leq 1$.*

Theorem 2.60. *If a metric space is path connected then it is connected.*

Proof. Suppose $X = A \cup B$ is a separation of X. Let $f : [a, b] \to X$ be any path in X. Being the continuous image of a connected set, the set $f([a, b])$ is connected, so that it lies entirely in either A or B. Therefore, there is no path in X joining a point of A to a point of B, contrary to the assumption that X is path connected. □

Definition 2.44. *We define an equivalence relation on the space X by defining $x \sim y$ if there is a path in X from x to y. The equivalence classes are called the path components of X.*

We now have the following theorem whose proof is immediate:

Theorem 2.61. *The path components of X are path-connected disjoint subsets of X whose union is X, such that each path-connected subset of X intersects only one of them.*

Example 2.21. *The components of the subspace $Y = [-1, 0) \cup (0, 1]$ of the real line $\mathbb{R}$ are the two sets $[-1, 0)$ and $(0, 1]$. These are also the path components of Y.*

Definition 2.45. *A subset E of $\mathbb{C}$ is said to be convex if for every $z, w \in E$, the line segment $[z, w] = \{tw + (1 - t)z : 0 \leq t \leq 1\}$ is contained in E. We say $G \subset \mathbb{C}$ is star-shaped at $z \in G$ if the line segment $[z, w] \subset G$ for each $w \in G$.*

Definition 2.46. *A metric space X is said to be locally path connected at x if for every neighborhood U of x, there is a path-connected neighborhood V of x contained in U. If X is locally path connected at each of its points, then it is said to be locally path connected.*

Example 2.22. *(1) Any disk $B(a,r)$ in $\mathbb{C}$ is convex. (2) $E \subset \mathbb{C}$ is convex if and only if it is star-shaped at each of its points. (3) Any star-shaped subset of $\mathbb{C}$ is path connected and hence connected.*

Suppose the set G is star shaped at $a \in G$. Let z and w be two points in G. Then $[z,a,w] = [z,a] \cup [a,w]$ is a polygon in G connecting z and w. Hence G is connected.

Another typical application of connectedness is the following.

Proposition 2.6. *Let U be an open subset of $\mathbb{C}$. Then U is connected if and only if it is path connected.*

Proof. Fix $x \in U$. Consider $E = \{y \in U : \text{there is a path from } x \text{ to } y\}$. Then E is open in U: if $y \in E$, choose $r > 0$ such that $B(y,r) \subset U$. If $z \in B(y,r)$, then y and z are connected by the line segment $[y,z]$. Since x and y are path connected, y and z are path connected hence x and z are path connected. Thus $z \in E$. Hence $B(y,r) \subset E$. Now E is closed in U. For, if $a \in U$ is a limit point of E and if $r > 0$ is such that $B(a,r) \subset U$, then there exists $z \in E \cap B(a,r)$ such that $z \neq a$. Now, x is path connected to z (as $z \in E$)and z is path connected to a (as $B(a,r)$ is convex) via, the line segment $[z,a]$. Hence $a \in E$.

Since $x \in E$, E is a nonempty subset of U which is open and closed in U. Hence $E = U$. □

2.6 Exercises

1. Sketch the following sets and state which of them are open and which are not.

(a)	$\{z \in \mathbb{C} : 1 < \lvert z\rvert < 2\}$	(b)	$\{z \in \mathbb{C} : \lvert z\rvert < 1 \text{ and } \mathrm{Re} z \geq 0\}$
(c)	$\{z \in \mathbb{C} : z^{20} = 1\}$	(d)	$\{z \in \mathbb{C} : \mathrm{Re} z > 0\}$
(e)	$\{z \in \mathbb{C} : \lvert z + z^2\rvert < 1\}$	(f)	$\{z \in \mathbb{C} : 1 < \lvert z - 2\rvert < 3, \mathrm{Re} z \geq 0\}$
(g)	$\{z \in \mathbb{C} : \mathrm{Im} z = \mathrm{Re}\, z\}$	(h)	$\{z \in \mathbb{C} : \lvert\mathrm{Re}\, z\rvert > 2\}$
(i)	$\{z \in \mathbb{C} : z = \frac{1}{n}, n \in \mathbb{N}\}$	(j)	$\mathbb{N}$
(k)	$\{z \in \mathbb{C} : \mathrm{Re}\, z \in Q\}$	(l)	$\{z \in \mathbb{C} : (\mathrm{Re}\, z)(\mathrm{Im}\, z) = 0\}$
(m)	$\{z \in \mathbb{C} : \mathrm{Im}\, z^2 < 0\}$	(n)	$\{z \in \mathbb{C} : \arg(z - 1 - i) = \frac{\pi}{3}\}$
(p)	$\{z \in \mathbb{C} : \frac{\pi}{6} < \arg z < \frac{\pi}{3}\}$		

2. Give an example of an infinite set which does not have a limit point.

3. Show that the set $\{z \in \mathbb{C} : \mathrm{Re} z^3 = \mathrm{Im} z^2\}$ is closed in $\mathbb{C}$.

4. For each of the following subsets of $\mathbb{C}$ state whether the set is open, closed or neither.

(i)	$\{z \in \mathbb{C} : z = x + iy, x, y \in Q\}$	(ii)	$\{z \in \mathbb{C} : \text{Re}z^2 = 4\}$
(iii)	$\{z \in \mathbb{C} : \text{Re}z^2 = 4\}$	(iv)	$\{z \in \mathbb{C} : \|z\|^2 > \text{Im}z\}$
(v)	$\{z \in \mathbb{C} : 3 \leq \text{Re}z \leq 7\}$	(vi)	$\{z \in \mathbb{C} : 4 \leq \|z - 5\| \leq 7\}$

5. Let (X, d) be a metric space. Show that if A is closed in Y and Y is closed in X then A is closed in X.

6. Suppose X is a metric space with metric d. Show that if U is open in X and A is closed in X, then $U - A$ is open in X and $A - U$ is closed in X.

7. Let A, B and A_α denote subsets of a metric space X. Show that (a)$\overline{A \cup B} = \overline{A} \cup \overline{B}$, (b) $\overline{\cup A_\alpha} \supset \cup \overline{A_\alpha}$; give an example where equality fails.

8. Let $A_n = \{z \in \mathbb{C} : |z| < 1 + \frac{1}{n}\}, n = 1, 2, 3, \cdots$. Show that $\cap_{n=1}^{\infty} A_n$ is not open.

9. A set is said to be discrete if all its points are isolated. Show that a discrete set in $\mathbb{R}$ or $\mathbb{C}$ is countable.

10. Show that a discrete set in a separable metric space is countable.

11. Let X be a set ; let $\mathcal{B}$ be a basis for a topology $\mathcal{T}$ on X. Show that $\mathcal{T}$ equals the collection of all unions of elements of $\mathcal{B}$.

12. Let X be a set. Let d and d' be two metrics on X. The two metrics d and d' are said to be equivalent if there exist $C > 0$ and $M > 0$ such that $Cd(x, y) \leq d'(x, y) \leq Md(x, y)$. Show that in $\mathbb{R}^n$ the two metrics d and ρ defined as $d(x, y) = (\sum_{i=1}^{n} |x_i - y_i|^2)^{\frac{1}{2}}$, $\rho(x, y) = \max\limits_{1 \leq i \leq n} \{|x_i - y_i|\}$ are equivalent if $x = (x_1, x_2, \cdots, x_n)$ and $y = (y_1, y_2, \cdots, y_n)$ belong to $\mathbb{R}^n$.

13. Let X and Ω be metric spaces and suppose that $f : X \longrightarrow \Omega$ is one-one and onto. Show that f is an open map if and only if f is a closed map.

14. Let $p : \mathbb{C} \longrightarrow \mathbb{R}$ be defined by $p(z) = \text{Re}z$; show that p is an open map but is not a closed map.

15. Suppose X and Y are two metric spaces. If A is closed in X and B is closed in Y then show that $A \times B$ is closed in $X \times Y$.

16. Let $A \subset X$ and $B \subset Y$ where X and Y are metric spaces. Show that in the space $X \times Y$, $\overline{A \times B} = \overline{A} \times \overline{B}$.

17. Show that the set Q of rational numbers is not connected in $\mathbb{R}$.

18. Which of the following subsets of $\mathbb{C}$ are connected ?

(i)	$A = \{z \in \mathbb{C} : \|z\| \leq 1\}$	(ii)	$B = \{a + ib \in \mathbb{C} : a, b \in Q\}$
(iii)	$C = \{z \in \mathbb{C} : 1 < \|z\| < 2\}$	(iv)	$D = \{z \in \mathbb{C} : 1 < \text{Re}z < 2\}$
(v)	$E = \mathbb{C} - \{a, b\}, a, b \in \mathbb{C}$		

19. Let X be a metric space and $f_n, g_n : X \to \mathbb{C}$ be such that f_n (respectively g_n) converge uniformly on X to f (respectively g). Assume that f and g are bounded on X. Show that $f_n g_n$ converges to fg uniformly on X.

20. Find out which of the following subsets of $\mathbb{C}$ are compact and which are not.

(i)	$\{z \in \mathbb{C} : \lvert z-3\rvert \leq 5\}$	(ii)	$\{z \in \mathbb{C} : \lvert z-4\rvert \geq 4\}$
(iii)	$\{z \in \mathbb{C} : \lvert z\rvert = 2\}$	(iv)	$\{z \in \mathbb{C} : 0 < \lvert z\rvert \leq 1\}$
(v)	$\{z \in \mathbb{C} : \lvert z\rvert = 1, z \neq 1\}$	(vi)	$\{z \in \mathbb{C} : \lvert z-2\rvert > 3\}$
(vii)	$\{z \in \mathbb{C} : 0 < \arg z < \pi, \lvert z\rvert \leq 1\}$.		

21. Give three examples of metric spaces that are not complete.

22. Show that the closure of a totally bounded set is bounded.

23. Let A and B be disjoint compact subsets of the metric space X. Show that there exist disjoint open sets U and V containing A and B respectively.

24. Show that every compact subset of a metric space is bounded and closed.

25. Show that the unit sphere S^{n-1} and the closed unit ball B^n in $\mathbb{R}^n$ are compact.

26. Prove that the set $A = \{x \times \frac{1}{x} : 0 < x \leq 1\}$ is closed in $\mathbb{R}^2$ but not compact.

27. Show that the set $S = \{x \times \sin\frac{1}{x} : 0 < x \leq 1\}$ is bounded in $\mathbb{R}^2$ but it is not compact.

28. Let (X, d) be a metric space. Show that X is totally bounded under d if and only if it is totally bounded under $\bar{d}(x,y) = \min\{d(x,y), 1\}$.

29. Suppose X and Y are two metric spaces and $f : X \to Y$ is continuous. If X is compact show that f is a closed map.

30. Give an example of a closed, bounded subset of a metric space which is not compact.

31. Show that the union of a finite number of compact sets is compact in a metric space X.

32. Show that $I = [0,1]$ with the usual topology and I^2 (with product topology) cannot be homeomorphic.

33. If X and Y are compact metric spaces then show that $X \times Y$ is compact in the product topology.

34. Let X be a metric space with metric d. Let $f, g : X \to \mathbb{R}$ be continuous. (a) Show that the set $\{x : f(x) \leq g(x)\}$ is closed in X. (b) Let $h : X \to \mathbb{R}$ be the function $h(x) = \min\{f(x), g(x)\}$. Show that h is continuous.

35. Define $f_n : [0,1] \to \mathbb{R}$ by the equation $f_n(x) = x^n$. Show that the sequence $(f_n(x))$ converges for each $x \in [0,1]$ but that the sequence (f_n) does not converge uniformly.

36. Let (X, d) be a metric space, $\Omega \subset \mathbb{C}$. Suppose $f : X \to \Omega$ is uniformly continuous. Show that if $\{x_n\}$ is a Cauchy sequence in X then $\{f(x_n)\}$ is a Cauchy sequence in Ω.

37. Show that $\mathbb{R}^n$ and $\mathbb{R}$ are not homeomorphic if $n > 1$.

38. Show that the intervals $[a, b]$ and $[c, d]$ in $\mathbb{R}$ are homeomorphic.

39. Find a bijective, continuous map from the open interval $(-1, 1)$ into $\mathbb{R}$ which is also a homeomorphism.

40. Show that $\mathbb{R}^n$ and $\mathbb{R}^2$ are not homeomorphic if $n > 2$.

41. Assume that $\mathbb{R}$ is uncountable. Show that if A is a countable subset of $\mathbb{R}^2$, then $\mathbb{R}^2 - A$ is path connected.(Hint: How many lines are there passing through a given point of $\mathbb{R}^2$?)

42. If X, Y, A are metric spaces and $f : A \to X \times Y$ is given by the equation $f(a) = (f_1(a), f_2(a))$ then show that f is continuous if and only if the functions $f_1 : A \to X$ and $f_2 : A \to Y$ are continuous.

43. Let X be a topological space and $A \subset X$. If C is a connected subset of X with nonempty intersection with A as well as $X - A$, show that it has also nonempty intersection with the boundary of A.

44. Show that a locally path connected space is path connected if and only if it is connected.

Chapter 3

Analysis on $\mathbb{C}$

3.1 Sequences of complex numbers

We assume the reader is familiar with the analysis on $\mathbb{R}$. The complex plane $\mathbb{C}$ as a set coincides with the Euclidean plane $\mathbb{R}^2$ and $\mathbb{R}^2$ becomes a metric space if we endow it with the Euclidean metric. Hence if we let $d(z, w) = |z - w|$, then d is a metric on $\mathbb{C}$. Thus, as we have already seen, all the concepts of metric spaces such as sequences, their convergence, continuity, open and closed sets, compactness and connectedness can be talked about in $\mathbb{C}$. The present chapter deals with analysis on $\mathbb{C}$ which is primarily concerned with limit processes and continuity. Based on the concept of distance, there are certain topological properties of $\mathbb{R}$ or $\mathbb{C}$, which are basis for the study of analysis on $\mathbb{C}$ and one needs to develop a thorough insight into the analytical structure of $\mathbb{C}$.

For $z \in \mathbb{C}$ and $r > 0$, we let $B(z, r) = \{w \in \mathbb{C} : |w - z| < r\}$ and $\overline{B(z, r)} = \{w \in \mathbb{C} : |w - z| \leq r\}$. They are respectively called the open disk and closed disk with center at z and radius r.

Definition 3.1. *A sequence in $\mathbb{C}$ (a complex sequence) is a function $f : \mathbb{N} \longrightarrow \mathbb{C}$ and is written as (z_n) where $z_n = f(n)$, the nth term of the sequence. As in practice, we shall use (z_n) to denote the sequence f. Given a sequence (z_n), a subsequence is defined as in the real case: A subsequence of a sequence is a sequence whose terms are chosen from the sequence and they are arranged in the same order as their relative order in the original sequence.*

Even though $\mathbb{C}$ is not an ordered field, it is possible to define bounded subset of $\mathbb{C}$ by using the concept of modulus. A subset D of $\mathbb{C}$ is called bounded if there exists $K > 0$ such that $|z| \leq K$ for all $z \in D$. A complex sequence (z_n) is called bounded if its range $\{z_1, z_2, z_3, \cdots\}$ is a bounded subset of $\mathbb{C}$, i.e. if there exists a constant $K > 0$ such that $|z_n| \leq K$ for all $n \in \mathbb{N}$.

Definition 3.2. *We say a complex sequence (z_n) converges to $z \in \mathbb{C}$ if given $\epsilon > 0$, there exists $n_0 \in \mathbb{N}$ depending on ϵ, such that for $n > n_0$, we have $|z_n - z| < \epsilon$. That is, $z_n \in B(z, \epsilon)$ for all but finitely many n. We write this as*

$\lim_{n\to\infty} z_n = z$ *or* $z_n \to z$ *and we call this number* z *as the limit of the sequence* (z_n).

Remark 3.1. Notice that we cannot rewrite the above statement as $z - \epsilon < z_n < z + \epsilon$, as we use to do in case of real z_n. Further observe that every convergent sequence is bounded: If $z_n \to z$ then given $\epsilon > 0$, there exists $n_0 \in \mathbb{N}$ such that for $n > n_0$, we have $|\,|z_n| - |z|\,| \leq |z_n - z| < \epsilon$. Hence $|z_n| < |z| + \epsilon$ for $n > n_0$. Let $K = \max\{|z_1|, \cdots, |z_{n_0}|, |z| + \epsilon\}$. Then it can be verified that $|z_n| \leq K$ for all n.

It is also easy to see that if $z_n \to z$ and $z_n \to w$ then $z = w$. This is so because given $\epsilon > 0$, $|z - w| \leq |z - z_n| + |z_n - w| < 2\epsilon$, for n sufficiently large.

Remark 3.2. If the sequence (z_n) does not converge to any $z \in \mathbb{C}$, then it is said to diverge. Further notice that the sequence (z_n) does not converge to $z \in \mathbb{C}$ if there exists $\epsilon > 0$ such that $|z_n - z| \geq \epsilon$ infinitely often.

We shall show below that convergence of the sequence $(z_n) = (x_n + iy_n)$ is equivalent to the simultaneous convergence of the sequence $(\text{Re}\, z_n) = (x_n)$ and $(\text{Im}\, z_n) = (y_n)$.

Proposition 3.1. *Let* (z_n) *be a sequence in* $\mathbb{C}$. *Then* $z_n \to z$ *if and only if* $Rez_n \to Rez$ *and* $Imz_n \to Imz$.

Proof. Let $z_n \to z$. Then $|\text{Re}z_n - \text{Re}z| = |\text{Re}(z_n - z)| \leq |z_n - z|$ from which the convergence of $\text{Re}z_n \to \text{Re}z$ follows. Conversely, if $\text{Re}z_n \to \text{Re}z$ and $\text{Im}z_n \to \text{Im}z$, then the inequality

$$|z_n - z| \leq |\text{Re}z_n - \text{Re}z| + |\text{Im}z_n - \text{Im}z|$$

shows that $z_n \to z$. □

Example 3.1. *(i) Let* $(z_n) = (i^n) = (i, -1, -i, 1, i, \cdots)$. *The sequence is bounded, divergent and has finite range. The set of cluster points of* (z_n) *is* $\{1, i, -i, -1\}$.

(ii) Let $z_n = \frac{i^n}{n}$ *is bounded and is convergent to the value 0.*

(iii) Let $(z_n) = (1 + \frac{1}{n}) + i(\frac{n-1}{n})$ *is bounded, has infinite range and converges to* $1 + i$.

(iv) $z_n = (1 + \frac{1}{n})^n + i(1 + \frac{3}{n})^n$ *is bounded, has infinite range and converges to* $e + ie^3$. *In fact* $Rez_n \to e$ *and* $Imz_n \to e^3$. *So* $\lim z_n = e + ie^3$.

(v) In (i) we have already seen that the sequence $(z_n) = (i^n)$ *is divergent. In this case, we have*

$$i^n = \begin{cases} 1 & if \quad n \equiv 0 (mod\ 4) \\ i & if \quad n \equiv 1 (mod\ 4) \\ -1 & if \quad n \equiv 2 (mod\ 4) \\ -i & if \quad n \equiv 3 (mod\ 4). \end{cases}$$

Hence, looking at the real and imaginary parts of i^n, we see that neither of these real sequences converges.

(vi) Let $z_n = \frac{(1+i)^n}{n}$. Notice that the limit of $|z_n|$ cannot exist and the limit of z_n cannot exist. That is, the sequence (z_n) diverges.

We have the following theorem on algebra of limits whose proofs are exactly analogous to the results in real theory and hence the proofs are left as an exercise to the reader.

Theorem 3.1. *Let $(z_n), (w_n)$ be sequences in $\mathbb{C}$. Assume that $\lim z_n = z$ and $\lim w_n = w$. Then the following are true:*

1. $\lim_{n\to\infty} (z_n + w_n) = z + w$.
2. $\lim_{n\to\infty} z_n w_n = zw$.
3. *If $\alpha \in \mathbb{C}$, then* $\lim_{n\to\infty} \alpha z_n = \alpha z$.
4. *The set of convergent sequence in $\mathbb{C}$ is a vector space over $\mathbb{C}$ with the vector addition and scalar multiplication defined in (i) and (iii).*
5. *If $w \neq 0$, then there exists $N \in \mathbb{N}$ such that for $n \geq N$, $w_n \neq 0$.*
6. *Let $w \neq 0$. Let N be as in (v). Then* $\lim_{n\to\infty} \left(\frac{z_n}{w_n}\right) = \frac{z}{w}$.

Remark 3.3. Notice that if $z_n \to z$ and $|z_n - w_n| \to 0$ then the sequence (w_n) converges to z. This happens because by triangle inequality

$$|w_n - z| \leq |w_n - z_n| + |z_n - z|.$$

Definition 3.3. *The sequence (z_n) is said to be Cauchy if given $\epsilon > 0$, there exists $n_0 \in \mathbb{N}$ such that $|z_n - z_m| < \epsilon$ for all $n, m \geq n_0$.*

Definition 3.4. *A subset D of $\mathbb{C}$ is said to be complete (or more appropriately Cauchy complete) if every Cauchy sequence in D is convergent to value in D.*

Theorem 3.2. *A sequence (z_n) in $\mathbb{C}$ is Cauchy if and only if it is convergent.*

Proof. Let $z_n \in \mathbb{C}$ be Cauchy. Let $z_n = x_n + iy_n$ where x_n, $y_n \in \mathbb{R}$. Now $z_n - z_m = (x_n - x_m) + i(y_n - y_m)$. Hence $|x_n - x_m| \leq |z_n - z_m|$, $|y_n - y_m| \leq |z_n - z_m|$. It follows from the above inequalities that (x_n) and (y_n) are Cauchy in $\mathbb{R}$ as (z_n) is Cauchy in $\mathbb{C}$. By Cauchy's completeness principle for $\mathbb{R}$, Example 2.9 there exists $x \in \mathbb{R}$, $y \in \mathbb{R}$ such that $x_n \to x$ and $y_n \to y$. Now define $z = x + iy$. Hence, by triangle law, $|z_n - z| = |(x_n - x) + i(y_n - y)| \leq |x_n - x| + |y_n - y| \to 0$ as $n \to \infty$. This proves that $\mathbb{C}$ is complete. If $z_n \to l$, it follows from the inequality: $|z_n - z_m| \leq |z_n - l| + |z_m - l|$ that (z_n) is Cauchy. □

In the following theorem we shall show that a convergent subsequence (z_{n_k}) may be chosen from any infinite bounded sequence (z_n).

Theorem 3.3. *(1) If (z_{n_k}) is a subsequence of (z_n) and $\lim_{n\to\infty} z_n = z$, then $\lim_{k\to\infty} z_{n_k} = z$. (2) (Bolzano-Weierstrass Theorem) Let (z_n) be a bounded sequence in $\mathbb{C}$. Then there is a convergent subsequence (z_{n_k}).*

Proof. The assertion (1) follows from the triangle inequality. To prove (2), notice that $(x_n = \text{Re}z_n)$ and $(y_n = \text{Im}z_n)$ are bounded sequences in $\mathbb{R}$. This is so because $|x_n| \leq |z_n|, |y_n| \leq |z_n|$ and (z_n) is a bounded sequence in $\mathbb{C}$. By Bolzano-Weierstrass theorem for real sequences, there is a subsequence (x_{n_k}) which is convergent, say, to x in $\mathbb{R}$. We once again appeal to Bolzano-Weierstrass theorem to the bounded real sequence (y_{n_k}) to find a subsequence $(y_{n_{k_r}})$, converging to, say y. Then the subsequence $(z_{n_{k_r}})$ is convergent to $x + iy$. □

We have already seen that a complex sequence need not converge at all. But if it does converge, it converges to only one point. We shall show below that if a bounded sequence (z_n) diverges then there exist two convergent subsequences of (z_n) with distinct limits.

Proposition 3.2. *Let (z_n) be a bounded sequence in $\mathbb{C}$ which is not convergent. Then there exist two convergent subsequences with different limits.*

Proof. By Bolzano-Weierstrass theorem (Theorem 3.3), there exists (z_{n_k}), a convergent subsequence of (z_n) with limit, say α. Since (z_n) is not convergent, in particular not to α, there exists $\epsilon > 0$ such that given any r, there exists $n_r \geq r$, $n_r \geq n_{r-1}$ and $|z_{n_r} - \alpha| \geq \epsilon$. Now (z_{n_r}) is a bounded sequence and hence it has a convergent subsequence, say $(z_{n_{r_s}})$ with limit β. This is a subsequence of (z_n) and by construction, $\alpha \neq \beta$. □

The following is an immediate corollary of the proposition.

Corollary 3.1. *Let (z_n) be a bounded sequence in $\mathbb{C}$. If there exists $z \in \mathbb{C}$ such that every convergent subsequence of (z_n) has the same limit z, then $\lim\limits_{n\to\infty} z_n = z$.*

Remark 3.4. It is not difficult to verify that if a subsequence (z_{n_k}) of a Cauchy sequence (z_n) converges to $z \in \mathbb{C}$ then the sequence (z_n) itself converges to $z \in \mathbb{C}$.

3.2 Series of complex numbers

An infinite series of complex numbers is of the form $\sum_{n=1}^{\infty} a_n$. One way to assign a meaning to an infinite sum is through convergence of the sequence (s_n) where

$$s_n = \sum_{k=1}^{n} a_k = a_1 + a_2 + \cdots + a_n$$

which is called nth partial sum of the series.

We say that the series is convergent if the sequence (s_n) of its partial sums $s_n = \sum_{k=1}^{n} a_k$ is convergent. If $\alpha = \lim s_n$ we then write $\sum_{k=1}^{\infty} a_k = \alpha$ and call α the sum of the series.

Example 3.2. *(**Geometric series**) Let $z \in \mathbb{C}$ and $|z| < 1$. Consider the infinite series $\sum_{n=0}^{\infty} z^n$. It is called the geometric series. It is not difficult to see that the series converges to $\alpha = \frac{1}{1-z}$ for $|z| < 1$. Its n-th partial sum s_n is given by*

$$s_n = \sum_{k=0}^{n} z^k = \frac{1 - z^{n+1}}{1 - z}.$$

Now, $|\alpha - s_n| = \left|\frac{z^{n+1}}{1-z}\right| \leq \frac{|z|^{n+1}}{1-|z|}$ which converges to 0 as $n \to \infty$. If $|z| > 1$, the n-th term of the series does not tend to 0; so the series cannot be convergent in this case.

Our treatment of sequences in the earlier section allows us to quickly obtain the basic principles of convergence of infinite series. In fact every theorem on sequences can be restated in terms of series by putting $a_0 = s_0$, $a_n = s_n - s_{n-1}$ for $n \geq 1$, and vice-versa. Sometimes it is useful to consider both the forms.

Theorem 3.4. *Let $\sum_{n=1}^{\infty} a_n$ and $\sum_{n=1}^{\infty} b_n$ be series of complex numbers. Assume that $\sum_{n=1}^{\infty} a_n = \alpha$ and $\sum_{n=1}^{\infty} b_n = \beta$. Then*

(1) For any $\lambda \in \mathbb{C}$ and $\mu \in \mathbb{C}$, we have

$$\sum_{n=1}^{\infty} \lambda a_n + \sum_{n=1}^{\infty} \mu b_n = \lambda\alpha + \mu\beta.$$

(2) The series $\sum a_n$ converges if and only if, for every $\epsilon > 0$, there exists $n_0 \in \mathbb{N}$ such that, for $m \geq n > n_0$,

$$\left|\sum_{k=n}^{m} a_k\right| < \epsilon.$$

(3) If $\sum_n |a_n|$ is convergent then $\sum_n a_n$ is convergent but the converse is not necessarily true.

(4) If $\sum a_n$ converges, then a necessary but not sufficient condition is that $\lim a_n = 0$.

(5) Let $a_n \in \mathbb{R}$ and $a_n \geq 0$. Then the series $\sum a_n$ converges if and only if the partial sums form a bounded sequence.

(6) $\sum_n a_n$ is convergent if and only if $\sum_n Re a_n$ and $\sum_n Im a_n$ is convergent. Furthermore, the sum of the infinite series $\sum_n a_n$ is $x + iy$ if x and y are respectively the sums of $\sum Re a_n$ and $\sum Im a_n$.

(7) Let $\sum a_n$ converges with sum α. Then $\sum \overline{a_n}$ converges and its sum is $\overline{\alpha}$.

Proof. Since $\mathbb{C}$ is complete, the series $\sum a_n$ converges if and only if the sequence of partial sums s_m is a Cauchy sequence. That is, if and only if for every $\epsilon > 0$, there exists $n_0 \in \mathbb{N}$ such that, for $m \geq n > n_0$, $|\sum_{k=n}^{m} a_k| = |s_m - s_{n-1}| < \epsilon$. This proves (2).

To prove (3), let s_n and σ_n denote the partial sums of $\sum a_n$ and $\sum |a_n|$ respectively. Then for $n > m$,

$$|s_n - s_m| = \left| \sum_{k=m+1}^{n} a_k \right| \leq \sum_{k=m+1}^{n} |a_k| = \sigma_n - \sigma_m,$$

which converges to 0, as (σ_n) is convergent. Hence (s_n) is Cauchy in $\mathbb{C}$. Now apply Theorem 3.2.The series $\sum_{n=1}^{\infty} \frac{(-1)^n}{n}$ is convergent (see Example 3.11) but $\sum_{n=1}^{\infty} \frac{1}{n}$ is not convergent (see Example 3.8).

(4) follows from (2) by taking $m = n$. In fact, we obtain $|a_n| < \epsilon$ $(n > n_0)$. To show the converse take the series $\sum_{n=1}^{\infty} \frac{1}{n}$ which is divergent though $a_n = \frac{1}{n} \to 0$. Since $a_n \geq 0$ implies that (s_n) is monotonic non-decreasing, the result (5) follows from the fact that monotonic increasing sequence bounded above is convergent.

□

Example 3.3. *(i) The series $\sum_{k=1}^{\infty} k = \infty$; for $s_n = \sum_{k=1}^{n} k = \frac{n(n+1)}{2} \to \infty$ as $n \to \infty$.*

(ii) The series $\sum_{k=0}^{\infty} (-1)^k$ is divergent; in fact in the above sense it has no meaning; for $(s_n) = (1, 0, 1, 0, 1, 0, \cdots)$ does not have any limit; it oscillates between finite bounds.

(iii) $\sum_{n=1}^{\infty} \frac{1}{n(n+1)} = 1$, since $s_n = \sum_{k=1}^{n} \frac{1}{k(k+1)} = \sum_{k=1}^{n} (\frac{1}{k} - \frac{1}{k+1}) = 1 - \frac{1}{n+1}$ converges to 1 as $n \to \infty$.

(iv) The series $\frac{1}{3} + \frac{2}{5} + \frac{3}{7} + \cdots + \frac{n}{2n+1} + \cdots$ diverges, since the nth term $a_n = \frac{n}{2n+1} \to \frac{1}{2}$ as $n \to \infty$.

(v) The series $\sum_{n=1}^{\infty} \frac{1}{n} \left(\frac{\cos \pi}{n} + i \frac{\sin \pi}{n} \right)$ converges.

Remark 3.5. We may remark that the statement

$$\sum (a_n + b_n) = \sum a_n + \sum b_n$$

is valid if $\sum a_n$ and $\sum b_n$ are both convergent. It may so happen that the series $\sum (a_n + b_n)$ may converge even when both $\sum a_n$ and $\sum b_n$ are divergent. For example take $a_n = \frac{1}{n}$, $b_n = -\frac{1}{n+1}$ (see Example 3.3 (iii) above).

It may further be noticed that if $n_0 \in \mathbb{N}$, then both the series

$$\sum_{k > n_0} a_k, \sum_{k=0}^{\infty} a_k$$

converge or diverge together (a consequence of Theorem 3.4), though their sum whenever it exists may be different.

Example 3.4. *(i) Consider the series* $\sum_{n=1}^{\infty}\left(\frac{1}{n^2}+\frac{i}{n}\right)$. *The series will converge if and only if both* $\sum_{n=1}^{\infty}\frac{1}{n^2}$ *and* $\sum_{n=1}^{\infty}\frac{i}{n}$ *converge. Since* $\sum_{n=1}^{\infty}\frac{1}{n}$ *diverges,* $\sum_{n=1}^{\infty}\left(\frac{1}{n^2}+\frac{i}{n}\right)$ *also diverges.*

(ii) The series $\sum_{n=1}^{\infty}\frac{1}{2^n}\left(\cos((\frac{\pi}{e})^n)+i\sin((\frac{\pi}{e})^n)\right)$ *is absolutely convergent.*

Since $\left|\frac{1}{2^n}\left(\cos((\frac{\pi}{e})^n)+i\sin((\frac{\pi}{e})^n)\right)\right|=\frac{1}{2^n}$ *for all* n *and* $\sum_{n=1}^{\infty}\frac{1}{2^n}$ *converges, the claim is verified.*

In Theorem 3.4, it is established that if (z_n) is a sequence of complex numbers then $\sum z_n$ is convergent if and only if $\sum \text{Re } z_n$ and $\sum \text{Im } z_n$ are convergent. Hence all the tests used to determine the convergence or divergence of a series whose terms are real numbers can be applied to the series $\sum \text{Re } z_n$ and $\sum \text{Im } z_n$ which will in turn determine the convergence or divergence of $\sum z_n$.

Theorem 3.5. ***(Comparison Test)***

Let $\sum a_n$ *and* $\sum b_n$ *be series of positive reals. Assume that* $a_n \leq b_n$ *for all* n. *Then (i) if* $\sum b_n$ *is convergent, then so is* $\sum a_n$, *and (ii) if* $\sum a_n$ *is divergent, so is* $\sum b_n$.

Proof. (i) Since $\sum b_n$ is convergent, given $\epsilon > 0$, there exists $n_0 \in \mathbb{N}$ such that for $m \geq n > n_0$, $\sum_{k=n}^{m} b_k < \epsilon$ (by Theorem 3.4). Hence by hypothesis

$$\sum_{k=n}^{m} a_k \leq \sum_{k=n}^{m} b_k < \epsilon$$

for $m \geq n > n_0$. This proves (i).

To prove (ii) suppose $\sum b_n$ converges, Then by (i) $\sum a_n$ should converge and this is a contradiction.

Note that if $\sum a_n$ is divergent then its partial sums form an increasing unbounded sequence. □

Example 3.5. ***(Harmonic series)***

The series $\sum_{n=1}^{\infty}\frac{1}{n^\alpha}$ *is called the harmonic series. Observe that*

$$\frac{2^k}{2^{\alpha(k+1)}} < \sum_{2^k+1}^{2^{k+1}} \frac{1}{n^\alpha} < \frac{2^k}{2^{\alpha k}}.$$

Now the geometric series $\sum_k \frac{2^k}{2^{\alpha(k+1)}}$ *is divergent if* $\alpha \leq 1$ *and the geometric series* $\sum_k \frac{2^k}{2^{\alpha k}}$ *is convergent if* $\alpha > 1$. *See Example 3.2 in this section. Hence the series* $\sum_{n=1}^{\infty}\frac{1}{n^\alpha}$ *is convergent if* $\alpha > 1$ *and is divergent if* $\alpha \leq 1$.

Corollary 3.2. *Suppose there exists* $n_0 \in \mathbb{N}$ *such that*

$$\frac{a_{n+1}}{a_n} \leq \frac{b_{n+1}}{b_n}(n > n_0)$$

then convergence of $\sum b_n$ *implies convergence of* $\sum a_n$. *If on the other hand*

$$\frac{a_{n+1}}{a_n} \geq \frac{b_{n+1}}{b_n}(n > n_0)$$

then divergence of $\sum b_n$ *to* ∞ *implies that* $\sum a_n$ *is divergent to* ∞.

Proof. For $n \geq n_0$,

$$\begin{aligned} a_n &= \frac{a_n}{a_{n-1}}\frac{a_{n-1}}{a_{n-2}}\cdots\frac{a_{n_0+1}}{a_{n_0}}.a_{n_0} \\ &\leq \frac{b_n}{b_{n-1}}\frac{b_{n-1}}{b_{n-2}}\cdots\frac{b_{n_0+1}}{b_{n_0}}a_{n_0} \\ &= \frac{a_{n_0}}{b_{n_0}}b_n. \end{aligned}$$

Hence the first part of the corollary follows from the comparison test. The proof of the second part is similar. □

Corollary 3.3. *If* $0 < \underline{\lim}_{n\to\infty}(\frac{a_n}{b_n}) \leq \overline{\lim_{n\to\infty}}(\frac{a_n}{b_n}) < \infty$ *then* $\sum a_n$ *and* $\sum b_n$ *converge or diverge together.*

Proof. Let $\lambda = \overline{\lim}(\frac{a_n}{b_n})$, $\mu = \underline{\lim}(\frac{a_n}{b_n})$ Then given $\epsilon > 0$, there exists $n_0 \in \mathbb{N}$ such that for $n > n_0$,

$$\frac{a_n}{b_n} < \lambda + \epsilon, \frac{a_n}{b_n} > \mu - \epsilon.$$

Since $\mu > 0$, we may choose ϵ such that $\mu - \epsilon > 0$. Now for $n > n_0$,

$$0 < \mu - \epsilon < \frac{a_n}{b_n} < \lambda + \epsilon.$$

Hence the result follows by the comparison test. □

Example 3.6. *The series* $\sum_{n=1}^{\infty} \frac{n+5}{n(n+1)\sqrt{n+2}}$ *converges.*

Solution: Let $a_n = \frac{n+5}{n(n+1)\sqrt{n+2}}, b_n = \frac{1}{n^{\frac{3}{2}}}$. Then $\frac{a_n}{b_n} = \frac{\sqrt{n}(n+5)}{(n+1)\sqrt{(n+2)}} \to 1$ as $n \to \infty$. Since $\sum \frac{1}{n^{\frac{3}{2}}}$ is convergent it follows from Corollary 3.3 that the given series is convergent.

The following test due to Cauchy determines the convergence of a series by considering the nature of a selected number of terms which constitute a condensed form of the original series.

Theorem 3.6. ***(Condensation test):*** *Suppose that* $a_n \geq a_{n+1} \geq 0$. *The series* $\sum a_n$ *converges if and only if the series* $\sum_{k=0}^{\infty} 2^k a_{2^k} = a_1 + 2a_2 + 4a_4 + 8a_8 + \cdots$ *converges.*

Proof. Let $s_n = \sum_{k=0}^{n} a_k$ and

$$t_k = a_1 + 2a_2 + \cdots + 2^k a_{2^k}.$$

For $n < 2^k$,

$$\begin{aligned} s_n &\leq a_1 + a_2 + \cdots + a_{2^k} \\ &\leq a_1 + (a_2 + a_3) + \cdots + (a_{2^k} + a_{2^k+1} + \cdots + a_{2^{k+1}-1}) + \cdots \\ &\leq a_1 + 2a_2 + 4a_4 + \cdots + 2^k a_{2^k} = t_k \ \ (\text{ as } a_k \geq a_{k+1}). \end{aligned} \tag{3.1}$$

For $n > 2^k$,

$$\begin{aligned} s_n &\geq a_1 + a_2 + (a_3 + a_4) + \cdots + (a_{2^{k-1}+1} + \cdots + a_{2^k}) \\ &\geq \frac{1}{2}a_1 + a_2 + 2a_4 + \cdots + 2^{k-1}a_{2^k} = \frac{1}{2}t_k. \end{aligned} \tag{3.2}$$

It follows from (3.1) and (3.2) that (s_n) and (t_n) are either both bounded or both unbounded. But we know that a series of positive terms is convergent if and only if it is bounded. Now the result follows from this. □

Example 3.7. *The series* $\sum_{n=2}^{\infty} \frac{1}{n(\log n)^{\alpha}}$ *converges if* $\alpha > 1$ *and diverges if* $\alpha \leq 1$.

Solution: Since $\frac{1}{n(\log n)^{\alpha}}$ is decreasing for all $\alpha \in \mathbb{R}$, by Theorem 3.6 the series converges if and only if $\sum_{k=2}^{\infty} 2^k \frac{1}{2^k(\log 2^k)^{\alpha}} = \frac{1}{(\log 2)^{\alpha}} \sum_{k=2}^{\infty} \frac{1}{k^{\alpha}}$ converges, but this is so if and only if $\alpha > 1$.

Remark 3.6. As is evident in the previous example, Cauchy's condensation test is particularly well adopted to suit the series that involves logarithm.

Example 3.8. *We again consider the series* $\sum_{n=1}^{\infty} \frac{1}{n^{\alpha}} = 1 + \frac{1}{2^{\alpha}} + \frac{1}{3^{\alpha}} + \cdots$. *It has already been verified that the series converges if* $\alpha > 1$, *diverges if* $\alpha \leq 1$. *We shall verify the same thing using Cauchy's condensation test.*

Solution: Since $\frac{1}{n^{\alpha}}$ $(\alpha \leq 0)$ does not converge to 0, it follows that the series diverges. If $\alpha > 0$, $\frac{1}{n^{\alpha}}$ is monotonically decreasing and hence by Theorem 3.6, $\sum \frac{1}{n^{\alpha}}$ converges if and only if $\sum 2^k \frac{1}{2^{k\alpha}} = \sum (\frac{1}{2^{\alpha-1}})^k$ converges. But this is a geometric series which converges if and only if $\frac{1}{2^{\alpha-1}} < 1$ and this is so if and only if $\alpha > 1$.

Theorem 3.7. ***(Cauchy's root test):*** *Let* (a_n) *be a sequence of real numbers.*

(a) If $\limsup a_n^{\frac{1}{n}} < 1$, $\sum a_n$ *converges.*

(b) If $\limsup a_n^{\frac{1}{n}} > 1$, $\sum a_n$ *diverges.*

(c) If $\limsup a_n^{\frac{1}{n}} = 1$, *no conclusion is obtained; that is, the series may converge or diverge.*

We know that if a sequence is convergent, its lim sup *and* lim inf *coincide with limit. Hence if* $(a_n)^{\frac{1}{n}}$ *converges, the root test takes a simpler form and holds with* $\lim a_n^{\frac{1}{n}}$ *in place of* $\limsup a_n^{\frac{1}{n}}$.

Proof. (a) Let $l = \limsup a_n^{\frac{1}{n}}$ and let $l < 1$. Choose δ such that $l < \delta < 1$. Then there exists $n_0 \in \mathbb{N}$ such that for $n > n_0$, $a_n^{\frac{1}{n}} < \delta$, i.e., $a_n < \delta^n$.

Since $0 < \delta < 1$, $\sum \delta^n$ converges. Hence, by the comparison test $\sum a_n$ converges.

(b) Let $l > 1$. This implies $a_n^{\frac{1}{n}} > 1$ for infinitely many values of n. Hence a_n does not converge to 0 and so $\sum a_n$ is divergent.

(c) The series $\sum_{n=1}^{\infty} \frac{1}{n}$ diverges even though $\limsup a_n^{\frac{1}{n}} = \limsup(\frac{1}{n})^{\frac{1}{n}} = 1$; also the series $\sum \frac{1}{n^2}$ converges for which $\lim(\frac{1}{n^2})^{\frac{1}{n}} = 1$. □

Before we discuss about the ratio test, we prove a lemma.

Lemma 3.1. *Let* $a_n > 0$ *for all* n. *Then* $\liminf(\frac{a_{n+1}}{a_n}) \leq \liminf a_n^{\frac{1}{n}} \leq \limsup a_n^{\frac{1}{n}} \leq \limsup(\frac{a_{n+1}}{a_n})$.

Proof. The middle inequality is obvious. The first is similar to the third, and so we prove the third inequality. Let $l = \limsup(\frac{a_{n+1}}{a_n})$. Take $l < \beta$. Then, there exists $n_0 \in \mathbb{N}$ such that $\frac{a_{n+1}}{a_n} < \beta$ $(n \geq n_0)$. Hence by repeated applications of the above inequality, we have $a_n < \beta^{n-n_0} a_{n_0}$ $(n \geq n_0)$. So, $\limsup a_n^{\frac{1}{n}} \leq \beta \limsup(a_{n_0}\beta^{-n_0})^{\frac{1}{n}} = \beta$. Since this is true for any $\beta > l$, it follows that $\limsup a_n^{\frac{1}{n}} \leq l$. □

Theorem 3.8. ***(D' Alembert's ratio test):*** *Let* $a_n > 0$. *Then (a) if* $\limsup(\frac{a_{n+1}}{a_n}) < 1$, $\sum a_n$ *converges.*

(b) if $\liminf(\frac{a_{n+1}}{a_n}) > 1$, $\sum a_n$ *diverges.*

(c) if $\liminf(\frac{a_{n+1}}{a_n}) \leq 1 \leq \limsup(\frac{a_{n+1}}{a_n})$, *no conclusion is obtained.*

Proof. Suppose that $\limsup(\frac{a_{n+1}}{a_n}) < 1$. By the lemma above, $\limsup a_n^{\frac{1}{n}} < 1$ and hence by the root test $\sum a_n$ converges. Similarly, $\liminf(\frac{a_{n+1}}{a_n}) > 1$ implies that $\limsup a_n^{\frac{1}{n}} > 1$, and so $\sum a_n$ diverges. To explain (c), consider the examples $\sum_{n=1}^{\infty} \frac{1}{n}$ and $\sum_{n=1}^{\infty} \frac{1}{n^2}$. □

Remark 3.7. Since we are able to deduce the ratio test from the root test, it is proved that the latter is superior to the former. In fact, whenever the ratio test proves convergence, the root test does so, but the converse is not true. There are series for which the root test proves convergence but for which the ratio test gives no information at all. But for certain kinds of series, the ratio test is as effective as the root test.

Corollary 3.4. *Let* $a_n > 0$ *for all* n *and let* $\lim(\frac{a_{n+1}}{a_n})$ *exist or diverge to* ∞. *Then both the root test and ratio test coincide, i.e., (a)* $\lim a_n^{\frac{1}{n}} = \lim(\frac{a_{n+1}}{a_n})$. *(b)* $\sum a_n$ *converges if* $\lim(\frac{a_{n+1}}{a_n}) < 1$, *diverges if* $\lim(\frac{a_{n+1}}{a_n}) > 1$ *and no information is obtained if* $\lim(\frac{a_{n+1}}{a_n}) = 1$.

Proof. In the case $\lim(\frac{a_{n+1}}{a_n})$ exists or diverges to ∞, the result (a) follows from Lemma 3.1. The result (b) follows from either the root or the ratio test. □

Definition 3.5. *Let (a_n) be a sequence of complex numbers. The series $\sum a_n$ is said to be absolutely convergent if $\sum |a_n|$ is convergent.*

Theorem 3.9. *Let $a_n \in \mathbb{C}$. The series $\sum |a_n|$ is convergent if and only if $\sum |Rea_n|$ and $\sum |Ima_n|$ are convergent.*

Proof. Let $a_n = b_n + ic_n$ where $b_n, c_n \in \mathbb{R}$. Let $A_k = \sum_{n=0}^{k} a_n$, $B_k = \sum_{n=0}^{k} b_n$, $C_k = \sum_{n=0}^{k} c_n$. Then $A_k = B_k + iC_k$. Since $|B_k| \leq |B_k + iC_k| \leq |B_k| + |C_k|$ and $|C_k| \leq |B_k + iC_k| \leq |B_k| + |C_k|$. The result is now evident. □

Example 3.9. *Using ratio test it is easy to verify that the series*

(i) $\sum_{n=1}^{\infty} \frac{n}{(2i)^n}$ *converges absolutely.*

(ii) $\sum_{n=1}^{\infty} \frac{n!}{(in)^n}$ *converges absolutely.*

(iii) *The series* $\sum_{n=1}^{\infty} \left(\frac{\cos n}{n} + i\frac{\sin n}{n}\right)$ *is convergent but does not converge absolutely.*

(iv) *The series* $\sum_{n=1}^{\infty} \left(\frac{\cos n\phi}{n} + i\frac{\sin n\phi}{n}\right)$ *diverges for* $\phi = 2k\pi, k \in \mathbb{Z}$. *The series converges for* $\phi \neq 2k\pi, k \in \mathbb{Z}$ *but it does not converge absolutely.*

(v) *The series* $\sum_{n=1}^{\infty} \left(\frac{\cos n}{n^2} + i\frac{\sin n}{n^2}\right)$ *converges absolutely.*

Example 3.10. ***(Exponential series)*** *We show that $\sum_{n=0}^{\infty} \frac{z^n}{n!}$ is absolutely convergent for any $z \in \mathbb{C}$.*

Given z, we select $N \in \mathbb{N}$ such that $|z| < \frac{N}{2}$. Then for $k \in \mathbb{N}$, one shows easily that $|\frac{z^{N+k}}{(N+k)!}| \leq \frac{1}{2^k}\frac{|z|^N}{N!}$. Hence we have $\sum_{n\geq N} |\frac{z^n}{n!}| \leq \frac{|z|^N}{N!}\sum_{n\geq N} \frac{1}{2^n} \leq 2\frac{|z|^N}{N!}$. By the comparison test the result follows. The value of the series is denoted by $\exp(z)$ *and the series is called the exponential series.*

In the case of an infinite series, its sum (if it exists) is effected by the rearrangement of terms and a divergent series may become convergent upon rearrangement. But if $\sum a_n$ is absolutely convergent, any rearrangement of $\sum a_n$ has the same sum. The proof is left as an exercise to the reader.

Conditionally convergent series

Definition 3.6. *A series $\sum a_n$ is said to be conditionally convergent if $\sum a_n$ is convergent but $\sum |a_n|$ is divergent.*

We now state the techniques of determining the convergence of conditionally convergent series.

Theorem 3.10. *Let (c_n) be a monotone decreasing sequence of nonnegative reals with $\lim c_n = 0$. Let (b_n) be a bounded sequence in $\mathbb{C}$. Then $\sum_n (b_{n+1} - b_n)c_n$ is convergent.*

Proof. Let s_n be the nth partial sum of the series. We have for $n \geq m$, $s_n - s_m = \sum_{k=m+1}^{n}(b_{k+1} - b_k)c_k = b_{n+1}c_n + \sum_{k=m+1}^{n} b_k(c_{k-1} - c_k) - b_m c_m$. Hence taking modulus, we get

$$|s_n - s_m| \leq M\left(c_n + \sum_{k=m+1}^{n}(c_{k-1} - c_k) + c_m\right) \leq 2Mc_m,$$

which shows that (s_n) is Cauchy. □

Corollary 3.5. ***(Alternating series test)****: Let $\sum_n (-1)^{n+1}x_n$ be an infinite series with $x_n \geq 0$ and $\lim_{n\to\infty} x_n = 0$. Then $\sum_n (-1)^{n+1}x_n$ is convergent.*

Proof. Take $b_n = \frac{(-1)^n}{2}$ in the theorem. □

Example 3.11. $\sum_{n=1}^{\infty} \frac{(-1)^n}{n}$ *is convergent. In fact the above series is conditionally convergent by the alternating series test.*

Multiplication of series

Multiplication of series can be defined in various ways. The term-by-term multiplication of terms of the series $\sum a_n$ and $\sum b_n$ gives a series $\sum a_n b_n$ which is called the Hadamard product. The convergence of $\sum a_n b_n$ can be tested by using the following techniques. First we prove a lemma of general interest.

Lemma 3.2. *(Abel's summation formula): Let $0 \leq m \leq n, s_n = a_0 + \cdots + a_n, a_{-1} = 0$. Then*

$$\sum_{k=m}^{n} a_k b_k = \sum_{k=m}^{n-1} s_k(b_k - b_{k+1}) + s_n b_n - s_{m-1}b_m.$$

Proof.

$$\begin{aligned}\sum_{k=m}^{n} a_k b_k &= \sum_{k=m}^{n}(s_k - s_{k-1})b_k \\ &= \sum_{k=m}^{n} s_k b_k - \sum_{k=m-1}^{n-1} s_k b_{k+1} \\ &= \sum_{k=m}^{n-1} s_k b_k + s_n b_n - \sum_{k=m}^{n-1} s_k b_{k+1} - s_{m-1}b_m.\end{aligned}$$

□

Theorem 3.11. ***(Dedekind's test)****: Let $a_k, b_k \in \mathbb{C}$ and let*

(i) s_n be bounded.

(ii) $\sum |b_k - b_{k+1}|$ be convergent.

(iii) $b_n \to 0$ as $n \to \infty$.

Then $\sum a_k b_k$ converges and $\sum a_k b_k = \sum s_k(b_k - b_{k+1})$.

Proof. By Abel's formula

$$\sum_{k=0}^{n} a_k b_k = \sum_{k=0}^{n-1} s_k(b_k - b_{k+1}) + s_n b_n. \tag{3.3}$$

Now by hypotheses (i) and (iii), $s_n b_n \to 0$ as $n \to \infty$. By hypotheses (i) and (ii), $\sum |s_k(b_k - b_{k+1})| \leq M \sum |b_k - b_{k+1}|$ where M is a positive constant such that $|s_n| \leq M$. Now by comparison test $\sum s_k(b_k - b_{k+1})$ is absolutely convergent.Now taking the limit as $n \to \infty$ in expression (3.3) we obtain the result. □

Theorem 3.12. *Let $a_k, b_k \in \mathbb{C}$ and let*
(i) $\sum a_n$ be convergent.
(ii) $\sum |b_n - b_{n+1}|$ also be convergent.
Then $\sum a_n b_n$ is convergent.

Proof. since $\sum a_n$ is convergent, s_n is bounded. Hence the series $\sum s_k(b_k - b_{k+1})$ is absolutely convergent by hypothesis (ii). Also,

$$b_n = b_0 - \sum_{k=0}^{n-1} (b_k - b_{k+1}).$$

Hence, as $\sum (b_k - b_{k+1})$ is convergent, $\lim b_n$ exists. Now it follows from Abel's summation formula that

$$\sum_{k=0}^{\infty} a_k b_k = \sum_{k=0}^{\infty} s_k(b_k - b_{k+1}) + (\lim s_n)(\lim b_n).$$

□

We can also define multiplication of two series in a way different from this.

Given a_n and b_n we define (c_n), the Cauchy product of (a_n) and (b_n) by $c_n = a_0 b_n + a_1 b_{n-1} + \cdots + a_n b_0 = \sum_{k=0}^{n} a_k b_{n-k}$, and call $\sum c_n$ as the Cauchy product or convolution product of the series. The following example shows that the Cauchy product of two convergent series may diverge.

Example 3.12. *The series $\sum_{n=0}^{\infty} (-1)^n \frac{1}{\sqrt{n+1}}$ is conditionally convergent but its Cauchy product with itself is divergent.*

Solution: If $a_n = (-1)^n \frac{1}{\sqrt{n+1}}$, then the Cauchy product with itself is given by

$$c_n = (-1)^n \sum_{k=0}^{n} \frac{1}{\sqrt{(n-k+1)}\sqrt{k+1}}.$$

Hence $|c_n| = \sum_{k=0}^{n} \frac{1}{\sqrt{(n-k+1)}\sqrt{k+1}} \geq \frac{1}{(n+1)} \sum_{k=0}^{n} 1 = 1$ and $\sum c_n$ diverges.

But if one or both series are absolutely convergent, the convergence of the product series is given by the following two theorems. Before we prove the next theorem, we need the following notation and simplification.

Let $A_n = \sum_{k=0}^{n} a_k$, $B_n = \sum_{k=0}^{n} b_k$, $C_n = \sum_{k=0}^{n} c_k$, and $A_n^* = \sum_{k=0}^{n} |a_k|$, $B_n^* = \sum_{k=0}^{n} |b_k|$, $C_n^* = \sum_{k=0}^{n} |c_k|$. Then

$$\begin{aligned} C_n &= \textstyle\sum_{k=0}^{n} c_k \\ &= \textstyle\sum_{k=0}^{n} (a_k b_0 + \cdots + a_0 b_k) \\ &= a_0 b_0 + (a_1 b_0 + a_0 b_1) + \cdots + (a_n b_0 + \cdots + a_0 b_n) \\ &= a_0(b_0 + b_1 + \cdots + b_n) + a_1(b_0 + b_1 + \cdots + b_{n-1}) + \cdots + a_n b_0. \end{aligned}$$

Thus

$$C_n = a_0 B_n + a_1 B_{n-1} + \cdots + a_n B_0. \tag{3.4}$$

Hence

$$C_n^* \le |a_0| B_n^* + |a_1| B_{n-1}^* + \cdots + |a_n| B_0^*. \tag{3.5}$$

Now we use the above relations to prove the following theorems:

Theorem 3.13. ***(Mertens' theorem):*** *Let $\sum a_n$ be absolutely convergent and $\sum b_n$ be convergent (not necessarily absolutely). Then the Cauchy product series $\sum c_n$ is convergent and its sum is given by $\sum c_n = (\sum a_n)(\sum b_n)$.*

Proof. Let $A = \sum a_n, B = \sum b_n$ and let $D_n = B - B_n$. $C_n = \sum_{k=0}^{n} c_k = a_0 B_n + a_1 B_{n-1} + \cdots + a_n B_0 = a_0(-B + B_n) + a_1(-B + B_{n-1}) + \cdots + a_n(-B + B_0) + B(\sum_{k=1}^{n} a_k) = A_n B - R_n$ where $R_n = a_0 D_n + a_1 D_{n-1} + \cdots + a_n D_0$. Let $\alpha := \sum_n |a_n|$. Since $D_n \to 0, (D_n)$ is bounded, say by $D : |D_n| \le D$ for all n. Given $\epsilon > 0$, there exists $N \in \mathbb{N}$ such that $\sum_{n \ge N} |a_n| < \epsilon$ and $|D_n| \le \epsilon$ for all $n \ge N$. For all $n \ge 2N$, we have $|R_n| \le (|a_0| + \cdots + |a_{n-N}|)\epsilon + (|a_{n-N+1}| + \cdots + |a_n|)D \le (\alpha + D)\epsilon$. Hence $R_n \to 0$. Since $A_n B \to AB$, the result follows. □

Example 3.13. *We know that $\sum_{n=0}^{\infty} z^n = \frac{1}{1-z}$ for $|z| < 1$. If we take $a_n = z^n = b_n$ in the theorem, we get $\sum_{n=1}^{\infty} n z^{n-1} = \frac{1}{(1-z)^2}$ for $|z| < 1$.*

Theorem 3.14. *Let $\sum a_k$ and $\sum b_k$ be absolutely convergent. Then their Cauchy product $\sum c_n$ is absolutely convergent and its sum is given by $\sum c_k = (\sum a_k)(\sum b_k)$.*

Proof. Let $\alpha = \sum_{k=0}^{\infty} |a_k|$ and $\beta = \sum_{k=0}^{\infty} |b_k|$. since $B_n^* = \sum_{k=0}^{n} |b_k| \le \beta$, $A_n^* \le \alpha$, it follows from (3.5) that $|C_n^*| = |\sum_{k=0}^{n} B_{n-k}^* a_k| \le \beta \sum_{k=0}^{n} |a_k| \le \alpha\beta$. Since C_n^* is monotonic increasing and bounded above by $\alpha\beta$, it follows that $\lim\limits_{n\to\infty} C_n^*$ exists in $\mathbb{R}$ and this proves that $\sum |c_n|$ is convergent. Now it follows from Theorem 3.13 that $\sum c_n = (\sum a_n)(\sum b_n)$. □

We state without proof another multiplication theorem whose proof is based on Abel's limit theorem.

Theorem 3.15. *Let $\sum a_n, \sum b_n$ and $\sum c_n$ be convergent where c_n is given by the Cauchy product $c_n = a_0 b_n + a_1 b_{n-1} + \cdots + a_n b_0$. Then*

$$\sum_{n=0}^{\infty} c_n = (\sum_{n=0}^{\infty} a_n)(\sum_{n=0}^{\infty} b_n).$$

Example 3.14. *Consider the series $\sum a_n$ where $a_n = \frac{(-1)^n}{n+1}$. We shall show that the Cauchy product of $\sum a_n$ with itself is the series $\sum c_n$ where $c_n = \frac{2(-1)^n}{n+2}(1 + \frac{1}{2} + \frac{1}{3} + \cdots + \frac{1}{n+1})$. The series $\sum c_n$ is convergent.*

Solution: The Cauchy product $c_n = \sum_{k=0}^{n} a_{n-k} a_k = \sum_{k=0}^{n} \frac{(-1)^{n-k}}{n+1-k} \frac{(-1)^k}{k+1}$ $= \frac{(-1)^n}{n+2} \sum_{k=0}^{n} (\frac{1}{n+1-k} + \frac{1}{k+1}) = \frac{2(-1)^n}{n+2} \sum_{k=0}^{n} \frac{1}{k+1}$. But since $d_n = \frac{1}{n+2} \sum_{k=0}^{n} \frac{1}{k+1} \to 0$ as $n \to \infty$, the series $\sum c_n = 2\sum(-1)^n d_n$ converges by alternating series test. The fact that $d_n \to 0$ follows from the following consideration. Since $\sum_{k=0}^{n} \frac{1}{k+1} - \log(n+1) \to \nu$ (Euler's constant) and $\frac{\log(n+1)}{n+2} \to 0$ it follows that $d_n = \frac{1}{n+2} [\sum_{k=0}^{n} \frac{1}{k+1} - \log(n+1)] + \frac{\log(n+1)}{n+2} \to 0$ as $n \to \infty$.

3.3 Power series

Definition 3.7. *Given a sequence (a_n) of complex numbers, the series $\sum a_n z^n$ is called a power series. The numbers a_n are called the coefficients of the series; z is a complex number.*

The terms of a power series are constant multiples of powers of z. The nature of the series depends on the complex number z as well as the sequence (a_n). More specifically, with every power series there is associated a circle, the circle of convergence such that the power series converges if z is in the interior of the circle and diverges if z is in the exterior.

A little more generally we may consider series $\sum_{n=0}^{\infty} a_n (z-a)^n$ which are power series with respect to center $a \in \mathbb{C}$, but the difference is so slight that we need not do so in a formal manner.

Example 3.15. *Consider the three power series:*
(1) $\sum_{n=1}^{\infty} n^n z^n$, (2) $\sum_{n=0}^{\infty} z^n$ and (3) $\sum_{n=0}^{\infty} \frac{z^n}{n!}$.

We claim that if $z \neq 0$ then the first series does not converge. For, if $z \neq 0$ is a complex number, choose $N \in \mathbb{N}$ so that $|z| > \frac{1}{N}$. Then for all $n \geq N$, we have $|(nz)^n| > 1$ and hence the series is not convergent.

We have already seen that the second series converges absolutely for all z with $|z| < 1$ whereas the third series converges absolutely for all $z \in \mathbb{C}$.

In what follows we shall show that once $\sum a_n z^n$ converges for a certain z_0 it is convergent for all z in the open disk $\{z \in \mathbb{C} : |z| < |z_0|\}$.

Theorem 3.16. *If $\sum a_n z_0^n$ is convergent, then $\sum a_n z^n$ is absolutely convergent for $|z| < |z_0|$.*

Proof. Since $\sum a_n z_0^n$ is convergent, we can find $n_0 \in \mathbb{N}$ such that $|a_n z_0^n| < 1$, for $n > n_0$. Hence for $n > n_0$, $|a_n z^n| = |a_n z_0^n \frac{z^n}{z_0^n}| < |\frac{z}{z_0}|^n$. But since $|\frac{z}{z_0}| < 1$, it follows that the geometric series $\sum |\frac{z}{z_0}|^n$ is convergent, and hence by the comparison test $\sum |a_n z^n|$ is convergent for $|z| < |z_0|$. □

Definition 3.8. *Given the power series $\sum a_n z^n$, the number*

$$R = \sup\{|z| : \sum a_n z^n \text{ is convergent}\}$$

is called the radius of convergence of the power series and $|z| = R$ is called the circle of convergence of the power series and $B(0, R)$ is called the disk of convergence of the power series. Note that by definition R may be 0 or ∞ or any real number satisfying $0 < R < \infty$.

The following theorem brings out the importance of the radius of convergence.

Theorem 3.17. *(a) If R is the finite radius of convergence of the power series $\sum a_n z^n$, the series converges for $|z| < R$, diverges for $|z| > R$ and no conclusion is obtained on the circle of convergence.*

(b) If $0 < \limsup |a_n|^{\frac{1}{n}} < \infty$, then the radius of convergence is given by the formula

$$\frac{1}{R} = \limsup |a_n|^{\frac{1}{n}}.$$

(c) If $\limsup |a_n|^{\frac{1}{n}} = 0$, then $\sum a_n z^n$ is convergent for all z (and in this case the radius of convergence of the power series is interpreted as ∞).

(d) If $\limsup |a_n|^{\frac{1}{n}} = \infty$, then $\sum a_n z^n$ converges only for $z = 0$ (and in this case the radius of convergence of the power series is interpreted as 0).

(e) If $a_n \neq 0$ for all n and $\lim |\frac{a_n}{a_{n+1}}|$ exists including divergence to ∞, then R is given by $R = \lim_{n\to\infty} \left|\frac{a_n}{a_{n+1}}\right|$.

Proof. (a) Suppose that $|z| < R$. Choose δ such that $|z| < \delta < R$. Then from definition of radius of convergence, $\sum a_n \delta^n$ is absolutely convergent. By the comparison test $\sum a_n z^n$ is convergent for $|z| < R$.

Let $|z_0| > R$. If $\sum a_n z_0^n$ is convergent, then $|z_0| \leq R$ which is a contradiction. Hence $\sum a_n z_0^n$ is divergent for $|z_0| > R$.

(b) By Cauchy's root test, the power series is absolutely convergent if $\limsup |a_n z^n|^{\frac{1}{n}} = |z| \limsup |a_n|^{\frac{1}{n}} < 1$. If $\lim |a_n|^{\frac{1}{n}}$ is finite, then the above expression can be rewritten as $|z| < \frac{1}{\limsup |a_n|^{\frac{1}{n}}}$. Similarly the power series is divergent if $|z| > \frac{1}{\limsup |a_n|^{\frac{1}{n}}}$. This proves that $R = \frac{1}{\limsup |a_n|^{\frac{1}{n}}}$.

(c) For every $z \in \mathbb{C}$, $\limsup |a_n z^n|^{\frac{1}{n}} = |z| \limsup |a_n|^{\frac{1}{n}} = 0$ and so by Cauchy's root test the series converges for all $z \in \mathbb{C}$.

(d) For $z \neq 0$, $\limsup |a_n z^n| = |z| \limsup |a_n|^{\frac{1}{n}} = \infty$ and so by Cauchy's root test $\sum a_n z^n$ diverges for all $z \in \mathbb{C} - \{0\}$.

(e) This follows from Lemma 3.1. □

Example 3.16. *Consider the series (i) $\sum_{n=0}^{\infty} n! z^n$; (ii) $\sum_{n=0}^{\infty} z^{n!}$; (iii) $\sum_{n=0}^{\infty} \frac{z^{3n}}{2^n}$.*

Solution: (i) Applying the ratio test we obtain that the radius of convergence of the first series is $R = \lim \frac{n!}{(n+1)!} = 0$ and so the series converges only at $z = 0$.

(ii) The series $\sum_{n=0}^{\infty} z^{n!} = z + z + z^2 + z^6 + z^{24} + \cdots$ has many vanishing coefficients (that is, the series has many gaps). In fact we have $a_0 = 0, a_1 = 2, a_2 = 1, a_3 = a_4 = a_5 = 0, a_6 = 1, a_7 = a_8 = a_9 = \cdots = a_{23} = 0, a_{24} = 1, \cdots$. So applying root test we obtain $R = \frac{1}{\overline{\lim}|a_n|^{\frac{1}{n}}} = 1$. Hence the series converges for $|z| < 1$; diverges for $|z| > 1$. Also the series diverges on $|z| = 1$ as the nth term does not converge to 0.

(iii) The coefficients of the series $\sum_{n=0}^{\infty} a_n z^n = \sum_{n=0}^{\infty} \frac{z^{3n}}{2^n}$ are as follows: $a_0 = 1, a_1 = a_2 = 0, a_3 = \frac{1}{2}, a_4 = a_5 = 0, a_6 = \frac{1}{4}$ and more generally $a_{3n} = \frac{1}{2^n}, a_{3n+1} = a_{3n+2} = 0$. By the root test, the radius of convergence is given by $R = \frac{1}{\overline{\lim}|a_n|^{\frac{1}{n}}} = \frac{1}{\overline{\lim}|a_{3n}|^{\frac{1}{3n}}} = \frac{1}{\overline{\lim}(\frac{1}{2^n})^{\frac{1}{3n}}} = 2^{\frac{1}{3}}$. Thus the series converges absolutely for $|z| < 2^{\frac{1}{3}}$, diverges for $|z| > 2^{\frac{1}{3}}$. The series also diverges at $|z| = 2^{\frac{1}{3}}$ (circle of convergence), since the nth term does not go to 0, as $\frac{|z|^{3n}}{2^n} = \frac{2^n}{2^n} = 1$.

Example 3.17. *Suppose we know that the radius of convergence of the power series $\sum_{n=0}^{\infty} a_n z^n$ is R $(0 < R < \infty)$. From this we can also evaluate the radii of convergence of many other power series. For example the radius of convergence of the power series $\sum_{n=0}^{\infty} n^m a_n z^n$ is equal to $\frac{1}{\overline{\lim}_{n\to\infty}|n^m a_n|^{\frac{1}{n}}} = \frac{1}{\overline{\lim}_{n\to\infty}|n^m|^{\frac{1}{n}}} \cdot \frac{1}{\overline{\lim}_{n\to\infty}|a_n|^{\frac{1}{n}}} = 1 \cdot R = R$ since $\lim_{n\to\infty} n^{\frac{1}{n}} = 1$.*

Similarly the radius of convergence of the power series $\sum_{n=0}^{\infty}(2^n - 1)a_n z^n$ is

$$\frac{1}{\overline{\lim}_{n\to\infty}|(2^n-1)a_n|^{\frac{1}{n}}} = \frac{1}{\overline{\lim}_{n\to\infty}|(2^n-1)|^{\frac{1}{n}}} \cdot \frac{1}{\overline{\lim}_{n\to\infty}|a_n|^{\frac{1}{n}}} = \frac{R}{2}$$

since

$$\lim_{n\to\infty}(2^n-1)^{\frac{1}{n}} = \lim_{n\to\infty} e^{\frac{1}{n}\log(2^n-1)} = \lim_{n\to\infty} e^{\frac{1}{n}\log 2^n(1-\frac{1}{2^n})}$$

$$= \lim_{n\to\infty} e^{\frac{1}{n}[\log 2^n + \log(1-\frac{1}{2^n})]} = \lim_{n\to\infty} e^{\log 2} \cdot e^{\frac{1}{n}\log(1-\frac{1}{2^n})} = 2.$$

Theorem 3.18. *The radius of convergence of $\sum_{n=0}^{\infty} a_n z^n$ is $\liminf |a_n|^{-\frac{1}{n}}$, where $0^{-\frac{1}{n}}$ is defined to be ∞.*

Proof. Let ρ be the radius of convergence of the power series. Let $0 < \lambda < \alpha := \liminf |a_n|^{-\frac{1}{n}}$. There exists $N \in \mathbb{N}$ such that for all $n \geq N$, we have $\lambda < |a_n|^{-\frac{1}{n}}$ or $|a_n|\lambda^n < 1$. (Note that this inequality certainly holds if $a_n = 0$.) In particular, $\{|a_n|\lambda^n\}$ is bounded, say, by M. Let z be such that $|z| < \lambda$. Then, $|a_n z^n| = |a_n||z|^n \leq M\lambda^{-n}|z|^n = M(\frac{|z|}{\lambda})^n$. Thus the series $\sum a_n z^n$ is dominated by the convergent geometric series $\sum M(\frac{|z|}{\lambda})^n$ and hence is convergent. Thus, $\lambda \leq \rho$. Since λ is an arbitrary real number such that $\lambda \leq \alpha$, it follows that $\alpha \leq \rho$.

Now assume that $\mu > \alpha$. Then for infinitely many n, $|a_n|^{-\frac{1}{n}} < \mu$ or $|a_n|^{\frac{1}{n}}\mu > 1$. Since the nth terms do not approach zero, $\sum c_n \mu^n$ diverges. Hence $\mu \geq \rho$. Therefore, $\alpha = \inf\{\mu : \mu > \alpha\} \geq \rho$. We have shown that $\rho = \alpha$. □

If the power series $\sum a_n z^n$ has a radius of convergence r and $|a| < r$, then writing $z^n = (z - a + a)^n = \sum_{k=0}^{n} \binom{n}{k} a^k (z-a)^{n-k}$ and rearranging the terms in powers of $(z - a)$, the series $\sum a_n z^n$ can be written in the form

$$\sum_{n=0}^{\infty} b_n (z-a)^n. \tag{3.6}$$

Note that by absolute convergence of the power series $\sum a_n z^n$ for $|z| < r$, any arrangement of this series does not affect the sum. The following theorem determines the radius of convergence of the rearranged series (3.6).

Theorem 3.19. *If $\sum a_n z^n$ has a radius of convergence r and $|a| < r$, then the rearranged series (3.6) has a radius of convergence $\geq r - |a|$.*

Proof. Let $z_0 \in \{z : |z-a| = r-|a|\} \cap \{z : |z| < r\}$. Hence $\sum a_n z_0^n$ is absolutely convergent. So the series (3.6) is absolutely convergent for $|z - a| < |z_0 - a| = r - |a|$. This proves that the radius of convergence of (3.6) is at least $r - |a|$. □

Example 3.18. *The power series $\sum_{n=0}^{\infty} \frac{z^n}{(n+1)^2}$ is absolutely convergent on its circle of convergence; in this case $r = 1$ and $\sum \frac{1}{(n+1)^2}$ is convergent.*

But there are series which do not converge at all on the circle of convergence.

Example 3.19. *The power series $\sum_{n=0}^{\infty} z^n$ has radius of convergence* 1 *and does not converge for $|z| = 1$. the series diverges for all z on the circle of convergence, as the nth terms do not approach* 0.

There are series that converge conditionally at all points on the circle of convergence.

Example 3.20. *Consider the power series $\sum_{n=1}^{\infty} \frac{(-1)^n z^{2n}}{2n}$. The radius of convergence of the power series is easily seen to be* 1. *So it is absolutely convergent for $|z| < 1$; divergent for $|z| > 1$. If $z = 1$ or $z = -1$, the series reduces to $\sum \frac{(-1)^n}{2n}$ which is convergent (conditionally convergent). If $|z| = 1$, we make use of Dedekind's test (Theorem 3.11). Take $b_n = \frac{1}{2n}$, $a_n = (-1)^n z^{2n}$.*

Now for all $|z| = 1$, $|a_0 + a_1 + \cdots + a_n| = |-z^2 + z^4 - z^6 + \cdots + (-1)^n z^{2n}| = \left|\frac{1-(-1)^{n+1}z^{2(n+1)}}{1+z^2}\right| \leq \frac{2}{|1+z^2|}$.

Hence a_n *has bounded partial sums, and by Dedekind's test, the series is conditionally convergent at all points on the circle of convergence* $|z| = 1$.

Further there are series that converge at all points except one, two or some points on the circle of convergence. Before we give examples of such power series we prove the following theorem which gives a test for conditional convergence on the circle of convergence.

Theorem 3.20. *Let the power series* $\sum b_n z^n$ *have radius of convergence* 1. *Let* b_n *be monotonic and converge to* 0. *Then the power series converges for* $|z| = 1, z \neq 1$.

Proof. Let $a_n = z^n$. Then

$$s_n = 1 + z + z^2 + \cdots + z^n = \frac{1 - z^{n+1}}{1 - z}, z \neq 1.$$

Hence for $|z| = 1, z \neq 1$,

$$|s_n| \leq \frac{1 + |z|^{n+1}}{|1 - z|} \leq \frac{2}{|1 - z|}, \tag{3.7}$$

so that s_n is bounded. Since as b_n is monotonic and $b_n \to 0$, $\sum |b_n - b_{n+1}|$ converges and so by Dedekind's test $\sum b_n z^n$ converges for $|z| = 1, z \neq 1$. □

Remark 3.8. On the circle of convergence, $z = \cos\theta + i\sin\theta$. Hence by De Moivre's formula, $z^n = \cos n\theta + i \sin n\theta$. So the series

$$\sum b_n z^n = \sum b_n \cos n\theta + i \sum b_n \sin n\theta$$

converges for $|z| = 1, z \neq 1$. Thus the series $\sum b_n \cos n\theta$ and $\sum b_n \sin n\theta$ converge for these values of z. Note that if $\theta \neq 2k\pi$ then $z = \cos\theta + i\sin\theta \neq 1$. Further, if $\theta = 2k\pi$, the sine series reduces to a series of 0 and cosine series reduces to $\sum b_n$ whose divergence or convergence is not known and this may be ascertained from other consideration.

Example 3.21. *Consider the power series* $\sum_{n=0}^{\infty} \frac{z^n}{(n+1)^\alpha}$ $(\alpha \in \mathbb{R})$. *The radius of convergence is given by* $R = \lim_{n\to\infty} \left(\frac{n+2}{n+1}\right)^\alpha = 1$ *for all values of* $\alpha \in \mathbb{R}$. *Hence the series is convergent for* $|z| < 1$; *divergent for* $|z| > 1$. *If* $\alpha > 1$, *the series* $\sum \frac{1}{(n+1)^\alpha}$ *is convergent and hence* $\sum \frac{z^n}{(n+1)^\alpha}$ *is absolutely convergent at all points on the circle of convergence* $|z| = 1$. *If* $0 < \alpha < 1$, *the series clearly diverges to* ∞ *at* $z = 1$. *But if* $0 < \alpha < 1$ *and* $|z| = 1, z \neq 1$ *then the series is conditionally convergent by Theorem 3.20.*

Example 3.22. *Consider the series* $\sum_{n=1}^{\infty} \frac{z^{2n}}{2n}$. *The radius of convergence of the power series is* 1. *So the series is absolutely convergent for* $|z| < 1$; *divergent for* $|z| > 1$. *On* $|z| = 1$, *the series is not absolutely convergent as* $\sum_{n=1}^{\infty} \frac{1}{2n} = \infty$. *If* $z^2 = 1$, *i.e., if* $z = 1$ *or* $z = -1$ *then the series diverges. But if* $|z| = 1$ *and* $z^2 \neq 1$, *we apply Theorem 3.20 to show that it is conditionally convergent.*

Example 3.23. *We describe below the behavior of some of the power series on the circle of convergence.*

(i) *From the ratio test it follows that the radius of convergence of the power series* $\sum_{n=1}^{\infty} \frac{z^n}{n}$ *is equal to* 1. *On the circle of convergence the series diverges when* $z = 1$ *and converges when* $z = -1$. *In fact the series converges non absolutely at all points on the circle of convergence except at* $z = 1$.

(ii) *The radius of convergence of the power series* $\sum_{n=0}^{\infty} z^n$ *is* 1 *and it diverges at all points of the circle of convergence.*

(iii) *The radius of convergence of the power series* $\sum_{n=0}^{\infty} \frac{z^n}{n^2}$ *is* 1 *and the series converges absolutely at all points on the circle of convergence.*

(iv) *The series* $\sum_{n=1}^{\infty} \frac{(-1)^n}{n} z^n$ *has radius of convergence* 1 *and it converges non absolutely at all points on the circle of convergence except at* $z = -1$.

(v) *The series* $\sum_{n=1}^{\infty} \frac{z^{pn}}{n}$ *(p a natural number) has radius of convergence* 1 *and it converges non absolutely at all points on the circle of convergence except at* $z = e^{\frac{2k\pi i}{p}}$ $(k = 0, 1, \cdots, p-1)$.

(vi) *The series* $\sum_{n=2}^{\infty} \frac{(-1)^n}{\ln n} z^{3n-1}$ *has radius of convergence* 1 *and it converges non absolutely at all points on the circle of convergence except at* $z = \frac{1 \pm i\sqrt{3}}{2}$ *and* $z = -1$.

(vii) *The series* $\sum_{n=1}^{\infty} \frac{z^{n!}}{n^2}$ *has radius of convergence* 1 *and it converges absolutely at all points on the circle of convergence.*

3.4 Continuity

Having introduced the complex number system, sequence and series of complex numbers, we proceed to the development of the theory of functions of a complex variable, beginning with the notion of continuity.

Definition 3.9. ***(Limits of functions)*** *Let* Ω *be an open subset of* $\mathbb{C}$ *and let* $a \in \mathbb{C}$ *be a limit point of* Ω, *not necessarily in* Ω. *Let* f *be a function* Ω *to* $\mathbb{C}$. *The function* f *is said to have the limit* L *as* z *tends to* a, *written as*

$$\lim_{z \to a} f(z) = L, \tag{3.8}$$

if and only if the following holds: For every $\epsilon > 0$ *there exists a number* $\delta > 0$ *with the property that* $|f(z) - L| < \epsilon$ *for all values of* z *such that* $|z - a| < \delta$ *and* $z \neq a$.

In the complex variable case there are an infinity of directions in which a variable can approach a point a. In the real case, however there are only two avenues of approach. Continuity, for example, of a function defined on $\mathbb{R}$ can be discussed in terms of right and left continuity; this is far from the case for functions of a complex variable.

Condition (3.8) is equivalent to

$$\lim_{z\to a}\overline{f(z)} = \overline{L}. \tag{3.9}$$

From (3.8) and (3.9) we obtain

$$\lim_{z\to a}\mathrm{Re}f(z) = \mathrm{Re}L \tag{3.10}$$

$$\lim_{z\to a}\mathrm{Im}f(z) = \mathrm{Im}L. \tag{3.11}$$

Conversely (3.8) is a consequence of (3.10) and (3.11).

The well-known results concerning the limit of a sum, a product, and a quotient of functions continue to hold in the complex case.

Theorem 3.21. *Let Ω be an open subset of $\mathbb{C}$ and f, g are functions from Ω into $\mathbb{C}$. Let $a \in \Omega$. If $\lim_{z\to a}f(z), \lim_{z\to a}g(z)$ exist then the following hold:*

(1) $\lim_{z\to a}(f(z)+g(z)) = \lim_{z\to a}f(z) + \lim_{z\to a}g(z)$;

(2) $\lim_{z\to a}(f(z)\cdot g(z)) = \lim_{z\to a}f(z)\cdot\lim_{z\to a}g(z)$;

(3) $\lim_{z\to a}\dfrac{f(z)}{g(z)} = \dfrac{\lim_{z\to a}f(z)}{\lim_{z\to a}g(z)}$ *provided that* $\lim_{z\to a}g(z) \neq 0$;

(4) $\lim_{z\to a}(\alpha f(z)) = \alpha\lim_{z\to a}f(z), \alpha \in \mathbb{C}$.

Example 3.24. *The following limit can easily be evaluated.*

(i) $$\lim_{z\to i}\frac{z^4-1}{z-i} = -4i;$$

(ii) $$\lim_{z\to -2i}\frac{(2z+3)(z-1)}{z^2-2z+4} = -\tfrac{1}{2}+\tfrac{11}{4}i.$$

Division by 0 is impossible in $\mathbb{C}$. The function $f(z) = \frac{1}{z}$ is undefined at $z = 0$. We can avoid this disadvantage by adding the point ∞ (called the point at infinity) to $\mathbb{C}$. Let $\mathbb{C}_\infty = \mathbb{C}\cup\{\infty\}$, the extended complex plane. Now the transformation $w = \frac{1}{z}$ maps the point $z = 0$ to $w = \infty$. Further to consider the behavior of $f(z)$ at $z = \infty$, it suffices to examine the behavior of $f(\frac{1}{w})$ at $w = 0$.

Definition 3.10. *Let $f : \mathbb{C} \to \mathbb{C}$ be a function. We say $\lim_{z\to\infty} f(z) = l$ if given $\epsilon > 0$ there exists an $R > 0$ such that for all z with $|z| > R$, we have*

$|f(z) - l| < \epsilon$. *We say that* $\lim_{z \to z_0} f(z) = \infty$ *or* $f(z)$ *approaches infinity as* z *approaches* z_0 *if for any* $N > 0$, *we can find* $\delta > 0$ *such that* $|f(z)| > N$ *whenever* $0 < |z - z_0| < \delta$.

Example 3.25. $\lim_{z \to \infty} \dfrac{1}{z^n} = 0$ *for any* $n \in \mathbb{N}$.

Definition 3.11. ***(Continuity)****: Let* Ω *be an open subset of* $\mathbb{C}$. *A function* $f : \Omega \to \mathbb{C}$ *is continuous at* $z_0 \in \Omega$ *if for any given* $\epsilon > 0$, *there exists a* $\delta > 0$ *such that* $|f(z) - f(z_0)| < \epsilon$ *whenever* $|z - z_0| < \delta$. *In other words,* f *is continuous at* z_0 *if and only if* $\lim_{z \to z_0} f(z) = f(z_0)$. *If* f *is continuous at each point of* Ω *then* f *is said to be a continuous function from* Ω *into* $\mathbb{C}$.

Continuity of a function of complex variable is more difficult to achieve than that of a function of real variable. In fact z can approach z_0 through various paths and in different paths it might take different limits. For example:

(i) let $f(z) = \frac{xy^2}{x^2+y^4}$ if $z \neq 0$ and is equal to 0 if $z = 0$ where $z = x + iy, x$, $y \in \mathbb{R}$. If $z \to 0$ along the line $y = mx (m \neq 0)$, then $\lim_{z \to 0} f(z) = 0$. But if $z \to 0$ along the curve $y^2 = x$, then $f(z) = \frac{y^4}{y^4+y^4} = \frac{1}{2}$ and $f(z)$ does not tend to zero.

(ii) $\lim_{z \to 0} \dfrac{\bar{z}}{z}$ does not exist. If the limit is to exist it must be independent of the manner in which z approaches the point 0. Let $z \to 0$ along the x-axis. Then $y = 0$, and $z = x + iy = x$ and $\bar{z} = x - iy = x$, so that the required limit is $\lim_{x \to 0} \dfrac{x}{x} = 1$. Let $z \to 0$ along the y-axis. Then $x = 0$, and $z = x + iy = iy$ and $\bar{z} = x - iy = -iy$, so that the required limit is $\lim_{y \to 0} \dfrac{-iy}{iy} = -1$. Since the two approaches do not give the same value, the limit does not exist.

Example 3.26. *1. The function* $f : \mathbb{C} \to \mathbb{C}$ *given by* $f(z) = |z|$ *is continuous. It follows from the inequality that* $|\,|z| - |z_0|\,| \leq |z - z_0|$.

2. The function $f : \mathbb{C} \to \mathbb{C}$ *is continuous if and only if* $\bar{f} : \mathbb{C} \to \mathbb{C}$ *is continuous where* $\bar{f}(z) = \overline{f(z)}$. *This follows from the fact that*

$$|\bar{f}(z) - \bar{f}(z_0)| = |\overline{f(z)} - \overline{f(z_0)}| = |f(z) - f(z_0)|.$$

3. Let (X, d_X) *be a metric space. Fix* $a \in X$. *Define* $f : X \to \mathbb{R}$ *by setting* $f(x) = d_X(a, x)$. *We shall show that* f *is continuous. Let* $\epsilon > 0$ *and let* $y \in X$. *Notice that*

$$\begin{aligned} d_X(a,x) &\leq d_X(a,y) + d_X(y,x); \\ d_X(a,y) &\leq d_X(a,x) + d_X(x,y). \end{aligned}$$

Thus

$$\begin{aligned} d_X(a,x) - d_X(a,y) &\leq d_X(y,x) = d_X(x,y); \\ d_X(a,y) - d_X(a,x) &\leq d_X(x,y). \end{aligned}$$

Hence

$$|d_X(a,x) - d_X(a,y)| \leq d_X(x,y). \tag{3.12}$$

Let $\delta = \epsilon$. Now whenever $d_X(x,y) < \delta$, the inequality (3.12) implies $|f(x) - f(y)| < \epsilon$. Thus f is continuous at y.

Remark 3.9. Let $f : \mathbb{C} \to \mathbb{C}$ be given. Write $f(z) = u(z) + iv(z)$. Then u and v are called the real and imaginary parts of f. It is not so difficult to verify that f is continuous if and only if the real valued functions $u, v : \mathbb{C} \to \mathbb{R}$ are continuous. Thus f is continuous in an open set G if and only if Ref and Imf are continuous in G.

Theorem 3.22. *Let $G \subseteq \mathbb{C}$. Let $f, g : G \to \mathbb{C}$ be continuous at $z \in G$. Then the following holds:*

(1) The functions $f + g, fg, \alpha f$ are continuous at $z \in G$ for any $\alpha \in \mathbb{C}$.

(2) If $g(w) \neq 0$ for all $w \in B(z,r)$ for some $r > 0$ such that $B(z,r) \subset G$ then $\frac{f}{g}$ is defined on $B(z,r)$ and $\frac{f}{g} : B(z,r) \to \mathbb{C}$ is continuous at z.

Proof. The proof follows from the Theorem (theorem 3.21) on algebra of limits. □

An immediate application of Theorem 3.22 is the following.

Corollary 3.6. *Let f be a polynomial with complex coefficients and $f(z) = a_0 + a_1 z + a_2 z^2 + \cdots + a_n z^n, a_k \in \mathbb{C}$ for $0 \leq k \leq n$. Then $f : \mathbb{C} \to \mathbb{C}$ is continuous.*

Remark 3.10. Let f be a rational function in an open set $G \subset \mathbb{C}$, i.e., suppose $f(z) = \frac{p(z)}{q(z)}$ where p and q are polynomials on G. Notice that f is continuous at each point of the domain where the denominator q does not vanish.

Lemma 3.3. *Let $f : (X,d) \to \mathbb{C}$ be continuous at $a \in X$. Assume that $f(a) \neq 0$. Then there exists $r > 0$ such that $|f(x)| > \frac{|f(a)|}{2} > 0$ for all $x \in B(a,r)$. (That is, continuous functions do not change their values abruptly.)*

Proof. Since f is continuous at a, hence for $\epsilon = \frac{|f(a)|}{2} > 0$ there exists $\delta > 0$, such that whenever $d(x,a) < \delta$, we have

$$\big|\, |f(x)| - |f(a)| \,\big| \leq |f(x) - f(a)| < \epsilon.$$

Thus $|f(a)| - \epsilon < |f(x)| < |f(a)| + \epsilon$ whenever $d(x,a) < \delta$. It follows therefore that $|f(x)| > \frac{|f(a)|}{2} > 0$ for all $x \in B(a,\delta)$. □

Example 3.27. *Recall that $exp : \mathbb{C} \to \mathbb{C}$ is defined by $exp(z) = \sum_{n=0}^{\infty} \frac{z^n}{n!}$. We claim that exp is a continuous function on $\mathbb{C}$. Let $z \in \mathbb{C}$ and let $\epsilon > 0$ be given. Choose δ such that $0 < \delta < 1$ and $\delta exp(1 + |z|) < \epsilon$. Let w be such that $|w - z| < \delta$. Then since $|w| - |z| < \delta$, we have $|w| < \delta + |z| < 1 + |z|$. Now*

$$
\begin{aligned}
|exp(z) - exp(w)| &\leq \sum_{n=1}^{\infty} \frac{|z^n - w^n|}{n!} \\
&\leq |z-w| \sum_{n=1}^{\infty} \frac{|z|^{n-1} + |z|^{n-2}|w| + \cdots + |w|^{n-1}}{n!} \\
&\leq |z-w| \sum_{n=1}^{\infty} \frac{n \; max\{|z|^{n-1}, |w|^{n-1}\}}{n!} \\
&\leq |z-w| \sum_{n=1}^{\infty} \frac{(1+|z|)^{n-1}}{(n-1)!} \\
&= |z-w| exp(1+|z|) \\
&< \delta exp(1+|z|) \\
&< \epsilon.
\end{aligned}
$$

Since $z \in \mathbb{C}$ is arbitrary, exp *is continuous on $\mathbb{C}$.*

Theorem 3.23. *Any power series function $f(z) = \sum_{n=0}^{\infty} a_n z^n$ is continuous on its disk of convergence.*

Proof. Let $R > 0$ be the radius of convergence and $|z| < R$. Choose $\delta > 0$ such that $|z| + 2\delta < R$. If $|h| < \delta$, from the identity, $z^n - w^n = (z-w)[\sum_{k=0}^{n-1} z^k w^{n-k-1}]$, we obtain the inequality

$$
\begin{aligned}
|(z+h)^n - z^n| &= |h| \left|\sum_{k=0}^{n-1} z^k (z+h)^{n-k-1}\right| \\
&\leq |h| \sum_{k=0}^{n-1} |z|^k |z+h|^{n-k-1} \\
&\leq |h| \sum_{k=0}^{n-1} |z|^k \, (|z| + |h|)^{n-k-1}.
\end{aligned}
$$

Hence

$$|(z+h)^n - z^n| \leq |h| \; n \; (|z| + \delta)^{n-1}. \tag{3.13}$$

From the binomial expansion for $[(|z| + \delta) + \delta]^n$ we obtain

$$(|z| + 2\delta)^n \geq n(|z| + \delta)^{n-1}\delta. \tag{3.14}$$

We shall use these inequalities below to show that f is continuous in $B(0, R)$. Since $|z| + 2\delta < R$, hence $M = \sum_{n=1}^{\infty} |a_n|(|z| + 2\delta)^n < \infty$. Thus we obtain

$$
\begin{aligned}
|f(z+h) - f(z)| &\leq \sum_{n=1}^{\infty} |a_n||(z+h)^n - z^n| \\
&\leq \sum_{n=1}^{\infty} |a_n| n |h| (|z| + \delta)^{n-1} \quad \text{(from equation (3.13))} \\
&\leq |h| \sum_{n=1}^{\infty} |a_n| \frac{(|z| + 2\delta)^n}{\delta} \quad \text{(from equation (3.14))} \\
&\leq \frac{|h|}{\delta} M.
\end{aligned}
$$

This implies $\lim_{h \to 0} |f(z+h) - f(z)| = 0$. □

Definition 3.12. *Let X be a metric space. Given a series $\sum f_n$ of functions $f_n : X \to \mathbb{C}$, we say that $\sum f_n$ is uniformly convergent on X if the sequence of partial sums $s_n = \sum_{k=1}^{n} f_k$ converges uniformly on X to a function $f : X \to \mathbb{C}$.*

Example 3.28. *The series $\sum_{n=1}^{\infty} z^n(1-z)$, converges for $|z|<1$ and the sum is z. The convergence is uniform on $|z| \le \frac{1}{2}$. The nth partial sum $s_n(z) = z(1-z)+z^2(1-z)+\cdots+z^n(1-z) = z-z^{n+1}$. Now $|s_n(z)-z| = |-z^{n+1}| = |z|^{n+1} < \epsilon$ for $(n+1)\ln|z| < \ln\epsilon$, that is, for $n > \frac{\ln\epsilon}{\ln|z|} - 1$ if $z \neq 0$. If $z=0$, $s_n(0)=0$ and $|s_n(0)-0| < \epsilon$ for all n. Hence $\lim_{n\to\infty} s_n(z) = z$. If $|z| \le \frac{1}{2}$, the largest value of $\frac{\ln\epsilon}{\ln|z|} - 1$ occurs when $|z| = \frac{1}{2}$ and is given by $\frac{\ln\epsilon}{\ln\frac{1}{2}} - 1 = N$. It follows that $|s_n(z)-z| < \epsilon$ for all $n > N$ where N depends only on ϵ and not on the particular z in $|z| \le \frac{1}{2}$. Thus the series converges uniformly to z on $|z| \le \frac{1}{2}$.*

Proposition 3.3. *Let X be a metric space. If $f_n : X \to \mathbb{C}$ are continuous and $\sum_n f_n$ is uniformly convergent on X to f then f is continuous.*

The proof follows from Definition 3.12 and Theorem 2.58.

The most useful sufficient condition for uniform convergence of a series of functions is Weierstrass-M-test.

Theorem 3.24. ***(Weierstrass-M-test):*** *Let M_n be a sequence of nonnegative real numbers such that $\sum M_n$ is convergent. Let X be a metric space and $f_n : X \to \mathbb{C}$ be such that $|f_n(x)| \le M_n$ for all $n \in \mathbb{N}$ and for every $x \in X$. Then the series $\sum f_n$ is uniformly convergent on X.*

Proof. The numerical series $\sum f_n(x)$ is absolutely convergent which can be verified by comparing it with the series $\sum M_n$. Let $f(x) = \sum f_n(x)$. We wish to show that $\sum f_n$ is uniformly convergent to f on X.
Let $s_n(x) = \sum_{k=1}^{n} f_k(x)$. Then $|s_n(x) - s_m(x)| \le \sum_{k=m+1}^{n} M_k$ for each x. The series $\sum_{k=1}^{\infty} M_k$ converges, hence $(s_n(x))$ is a Cauchy sequence in $\mathbb{C}$. Since $\mathbb{C}$ is complete there is a number $\zeta \in \mathbb{C}$ with $\zeta = \lim s_n(x) = \sum_{n=1}^{\infty} f_n(x)$. Notice that ζ depends on x. Define $f : X \to \mathbb{C}$ as $f(x) = \sum_{n=1}^{\infty} f_n(x)$.

Given $\epsilon > 0$, choose $N \in \mathbb{N}$ so that $\sum_{n>N} M_n < \epsilon$. For all $n > N$, $|\sum_{k=N+1}^{n} f_k(x)| \le \sum_{k=N+1}^{n} M_k < \epsilon$. Hence $\left|\sum_{k>N} f_k(x)\right| < \epsilon$. Let $m \ge N$. Then $|f(x) - s_m(x)| = \left|\sum_{n=m+1}^{\infty} f_n(x)\right| < \epsilon$. □

Theorem 3.25. *Let $\sum_{n=0}^{\infty} a_n z^n$ has radius of convergence $R, 0 < R \le \infty$. Then for any r with $0 \le r < R$, the power series is uniformly convergent in $|z| \le r$.*

Proof. Since the radius of convergence of the power series $\sum_{n=0}^{\infty} a_n z^n$ is R and $0 \le r < R$, there exists z_0 such that $|z_0| > r$ and $\sum_{n=0}^{\infty} a_n z_0^n$ is convergent. Hence $a_n z_0^n \to 0$. Thus the sequence $\{a_n z_0^n\}$is bounded, say, by $M : |a_n z_0^n| \le M$. Now, if $|z| \le r$, then $|a_n z^n| \le |a_n| r^n \le |a_n z_0^n| (\frac{r}{|z_0|})^n \le M(\frac{r}{|z_0|})^n$. But the geometric series $\sum M(\frac{r}{|z_0|})^n$ is convergent. □

Remark 3.11. Thus using Proposition 3.3 and Theorem 3.25 we can also conclude that any power series with positive radius of convergence is continuous in its disk of convergence.

Corollary 3.7. *(Uniqueness theorem for power series): Let* $f(z) = \sum_{n=0}^{\infty} a_n z^n$ *and* $g(z) = \sum_{n=0}^{\infty} b_n z^n$ *for all* z *such that* $|z| \leq R, R > 0$. *Assume that there exists a sequence* (z_n) *in* $B(0, R)$ *such that (1)* $z_n \neq 0$ *for all* n, *(2)* $\lim_{n\to\infty} z_n = 0$ *and (3)* $f(z_n) = g(z_n)$ *for all* n. *Then* $a_n = b_n$ *for all* n.

Proof. Since $f(z_n) = g(z_n)$ for all n and f, g are continuous (Theorem 3.23) at 0, $f(0) = \lim f(z_n) = \lim g(z_n) = g(0)$ and hence $a_0 = b_0$. We shall prove the result by using principle of mathematical induction. By induction hypothesis, we have $a_k = b_k$ for $0 \leq k \leq n$. If we let $\phi(z) = a_{n+1} + a_{n+2}z + \cdots$ and $\psi(z) = b_{n+1} + b_{n+2}z + \cdots$, then ϕ and ψ are continuous on $B(0, R)$. Note that $\phi(z) = \frac{1}{z^{n+1}}[f(z) - \sum_{k=0}^{n} a_k z^k]$ for $z \neq 0$ and $|z| < R$. Similarly $\psi(z) = \frac{1}{z^{n+1}}[g(z) - \sum_{k=0}^{n} b_k z^k]$ for $z \neq 0$ and $|z| < R$. Consequently, $\phi(z_r) = \psi(z_r)$ for all $r \in \mathbb{N}$. Hence we have $\phi(0) = \lim \phi(z_r) = \lim \psi(z_r) = \psi(0)$. Since $\phi(0) = a_{n+1}$ and $\psi(0) = b_{n+1}$, the result follows. □

Remark 3.12. Notice that the above corollary says that if two power series with radius of convergence $\geq R$ agree on a set $\{z_n\}, z_n \in B(0, R)$ with a limit point $z \in B(0, R)$, then they agree on all of $B(0, R)$.

Let G be a region in $\mathbb{C}$ and f be a complex valued function on G. Then f is uniformly continuous (Definition 2.35) on G if given $\epsilon > 0$, there exists $\delta > 0$ (depending only on ϵ) such that $|f(z_1) - f(z_2)| < \epsilon$ whenever $|z_1 - z_2| < \delta$ where z_1, z_2 are any two points in G.

Example 3.29. *The function* $f(z) = z^2$ *is uniformly continuous in* $\mathbb{D} = \{z \in \mathbb{C} : |z| < 1\}$.

Solution: We must show that given any $\epsilon > 0$, we can find $\delta > 0$ such that $|z^2 - z_0^2| < \epsilon$ when $|z - z_0| < \delta$, where δ depends only on ϵ and not on the choice of the point $z_0 \in \mathbb{D}$.

If z and z_0 are any two points in $\mathbb{D}$, then

$$|z^2 - z_0^2| = |z + z_0||z - z_0| \leq \{|z| + |z_0|\}|z - z_0| < 2|z - z_0|.$$

Thus if $|z - z_0| < \delta$, it follows that $|z^2 - z_0^2| < 2\delta$. Choosing $\delta = \frac{\epsilon}{2}$, we see that $|z^2 - z_0^2| < \epsilon$ when $|z - z_0| < \delta$, where δ depends only on ϵ and not on z_0. Hence $f(z) = z^2$ is uniformly continuous in $\mathbb{D}$.

Example 3.30. *The function* $f(z) = \frac{1}{z}$ *is not uniformly continuous in* $\mathbb{D} = \{z \in \mathbb{C} : |z| < 1\}$.

Solution: Suppose that f is uniformly continuous in $\mathbb{D}$. Then for any $\epsilon > 0$, we should be able to find δ, say between 0 and 1, such that $|f(z) - f(z_0)| < \epsilon$ when $|z - z_0| < \delta$ for all z and z_0 in $\mathbb{D}$. Let $z = \delta$ and $z_0 = \frac{\delta}{1+\epsilon}$. Then $|z - z_0| = |\delta - \frac{\delta}{1+\epsilon}| = \frac{\delta\epsilon}{1+\epsilon} < \delta$. However, $|\frac{1}{z} - \frac{1}{z_0}| = |\frac{1}{\delta} - \frac{1+\epsilon}{\delta}| = \frac{\epsilon}{\delta} > \epsilon$ since $0 < \delta < 1$. Thus we have a contradiction, and it follows that $f(z) = \frac{1}{z}$ cannot be uniformly continuous in $\mathbb{D} = \{z \in \mathbb{C} : |z| < 1\}$.

3.5 Continuous choice of an argument

Theorem 3.26. *The function $\theta : \mathbb{C}^* \to \mathbb{R}$ given by $\theta(z) = Arg(z)$ is continuous on $U = \mathbb{C}/(-\infty, 0]$.*

Proof. The principal argument $\text{Arg} : \mathbb{C}^* = \mathbb{C} - \{0\} \to \mathbb{R}$ was defined by

$$\text{Arg}z = \begin{cases} \sin^{-1}(\frac{y}{|z|}) & \text{if } \ x \geq 0; \\ \pi - \sin^{-1}(\frac{y}{|z|}) & \text{if } \ x < 0 \text{ and } \ y \geq 0; \\ -\pi - \sin^{-1}(\frac{y}{|z|}) & \text{if } \ x < 0 \text{ and } y < 0. \end{cases}$$

where $\sin : [-\frac{\pi}{2}, \frac{\pi}{2}] \to [-1, 1]$ is a strictly increasing one-one, onto function. Its inverse $\sin^{-1} : [-1, 1] \to [-\frac{\pi}{2}, \frac{\pi}{2}]$ is therefore bijective and continuous. We let $\phi : \mathbb{C}^* \to [-\frac{\pi}{2}, \frac{\pi}{2}]$ be defined by $\phi(z) = \sin^{-1}(\frac{y}{|z|})$. Then ϕ is continuous. Continuity of ϕ implies that of θ on the three open sets $U_1 = \{z \in \mathbb{C} : \text{Re}z > 0\}, U_2 = \{z \in U : \text{Re}z < 0, \text{Im}z > 0\}$ and $U_3 = \{z \in U : \text{Re}z < 0, \text{Im}z < 0\}$. We need only verify the continuity of θ at points of the imaginary axis $\{z : \text{Re}z = 0\}$. Let $z_0 = \text{Im}z_0 > 0$. Let $z = x + iy \in U$. If $x \geq 0$, we observe that $|\theta(z) - \theta(z_0)| = |\phi(z) - \phi(z_0)|$. If $x \leq 0$, since $\phi(z_0) = \frac{\pi}{2}$, we have

$$|\theta(z) - \theta(z_0)| = |\pi - \phi(z) - \frac{\pi}{2}| = |\frac{\pi}{2} - \phi(z)| = |\phi(z) - \phi(z_0)|.$$

Hence $|\theta(z) - \theta(z_0)| = |\phi(z) - \phi(z_0)|$ holds for all $z \in U$ with $\text{Im}z > 0$. Since ϕ is continuous at z_0 this implies that θ is continuous at z_0. The case when $\text{Im}z_0 < 0$ is treated similarly. □

Remark 3.13. It is easy to see that θ is not continuous at any negative real. In fact, one cannot find a continuous argument on all of $\mathbb{C}^*$. This is the content of the next theorem.

Theorem 3.27. *There is no continuous map $\theta : \mathbb{C}^* \to \mathbb{R}$ such that $z = |z|(cos\theta(z) + isin\theta(z))$ for $z \in \mathbb{C}^*$. That is, it is impossible to make a continuous choice $\theta(z) \in arg(z)$ on $\mathbb{C}^*$.*

Proof. Let $g(t) = \text{cos}t + i\text{sin}t$. Assuming such a θ exists, consider $f : [0, 2\pi] \to \mathbb{R}$ defined by $f(t) = [\frac{\theta(g(t)) + \theta(g(-t))}{2\pi}]$. Then f is a real valued continuous function on $[0, 2\pi]$. The value $2\pi f(t)$ is a choice of $\arg(g(t)g(-t))$ and hence of $\arg(1)$. Thus f is an integer valued continuous function on the interval $[0, 2\pi]$. By intermediate value theorem, it is a constant. In particular, $f(0) = f(\pi)$. This implies that $[\frac{\theta(1)+\theta(1)}{2\pi}] = [\frac{\theta(-1)+\theta(-1)}{2\pi}]$ or, $\theta(1) = \theta(-1)$, which is impossible as the sets $\arg(1)$ and $\arg(-1)$ are disjoint. □

3.6 Exercises

1. Discuss the convergence of the following sequences whose n^{th} terms are

(i)	$\frac{i^n}{n^2}$	(ii)	$\frac{1}{n}+i^n$	(iii)	$(-1)^n+i(\frac{n+1}{n})$
(iv)	$\frac{n^2 i}{(n^2+1)}$	(v)	$\frac{in!}{n^n}$	(vi)	$\cos n\pi+i$
(vii)	$\cos(\frac{n\pi}{2})+i^n$	(viii)	$i^n\sin(\frac{n\pi}{4})$	(ix)	$z^n, \|z\|<1$
(x)	$\frac{n}{n+i}$	(xi)	$(1+\frac{4}{n})^n+\frac{7ni}{(n+4)}$	(xii)	$\sqrt{n+1}-\sqrt{n}+3i$
(xiii)	$\frac{1}{n}+\frac{i}{n^2}$.				

2. Show that if $\lim z_n = z$ then $\lim |z_n| = |z|$. Is the converse true?

3. Show that $\lim z_n = 0$ if and only if $\lim |z_n| = 0$.

4. Let $z \in \mathbb{C}$ be such that $z \neq 1$ but $|z| = 1$. Then prove that $\lim\limits_{n\to\infty} \dfrac{\sum_{k=1}^{n} z^k}{n} = 0$.

5. Let (z_n) be bounded. Prove that $(z_n) = (x_n + iy_n)$ is convergent in each of the following cases. (a) (x_n) is monotonic increasing, (y_n) is monotonic decreasing (b) Both (x_n) and (y_n) are monotonic increasing or decreasing (c) (x_n) is monotonic decreasing, (y_n) is monotonic increasing.

6. Let $z \in \mathbb{C}$ and $|z| < 1$. Show that z^n converges to the value 0.

7. Show that $(1 + \frac{i}{n})^n$ converges.

8. Give an example of a complex sequence $z_n = x_n + iy_n$ which is
   (i) bounded and divergent (ii) unbounded
   (iii) convergent to $1+i$ (iv) convergent to 1
   (v) bounded, (x_n) increasing, (y_n) decreasing
   (vi) unbounded with (x_n) decreasing, (y_n) increasing
   (vii) divergent but $(|z_n|)$ convergent.

9. Prove that if $l \geq 0, z_n \to l$, then
   (i) $|z_n| \to l$ (ii) $\alpha z_n \to \alpha l, \alpha \in \mathbb{C}$
   (iii) $\frac{1}{z_n} \to \frac{1}{l}$ if $l \neq 0$ (iv) $\frac{z_n}{1+z_n} \to \frac{l}{l+1}$ if $l+1 \neq 0$
   (v) $(1+z_n)^2 \to (1+l)^2$.

10. Show that every Cauchy sequence in a metric space is bounded.

11. Let a_n be nonnegative reals and assume that $\sum a_n$ is convergent. Then its sum s is given by $s = \sup\{s_n : n \in \mathbb{N}\}$.

12. Prove that $\sum_{n=0}^{\infty} z^n$ and $\sum_{n=1}^{\infty} \frac{z^n}{n^2}$ both diverge if $|z| > 1$.

13. Show that $\sum_{n=1}^{\infty} \frac{1}{2^k}(1+i)^{2k}$ is divergent.

14. If $\sum a_n$ is absolutely convergent then show that any rearrangement of $\sum a_n$ has the same sum.

15. If $\lim\limits_{n\to\infty} z_n = A$, prove that $\lim\limits_{n\to\infty} \dfrac{1}{n}(z_1 + z_2 + \cdots + z_n) = A$.

16. Discuss about the convergence and uniform convergence of the sequence $\{nz^n\}_{n=1}^{\infty}$.

17. The radius of convergence of $\sum_{n=0}^{\infty} a_n z^n$ is R $(0 < R < \infty)$. Determine the radii of convergence of the following series:

(i) $\sum_{n=0}^{\infty} \frac{a_n}{n!} z^n$ (ii) $\sum_{n=1}^{\infty} n^n a_n z^n$

(iii) $\sum_{n=0}^{\infty} a_n^k z^n$ (iv) $\sum_{n=0}^{\infty} (1 + z_0^n) a_n z^n$.

18. Suppose $\sum_{n=1}^{\infty} a_n z^n$ has radius of convergence $R \neq 0, R \neq \infty$. Find the radius of convergence of

(i) $\sum_{n=1}^{\infty} a_n z^{n^2}$ (ii) $\sum_{n=1}^{\infty} a_n z^{3n}$ (iii) $\sum_{n=1}^{\infty} 2^n a_n z^n$

(iv) $\sum_{n=1}^{\infty} a_n^2 z^n$ (v) $\sum_{n=1}^{\infty} a_n z^{2n}$ (vi) $\sum_{n=1}^{\infty} n^2 a_n z^n$.

19. If $\sum a_n z^n$ and $\sum b_n z^n$ have radius of convergence R_1 and R_2 then show that

(i) the radius of convergence of $\sum_{n=0}^{\infty} a_n b_n z^n$ is greater than or equal to $R_1 R_2$.

(ii) the radius of convergence of $\sum_{n=0}^{\infty} (a_n \pm b_n) z^n$ is greater than or equal to $\min\{R_1, R_2\}$.

(iii) the radius of convergence of $\sum_{n=0}^{\infty} \frac{a_n}{b_n} z^n$ is less than or equal to $\frac{R_1}{R_2}$.

20. For what values of z is $\sum_{n=0}^{\infty} \left(\frac{z}{1+z}\right)^n$ and $\sum_{n=0}^{\infty} \frac{z^n}{1+z^{2n}}$ are convergent?

21. Find the radius of convergence of the following power series.

(i) $\displaystyle\sum_{n=1}^{\infty} \frac{(\log n)^p}{n} z^n$ (ii) $\displaystyle\sum_{n=1}^{\infty} \frac{n}{2^n} z^n$ (iii) $\displaystyle\sum_{n=1}^{\infty} \frac{(-1)^n}{n} z^{n(n+1)}$

(iv) $\displaystyle\sum_{n=0}^{\infty} n! z^{n!}$ (v) $\displaystyle\sum_{n=0}^{\infty} n^n z^{n^2}$ (vi) $\displaystyle\sum_{n=1}^{\infty} n z^{n-1}$

(vii) $\displaystyle\sum_{n=0}^{\infty} z^{n(n+1)}$ (viii) $\displaystyle\sum_{n=0}^{\infty} z^{2^n}$ (ix) $\displaystyle\sum_{n=2}^{\infty} \frac{n(n-1)}{2} z^{n-2}$

(x) $\displaystyle\sum_{n=1}^{\infty} \frac{n!}{n^n} z^n$ (xi) $\displaystyle\sum_{n=3}^{\infty} \frac{z^n}{n \log(n-1)}$ (xii) $\displaystyle\sum_{n=1}^{\infty} z^{n^2}$

(xiii) $\displaystyle\sum_{n=0}^{\infty} \frac{(n!)^3}{(3n)!} z^{3n}$ (xiv) $\displaystyle\sum_{n=0}^{\infty} 2^n z^{n!}$ (xv) $\displaystyle\sum_{n=0}^{\infty} [3 + (-1)^n]^n z^n$

(xvi) $\displaystyle\sum_{n=0}^{\infty} 3^{n^2} z^n$ (xvii) $\displaystyle\sum_{n=0}^{\infty} 4^n z^n$ (xviii) $\displaystyle\sum_{n=0}^{\infty} \cos in \cdot z^n$

(xix) $\displaystyle\sum_{n=0}^{\infty} (n + a^n) z^n$ (xx) $\displaystyle\sum_{n=0}^{\infty} n! z^{n^2}$ (xxi) $\displaystyle\sum_{n=0}^{\infty} (-1)^n z^n$

(xxii) $\displaystyle\sum_{n=0}^{\infty} \left(\frac{1}{2}\right)^{n^2} z^n$.

22. Show that $(\sum_{n=0}^{\infty} \frac{z^n}{n!})(\sum_{n=0}^{\infty} \frac{(-z)^n}{n!}) = 1$ for all z. In particular, $\exp(z) \neq 0$ for any $z \in \mathbb{C}$. More generally, show that the Cauchy product of the series for $\exp(z)$ and $\exp(w)$ converges to $\exp(z+w)$.

23. Sum the following series for $|z| < 1$:

    (i) $\sum_{n=1}^{\infty} nz^n$ (ii) $\sum_{n=1}^{\infty} \frac{z^n}{n}$

    (iii) $\sum_{n=0}^{\infty} \frac{z^{2n+1}}{2n+1}$ (iv) $\sum_{n=1}^{\infty} (-1)^{n+1} \frac{z^n}{n}$.

24. Show that the set $\{z \in \mathbb{C} : |\exp(z)| < 1\}$ is an open set.

25. Suppose that $\sum_{n=0}^{\infty} a_n z^n$ has radius of convergence $R > 0$, and set $f(z) = \sum_{n=0}^{\infty} a_n z^n$ for $|z| < R$.

    (a) If $g(z) = f(z) + f(-z)$, find the value of $g^{(k)}(0)$ for each k (in terms of the a_n).

    (b) Deduce that if f is an odd function (i.e., $f(-z) = -f(z)$), then $a_n = 0$ for all even values of n.

26. Show that the function

$$f(z) = \begin{cases} z\frac{\text{Re } z}{|z|} & \text{if } z \neq 0; \\ 0 & \text{if } z = 0 \end{cases}$$

    is continuous at $z = 0$.

27. Let

$$f(z) = \begin{cases} 1 & \text{if } |z| \text{ is rational ;} \\ 0 & \text{if } |z| \text{ is irrational .} \end{cases}$$

    Prove that $f(z)$ is discontinuous at all values of z.

28. Prove that

$$f(z) = \begin{cases} z^2 & \text{if } z \neq a; \\ 0 & \text{if } z = a \end{cases}$$

    where $a \neq 0$, is not continuous at $z = a$.

29. Which of the following functions are uniformly continuous on the open unit disk $\mathbb{D} = \{z \in \mathbb{C} : |z| < 1\}$: (i) $\frac{1}{1-z}$; (ii) $\frac{1}{1+z^2}$.

Chapter 4

The Exponential and Trigonometric Functions

We have considered in the previous chapters such elementary functions as polynomials and rational functions. In this chapter, we study elementary functions such as the exponential, logarithmic, trigonometric and hyperbolic functions.

4.1 Elementary properties of exponential function

We define the exponential function $exp : \mathbb{C} \longrightarrow \mathbb{C}$ as $exp(z) = \sum_{n=0}^{\infty} \frac{z^n}{n!}$. This is the most important function in Mathematics. From ratio test and comparison test it follows that the power series converges for all $z \in \mathbb{C}$. It converges absolutely for every z and converges uniformly on every bounded subset of the complex plane.Thus the mapping exp is well defined. We define the number e to be $exp(1)$, and shall usually replace $\mathbf{exp(z)}$ by the customary shorter expression $\mathbf{e^z}$. Note that $e^0 = exp(0) = 1$. Below we list out some of the properties of exponential function.

Theorem 4.1. *The exponential function $exp : \mathbb{C} \rightarrow \mathbb{C}$ has the following properties:*

(1) $exp(z + w) = exp(z)exp(w)$, for all $z, w \in \mathbb{C}$.

(2) exp, when restricted to $\mathbb{R}$, is positive and monotonically increasing: $exp(x) \geq 0$ for $x \in \mathbb{R}$ and if x and y are real numbers with $x < y$ then $exp(x) < exp(y)$. Further $e^x \rightarrow \infty$ as $x \rightarrow \infty$ and $e^x \rightarrow 0$ as $x \rightarrow -\infty$.

(3) exp is continuous on the whole of $\mathbb{C}$.

(4) For every complex number z we have $exp(z) \neq 0$ and $exp(-z) = \frac{1}{exp(z)}$.

(5) $exp(\bar{z}) = \overline{exp(z)}$ for all $z \in \mathbb{C}$.

(6) *If x is real then $|exp(ix)| = 1$. That is, the mapping $t \mapsto e^{it}$ maps the real axis onto the unit circle.*

(7) $|exp(z)| = exp(Re(z)) \leq exp(|z|)$ *for* $z \in \mathbb{C}$.

(8) *The map $f : \mathbb{R} \to \mathbb{R}_+ = \{x \in \mathbb{R} : x > 0\}$ defined by $f(y) = \exp(y) = e^y$ is one-one and onto. That is, for any $\alpha \in \mathbb{R}$ with $\alpha > 0$, there is a unique $x \in \mathbb{R}$ such that $exp(x) = \alpha$.*

(9) *For any $n \in \mathbb{N}$, $\frac{exp(x)}{x^n} \to \infty$ and $x^n exp(-x) \to 0$ as $x \to \infty$. That is, e^x grows faster than any polynomial.*

Proof. The proof is as follows.

(1) The series for both $\exp(z)$ and $\exp(w)$ are absolutely convergent and hence their Cauchy product converges to $\exp(z)\exp(w)$. Thus

$$\begin{aligned} \exp(z)\exp(w) &= \textstyle\sum_{n=0}^{\infty} \frac{1}{n!} \sum_{k=0}^{n} \frac{n!}{k!(n-k)!} z^k w^{n-k} \\ &= \textstyle\sum_{n=0}^{\infty} \frac{(z+w)^n}{n!} \\ &= \exp(z+w). \end{aligned}$$

(2) If $x \geq 0$, then notice that $\exp(x) = \sup_n \{\sum_{k=0}^{n} \frac{x^k}{k!}\} > 1 + x > x$. From (1), we have $\exp(x)\exp(-x) = \exp(0) = 1$. Since $\exp(x) > x > 0$ we infer that $\exp(-x) > 0$. Also, if $x < y$, then $\exp(y) = \exp((y-x)+x) = \exp(y-x)\exp(x) > (1+(y-x))\exp(x) > \exp(x)$. Hence exp is increasing in $\mathbb{R}$ and if $x \to \infty, e^x \to \infty$. Therefore, $e^x = \frac{1}{e^{-x}} \to 0$ as $x \to -\infty$.

(3) We have already shown that exp is continuous on $\mathbb{C}$. See Example 3.27.

(4) From (1), $\exp(z)\exp(-z) = \exp(0) = 1$ and (4) follows.

(5) It follows from the fact that $\sum z_n$ converges and its sum is z if and only if $\sum \bar{z}_n$ converges and its sum is $\bar{z}$.

(6) For $x \in \mathbb{R}$, we have $|\exp(ix)|^2 = \exp(ix)\overline{\exp(ix)} = \exp(ix)\exp(-ix) = \exp(0) = 1$.

(7) If $z = x + iy$, then $\exp(z) = \exp(x+iy) = \exp(x)\exp(iy)$ so that (7) follows from (6) and (2).

(8) From the proof of (2), $\exp(\alpha) > \alpha$ and hence $\exp(-\frac{1}{\alpha}) = \frac{1}{\exp(\frac{1}{\alpha})} < \frac{1}{(\frac{1}{\alpha})} = \alpha < \exp(\alpha)$. Since exp is continuous on $[-\frac{1}{\alpha}, \alpha]$ the result follows from intermediate value theorem.

(9) Observe that, for $x > 0, \exp(x) > \frac{x^{n+1}}{(n+1)!}$ so that $x^{-n}\exp(x) > \frac{x}{(n+1)!}$.

□

Let $z = x + iy, x, y \in \mathbb{R}$. Since $|e^z| = e^x$, the map $z \mapsto e^z$ sends the vertical line $x = x_0$ onto the circle with center 0 and radius e^{x_0}. As a point moves along the vertical line in the upward direction, its image point on the circle moves in the counterclockwise direction, making one complete circuit each time its pre-image traverses an interval of length 2π. Since $\arg e^z = y$, the map $z \mapsto e^z$ maps the horizontal line $y = y_0$ onto the open ray emanating from the origin that makes the angle y_0 with the positive real axis. As a point moves toward the right on the horizontal line, its image point on the ray moves away from the origin. Thus, the map $z \mapsto e^z$ sends the coordinate grid for rectangular coordinates to the coordinate grid for polar coordinates. Its range is $\mathbb{C}/\{0\}$, and every point in the range has infinitely many pre-images, any two of which differ by an integer multiple of $2\pi i$. The restriction of the map $z \mapsto e^z$ to an open horizontal strip of width at most 2π is one-to-one, and the map sends the strip to the interior of an angle with vertex at the origin. For example, the strip $-\frac{\pi}{2} < y < \frac{\pi}{2}$ is sent onto the right half plane. Notice that for $x \in \mathbb{R}$,

$$\begin{aligned} e^{ix} &= \textstyle\sum_{n=0}^{\infty} \frac{(ix)^n}{n!} \\ &= 1 - \tfrac{x^2}{2!} + \tfrac{x^4}{4!} + \cdots + \tfrac{(-1)^n}{(2n)!}x^{2n} + i\left(x - \tfrac{x^3}{3!} + \tfrac{x^5}{5!} - \cdots + \tfrac{(-1)^n}{(2n+1)!}x^{2n+1}\right) \\ &= \cos x + i \sin x. \end{aligned}$$

The identity $e^{ix} = \cos x + i \sin x$ is called the **Euler's identity**. Thus $e^{-ix} = \cos x - i \sin x$ and it follows that

$$\cos x = \frac{e^{ix} + e^{-ix}}{2}; \sin x = \frac{e^{ix} - e^{-ix}}{2i}.$$

4.2 Trigonometric functions

Definition 4.1. *We define*

$$cos(z) = \frac{exp(iz) + exp(-iz)}{2}$$

and

$$sin(z) = \frac{exp(iz) - exp(-iz)}{2i}$$

for $z \in \mathbb{C}$.

For brevity, we write $\cos z$ for $\cos(z)$ etc. It follows from the definition of $\sin z$ and $\cos z$ that for all $z \in \mathbb{C}$,

$$\cos z = 1 - \frac{z^2}{2!} + \frac{z^4}{4!} - \cdots = \sum_{n=0}^{\infty} \frac{(-1)^n z^{2n}}{(2n)!};$$

$$\sin z = z - \frac{z^3}{3!} + \frac{z^5}{5!} - \cdots = \sum_{n=0}^{\infty} \frac{(-1)^n z^{2n+1}}{(2n+1)!}.$$

The radius of convergence of each of the series is infinite.

Theorem 4.2. *For all $z, w \in \mathbb{C}$, the following holds:*

(1) The trigonometric functions cos and sin are continuous on the whole of $\mathbb{C}$.

(2) $exp(iz) = cosz + isinz$.

(3) $cos^2 z + sin^2 z = 1$.

(4) $cos(z + w) = cosz \cdot cosw - sinz \cdot sinw$.

(5) $sin(z + w) = sinz \cdot cosw + cosz \cdot sinw$.

(6) $cosz = \sum_{n=0}^{\infty} \frac{(-1)^n z^{2n}}{(2n)!}$.

(7) $sinz = \sum_{n=0}^{\infty} \frac{(-1)^n z^{2n+1}}{(2n+1)!}$.

(8) $(\cos z + i \sin z)^n = \cos nz + i \sin nz$ *for all positive integers* n.

(9) For each $z \in \mathbb{C}, \cos(-z) = \cos z$ *and* $\sin(-z) = -\sin z$. *That is,* $\cos z$ *is an even function and* $\sin z$ *is an odd function.*

(10) $e^{-iz} = \cos z - i \sin z$.

(11) For $z \in \mathbb{C}, \cos z = \cos(x + iy) = \cos x \cosh y - i \sin x \sinh y$ *and* $\sin z = \sin(x + iy) = \sin x \cosh y + i \cos x \sinh y$.

Proof. Using (2), we get

$$\begin{aligned} 1 &= \exp(iz)\exp(-iz) \\ &= (\cos z + i\sin z)(\cos z - i\sin z) \\ &= \cos^2 z + \sin^2 z. \end{aligned}$$

This proves (3).

To prove (4) and (5), note that

$$\begin{aligned} &\exp(i(z + w)) \pm \exp(-i(z + w)) \\ &= \exp(iz)\exp(iw) \pm \exp(-iz)\exp(-iw) \\ &= (\cos z + i\sin z)(\cos w + i\sin w) \pm (\cos z - i\sin z)(\cos w - i\sin w). \end{aligned}$$

□

The formula in (11) separates the real and imaginary parts of $\cos z$ and $\sin z$. Using the formula (11) one can also solve the following types of equation. Suppose we wish to find all values of z such that $\sin z = 4$. One can readily verify that $z = \frac{4n+1}{2}\pi + i\cosh^{-1} 4, n \in \mathbb{Z}$. Similarly we may now establish that

$$\begin{aligned} \sin^2 z &= \tfrac{1}{2}(1 - \cos 2z); \\ \cos^2 z &= \tfrac{1}{2}(1 + \cos 2z); \\ \sin 2z &= 2 \sin z \cos z. \end{aligned}$$

Let E be the set of all complex numbers which are not odd integral multiples of $\frac{\pi}{2}$, and let G be the set of all complex numbers which are not integral multiples of π. We define $\tan z, \cot z, secz$ and $\csc z$ as follows:

$$\tan z = \frac{\sin z}{\cos z}, \sec z = \frac{1}{\cos z}$$

for each $z \in E$;

$$\cot z = \frac{\cos z}{\sin z}, \csc z = \frac{1}{\sin z}$$

for each $z \in G$. We may now establish that $1 + \tan^2 z = \sec^2 z$ on E and $1 + \cot^2 z = \csc^2 z$ on G.

4.3 Periodicity

In this section we shall establish that $\sin z, \cos z$ and e^z are periodic functions.

Let $G \subset \mathbb{C}$ and f is a function from G into $\mathbb{C}$. The point $w \in G$ is a **zero** of the function f if $f(w) = 0$.

Proposition 4.1. $\cos t$ *has at least one zero in* $(0, 2)$.

Proof. Note that

$$\cos t = 1 - \frac{t^2}{2!} + \frac{t^4}{4!} - \frac{t^6}{6!} + \cdots \tag{4.1}$$

Take $t = 2$. The terms of the series (4.1) then decrease in absolute value (except for the first one) and their signs alternate. Hence $\cos 2$ is less than the sum of the first three terms of (4.1), with $t = 2$; thus $\cos 2 < -\frac{1}{3}$. Since $\cos 0 = 1$ and cos is a continuous real function on the real axis we conclude that there is a smallest positive number t_0 for which $\cos t_0 = 0$. □

Lemma 4.1. *The zeros of* $\sin z$ *lie on the real axis only. That is, if* $\sin z = 0$ *then* $z \in \mathbb{R}$.

Proof. If $\sin z = 0$, then $\exp(iz) - \exp(-iz) = 0$ so that $\exp(2iz) = 1$. Hence $\exp(-2y) = |\exp(2iz)| = 1$. Since exp is strictly increasing on $\mathbb{R}$, this means $y = 0$. □

Lemma 4.2. *There exists* $t \in [0, 1]$ *such that* $\sin t = \cos t$.

Proof. If $0 < x < 2$, we have

$$\begin{aligned}
|\sin x - x| &\leq \tfrac{x^3}{3!} + \tfrac{x^5}{5!} + \cdots \\
&\leq \tfrac{x^3}{3!}\left(1 + \tfrac{x^2}{20} + (\tfrac{x^2}{20})^2 + \cdots\right) \\
&\leq \tfrac{x^3}{3!}\left[1 + \tfrac{1}{5} + (\tfrac{1}{5})^2 + \cdots\right] \\
&< \tfrac{x^3}{4}.
\end{aligned}$$

Hence we find that $\sin x > x - \frac{x^3}{4} > \frac{x(4-x^2)}{4} > 0, 0 < x < 2$.

Clearly $\sin 1 > \frac{3}{4}$ and hence $\cos 1 < \frac{3}{4}$. Thus $\sin 1 - \cos 1 > 0$ and $\sin 0 - \cos 0 = 0 - 1 = -1 < 0$. Apply intermediate value theorem to sin − cos on [0,1] to assert the existence of a $t \in [0, 1]$ such that $\cos t = \sin t$. □

Lemma 4.3. *There exists* $t \in (0, 4)$ *such that* $\sin t = 0$.

Proof. Let $t \in (0, 1)$ be such that $\cos t = \sin t$. Hence $\cos 2t = \cos^2 t - \sin^2 t = 0$ and hence $\sin 4t = 2\cos 2t \sin 2t = 0$. Hence sin has at least one zero in (0, 4) (as $0 < 4t < 4$ if $t \in (0, 1)$). □

Definition 4.2. *Define* $\pi = \inf\{x > 0 : \sin x = 0\}$.

In course of proof of Lemma 4.2 and Lemma 4.3, we have shown that $\sin x > 0$ if $0 < x < 2$ and sin has at least one zero in $(0, 4)$. Hence $2 < \pi < 4$. For each $n \in \mathbb{N}$, $\pi < \pi + \frac{1}{n}$ and hence there exists x_n with $\pi < x_n < \pi + \frac{1}{n}$ such that $\sin(x_n) = 0$. This follows from the definition of infimum or g.l.b.

Since $x_n \to \pi$ and sin is continuous, we conclude $\sin\pi = 0$.

The following observations can be made about the sin function.

(1) $\sin\frac{\pi}{2} = 1$. For, $\sin\pi = 2\sin\frac{\pi}{2}\cos\frac{\pi}{2}$. Since $0 < \frac{\pi}{2} < 2, \sin\frac{\pi}{2} > 0$. Hence $\cos\frac{\pi}{2} = 0$ as $\sin\pi = 0$. Since $\cos^2\frac{\pi}{2} + \sin^2\frac{\pi}{2} = 1$, the result follows. Thus we obtain (since $\cos\frac{\pi}{2} = 0, \sin\frac{\pi}{2} = 1$) $e^{\frac{\pi i}{2}} = i$. It follows therefore that $e^{\pi i} = i^2 = -1, e^{2\pi i} = (-1)^2 = 1$ and that $e^{2\pi i n} = 1$ for every integer n.

(2) $\sin x > 0$ for $x \in (0, \pi)$. For, $\sin x > 0$ for $x \in (0, \frac{\pi}{2})$. Observe that $\sin(x + \frac{\pi}{2}) = \sin(-x + \frac{\pi}{2})$. This follows from the fact that cos is an even function and $\cos\frac{\pi}{2} = 0$. (This also follows if we expand both sides using the addition formula for sin.) If $x \in (0, \frac{\pi}{2})$, the right side of this identity is positive.

    One can also prove this using the definition of π and continuity of sin. If $\sin x \leq 0$ for $x \in (0, \pi)$, then $\sin y = 0$ for some $y \in (0, \pi)$ by intermediate value theorem. This violates the definition of π.

(3) sin is increasing on $[0, \frac{\pi}{2}]$. If $0 < x < y < \frac{\pi}{2}$, then $0 < \sin(y+x)\sin(y-x) = \sin^2 y - \sin^2 x$. Since $\sin x > 0$ for $x \in (0, \pi)$, the result follows.

(4) For $z \in \mathbb{C}$, we have $\sin(z + \pi) = -\sin z$ and hence $\sin(z + 2\pi) = \sin z$. This follows from the fact that $\cos(\pi) = \cos(2\frac{\pi}{2}) = -1$ and the addition formula for sin function. Note also that $\cos(z + 2\pi) = \cos z$ for all $z \in \mathbb{C}$.

Theorem 4.3. *The following hold.*

(1) $\{z \in \mathbb{C} : \sin z = 0\} = \{n\pi : n \in \mathbb{Z}\}$.

(2) $\{z \in \mathbb{C} : \cos z = 0\} = \{(n + \frac{1}{2})\pi : n \in \mathbb{Z}\}$.

(3) $\exp z = 1$ *if and only if* $z = 2\pi i n$ *for some* $n \in \mathbb{Z}$. *Hence* $\exp z = \exp w$ *if and only if* $z - w = 2\pi i k$ *for some* $k \in \mathbb{Z}$.

Proof. We know that if sin$z = 0$ then $z \in \mathbb{R}$. Since $\sin(-x) = -\sin(x)$ and sin is strictly increasing in $[0, \frac{\pi}{2}]$, the only zeros of sin in $[-\pi, \pi]$ are $-\pi, 0, \pi$. Let $J = (-\pi, \pi]$. Then $\mathbb{R} = \cup_{n\in\mathbb{Z}}(J + 2n\pi)$ is a disjoint union. If $x \in \mathbb{R}$ is such that sin$x = 0$, then there exists $x_0 \in J$ and $n \in \mathbb{Z}$ such that $x = x_0 + 2n\pi$. Now, $0 = \sin x = \sin(x_0 + 2n\pi) = \sin x_0$. We hence conclude that x_0 is either 0 or π so that $x = 2n\pi$ or $(2n+1)\pi$. The assertion (1) is thus proved. The assertion (2) is a consequence of $\sin(z + \frac{\pi}{2}) = \cos z$. To prove (3), let $z = x + iy$ be such that exp$z = 1$. Then $|\exp z| = \exp x = 1$ so that $x = 0$. Thus $e^{iy} = 1$. Hence cos$y = 1$ and sin$y = 0$. From (1) and (2), we infer that $y = 2n\pi$. □

An alternate way to show that $e^{iy} \neq 1$ if $0 < y < 2\pi$, is as follows. Suppose $0 < y < 2\pi$ and $e^{\frac{iy}{4}} = u + iv$ (u and v real). Since $0 < \frac{y}{4} < \frac{\pi}{2}$, we have $u > 0$ and $v > 0$. Also

$$e^{iy} = (u + iv)^4 = u^4 - 6u^2v^2 + v^4 + 4iuv(u^2 - v^2). \tag{4.2}$$

The right side of (4.2) is real only if $u^2 = v^2$; since $u^2 + v^2 = 1$, this happens only when $u^2 = v^2 = \frac{1}{2}$, and then (4.2) shows that $e^{iy} = -1 \neq 1$.

Definition 4.3. *Let $f : \mathbb{C} \to \mathbb{C}$ be given. A number $p \in \mathbb{C}$ is called a period of f if for every $z \in \mathbb{C}$, we have $f(z + p) = f(z)$. If f has a nonzero period p, then f is said to be periodic. Note that if p is a period of f, so is np for any $n \in \mathbb{Z}$.*

It follows from the above discussions that exp, sin and cos are periodic with the set of periods $\{2\pi in : n \in \mathbb{Z}\}, \{2\pi n : n \in \mathbb{Z}\}$ and $\{2\pi n : n \in \mathbb{Z}\}$ respectively.

Theorem 4.4. *The mapping $t \mapsto e^{it}$ maps the real axis onto the unit circle.*

Proof. The function $f(t) = e^{it}$ maps the real axis into the unit circle. Fix w so that $|w| = 1$. We shall show that $w = e^{it}$ for some real t. Let $w = u + iv$, u and v real, and suppose that $u \geq 0$ and $v \geq 0$. Since $u \leq 1$, there exists a $t, 0 \leq t \leq \frac{\pi}{2}$, such that $\cos t = u$. This happens because $\cos 0 = 1$ and $\cos\frac{\pi}{2} = 0$. Then $\sin^2 t = 1 - u^2 = v^2$, and since $\sin t \geq 0$ if $0 \leq t \leq \frac{\pi}{2}$, we have $\sin t = v$. Thus $w = e^{it}$. If $u < 0$ and $v \geq 0$, the preceding conditions are satisfied by $-iw$. Hence $-iw = e^{it}$ for some real t, and $w = e^{i(t+\frac{\pi}{2})}$. Finally, if $v < 0$, the preceding two cases show that $-w = e^{it}$ for some real t, hence $w = e^{i(t+\pi)}$. This completes the proof. □

Theorem 4.5. *If w is a complex number and $w \neq 0$, then $w = e^z$ for some $z \in \mathbb{C}$.*

Proof. If $w \neq 0$, put $\alpha = \frac{w}{|w|}$. Then $w = |w|\alpha$. Thus there is a real x such that $|w| = e^x$. Since $|\alpha| = 1$, by Theorem 4.4, there exists $y \in \mathbb{R}$ such that $\alpha = e^{iy}$. Hence $w = e^{x+iy}$. □

Let $w = f(z) = e^z = e^x e^{iy}, z \in \mathbb{C}$. Let $\rho = |w|$ and $\phi = \arg w = y$. We have already noted that the line $x = a$ is mapped by e^z onto the circle $|w| = e^a$ and the line $y = b$ is mapped by e^z onto the ray $\phi = b$ with $\rho > 0$. In particular, the x-axis ($y = 0$) is mapped onto the ray $\phi = 0$ with $\rho > 0$ while the line $y = \pi$ is mapped onto the ray $\phi = \pi$ with $\rho > 0$. Now as b increases from 0 to π, the ray $\phi = b$ rotates through the upper half plane. This means that the horizontal strip $-\infty < x < +\infty$ and $0 \leq y \leq \pi$ is mapped onto the upper half plane (with the origin deleted) $\rho > 0$ and $0 \leq \phi \leq \pi$. Similarly, the strip $-\infty < x < +\infty$ and $-\pi < y < 0$ is mapped onto the lower half plane $\rho > 0$ and $-\pi < \phi < 0$. Thus f maps the strip $\{z \in \mathbb{C} : -\pi < y \leq \pi\}$ onto $\mathbb{C} - \{0\}$. Now for each $k \in \mathbb{Z}$, we have $w = f(z + 2k\pi i) = e^z e^{2k\pi i} = e^z = f(z), z \in \mathbb{C}$. Thus for each $k \in \mathbb{Z}$, the strip $\{z \in \mathbb{C} : (2k-1)\pi < y \leq (2k+1)\pi\}$ is mapped onto $\mathbb{C} - \{0\}$. For this reason, the strip $\{z \in \mathbb{C} : -\pi < y \leq \pi\}$ is called **the fundamental strip**. Suppose

$$\begin{aligned} A &= \{(x, y) : 0 \leq x < +\infty \text{ and } 0 \leq y \leq \pi\}; \\ B &= \{(x, y) : -\infty < x < 0 \text{ and } 0 \leq y \leq \pi\}. \end{aligned}$$

It is not difficult to check that

$$\begin{aligned} f(A) &= \{w \in \mathbb{C} : |w| \geq 1 \text{ and Im } w \geq 0\}; \\ f(B) &= \{w \in \mathbb{C} : |w| < 1, \text{Im } w \geq 0 \text{ and } w \neq 0\}. \end{aligned}$$

4.4 Argument of a nonzero complex number

Definition 4.4. *Let $z \in \mathbb{C}$ be nonzero. We define* $\arg(z)$ *to be the set* $\{t \in \mathbb{R} : z = |z| exp(it)\} = \{t \in \mathbb{R} : z = |z| e^{it}\}$.

Proposition 4.2. *Let $z \in \mathbb{C}$ be nonzero. Then there exists a unique $\theta \in [0, 2\pi)$ such that $z = |z| exp(i\theta)$. Also,* $\arg(z) = \{\theta + 2\pi n : n \in \mathbb{Z}\}$.

Proof. Let $\frac{z}{|z|} = x + iy$ so that $x^2 + y^2 = 1$. We consider four cases corresponding to four quadrants.

(1) Let $x \geq 0$ and $y \geq 0$. By intermediate value theorem, there exists $t \in [0, \frac{\pi}{2}]$ such that $\cos t = x$. Then $\sin^2 t = y^2$. Since $\sin t \geq 0$ on $[0, \frac{\pi}{2}]$, we see that $y = \sin t$.

(2) Let $x \geq 0$ and $y \geq 0$. Then $x + iy = i(y - ix)$. By (1), there is a $t \in [0, \frac{\pi}{2}]$ such that $y - ix = \exp(it)$. But then $x + iy = i(y - ix) = \exp(i\frac{\pi}{2})\exp(it) = \exp(i(t + \frac{\pi}{2}))$.

The other cases are treated similarly. If $z = |z|\exp(is) = |z|\exp(it)$ for $s, t \in [0, 2\pi)$, then $\exp(i(s - t)) = 1$ so that $s - t$ is an integral multiple of 2π. Hence, $s = t$. □

Example 4.1. *(i) Let $z = x + iy \in \mathbb{C}$. Then $expz = exp(x)exp(iy)$. Since $|expz| = exp(x)$, we see that* $\arg(exp(z)) = \{y + 2n\pi : n \in \mathbb{Z}\}$.

(ii) Let $z = x + iy$ *be in the first quadrant. Let* $\theta \in \arg(z)$*. Then* $x + iy = |z|(cos\theta + isin\theta)$ *so that* $cos\theta = \frac{x}{|z|}$ *and* $sin\theta = \frac{y}{|z|}$.

Example 4.2. *If* $z_1, z_2 \in \mathbb{C}, z_1 z_2 \neq 0$ *then* $Arg(z_1 z_2) - (Argz_1 + Argz_2) = 2n\pi$ *where* n *has one of the values* $0, 1$ *or* -1.

Solution: Let $z_j = |z_j|(\cos\theta_j + i\sin\theta_j), j = 1, 2$ where $-\pi < \theta_j = \text{Arg}z_j \leq \pi$. Then $z_1 z_2 = |z_1||z_2|(\cos(\theta_1+\theta_2) + i\sin(\theta_1+\theta_2)) = |z_1||z_2|(\cos(\theta_1+\theta_2+2k\pi) + i\sin(\theta_1+\theta_2+2k\pi))$ for each $k \in \mathbb{Z}$. Since $-2\pi < \theta_1 + \theta_2 \leq 2\pi$, hence $\theta_1 + \theta_2$ is in one of the intervals $(-2\pi, -\pi], (-\pi, \pi]$ or $(\pi, 2\pi]$. If $\theta_1 + \theta_2$ is in $(-\pi, \pi]$, then $\text{Arg}(z_1 z_2) = \theta_1 + \theta_2 = (\text{Arg}z_1 + \text{Arg}z_2)$. That is, $n = 0$. If $\theta_1 + \theta_2$ is in $(\pi, 2\pi]$, then $\pi < \theta_1 + \theta_2 \leq 2\pi$. Hence $-\pi < \theta_1 + \theta_2 - 2\pi \leq 0 \leq \pi$ and therefore $\text{Arg}(z_1 z_2) = \theta_1 + \theta_2 = \text{Arg}z_1 + \text{Arg}z_2 - 2\pi$. That is, $n = -1$. Similarly if $\theta_1+\theta_2$ is in $(-2\pi, -\pi]$, then one can show that $\text{Arg}(z_1 z_2) = \theta_1 + \theta_2 = \text{Arg}z_1 + \text{Arg}z_2 + 2\pi$. That is, $n = 1$.

We have already proved a result which says that we cannot assign argument on $\mathbb{C}^*$ in a continuous manner. However, if we are ready to ignore a closed half-line we can get a continuous choice θ of arg . Let $\alpha \in \mathbb{R}$ and let $L_\alpha = \{te^{i\alpha} : t \geq 0\}$. Note that $z \in \mathbb{C}/L_\alpha$ if and only if $\alpha \notin \arg(z)$. For $z \notin L_\alpha$, there is a unique choice of $\arg(z)$ in the interval $(\alpha, \alpha + 2\pi)$. We denote this choice by $\text{Arg}_\alpha : \mathbb{C}/L_\alpha \to (\alpha, \alpha + 2\pi)$.

Theorem 4.6. Arg_α *is continuous on* $\mathbb{C}/L_\alpha$.

Proof. Let $\theta = \text{Arg}_0$. Then $\mathbb{C}/L_0 = H_0 \cup H_1 \cup H_2$ where $H_0 = \{x + iy : y > 0\}, H_1 = \{x + iy : x < 0\}$ and $H_2 = \{x + iy : y < 0\}$. We show that θ is continuous on each of the open half-planes $H_i, 0 \leq i \leq 2$. If $z \in H_0$, then $\text{Im}z = |z|\text{sin}(\theta(z))$. So, $\sin(\theta(z)) > 0$. Since θ takes values only in $(0, 2\pi)$, this means that $\theta(z) \in (0, \pi)$ for such z. Now, $\cos : (0, \pi) \to (-1, 1)$ is strictly decreasing. Hence it has a continuous inverse $\cos^{-1} : (-1, 1) \to (0, \pi)$. Hence $\theta(z) = \cos^{-1}(\cos(\theta(z))) = \cos^{-1}\left(\frac{\text{Re}(z)}{|z|}\right), z \in H_0$.

Since the RHS is continuous, it follows that θ is continuous on H_0. Similar reasoning establishes the continuity of θ on H_1 and H_2. For the general case consider the map $f(z) = z\exp(-i\alpha)$ which maps $\mathbb{C}/L_\alpha$ onto $\mathbb{C}/L_0$. Then one shows that $\text{Arg}_\alpha(z) = \alpha + \theta(f(z))$. □

Proposition 4.3. *Let* $\theta : U = \mathbb{C}/L_\alpha \to \mathbb{R}$ *be a continuous choice of the argument, i.e., for each* $z \in \mathbb{C}/L_\alpha$*, we have* $z = |z|exp(i\theta(z))$*. Then there exists an integer* n *such that* $\theta - Arg_\alpha = 2\pi n$.

Proof. Note that U is star shaped and hence path connected whence connected. Now for each $z \in U$, $\theta(z) - \text{Arg}_\alpha(z) \in \{2\pi n : n \in \mathbb{Z}\}$. Since this latter is not connected, the continuous function $\theta - \text{Arg}_\alpha$ must be a constant. □

Remark 4.1. This proposition yields another proof of impossibility of a continuous choice of the argument on $\mathbb{C}^*$. For, if θ is one such then it must coincide with $\theta_0 + 2n\pi$ on $\mathbb{C}/L_0$ for some n. But then we can easily show that $\theta_0 + 2n\pi$ does not extend continuously to $\mathbb{C}$.

4.5 Hyperbolic functions

The hyperbolic functions cosh and sinh are defined as follows:

$$\begin{aligned}\cosh(z) &= \tfrac{[\exp(z)+\exp(-z)]}{2}, z \in \mathbb{C}\\ \sinh(z) &= \tfrac{[\exp(z)-\exp(-z)]}{2}, z \in \mathbb{C}.\end{aligned}$$

Proposition 4.4. *The following identities hold.*

(1) $\cosh(z) = cos(iz)$ *and* $\sinh(z) = -isin(iz)$ *for all* $z \in \mathbb{C}$. *Hence* $\cosh^2(z) - sinh^2(z) = 1$.

(2) $\cosh(z) = \sum_{n=0}^{\infty} \frac{z^{2n}}{(2n)!}$ *and* $sinh(z) = \sum_{n=0}^{\infty} \frac{z^{2n+1}}{(2n+1)!}$ *for* $z \in \mathbb{C}$.

(3) cosh *and sinh are positive and strictly increasing on the set of positive reals.*

(4) $\cosh(-z) = coshz$ *and* $sinh(-z) = -sinh(z)$ *for* $z \in \mathbb{C}$.

(5) $$\begin{aligned}\cosh(x+iy) &= \cos(x)\cosh(y) - i\sin(x)\sinh(y)\\ \sinh(x+iy) &= \sin(x)\cosh(y) + i\cos(x)\sinh(y).\end{aligned}$$

(6) For $z = x + iy \in \mathbb{C}$, *we have*

$$\begin{aligned}sinh(y) &\le |cos(x \pm iy)| \le cosh(y)\\ sinh(y) &\le |sin(x \pm iy)| \le cosh(y).\end{aligned}$$

(7) $$\begin{aligned}\sinh(z_1 + z_2) &= \sinh z_1 \cdot \cosh z_2 + \cosh z_1 \cdot \sinh z_2;\\ \cosh(z_1 + z_2) &= \cosh z_1 \cdot \cosh z_2 + \sinh z_1 \cdot \sinh z_2.\end{aligned}$$

(8) $\sinh 2z = 2\sinh z \cosh z$.

(9) $$\begin{aligned}|\sinh z|^2 &= \sinh^2 x + \sin^2 y = \cosh^2 x - \cos^2 y;\\ |\cosh z|^2 &= \sinh^2 x + \cos^2 y = \cosh^2 x - \sin^2 y.\end{aligned}$$

Proof. We indicate a proof for a part of (6). We have

$$\cos(x+iy) = \frac{[\exp(-y+ix) + \exp(y-ix)]}{2}.$$

Now use triangle inequalities to estimate below and above using the fact that $|\exp(u+iv)| = \exp(u)$. □

The zeros of the function $\sinh z$ are the solutions for z of the equation $e^z = e^{-z}$, which can be rewritten as $e^{2z} = 1$. Thus the zeros of $\sinh z$ are the numbers $n\pi i$ with n an integer. Similarly the zeros of $\cosh z$ are the numbers $(n + \frac{1}{2})\pi i$ with n an integer.

For each complex number z which is not of the form $\frac{2n+1}{2}\pi i$ where $n \in \mathbb{Z}$, define

$$\tanh z = \frac{\sinh z}{\cosh z} \text{ and } \mathrm{sech} z = \frac{1}{\cosh z}.$$

Also, for each $z \in \mathbb{C}$ such that $z \neq n\pi i$ where n is an integer define

$$\coth z = \frac{\cosh z}{\sinh z} \text{ and } \operatorname{csch} z = \frac{1}{\sinh z}.$$

4.6 Logarithms

We know already that $\exp : \mathbb{R} \to \mathbb{R}_+$ is a bijective function and hence its inverse exists. The inverse map $\exp^{-1} : \mathbb{R}_+ \to \mathbb{R}$ is called the logarithm function. Thus, we have a function $\log : \mathbb{R}_+ \to \mathbb{R}$ defined by $\exp(\log(x)) = x$ for all $x \in \mathbb{R}_+$. Since exp is a continuous increasing function, log is continuous and increasing.

Now one has to determine the inverse of $\exp : \mathbb{C} \to \mathbb{C} - \{0\}$. Since $e^{z+2k\pi i} = e^z.e^{2k\pi i} = e^z, k \in \mathbb{Z}$, it is obvious that e^z is not one-to-one. But this is onto. To prove this, take a nonzero complex number $a + ib$. We can write this as $a + ib = r(\cos\theta + i\sin\theta) = re^{i\theta}$ where $r > 0$. Since we know that e^x is onto from $\mathbb{R}$ to $\mathbb{R}_+$, we can find $x \in \mathbb{R}$ such that $r = e^x$ and so we get

$$a + ib = re^{i\theta} = e^x e^{i\theta} = e^{x+i\theta}.$$

This shows that $\exp : \mathbb{C} \to \mathbb{C} - \{0\}$ is an onto map.

Definition 4.5. *Given $z \in \mathbb{C}^*$, any complex number w such that $exp(w) = z$ is called a logarithm of z. If w is a logarithm of z then we shall write $w = \log z$. We let $\mathcal{L}og(z)$ stand for the set of all logarithms of a nonzero $z \in \mathbb{C}$. Since $e^w \neq 0$ for any w we cannot define $\mathcal{L}og(0)$.*

Example 4.3. *If $z \in \mathbb{C}^*$, then $w = log(|z|) + it$ where $t \in \arg(z)$ is a logarithm of z. Furthermore, any logarithm of z is of this form.*

Remark 4.2. Note that for $x > 0$, $\log(x) \in \mathcal{L}og(x)$. Thus, even here, $\mathcal{L}og(x)$ is the infinite set $\{\log(x) + 2n\pi i : n \in \mathbb{Z}\}$.

Recall the definitions $L_\alpha = \{t\exp(i\alpha) : t \geq 0\}$ for $\alpha \in \mathbb{R}$ and the argument $\operatorname{Arg}_\alpha : \mathbb{C}/L_\alpha \to (\alpha, \alpha + 2\pi)$.

We define a logarithm $\log_\alpha : \mathbb{C}/L_\alpha \to \mathbb{C}$ by setting $\log_\alpha(z) = \log(|z|) + i\operatorname{Arg}_\alpha(z)$, $z \in \mathbb{C}/L_\alpha$. Then $\log_\alpha$ is continuous on its domain. Since exp maps $\{z = x + iy : \alpha < y < \alpha + 2\pi\}$ onto $\mathbb{C}/L_\alpha$, $\log_\alpha$ is its continuous inverse.

Power functions

Definition 4.6. *Given $x > 0$ and $\lambda \in \mathbb{C}$, we define $x^\lambda = exp(\lambda log(x))$.*

Note that the following laws of exponents are valid:
(1) $a^z a^w = a^{z+w}$ for $a > 0$ and $z, w \in \mathbb{C}$.
(2) $(ab)^z = a^z b^z$ for $a > 0$, $b > 0$ and $z \in \mathbb{C}$.
In analogy with the above definition, we make the following.

Definition 4.7. *For $\lambda \in \mathbb{C}$ and $z \neq 0$, we define z^λ by setting $z^\lambda = exp(\lambda \mathcal{L}og(z))$, a subset of $\mathbb{C}$.*

Example 4.4. *Let us find i^i. Since $\arg(i) = \{\frac{\pi}{2} + 2n\pi : n \in \mathbb{Z}\}$, we have $\mathcal{L}og(i) = log(1) + i\arg(i) = i\arg(i)$. Hence $i^i = \{exp(i[i(2n\pi + \frac{\pi}{2})]) : n \in \mathbb{Z}\} = \{exp(-(2n\pi + \frac{\pi}{2})) : n \in \mathbb{Z}\}$. This set is a subset of $\mathbb{R}$. Similarly $(-1)^i = \{exp(-(2n+1)\pi i) : n \in \mathbb{Z}\}$.*

Thus if z, w_1 and w_2 are complex numbers and $z \neq 0$, then $z^{w_1} z^{w_2} = z^{w_1+w_2}$. Further, if z_1, z_2 and w are complex numbers and $z_1 z_2 \neq 0$, then $(z_1 z_2)^w = z_1^w z_2^w e^{2n\pi i w}$ for some $n = 0, \pm 1$. One can also verify the following: Let $z, w, a \in \mathbb{C}$ and $z \neq 0$. Then

(i) $\text{Log} z^w = w\text{Log} z + 2n\pi i$ where $n \in \mathbb{Z}$ is such that $-\pi < \text{Im}\ (w\text{Log} z) + 2n\pi \leq \pi$;

(ii) $(z^w)^a = z^{aw} e^{2n\pi i a}$ where $n \in \mathbb{Z}$ is such that $-\pi < \text{Im}\ (w\text{Log} z) + 2n\pi \leq \pi$.

For completeness we shall define 0^c to be 0 if $c \in \mathbb{C}$ and $c \neq 0$ and 0^c to be 1 if $c = 0$.

To get a single valued function, we proceed as follows. For $\lambda \in \mathbb{C}$ and $z \neq 0$, we define $P_\alpha^\lambda(z) = \exp(\lambda \log_\alpha(z))$, for $z \in \mathbb{C}/L_\alpha$. It is immediate from the definition that P_α^λ is continuous on its domain.

Remark 4.3. Why cannot we work with the set valued maps z^λ? Consider $f(z) = z^{\frac{1}{2}}$. Then $f + f = 2f$. For, at $z = 1$, the LHS is $\{-2, 0, 2\}$ while the RHS is $\{\pm 2\}$.

Thus to summarize, if a is a complex number, then by a logarithm of a one means any complex number b such that $e^b = a$. If $a = 0$ there is no such b, but otherwise there are infinitely many, any two of which differ by an integer multiple of $2\pi i$. For example, the logarithms of the number 1 are the numbers $2\pi i n, n \in \mathbb{Z}$, the logarithms of the number -1 are the numbers $(2n+1)\pi i, n \in \mathbb{Z}$, and the logarithms of the number i are the numbers $(2n + \frac{1}{2})\pi i, n \in \mathbb{Z}$.

Since $|e^b| = e^{\text{Re} b}$ and $\arg e^b = \text{Im} b$, the number b is a logarithm of the nonzero number a if and only if $e^{\text{Re} b} = |a|$ and $\text{Im} b = \arg a$. Thus, letting *log* denote the natural logarithm function of calculus, we see that the logarithms of a are the numbers $log|a| + i\arg a$. If b is a logarithm of a, we shall write $b = \log a$. The particular logarithm of a whose imaginary part is in the interval $(-\pi, \pi]$; corresponding to the principal value of $\arg a$, will be denoted by $\text{Log} a = log|a| + i\text{Arg} a$.

Example 4.5. *We shall evaluate each of the following principal logarithms.*

(i)	*Log i*	$=$	$\frac{\pi i}{2}$
(ii)	*Log*(-1)	$=$	πi
(iii)	*Log*$(-1-i)$	$=$	$\frac{1}{2}\log 2 - \frac{3\pi i}{4}$
(iv)	*Log*$(1-i)$	$=$	$\frac{1}{2}\log 2 - \frac{\pi i}{4}$.

Example 4.6. *The principal value of* $(1-i)^i$ *is* $e^{\frac{\pi}{4}}[\cos(\frac{1}{2}\log 2)+i\sin(\frac{1}{2}\log 2)]$ *and the principal value of* i^{-i} *is* $e^{\frac{\pi}{2}}$.

4.7 Continuous logarithms

Let ϕ be a continuous, nowhere vanishing, complex-valued function on the subinterval $[a,b]$ of $\mathbb{R}$. Then there is a continuous function ψ on $[a,b]$ such that $\phi = e^{\psi}$. The function ψ is unique to within addition of a constant integer multiple of $2\pi i$. As for uniqueness, if ψ_1 and ψ_2 are two functions with the required properties, then the function $\frac{1}{2\pi i}(\psi_1 - \psi_2)$ is continuous and integer-valued, hence constant since $[a,b]$ is connected.

We establish the existence of ψ first for the case where the range of ϕ lies in an open half-plane H bounded by a line through the origin. In that case, there is a branch l of $\log z$ in H. The composite function $\psi = l \circ \phi$ then has the required properties. Note that, by suitably choosing the branch l of $\log z$, we can arrange to make $\psi(a)$ equal to any preassigned value of $\log\phi(a)$.

To establish the existence of ψ in the general case, we partition $[a,b]$ into subintervals $[t_0,t_1],[t_1,t_2].\cdots,[t_{p-1},t_p]$, where $a=t_0<t_1<\cdots<t_p=b$, such that the range of ϕ on each subinterval lies in an open half-plane bounded by a line through the origin. This is easily done on the basis of the continuity of ϕ and the compactness of $[a,b]$.

By the special case already treated, there is a continuous function ψ_1 on $[t_0,t_1]$ such that $\phi = e^{\psi_1}$ on that interval. By the same special case, there is a continuous function ψ_2 on $[t_1,t_2]$ such that $\phi = e^{\psi_2}$ on that interval, and such that $\psi_2(t_1)=\psi_1(t_1)$. Continuing in this way, we obtain on each interval $[t_{j-1},t_j]$ a continuous function ψ_j such that $\phi = e^{\psi_j}$ there, and $\psi_j(t_{j-1}) = \psi_{j-1}(t_{j-1})$. The function ψ on $[a,b]$ that equals ψ_j on $[t_{j-1},t_j]$ $(j=1,\cdots,p)$ then has the required properties.

4.8 Branches of arg z and log z

Let Ω be an open connected subset of $\mathbb{C}$ not containing the origin. By a branch of $\arg z$ in Ω is meant a continuous function α in Ω such that, for each z in Ω, the value $\alpha(z)$ is a value of $\arg z$. By a branch of $\log z$ in Ω is meant a continuous function l in Ω such that, for each z in Ω, the value $l(z)$ is a logarithm of z. For a simple example, we can take for Ω the region one obtains by removing from $\mathbb{C}$ the interval $(-\infty,0]$ on the real axis. In this region the functions $\alpha(z) = \mathrm{Arg} z$ and $l(z) = \mathrm{Log} z$ are branches of $\arg z$ and $\log z$, respectively.

As we shall see, depending on the shape of Ω, there may or may not exist branches of $\arg z$ and $\log z$ in Ω. However, if there is a branch α of $\arg z$ in Ω then there is a branch of $\log z$, namely the function $l(z) = \log|z| + i\alpha(z)$. Similarly, if there is a branch l of $\log z$ in Ω then there is a branch of $\arg z$, namely, the function $\alpha(z) = \text{Im}l(z)$. Thus if Ω is a region and $\Omega \subset \mathbb{C}/\{0\}$, a function $f : \Omega \longrightarrow \mathbb{C}$ is called a branch of a logarithm if $exp(f(z)) = z, z \in \Omega$. If we consider the upper half plane $\{z \in \mathbb{C} : z = re^{i\theta}, 0 < \theta < \pi, r > 0\}$, then we may define a branch of the logarithm as $logz = log(re^{i\theta}) = logr + i\theta$ where $0 < \theta < \pi$. But we could also define $logre^{i\theta} = logr + i\theta + 2\pi ik, k \in \mathbb{Z}$. It says that there could be several branches of logarithm. Thus every complex number other than zero has infinitely many logarithms which differ from each other by multiples of $2\pi i$.

If α is a branch of $\arg z$ in Ω then $\alpha + 2\pi n$ is another branch for any integer n. On the other hand, if α_1 and α_2 are two branches of arg$z$ in Ω, then the function $\frac{\alpha_1 - \alpha_2}{2\pi}$ is continuous in Ω and takes only integer values, so it must be constant because by assumption Ω is connected. Thus, if there is a branch of $\arg z$ in Ω, then any two branches differ by a constant integer multiple of 2π. Similarly, if there is a branch l of $\log z$ in Ω, then $l + 2\pi in$ is another branch for every integer n, and every branch has this form.

An example of a region that carries no branch of arg$z$ is the following:

There is no branch of $\arg z$ in the region $0 < |z| < 1$.

On the other hand, if G is any open disk that excludes the origin, then G does carry such a branch. In fact suppose θ_0 is a particular value of the argument of the center of the disk. Then the disk contained in the half-plane $\{re^{i\theta} : r > 0, \theta_0 - \frac{\pi}{2} < \theta < \theta_0 + \frac{\pi}{2}\}$, and we can obtain the desired branch α by letting $\alpha(z)$, for z in G, equal the value of arg$z$ in the interval $(\theta_0 - \frac{\pi}{2}, \theta_0 + \frac{\pi}{2})$. One can also verify that since $e^{z_1+z_2} = e^{z_1}e^{z_2}$ the following holds:

$$\begin{aligned} \log(z_1 z_2) &= \log z_1 + \log z_2, \\ \arg(z_1 z_2) &= \arg z_1 + \arg z_2. \end{aligned}$$

We omit an additive constant $2k\pi i, k = 0, \pm 1, \pm 2, \cdots$ in the logarithm.

4.9 Exercises

1. Prove or disprove that $\lim_{z \to \infty} |e^{-z}| = 0$.

2. Evaluate $\lim_{z \to a} \dfrac{\cos z - \cos a}{z - a}$.

3. For $z = x + iy$, prove that

   (i) $|\sin \pi z|^2 = \cosh^2 \pi y - \cos^2 \pi x$

   (ii) $|\cos z|^2 = \sinh^2 y + \cos^2 x$

4. Find the values of $\sin i, \cos i, \tan(1 + i)$.

5. Determine all values of 2^i.

6. For what values of z is e^z equal to $i, 1 + 2i$?

7. Find the values of e^z for $z = \frac{2}{3}\pi i$.

8. Find the values of $\mathcal{L}og(-1), \mathcal{L}og(i)$ and $\mathcal{L}og(exp(z))$.

9. Find all solutions of $z^{1-i} = 4$.

10. Prove the following:

    (i) $\lim\limits_{z \to 0} \dfrac{\sin z}{z} = 1$.

    (ii) $\lim\limits_{z \to 0} \dfrac{\sinh z}{iz} = -i$.

11. Show that the series $\sum_{n=1}^{\infty} \frac{\cos in}{2^n}$ diverges.

12. Show that $\sum_{n=1}^{\infty} \frac{e^{in}}{n^2}$ converges absolutely.

13. Prove that any function with three or more independent periods is a constant.

14. Express $\cos^2 z$ as a power series in z, giving all terms up to z^6.

15. Show that the functions (i) e^z; (ii) $\sin z$; (iii) $\cos z$ have fixed points. See Example 18.4.

16. Show that for each $n \in \mathbb{Z}$, each of the following identities holds. (a) $\cos[z + (2n+1)\pi] = -\cos z$; (b) $\sin[z + (2n+1)\pi] = -\sin z$.

17. Solve the equation $e^z = a + ib$ for z where a and b are real and $ab \neq 0$.

18. Evaluate: $\lim\limits_{x \to -\infty} e^{2x+3iy}$.

19. (a) Show that for $z \in \mathbb{C}$, (i) $\sinh(z + \pi i) = -\sinh z$;

    (ii) $\cosh(z + \pi i) = -\cosh z$.

    (b) Prove that $\tanh(z + \pi i) = \tanh z$ for each $z \neq \frac{2n+1}{2}\pi i, n \in \mathbb{Z}$.

20. Prove that $|\sinh x| \leq |\cosh z| \leq \cosh x$.

21. Find all values of z such that $\cosh z = -2$.

22. If $z_1, z_2 \in \mathbb{C}, z_1 z_2 \neq 0$ then prove that $\text{Log} z_1 z_2 = \text{Log} z_1 + \text{Log} z_2 + 2n\pi i$ where n has one of the values $0, 1$ or -1.

23. Let $w = f(z) = z^m, 0 < m < 1$. Prove that f is a one-to-one mapping of the upper half plane $H = \{z \in \mathbb{C} : \text{ Im } z > 0\}$ onto the annular region $A = \{w \in \mathbb{C} : 0 < \text{Arg} w < m\pi\}$.

24. Let $f(z) = \text{Log} z$ and let $H = \{z \in \mathbb{C} : \text{ Im } z > 0\}$. Show that $f(H) = \{z \in \mathbb{C} : 0 < \text{ Im } w < \pi\}$.

25. If $z = e^{i\theta} \cos\theta$ with $0 < \theta < \frac{\pi}{2}$ and if n is a positive integer, then prove that $|1 - z| < |1 - z^n|$.

26. Let z be a complex number and $0 < |z| < 1$ then prove that

$$\frac{1}{4}|z| < |e^z - 1| < \frac{7}{4}|z|.$$

27. For $|z| < 1$, prove that

$$|\log(1+z)| \leq -\log(1-|z|).$$

28. Let z be a complex number and $|z| < 1$. Prove that

$$\frac{|z|}{1+|z|} \leq |\log(1+z)| \leq |z|\frac{1+|z|}{|1+z|}.$$

29. If n is a natural number and z any complex number, then prove that

$$\left|e^z - \left(1+\frac{z}{n}\right)^n\right| < \left|e^{|z|} - \left(1+\frac{|z|}{n}\right)^n\right| < e^{|z|}\frac{|z|^2}{2n}.$$

Chapter 5

Differentiation

5.1 Functions defined on $\mathbb{R}$

In this section we deal with differentiation of functions defined on a subset of $\mathbb{R}$ into $\mathbb{C}$.

Definition 5.1. *Let $f : [a,b] \to \mathbb{C}$ be given. We say f is differentiable at $t \in [a,b]$ if there exists an $\alpha \in \mathbb{C}$ such that given $\epsilon > 0$ there exists a $\delta > 0$ such that whenever $0 \neq h \in \mathbb{R}$, $|h| < \delta$ and $t + h \in [a,b]$ implies*

$$\left| \frac{f(t+h) - f(t)}{h} - \alpha \right| < \epsilon.$$

The number α is unique and is called the derivative of f at t. It is denoted by $f'(t)$. The function f is differentiable in $[a,b]$ if f is differentiable at each $t \in [a,b]$.

Let f be a function from $[a,b]$ into $\mathbb{C}$, $f = u + iv$ where $u = \mathrm{Re}f$ and $v = \mathrm{Im}f$. Then f is differentiable at $t \in [a,b]$ if and only if the real valued functions $u, v : [a,b] \to \mathbb{R}$ are differentiable at t. It can also be be verified that $f'(t) = u'(t) + iv'(t)$. Hence the existence of $f'(t)$ is equivalent to the simultaneous existence of $u'(t)$ and $v'(t)$. Thus the derivative of a complex valued function of a real variable can be reduced to the real case. Further, if f is differentiable at t, then it is continuous at t.

Remark 5.1. If $f : [a,b] \to \mathbb{C}$ is differentiable at $t \in [a,b]$, then we may think of f as a differentiable curve in the complex plane. That is, if $f(t) = u(t) + iv(t)$ then $(u(t), v(t)) \in \mathbb{R}^2$ and $f'(t) = (u'(t), v'(t))$ is thought of as the tangent vector to the curve at t. In physical terms, we may think of $f(t)$ as the position of a particle at time t and $f'(t)$ as the "velocity".

From the basic rules of differentiation noted for real valued functions one sees that the following functions are differentiable.

Example 5.1. *Let $z, w \in \mathbb{C}$. Let $\gamma : [0,1] \to \mathbb{C}$ be given by $\gamma(t) = tw + (1-t)z$. Then γ is the line segment $[z, w]$ joining z to w and for $t \in [0,1]$, $\gamma'(t) = w - z$ since $\lim_{h\to 0} \frac{\gamma(t+h)-\gamma(t)}{h} = (w - z)$. Hence γ is differentiable on $[0,1]$.*

Example 5.2. *Let $a \in \mathbb{C}, R > 0$. Consider the map $\sigma : [0, 2\pi] \to \mathbb{C}$ given by $\sigma(t) = a + R(\cos t + i \sin t)$. Then σ is the circle with center at a and radius R. It is not difficult to see that σ is differentiable on $[0, 2\pi]$ and $\sigma'(t) = R(-\sin t + i \cos t), t \in [0, 2\pi]$.*

When we turn to the mean value theorem, however, and to one of its consequences, namely, L'Hospital's rule, the situation changes. The next two examples will show that each of these results fails to be true for complex-valued functions.

Example 5.3. *Define for real x, $f(x) = e^{ix} = cosx + isinx$. Then $f(2\pi) - f(0) = 1 - 1 = 0$, but $f'(x) = ie^{ix}$ so that $|f'(x)| = 1$ for all real x.*

Example 5.4. *On the segment $(0,1)$, define $f(x) = x$ and $g(x) = x + x^2 e^{\frac{i}{x^2}}$. Since $|e^{it}| = 1$ for all real t, we see that*

$$\lim_{x\to 0} \frac{f(x)}{g(x)} = 1. \tag{5.1}$$

Next, for $0 < x < 1$, $g'(x) = 1 + \{2x - \frac{2i}{x}\} e^{\frac{i}{x^2}}$ and hence

$$|g'(x)| \geq \left| 2x - \frac{2i}{x} \right| - 1 \geq \frac{2}{x} - 1. \tag{5.2}$$

Thus $\left|\frac{f'(x)}{g'(x)}\right| = \frac{1}{|g'(x)|} \leq \frac{x}{2-x}$ and so

$$\lim_{x\to 0} \frac{f'(x)}{g'(x)} = 0. \tag{5.3}$$

From (5.1) and (5.3), one observes that L'Hospital's rule fails in this case. Note also that $g'(x) \neq 0$ on $(0,1)$, by (5.2).

However, there is a consequence of the mean value theorem which, for purposes of applications, is almost as useful as the mean value theorem for real-valued functions which states:

If f is a real continuous function on $[a, b]$ which is differentiable in (a, b), then there is a point $x \in (a, b)$ at which $f(b) - f(a) = (b - a)f'(x)$. From the theorem it follows that $|f(b) - f(a)| \leq (b - a) \sup_{a<x<b} |f'(x)|$.

Theorem 5.1. *Suppose f is a continuous function from $[a, b] \subset \mathbb{R}$ into $\mathbb{C}$ and f is differentiable in (a, b). Then there exists $x \in (a, b)$ such that*

$$|f(b) - f(a)| \leq (b - a)|f'(x)|.$$

Proof. Put $z = z_1 + iz_2 = f(b) - f(a)$, and $f(t) = f_1(t) + if_2(t)$. Define for $a \le t \le b$, $\phi(t) = z \cdot f(t) = z_1 f_1(t) + z_2 f_2(t)$. Then ϕ is a real-valued continuous function on $[a, b]$ which is differentiable in (a, b). The mean value theorem shows therefore that $\phi(b) - \phi(a) = (b - a)\phi'(x) = (b - a)(z \cdot f'(x)) = (b-a)(z_1 f_1'(t) + z_2 f_2'(t))$ for some $x \in (a, b)$. On the other hand, $\phi(b) - \phi(a) = z \cdot f(b) - z \cdot f(a) = z \cdot z = |z|^2$.

The Cauchy-Schwarz inequality now gives $|z|^2 = (b - a)|z \cdot f'(x)| \le (b-a)|z||f'(x)|$. Hence $|z| \le (b-a)|f'(x)|$, which is the desired conclusion. □

Notice that if $h : [a, b] \to \mathbb{C}$ be differentiable with $h'(t) = 0$ then h is a constant. This can be shown by invoking the result from real analysis to the real and imaginary parts of h.

Definition 5.2. *Given $f : [a, b] \to \mathbb{C}$ continuous, we define $\int_a^b f(t)dt = \int_a^b Re(f(t))dt + i\int_a^b Im(f(t))dt$. Here the integrals are Riemann and they exist since $Re(f)$ and $Im(f)$ are continuous.*

Theorem 5.2. ***(Fundamental theorem of calculus)***

(1) *Let $f : [a, b] \to \mathbb{C}$ be continuous. Define $F(x) = \int_a^x f(t)dt$. Then F is differentiable and $F'(x) = f(x)$ for $x \in [a, b]$.*

(2) *Let $G : [a, b] \to \mathbb{C}$ be differentiable with $g = G'$ is continuous. Then $\int_a^b g(t)dt = G(b) - G(a)$.*

Proof. The assertions follow by involving their real counterparts. □

5.2 Differentiable functions on $\mathbb{C}$

Definition 5.3. *Let $U \subset \mathbb{C}$ be open. A function $f : U \to \mathbb{C}$ is said to be differentiable at $z \in U$ if there exists an $\alpha \in \mathbb{C}$ such that given $\epsilon > 0$ there exists a $\delta > 0$ such that $|\frac{f(z+h)-f(z)}{h} - \alpha| < \epsilon$, for $0 < |h| < \delta$. The number α is unique and called the derivative of f at z. It is denoted by $f'(z)$. If f is differentiable at each $z \in U$, then f is said to be **holomorphic** on U. Let $H(U)$ denote the set of functions holomorphic on U. If $f \in H(\mathbb{C})$, then f is called an **entire function**.*

Suppose f is a real-valued function of a complex variable whose derivative exists at $z = a$. Then $f'(a)$ is on one side real, for it is the limit of the quotients

$$\frac{f(a+h) - f(a)}{h}$$

as h tends to zero through real values. On the other side it is also the limit of the quotients $\frac{f(a+ih)-f(a)}{ih}$ and as such purely imaginary. Therefore $f'(a)$ must be zero. Thus a real function of a complex variable either has the derivative zero, or else the derivative does not exist.

Remark 5.2. One can visualize a real-valued function of a real variable by means of its graph, which is a curve in $\mathbb{R}^2$. A complex-valued function of a complex variable also has a graph, but its graph is a two-dimensional object in the four-dimensional space $\mathbb{C} \times \mathbb{C}$, which is difficult to visualize. So if one wants to obtain a geometric picture of a holomorphic function, a more sensible approach is to think of the function as a map from the complex plane to itself, and to try to understand how the map deforms the plane; for example, how does it transform lines and circles?

We can restate the definition of differentiable functions in terms of linear approximation. The statement says that f is differentiable at z, with $f'(z) = d$, if and only if f is well approximated near z by the linear function $f(z) + d(w - z)$, in the sense that the remainder $R(w) = f(w) - f(z) - d(w - z)$ in the approximation is small compared to the distance of w from z.

Proposition 5.1. *Let $f : U \to \mathbb{C}$ be given and $z \in U$. Then f is differentiable at z if and only if there exists a function $f_1 : U \to \mathbb{C}$, continuous at z and such that $f(w) = f(z) + (w - z)f_1(w)$. Furthermore, $f'(z) = f_1(z)$.*

Proof. Let f be differentiable at z. Define $f_1(w) = \frac{f(w)-f(z)}{w-z}$ for $w \neq z$ and $f_1(z) = f'(z)$.

To check the continuity of f_1 at z, let $\epsilon > 0$ be given. Since f is differentiable at z, for this ϵ there exists a $\delta > 0$ such that if $|w - z| < \delta$, then $|\frac{f(w)-f(z)}{w-z} - f'(z)| < \epsilon$. Hence f_1 is continuous at z.

Conversely, if f_1 exists as specified, observe that $f_1(w)$ must equal $\frac{f(w)-f(z)}{w-z}$ for $w \neq z$. We claim that f is differentiable at z and $f'(z) = f_1(z)$. By continuity of f_1 at z, given $\epsilon > 0$ there exists a $\delta > 0$ such that if $0 < |w - z| < \delta$, then $|\frac{f(w)-f(z)}{w-z} - f_1(z)| < \epsilon$. □

Corollary 5.1. *If f is differentiable at a, then f is continuous at a.*

Proof. In fact, $\lim_{z\to a} |f(z) - f(a)| = \left[\lim_{z\to a} \frac{|f(z) - f(a)|}{|z - a|}\right] \left[\lim_{z\to a} |z - a|\right] = f'(a) \cdot 0 = 0$. □

We shall now consider only complex valued functions defined on a region $U \subseteq \mathbb{C}$. From the definition of the derivative it is immediate that a constant function is differentiable everywhere in $\mathbb{C}$ with derivative 0 and that the identity function (the function $f(z) = z$) is differentiable everywhere in $\mathbb{C}$ with derivative 1.

Example 5.5. *Let $f : \mathbb{C} \to \mathbb{C}$ be a constant. Then f is holomorphic on $\mathbb{C}$ and $f'(z) = 0$ for all $z \in \mathbb{C}$.*

Example 5.6. *Let $f : \mathbb{C} \to \mathbb{C}$ be given by $f(z) = z^n$, for $n \in \mathbb{N}$. Then f is holomorphic on $\mathbb{C}$ and $f'(z) = nz^{n-1}$. Notice that*

$$(z + h)^n - z^n = h\left(nz^{n-1} + \sum_{k\geq 2} \binom{n}{k} h^{k-1} z^{n-k}\right).$$

The right hand side is of the form $[(z+h)-z]f_1$, *with* f_1 *continuous. Further,* $\lim_{h\to 0}\frac{(z+h)^n-z^n}{h}=nz^{n-1}$.

Example 5.7. *Let* $f(z)=\bar{z}$, *for* $z\in\mathbb{C}$. *We claim that it is not differentiable at* $z\in\mathbb{C}$. *This follows from the following observations:*

$$\frac{f(z+h)-f(z)}{h}=\frac{\bar{z}+\bar{h}-\bar{z}}{h}=\begin{cases}\frac{h}{h}=1 & \text{if } h\in\mathbb{R};\\ \frac{-it}{it}=-1 & \text{if } h=it, t\in\mathbb{R}.\end{cases}$$

Example 5.8. *The function* $f(z)=|z|$ *is not differentiable at* 0. *See the remark following Definition 5.3.*

The following properties of differentiation are proved from the basic definition in exactly the same way as the corresponding properties in the theory of real valued functions of a real variable.

Theorem 5.3. *Let* U *be a region in* $\mathbb{C}$ *and* $f,g:U\to\mathbb{C}$ *be differentiable at* $z\in U$. *Let* $\alpha,\beta\in\mathbb{C}$. *Then* $\alpha f+\beta g, fg$ *and* $\frac{f}{g}$ *if* $g(z)\neq 0$ *are differentiable at* z *and we have*

$$\begin{aligned}(\alpha f+\beta g)'(z) &= \alpha f'(z)+\beta g'(z)\\ (fg)'(z) &= f'(z)g(z)+f(z)g'(z)\\ \left(\frac{f}{g}\right)'(z) &= \frac{f'(z)g(z)-f(z)g'(z)}{(g(z))^2}.\end{aligned}$$

Thus if f and g are holomorphic functions on U then $\alpha f+\beta g, fg$ and $\frac{f}{g}$ (if $g(z)\neq 0$ for all $z\in U$) are holomorphic functions on U for all $\alpha,\beta\in\mathbb{C}$.

Example 5.9. *Any polynomial function* $f(z)=\sum_{k=0}^{n}a_k z^k$ *is holomorphic on* $\mathbb{C}$ *and* $f'(z)=\sum_{k=1}^{n}ka_k z^{k-1}$.

Example 5.10. *The quotient rule and the differentiability of polynomials imply that a rational function is differentiable at every point where it is defined and that its derivative is a rational function. That is, any rational function* $r(z)=\frac{p(z)}{q(z)}$ *is holomorphic on the open set* G, *where* G *is the complement with respect to* $\mathbb{C}$ *of the set of zeros of* q.

It may be pointed out here that although the definition of the derivative of a complex valued function of a complex variable is formally the same as that of the derivative of a real valued function of a real variable, the concept holds surprises.

We have already verified that polynomials are holomorphic functions. Also the rational function (the quotient of two polynomials) with domain which does not include the zeros of the denominator is a holomorphic function. Are there any other examples of holomorphic functions? The answer is provided by the following theorem.

We shall show below that every power series represents a holomorphic function within its circle of convergence.

Theorem 5.4. *Let $f(z) = \sum_{n=0}^{\infty} a_n z^n$ be a power series with radius of convergence $R > 0$. Then f is differentiable on $B(0, R)$. Furthermore, $f'(z) = \sum_{n=1}^{\infty} n a_n z^{n-1}$ for $z \in B(0, R)$. That is, a power series can be differentiated term by term in the disk of convergence.*

Proof. For $|z| < R$, we shall write $f(z) = \sum_{k=0}^{\infty} a_k z^k = s_n(z) + R_n(z)$ where

$$\begin{aligned} s_n(z) &= a_0 + a_1 z + \cdots + a_{n-1} z^{n-1}, \\ R_n(z) &= \sum_{k=n}^{\infty} a_k z^k, \end{aligned}$$

$$\text{and also } f_1(z) = \sum_{n=1}^{\infty} n a_n z^{n-1} = \lim_{n\to\infty} s_n'(z).$$

We have to show that $f'(z) = f_1(z)$. Consider the identity

$$\frac{f(z) - f(z_0)}{z - z_0} - f_1(z_0) = \left(\frac{s_n(z) - s_n(z_0)}{z - z_0} - s_n'(z_0) \right)$$

$$+ (s_n'(z_0) - f_1(z_0)) + \left(\frac{R_n(z) - R_n(z_0)}{z - z_0} \right), \tag{5.4}$$

where we assume that $z \neq z_0$ and $|z|, |z_0| < \rho < R$. The last term can be rewritten as $\sum_{k=n}^{\infty} a_k (z^{k-1} + z^{k-2} z_0 + \cdots + z z_0^{k-2} + z_0^{k-1})$, and we conclude that

$$\left| \frac{R_n(z) - R_n(z_0)}{z - z_0} \right| \leq \sum_{k=n}^{\infty} k |a_k| \rho^{k-1}.$$

The expression on the right is the remainder term in a convergent series. Hence we can find n_0 such that $\left| \frac{R_n(z) - R_n(z_0)}{z - z_0} \right| < \frac{\epsilon}{3}$ for $n \geq n_0$. There is also an n_1 such that $|s_n'(z_0) - f_1(z_0)| < \frac{\epsilon}{3}$ for $n \geq n_1$. Choose a fixed $n \geq n_0, n_1$. By the definition of derivative we can find $\delta > 0$ such that $0 < |z - z_0| < \delta$ implies $\left| \frac{s_n(z) - s_n(z_0)}{z - z_0} - s_n'(z_0) \right| < \frac{\epsilon}{3}$. When all these inequalities are combined it follows by (5.4) that $\left| \frac{f(z) - f(z_0)}{z - z_0} - f_1(z_0) \right| < \epsilon$ when $0 < |z - z_0| < \delta$. We have proved that $f'(z_0)$ exists and equals $f_1(z_0)$. □

Theorem 5.5. *The radius of convergence of $\sum_{n=0}^{\infty} a_n z^n$ and $\sum_{n=1}^{\infty} n a_n z^{n-1}$ are the same.*

Proof. It is to observe that $\frac{1}{R} = \limsup_{n\to\infty} |a_n|^{\frac{1}{n}} = \limsup_{n\to\infty} |n a_n|^{\frac{1}{n}}$ as $\lim_{n\to\infty} n^{\frac{1}{n}} = 1$. The reason is as follows: Set $n^{\frac{1}{n}} = 1 + \delta_n$. Then $\delta_n > 0$, and by use of the binomial theorem $n = (1 + \delta_n)^n > 1 + \frac{1}{2} n(n-1) \delta_n^2$. This gives $\delta_n^2 < \frac{2}{n}$, and hence $\delta_n \to 0$ and therefore, $n^{\frac{1}{n}} \to 1$. □

Example 5.11. *Consider* $exp(z) = \sum_{n=0}^{\infty} \frac{z^n}{n!}$. *Then we have* $exp'(z) = exp(z)$ *for* $z \in \mathbb{C}$. *Hence,* $\frac{d}{dz} sinz = \frac{d}{dz}(\frac{e^{iz}-e^{-iz}}{2i}) = \frac{e^{iz}+e^{-iz}}{2} = cosz$ *and* $\frac{d}{dz} cosz = -sinz$ *for* $z \in \mathbb{C}$.

Example 5.12. $\frac{1}{1-z} = \sum_{n=0}^{\infty} z^n$ *if* $|z| < 1$, *so* $\frac{1}{(1-z)^2} = \sum_{n=1}^{\infty} nz^{n-1}$ *in* $|z| < 1$.

Example 5.13. *The series* $\sum_{n=1}^{\infty} \frac{\sin nx}{n^2}$ *is absolutely convergent for* $x \in \mathbb{R}$ *but differentiating the series term by term we obtain the series* $\sum_{n=1}^{\infty} \frac{\cos nx}{n}$ *which diverges at* $x = 0$. *Notice that Theorem 5.4 states a property of the power series only.*

Thus $e^z, \sin z$ and $\cos z$ are holomorphic in $\mathbb{C}$ and $\tan z, \cot z, \sec z, \csc z$ are holomorphic in those regions in $\mathbb{C}$ where their denominators are nonzero. Because the function e^z is holomorphic in $\mathbb{C}$, hence the hyperbolic functions $\sinh z$ and $\cosh z$ are holomorphic in $\mathbb{C}$. The functions $\tanh z, \coth z$ are holomorphic in those regions where their denominators does not vanish.

We shall prove in Chapter 7 that if $f : U \to \mathbb{C}$ is differentiable then f is infinitely differentiable in U. Thus although, as we have seen above, some of the basic properties of real and complex differentiability are formally identical, the repeated differentiability of holomorphic functions points to a glaring dissimilarity.

Definition 5.4. *If* $f : U \to \mathbb{C}$ *is differentiable and if* $g = f' : U \to \mathbb{C}$ *is differentiable at* $z \in U$, *we denote* $f''(z) = g'(z)$. *The function* f'' *is called the second derivative of* f *at* z. *More generally, we define inductively the n-th derivative of* f *at* z *by setting* $f^{(0)}(z) = f(z)$ *and* $f^{(n)}(z) = (f^{(n-1)})'(z)$, *the derivative of* $f^{(n-1)}$ *at* z. *In general, the first, second and third derivatives are denoted by* f', f'' *and* f''' *respectively. If* $f^{(n)}(z)$ *exists for all* $n \in \mathbb{N}$, *we say that* f *is infinitely differentiable.*

Proposition 5.2. *Any power series function* $f(z) = \sum a_n z^n$ *is infinitely differentiable on its disk of convergence. In fact, we have*

$$f^{(k)}(z) = \sum_{n=k}^{\infty} n(n-1)\cdots(n-k+1)a_n z^{n-k}.$$

In particular, we have $a_n = \frac{f^{(n)}(0)}{n!}$.

Proof. Repeated application of Theorem 5.4 yields the result. □

Definition 5.5. ***(Analytic functions):*** *Let* $U \subset \mathbb{C}$ *be open. A function* $f : U \to \mathbb{C}$ *is said to be analytic in* U *if for any* $z_0 \in U$, *there exists a sequence* (a_n) *of complex numbers and an* $r > 0$ *such that*

$$f(z) = \sum_{n=0}^{\infty} a_n (z - z_0)^n \tag{5.5}$$

for all $z \in B(z_0, r)$. *The sequence* (a_n) *may depend on* z_0.

It is obvious that the power series (5.5) has radius of convergence greater than or equal to r.

Example 5.14. *Let $U = \mathbb{C} - \{0\}$ and $f(z) = \frac{1}{z}$. Then for any $a \neq 0$, we have $f(z) = \frac{1}{z} = \frac{1}{z-a+a} = \frac{1}{a(1+\frac{z-a}{a})} = \frac{1}{a}\sum_{n=0}^{\infty}\left(-\frac{(z-a)}{a}\right)^n$ for all $z \in B(a, |a|)$. Thus, f is analytic in U.*

Example 5.15. *$exp : \mathbb{C} \to \mathbb{C}$ is analytic in $\mathbb{C}$.*

It follows from Theorem 5.4 that every analytic functions on a region U is holomorphic in U. In fact, analytic functions are infinitely differentiable.

Remark 5.3. It is a central result of Cauchy theory that the converse is also true. This is striking since this implies that any function which is once differentiable on an open set is infinitely differentiable. This should be contrasted with the existence of differentiable functions on intervals in $\mathbb{R}$ whose derivatives are not even continuous, leave alone differentiable. For example, consider the function $f(x) = x^2 \sin\frac{1}{x}(x \neq 0), f(0) = 0$. It is continuous and differentiable on $\mathbb{R}$. If $x \neq 0$, then $f'(x) = 2x\sin\frac{1}{x} - \cos\frac{1}{x}$. To calculate $f'(0)$, we calculate

$$\lim_{x\to 0}\frac{f(x)-f(0)}{x-0} = \lim_{x\to 0}\left(x\sin\frac{1}{x}\right) = 0.$$

Hence $f'(0) = 0$. Note further that f' is not continuous as $\cos\frac{1}{x}$ oscillates at $x = 0$ and f' is not differentiable at $x = 0$.

Definition 5.6. *A real or complex function $\psi(t)$ of a real variable t, defined on an interval $a < t < b$, is said to be real analytic (that is, analytic in the real sense) if, for every t_0 in the interval, the Taylor series*

$$\psi(t) = \psi(t_0) + \psi'(t_0)(t-t_0) + \frac{1}{2}\psi''(t_0)(t-t_0)^2 + \cdots$$

converges in some interval $(t_0 - \rho, t_0 + \rho), \rho > 0$.

Notice that if this is so then the series is also convergent for complex values of t, as long as $|t - t_0| < \rho$, and it represents an analytic function in that disk.

Lemma 5.1. *Given a double sequence $\{a_{ij}\}, i = 1, 2, 3, \cdots, j = 1, 2, 3, \cdots$, suppose that $\sum_{j=1}^{\infty}|a_{ij}| = b_i$ $(i = 1, 2, 3, \cdots)$ and $\sum b_i$ converges. Then $\sum_{i=1}^{\infty}\sum_{j=1}^{\infty}a_{ij} = \sum_{j=1}^{\infty}\sum_{i=1}^{\infty}a_{ij}$.*

Proof. Let E be a countable set, consisting of the points x_0, $x_1, x_2, \cdots$, and suppose $x_n \to x_0$ as $n \to \infty$. Define

$$f_i(x_0) = \sum_{j=1}^{\infty} a_{ij} \ (i = 1, 2, 3, \cdots) \tag{5.6}$$

$$f_i(x_n) = \sum_{j=1}^{n} a_{ij} \ (i, n = 1, 2, 3, \cdots) \tag{5.7}$$

$$\sum_{j=1}^{\infty} |a_{ij}| = b_i \ (i = 1, 2, 3, \cdots) \tag{5.8}$$

$$g(x) = \sum_{i=1}^{\infty} f_i(x) \ (x \in E). \tag{5.9}$$

Now, (5.6) and (5.7), together with (5.8), show that each f_i is continuous at x_0. Since $|f_i(x)| \leq b_i$ for $x \in E$, (5.9) converges uniformly, so that g is continuous at x_0. It follows that

$$\begin{aligned}
\sum_{i=1}^{\infty}\sum_{j=1}^{\infty} a_{ij} &= \sum_{i=1}^{\infty} f_i(x_0) \\
&= g(x_0) = \lim_{n\to\infty} g(x_n) \\
&= \lim_{n\to\infty} \sum_{i=1}^{\infty} f_i(x_n) \\
&= \lim_{n\to\infty} \sum_{i=1}^{\infty}\sum_{j=1}^{n} a_{ij} \\
&= \lim_{n\to\infty} \sum_{j=1}^{n}\sum_{i=1}^{\infty} a_{ij} = \sum_{j=1}^{\infty}\sum_{i=1}^{\infty} a_{ij}.
\end{aligned}$$

□

Let $f(z) = \sum_{n=0}^{\infty} a_n(z - z_0)^n$ be a power series with radius of convergence $R > 0$. Then f is analytic in $B(z_0, R)$. Note that this is not obvious and requires proof. In fact we are supposed to exhibit a power series representation of f in $B(a, r)$ for each $a \in B(z_0, R)$ and for some $r > 0$. That is, we shall show below every power series represents an analytic function within its circle of convergence.

Theorem 5.6. *Let $f(z) = \sum_{n=0}^{\infty} a_n z^n$ be a power series for $z \in B(0, R)$ with $R > 0$. Then f is analytic in $B(0, R)$.*

Proof. Let $a \in B(0, R)$. We have

$$\begin{aligned}
f(z) &= \textstyle\sum_{n=0}^{\infty} a_n(z - a + a)^n \\
&= \textstyle\sum_{n=0}^{\infty} a_n \sum_{k=0}^{n} \binom{n}{k}(z - a)^k a^{n-k} \\
&= \textstyle\sum_{k=0}^{\infty} \left[\sum_{n=k}^{\infty} \binom{n}{k} a_n a^{n-k}\right] (z - a)^k.
\end{aligned}$$

The only thing is to justify the last equality, i.e., to justify the interchange of the order of the sums. By Lemma 5.1, this is valid, if

$$\sum_{n=0}^{\infty}\sum_{k=0}^{n}\left|a_n\binom{n}{k}a^{n-k}(z-a)^k\right|$$

is convergent. But this series is nothing other than $\sum_{n=0}^{\infty}|a_n|\,(|z-a|+|a|)^n$ which is convergent if $|z-a|+|a|<R$. Thus if the series $f(z)=\sum_{n=0}^{\infty}a_nz^n$ converges in $|z|<R$ and if $a\in B(0,R)$, then f can be expanded in a power series about the point $z=a$ which converges in $|z-a|<R-|a|$, and $f(z)=\sum_{k=0}^{\infty}\frac{f^{(k)}(a)}{k!}(z-a)^k$ for $|z-a|<R-|a|$. The fact that $\sum_{n=k}^{\infty}\binom{n}{k}a_na^{n-k}=\frac{f^{(k)}(a)}{k!}$ follows from Proposition 5.2. □

We have already considered a few power series of interest defining the functions $e^z, \sin z, \cos z, \sinh z, \cosh z$ etc. It is important to note that they give rise to holomorphic and analytic functions.

Theorem 5.7. *(a) (L'Hospital's rule for holomorphic functions): Let f and g be holomorphic in a region $G\subseteq\mathbb{C}$ containing the point a. Suppose $f(a)=g(a)=0$ but $g'(a)\neq 0$. Then $\lim_{z\to a}\frac{f(z)}{g(z)}=\frac{f'(a)}{g'(a)}$. (b)(Chain rule): (i) Let $\gamma:[a,b]\to U\subset\mathbb{C}$ and $f:U\to\mathbb{C}$ be differentiable. Then the map $g=f\circ\gamma:[a,b]\to\mathbb{C}$ is differentiable and $g'(t)=f'(\gamma(t))\gamma'(t)$. (ii) Let U and V be open sets in $\mathbb{C}$. Let $f:U\to V$ be differentiable at $z\in U$ and $g:V\to\mathbb{C}$ be differentiable at $w=f(z)\in V$. Then $g\circ f:U\to\mathbb{C}$ is differentiable at z and $(g\circ f)'(z)=g'(w)f'(z)$.*

Proof. Let $\frac{f(z)-f(a)}{z-a}-f'(a)=\eta_1$ so that $f(z)=f(a)+f'(a)(z-a)+\eta_1(z-a)$. Then since f is holomorphic in G and $a\in G$, hence $\lim_{z\to a}\eta_1=\lim_{z\to a}\left\{\frac{f(z)-f(a)}{z-a}-f'(a)\right\}=f'(a)-f'(a)=0$.

Similarly, $g(z)=g(a)+g'(a)(z-a)+\eta_2(z-a)$ where $\eta_2\to 0$ as $z\to a$. Now using the fact that $f(a)=g(a)=0$, we obtain $f(z)=f'(a)(z-a)+\eta_1(z-a)$ and $g(z)=g'(a)(z-a)+\eta_2(z-a)$ where $\eta_i\to 0$ as $z\to a$ for $i=1,2$. Thus,

$$\lim_{z\to a}\frac{f(z)}{g(z)}=\lim_{z\to a}\frac{(f'(a)+\eta_1)(z-a)}{(g'(a)+\eta_2)(z-a)}=\frac{f'(a)}{g'(a)}.$$

For proof of (b)(i), observe that $(f\circ\gamma)(t+h)-(f\circ\gamma)(t)=[\gamma(t+h)-\gamma(t)]f_1(\gamma(t+h))$. As $h\to 0$, we see that $\frac{g(t+h)-g(t)}{h}=\frac{\gamma(t+h)-\gamma(t)}{h}f_1(\gamma(t+h))$ goes to $\gamma'(t)f'(\gamma(t))$. For proof of (b)(ii), notice that $g(z^*)=g(w)+(z^*-w)g_1(z^*)$, $f(\widetilde{w})=f(z)+(\widetilde{w}-z)f_1(\widetilde{w})$ so that $(g\circ f)(\widetilde{w})=g(f(\widetilde{w}))=g(w)+(f(\widetilde{w})-f(z))g_1(f(\widetilde{w}))=(g\circ f)(z)+(\widetilde{w}-z)f_1(\widetilde{w})g_1(f(\widetilde{w}))$. If we let $h_1(\widetilde{w})=f_1(\widetilde{w})g_1(f(\widetilde{w}))$, then h_1 is continuous at z and we see that $h=g\circ f$ is differentiable at z. □

Using Theorem 5.7, it is not difficult to evaluate the following limits:

$$\begin{aligned}
(i) \quad & \lim_{z\to 0} \frac{1-\cos z}{z^2} &=& \tfrac{1}{2};\\
(ii) \quad & \lim_{z\to 0} \frac{1-\cos z}{\sin z^2} &=& \tfrac{1}{2};\\
(iii) \quad & \lim_{z\to 0} (\cos z)^{\frac{1}{z^2}} &=& -\tfrac{1}{2};\\
(iv) \quad & \lim_{z\to i} \frac{z^{10}+1}{z^6+1} &=& \tfrac{5}{3}.
\end{aligned}$$

Lemma 5.2. *Let U be a region in $\mathbb{C}$ and $f : U \to \mathbb{C}$ be differentiable. Assume that $f'(z) = 0$ for all $z \in U$. Then f is locally constant and hence a constant function on the whole of U.*

Proof. Given $z \in U$, chose $r > 0$ such that $B(z,r) \subset U$. This is possible since U is open. Fix any $w \in B(z,r)$. Consider $\gamma(t) = tw + (1-t)z$ for $t \in [0,1]$. Then $\gamma(t) \in B(z,r)$ for $t \in [0,1]$ as $B(z,r)$ is convex. Then $g = f \circ \gamma : [0,1] \to \mathbb{C}$ is differentiable and $g'(t) = f'(\gamma(t))\gamma'(t) = 0 \cdot (w-z) = 0$. Hence g is a constant and $f(w) = g(1) = g(0) = f(z)$. The lemma follows from Theorem 5.1 and Theorem 2.59. □

Observe that the condition which allowed us to conclude that the function f is constant from the hypothesis in Lemma 5.2 is that of connectedness.

Example 5.16. *Let $U = B(0,1) \cup B(2,1)$. Let*

$$f(z) = \begin{cases} 0 & \text{if } \ z \in B(0,1); \\ 2 & \text{if } \ z \in B(2,1). \end{cases}$$

Then it is easy to check that f is differentiable on the open set U and $f'(z) = 0$. However, f is not a constant on U. Here U is not connected.

5.3 Cauchy-Riemann equations

Suppose that f is complex differentiable at z. We shall investigate the existence of the limit of the difference quotients? Suppose we want h to approach 0 either completely from the real axis or from the imaginary axis. That is, we want to look at $f(z+h) - f(z)$ as $h \in \mathbb{R}$ goes to 0. We have

$$\begin{aligned}
f'(z) &= \lim_{\substack{h\in\mathbb{R}\\ h\to 0}} \frac{f(z+h)-f(z)}{h}\\
&= \lim_{\substack{h\in\mathbb{R}\\ h\to 0}} \frac{[u(x+h,y)+iv(x+h,y)] - [u(x,y)+iv(x,y)]}{h}\\
&= u_x(x,y) + iv_x(x,y). \qquad (5.10)
\end{aligned}$$

We consider increments ih with $h \in \mathbb{R}$ and proceed in a similar way to get

$$\begin{aligned} f'(z) &= \lim_{\substack{h\in\mathbb{R}\\ h\to 0}} \frac{f(z+ih)-f(z)}{ih} \\ &= \lim_{\substack{h\in\mathbb{R}\\ h\to 0}} \frac{[u(x,y+h)+iv(x,y+h)]-[u(x,y)+iv(x,y)]}{ih} \\ &= \frac{1}{i}(u_y+iv_y) = -iu_y+v_y. \end{aligned} \tag{5.11}$$

From equation (5.10) and equation (5.11), we get the Cauchy-Riemann equations

$$u_x = v_y \text{ and } u_y = -v_x. \tag{5.12}$$

In a similar way, we can show that if f is complex differentiable at z, then

$$\frac{\partial f}{\partial x}(z) = -i\frac{\partial f}{\partial y}(z) = f'(z).$$

If we use the new variables z and $\bar{z}$ and write

$$\frac{\partial f}{\partial z} = \frac{\partial f}{\partial x}\frac{\partial x}{\partial z} + \frac{\partial f}{\partial y}\frac{\partial y}{\partial z}.$$

Thus $\frac{\partial f}{\partial z} = \frac{1}{2}[f_x - if_y]$. Again $\frac{\partial f}{\partial \bar{z}} = \frac{\partial f}{\partial x}\frac{\partial x}{\partial \bar{z}} + \frac{\partial f}{\partial y}\frac{\partial y}{\partial \bar{z}}$. Hence $\frac{\partial f}{\partial \bar{z}} = \frac{1}{2}[f_x + if_y]$.

Thus we have a host of Cauchy-Riemann differential equations:

$$f'(z) = \frac{\partial f}{\partial x} = -i\frac{\partial f}{\partial y}. \tag{5.13}$$

$$\frac{\partial f}{\partial \bar{z}} = 0. \tag{5.14}$$

$$\frac{\partial u}{\partial x} = \frac{\partial v}{\partial y}, \quad \frac{\partial u}{\partial y} = -\frac{\partial v}{\partial x}. \tag{5.15}$$

These are the Cauchy-Riemann equations which must be satisfied by the real and imaginary part of any holomorphic function. We remark that the existence of the four partial derivatives in (5.15) is implied by the existence of $f'(z)$. Using (5.15) we can write down four formally different expressions for $f'(z)$; the simplest is

$$f'(z) = \frac{\partial u}{\partial x} + i\frac{\partial v}{\partial x}.$$

Thus the quantity

$$\begin{aligned} |f'(z)|^2 &= \left(\frac{\partial u}{\partial x}\right)^2 + \left(\frac{\partial u}{\partial y}\right)^2 \\ &= \left(\frac{\partial u}{\partial x}\right)^2 + \left(\frac{\partial v}{\partial x}\right)^2 \\ &= \frac{\partial u}{\partial x}\frac{\partial v}{\partial y} - \frac{\partial u}{\partial y}\frac{\partial v}{\partial x}. \end{aligned}$$

The last expression shows that $|f'(z)|^2$ is the Jacobian of u and v with respect to x and y.

Example 5.17. *Let $f(z) = z^2 = (x+iy)^2$. Then f is holomorphic and here $u = x^2 - y^2, v = 2xy$. It can be checked that $u_x = v_y = 2x, u_y = -v_x = -2y$. Hence the real and imaginary part of f satisfy the Cauchy-Riemann equations.*

Example 5.18. *Let $f(z) = |z|^2 = x^2 + y^2$. Then $u = x^2 + y^2, v = 0$. Thus $u_x = 2x, v_y = 0, u_y = 2y, v_x = 0$. So the only place the Cauchy-Riemann equations are satisfied is $x = y = 0$. So f is holomorphic nowhere except possibly at 0. It is differentiable at 0 as $\frac{f(0+h)-f(0)}{h} = \frac{|h|^2}{h} \to 0$ as $h \to 0$. So $f'(0) = 0$.*

Remark 5.4. We have already seen that if $f(z) = u(x,y) + iv(x,y)$ is holomorphic then $u_x = v_y$ and $u_y = -v_x$. But the converse is false. We can find $f = u + iv$ that satisfies Cauchy-Riemann equation at $c = a + ib$ but f not even continuous there. For example, let

$$f(z) = \begin{cases} 0 & \text{if} \quad x = 0 \text{ or } y = 0 (\text{ or both }); \\ \frac{1}{z} & \qquad \text{otherwise.} \end{cases}$$

Example 5.19. *The function $f(z) = \sqrt{|xy|}$ is not differentiable at the origin but satisfies the Cauchy-Riemann equation there.*

Solution: Let $f(z) = \sqrt{|xy|} = u(x,y) + iv(x,y)$ where $u(x,y) = \sqrt{|xy|}$ and $v(x,y) = 0$. Then $f'(0) = \lim\limits_{z\to 0} \dfrac{f(z) - f(0)}{z - 0} = \lim\limits_{z \to 0} \dfrac{\sqrt{|xy|}}{x + iy}$. If $z \to 0$ along the line $y = mx$, we get $f'(0) = \lim\limits_{x\to 0} \dfrac{\sqrt{|mx^2|}}{(1+im)x} = \lim\limits_{x\to 0} \dfrac{\sqrt{|m|}x}{(1+im)x} = \dfrac{\sqrt{|m|}}{1+im}$. The limit is not unique since it depends on m. Hence the limit does not exist. Now

$$\left.\frac{\partial u}{\partial x}\right|_{(0,0)} = \lim_{h\to 0} \frac{u(0+h,0) - u(0,0)}{h} = \lim_{h\to 0} \frac{0}{h} = 0.$$

Similarly one can show that $\frac{\partial u}{\partial y} = \frac{\partial v}{\partial x} = \frac{\partial v}{\partial y} = 0$.

Remark 5.5. Unlike continuity, complex differentiability is not the same as being differentiable with respect to two real variables. However they are related.We shall show below that if u_x, u_y, v_x, v_y exist near c and are continuous at c, and satisfy the Cauchy-Riemann equation at c, then f is differentiable at c.

Lemma 5.3. *Let G be a region in $\mathbb{C}$. Let $u : G \to \mathbb{R}$ have continuous partial derivatives u_x and u_y in G. Fix $(x,y) \in G$. Then for all (ζ, η) near $(0,0)$ we have $u(x+\zeta, y+\eta) - u(x,y) = \zeta u_x(x,y) + \eta u_y(x,y) + \zeta\epsilon_1 + \eta\epsilon_2$, where $\epsilon_i \to 0$ as $(\zeta,\eta) \to 0, i = 1,2$.*

Proof. We shall use mean value theorem of calculus:

$$\begin{aligned}
&u(x+\zeta,y+\eta)-u(x,y)\\
&= u(x+\zeta,y+\eta)-u(x,y+\eta)+u(x,y+\eta)-u(x,y)\\
&= \zeta u_x(x+\theta_1\zeta,y+\eta)+\eta u_y(x,y+\theta_2\eta)\\
&= \zeta[u_x(x,y)+(u_x(x+\theta_1\zeta,y+\eta)-u_x(x,y))]\\
&\quad +\eta[u_y(x,y)+(u_y(x,y+\theta_2\eta)-u_y(x,y))]\\
&= \zeta u_x(x,y)+\eta u_y(x,y)+\zeta\epsilon_1+\eta\epsilon_2
\end{aligned}$$

where $\epsilon_i \to 0$ as $(\zeta,\eta)\to 0, i=1,2$. □

Proposition 5.3. *Let $f:\Omega\to\mathbb{C}$. Let $f=u+iv$. Assume that the partial derivatives u_x, u_y, v_x and v_y exist and are continuous and further that the Cauchy-Riemann equations hold. Then f is differentiable on Ω. In fact $f'=u_x+iv_x$ on Ω.*

Proof. We use the lemma for u and v. Let $h=\zeta+i\eta$. Then $f(z+h)-f(z)$

$$\begin{aligned}
&= [u(x+\zeta,y+\eta)+iv(x+\zeta,y+\eta)]-[u(x,y)+iv(x,y)]\\
&= \zeta u_x+\eta u_y+\zeta\epsilon_1+\eta\epsilon_2+i[\zeta v_x+\eta v_y+\zeta\epsilon_3+\eta\epsilon_4]\\
&= h(u_x+iv_x)+(\zeta\epsilon_1+\eta\epsilon_2)+i(\zeta\epsilon_3+\eta\epsilon_4).
\end{aligned}$$

In the above, we made use of Cauchy-Riemann equations. The result follows from this and the observation that $|\frac{\zeta}{h}|\le 1$ etc. □

Lemma 5.4. *In polar form the Cauchy-Riemann equations can be written as*

$$\frac{\partial u}{\partial r}=\frac{1}{r}\frac{\partial v}{\partial \theta},\ \frac{\partial v}{\partial r}=-\frac{1}{r}\frac{\partial u}{\partial \theta}.$$

Proof. We have $x=r\cos\theta, y=r\sin\theta$ or $r=\sqrt{x^2+y^2}, \theta=\tan^{-1}(\frac{y}{x})$. Then

$$\frac{\partial u}{\partial x}=\frac{\partial u}{\partial r}\frac{\partial r}{\partial x}+\frac{\partial u}{\partial \theta}\frac{\partial \theta}{\partial x}=\frac{\partial u}{\partial r}\left(\frac{x}{\sqrt{x^2+y^2}}\right)+\frac{\partial u}{\partial \theta}\left(\frac{-y}{x^2+y^2}\right)$$

$$=\frac{\partial u}{\partial r}\cos\theta-\frac{1}{r}\frac{\partial u}{\partial \theta}\sin\theta; \tag{5.16}$$

$$\frac{\partial u}{\partial y}=\frac{\partial u}{\partial r}\frac{\partial r}{\partial y}+\frac{\partial u}{\partial \theta}\frac{\partial \theta}{\partial y}=\frac{\partial u}{\partial r}\left(\frac{y}{\sqrt{x^2+y^2}}\right)+\frac{\partial u}{\partial \theta}\left(\frac{x}{x^2+y^2}\right)$$

$$=\frac{\partial u}{\partial r}\sin\theta+\frac{1}{r}\frac{\partial u}{\partial \theta}\cos\theta. \tag{5.17}$$

Similarly,

$$\frac{\partial v}{\partial x}=\frac{\partial v}{\partial r}\frac{\partial r}{\partial x}+\frac{\partial v}{\partial \theta}\frac{\partial \theta}{\partial x}=\frac{\partial v}{\partial r}\cos\theta-\frac{1}{r}\frac{\partial v}{\partial \theta}\sin\theta \tag{5.18}$$

and

$$\frac{\partial v}{\partial y} = \frac{\partial v}{\partial r}\frac{\partial r}{\partial y} + \frac{\partial v}{\partial \theta}\frac{\partial \theta}{\partial y} = \frac{\partial v}{\partial r}\sin\theta + \frac{1}{r}\frac{\partial v}{\partial \theta}\cos\theta. \tag{5.19}$$

From the Cauchy-Riemann equation $\frac{\partial u}{\partial x} = \frac{\partial v}{\partial y}$ we have, using (5.16) and (5.19),

$$\left(\frac{\partial u}{\partial r} - \frac{1}{r}\frac{\partial v}{\partial \theta}\right)\cos\theta - \left(\frac{\partial v}{\partial r} + \frac{1}{r}\frac{\partial u}{\partial \theta}\right)\sin\theta = 0. \tag{5.20}$$

From the Cauchy-Riemann equation $\frac{\partial u}{\partial y} = -\frac{\partial v}{\partial x}$ we have, using (5.17) and (5.18),

$$\left(\frac{\partial u}{\partial r} - \frac{1}{r}\frac{\partial v}{\partial \theta}\right)\sin\theta + \left(\frac{\partial v}{\partial r} + \frac{1}{r}\frac{\partial u}{\partial \theta}\right)\cos\theta = 0. \tag{5.21}$$

Multiplying (5.20) by $\cos\theta$, (5.21) by $\sin\theta$ and adding yields $\frac{\partial u}{\partial r} - \frac{1}{r}\frac{\partial v}{\partial \theta} = 0$ or $\frac{\partial u}{\partial r} = \frac{1}{r}\frac{\partial v}{\partial \theta}$. Multiplying (5.20) by $-\sin\theta$, (5.21) by $\cos\theta$ and adding yields $\frac{\partial v}{\partial r} + \frac{1}{r}\frac{\partial u}{\partial \theta} = 0$ or $\frac{\partial v}{\partial r} = -\frac{1}{r}\frac{\partial u}{\partial \theta}$.

□

Example 5.20. *If $f = u + iv$ is holomorphic in a region G then*

$$\left(\frac{\partial^2}{\partial x^2} + \frac{\partial^2}{\partial y^2}\right)|f(z)|^p = p^2|f(z)|^{p-2}|f'(z)|^2.$$

Solution: Let $\Phi(z) = |f(z)|^p = (u^2+v^2)^{\frac{p}{2}}$. Then

$$\frac{\partial \Phi}{\partial x} = \frac{p}{2}(u^2+v^2)^{\frac{p}{2}-1}\left(2u\frac{\partial u}{\partial x} + 2v\frac{\partial v}{\partial x}\right) = p(u^2+v^2)^{\frac{p}{2}-1}\left(u\frac{\partial u}{\partial x} + v\frac{\partial v}{\partial x}\right).$$

Thus

$$\begin{aligned}
\frac{\partial^2 \Phi}{\partial x^2} &= p\left(\tfrac{p}{2}-1\right)(u^2+v^2)^{\frac{p}{2}-2}\left(2u\tfrac{\partial u}{\partial x} + 2v\tfrac{\partial v}{\partial x}\right)\left(u\tfrac{\partial u}{\partial x} + v\tfrac{\partial v}{\partial x}\right)\\
&\quad + p(u^2+v^2)^{\frac{p}{2}-1}\left(\left(\tfrac{\partial u}{\partial x}\right)^2 + \left(\tfrac{\partial v}{\partial x}\right)^2 + u\tfrac{\partial^2 u}{\partial x^2} + v\tfrac{\partial^2 v}{\partial x^2}\right)\\
&= p(p-2)(u^2+v^2)^{\frac{p}{2}-2}\left(u\tfrac{\partial u}{\partial x} + v\tfrac{\partial v}{\partial x}\right)^2\\
&\quad + p(u^2+v^2)^{\frac{p}{2}-1}\left(\left(\tfrac{\partial u}{\partial x}\right)^2 + \left(\tfrac{\partial v}{\partial x}\right)^2 + u\tfrac{\partial^2 u}{\partial x^2} + v\tfrac{\partial^2 v}{\partial x^2}\right).
\end{aligned}$$

Similarly,

$$\begin{aligned}
\frac{\partial^2 \Phi}{\partial y^2} &= p(p-2)(u^2+v^2)^{\frac{p}{2}-2}\left(u\tfrac{\partial u}{\partial y} + v\tfrac{\partial v}{\partial y}\right)^2\\
&\quad + p(u^2+v^2)^{\frac{p}{2}-1}\left(\left(\tfrac{\partial u}{\partial y}\right)^2 + \left(\tfrac{\partial v}{\partial y}\right)^2 + u\tfrac{\partial^2 u}{\partial y^2} + v\tfrac{\partial^2 v}{\partial y^2}\right).
\end{aligned}$$

Hence

$$\begin{aligned}\frac{\partial^2\Phi}{\partial x^2}+\frac{\partial^2\Phi}{\partial y^2} &= p(p-2)(u^2+v^2)^{\frac{p}{2}-2}\left[u^2\left(\frac{\partial u}{\partial x}\right)^2+v^2\left(\frac{\partial v}{\partial x}\right)^2\right.\\ &\quad \left.+\,2uv\frac{\partial u}{\partial x}\frac{\partial v}{\partial x}+u^2\left(\frac{\partial u}{\partial y}\right)^2+v^2\left(\frac{\partial v}{\partial y}\right)^2+2uv\frac{\partial u}{\partial y}\frac{\partial v}{\partial y}\right]\\ &\quad +\,p(u^2+v^2)^{\frac{p}{2}-1}\left[2\left(\left(\frac{\partial u}{\partial x}\right)^2+\left(\frac{\partial v}{\partial x}\right)^2\right)+u\cdot 0+v\cdot 0\right]\\ &= p(p-2)(u^2+v^2)^{\frac{p}{2}-2}\left[(u^2+v^2)\left(\left(\frac{\partial u}{\partial x}\right)^2+\left(\frac{\partial v}{\partial x}\right)^2\right)\right]\\ &\quad +\,p(u^2+v^2)^{\frac{p}{2}-1}\left[2\left(\left(\frac{\partial u}{\partial x}\right)^2+\left(\frac{\partial v}{\partial x}\right)^2\right)\right]\\ &= p(p-2)(u^2+v^2)^{\frac{p}{2}-1}|f'(z)|^2+p(u^2+v^2)^{\frac{p}{2}-1}[2|f'(z)|^2]\\ &= p^2|f(z)|^{p-2}|f'(z)|^2.\end{aligned}$$

Theorem 5.8. *A holomorphic function f in a region Ω whose derivative vanishes identically is a constant. Further if either the $Ref, Imf, |f|$ or $\arg f$ is constant then f is a constant function on Ω.*

Proof. We have already proved that if $f'(z) = 0$ for all $z \in \Omega$ then f is a constant. We further repeat that this is so because the vanishing of the derivative implies that u_x, u_y, v_x and v_y are all zero. Thus u and v are constant on any line segment in Ω which is parallel to one of the coordinate axes. Further any two points in a region can be joined within the region by a polygon whose sides are parallel to the axes. Hence we conclude that $u+iv = f$ is a constant. If u or v is constant, $f'(z) = \frac{\partial u}{\partial x} - i\frac{\partial u}{\partial y} = \frac{\partial v}{\partial y} + i\frac{\partial v}{\partial x} = 0$, and hence $f(z)$ must be constant. If u^2+v^2 is constant, we obtain

$$u\frac{\partial u}{\partial x}+v\frac{\partial v}{\partial x}=0$$

and

$$u\frac{\partial u}{\partial y}+v\frac{\partial v}{\partial y}=-u\frac{\partial v}{\partial x}+v\frac{\partial u}{\partial x}=0.$$

These equations permit the conclusion $\frac{\partial u}{\partial x} = \frac{\partial v}{\partial x} = 0$ unless the determinant u^2+v^2 vanishes. But if $u^2+v^2 = 0$ at a single point it is constantly zero and $f(z)$ vanishes identically. Hence $f(z)$ in any case a constant. Finally, if $\arg f(z)$ is constant, we can set $u = kv$ with constant k (unless v is identically zero). But $u - kv$ is the real part of $(1+ik)f$, and we conclude again that f must reduce to a constant. □

Note that in the above result connectedness is necessary.

Example 5.21. *Let* $U = \{z \in \mathbb{C} : |z| < 1\} \cup \{z \in \mathbb{C} : |z| > 1\}$. *Define*

$$f(z) = \begin{cases} 0 & \text{if } |z| < 1; \\ 1 & \text{if } |z| > 1. \end{cases}$$

Then $f' = 0, Im\ f = 0$. *Further, Re* $f, |f|$ *are locally constants but* f *is not a constant function on* U.

Example 5.22. *Let* G *be a region and suppose that* $f : G \to \mathbb{C}$ *is holomorphic such that* $f(G)$ *is a subset of a circle. Then* f *is a constant.*

Solution: Suppose $f(G) \subset \{w \in \mathbb{C} : |w - a| = r\}$. Hence $|f(z) - a| = r$ for all $z \in G$. Let $g(z) = f(z) - a$. Then g is holomorphic and $|g(z)| = r$, i.e., $|g|$ is constant in G. Thus g is constant by Theorem 5.8 and hence $f(z) = a + d$ for some constant d and for all $z \in G$.

5.4 Inverse functions

Theorem 5.9. ***(Differentiability of inverse functions):*** *Let* U *and* V *be open in* $\mathbb{C}$. *Let* $f : U \to V$ *be a bijection with* $g = f^{-1}$. *Let* $a \in U$ *and* $b = f(a)$. *Assume that (1)* f *is differentiable at* a *and* $f'(a) \neq 0$ *and (2)* g *is continuous at* b. *Then* g *is differentiable at* b *and* $g'(b) = (f'(a))^{-1}$.

Proof. Let $z \in U$. Since f is differentiable at a, hence $f(z) - f(a) = (z - a)f_1(z)$. Let $w = f(z)$. Then $g(w) = z$. Thus if we substitute $g(w)$ for z in the above equation we obtain $w - b = [g(w) - g(b)]f_1(g(w))$.

Hence $g(w) - g(b) = (w - b)\frac{1}{f_1(g(w))}$. Since $f_1 \circ g$ is continuous and nonzero at b, the result follows. □

We recall that $\log_\alpha : \mathbb{C}/L_\alpha \to \mathbb{C}$ is defined by setting

$$\log_\alpha(z) = \log|z| + i\text{Arg}_\alpha(z), z \in \mathbb{C}/L_\alpha.$$

Then $\log_\alpha$ is continuous on its domain. Since exp maps $\{z = x + iy : y \in (\alpha, \alpha + 2\pi)\}$ onto $\mathbb{C}/L_\alpha$, $\log_\alpha$ is its inverse.

The following is an immediate application of Theorem 5.9.

Theorem 5.10. log_α *is holomorphic and* $log'_\alpha(z) = \frac{1}{z}$ *for* $z \in \mathbb{C}/L_\alpha$.

Proof. Let $g = \log_\alpha$ and $f(z) = \exp(z)$ for z such that $\text{Im}(z) \in (\alpha, \alpha + 2\pi)$. Let $f(z) = w$. Then $g'(w) = \frac{1}{f'(z)} = \frac{1}{\exp(z)} = \frac{1}{w}$. □

Example 5.23. *If* $U = B(1 + i, 1)$ *and if we take as* g *the restriction of any one of* log_0, $log_{\frac{\pi}{2}}$, log_π *and* $log_{\frac{3\pi}{2}}$ *to* U, *then* g *is a logarithm on* U.

Theorem 5.11. *If* $|z| < 1$, *then* $log_\pi(1 + z) = \sum_{n=1}^{\infty}(-1)^{n-1}\frac{z^n}{n}$.

Proof. Let $f(z) = \log_\pi(1+z)$ and $g(z) = \sum_{n=1}^{\infty}(-1)^{n-1}\frac{z^n}{n}$ for $|z| < 1$. Then by Theorem 5.10 , $f'(z) = \frac{1}{1+z}$ and $g'(z) = \sum_{n=1}^{\infty}(-1)^{n-1}z^{n-1} = \frac{1}{1+z}$ by Theorem 5.4. Hence $f - g$ is a constant on $B(0,1)$. Since $f(0) = 0 = g(0)$, the result follows. □

Let U be an open subset of $\mathbb{C}$. We say a continuous function $g : U \to \mathbb{C}$ is a logarithm of z on U if we have $\exp(g(z)) = z$ for all $z \in U$. It can be verified that if g is a logarithm on U, then $g'(z) = \frac{1}{z}$ for $z \in U$.

Theorem 5.12. *If l is a branch of $\log z$ in the open connected set $G \subset \mathbb{C}$, then l is holomorphic, and $l'(z) = \frac{1}{z}$.*

Proof. We shall verify that l is holomorphic by checking that l satisfies the Cauchy-Riemann equations in polar coordinates. Let $u = \mathrm{Re} l$ and $v = \mathrm{Im} l$. The Cauchy-Riemann equations in polar coordinates are

$$r\frac{\partial u}{\partial r} = \frac{\partial v}{\partial \theta}, \frac{\partial u}{\partial \theta} = -r\frac{\partial v}{\partial r}.$$

Since l is a branch of $\log z$ we have $u(z) = \log|z|$ and $v(z) = \arg z$, from which it is immediate that

$$\frac{\partial u}{\partial r} = \frac{1}{r}, \frac{\partial u}{\partial \theta} = 0 = \frac{\partial v}{\partial r}, \frac{\partial v}{\partial \theta} = 1.$$

The Cauchy-Riemann equations are thus satisfied, and hence l is holomorphic by Proposition 5.3. Moreover, we have $l' = \frac{\partial u}{\partial x} + i\frac{\partial v}{\partial x}$. Using the chain rule, we obtain

$$\frac{\partial u}{\partial x} = \frac{\partial u}{\partial r}\frac{\partial r}{\partial x} + \frac{\partial u}{\partial \theta}\frac{\partial \theta}{\partial x} = \frac{x}{r}\frac{\partial u}{\partial r} - \frac{y}{r^2}\frac{\partial u}{\partial \theta} = \frac{x}{r^2}.$$

$$\frac{\partial v}{\partial x} = \frac{\partial v}{\partial r}\frac{\partial r}{\partial x} + \frac{\partial v}{\partial \theta}\frac{\partial \theta}{\partial x} = \frac{x}{r}\frac{\partial v}{\partial r} - \frac{y}{r^2}\frac{\partial v}{\partial \theta} = \frac{-y}{r^2}.$$

Hence $l'(z) = \frac{x-iy}{r^2} = \frac{1}{x+iy} = \frac{1}{z}$. □

5.5 Logarithms of holomorphic functions

Let Ω be an open connected subset of $\mathbb{C}$, and let f be a holomorphic function in Ω that does not assume the value 0. By a branch of $\log f$ in Ω one means a continuous function g in Ω such that $f = e^g$, in other words, such that, for each z in Ω, the value $g(z)$ is a logarithm of $f(z)$. The special case of the function $f(z) = z$ is the one discussed earlier. Note if there is a branch of $\log f$ in Ω, then one obtains the most general such branch by adding a constant integer multiple of $2\pi i$ to any particular one.

If it happens that the range of f is contained in a region in which there is a branch l of $\log z$, then the composite function $l \circ f$ is a branch of $\log f$; moreover, being a composite function of two holomorphic functions it is holomorphic, and by the chain rule its derivative equals $\frac{f'}{f}$.

Whether or not there is a branch of log$f$ in Ω; such branches exist locally, in the following sense. Fix a point z_0 in Ω. Since by assumption $f(z_0) \neq 0$, there is by the continuity of f an open disk D with center z_0 such that $f(D)$ is contained in an open disk D' that excludes the origin. Then there is a branch l of log$z$ in D' and by composing l with f we obtain a branch of log$f$ in D. In case there happens to be a branch g of log$f$ in all of Ω, the functions g and $l \circ f$ differ in D by a constant integer multiple of $2\pi i$. In fact, we can arrange to have $g = l \circ f$ by choosing for l the particular branch of log$z$ in D' satisfying $l(f(z_0)) = g(z_0)$. Since $l \circ f$ is a holomorphic function with derivative $\frac{f'}{f}$, we can draw the following conclusion: If there is a branch of log$f$ in Ω, then any such branch is a holomorphic function whose derivative is $\frac{f'}{f}$. Whether or not there is a branch of log$f$ in Ω, one refers to the function $\frac{f'}{f}$ as the logarithmic derivative of f.

Although, a branch of log$f$ in Ω, if it exists, is locally the composite of f with a branch of log$z$, it may not be globally such a composite. A simple example is provided by the function $f(z) = e^z$ in $\mathbb{C}$. The function $g(z) = z$ is a branch of log$f$, yet the range of f is $\mathbb{C}/\{0\}$, in which there is no branch of log$z$.

This concept of branches gives a way of dealing with inverses of functions that are not one-to-one. This happens also in the theory of functions of a real variable. The function $f(x) = x^2$ from $\mathbb{R}$ onto $[0, \infty)$ is not one-one. But if we restrict the domain to the set of positive real numbers (or negative real numbers) the function f is one-one. Similarly, the sine function on $\mathbb{R}$ has range $[-1, 1]$ and it is not one-one. But if we restrict the domain of sine function to the interval $[-\frac{\pi}{2}, \frac{\pi}{2}]$, the function becomes one-one. In this case the so-called inverse of sine function, the function arc sin can be defined from the interval $[-1, 1]$ to the interval $[-\frac{\pi}{2}, \frac{\pi}{2}]$. Notice that this is only one of infinitely many branches of the inverse of the sine function.

Suppose f is a holomorphic function in the region $G \subset \mathbb{C}$. If f is one-to-one (or univalent) then the inverse of f is the function g defined in the region $f(G)$ by the rule $g(f(z)) = z$. If the holomorphic function f, defined in the region G is not univalent, then it does not possess an inverse function in the strict sense. But it is still possible to speak of branches of the inverse of f. Let H be a region in G on which f is univalent; the inverse of the restriction of f to H is then called a branch of the inverse of f. Starting with the function $f(z) = e^z$ one obtains in this way branches of $\log z$; starting with the function $f(z) = z^n$ one obtains the branches of $z^{\frac{1}{n}}$. If we start with trigonometric functions, then we shall be able to define branches of the corresponding inverse functions. Because of the relation between the trigonometric functions and the exponential function, those branches can be related to the logarithm function.

Consider the cosine function. The branches of its inverse are referred to as the branches of $\arccos w$. Let us fix a point w in the domain of such a branch and ask what possible values the branch can have at w. If z is such a value, then z solves the equation $\cos z = w$; in other words, $\frac{e^{iz}+e^{-iz}}{2} = w$, which can be rewritten as $e^{2iz} - 2we^{iz} + 1 = 0$. The roots of the quadratic equation are $e^{iz} = w \pm \sqrt{w^2-1}$. The possible values of z are therefore expressed by $z = \frac{1}{i}\log(w \pm \sqrt{w^2-1})$. Thus the branches of $\arccos w$ are just the branches of $\frac{1}{i}\log(w \pm \sqrt{w^2-1}) = \pm i \log(w + \sqrt{w^2-1})$.

For each $z \in \mathbb{C}$, there are infinitely many values of w such that $z = \sin w$. To find these values of w (for a given z), we write $z = \sin w = \frac{e^{iw}-e^{-iw}}{2i} = \frac{e^{2iw}-1}{2ie^{iw}} = \frac{p^2-1}{2ip}$ where $p = e^{iw}$. Thus $p^2 - 2izp - 1 = 0$. Solving for p, $e^{iw} = p = iz \pm \sqrt{1-z^2} = iz \pm i\sqrt{z^2-1}$, where $\sqrt{z^2-1}$ is one of the two values of $(z^2-1)^{\frac{1}{2}}$. Thus $iw = \log i[z \pm \sqrt{z^2-1}]$. Now for each $z \in \mathbb{C}$, there are infinitely many values of w such that $w = -i\log[i(z \pm \sqrt{z^2-1})] = -i[\log i + \log(z \pm \sqrt{z^2-1})] = -i\log i \mp i\log(z + \sqrt{z^2-1})$. Each value of w is a complex number such that $\sin w = z$. For a fixed number z, each number w given by the formula above is denoted by $\sin^{-1} z$ or $\arcsin z$. One of the infinitely many branches of w is singled out and called the principal arcsine of z and is denoted by $\sin^{-1} z$. This branch is the following: $\sin^{-1} z = \frac{\pi}{2} + i\text{Log}(z + \sqrt{z^2-1})$.

Similarly one can define branches of the inverse hyperbolic sine and cosine of z. Let $z = \cosh iw = \cos w$. Thus $w = \cos^{-1} z$ and $iw = \cosh^{-1} z$. Hence for each $z \in \mathbb{C}$, $\cosh^{-1} z = i\cos^{-1} z = i[-i\log(z + (z^2-1)^{\frac{1}{2}})] = \log[z + (z^2-1)^{\frac{1}{2}}]$. Similarly, let $z = \sinh iw = i\sin w$. Then $w = \sin^{-1}(-iz)$ and $iw = \sinh^{-1} z$. Hence for each $z \in \mathbb{C}$, $\sinh^{-1} z = i\sin^{-1}(-iz) = i[-i\log i(-iz + (-z^2-1)^{\frac{1}{2}})] = \log[z + (z^2+1)^{\frac{1}{2}}]$ since $(-1)^{\frac{1}{2}} = \pm i$. Similarly one can show that $\tanh^{-1} z = \frac{1}{2}\log\frac{1+z}{1-z}$ for each $z \neq \pm 1$. Use the formula $z = \tanh iw = i\tan w$. We can also define branches of inverse of other hyperbolic functions.

This happens only in the sense that both sides represent the same infinite set of complex numbers. If we want to compare a value on the left with a value on the right, then we can merely assert that they differ by a multiple of $2\pi i$ (or 2π).

Let G be a region in $\mathbb{C}$ and $f : G \longrightarrow \mathbb{C}$ be holomorphic. Suppose $f(z) \neq 0$ for all $z \in G$. Let n be a positive integer. By a branch of $f^{\frac{1}{n}}$ in G one means a continuous function h in G such that, for each z in G, the value $h(z)$ is an n-th root of $f(z)$. Notice that for $n > 1$, there is no branch of $z^{\frac{1}{n}}$ in the region $0 < |z| < 1$. It is for this reason we assumed that $f(z) \neq 0$ for all $z \in G$. To define branches of $f(z)^{\frac{1}{n}}$ we merely write it in the form $exp((\frac{1}{n})\log f(z))$. Thus the discussion of branches of $f^{\frac{1}{n}}$ parallels that of branches of $\log f$. Moreover, if there is a branch of $f^{\frac{1}{n}}$ in G then there are precisely n such branches, each obtainable from any particular one through multiplication by an n-th root of 1. If there is a branch g of $\log f$ in G, then the function $e^{\frac{g}{n}}$ is a branch of $f^{\frac{1}{n}}$ in G. If h is a branch of $f^{\frac{1}{n}}$ then h is holomorphic and $\frac{h'}{h} = \frac{f'}{nf}$.

5.6 Exercises

1. Suppose f and g are complex differentiable functions on $(0,1)$, $f(x) \to 0$, $g(x) \to 0$, $f'(x) \to A$, $g'(x) \to B$ as $x \to 0$, where A and B are complex numbers, $B \neq 0$. Prove that $\frac{A}{B} = \lim_{x \to 0} \dfrac{f(x)}{g(x)}$. Hint: $\frac{f(x)}{g(x)} = \{\frac{f(x)}{x} - A\}\frac{x}{g(x)} + A\frac{x}{g(x)}$. Apply L'Hospital's rule for real-valued functions to the real and imaginary parts of $\frac{f(x)}{x}$ and $\frac{g(x)}{x}$.

2. If $f(z) = z\text{Re } z$ then state where $f'(z)$ exists and find $f'(z)$ at these points.

3. Show that $f(z) = |z|^2$ is differentiable at $z = 0$ and nowhere else.

4. Show that $f(z) = x^2 - iy^3$ is differentiable only at $z = 0$.

5. (a) At which points the following functions are differentiable:
   (i) $f(z) = \bar{z}^2$; (ii) $g(z) = \text{Re}(z)$; (iii) $h(z) = \text{Im}(z)$.

   (b) Express each of the following functions of $z = x + iy$ in the form $u(x,y) + iv(x,y)$ and use the Cauchy-Riemann equation to determine the set of points at which it is holomorphic: (i) $\bar{z}, z \in \mathbb{C}$; (ii) $\arg z(\text{ Re } z > 0)$.

   (c) State appropriate theorems showing why the following functions are entire: (i) $3iz^2 + 4 + \frac{1}{e^z}$; (ii) $e^{\sin z}$; (iii) $\cosh e^z$.

   (d) Show that $\sin \bar{z}$ and $\cos \bar{z}$ are not analytic anywhere.

6. Show that $\lim_{z \to 0} \dfrac{e^{iz} - 1}{z} = i$. Hint: Recall that $\frac{f(z) - f(0)}{z} \to f'(0)$ as $z \to 0$.

7. If $0 \leq r < 1$, then show that $nr^n \to 0$ as $n \to \infty$. Hint: Write $r = \frac{1}{1+h}$ for $h > 0$. By binomial theorem, $(1+h)^n \geq \binom{n}{2}h^2$.

8. Let $f : [0, 2\pi] \to \mathbb{C}$ be given by $f(t) = e^{int}$ for some $n \in \mathbb{Z}$. Compute $\int_0^{2\pi} f(t)dt$.

9. Let $f(z) = \sum_{n=0}^{\infty} a_n z^n$ with $R > 0$ as its radius of convergence. Assume that $f' = f$ on $B(0,R)$ and that $f(0) = 1$. Find a_n explicitly and hence f.

10. Let $f(z) = \sum_{n=0}^{\infty} \frac{z^{2n}}{(n!)^2}$. Show that f satisfies the differential equation $z^2 f''(z) + zf'(z) = 4f(z), z \in \mathbb{C}$.

11. For any $z \in \mathbb{C}$, show that $n\log_\pi(1 + \frac{z}{n})$ is defined for all large values of n and that it tends to z as $n \to \infty$. Hence deduce that $(1 + \frac{z}{n})^n \to \exp(z)$ as $n \to \infty$.

12. Let $f = u + iv$ be holomorphic in a region G, where u and v are real valued. Suppose there are real constants a, b and c such that $a^2 + b^2 \neq 0$ and $au + bv = c$ in G. Show that f is constant in G.

13. Can there be a nonconstant holomorphic function with constant absolute value? Justify your answer.

14. Find a holomorphic function whose real part is $x^3 - 3xy^2$. Can there be a holomorphic function whose real part is x^2?

15. If an entire function f is real on the real axis and imaginary on the imaginary axis, then show that it is an odd function. Hint: Let $f(z) = \sum_{n=0}^{\infty} a_n z^n$ be the Taylor series expansion of f around origin. Since $f(\mathbb{R}) \subset \mathbb{R}$, each a_n is real. Further, since $f(i\mathbb{R}) \subset i\mathbb{R}$, $a_n = 0$ for n even.

16. If E is any straight line in $\mathbb{C}$, and if a holomorphic function f on a region U takes values in E, then show that f is a constant.

17. Let $f : \mathbb{C} \to \mathbb{C}$ be a function. Suppose $g(z) = \overline{f(\bar{z})}$ for all $z \in \mathbb{C}$. Show that f and g are simultaneously holomorphic.

18. Let U be a region in $\mathbb{C}$. If f and $\bar{f}$ are both holomorphic in U, then show that f is constant.

19. If the range of a holomorphic function f defined on a region G lies in either a straight line or a circle, then prove that f is a constant.

20. If f is holomorphic in a region G and $z = x + iy$ then prove that

$$\left(\frac{\partial^2}{\partial x^2} + \frac{\partial^2}{\partial y^2}\right) |\mathrm{Re} f(z)|^p = p(p-1)|\mathrm{Re} f(z)|^{p-2}|f'(z)|^2.$$

21. Prove (assuming equality of second-order mixed partial derivatives) that

$$\frac{\partial^2}{\partial \bar{z} \partial z} = \frac{1}{4}\left(\frac{\partial^2}{\partial x^2} + \frac{\partial^2}{\partial y^2}\right).$$

22. Let $u(x, y) = x^3 - 3xy^2 + 1$. Verify that $\frac{\partial^2 u}{\partial x^2} + \frac{\partial^2 u}{\partial y^2} = 0$. Find a holomorphic function f such that u is the real part of f.

Chapter 6

Integration

6.1 Integration of functions from $\mathbb{R}$ to $\mathbb{C}$

We assume the reader is familiar with the theory of definite integrals of real valued continuous functions. In this section we shall define integrals of complex valued functions defined on a closed interval $[a, b] \subset \mathbb{R}$ in terms of integrals of real valued functions. This will save us from repeating existence proofs which are essentially the same as in the real case.

Definition 6.1. *(Piecewise continuous functions): Let $[a, b] \subseteq \mathbb{R}$ and let ϕ be a complex valued function on $[a, b]$. The function ϕ is said to be piecewise continuous if it is continuous at all but finitely many points of $[a, b]$ and has finite one-sided limits at each point of discontinuity.*

Let $I = [a, b] \subseteq \mathbb{R}$ and $f : I \to \mathbb{C}$ be piecewise continuous. Then we define the integral of f by $\int_a^b f(t)dt = \int_a^b \text{Re}f(t)dt + i\int_a^b \text{Im}f(t)dt$. Since $\text{Re}f$ and $\text{Im}f$ are piecewise continuous on $[a, b]$ hence the Riemann integral of these functions are defined over $[a, b]$.

Proposition 6.1. *If ϕ_1 and ϕ_2 are piecewise continuous complex-valued functions on $[a, b]$ and c_1 and c_2 are complex constants then*

$$\int_a^b [c_1\phi_1(t) + c_2\phi_2(t)]dt = c_1 \int_a^b \phi_1(t)dt + c_2 \int_a^b \phi_2(t)dt.$$

Proof. It follows from the linearity of the Riemann integral for real-valued functions. □

It is to be noted that one can use the Riemann integral under less restrictive conditions than the piecewise continuity of the function to be integrated.

Definition 6.2. *The function ϕ is said to be piecewise C^1 (piecewise smooth) if it is continuous, it is differentiable at all but finitely many points of $[a, b]$, its derivative is continuous at each point where it exists and its derivative has finite one-sided limits at each of the finitely many points where it fails to exist.*

Theorem 6.1. *(Fundamental theorem of calculus)*

(1) Let $f : [a, b] \to \mathbb{C}$ be piecewise continuous. Define $F(x) = \int_a^x f(t)dt$. Then F is piecewise C^1 and $F'(x) = f(x)$ for $x \in [a, b]$.

(2) Let $G : [a, b] \to \mathbb{C}$ be piecewise C^1 with $G' = g$ continuous. Then $\int_a^b g(t)dt = G(b) - G(a)$.

Proof. To prove (1), notice that if $f(t) = u(t) + iv(t)$ then $F(x) = \int_a^x u(t)dt + i\int_a^x v(t)dt$. Hence $F'(x) = u(x) + iv(x) = f(x)$ using the fundamental theorem of calculus for the real indefinite integrals. For the proof of (2), let $H(x) = \int_a^x g(t)dt$. Hence $H'(x) = g(x)$ by (1). Hence $(H - G)' = 0$. Thus $H - G$ is a constant on $[a, b]$ whence $H(b) - G(b) = H(a) - G(a)$. □

The second part of Theorem 6.1 can also be proved as follows: Let $G(x) = u(x) + iv(x)$. Then $G'(x) = u'(x) + iv'(x) = g(x)$. Now it follows from the fundamental theorem of calculus for the real indefinite integral that $\int_a^b g(t)dt = \int_a^b u'(t)dt + i\int_a^b v'(t)dt = u(b) - u(a) + i(v(b) - v(a)) = G(b) - G(a)$.

Corollary 6.1. *(Integration by parts): Let $f, g : [a, b] \to \mathbb{C}$ be piecewise C^1. Then $\int_a^b f(t)g'(t)dt = [f(t)g(t)]_a^b - \int_a^b f'(t)g(t)dt$.*

Proof. We already know the result if f and g are real-valued functions. The complex-valued case follows from this by writing $f(t) = u(t) + iv(t)$ and $g(t) = u_1(t) + iv_1(t)$ where u, u_1, v, v_1 are real-valued functions. Let $\phi(t) = f(t)g(t)$. Then $\phi'(t) = f'(t)g(t) + g'(t)f(t)$ and ϕ' is piecewise continuous. Hence ϕ' is integrable. Thus by fundamental theorem of calculus,

$$\begin{aligned} \int_a^b f'(t)g(t)dt + \int_a^b g'(t)f(t)dt &= \int_a^b \phi'(t)dt \\ &= \phi(b) - \phi(a) \\ &= f(b)g(b) - f(a)g(a). \end{aligned}$$

□

Proposition 6.2. *(Change of variable): Let $h : [c, d] \longrightarrow \mathbb{R}$ be continuously differentiable and $f : [a, b] \longrightarrow \mathbb{R}$ be continuous. Assume that $h([c, d]) \subset [a, b]$. Then*

$$\int_c^d (f \circ h)(s)h'(s)ds = \int_{h(c)}^{h(d)} f(t)dt.$$

Proof. Consider $\phi(s) = \int_{h(c)}^{h(s)} f(t)dt$. Then ϕ is the composition of the functions $s \mapsto h(s)$ and $x \mapsto \int_{h(c)}^x f(t)dt$. Hence $\phi'(s) = f(h(s))h'(s)$ by chain rule and Theorem 6.1. Again by Theorem 6.1, we have

$$\int_{h(c)}^{h(d)} f(t)dt = \phi(d) - \phi(c) = \int_c^d \phi'(s)ds = \int_c^d f(h(s))h'(s)ds.$$

□

Proposition 6.3. *If f is a piecewise continuous complex-valued function on the interval $[a,b]$, then* $\left|\int_a^b f(t)dt\right| \leq \int_a^b |f(t)|\,dt$.

Proof. If $\int_a^b f(t)dt = 0$, then there is nothing to prove. So assume $\int_a^b f(t)dt \neq 0$, and let $\alpha = \frac{|\int_a^b f(t)dt|}{\int_a^b f(t)dt}$. Then $\left|\int_a^b f(t)dt\right| = \alpha \int_a^b f(t)dt = \int_a^b \alpha f(t)dt = \mathrm{Re}\int_a^b \alpha f(t)dt = \int_a^b \mathrm{Re}[\alpha f(t)]dt \leq \int_a^b |\alpha f(t)|dt = \int_a^b |f(t)|dt$. □

Proposition 6.4. *If $f_n : [a,b] \to \mathbb{C}$ are continuous and they converge uniformly to a function f, then f is necessarily continuous. Moreover, if $M_n = \sup_{x\in[a,b]} |f_n(x) - f(x)|$ then $\lim_{n\to\infty} M_n = 0$ and we have*

$$\lim_{n\to\infty} \int_a^b f_n(t)dt = \int_a^b f(t)dt. \tag{6.1}$$

Proof. The first part of the statement follows from the real case. For the proof of (6.1), note that

$$\begin{aligned}\left|\int_a^b [f_n(t) - f(t)]dt\right| &\leq \int_a^b |f_n(t) - f(t)|\,dt \\ &\leq \int_a^b M_n dt \leq M_n(b-a) < \epsilon(b-a)\end{aligned}$$

if n is sufficiently large. □

Example 6.1. *Let f be a continuous complex valued function on $[0,1]$, and define the function g by $g(z) = \int_0^1 f(t)e^{tz}dt, z \in \mathbb{C}$. Then g is holomorphic in the entire complex plane.*

Solution 1: Let $z \in \mathbb{C}$. We have $g(z) = \int_0^1 \sum_{n=0}^{\infty} f(t)\frac{t^n z^n}{n!}dt$. Since f is bounded, this series converges uniformly in t, so we can change the order of summation and get $g(z) = \sum_{n=0}^{\infty}(\frac{1}{n!}\int_0^1 f(t)t^n dt)z^n = \sum_{n=0}^{\infty} \zeta_n z^n$ where $\zeta_n = \frac{1}{n!}\int_0^1 f(t)t^n dt$. We have $|\zeta_n| \leq \frac{1}{n!}\int_0^1 |f(t)|dt$ so the radius of convergence of the series of g is ∞.

Solution 2: Let $z_0 \in \mathbb{C}$. We have

$$\frac{g(z) - g(z_0)}{z - z_0} = \int_0^1 f(t)\frac{e^{tz} - e^{tz_0}}{z - z_0}dt.$$

From the power series expansion $e^{t(z-z_0)} = \sum_{n=0}^{\infty} \frac{t^n(z-z_0)^n}{n!}$, one gets

$$\frac{e^{tz} - e^{tz_0}}{z - z_0} = te^{tz_0} + O(z - z_0)$$

uniformly on $0 \leq t \leq 1$, when $z \to z_0$. Thus, as $z \to z_0$, the integrand in the integral above converges uniformly on $[0,1]$ to $tf(t)e^{tz_0}$, and one can pass to the limit under the integral sign to get $\lim_{z\to z_0} \frac{g(z)-g(z_0)}{z-z_0} = \int_0^1 tf(t)e^{tz_0}dt$, proving that g is differentiable at z_0.

Example 6.2. *Let f be a continuous real valued function on $[0,1]$, and let the function h in the complex plane be defined by $h(z) = \int_0^1 f(t)\cos(zt)dt$. Then h is holomorphic in the entire plane and h is the zero function only if f is the zero function.*

Solution: The Maclaurin series for $\cos z$ is $\sum_{n=0}^{\infty} \frac{(-1)^n z^{2n}}{(2n)!}$, and it converges uniformly on compact sets. Hence, for fixed z,

$$f(t)\cos(zt) = \sum_{n=0}^{\infty} \frac{(-1)^n f(t)t^{2n}z^{2n}}{(2n)!}$$

with the series converging uniformly on $[0,1]$. We can therefore, interchange the order of integration and summation to get

$$h(z) = \sum_{n=0}^{\infty} \frac{(-1)^n}{(2n)!}\left(\int_0^1 t^{2n}f(t)dt\right)z^{2n}.$$

In other words, h has the power series representation $h(z) = \sum_{n=0}^{\infty} c_{2n}z^{2n}$ with $c_{2n} = \frac{(-1)^n}{(2n)!}\int_0^1 t^{2n}f(t)dt$. Since h is given by a convergent power series, it is holomorphic. Now suppose h is the zero function. By the first part, $\int_0^1 t^{2n}f(t)dt = 0$ for $n = 0,1,2,\ldots$. Hence, if p is any polynomial, then $\int_0^1 p(t^2)f(t)dt = 0$. By the Stone-Weierstrass approximation theorem, there is a sequence (p_k) of polynomials such that $p_k(t) \to f(\sqrt{t})$ uniformly on $[0,1]$. Then $p_k(t^2) \to f(t)$ uniformly on $[0,1]$, so

$$\int_0^1 (f(t))^2dt = \lim_{k\to\infty}\int_0^1 p_k(t^2)f(t)dt = 0$$

implying that $f \equiv 0$.

Theorem 6.2. *(Leibniz rule): Let $f : [a,b]\times[c,d] \to \mathbb{C}$ be a continuous function. Assume that $\frac{\partial f}{\partial x}(t,x)$ exists for all $(t,x) \in [a,b]\times[c,d]$ and is continuous. Then we have*

$$\frac{d}{dx}\left(\int_a^b f(t,x)dt\right) = \int_a^b \frac{\partial f}{\partial x}(t,x)dt.$$

Proof. Without loss of generality we may assume that f is real-valued (since what we are doing can be done for real and imaginary parts separately). Let $g(x) = \int_a^b f(t,x)dt$. Consider

$$\begin{aligned}\left|\frac{g(x+h)-g(x)}{h} - \int_a^b \frac{\partial f}{\partial x}(t,x)dt\right| &= \left|\int_a^b \left[\frac{f(t,x+h)-f(t,x)}{h} - \frac{\partial f}{\partial x}(t,x)\right]dt\right| \\ &= \left|\int_a^b \left[\frac{\partial f}{\partial x}(t,c_{t,h}) - \frac{\partial f}{\partial x}(t,x)\right]dt\right|\end{aligned}$$

for some $c_{t,h} \in (x, x+h)$ by mean value theorem. Since $\frac{\partial f}{\partial x}$ is continuous, and $[a,b] \times [c,d]$ is compact hence $\frac{\partial f}{\partial x}$ is uniformly continuous on $[a,b] \times [c,d]$. Thus given $\epsilon > 0$, we choose $\delta > 0$ such that

$$\left|\frac{\partial f}{\partial x}(t, x_0) - \frac{\partial f}{\partial x}(t, x)\right| < \frac{\epsilon}{b-a},$$

if $|t - t_C| < \delta$ and $|x - x_0| < \delta$. Then for $|h| < \delta$, we have

$$\left|\frac{g(x+h) - g(x)}{h} - \int_a^b \frac{\partial f}{\partial x}(t, x)dt\right|$$

$$\leq \int_a^b \left|\frac{\partial f}{\partial x}(t, c_{t,h}) - \frac{\partial f}{\partial x}(t, x)\right| dt \leq \frac{\epsilon}{b-a}(b-a) = \epsilon.$$

□

Example 6.3. *We shall show that* $\int_0^{2\pi} \frac{e^{it}}{e^{it}-z}dt = 2\pi$ *if* $|z| < 1$. *Let* $f(t,x) = \frac{e^{it}}{e^{it}-xz}$ *for* $0 \leq x \leq 1, 0 \leq t \leq 2\pi$. *The function* f *is continuously differentiable in* $|z| < 1$. *Therefore* $\phi(x) = \int_0^{2\pi} f(t,x)dt$ *is continuously differentiable and* $\phi(0) = 2\pi$. *If it can be shown now that* ϕ *is a constant then we shall obtain* $\phi(1) = 2\pi$. *By Lebniz's rule,* $\phi'(x) = \int_0^{2\pi} \frac{ze^{it}}{(e^{it}-xz)^2}dt$. *But for* x *fixed, if* $\psi(t) = \frac{zi}{e^{it}-xz}$ *then* $\psi'(t) = \frac{-zi(ie^{it})}{(e^{it}-xz)^2} = \frac{ze^{it}}{(e^{it}-xz)^2}$. *Hence* $\phi'(x) = \psi(2\pi) - \psi(0) = 0$, *so* ϕ *must be a constant.*

6.2 Riemann-Stieltjes integral

Definition 6.3. *A function* $\gamma : [a,b] \longrightarrow \mathbb{C}$ *is of bounded variation if there is a constant* $M > 0$ *such that for any partition* $P = \{a = t_0 < t_1 < \cdots < t_n = b\}$ *of* $[a,b]$

$$v(\gamma, P) = \sum_{k=1}^{m} |\gamma(t_k) - \gamma(t_{k-1})| \leq M.$$

The total variation of $\gamma, V(\gamma)$, *is defined by*

$$V(\gamma) = \sup\{v(\gamma, P) : P \text{ a partition of } [a,b]\}.$$

Notice that $V(\gamma) \leq M < \infty$.

Theorem 6.3. *The following holds:*

(1) The function $\gamma : [a,b] \longrightarrow \mathbb{C}$ *is of bounded variation if and only if* $Re\gamma$ *and* $Im\gamma$ *are of bounded variation.*

(2) If $\gamma : [a,b] \longrightarrow \mathbb{R}$ *is non-decreasing then* γ *is of bounded variation and* $V(\gamma) = \gamma(b) - \gamma(a)$.

(3) If $\gamma, \sigma : [a,b] \longrightarrow \mathbb{C}$ are of bounded variation, $\alpha, \beta \in \mathbb{C}$ then $\alpha\gamma + \beta\sigma$ are of bounded variation.

Proof. The proof follows from the definition of a function of bounded variation. □

Proposition 6.5. *If $\gamma : [a,b] \longrightarrow \mathbb{C}$ is piecewise smooth then γ is of bounded variation and $V(\gamma) = \int_a^b |\gamma'(t)|dt$.*

Proof. Without loss of generality assume that γ is smooth. Hence γ' is continuous. Let $P = \{a = t_0 < t_1 < \cdots < t_m = b\}$ is a partition of $[a,b]$. Then

$$\begin{aligned} v(\gamma, P) &= \textstyle\sum_{k=1}^m |\gamma(t_k) - \gamma(t_{k-1})| \\ &= \textstyle\sum_{k=1}^m |\int_{t_{k-1}}^{t_k} \gamma'(t)dt| \\ &\leq \textstyle\sum_{k=1}^m \int_{t_{k-1}}^{t_k} |\gamma'(t)|dt = \int_a^b |\gamma'(t)|dt. \end{aligned}$$

Hence

$$V(\gamma) \leq \int_a^b |\gamma'(t)|dt. \tag{6.2}$$

The closed interval $[a,b]$ is compact, γ' is continuous on $[a,b]$. Hence γ' is uniformly continuous on $[a,b]$. So given $\epsilon > 0$, there exists $\delta_1 > 0$ such that $|s-t| < \delta_1$ implies $|\gamma'(s) - \gamma'(t)| < \epsilon$. Also from the definition of Riemann integration it follows that there exists $\delta_2 > 0$ such that if $P = \{a = t_0 < t_1 < \cdots < t_m = b\}$ and $\|P\| = \max\limits_{1\leq k\leq m} |t_k - t_{k-1}| < \delta_2$ then

$$\left| \int_a^b |\gamma'(t)|dt - \sum_{k=1}^m |\gamma'(\tau_k)|(t_k - t_{k-1}) \right| < \epsilon$$

where τ_k is any point in $[t_{k-1}, t_k]$. Hence

$$\begin{aligned} \textstyle\int_a^b |\gamma'(t)|dt &\leq \epsilon + \textstyle\sum_{k=1}^m |\gamma'(\tau_k)|(t_k - t_{k-1}) \\ &= \epsilon + \textstyle\sum_{k=1}^m \left| \int_{t_{k-1}}^{t_k} \gamma'(\tau_k)dt \right| \\ &\leq \epsilon + \textstyle\sum_{k=1}^m \left| \int_{t_{k-1}}^{t_k} [\gamma'(\tau_k) - \gamma'(t)]\, dt \right| + \sum_{k=1}^m \left| \int_{t_{k-1}}^{t_k} \gamma'(t)dt \right|. \end{aligned}$$

If $\|P\| < \delta = \min\{\delta_1, \delta_2\}$, then $|\gamma'(\tau_k) - \gamma'(t)| < \epsilon$ for t in $[t_{k-1}, t_k]$ and

$$\begin{aligned} \textstyle\int_a^b |\gamma'(t)|dt &\leq \epsilon + \epsilon(b-a) + \textstyle\sum_{k=1}^m |\gamma(t_k) - \gamma(t_{k-1})| \\ &= \epsilon + \epsilon(b-a) + v(\gamma, P) \leq \epsilon + \epsilon(b-a) + V(\gamma). \end{aligned}$$

Letting $\epsilon \longrightarrow 0^+$, we obtain

$$\int_a^b |\gamma'(t)|dt \leq V(\gamma). \tag{6.3}$$

From (6.2) and (6.3), the result follows. □

Definition 6.4. *Let $\gamma : [a, b] \longrightarrow \mathbb{C}$ be a function of bounded variation and suppose that $f : [a, b] \longrightarrow \mathbb{C}$ is continuous. If there is a complex number I such that for every $\epsilon > 0$ there is a $\delta > 0$ such that when $P = \{t_0 < t_1 < \cdots < t_m\}$ is a partition of $[a, b]$ with $\|P\| = \max_{1 \le k \le m} (t_k - t_{k-1}) < \delta$ then*

$$\left| I - \sum_{k=1}^{m} f(\tau_k)[\gamma(t_k) - \gamma(t_{k-1})] \right| < \epsilon$$

for whatever choice of points $\tau_k, t_{k-1} \le \tau_k \le t_k$ then we shall say the Riemann-Stieltjes integral of f with respect to γ over $[a, b]$ exists and we write

$$I = \int_a^b f d\gamma = \int_a^b f(t) d\gamma(t).$$

The number I in the theorem above is called the integral of f with respect to γ over $[a, b]$. In fact one can show that with this background a complex number I always exists. That is, if $f : [a, b] \longrightarrow \mathbb{C}$ is continuous and $\gamma : [a, b] \longrightarrow \mathbb{C}$ be a function of bounded variation then f is Riemann-Stieltjes integrable with respect to γ.

The next result follows from Definition 6.4.

Proposition 6.6. *Let f and g be continuous on $[a, b]$ and let γ and σ be functions of bounded variation on $[a, b]$. Then for $\alpha, \beta \in \mathbb{C}$,*

(1) $\int_a^b (\alpha f + \beta g) d\gamma = \alpha \int_a^b f d\gamma + \beta \int_a^b g d\gamma.$

(2) $\int_a^b f d(\alpha\gamma + \beta\sigma) = \alpha \int_a^b f d\gamma + \beta \int_a^b f d\sigma$

(3) $\int_a^b f d\gamma = \sum_{k=1}^{n} \int_{t_{k-1}}^{t_k} f d\gamma$ *if* $a = t_0 < t_1 < \cdots < t_n = b.$

Definition 6.5. *A continuous function $\gamma : [a, b] \longrightarrow \mathbb{C}$ is said to be rectifiable if γ is a function of bounded variation. To say that γ is rectifiable is to say that γ has finite length and its length $L(\gamma) = V(\gamma)$. In Proposition 6.5 it is shown that if γ is piecewise smooth then γ is rectifiable and its length is $\int_a^b |\gamma'(t)| \, dt$.*

6.3 Path integrals

Definition 6.6. *(Curve): Let G be a region in $\mathbb{C}$. Given points z and w in G, a curve in G from z to w is a continuous map $\gamma : [a, b] \to G$ of some closed interval in the real line into G, such that $\gamma(a) = z$ and $\gamma(b) = w$.*

A curve $\gamma : [a, b] \longrightarrow G$ is said to be closed if $\gamma(a) = \gamma(b)$. A curve γ in G is said to be piecewise smooth on $[a, b]$ if γ is piecewise C^1. Henceforth a path in G is a piecewise smooth map from an interval $[a, b]$ to G. The curve γ is called a smooth path if γ is continuously differentiable on its domain. The word "smooth" implies that the motion of a point which traces the curve has *no abrupt change in direction. We also say that* $\gamma(a)$*is the initial point and* $\gamma(b)$ *the final (terminal) point, of the path* γ*. We say that* γ *is a closed path if the initial and terminal points are the same.*

If $\gamma : [a, b] \to G$ is a path, we let $[\gamma]$ denote its image: $[\gamma] = \{\gamma(t) : a \le t \le b\}$ and $[\gamma]$ is called the trace of the curve γ and is not to be confused with the curve. Below are some examples of paths.

Example 6.4. *Let* G *be a region and* $p \in G$*. Define* $\gamma : [a, b] \to G$ *such that* $\gamma(t) = p$ *for* $a \le t \le b$*. Note* γ *is continuous and it is called a constant path in* G *or null point in* G*.*

Example 6.5. *Consider the open ball* $B(0, r) \subseteq \mathbb{C}$ *and let* $p, q \in B(0, r), p \neq q$*. Define* $\gamma : [0, 1] \to B(0, r)$ *such that* $\gamma(t) = (1 - t)p + tq, 0 \le t \le 1$*. Then* γ *is a path from* p *to* q *in* $B(0, r)$*. We denote this path by* $[p, q]$*. In fact the path* γ *is the line segment from* p *to* q*, the direction of the parametrization being from* p *to* q*.*

Example 6.6. *Let* $a \in G$ *and* $r > 0$ *be such that* $\overline{B(0, r)} \subset G$ *where* G *is a region. Then define* $\gamma : [0, 2\pi] \to G$ *such that* $\gamma(t) = a + re^{it}$*,* $0 \le t \le 2\pi$*. Then* γ *is a closed path as* γ *is continuous and* $\gamma(0) = \gamma(2\pi)$*.* γ *is continuously differentiable in* $[0, 2\pi]$*. We denote this path by* $S(a, r)$ *and it is a parametrization of the circle with center at* a *and radius* r *(called the circular path), the circle being traversed once in the counterclockwise direction as* t *moves from the initial to the terminal point of the parameter interval* $[0, 2\pi]$*.*

Example 6.7. *Let the notation be as above, but, let* $\sigma(t) = a + re^{-it}$ *and* $\tau(t) = a + re^{2it}$*. The images of* σ*,* τ *and* γ *are the same, namely, the circle centered at* a *and radius* r*. But, however, these maps are different and hence as paths they are different. The path* σ *describes a circle which is traversed once in the clockwise direction and the path* τ *describes a circle which is traversed twice in the clockwise direction.*

Example 6.8. *(i) Let*

$$\gamma(t) = \begin{cases} t & \text{if } 0 \le t \le 1; \\ 1 + i(t-1) & \text{if } 1 \le t \le 2; \\ i + (3-t) & \text{if } 2 \le t \le 3; \\ (4-t)i & \text{if } 3 \le t \le 4. \end{cases}$$

Then γ *is piecewise smooth but not a smooth path.*

(ii) Let $G = \mathbb{C}$ *and*

$$\gamma(t) = \begin{cases} t & \text{if } 0 \le t \le 1; \\ 1 + i(t-1) & \text{if } 1 \le t \le 2. \end{cases}$$

Then γ *is a piecewise smooth path but not a smooth path.*

Example 6.9. *(i) Let $\gamma(t) = t^3 + i|t|^2$ on $[-1, 1]$. This is a parametrization of the square with vertices $0, 1, 1+i, i$. The square is traversed once in the counterclockwise direction and γ is not smooth.*

(ii) Consider the map

$$\gamma(t) = \begin{cases} e^{i(1-t)\pi} & \text{if } 0 \leq t \leq 1; \\ t & \text{if } 1 \leq t \leq 2. \end{cases}$$

γ is piecewise smooth but not a smooth path.

Example 6.10. *Let $\gamma_k : [a_k, b_k] \to \mathbb{C}$ be paths such that $\gamma_1(b_1) = \gamma_2(a_2)$, $k = 1, 2$. That is, the terminal point of γ_1 is the initial point of γ_2. we define a new path $\gamma : [a_1, b_1 + (b_2 - a_2)] \to \mathbb{C}$ by setting*

$$\gamma(t) = \begin{cases} \gamma_1(t) & \text{if } t \in [a_1, b_1]; \\ \gamma_2(a_2 + t - b_1) & \text{if } t \in [b_1, b_1 + (b_2 - a_2)]. \end{cases}$$

In this case we write $\gamma = \gamma_1 + \gamma_2$.

Definition 6.7. *Let $G \subseteq \mathbb{C}$ be a region and let γ be a smooth path in G. Let $f : [\gamma] \to \mathbb{C}$ be continuous. We define the line integral or path integral of f over γ by setting*

$$\int_\gamma f(z)dz = \int_a^b f(\gamma(t))\gamma'(t)dt. \tag{6.4}$$

Now suppose $\gamma : [a, b] \to G$ is a piecewise smooth path. That is γ is continuous and there exist a finite number of points $a = t_0 < t_1 < \cdots < t_n = b$ such that γ is continuously differentiable on the closed intervals $[t_j, t_{j+1}]$ for $0 \leq j \leq n-1$. (The right and left side derivatives at t_j may not be the same but they are all finite). Using the above notation, we set

$$\int_\gamma f(z)dz = \sum_{j=0}^{n-1} \int_{t_j}^{t_{j+1}} f(\gamma(t))\gamma'(t)dt.$$

Note that the integrals on the RHS make sense, since γ' exists and is continuous on $[t_j, t_{j+1}]$. The value $\int_\gamma f$ is called the integral (line integral) of f over the path γ.

Note also that if $\gamma(t) = t, t \in [a, b]$, then $\int_\gamma f(z)dz = \int_a^b f(t)dt$.

Example 6.11. *(i) We shall compute $\int_\gamma \bar{\mathbf{z}}dz$ from $z = 0$ to $z = 4 + 2i$ along the curve γ given by $z = \gamma(t) = t^2 + it$. The points $z = 0$ and $z = 4 + 2i$ on γ correspond to $t = 0$ and $t = 2$ respectively. Then $\int_\gamma \bar{\mathbf{z}}dz$ $= \int_{t=0}^2 \overline{(t^2 + it)}(2t + i)dt = 10 - \frac{8i}{3}$.*

(ii) We compute $\int_\gamma cos z dz$ where γ is the line segment from $-\frac{\pi}{2} + i$ to $\pi + i$. We have $\gamma(t) = (1-t)[i - \frac{\pi}{2}] + t(i + \pi) = \frac{3\pi}{2}t - \frac{\pi}{2} + i$. Also, $cos(x + iy) = cos x \cdot cosh y - i sin x \cdot sinh y$. Hence,

$$\begin{aligned}\int_\gamma cos z dz &= \int_0^1 \left[cos\left(\tfrac{3\pi t}{2}-\tfrac{\pi}{2}\right)cosh(1) - isin\left(\tfrac{3\pi t}{2}-\tfrac{\pi}{2}\right)sinh(1)\right]\tfrac{3\pi}{2}dt \\ &= cosh(1)sin\left(\tfrac{3\pi t}{2}-\tfrac{\pi}{2}\right) + isinh(1)cos\left(\tfrac{3\pi t}{2}-\tfrac{\pi}{2}\right) \\ &= cosh(1) - isinh(1).\end{aligned}$$

Theorem 6.4. *Let G be a region and $\gamma : [a,b] \to G$ is a path in G. Suppose $f : G \to \mathbb{C}$ is continuous and that there exists a function $F : G \to \mathbb{C}$ such that $F' = f$(the function F is called the primitive of f in G) on G. Then $\int_\gamma f(z)dz = F(\gamma(b)) - F(\gamma(a))$.*

Proof. Using chain rule and fundamental theorem of calculus, we get

$$\begin{aligned}\int_\gamma f(z)dz &= \int_a^b f(\gamma(t))\gamma'(t)dt \\ &= \int_a^b F'(\gamma(t))\gamma'(t)dt \\ &= \int_a^b (F\circ\gamma)'(t)dt \\ &= F(\gamma(b)) - F(\gamma(a)).\end{aligned}$$

□

Corollary 6.2. *If $f : G \to \mathbb{C}$ is continuous and has a primitive in G, then $\int_\gamma f = 0$ for any closed path γ in G.*

Example 6.12. *Compute $\int_\gamma (z-a)^n dz$, where $\gamma(t) = a + re^{it}, 0 \le t \le 2\pi$. Here $\gamma'(t) = rie^{it}$ and $\int_\gamma (z-a)^n dz = \int_0^{2\pi}(\gamma(t)-a)^n\gamma'(t)dt = \int_0^{2\pi}(re^{it})^n ire^{it}dt$ $= r^{n+1}\int_0^{2\pi} ie^{i(n+1)t}dt$. Since $\frac{1}{n+1}e^{i(n+1)t}$ is a primitive of $ie^{i(n+1)t}$ if $n \neq -1$, hence we get*

$$\int_\gamma (z-a)^n dz = \begin{cases} 0 & if \;\; n \neq -1; \\ 2\pi i & if \;\; n = -1. \end{cases}$$

Example 6.13. *We shall evaluate $\int_\gamma \frac{dz}{z-w}$ where γ is the boundary of a square with corners $w \pm l \pm li$. Thus $\gamma = \gamma_1 + \gamma_2 + \gamma_3 + \gamma_4$ where $\gamma_1, \gamma_2, \gamma_3, \gamma_4$ are the lines as in Fig. 6.1. So $\int_\gamma \frac{dz}{z-w} = \sum_{i=1}^4 \int_{\gamma_i}\frac{dz}{z-w}$. First we shall compute $\int_{\gamma_1}\frac{dz}{z-w}$ where $\gamma_1(t) = w + l + it, -l \le t \le l$.*

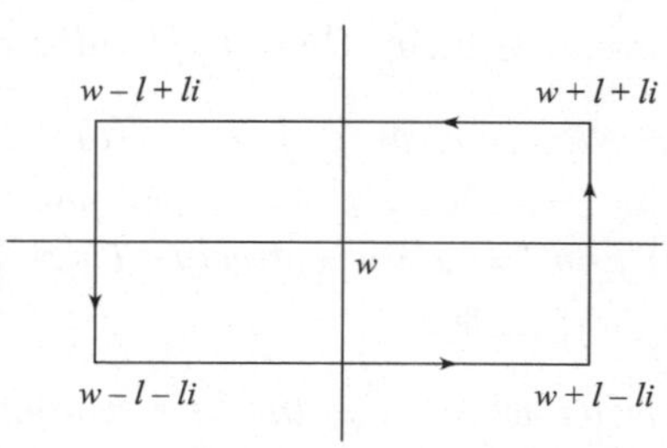

Fig. 6.1: The square with corners $w \pm l \pm li$

$$\begin{aligned}
\int_{\gamma_1} \frac{dz}{z-w} &= \int_{-l}^{l} \frac{\gamma_1'(t)}{\gamma_1(t)-w} dt &&= \int_{-l}^{l} \frac{i}{l+it} dt \\
&= \int_{-l}^{l} \frac{i(l-it)}{l^2+t^2} idt &&= \int_{-l}^{l} \frac{t}{l^2+t^2} dt + i\int_{-l}^{l} \frac{l}{l^2+t^2} dt \\
&= 0 + i\left[\tan^{-1} \tfrac{t}{l}\right]_{-l}^{l} &&= i\tfrac{\pi}{2}.
\end{aligned}$$

Similarly, one can show that $\int_{\gamma_i} \frac{dz}{z-w} = i\frac{\pi}{2}$ *for* $i = 2, 3, 4$. *Thus* $\int_{\gamma} \frac{dz}{z-w} = 2\pi i$.

Definition 6.8. *Let* $\gamma : [a, b] \to \mathbb{C}$ *be a path. Then* $\widetilde{\gamma} : [a, b] \to \mathbb{C}$ *defined by* $\widetilde{\gamma}(t) = \gamma(a + b - t)$ *is said to be the reverse path or opposite path of* γ *and its initial point is* $\gamma(b)$ *and final point is* $\gamma(a)$ *(i.e.,* $\widetilde{\gamma}$ *is a path from* $\gamma(b)$ *to* $\gamma(a)$*).*

If $\gamma : [a, b] \to \mathbb{C}$ is a rectifiable path then denote by $-\gamma$ the path defined by $(-\gamma)(t) = \gamma(-t)$ for $-b \leq t \leq -a$. Notice also that γ and $-\gamma$ have the same but opposite end points. The path $\widetilde{\gamma}$ is just a reparametrization of $-\gamma$ (see definition 6.9). In fact $\widetilde{\gamma}(t) = (-\gamma)(t-b-a) = \gamma(a+b-t), a \leq t \leq b$. Another notation for this is γ^{-1}. So we shall call the paths $\widetilde{\gamma}, -\gamma, \gamma^{-1}$ as the reverse path of γ. Also if $c \in \mathbb{C}$, let $\gamma + c$ denote the curve defined by $(\gamma + c)(t) = \gamma(t) + c$.

Example 6.14. *If* $\gamma : [0, 1] \to \mathbb{C}$ *is defined by* $\gamma(t) = (1 - t)z + tw$ *then* $\widetilde{\gamma}(t) = \gamma(1 - t), 0 \leq t \leq 1$ *is the line segment* $[w, z]$ *from* w *to* z. *Here* $\widetilde{\gamma}$ *is the reverse path of* γ *in* $\mathbb{C}$.

Example 6.15. *Let* $\gamma(t) = a + re^{it}, 0 \leq t \leq 2\pi$. *Then* γ *is the circular path from* $a+r$ *to* $a+r$ *in the counter-clockwise direction. Its opposite path is given by* $\widetilde{\gamma}(t) = a + re^{-it}, 0 \leq t \leq 2\pi$, *which is the circle traversed in the clockwise direction.*

Proposition 6.7. *Let* γ *be a path and let* $f : [\gamma] \to \mathbb{C}$ *be a continuous function. Then*

(i) $\int_{\widetilde{\gamma}} f dz = -\int_{\gamma} f dz$

(ii) $\int_{\gamma} f = -\int_{-\gamma} f$

(iii) *If* $c \in \mathbb{C}$ *then* $\int_{\gamma} f(z)dz = \int_{\gamma+c} f(z - c)dz$.

Proof. We note first that $\widetilde{\gamma}'(t) = -\gamma'(a + b - t)$. We compute $\int_{\widetilde{\gamma}} f dz = \int_a^b f(\widetilde{\gamma}(t))\widetilde{\gamma}'(t)dt = \int_a^b f(\gamma(a+b-t))(-\gamma'(a+b-t))dt = \int_b^a f(\gamma(u))\gamma'(u)du = -\int_{\gamma} f dz$ where $u = a + b - t$. □

Definition 6.9. *Let* $\gamma : [a, b] \to \mathbb{C}$ *and* $\sigma : [c, d] \to \mathbb{C}$ *be paths. We say that* σ *is a reparametrization of* γ *if there exists a bijective continuously differentiable function* $h : [c, d] \to [a, b]$ *such that* $h' > 0$ *and* $\sigma(s) = (\gamma \circ h)(s)$ *for* $s \in [c, d]$.

Theorem 6.5. *If the path $\sigma : [c, d] \to \mathbb{C}$ is reparametrization of the path $\gamma : [a, b] \to \mathbb{C}$ as above then $[\gamma] = [\sigma]$. Further if $f : [\gamma] \to \mathbb{C}$ is continuous, then $\int_\sigma f = \int_\gamma f$.*

Proof. Note that the set $\{\gamma(t) : a \leq t \leq b\} = \{(\gamma \circ h)(s) : c \leq s \leq d\}$ as h is bijective and increasing from $[c, d]$ onto $[a, b]$. Now $\int_\gamma f(z)dz = \int_a^b f(\gamma(t))\gamma'(t)dt$ and $\int_\sigma f(z)dz = \int_c^d f(\gamma(h(s)))\gamma'(h(s))h'(s)ds$.

The two integrals on the right side are equal by the change of variables formula. Thus, the curve γ and its reparametrizations are indistinguishable for purposes of integration. □

If $\gamma_1 : [a, b] \longrightarrow \mathbb{C}$ and $\gamma_2 : [c, d] \longrightarrow \mathbb{C}$ are two continuous functions such that $\gamma_1(b) = \gamma_2(c)$ then we can reparametrize γ_2 as $\widehat{\gamma_2} : [b, d + b - c] \longrightarrow \mathbb{C}$ by $\widehat{\gamma_2}(s) = \gamma_2(c - b + s)$. Then $\gamma_1(b) = \widehat{\gamma_2}(b)$. So we can form $\gamma_1 + \widehat{\gamma_2}$, often written simply as $\gamma_1 + \gamma_2$.

Suppose γ and σ are paths defined on $[0, 1]$ with $\gamma(1) = \sigma(0)$. Then $\gamma + \sigma$ is the curve $(\gamma + \sigma)(t) = \gamma(2t)$ if $0 \leq t \leq \frac{1}{2}$ and $(\gamma + \sigma)(t) = \sigma(1 - 2t)$ if $\frac{1}{2} \leq t \leq 1$.

Proposition 6.8. *Let G be a region and let $f : G \to \mathbb{C}$ be a continuous function with a primitive F in G. Let γ_1 and γ_2 be two paths in G having the same initial and terminal points z and w respectively. Then $\int_{\gamma_1} f = F(w) - F(z) = \int_{\gamma_2} f$. Thus, in this case $\int_\sigma f$ depends only on the end points of σ for any path σ in G.*

Proof. Let $\gamma = \gamma_1 + \widetilde{\gamma_2}$. Since γ is closed, by Corollary 6.2, $\int_\gamma f = 0$. Hence $\int_\gamma f = \int_{\gamma_1} f - \int_{\gamma_2} f$. □

Definition 6.10. *If $\gamma : [a, b] \to \mathbb{C}$ is a smooth path, then the length $L(\gamma)$ is defined by $L(\gamma) = \int_a^b |\gamma'(t)|dt$. If γ is a (piecewise smooth) path we define $L(\gamma) = \sum_{j=0}^{n-1} \int_{t_j}^{t_{j+1}} |\gamma'(t)|dt$ where $t_0, t_1, \cdots, t_n$ are the finitely many points where the path γ is not differentiable. Note since γ is piecewise smooth, hence $L(\gamma) < \infty$ and γ is rectifiable.*

Example 6.16. *If γ is the line segment $[z, w]$, then $L(\gamma) = |w - z|$.*

Example 6.17. *If γ is the circular path $\gamma(t) = a + re^{it}, 0 \leq t \leq 2\pi$ then $L(\gamma) = 2\pi r$.*

Note that if the path σ is a reparametrization of the path γ then also $L(\gamma) = L(\sigma)$, i.e., $L(\gamma)$ remains invariant under reparametrization. This follows from Theorem 6.5 . Further if $\gamma = \gamma_1 + \gamma_2$ then $L(\gamma) = L(\gamma_1) + L(\gamma_2)$.

Theorem 6.6. *(Estimation theorem): Let $\gamma : [a, b] \to \mathbb{C}$ be a path. Suppose $f : [\gamma] \to \mathbb{C}$ be continuous and there exist a constant $M > 0$ such that $|f(\gamma(t))| \leq M$ for all $t \in [a, b]$. Then $\left|\int_\gamma f(z)dz\right| \leq ML(\gamma)$.*

Proof. The inequality in Theorem 6.6 is sometimes referred to as **ML inequality** and it can be verified as follows:

$$\begin{aligned}\left|\int_\gamma f(z)dz\right| &\leq \left|\int_a^b f(\gamma(t))\gamma'(t)dt\right| \\ &\leq \int_a^b |f(\gamma(t))|\,|\gamma'(t)|\,dt \\ &\leq M\int_a^b |\gamma'(t)|\,dt \\ &= ML(\gamma).\end{aligned}$$

□

Proposition 6.9. *Let G be a region and let $f : G \to \mathbb{C}$ be continuous. Then f has a primitive in G if and only if $\int_\gamma f = 0$ for any closed path γ in G.*

Proof. One way is already proved in Corollary 6.2. For the converse, let $z_0 \in G$ be fixed. We define $F(z) = \int_\gamma f(z)dz$ where γ is any path from z_0 to z. Since G is open and connected therefore G is path connected and so there exist paths from z_0 to z in G. We claim that the value of the integral $\int_\gamma f(z)dz$ is independent of the choice of the path and hence well-defined. If σ is another such path from z_0 to z then let $\tau = \gamma + \tilde{\sigma}$. Now τ is a closed path and hence $\int_\tau f(z)dz = 0$. Hence $\int_\gamma f(z)dz = \int_\sigma f(z)dz$. Now if $|h|$ is sufficiently small, then

$$\left|\frac{F(z+h)-F(z)}{h} - f(z)\right|$$

$$= \frac{1}{|h|}\left|\int_{[z,z+h]} [f(w)-f(z)]dw\right| \leq \frac{1}{|h|}M|h| = M$$

where $M = \max\{|f(w)-f(z)| : w \in [z, z+h]\}$.The right side of the inequality tends to zero as $|h| \to 0$ by continuity of f at z. □

Example 6.18. *Let $U = \mathbb{C}\setminus\{w\}$ and γ be the circle with center at w and radius r, $f(z) = (z-w)^n$, if $n \neq -1$. Then $f = F'$, with $F(z) = \frac{(z-w)^{n+1}}{n+1}$. So $\int_\gamma f(z)dz = 0$ by Proposition 6.9 . But if $n = -1$, then $\int_\gamma f(z)dz = 2\pi i$, so there is no analytic function F on $\mathbb{C}\setminus\{w\}$ with $F'(z) = \frac{1}{z-w}$.*

Example 6.19. *The Fundamental theorem of Calculus says that each continuous function has a primitive. This is far from being true for functions of a complex variable. For example, let $f(z) = |z|^2 = x^2+y^2 = u(x,y)+iv(x,y)$. If F is a primitive of f then F is holomorphic. So if $F = U+iV$ then $x^2+y^2 = F'(x+iy)$. Now, using the Cauchy-Riemann equations, $\frac{\partial U}{\partial x} = \frac{\partial V}{\partial y} = x^2+y^2$ and $\frac{\partial U}{\partial y} = \frac{\partial V}{\partial x} = 0$. But $\frac{\partial U}{\partial y} = 0$ implies that $U(x,y) = u(x)$ for some differentiable function u. But this gives $x^2+y^2 = \frac{\partial U}{\partial x} = u'(x)$, a contradiction.*

Proposition 6.10. *Let $\gamma : [a,b] \to \mathbb{C}$ be a path. Let $f_n : [\gamma] \to \mathbb{C}$ be continuous, $n \in \mathbb{N}$ and that f_n converge uniformly on $[\gamma]$ to a function $f : [\gamma] \to \mathbb{C}$. Then*

$$\lim_{n\to\infty}\int_\gamma f_n(z)dz = \int_\gamma f(z)dz.$$

Proof. The function f is continuous on $[\gamma]$ as the sequence $\{f_n\}$ converges uniformly to f and because each f_n is continuous on $[\gamma]$. Now let $\epsilon > 0$; then there is an integer N such that $|f_n(w) - f(w)| < \frac{\epsilon}{L(\gamma)}$ for all w on $[\gamma]$ and $n \geq N$. Hence

$$\begin{aligned} \left|\int_\gamma [f(z) - f_n(z)]dz\right| &= \left|\int_a^b [f(\gamma(t)) - f_n(\gamma(t))]\gamma'(t)dt\right| \\ &\leq \frac{\epsilon}{L(\gamma)} \int_a^b |\gamma'(t)|\, dt \\ &< \frac{\epsilon}{L(\gamma)} L(\gamma) \\ &= \epsilon \end{aligned}$$

if n is sufficiently large. □

Corollary 6.3. *(Termwise integration of series): Let $U \subset \mathbb{C}$ be open, $\gamma : [a, b] \longrightarrow \mathbb{C}$ be a path. Suppose $f_k, k \geq 0$ and g are continuous functions from U into $\mathbb{C}$.*

(i) If $\sum_{k=0}^{\infty} f_k(z)$ is uniformly convergent on U to some function $g(z)$ on U then $\int_\gamma g(z)dz = \int_\gamma (\sum_{k=0}^{\infty} f_k(z))dz = \sum_{k=0}^{\infty} \int_\gamma f_k(z)dz$.

(ii) If for each $z \in [\gamma]$, $\sum_{k=0}^{\infty} f_k(z)$ converges to $g(z)$ and there exist constants M_k such that $|f_k(z)| \leq M_k$ for all $z \in [\gamma]$ and $\sum_{k=0}^{\infty} M_k < \infty$ then $\sum_{k=0}^{\infty}(\int_\gamma f_k(z)dz)$ converges to $\int_\gamma g(z)dz$.

Proof. For proof of (i), let $s_n(z) = \sum_{k=0}^{n} f_k(z)$. Then $s_n(z)$ is a continuous function for all $n = 1, 2, \ldots$. By Definition 3.12, $\{s_n(z)\}$ converges to $g(z)$ uniformly on U. By Proposition 6.10, we obtain

$$\int_\gamma (\sum_{k=0}^{\infty} f_k(z))dz = \int_\gamma g(z)dz = \lim_{n\to\infty} \int_\gamma s_n(z)dz = \sum_{k=0}^{\infty} \int_\gamma f_k(z)dz.$$

The proof of (ii) follows from the Weierstrass M test and from (i) above. Without assuming these, one can also simply verify as follows. Let $M = \sum_{k=0}^{\infty} M_k$, then

$$\begin{aligned} \left|g(z) - \sum_{k=0}^{n} f_k(z)\right| &= |\textstyle\sum_{k=n+1}^{\infty} f_k(z)| \\ &\leq \sum_{k=n+1}^{\infty} |f_k(z)| \\ &\leq \sum_{k=n+1}^{\infty} M_k \\ &= M - \sum_{k=0}^{n} M_k. \end{aligned}$$

Hence

$$\begin{aligned}\left|\int_\gamma g(z)dz - \sum_{k=0}^{n}\int_\gamma f_k(z)dz\right| &= \left|\int_\gamma (g(z) - \sum_{k=0}^{n} f_k(z))dz\right| \\ &\le (M - \sum_{k=0}^{n} M_k)L(\gamma) \to 0\end{aligned}$$

as $n \to \infty$. That is,

$$\sum_{k=0}^{\infty}\int_\gamma f_k(z)dz = \int_\gamma g(z)dz.$$

□

Let $\gamma : [a, b] \to \mathbb{C}$ be a piecewise C^1 curve. A complex-valued function f is said to be defined on γ if it is defined on a set that contains the range of γ (trace of γ). In that case, the composition function $f \circ \gamma$ is defined. In particular, if f is continuous on $[\gamma]$, then $f \circ \gamma$ is continuous on $[a, b]$.

The integral just defined has the linearity property. If f and g are two continuous functions defined on γ, and if c_1 and c_2 are complex constants, then $\int_\gamma (c_1 f(z) + c_2 g(z))dz = c_1 \int_\gamma f(z)dz + c_2 \int_\gamma g(z)dz$. Moreover, if $a < c < b$, and if $\gamma_1 = \gamma\mid_{[a,c]}$ and $\gamma_2 = \gamma\mid_{[c,b]}$, then $\int_\gamma f(z)dz = \int_{\gamma_1} f(z)dz + \int_{\gamma_2} f(z)dz$. This is a trivial consequence of the corresponding property of the Riemann integral. More generally, we can subdivide a curve γ into a finite number of subcurves. A subdivision can be indicated by a symbolic equation $\gamma = \gamma_1 + \gamma_2 + \cdots + \gamma_n$ and the corresponding integrals satisfy the relation

$$\int_{\gamma_1+\gamma_2+\cdots+\gamma_n} f(z)dz = \int_{\gamma_1} f(z)dz + \int_{\gamma_2} f(z)dz + \cdots + \int_{\gamma_n} f(z)dz.$$

Proposition 6.11. *Let $f(z) = \sum_{n=0}^{\infty} a_n(z - z_0)^n$ on $B(z_0, R)$. Then for any $0 < r < R$ and $\gamma(t) = z_0 + re^{it}, 0 \le t \le 2\pi$,*

$$\frac{1}{n!}f^{(n)}(z_0) = a_n = \frac{1}{2\pi i}\int_\gamma \frac{f(z)}{(z - z_0)^{n+1}}dz, n \in \mathbb{Z}_+.$$

Proof. The integral on the right side of the above equation is

$$\sum_{k=0}^{\infty} a_k \frac{1}{2\pi i}\int_\gamma (z - z_0)^{k-n-1}dz = a_n,$$

by Example 6.12. Since $a_n = \frac{f^{(n)}(z_0)}{n!}$, the result follows. □

Corollary 6.4. *Let $M(r) = \sup\{|f(z)| : z \in [\gamma]\}$. Then*

$$n!|a_n| = |f^{(n)}(z_0)| \le \frac{M(r)}{r^n}, (0 < r < R).$$

This inequality is called Cauchy's inequality.

Proof. The result follows from Proposition 6.3 and Proposition 6.11. □

Proposition 6.12. *Let $\gamma(t) = z_0 + re^{it}$ for $0 \le t \le 2\pi$. Then*

$$\int_\gamma \frac{dz}{(z-a)} = \begin{cases} 2\pi i & \text{if } a \in B(z_0,r); \\ 0 & \text{if } a \notin \overline{B(z_0,r)}. \end{cases}$$

Proof. The integrand $\frac{1}{z-a} = \frac{1}{(z-z_0)-(a-z_0)} = \frac{1}{(z-z_0)[1-\frac{a-z_0}{z-z_0}]} = \frac{1}{(z-z_0)}\frac{1}{(1-w)}$ where $w = \frac{a-z_0}{z-z_0}$.

If $a \in B(z_0,r)$, then $|w| < 1$ and we can substitute the geometric series expansion for $\frac{1}{(1-w)}$ in the integral. Since the convergence is uniform on $[\gamma]$ we have by Corollary 6.3 and Example 6.12, $\int_\gamma \frac{dz}{(z-a)} = \int_\gamma \frac{1}{(z-z_0)}\frac{1}{(1-w)}dz = \int_\gamma \sum_{n=0}^\infty \frac{w^n}{z-z_0}dz = \int_\gamma \sum_{n=0}^\infty \frac{(a-z_0)^n}{(z-z_0)^{n+1}}dz = \sum_{n=0}^\infty (a-z_0)^n \int_\gamma \frac{1}{(z-z_0)^{n+1}}dz = 2\pi i$. If $a \notin \overline{B(z_0,r)}$, then write $\frac{1}{z-a} = -\frac{1}{(a-z_0)}\frac{1}{(1-w)}$, where $w = \frac{z-z_0}{a-z_0}$ and complete the argument. □

Theorem 6.7. *Let γ be any path and $f : [\gamma] \to \mathbb{C}$ be continuous. Then $F(z) = \int_\gamma \frac{f(w)}{w-z}dw$ is differentiable on $\mathbb{C} \setminus [\gamma]$. In fact, F is locally a power series, i.e., analytic in $\mathbb{C} - [\gamma]$.*

Proof. Notice that $[\gamma]$ is compact and $\mathbb{C} \setminus [\gamma]$ is an open set. If $z_0 \notin [\gamma]$, choose $r > 0$ such that $B(z_0,r) \cap [\gamma] = \emptyset$. This follows from Corollary 2.7. If $z \in B(z_0,r)$, choose ρ such that $0 < \rho < r$ and $z \in B(z_0,\rho)$. Then $|\frac{z-z_0}{w-z_0}| < 1$ for $w \in [\gamma]$. Thus the series $\frac{f(w)}{w-z} = \frac{f(w)}{w-z_0}\sum_{n=0}^\infty (\frac{z-z_0}{w-z_0})^n$ of functions in w for fixed z and z_0 converge uniformly on $[\gamma]$. Therefore, $F(z) = \sum_{n=0}^\infty a_n(z-z_0)^n$, where $a_n = \int_\gamma \frac{f(w)}{(w-z_0)^{n+1}}dw$. Thus F is locally a power series and by Theorem 5.4, F is differentiable on $\mathbb{C} - [\gamma]$. □

Proposition 6.13. *Let $\gamma(t) = a + re^{it}, 0 \le t \le 2\pi$. Let $g : [\gamma] \to \mathbb{C}$ be continuous and $n \in \mathbb{Z}_+$. Let $\phi(z) = \int_\gamma \frac{g(w)}{(w-z)^n}dw, z \in B(a,r)$. Then ϕ is holomorphic in $B(a,r)$ and we have $\phi'(z) = n\int_\gamma \frac{g(w)}{(w-z)^{n+1}}dw, z \in B(a,r)$.*

Proof. Let $z_0 \in B(a,r)$ and choose $\rho < r$ such that $z_0 \in B(a,\rho)$. Let $z_1 \in B(a,\rho)$. We have

$$\begin{aligned} &\left|\frac{1}{(w-z_1)^n} - \frac{1}{(w-z_0)^n}\right| \\ &= \left|\frac{z_1 - z_0}{(w-z_0)^n(w-z_1)^n}\right| \\ &\times \left|(w-z_0)^{n-1} + (w-z_0)^{n-2}(w-z_1) + \cdots + (w-z_1)^{n-1}\right| \\ &\le \frac{|z_1-z_0|}{(r-\rho)^{2n}}n(2r)^{n-1}. \end{aligned}$$

Since $[\gamma]$ is compact, by Corollary 2.13, we can find $M < \infty$ such that $|g(w)| \leq M$ for $w \in [\gamma]$.

Hence

$$|\phi(z_1) - \phi(z_0)| = \left| \int_\gamma [\frac{g(w)}{(w-z_1)^n} - \frac{g(w)}{(w-z_0)^n}] dw \right|$$
$$\leq 2\pi r M \frac{n(2r)^{n-1}}{(r-\rho)^{2n}} |z_1 - z_0|.$$

Thus ϕ is continuous at z_0. Now

$$\frac{\phi(z) - \phi(z_0)}{z - z_0} = \sum_{j=1}^{n} \int_\gamma \frac{g(w)}{(w-z_0)^{n-j+1}(w-z)^j} dw.$$

Each integrand can be written as $\frac{g_j}{(w-z)^j}$ where g_j is continuous on $[\gamma]$. Replacing g by g_j, it follows from the first part of the proof that each of the summand integrals ϕ_j is continuous. Taking limit $z \longrightarrow z_0$, we obtain $\phi'(z_0) = n \int_\gamma \frac{g(w)}{(w-z_0)^{n+1}} dw$. Hence ϕ is differentiable at z_0 and since the choice of z_0 in $B(a, r)$ is arbitrary, the result follows. □

Lemma 6.1. *If G is open in $\mathbb{C}$, $\gamma : [a, b] \longrightarrow G$ is a rectifiable curve, and $f : G \longrightarrow \mathbb{C}$ is continuous then for every $\epsilon > 0$ there is a polygonal path Γ in G such that $\Gamma(a) = \gamma(a), \Gamma(b) = \gamma(b)$, and $|\int_\gamma f - \int_\Gamma f| < \epsilon$.*

Proof. Case I: Suppose $G = B(c, r)$, an open disk in $\mathbb{C}$. Notice that $[\gamma]$ is compact and ∂G is closed in $\mathbb{C}$. Hence by Theorem 2.57, $d = \text{dist}([\gamma], \partial G) > 0$. Let $\rho = r - \frac{d}{2}$. Then $[\gamma] \subset B(c, \rho)$ and $\overline{B(c, \rho)} \subset G$. The function f is continuous on $\overline{B(c, \rho)}$ which is compact. Hence f is uniformly continuous on $\overline{B(c, \rho)}$. Without loss of generality we shall assume f is uniformly continuous on G. Suppose $\epsilon > 0$ is given. Choose $\delta > 0$ such that $|f(z) - f(w)| < \epsilon$ whenever $|z - w| < \delta$. The function γ is uniformly continuous on $[a, b]$. So there is a partition $\{t_0 < t_1 < \cdots < t_n\}$ of $[a, b]$ such that

$$|\gamma(s) - \gamma(t)| < \delta \tag{6.5}$$

if $t_{k-1} \leq s, t \leq t_k$; and such that for $t_{k-1} \leq \tau_k \leq t_k$ we have

$$\left| \int_\gamma f - \sum_{k=1}^{n} f(\gamma(\tau_k))[\gamma(t_k) - \gamma(t_{k-1})] \right| < \epsilon. \tag{6.6}$$

Define $\Gamma : [a, b] \longrightarrow \mathbb{C}$ by

$$\Gamma(t) = \frac{1}{t_k - t_{k-1}} [(t_k - t)\gamma(t_{k-1}) + (t - t_{k-1})\gamma(t_k)]$$

if $t_{k-1} \leq t \leq t_k$. So on $[t_{k-1}, t_k], \Gamma(t)$ traces out the straight line segment from $\gamma(t_{k-1})$ to $\gamma(t_k)$; that is, Γ is a polygonal path in G. From (6.5),

$$|\Gamma(t) - \gamma(\tau_k)| < \delta \text{ for } t_{k-1} \leq t \leq t_k. \tag{6.7}$$

Since $\int_\Gamma f = \int_a^b f(\Gamma(t))\Gamma'(t)dt$ it follows that

$$\int_\Gamma f = \sum_{k=1}^{n} \frac{\gamma(t_k) - \gamma(t_{k-1})}{t_k - t_{k-1}} \int_{t_{k-1}}^{t_k} f(\Gamma(t))dt.$$

Using (6.6) we obtain

$$\left|\int_\gamma f - \int_\Gamma f\right| \le \epsilon + \left|\sum_{k=1}^{n} f(\gamma(\tau_k))[\gamma(t_k) - \gamma(t_{k-1})] - \int_\Gamma f\right|$$

$$\le \epsilon + \sum_{k=1}^{n} \frac{|\gamma(t_k) - \gamma(t_{k-1})|}{t_k - t_{k-1}} \int_{t_{k-1}}^{t_k} |f(\Gamma(t)) - f(\gamma(\tau_k))|\, dt.$$

Applying (6.7) to the integrand gives

$$\left|\int_\gamma f - \int_\Gamma f\right| \le \epsilon + \epsilon \sum_{k=1}^{n} |\gamma(t_k) - \gamma(t_{k-1})|$$

$$\le \epsilon(1 + V(\gamma)).$$

The proof of case I now follows.

Case II: G is arbitrary. Since $[\gamma]$ is compact there is a number r with $0 < r < \text{dist}([\gamma], \partial G)$. Choose $\delta > 0$ such that $|\gamma(s) - \gamma(t)| < r$ when $|s - t| < \delta$. If $P = \{t_0 < t_1 < \cdots < t_n\}$ is a partition of $[a, b]$ with $\|P\| < \delta$ then $|\gamma(t) - \gamma(t_{k-1})| < r$ for $t_{k-1} \le t \le t_k$. That is if $\gamma_k : [t_{k-1}, t_k] \longrightarrow G$ is defined by $\gamma_k(t) = \gamma(t)$ then $[\gamma_k] \subset B(\gamma(t_{k-1}), r)$ for $1 \le k \le n$. By case-I there is a polygonal path $\Gamma_k : [t_{k-1}, t_k] \longrightarrow B(\gamma(t_{k-1}), r)$ such that $\Gamma_k(t_{k-1}) = \gamma(t_{k-1}), \Gamma_k(t_k) = \gamma(t_k)$, and $\left|\int_{\gamma_k} f - \int_{\Gamma_k} f\right| < \frac{\epsilon}{n}$. If $\Gamma(t) = \Gamma_k(t)$ on $[t_{k-1}, t_k]$ then Γ has the required properties. □

The next result is the analogue of the Fundamental theorem of calculus for line integrals.

Theorem 6.8. *Let G be open in $\mathbb{C}$ and let γ be a rectifiable curve in G with initial and final points α and β respectively. If $f : G \longrightarrow \mathbb{C}$ is a continuous function with a primitive $F : G \longrightarrow \mathbb{C}$, then $\int_\gamma f = F(\beta) - F(\alpha)$.*

Proof. If $\gamma : [a, b] \longrightarrow \mathbb{C}$ is piecewise smooth then the result follows from Theorem 6.4. For the general case let $\epsilon > 0$. By Lemma 6.1, there is a polygonal path Γ from α to β such that $\left|\int_\gamma f - \int_\Gamma f\right| < \epsilon$. But Γ is piecewise smooth. So $\int_\Gamma f = F(\beta) - F(\alpha)$. Hence $\left|\int_\gamma f - [F(\beta) - F(\alpha)]\right| < \epsilon$. Since $\epsilon > 0$ is arbitrary, the desired equality follows. □

Corollary 6.5. *Let G be open in $\mathbb{C}$ and let γ be a closed rectifiable curve in G. If $f : G \longrightarrow \mathbb{C}$ is a continuous function with a primitive $F : G \longrightarrow \mathbb{C}$, then $\int_\gamma f = 0$.*

6.4 Fubini's theorem

In this section our main goal is to prove Fubini's theorem for complex valued functions defined on compact subsets of $\mathbb{C} \times \mathbb{C}$. The organization of the section is as follows: We first define double integral of real valued functions defined on a rectangle contained in $\mathbb{R}^2$ and we make a quick review of the results on existence of double integrals and the properties of double integrals. We shall not prove them here. For details refer any standard book in real analysis. Then we shall proceed to prove Fubini's theorem for real valued functions defined on a rectangle contained in $\mathbb{C}$ and the generalization of the theorem to subsets of $\mathbb{C} \times \mathbb{C}$.

Let f be a real-valued function defined and bounded on the rectangle

$$E = \{(x, y) \in \mathbb{R}^2 : a \leq x \leq b \text{ and } c \leq y \leq d\}.$$

Construct a partition $\mathcal{P}$ of E as follows. Draw lines through E parallel to the y-axis $x = x_j$ for $j = 0, 1, 2, \ldots, n$ and parallel to the x-axis $y = y_k$ for $k = 0, 1, 2, \ldots, m$, with $a = x_0 < x_1 < \ldots < x_{n-1} < x_n = b$ and $c = y_0 < y_1 < y_2 < \ldots < y_{m-1} < y_m = d$. Denote the subrectangles

$$E_{jk} = \{(x, y) \in E : x_{j-1} \leq x \leq x_j \text{ and } y_{k-1} \leq y \leq y_k\}$$

for each $j = 1, 2, \ldots, n$ and $k = 1, 2, \ldots, m$. Let

$$\mathcal{P} = \{E_{jk} : j = 1, 2, \ldots, n \text{ and } k = 1, 2, \ldots, m\}.$$

Let

$$M = \sup_{(x,y)\in E} f(x, y) \text{ and } m = \inf_{(x,y)\in E} f(x, y).$$

Let

$$M_{jk} = \sup_{(x,y)\in E_{jk}} f(x, y) \text{ and } m_{jk} = \inf_{(x,y)\in E_{jk}} f(x, y).$$

The upper sum and lower sum are defined by $U(f, \mathcal{P}) = \sum_{j,k} M_{jk} A(E_{jk})$ and $L(f, \mathcal{P}) = \sum_{j,k} m_{jk} A(E_{jk})$ where $A(E_{jk})$ is the area of the rectangle E_{jk}. It follows from the definition that $mA(E) \leq L(f, \mathcal{P}) \leq U(f, \mathcal{P}) \leq MA(E)$.

We now define the upper double integral of f over E by the expression

$$\overline{\iint_E} f(x, y)\,dxdy = \inf_{\mathcal{P}} U(f, \mathcal{P}).$$

Similarly, we define the lower double integral of f over E by

$$\underline{\iint_E} f(x, y)\,dxdy = \sup_{\mathcal{P}} L(f, \mathcal{P}).$$

Notice that both integrals exist because of boundedness of f on E.

Definition 6.11. *A bounded function $f : E \to \mathbb{R}$ is said to be integrable provided*

$$\underline{\iint_E} f(x,y)dxdy = \overline{\iint_E} f(x,y)dxdy.$$

In this case we define the double integral of f over E by

$$\underline{\iint_E} f(x,y)dxdy = \overline{\iint_E} f(x,y)dxdy = \iint_E f(x,y)dxdy.$$

The following holds:

Theorem 6.9. *Let f be bounded on the rectangle E. Then f is integrable over E if and only if given $\epsilon > 0$, there is a partition $\mathcal{P}$ of E such that $U(f,\mathcal{P}) - L(f,\mathcal{P}) < \epsilon$.*

Example 6.20. *Let*

$$E = \{(x,y) \in \mathbb{R}^2 : 0 \le x \le 1, 0 \le y \le 1\}.$$

Define $f(x,y) = 1$ if x and y are rational numbers in $[0,1]$ and $f(x,y) = 0$ for all other points $(x,y) \in E$. Then f is not integrable on E. That is, $\iint_E f(x,y)dxdy$ does not exist.

Example 6.21. *Let*

$$E = \{(x,y) \in \mathbb{R}^2 : 0 \le x \le 1, 0 \le y \le 1\}.$$

Define f by

$$f(x,y) = \begin{cases} 1 & \text{if } x = 1; \\ 0 & \text{if } 0 \le x < 1. \end{cases}$$

Then one can verify that f is integrable on E and $\iint_E f(x,y)dxdy = 0$.

We define r_{jk} to the length of a diagonal of E_{jk} and we define the norm of a partition $\mathcal{P}$ of E to be

$$\|\mathcal{P}\| = \max\{r_{jk} : 1 \le j \le n, 1 \le k \le m\}.$$

Let $\mathcal{P}$ be any partition of E and choose $\eta_{jk} \in E_{jk}$ for each $j = 1, 2, \dots, n$; $k = 1, 2, \dots, m$. We form the Riemann sum $S(f,\mathcal{P},\eta) = \sum_{j,k} f(\eta_{jk})A(E_{jk})$.

Notice that for each choice of the points $\eta = \eta_{jk}$,

$$L(f,\mathcal{P}) \le S(f,\mathcal{P},\eta) \le U(f,\mathcal{P}).$$

The following theorem is the analog of the theorem on Riemann integration of f defined on $[a,b]$.

Theorem 6.10. *The double integral of f on E exists and equals I if and only if given $\epsilon > 0$, there is a $\delta > 0$ such that $|S(f, \mathcal{P}, \eta) - f| < \epsilon$ for every Riemann sum $S(f, \mathcal{P}, \eta)$ with $\|\mathcal{P}\| < \delta$.*

Example 6.22. *Let f be a continuous function defined on $E = [a, b] \times [c, d] \subset \mathbb{R}^2$. Then f is integrable. It can be seen as follows: Since f is continuous on the compact set E, it follows that f is uniformly continuous on E (see Theorem 2.51). Thus by the uniform continuity of f on E, given $\frac{\epsilon}{(b-a)(d-c)} > 0$, there exists a $\delta > 0$ such that*

$$|f(x_1, y_1) - f(x_2, y_2)| < \frac{\epsilon}{(b-a)(d-c)}$$

whenever $|x_1 - x_2| < \delta$ and $|y_1 - y_2| < \delta$.

Let $\mathcal{P}$ be any partition of E with $\|\mathcal{P}\| < \delta$. Since continuous function on a compact set attains its minimum and maximum (corollary 2.12) hence there exist points $\mu_{jk} = (x_{jk}, y_{jk}), \xi_{jk} = (x'_{jk}, y'_{jk})$ in E_{jk} such that $f(\mu_{jk}) = M_{jk}$ and $f(\xi_{jk}) = m_{jk}$ and $|x_{jk} - x'_{jk}| < \delta, |y_{jk} - y'_{jk}| < \delta$. Hence $M_{jk} - m_{jk} = f(\mu_{jk}) - f(\xi_{jk}) < \frac{\epsilon}{(b-a)(d-c)}, j = 1, 2, \ldots, n; k = 1, 2, \ldots, m$. Hence $U(f, \mathcal{P}) - L(f, \mathcal{P}) = \sum_{j,k}(M_{jk} - m_{jk})A(E_{jk}) < \frac{\epsilon}{(b-a)(d-c)} \sum_{j,k} A(E_{jk}) = \epsilon$ whenever $\mathcal{P}$ is a partition of E with $\|\mathcal{P}\| < \delta$. This implies that f is integrable on E.

The following are some properties of double integrals which can be verified easily.

Let f, g be integrable functions on a rectangle $E \subset \mathbb{R}^2$. Then

(a)

$$\int\int_E kf(x, y)dxdy = k\int\int_E f(x, y)dxdy$$

for any real number k.

(b)

$$\int\int_E (f(x, y) + g(x, y))dxdy = \int\int_E f(x, y)dxdy + \int\int_E g(x, y)dxdy.$$

(c) If f is integrable on a rectangle $E \subset \mathbb{R}^2$, then $|f|$ is integrable on E and

$$\left|\int\int_E f(x, y)dxdy\right| \leq \int\int_E |f(x, y)|dxdy.$$

Let f be bounded on the rectangle E given by $E = \{(x, y) \in \mathbb{R}^2 : a \leq x \leq b, c \leq y \leq d\}$. If we fix $x \in [a, b]$ then $f(x, y)$ can be thought of as a function of the single real variable $y \in [c, d]$. Define then a function on $[a, b]$ by $g(x) = \int_c^d f(x, y)dy$. If g is integrable on $[a, b]$ then

$$\int_a^b g(x)dx = \int_a^b \left[\int_c^d f(x, y)dy\right] dx$$

is called an iterated integral. We write this integral as $\int_a^b \int_c^d f(x,y)dydx = \int\int_E f(x,y)dydx$.

We now state and prove a very important theorem which is known as Fubini's theorem. This theorem allows us to evaluate double integrals by evaluating an iterated integral.

Theorem 6.11. *(Fubini' theorem for real valued functions): Let f be a bounded function defined on $E = [a,b] \times [c,d]$. The following holds:*

(i) Let $f : E \to \mathbb{R}$ be continuous. Then

$$\int\int_E f(x,y)dxdy = \int_a^b \left[\int_c^d f(x,y)dy\right] dx = \int_c^d \left[\int_a^b f(x,y)dx\right] dy.$$

(ii) Suppose f is integrable and the function $f_x : [c,d] \to \mathbb{R}$ defined by $f_x(y) = f(x,y)$ is integrable for each fixed $x \in [a,b]$. Then

$$\int\int_E f(x,y)dxdy = \int_a^b \left[\int_c^d f(x,y)dy\right] dx.$$

Similarly, assuming that $\int_a^b f(x,y)dx$ exists for each $y \in [c,d]$ we obtain

$$\int\int_E f(x,y)dxdy = \int_c^d \left[\int_a^b f(x,y)dx\right] dy.$$

Proof. We need only to prove (ii) as (i) is a special case of (ii). Let $g : [a,b] \to \mathbb{R}$ be the function defined by $g(x) = \int_c^d f(x,y)dy$. We shall show that g is integrable over $[a,b]$ and that $\displaystyle\int\int_E f(x,y)dxdy = \int_a^b g(x)dx$.

Since f is integrable on E, there is a $\delta > 0$ such that whenever $\mathcal{P}$ is a partition of E with $\|\mathcal{P}\| < \delta$, we have $U(f,\mathcal{P}) - L(f,\mathcal{P}) < \epsilon$. Let $a = x_0 < x_1 < x_2 < \cdots < x_{i-1} < x_i < \cdots < x_n = b$ be any partition of $[a,b]$ such that $\max\limits_{1\le j\le n} |x_j - x_{j-1}| < \dfrac{\delta}{2}$. Let $\epsilon_i \in [x_{i-1}, x_i]$ for $i = 1, 2, \ldots, n$ be arbitrary. Let $c = y_0 < y_1 < y_2 < \cdots < y_{i-1} < y_i < \cdots < y_m = d$ be a partition of $[c,d]$ such that $\max\limits_{1\le j\le m} |y_j - y_{j-1}| < \dfrac{\delta}{2}$. This implies the partition $\mathcal{P} = (E_{ij})$ of E with $E_{ij} = [x_{i-1}, x_i] \times [y_{j-1}, y_j], 1 \le i \le n; 1 \le j \le m$ satisfies $\|\mathcal{P}\| < \delta$.

For $i = 1, 2, \ldots, n$, let q_{ij} be the mean value of $f(\xi_i, y)$ on $[y_{j-1}, y_j]$, $j = 1, 2, \ldots, m$; that is,

$$q_{ij} = \frac{1}{\triangle y_j} \int_{y_{j-1}}^{y_j} f(\xi_i, y)dy.$$

Notice that q_{ij} depends on the intermediate points ξ_i as well as on i and j.

Then

$$\begin{aligned}\inf_{(x,y)\in E_{ij}} f(x,y) &= m_{ij} \\ &\le \inf_{y\in[y_{j-1},y_j]} f(\xi_i,y) \\ &\le q_{ij} \\ &\le \sup_{y\in[y_{j-1},y_j]} f(\xi_i,y) \\ &\le M_{ij} \\ &= \sup_{(x,y)\in E_{ij}} f(x,y).\end{aligned}$$

Therefore,

$$\begin{aligned}L(f,\mathcal{P}) &= \sum_{i,j} m_{ij}A(E_{ij}) \\ &\le \sum_{i,j} q_{ij}A(E_{ij}) \\ &\le \sum_{i,j} M_{ij}A(E_{ij}) \\ &= U(f,\mathcal{P}).\end{aligned}$$

Now

$$g(\xi_i) = \int_c^d f(\xi_i,y)dy$$

for $i = 1, 2, \ldots, n$ and so

$$\begin{aligned}\textstyle\sum_{i=1}^n g(\xi_i)\triangle x_i &= \textstyle\sum_{i=1}^n \left(\int_c^d f(\xi_i,y)dy\right)\triangle x_i \\ &= \textstyle\sum_{i=1}^n \left(\sum_{j=1}^m \int_{y_{j-1}}^{y_j} f(\xi_i,y)dy\right)\triangle x_i \\ &= \textstyle\sum_{i=1}^n \left(\sum_{j=1}^m q_{ij}\triangle y_j\right)\triangle x_i \\ &= \textstyle\sum_{i=1}^n \sum_{j=1}^m q_{ij}\triangle y_j\triangle x_i \\ &= \textstyle\sum_{i,j} q_{ij}A(E_{ij}).\end{aligned}$$

Consequently,

$$L(f,\mathcal{P}) \le \sum_{i=1}^n g(\xi_i)\triangle x_i \le U(f,\mathcal{P}).$$

Since $L(f,\mathcal{P}) \le \int\int_E f(x,y)dxdy \le U(f,\mathcal{P})$ and if $\|\mathcal{P}\| < \delta$, it follows that $U(f,\mathcal{P}) - L(f,\mathcal{P}) < \epsilon$. Since f is integrable over E, the above inequality shows that g is integrable and $\int_c^d g(x)dx = \int\int_E f(x,y)dxdy = \int_a^b \left(\int_c^d f(x,y)dy\right)dx$. Similarly, one can show that if $\int_a^b f(x,y)dx$ exists for each y then

$$\int\int_E f(x,y)dxdy = \int_c^d \left(\int_a^b f(x,y)dx\right)dy.$$

□

Theorem 6.12. *Let $\gamma : [a,b] \to \mathbb{C}$ be a piecewise smooth curve in $\mathbb{C}$ and let G be a region in $\mathbb{C}$. Let $F(z,w)$ be a complex valued function such that if w is any fixed point on $[\gamma]$, then F is holomorphic on G and such that if z is fixed in G, then $F(z,w)$ is a piecewise continuous on $[\gamma]$. Let $g(z) = \int_\gamma F(z,w)dw$ for each z in G.*

If for each $z_0 \in G$,

$$\frac{F(z,w) - F(z_0,w)}{z - z_0} \to F_z(z_0,w)$$

uniformly on $[\gamma]$ (that is, for each $\epsilon' > 0$, there exists a $\delta > 0$, independent of w such that

$$\left|\frac{F(z,w) - F(z_0,w)}{z - z_0} - F_z(z_0,w)\right| < \epsilon' \tag{6.8}$$

for each w on $[\gamma]$ whenever $z \in G$ and $0 < |z - z_0| < \delta$) then $g \in H(G)$ and $g'(z_0) = \int_\gamma F_z(z_0,w)dw$, where $F_z(z_0,w) = \frac{d}{dz}F(z,w)|_{z=z_0}$ for each fixed w on $[\gamma]$.

Proof. Let $\epsilon > 0$ and let $L = L(\gamma)$, the length of γ. Now

$$\begin{aligned} \left|\tfrac{g(z)-g(z_0)}{z-z_0} - \int_\gamma F_z(z_0,w)dw\right| &= \left|\int_\gamma \left(\tfrac{F(z,w)-F(z_0,w)}{z-z_0} - F_z(z_0,w)\right) dw\right| \\ &< L\epsilon' \\ &= \tfrac{L\epsilon}{2L} < \epsilon \end{aligned}$$

if $z \in G$ and $0 < |z - z_0| < \delta$ where δ is chosen so that (6.8) holds with $\epsilon' = \frac{\epsilon}{2L}$. The result follows. □

Example 6.23. *Let h be piecewise continuous on the counter clockwise oriented circle $\gamma = \{z \in \mathbb{C} : |z| = r_0\}$ and*

$$g(z) = \int_\gamma \frac{w+z}{w-z} h(w)dw$$

for each $z \in A$ where $A = \{z \in \mathbb{C} : |z| < r_0\}$. Then using Theorem 6.12, we shall show that g is holomorphic on A. Let $F(z,w) = \frac{w+z}{w-z}h(w)$. Then

$$\begin{aligned} \left|\tfrac{F(z,w)-F(z_0,w)}{z-z_0} - \tfrac{2w}{(w-z_0)^2}h(w)\right| &= \tfrac{2|wh(w)|}{|w-z_0|^2}\tfrac{|z-z_0|}{|w-z|} \\ &\leq M\tfrac{|z-z_0|}{r_0-|z|}, \end{aligned}$$

where M is the maximum of $\frac{2|wh(w)|}{|r_0-|z_0||^2}$ on γ. Thus g is holomorphic on A.

Let $\gamma : [a,b] \to \mathbb{C}$ be a piecewise smooth curve in $\mathbb{C}$ and let G be a region in $\mathbb{C}$. Let $g(z) = \int_\gamma f(z,w)dw$, where for each fixed w on $[\gamma]$, the function f is analytic on G. We shall find certain conditions under which g is holomorphic on G and $g'(z) = \int_\gamma f_z(z,w)dw$, where $f_z(z,w) = \frac{\partial}{\partial z}f(z,w)$ for each fixed w.

Theorem 6.13. *Let f be continuous on $G \times [\gamma]$. Then $g(z) = \int_\gamma f(z,w)dw$, $z \in G$ is continuous on G.*

Proof. Let $a \in G$ and let $\epsilon > 0$. Since f is continuous on $G \times [\gamma]$, for each w' on $[\gamma]$, let $\delta_{w'} > 0$ such that

$$|f(z,w) - f(a,w')| < \frac{\epsilon}{2L}$$

if $(z,w) \in G \times [\gamma], |z-a| < \delta_{w'}$ and $|w - w'| < \delta_{w'}$ where L is the length of γ. Now $[\gamma]$ is compact. Let $\{B(w_1, \delta_{w_1}), B(w_2, \delta_{w_2}), \ldots , B(w_n, \delta_{w_n})\}$ be a finite subcover of $\{B(w', \delta_{w'}) : w' \in [\gamma]\}$ for $[\gamma]$.

Let $\delta = \min\{\delta_{w_1}, \delta_{w_2}, \ldots , \delta_{w_n}\}$. Now let $w \in [\gamma]$. Then $w \in B(w_j, \delta_{w_j})$ for some $j \in \{1, 2, \ldots , n\}$. Suppose $z \in G$ is such that $|z - a| < \delta \leq \delta_{w_j}$. Then

$$\begin{aligned} |f(z,w) - f(a,w)| &\leq |f(z,w) - f(a,w_j)| + |f(a,w_j) - f(a,w)| \\ &< \tfrac{\epsilon}{2L} + \tfrac{\epsilon}{2L} \\ &= \tfrac{\epsilon}{L}. \end{aligned}$$

Thus we have shown that there is some $\delta > 0$ such that if $z \in G$ and $|z-a| < \delta$, then $|f(z,w) - f(a,w)| < \frac{\epsilon}{L}$ holds for each w on $[\gamma]$ (where δ is independent of w). Now for each $z \in G$, such that $|z - a| < \delta$, we have $|g(z) - g(a)| = \left|\int_\gamma [f(z,w) - f(a,w)]dw\right| < L \cdot \frac{\epsilon}{L} = \epsilon$. Since a is an arbitrary element of G, hence g is continuous on G. □

Remark 6.1. From Theorem 6.13, it follows that

$$\begin{aligned} \lim_{z\to a} \int_\gamma f(z,w)dw &= \lim_{z\to a} g(z) = g(a) \\ &= \int_\gamma f(a,w)dw = \int_\gamma [\lim_{z\to a} f(z,w)]dw. \end{aligned}$$

Theorem 6.14. *(Fubini's theorem): Let $\gamma : [a,b] \to \mathbb{C}$ and $\sigma : [c,d] \to \mathbb{C}$ be two piecewise smooth curves in $\mathbb{C}$. Suppose F is continuous on $[\gamma] \times [\sigma]$. Then*

$$\int_\gamma \left[\int_\sigma F(z,w)dw\right] dz = \int_\sigma \left[\int_\gamma F(z,w)dz\right] dw.$$

Proof. First suppose γ and σ are smooth. Then

$$\begin{aligned} \int_\gamma \left[\int_\sigma F(z,w)dw\right] dz &= \int_a^b \left[\int_c^d F(\gamma(s), \sigma(t))\sigma'(t)dt\right] \gamma'(s)ds \\ &= \int_c^d \left[\int_a^b F(\gamma(s), \sigma(t))\gamma'(s)ds\right] \sigma'(t)dt \\ &= \int_\sigma \left[\int_\gamma F(z,w)dz\right] dw. \end{aligned}$$

If γ and σ are piecewise smooth, then the intervals $[a,b]$ and $[c,d]$ can be broken up into subintervals on which $\gamma'(s)$ and $\sigma'(t)$ are continuous respectively. □

Theorem 6.15. *Let $\gamma : [a,b] \to \mathbb{C}$ be a piecewise smooth curve in $\mathbb{C}$ and G be a region in $\mathbb{C}$. Suppose $f(z,w)$ is continuous on $G \times [\gamma]$. If for each fixed w on $[\gamma]$, the function $f(z,w)$ (as a function of z) is analytic in G and if $g(z) = \int_\gamma f(z,w)dw$ for each $z \in G$, then g is holomorphic on G and $g'(z) = \int_\gamma f_z(z,w)dw$ for each $z \in G$.*

Proof. Let $z_0 \in G$. Since G is open, there exists $r > 0$ such that $\overline{B(z_0,r)} \subset G$. Let $\sigma = \{z \in \mathbb{C} : |z - z_0| = r\}$, positively oriented. If $z \in B(z_0,r)$, by Fubini's theorem (Theorem 6.14),

$$\begin{aligned} g(z) &= \int_\gamma f(z,w)dw = \int_\gamma \tfrac{1}{2\pi i}\left[\int_\sigma \tfrac{f(\widetilde{w},w)}{\widetilde{w}-z}d\widetilde{w}\right]dw \\ &= \tfrac{1}{2\pi i}\int_\sigma \left[\int_\gamma f(\widetilde{w},w)dw\right]\tfrac{d\widetilde{w}}{\widetilde{w}-z} \\ &= \tfrac{1}{2\pi i}\int_\sigma \tfrac{g(\widetilde{w})}{\widetilde{w}-z}d\widetilde{w}. \end{aligned}$$

Now g is continuous on G by Theorem 6.13. Thus by Proposition 6.13 and Fubini's theorem (Theorem 6.14), the function g is holomorphic on $B(z_0,r)$ and for each $z \in B(z_0,r)$,

$$\begin{aligned} g'(z) &= \tfrac{1}{2\pi i}\int_\sigma \tfrac{g(\widetilde{w})}{(\widetilde{w}-z)^2}d\widetilde{w} \\ &= \tfrac{1}{2\pi i}\int_\sigma \left[\int_\gamma f(\widetilde{w},w)dw\right]\tfrac{d\widetilde{w}}{(\widetilde{w}-z)^2} \\ &= \int_\gamma \left[\tfrac{1}{2\pi i}\int_\sigma \tfrac{f(\widetilde{w},w)}{(\widetilde{w}-z)^2}d\widetilde{w}\right]dw \\ &= \int_\gamma f_z(z,w)dw. \end{aligned}$$

Since z_0 is arbitrary, the theorem follows. □

Theorem 6.16. *Let F be continuous on $G \times [a,b]$, where G is a region in $\mathbb{C}$. If $z_0 \in G$, then*

$$\lim_{z\to z_0}\int_a^b F(z,t)dt = \int_a^b F(z_0,t)dt.$$

Proof. Define $\gamma : [a,b] \to \mathbb{C}$ as $\gamma(t) = t$. Then γ is a smooth curve in $\mathbb{C}$. In fact $[\gamma] = \{\gamma(t) : a \le t \le b\} = [a,b]$. Now

$$\lim_{z\to z_0}\int_a^b F(z,t)dt = \lim_{z\to z_0}\int_\gamma F(z,w)dw = \int_\gamma F(z_0,w)dw = \int_a^b F(z_0,t)dt$$

by Theorem 6.13 and Remark 6.1. □

The following is extension of Leibniz rule stated in Theorem 6.2.

Theorem 6.17. *Let $f : S \to \mathbb{C}$ be continuous where $S = \{(a,x) : c \le a \le d$ and $b \le x < +\infty\}$. Assume that $f_a(a,x) \in C(S)$, the space of continuous complex* *valued functions on* S*, and let* $\int_b^\infty f_a(a,x)dx$ *converges uniformly on* $[c,d]$*. If* $I(a) = \int_b^\infty f(a,x)dx$ *for each* $a \in [c,d]$ *then* $I'(a) = \int_b^\infty f_a(a,x)dx$ *for each* $a \in [c,d]$*.*

For proof of Theorem 6.17, see any standard text in real analysis. Theorem 6.17 is very often used in evaluating certain improper integrals (see Chapter 12). One can also prove the following generalization of Fubini's theorem to infinite intervals.

Let $f(x,y)$ be a real-valued function defined on the semi-infinite strip $U = \{(x,y) : a \leq x \leq b \text{ and } c \leq y < +\infty\}$. Let $\int_c^\infty f(x,y)dy = g(x)$ for each $x \in [a,b]$. Then we say that $\int_c^\infty f(x,y)dy$ converges uniformly to g on $[a,b]$ if and only if for each $\epsilon > 0$, there is a number r (independent of x) such that if $d \geq r$, then

$$\left| g(x) - \int_c^d f(x,y)dy \right| = \left| \int_d^\infty f(x,y)dy \right| < \epsilon$$

on $[a,b]$.

Theorem 6.18. *Let* f *be a real valued continuous function on* U *and suppose* $\int_c^\infty f(x,y)dy$ *converges uniformly on* $[a,b]$ *to* $g(x)$*. Then* g *is continuous on* $[a,b]$ *and*

$$\int_a^b \left[\int_c^\infty f(x,y)dy \right] dx = \int_c^\infty \left[\int_a^b f(x,y)dx \right] dy.$$

Proof. Let $x_0 \in [a,b]$ and let $\epsilon > 0$. By hypothesis, $\int_c^\infty f(x,y)dy$ converges uniformly on $[a,b]$. Let $d \geq r > c$ such that $\left|\int_d^\infty f(x,y)dy\right| < \frac{\epsilon}{3}$ on $[a,b]$. Let

$$S = \{(x,y) : a \leq x \leq b \text{ and } c \leq y \leq d\}.$$

The set S is closed and bounded and therefore compact. By Theorem 2.51, f is uniformly continuous on S. Thus we let $\delta > 0$ such that if (x,y) and (t,y) are in S and if $|x-t| < \delta$, then $|f(x,y) - f(t,y)| < \frac{\epsilon}{3(d-c)}$. Now for each $x \in [a,b]$ such that $|x - x_0| < \delta$, we have

$$\begin{aligned} & |g(x) - g(x_0)| \\ \leq\ & \left|\int_d^\infty f(x,y)dy\right| + \left|\int_c^d [f(x,y) - f(x_0,y)]dy\right| + \left|\int_d^\infty f(x_0,y)dy\right| \\ <\ & \tfrac{\epsilon}{3} + \tfrac{\epsilon}{3} + \tfrac{\epsilon}{3} = \epsilon. \end{aligned}$$

Thus g is continuous on $[a,b]$. To show that the integrals actually commute, let $\epsilon > 0$. Then there exist $r \in \mathbb{R}$ such that if $d \geq r$,

$$\left| g(x) - \int_c^d f(x,y)dy \right| < \tfrac{\epsilon}{b-a}$$

on $[a, b]$. Now for each $d > r$,

$$\begin{aligned} \left|\int_a^b g(x)dx - \int_c^d \left[\int_a^b f(x,y)dx\right] dy\right| &= \left|\int_a^b g(x)dx - \int_a^b \left[\int_c^d f(x,y)dy\right] dx\right| \\ &= \left|\int_a^b \left[g(x) - \int_c^d f(x,y)dy\right] dx\right| \\ &\le \int_a^b \left|g(x) - \int_c^d f(x,y)dy\right| dx \\ &< \tfrac{\epsilon}{b-a}(b-a) = \epsilon. \end{aligned}$$

But this means that $\int_a^b g(x)dx = \int_c^\infty [\int_a^b f(x,y)dx]dy$. □

6.5 Exercises

1. Show that if $\gamma : [a, b] \to \mathbb{C}$ is a Lipschitz function then γ is of bounded variation.

2. Let $\gamma(t) = e^{\frac{-1+i}{t}}$ for $0 < t \le 1$ and $\gamma(0) = 0$. Show that γ is a rectifiable path and find $V(\gamma)$.

3. Sketch the following contours:

   (a) $\gamma_1(t) = 2 + 2e^{2\pi it} \quad (0 \le t \le 1)$;

   (b) $\gamma_2(t) = \begin{cases} it & \text{if } 0 \le t \le 1, \\ t - 1 + i & \text{if } 1 \le t \le 2; \end{cases}$

   (c) $\gamma_3(t) = i + e^{-it}, \quad (0 \le t \le \frac{\pi}{2})$.

   Using these contours $\gamma_1, \gamma_2, \gamma_3$ and Theorem 6.8 (fundamental theorem of calculus for contour integrals) evaluate: (i) $\int_{\gamma_1} zdz$; (ii) $\int_{\gamma_2} \cos(\pi z)dz$; (iii) $\int_{\gamma_3} \frac{dz}{(z-i)^2}$.

4. Compute the following path integrals $\int_\gamma f(z)dz$:

   (i) $f(z) = |z|^2$ and γ is the line segment from 2 to $3 + i$.
   (ii) $f(z) = \bar{z}$ and γ is the semicircle from 1 to -1 passing through i.
   (iii) $f(z) = \text{Re}z$ and γ is the line segment from 1 to $-i$.
   (iv) $f(z) = \bar{z}$ and γ is the semicircle from 1 to -1 passing through $-i$.

5. Let γ and σ be the two polygons $[1, i]$ and $[1, 1 + i, i]$. Express γ and σ as paths and calculate $\int_\gamma f$ and $\int_\sigma f$ where $f(z) = |z|^2$.

6. Evaluate the following integrals:

   (i) $\int_{\gamma_1} \frac{dz}{z-2}$ where $\gamma_1(t) = 2 + 2e^{2\pi it} \quad (0 \le t \le 1)$;

   (ii) $\int_{\gamma_2} \bar{z}dz$ where

$$\gamma_2(t) = \begin{cases} it & \text{if } 0 \le t \le 1, \\ t - 1 + i & \text{if } 1 \le t \le 2; \end{cases}$$

(iii) $\int_{\gamma_3} \frac{dz}{(z-i)^2}$ where $\gamma_3(t) = i + e^{-it}, \quad (0 \le t \le \frac{\pi}{2})$.

7. Set $\gamma(t) = 3e^{it} \quad (0 \le t \le \pi)$. Use the estimation theorem to show that

$$\left| \int_\gamma \frac{e^z}{z-1} dz \right| \le \frac{3}{2}\pi e^3.$$

8. Establish the following:

(i)

$$\left| \int_\gamma \frac{dz}{z^2+4} \right| \le \frac{\pi R}{(R^2-4)}$$

where $\gamma(t) = Re^{it}$ for $0 \le t \le \pi$ and $R > 2$.

(ii)

$$\left| \int_\gamma \frac{e^z}{z} dz \right| \le 2\pi e$$

where γ is the unit circle.

9. Let $f : [a, b] \to \mathbb{C}$ be continuous. Then prove that $F(z) = \int_a^b f(t)e^{-tz} dt$ is holomorphic on $\mathbb{C}$.

10. Let p be a polynomial in z. Let γ be any circle $|z - z_0| = r$ with standard parametrization. Then show that for $a \notin [\gamma]$,

$$\frac{1}{2\pi i} \int_\gamma \frac{p(z)}{z-a} dz = \begin{cases} p(a) & \text{if} \quad |a - z_0| < r; \\ 0 & \text{if} \quad |a - z_0| > r. \end{cases}$$

Hint: $p(z) - p(a)$ is divisible by $z - a$.

11. Show that if G is a region in $\mathbb{C}$ and $f : G \longrightarrow \mathbb{C}$ is holomorphic and γ is a rectifiable curve in G then $f \circ \gamma$ is also a rectifiable curve.

12. Evaluate:

(i) $\int_\gamma \frac{z^2+1}{z+1} dz$ where $\gamma(t) = 2e^{it}, 0 \le t \le 2\pi$;

(ii) $\int_\gamma \frac{\cos z}{z} dz$ where $\gamma(t) = e^{it}, 0 \le t \le 2\pi$;

(iii) $\frac{1}{2\pi i} \int_\gamma \frac{e^{az}}{z^2+1} dz$ where $\gamma(t) = 2e^{it}, 0 \le t \le 2\pi$;

(iv) $\int_0^{2\pi} \frac{d\theta}{2+\cos\theta}$;

(v) $\int_\gamma x dz$ where γ is the directed line segment from 0 to $1 + i$.

13. Compute:

(i) $\int_{|z|=1} |z-1||dz|$; (ii) $\int_{|z|=\rho} \frac{|dz|}{|z-a|^2}, |a| \ne \rho$; (iii) $\int_{|z|=r} x dz$;

(iv) $\int_{|z|=r} \frac{dz}{|z^2+1|}, r < 1$; (v) $\int_{|z|=r} \frac{dz}{|z^2+1|^2}, r < 1$.

14. If F and G are primitives of f on U, then prove that $F - G$ is locally constant in U.

15. Let $f : B(z_0, r) \to \mathbb{C}$ be continuous. Let $\gamma_\epsilon = z_0 + \epsilon e^{it}$ for $0 \leq t \leq 2\pi$. Show that
$$\lim_{\epsilon \to 0} \frac{1}{2\pi i} \int_\gamma \frac{f(z)}{z - z_0} dz = f(z_0).$$

Chapter 7

Simple Versions of Cauchy's Theorem and Consequences

7.1 Cauchy theorem for triangles

One of the important theorem of complex function theory is the Cauchy's theorem. In the simplest form, it states that if γ is a closed curve and f is a holomorphic function in an open set $G \subseteq \mathbb{C}$ containing γ and its interior, then $\int_\gamma f(z)dz = 0$. Here the single curve γ can be replaced by a system of curves. A general version of Cauchy's theorem is presented in a later chapter as it requires considerable preparation. In this chapter we shall present two special cases, the case where γ is the boundary of a triangle and, on the basis of that, the case where the function f is holomorphic in a star-shaped open set containing γ. This will set enough background to draw some far-reaching conclusions concerning the behavior of holomorphic functions.

Let $\gamma(t) = x(t)+iy(t), a \le t \le b$ be a (piecewise smooth) path. If $f = u+iv$ then

$$\int_\gamma f(z)dz = \int_\gamma (udx - vdy) + i\int_\gamma (udy + vdx).$$

Thus we could have started by defining integrals of the form

$$\int_\gamma pdx + qdy, \tag{7.1}$$

where p and q are defined and continuous in a region G and γ varies in G. Integrals of the form (7.1) are often studied as functions of the arc γ. That is, if γ_1 and γ_2 have the same initial point and the same end point, we have $\int_{\gamma_1} pdx+qdy = \int_{\gamma_2} pdx+qdy$. Then this is equivalent to saying that the integral over any closed curve is zero. Indeed, if γ is a closed curve, then γ and $-\gamma$ have the same (opposite) end points, and if the integral depends only on the end points, we obtain $\int_\gamma pdx + qdy = \int_{-\gamma} pdx + qdy = -\int_\gamma pdx + qdy$ and consequently, $\int_\gamma pdx + qdy = 0$. Conversely, suppose the integral over any closed curve is zero. Now let γ_1 and γ_2 have the same end points. Then $\gamma_1 - \gamma_2$ is a closed curve and, it follows that $\int_{\gamma_1-\gamma_2} pdx + qdy = 0$. We have already presented a necessary and sufficient condition (Theorem 6.8) under which a line integral depends only on the end points. There are two versions of the proof. Since each are useful in evaluating line integrals we present the second version of the proof here.

Theorem 7.1. *The integral $\int_\gamma pdx + qdy$, defined in a region G depends only on the end points of γ if and only if there exists a function $U(x, y)$ in G with the partial derivatives $\frac{\partial U}{\partial x} = p$, $\frac{\partial U}{\partial y} = q$.*

Proof. For the sufficiency suppose there exists a function $U(x, y)$ in G such that $\frac{\partial U}{\partial x} = p$ and $\frac{\partial U}{\partial y} = q$. Now

$$\begin{aligned}\int_\gamma pdx + qdy &= \int_a^b \left(\frac{\partial U}{\partial x}x'(t) + \frac{\partial U}{\partial y}y'(t)\right) dt \\ &= \int_a^b \frac{d}{dt}U(x(t), y(t))dt \\ &= U(x(b), y(b)) - U(x(a), y(a)),\end{aligned}$$

and the value of this difference depends only on the end points.

To prove the necessity, we chose a fixed point $(x_0, y_0) \in G$, join it to (x, y) by a polygon γ, contained in G, whose sides are parallel to the coordinate axes and define a function by

$$U(x, y) = \int_\gamma pdx + qdy.$$

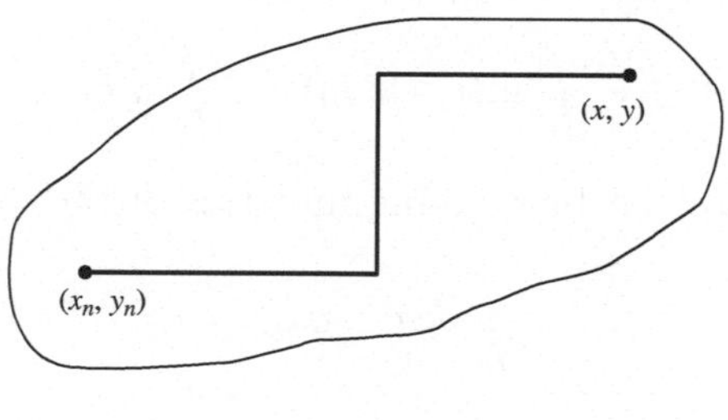

Fig. 7.1: The polygon γ in the region G

Since the integral depends only on the end points, the function is well defined. Moreover, if we chose the last segment of γ horizontal, we can keep y constant and let x vary without changing the other segments. On the last segment we can chose x for parameter and obtain

$$U(x, y) = \int^x p(x, y)dx + constant,$$

the lower limit of the integral being irrelevant. From this expression it follows at once that $\frac{\partial U}{\partial x} = p$. In the same way, by choosing the last segment vertical, we can show that $\frac{\partial U}{\partial y} = q$. □

Let $dU = (\frac{\partial U}{\partial x})dx + (\frac{\partial U}{\partial y})dy$. An expression $pdx + qdy$ which can be written in this form is an exact differential. Thus an integral $\int_\gamma pdx + qdy$ depends only on the end points if and only if the integrand is an exact differential. Here p, q and U can be either real-valued or complex-valued functions. Note that the function U, if it exists, is uniquely determined up to an additive constant, for if two functions have the same partial derivatives their difference must be constant.

When is $f(z)dz = f(z)dx + if(z)dy$ an exact differential? According to the definition there must exist a function $F(z)$ in G with the partial derivatives

$$\frac{\partial F(z)}{\partial x} = f(z),$$

$$\frac{\partial F(z)}{\partial y} = if(z).$$

If this is so, $F(z)$ satisfy the Cauchy-Riemann equation

$$\frac{\partial F}{\partial x} = -i\frac{\partial F}{\partial y};$$

since f is by assumption continuous (otherwise $\int_\gamma f dz$ would not be defined) $F(z)$ is analytic in G with the derivative $f(z)$. Thus we again prove Theorem 6.8 which states:

The integral $\int_\gamma f(z)dz$, with continuous f, depends only on the end points of γ if and only if f is the derivative of a holomorphic function in G.

We have already seen that if $f : G \to \mathbb{C}$ has a primitive F in G, then $\int_\gamma f(z)dz = 0$ for any closed path γ in G. Cauchy's theorem says that if f is holomorphic on G, then $\int_\gamma f(z)dz = 0$ under certain restrictions on G or on the path γ. In this chapter we shall only consider restrictions on G. As a by product, one shows that any $f \in H(G)$ has primitive locally, i.e. given $z \in G$, there exists $r > 0$ such that there exists a holomorphic function F on $B(z, r)$ such that $F' = f$ on $B(z, r)$.

In real analysis, we can find a primitive for any continuous function $f : [a, b] \to \mathbb{R}$ by setting $F(x) = \int_a^x f(t)dt$. This motivates the following approach to find primitives of continuous function $f : G \subset \mathbb{C} \to \mathbb{C} : F(z) = \int_{[z_0,z]} f(w)dw$ if there exists $z_0 \in G$ such that $[z_0, z] \subset G$ for all $z \in G$. (This means that G is star-shaped at z_0.)

If we want to show that F as defined above is differentiable at an arbitrary $z \in G$, observe in the following theorem that we need to verify the following:

For w near z,

$$\int_{[z_0,z]} f(v)dv + \int_{[z,w]} f(v)dv + \int_{[w,z_0]} f(v)dv = 0.$$

That is, loosely speaking, the integral of f over the boundary of a triangle is 0.

Definition 7.1. *If a, b, c are three (not necessarily non-collinear) points in $\mathbb{C}$, the triangle $\triangle(a, b, c) \equiv [a, b, c]$ is the set of points on the line segments of the form $[a, w]$, where $w \in [b, c]$. One easily sees that $\triangle(a, b, c) = \{w \in \mathbb{C} : w = t_1 a + t_2 b + t_3 c, 0 \leq t_i \leq 1, t_1 + t_2 + t_3 = 1\}$. The above description shows that $\triangle(a, b, c)$ is independent of the choice of a in the definition. One also shows that $\triangle(a, b, c)$ is the smallest convex set containing a, b, c.*

The points a, b, c are called the vertices of $[a, b, c]$ and the line segments $[a, b], [b, c], [c, a]$ are called its sides. The side $[b, c]$ is said to be opposite to the vertex a. If we let $\triangle = \triangle(a, b, c)$, the path $[a, b] + [b, c] + [c, a]$ is called the boundary of $\triangle$ and is denoted by $\partial\triangle$ or $T(a, b, c)$. Note that while in the definition of $\triangle(a, b, c)$ the order is not important, it is important for the definition of its boundary T as it gives the direction.

Theorem 7.2. *Let G be a convex open subset of $\mathbb{C}$ and let $f : G \to \mathbb{C}$ be a continuous function such that $\int_T f(z)dz = 0$ for the boundary T of any triangle $\triangle \subset G$. Then f has a primitive in G.*

Proof. To prove the theorem we fix a point z_0 in G and define the function g in G by

$$g(z) = \int_{[z_0,z]} f(w)dw.$$

We shall show that g is holomorphic and that $g' = f$. Fix a point z_1 in G. If z is any other point of G, then from the hypothesis it follows that

$$\int_{[z_0,z_1]} f(w)dw + \int_{[z_1,z]} f(w)dw + \int_{[z,z_0]} f(w)dw = 0.$$

The preceding equality can be written as

$$g(z) - g(z_1) = \int_{[z_1,z]} f(w)dw.$$

Consequently,

$$\frac{g(z) - g(z_1)}{z - z_1} - f(z_1) = \frac{1}{z - z_1} \int_{[z_1,z]} [f(w) - f(z_1)]dw.$$

Now fix a positive number ϵ. By the continuity of f, there is a positive number δ such that $|f(w) - f(z_1)| < \epsilon$ whenever $|w - z_1| < \delta$. Hence, if $|z - z_1| < \delta$, then the absolute value of the integrand in the preceding integral is bounded by ϵ, and we have

$$\begin{aligned}\left|\frac{g(z)-g(z_1)}{z-z_1}-f(z_1)\right| &= \frac{1}{|z-z_1|}\left|\int_{[z_1,z]}[f(w)-f(z_1)]dw\right| \\ &\leq \frac{1}{|z-z_1|}\epsilon|z-z_1| = \epsilon.\end{aligned}$$

This proves that

$$\lim_{z\to z_1}\frac{g(z)-g(z_1)}{z-z_1} = f(z_1),$$

in other words, g is differentiable at z_1 and $g'(z_1) = f(z_1)$. Now since z_1 is an arbitrary point of G, the result follows. □

Theorem 7.3. *Let G be star-shaped at $a \in G$ and let G be open. Further let f be a continuous complex valued function from G into $\mathbb{C}$ such that $\int_T f(z)dz = 0$ for the boundary T of any triangle $\triangle \subset G$. Then f has a primitive in G.*

Proof. Let $z_0 \in G$. Since G is open there exists $r > 0$ such that $B(z_0, r) \subset G$. If $z \in B(z_0, r)$, then $[z_0, z] \subset B(z_0, r) \subset G$. Now since G is star-shaped at $a \in G$, it follows that $[a, w] \subset G$ for any $w \in [z_0, z]$. We define

$$F(z) = \int_{[a,z]} f(w)dw.$$

Thus from the hypothesis it follows that

$$\int_{[a,z]} f(w)dw + \int_{[z,z_0]} f(w)dw + \int_{[z_0,a]} f(w)dw = 0.$$

That is,

$$F(z) = F(z_0) + \int_{[z_0,z]} f(w)dw.$$

Hence

$$\begin{aligned}\left|\frac{F(z)-F(z_0)}{z-z_0}-f(z_0)\right| &= \left|\frac{1}{z-z_0}\int_{[z_0,z]} f(w)dw - f(z_0)\right| \\ &\leq \frac{1}{|z-z_0|}\int_{[z_0,z]} |f(w)-f(z_0)|\,dw \\ &\leq |f(z)-f(z_0)|.\end{aligned}$$

Now since f is continuous at z_0 it follows that

$$\lim_{z\to z_0}\frac{F(z)-F(z_0)}{z-z_0} = f(z_0).$$

That is, $F'(z_0) = f(z_0)$. Since z_0 is an arbitrary point of G, we have $F' = f$. The result follows. □

Theorem 7.4. *(Cauchy's theorem for a triangle): Let G be any open subset of $\mathbb{C}$. Let $f \in H(G)$. Let $\triangle = [a, b, c] \subset G$ be a triangle and let T be its boundary. Then $\int_T f(z)dz = 0$.*

Proof. We shall assume that T has the counterclockwise orientation, and we let $I = \left|\int_T f(z)dz\right|$. Let a_1, b_1, c_1 be the mid points of sides opposite to the vertices a, b, c respectively. Consider the triangles $\triangle^1 = [a, c_1, b_1]$, $\triangle^2 = [c_1, b, a_1]$, $\triangle^3 = [a_1, c, b_1]$ and $\triangle^4 = [a_1, b_1, c_1]$. Let $J^j, j = 1, 2, 3, 4$ denote the boundary of $\triangle^1, \triangle^2, \triangle^3, \triangle^4$ respectively.

Then

$$\int_T f(z)dz = \sum_{j=1}^{4} \int_{T^j} f(z)dz.$$

Since $I \leq \sum_{j=1}^4 \left|\int_{T^j} f(z)dz\right|$, we see that there exists at least one j such that

$$\left|\int_{T^j} f(z)dz\right| \geq \frac{I}{4}.$$

Call the corresponding $\triangle^j$ as $\triangle_1$ and its boundary as T_1. Repeat the above process for this triangle. Proceeding this way, we get a sequence of triangles $(\triangle_n)$ with the following properties.

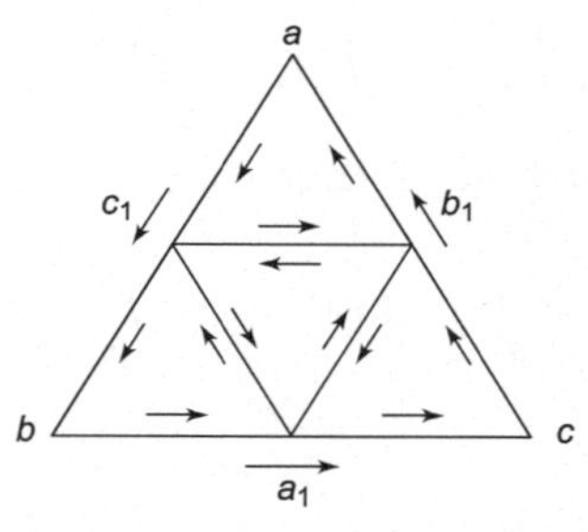

Fig. 7.2: The triangle $\triangle$

(1) The vertices of $\triangle_n$ are the midpoints of the sides of $\triangle_{n-1}$, so that for all $n \in \mathbb{N}$,
   (i) $L(T_n) = \frac{L(T_{n-1})}{2}$;
   (ii) $\text{diam}(\triangle_n) = \frac{\text{diam}(\triangle_{n-1})}{2}$ and $\triangle_n \subset \triangle_{n-1}$, where $\triangle_0 = \triangle$.

(2) $\left|\int_{T_n} f(z)dz\right| \geq \frac{1}{4}\left|\int_{T_{n-1}} f(z)dz\right|$ so that $\left|\int_{T_n} f(z)dz\right| \geq \frac{I}{4^n}$.

The sets $\triangle_n$ are compact, they decrease as n increases, and their diameters tend to 0 as n tends to infinity. Hence, $\cap_n \triangle_n \neq \emptyset$ and $\cap_n \triangle_n$ consists of a single point. Let $z_0 \in \cap_n \triangle_n$. Now $z_0 \in G$. Since f is differentiable at z_0, given any $\epsilon > 0$, there exists a $\delta > 0$ such that $|f(z) - f(z_0) - f'(z_0)(z - z_0)| \leq \epsilon|z - z_0|$ for $z \in B(z_0, \delta)$. Since $\text{diam}(\triangle_n) \to 0$, there exists n_0 such that

$\triangle_n \subset B(z_0, \delta)$ for $n \geq n_0$. Since the constant function $f(z_0)$ and the function $z \to f'(z_0)(z - z_0)$ have primitives in G, their integrals over T_n are 0. Hence we have

$$\int_{T_n} f(z)dz = \int_{T_n} [f(z) - f(z_0) - f'(z_0)(z - z_0)]dz.$$

Thus we have,

$$\begin{aligned}\left|\int_{T_n} f(z)dz\right| &\leq \int_{T_n} |f(z) - f(z_0) - f'(z_0)(z - z_0)|\, dz \\ &\leq \epsilon \int_{T_n} |z - z_0| dz \\ &\leq \epsilon \int_{T_n} \operatorname{diam}(\triangle_n) dz \\ &= \epsilon \operatorname{diam}(\triangle_n) L(T_n) \\ &\leq \epsilon \frac{\operatorname{diam}(\triangle)}{2^n} \frac{L(T)}{2^n}.\end{aligned}$$

Now

$$\left|\int_T f(z)dz\right| \leq 4^n \left|\int_{T_n} f(z)dz\right| \leq \epsilon \operatorname{diam}(\triangle) L(T).$$

Since ϵ is arbitrary, the result follows. □

If the function f is defined in the open subset G of $\mathbb{C}$, then by a primitive of f one means a holomorphic function in G whose derivative equals f. Let $C(G)$ be the space of all continuous complex valued functions on G.

Corollary 7.1. *Let G be a convex open subset of $\mathbb{C}$ and f a holomorphic function in G. Then f has a primitive in G and hence $\int_\gamma f(z)dz = 0$ for any closed path γ in G.*

Proof. Since f is holomorphic in G by Cauchy's theorem for a triangle (Theorem 7.4) we have $\int_T f(z)dz = 0$ for the boundary T of any triangle $\triangle \subset G$. Hence by Theorem 7.2 the function f has a primitive in G. By Corollary 6.5, $\int_\gamma f(z)dz = 0$ for any closed path (piecewise C^1 closed curve) γ in G. □

Corollary 7.2. *Let G be star-shaped and open in $\mathbb{C}$. Let $f \in H(G)$. Then f has a primitive in G and hence $\int_\gamma f(z)dz = 0$ for any closed path γ in G.*

Proof. By Theorem 7.4, we have $\int_T f(z)dz = 0$ for the boundary T of any triangle $\triangle \subset G$. Hence by Theorem 7.3, f has a primitive in G. By Corollary 6.5, $\int_\gamma f(z)dz = 0$. □

Theorem 7.5. *Let G be any open set, $p \in G$. Assume that $f : G \to \mathbb{C}$ is continuous and holomorphic on $G \setminus \{p\}$. Let $\triangle = [a, b, c]$ be a triangle in G with boundary T. Then $\int_T f(z)dz = 0$.*

Proof. If $p \notin \triangle$, the result follows from Theorem 7.4 . If $p \in \triangle$, let us assume that p is one of the vertices, say, a. If a, b, c are collinear, then the result is true even for a continuous f. If not, choose a point $z \in [a, b]$ and $w \in [a, c]$ very near to a. For example, choose z and w so that $d(a, z) = \frac{1}{n} = d(a, w)$. Then

$$\int_T f(\zeta)d\zeta = \int_{[a,z,w]} f(\zeta)d\zeta + \int_{[z,b,w]} f(\zeta)d\zeta + \int_{[b,c,w]} f(\zeta)d\zeta.$$

The last two integrals are 0, since they do not contain $p = a$. The first integral can be made arbitrarily small, since f is bounded on $\triangle$ (as $\triangle$ is compact and f is continuous). If p is in the interior, then by the argument of the last paragraph, the integrals over $[a, b, p]$, $[b, c, p]$ and $[c, a, p]$ are zero and their sum is $\int_T f(z)dz$. □

Corollary 7.3. *Let G be open, star-shaped and $f \in C(G)$ and $f \in H(G - \{p\})$. Then*

(i) f has a primitive in G and

(ii) $\int_\gamma f(z)dz = 0$ for every closed path γ in G.

Proof. The proof follows from Theorem 7.5, Theorem 7.3 and Corollary 6.5. □

Corollary 7.4. *Let $G \subset \mathbb{C}$ be any open set. If $f \in H(G)$, then f has local primitives.*

Proof. This follows from the fact that every open ball $B(z, r) \subset \mathbb{C}$ is star-shaped and every open set contains a neighborhood of each of its points. □

Example 7.1. *Let f be a holomorphic function on a disk D whose center is the point z_0. Assume that $|f'(z) - f'(z_0)| < |f'(z_0)|$ on $\mathbb{D}$. We shall show that f is one-to-one on D.*

Solution: Suppose $f(z_1) = f(z_2)$ and let γ be the segment connecting these two points. We have $0 = \int_\gamma f'(z)dz$. Hence,

$$\int_\gamma (f'(z) - f'(z_0))\, dz = -f'(z_0)(z_2 - z_1).$$

Taking absolute values, we get

$$|f'(z_0)||z_2 - z_1| \leq \int_\gamma |f'(z) - f'(z_0)||dz| < \int_\gamma |f'(z_0)||dz| = |f'(z_0)|\, |z_2 - z_1|,$$

which is absurd. Thus f is injective.

7.2 Cauchy integral formula

Theorem 7.6. *(Cauchy Integral Formula): Let G be an open set star-shaped at $p \in G$. Let $f \in H(G)$. Let γ be a closed path in G. Then for any $a \in G \setminus [\gamma]$, we have*

$$\frac{1}{2\pi i}\int_\gamma \frac{f(w)}{w-a}dw = f(a)\left[\frac{1}{2\pi i}\int_\gamma \frac{dw}{w-a}\right].$$

Proof. Let

$$g(w) = \begin{cases} \frac{f(w)-f(z)}{w-z} & \text{if} \quad w \in G \text{ and } w \neq a; \\ f'(a) & \text{if} \quad w = a. \end{cases}$$

Then g is continuous on G and $g \in H(G - \{a\})$. Then by Corollary 7.3, $\int_\gamma g(w)dw = 0$. But on γ, we have $g(w) = \frac{f(w)-f(a)}{w-a}$. Now $0 = \frac{1}{2\pi i}\int_\gamma g(w)dw = \frac{1}{2\pi i}\int_\gamma \frac{f(w)-f(a)}{w-a}dw = \frac{1}{2\pi i}\int_\gamma \frac{f(w)}{w-a}dw - f(a)[\frac{1}{2\pi i}\int_\gamma \frac{dw}{w-a}]$. □

Corollary 7.5. *Let G be any open set, $z_0 \in G, R > 0$ and $B(z_0, R) \subset G$. Let $f \in H(G)$. Let $\gamma(t) = z_0 + re^{it}, 0 < r < R$ and $0 \leq t \leq 2\pi$. Then we have*

$$f(z) = \frac{1}{2\pi i}\int_\gamma \frac{f(w)}{w-z}dw, z \in B(z_0, r).$$

Proof. Follows from Proposition 6.12 and Theorem 7.6. □

Example 7.2. *We shall evaluate $\int_{|z-2|=2} \frac{e^z}{z^2-9}dz$.*

Solution: Note

$$\begin{aligned} \int_{|z-2|=2} \frac{e^z}{z^2-9}dz &= \tfrac{1}{6}\int_{|z-2|=2} \frac{e^z}{z-3}dz - \frac{1}{6}\int_{|z-2|=2} \frac{e^z}{z+3}dz \\ &= \frac{1}{6}2\pi i e^3 - 0 \\ &= \frac{i\pi e^3}{3}. \end{aligned}$$

Example 7.3. *Evaluate $\int_0^{2\pi} e^{e^{i\theta}}d\theta$.*

Solution: By Cauchy's integral formula, we have

$$1 = e^0 = \frac{1}{2\pi i}\int_{|z|=1} \frac{e^z}{z}dz = \frac{1}{2\pi}\int_0^{2\pi} e^{e^{i\theta}}d\theta.$$

Thus

$$\int_0^{2\pi} e^{e^{i\theta}}d\theta = 2\pi.$$

Corollary 7.6. *Any holomorphic function f on an open set G is locally a power series, i.e., an analytic function. In particular, f' is also holomorphic in G.*

Proof. Let $z_0 \in G$ and $B(z_0, r) \subset G$ for some $r > 0$. Given $z \in B(z_0, r)$, choose ρ such that $0 < \rho < r$ and $z \in B(z_0, \rho)$. Let $\gamma(t) = z_0 + \rho e^{it}, 0 \leq t \leq 2\pi$. Then for $w \in [\gamma]$,

$$\begin{aligned} \frac{1}{w-z} &= \frac{1}{(w-z_0)+(z_0-z)} \\ &= \frac{1}{(w-z_0)\left[1-\frac{z-z_0}{w-z_0}\right]} \\ &= \sum_{n=0}^{\infty} \frac{(z-z_0)^n}{(w-z_0)^{n+1}} \end{aligned}$$

since $\left|\frac{z-z_0}{w-z_0}\right| < 1$. The above series is uniformly convergent on $[\gamma]$ for fixed z, z_0. Hence by Cauchy Integral formula, for $z \in B(z_0, \rho)$,

$$\begin{aligned} f(z) &= \frac{1}{2\pi i}\int_\gamma \frac{f(w)}{w-z}dw \\ &= \frac{1}{2\pi i}\int_\gamma \left(\sum_{n=0}^{\infty}\left[\frac{(z-z_0)^n}{(w-z_0)^{n+1}}\right] f(w)\right) dw \\ &= \sum_{n=0}^{\infty} a_n(z-z_0)^n \end{aligned}$$

where $a_n = \frac{1}{2\pi i}\int_\gamma \frac{f(w)}{(w-z_0)^{n+1}}dw$. □

Thus we have arrived at the heart of the Cauchy theory: The function $f : G \to \mathbb{C}$ is holomorphic if and only if it is analytic in G.

Remark 7.1. 1. It is important to observe that if $B(z_0, r) \subset G$, then we have a power series expansion $f(z) = \sum_{n=0}^{\infty} a_n(z-z_0)^n$ for $z \in B(z_0, r)$. In particular, the power series expansion of f at $z_0 \in G$ holds in the largest open disk (centered at z_0) contained in G. That is, in the disk $|z - z_0| < \text{dist}(z_0, \mathbb{C} \setminus G)$, the function f is represented by a power series of the form $\sum_{n=0}^{\infty} a_n(z-z_0)^n$. We shall show below that the $a_n's$ are the Taylor's coefficients of f.

2. From Corollary 7.6 it follows that if $f : G \to \mathbb{C}$ is differentiable on an open set G, then f is infinitely differentiable. That is, a holomorphic function is differentiable to all orders, and its real and imaginary parts have continuous partial derivatives to all orders. This should be contrasted with the fact that for any $k \in \mathbb{N}$, there exist functions from $\mathbb{R}$ to $\mathbb{R}$ which are k-times differentiable but not $k + 1$ times differentiable. There are well-known examples of continuous real-valued functions on $\mathbb{R}$ that are nowhere differentiable. An indefinite integral of such a function is differentiable everywhere while its derivative is

differentiable nowhere. By taking an n-fold indefinite integral, one can produce a function that is differentiable to order n yet whose nth derivative is nowhere differentiable. But this does not occur in the realm of complex differentiation.

On the other hand, there exists a real-valued function which is infinitely differentiable on all of $\mathbb{R}$ but does not have a power series expansion at the point $x = 0$: The standard example is $f(x) = e^{-\frac{1}{x^2}}$, for $x \neq 0$ and $f(0) = 0$. In fact f is continuous and infinitely differentiable at the origin and $f^{(n)}(0) = 0$ for all $n \in \mathbb{N}$. So if we could expand it in a power series using Taylor's theorem, then we would get a series of zeros which does not agree with $f(x)$ any where except at $x = 0$.

For example let

$$f(x) = \begin{cases} x^2 \sin \frac{1}{x} & \text{if} \quad x \neq 0; \\ 0 & \text{if} \quad x = 0 \end{cases}$$

where x is real. The function has a first derivative at all values of x for which $0 \leq x \leq 1$ but does not have a second derivative in $0 \leq x \leq 1$. The only place where there is any question as to existence of the first derivative is at $x = 0$. But at $x = 0$, the derivative is $\lim\limits_{\triangle x \to 0} \dfrac{f(0 + \triangle x) - f(0)}{\triangle x} = 0$ and so exists. Further

$$f'(x) = \begin{cases} 2x \sin \frac{1}{x} - \cos \frac{1}{x} & \text{if} \quad x \neq 0; \\ 0 & \text{if} \quad x = 0 \end{cases}$$

and the second derivative of f exists for all x such that $0 < x \leq 1$. At $x = 0$, notice that the second derivative does not exist.

3. The power series expansion of the holomorphic function $f : G \to \mathbb{C}$ around $z_0 \in G$ is valid in the largest possible disk $B(z_0, R) \subset G$. The only subtle point in the above proof is that the coefficients a_n, which apparently depend on the path γ (i.e., on ρ) are, in fact independent. This can be seen in two ways. The first is to recall that $a_n = \frac{f^{(n)}(z_0)}{n!}$ (Proposition 6.11). The second way is to invoke the uniqueness for power series. If

$$f(z) = \sum_{n=0}^{\infty} a_n(r)(z - z_0)^n = \sum_{n=0}^{\infty} a_n(\rho)(z - z_0)^n,$$

then they agree on $B(z_0, \rho)$ if $\rho < r$. Hence $a_n(\rho) = a_n(r)$ for all n.

Thus the coefficients a_n of the power series expansion of f around z_0 are unique. This follows from the uniqueness theorem for power series (Corollary 3.7) or the integral representation for a_n (Proposition 6.11).

For future reference we state the result on uniqueness of power series as a lemma.

Lemma 7.1. *Suppose that for some $\delta > 0$, $\sum_{n=0}^{\infty} a_n h^n = \sum_{n=0}^{\infty} b_n h^n$ for all $h \in \mathbb{C}$ such that $|h| < \delta$. Then $a_n = b_n$ for all n.*

Proof. Let $g(h) = \sum_{n=0}^{\infty} a_n h^n = \sum_{n=0}^{\infty} b_n h^n$ for $|h| < \delta$. By Proposition 6.11, $a_n = \frac{g^{(n)}(0)}{n!}$, also $b_n = \frac{g^{(n)}(0)}{n!}$. Hence $a_n = b_n$ for all n. □

Remark 7.2. The most important fall-out from this is the following: If $f \in H(\mathbb{C})$, then there exists a power series expansion of f around 0 with infinite radius of convergence: $f(z) = \sum_{n=0}^{\infty} a_n z^n$ for $z \in \mathbb{C}$.

Remark 7.3. The proof of Corollary 7.6 will not yield a single power series expansion for a holomorphic function f on a connected open set. This is not due to the inadequacy of the method. It is just not possible to do so. For instance, if we let $f(z) = \frac{1}{z}$ on $\mathbb{C}^*$ and $a \in \mathbb{C}^*$, we have

$$f(z) = \frac{1}{a} \sum_{n=0}^{\infty} \left(\frac{-(z-a)}{a} \right)^n$$

for all $z \in B(a, |a|)$. By the uniqueness of a_n's (Remark 7.1), this is the power series expansion of f around a. Clearly, its radius of convergence is $|a|$. Thus, it is impossible to find a single power series expansion for f on all of $\mathbb{C}^*$.

Theorem 7.7. *(Mean value property): Let $f \in H(G)$. Suppose $B(a, R) \subset G$ and $\gamma_r(t) = a + re^{it}, 0 < r < R$. Then*

$$f(a) = \frac{1}{2\pi} \int_0^{2\pi} f(a + re^{it}) dt.$$

Proof. By Cauchy Integral formula,

$$\begin{aligned} f(a) &= \tfrac{1}{2\pi i} \int_{\gamma_r} \frac{f(z)}{z-a} dz \\ &= \tfrac{1}{2\pi i} \int_0^{2\pi} \frac{f(a+re^{it})}{re^{it}} re^{it} i dt \\ &= \frac{1}{2\pi} \int_0^{2\pi} f(a + re^{it}) dt. \end{aligned}$$

□

The right side of the above equation may be considered as the "average" or "mean" of f on the circle γ_r. Thus the theorem says that the mean value of f on any circle γ_r centered at a (for $0 < r < R$) is $f(a)$.

Theorem 7.8. *(Cauchy's Integral formula for derivatives): Let $f \in H(G)$ and $B(a, R) \subset G$. Let $\gamma(t) = a + re^{it}$ for any r such that $0 < r < R$ and $0 \leq t \leq 2\pi$. We have*

$$f^{(n)}(a) = \frac{n!}{2\pi i} \int_\gamma \frac{f(w)}{(w-a)^{n+1}} dw.$$

Proof. Immediate consequence of Proposition 6.13 and Corollary 7.5. □

Example 7.4. *Evaluate $\int_0^{2\pi} e^{(e^{i\theta} - i\theta)} d\theta$.*

Solution: By Cauchy integral formula for derivatives, we have

$$\begin{aligned} \frac{d}{dz}e^z\bigg|_{z=0} &= \frac{1}{2\pi i}\int_{|z|=1}\frac{e^z}{z^2}dz \\ &= \frac{1}{2\pi}\int_0^{2\pi} e^{(e^{i\theta}-i\theta)}d\theta, \end{aligned}$$

therefore,

$$\int_0^{2\pi} e^{(e^{i\theta}-i\theta)}d\theta = 2\pi.$$

Theorem 7.9. *(Cauchy's estimate): Let f be holomorphic in $B(a,R)$ and suppose $|f(z)| \leq M$ for all z in $B(a,R)$. Then*

$$|f^{(n)}(a)| \leq \frac{n!M}{R^n}.$$

Proof. Let $\gamma(t) = a + re^{it}, 0 \leq t \leq 2\pi, 0 < r < R$. From Theorem 7.8, it follows that

$$\begin{aligned} |f^{(n)}(a)| &\leq \frac{n!}{2\pi}\int_\gamma \frac{|f(w)|}{|w-a|^{n+1}}|dw| \\ &\leq \frac{n!}{2\pi}\frac{M}{r^{n+1}}2\pi r \\ &= \frac{n!M}{r^n}. \end{aligned}$$

Since $r < R$ is arbitrary, the result follows by letting $r \to R^-$. □

Definition 7.2. *Let f be an infinitely differentiable function on an open set $G \subset \mathbb{C}$. The Taylor series of the function at $a \in G$ is defined by*

$$\sum_{n=0}^{\infty}\frac{f^{(n)}(a)}{n!}(z-a)^n.$$

Corollary 7.7. *Let $f \in H(G)$ and $B(a,R) \subset G$ where G is an open subset of $\mathbb{C}$. Then the Taylor series of f converges to f absolutely and uniformly on compact subsets of $B(a,R)$.*

Remark 7.4. The real analogue of this result is false. For instance, the function $f(x) = e^{-\frac{1}{x^2}}$ for $x \neq 0$ and $f(0) = 0$ is infinitely differentiable and $f^{(n)}(0) = 0$. Thus, the Taylor series of f at 0 is the identically zero power series, hence does not converge to f. Hence, we conclude that for any $r > 0$, there is no holomorphic function g on $B(0,r)$ such that $f = g$ on $(-r,r)$.

Corollary 7.8. *Let $f(z) = \sum_{n=0}^{\infty} a_n z^n$ and $g(z) = \sum_{n=0}^{\infty} b_n z^n$ be power series with radii of convergence R_1 and R_2 respectively. Let $h(z) = \sum_{n=0}^{\infty} c_n z^n$ where $c_n = \sum_{k=0}^{n} a_k b_{n-k}$. Then h has radius of convergence at least $R = min\{R_1, R_2\}$ and we have $h(z) = f(z)g(z)$.*

Proof. In $B(0,R)$ both f and g are holomorphic and we know that $a_n = \frac{f^{(n)}(0)}{n!}$ and $b_n = \frac{g^{(n)}(0)}{n!}$. The product function fg is holomorphic in $B(0,R)$. Hence, it is represented in $B(0,R)$ by its Taylor series $f(z)g(z) = \sum_{n=0}^{\infty} \alpha_n z^n$ where

$$\begin{aligned}\alpha_n &= \frac{(fg)^{(n)}(0)}{n!} \\ &= \sum_{r=0}^{n} \frac{1}{n!}\frac{n!}{r!(n-r)!} f^{(r)}(0) g^{(n-r)}(0) \\ &= \sum_{r=0}^{n} a_r b_{n-r} = c_n.\end{aligned}$$

(We had used the Leibniz's formula for the n-th derivative of a product.) □

Remark 7.5. The proper setting to study the power series (or analytic functions) is over the field of complex numbers. To illustrate this point, let us look at the function $f : \mathbb{R} \to \mathbb{R}$ defined by $f(x) = \frac{1}{1+x^2}$. This function is infinitely differentiable and its Taylor series at $0 \in \mathbb{R}$ is $\sum_{n=0}^{\infty}(-1)^n x^{2n}$ and it has radius of convergence 1. There is nothing in the behavior of f in $\mathbb{R}$ accounts for this restriction on the radius of convergence. However, if we try to extend the domain of the function to $\mathbb{C}$, we detect the reason for this. The points $\pm i$, where the function is not defined, lie on the unit circle. We leave it to the reader to think over this.

Example 7.5. *Let f be an entire function and suppose there is a constant M, an $R > 0$ and an integer $n \geq 1$ such that $|f(z)| \leq M|z|^n$ for $|z| > R$. Then f is a polynomial of degree less than or equal to $n - 1$.*

Solution: The function f is an entire function. Let $f(z) = \sum_{m=0}^{\infty} a_m z^m$ be the Taylor's series expansion of f around origin. Notice that the radius of convergence (of the Taylor series around origin) is ∞ and for $m \geq 0, a_m$ is given by

$$a_m = \frac{1}{2\pi}\int_0^{2\pi} \frac{f(re^{i\theta})}{(re^{i\theta})^{m+1}} d\theta, \text{ for all } r > 0.$$

Now for $m > n, r > R$ we have

$$\begin{aligned}|a_m| &\leq \tfrac{1}{2\pi}\int_0^{2\pi} \frac{|f(re^{i\theta})|}{r^{m+1}} d\theta \\ &\leq \tfrac{1}{2\pi}\int_0^{2\pi} \frac{M|re^{i\theta}|^n}{r^{m+1}} d\theta \\ &= \tfrac{1}{2\pi}\int_0^{2\pi} \frac{Mr^n}{r^{m+1}} d\theta \\ &= \frac{M}{r^{m+1-n}}.\end{aligned}$$

Hence $|a_m| \le \frac{M}{r^{m+1-n}}$. Taking limit $r \to \infty$, we get $|a_m| = 0$, (for $m+1 > n$). This implies $a_m = 0$, for $m+1 > n$. Thus $f(z) = \sum_{m=0}^{\infty} a_m z^m = \sum_{m=0}^{n-1} a_m z^m$. That is, $f(z)$ is a polynomial of degree $\le n-1$.

Example 7.6. *Let $f : \mathbb{C} \to \mathbb{C}$ be an entire function and let $a > 0$ and $b > 0$ be constants. If*
(a) $|f(z)| \le a\sqrt{|z|} + b$ for all z, then f is a constant.
(b) If $|f(z)| \le a|z|^{\frac{5}{2}} + b$ for all z, then f is a polynomial of degree at most 2.

Solution: (a) Using Cauchy's integral formula for derivatives, we get

$$|f^{(k)}(0)| \le \frac{k!}{2\pi} \int_{|z|=R} \left| \frac{f(z)}{z^{k+1}} \right| |dz|$$

$$\le \frac{k!}{2\pi R^{k+1}} \int_{|z|=R} |a\sqrt{|z|} + b||dz| = \frac{k!(a\sqrt{|z|} + b)}{R^k} = o(1)$$

as $R \to \infty$. So $f^{(k)}(0) = 0$ for $k \ge 1$, and f reduces to a constant.
(b) For $k \ge 3$, we get

$$\begin{aligned} |f^{(k)}(0)| &\le \tfrac{k!}{2\pi} \int_{|z|=R} \left| \frac{f(z)}{z^{k+1}} \right| |dz| \\ &\le \tfrac{k!}{2\pi R^{k+1}} \int_{|z|=R} |a\sqrt{|z|^5} + b||dz| \\ &= \frac{k!(a\sqrt{|z|^5} + b)}{R^k} = o(1) \end{aligned}$$

as $R \to \infty$. So $f^{(k)}(0) = 0$ for $k \ge 3$ and f reduces to a polynomial of degree at most 2.

Definition 7.3. *A sequence (f_n) of functions in an open set $\Omega \subset \mathbb{C}$ is said to converge to f uniformly on compact subsets of Ω if to every compact set $K \subset \Omega$ and to every $\epsilon > 0$ there corresponds an $N = N(K, \epsilon)$ such that $|f_n(z) - f(z)| < \epsilon$ for all $z \in K$ if $n > N$.*

For instance, the sequence (z^n) converges to 0 uniformly on compact subsets of $B(0,1)$, but the convergence is not uniform in $B(0,1)$.

If in a neighborhood of a point z a function f is the uniform limit of analytic functions, then f is analytic at z. (see below.) Therefore, if (f_n) is a sequence of analytic functions such that $\lim_{n\to\infty} f_n(z) = f(z)$ exists for each z in a region Ω, then f will be analytic if the convergence is "locally uniform", uniform in some neighborhood of each point.

Definition 7.4. *A sequence (f_n) converges locally uniformly on a region $\Omega \subset \mathbb{C}$ if for each closed disk C in Ω, $\lim_{n\to\infty} f_n(z) = f(z)$ exists uniformly on C.*

Consider again the sequence $f_n(z) = z^n$. If $0 < r_0 < 1$ then this sequence converges uniformly to 0 in the disk $|z| \leq r_0$, and so it converges locally uniformly to 0 in the disk $|z| < 1$. However, the sequence does not converge uniformly in the disk $|z| < 1$.

It is useful to note two conditions which are equivalent with local uniform convergence. The first comes from Cauchy's criterion. A sequence converges locally uniformly if for each closed disk C and $\epsilon > 0$ there exists an integer N such that

$$\max_{z \in C} |f_n(z) - f_m(z)| < \epsilon$$

for $m, n \geq N$.

The second equivalent condition is "uniform convergence on compact sets." If K is a compact subset of Ω, then $\lim_{n \to \infty} f_n(z) = f(z)$ should exist uniformly on K. This condition arises by covering K with (the interiors of) finitely many closed disks in Ω(using the definition of compactness).

Example 7.7. *Convergent power series give rise to examples of local uniform convergence. Recall that if $f(z) = \sum_{k=0}^{\infty} a_k(z - z_0)^k$ has radius of convergence $R > 0$, then for each $r < R$ the partial sums*

$$f_n(z) = \sum_{k=0}^{n} a_k(z - z_0)^k$$

converge uniformly for $|z - z_0| \leq r$. If C is a closed disk contained in the region $\Omega = \{z \in \mathbb{C} : |z - z_0| < R\}$, then for some $r < R$, C is also contained in $\{z \in \mathbb{C} : |z - z_0| \leq r\}$. Therefore, (f_n) converges uniformly on C, and since C is arbitrary, (f_n) converges locally uniformly on Ω.

If f and Ω are as above then it follows that f is holomorphic in Ω, and for each $m > 0$ the mth derivative of f is represented on Ω by the convergent power series

$$f^{(m)}(z) = \sum_{k=m}^{\infty} a_k(k(k-1)\cdots(k-m+1))(z - z_0)^{k-m}.$$

From the above discussion it also follows that $(f_n^{(m)})$ converges locally uniformly to $f^{(m)}$. Theorem 7.10 below proves that if (f_n) is a sequence of holomorphic functions defined on an open set G and (f_n) converges uniformly on compact subsets of G to a function f then f is holomorphic on G and for any $k \in \mathbb{N}$, $f_n^{(k)}$ also converges to $f^{(k)}$ uniformly on compact subsets of G.

Theorem 7.10. *Let G be open and $f_n \in H(G)$ for $n \in \mathbb{N}$. Assume that f_n converges uniformly on compact subsets of G to a function f. Then $f \in H(G)$. Furthermore, for any $k \in \mathbb{N}$, $f_n^{(k)}$ converges to $f^{(k)}$ uniformly on compact subsets of G.*

Proof. By Theorem 2.58, f is continuous on G. Choose $r > 0$ such that the closed ball $\overline{B(z_0, r)} \subset G$. Let $z \in B(z_0, r)$. Let $\gamma(t) = z_0 + re^{it}, 0 \leq t \leq 2\pi$. By Cauchy Integral formula,

$$f_n(z) = \frac{1}{2\pi i} \int_\gamma \frac{f_n(w)}{w - z} dw.$$

Since $[\gamma]$ is compact, $d(z, [\gamma]) > 0$ by Theorem 2.57. Hence $\frac{f_n(w)}{w-z} \to \frac{f(w)}{w-z}$ uniformly on $[\gamma]$. Hence $f(z) = \frac{1}{2\pi i} \int_\gamma \frac{f(w)}{w-z} dw$. Now, f is holomorphic on $B(z_0, r)$ either by Theorem 6.7 or by Proposition 6.13. Now, by Theorem 7.8

$$f_n^{(k)}(z) - f^{(k)}(z) = \frac{k!}{2\pi i} \int_\gamma \frac{f_n(w) - f(w)}{(w - z)^{k+1}} dw, \quad z \in B(z_0, r).$$

If $z \in \overline{B(z_0, \rho)}$ with $\rho < r$, then we have

$$|f_n^{(k)}(z) - f^{(k)}(z)| \leq \frac{k!}{2\pi} \max_{w \in \gamma} |f_n(w) - f(w)| \cdot \frac{2\pi r}{(r - \rho)^{k+1}} \to 0$$

as $n \to \infty$. Thus $f_n^{(k)}$ converges to $f^{(k)}$ uniformly on closed disks $\overline{B(z_0, r)}$ and hence uniformly on compact subsets of G. ☐

Compare this with the situation on the real line, where sequences of infinitely differentiable functions can converge uniformly to nowhere differentiable functions. The local uniform convergence on $\mathbb{R}$ of a sequence of real valued $\mathbb{C}^\infty$ functions does not imply even the differentiability let alone the indefinite differentiability of the limit function. In fact, by the classical Weierstrass approximation theorem, every continuous real-valued function on $\mathbb{R}$, even a nowhere differentiable one, is the local uniform limit of a sequence of polynomials.

7.3 Liouville's theorem

Definition 7.5. *If $f : \mathbb{C} \to \mathbb{C}$ is holomorphic then f is called an entire function.*

Example 7.8. *The complex valued functions e^z, $sinz$, $cosz$, $a_0 + a_1 z + \cdots + a_n z^n$, $a_0, a_1, \ldots, a_n \in \mathbb{C}$ are examples of entire functions.*

Remark 7.6. If g is an entire function then $f(z) = e^{g(z)}$ is entire and $f(z) \neq 0$ for all $z \in \mathbb{C}$. Conversely, if $f(z)$ is any entire function which is never zero, one can show that $f(z)$ is of the form $e^{g(z)}$. To this end we observe that the function $\frac{f'(z)}{f(z)}$ being holomorphic in the whole plane, is the derivative of an entire function $g(z)$ (Theorem 7.3 and Theorem 7.4). Now $\frac{d}{dz}(f(z)e^{-g(z)}) = 0$. Hence $f(z)$ is a constant multiple of $e^{g(z)}$, the constant can be absorbed in $g(z)$.

Theorem 7.11. *(Liouville's theorem): Let f be a bounded entire function. Then f is a constant.*

Proof. Suppose $|f(z)| \leq M$ for all $z \in \mathbb{C}$. Then by Cauchy's estimate, since f is holomorphic in any disk $B(z, R)$ we have, $|f'(z)| \leq \frac{M}{R}$. Since R is arbitrary, it follows that $f'(z) = 0$ for each z in $\mathbb{C}$. Hence f is a constant. □

We shall now prove the Fundamental theorem of algebra which says that the complex field is algebraically closed. That is, every nonconstant polynomial with complex coefficients has at least one complex zero.

Theorem 7.12. *(Fundamental theorem of algebra): If $P(z)$ is a nonconstant polynomial with complex coefficients then there is a complex number a with $P(a) = 0$.*

Proof. Let $P(z) = a_0 + a_1 z + \cdots + a_n z^n, a_n \neq 0$ be a polynomial of degree $n > 0$, with complex coefficients. Suppose $P(z) \neq 0$ for all z and let $f(z) = \frac{1}{P(z)}$. Then f is an entire function. Since $a_n \neq 0$ we can write, for $z \neq 0$,

$$P(z) = a_n z^n \left(1 + \frac{a_{n-1}}{a_n}\frac{1}{z} + \cdots + \frac{a_1}{a_n}\frac{1}{z^{n-1}} + \frac{a_0}{a_n}\frac{1}{z^n}\right).$$

Consequently,

$$|P(z)| \geq |a_n||z|^n \left(1 - \frac{|a_{n-1}|}{|a_n|}\frac{1}{|z|} - \cdots - \frac{|a_1|}{|a_n|}\frac{1}{|z|^{n-1}} - \frac{|a_0|}{|a_n|}\frac{1}{|z|^n}\right).$$

Clearly, as $r \to \infty$,

$$\left(\frac{|a_{n-1}|}{|a_n|}\frac{1}{r} + \cdots + \frac{|a_1|}{|a_n|}\frac{1}{r^{n-1}} + \frac{|a_0|}{|a_n|}\frac{1}{r^n}\right) \to 0.$$

That is for $\epsilon > 0$, in particular, $\epsilon = \frac{1}{2}$ there exists R such that for $r \geq R$,

$$\frac{|a_{n-1}|}{|a_n|}\frac{1}{r} + \cdots + \frac{|a_1|}{|a_n|}\frac{1}{r^{n-1}} + \frac{|a_0|}{|a_n|}\frac{1}{r^n} < \epsilon = \frac{1}{2}.$$

So for $|z| \geq R$,

$$|P(z)| \geq |a_n|R^n(1 - \epsilon) = |a_n|R^n \cdot \left(\frac{1}{2}\right).$$

Hence

$$|f(z)| \leq \frac{2}{|a_n|R^n}, \ |z| \geq R.$$

On the other hand, since $|f|$ is a continuous function, it is bounded on the compact set $|z| \leq R$. Thus f is a bounded entire function, and so it is constant by Liouville's theorem. This is the desired contradiction, because the polynomial P was assumed to be nonconstant. □

The proof just given is purely an existence proof; it does not provide one with a method of locating the roots of a polynomial. The problem of finding approximations to the roots of complex polynomials is a major theme in numerical analysis.

We shall say $a \in \mathbb{C}$ is a zero of the polynomial P of multiplicity (or order) $m \geq 1$, if $P^{(n)}(a) = 0$ for $n = 0, 1, 2, \ldots, m-1$, while $P^{(m)}(a) \neq 0$.

Corollary 7.9. *Any polynomial P of degree n has exactly n zeros in $\mathbb{C}$, counting multiplicities. (A zero of order m say, is counted as m zeros.)*

Proof. By the Fundamental theorem of algebra, there exists a $z_1 \in \mathbb{C}$ such that $P(z_1) = 0$. Consequently, there is a polynomial Q, of degree $n-1$, such that $P(z) = (z - z_1)Q(z)$. The proof is completed by induction on n. □

Definition 7.6. *(Algebraic and Transcendental Numbers) A number $z \in \mathbb{C}$ is called algebraic if it is the solution of an algebraic or polynomial equation with integral coefficients; otherwise it is called transcendental.*

Theorem 7.13. *(i) The set of algebraic numbers is countably infinite.*
(ii)The set of transcendental numbers is uncountable.

Proof. Let the algebraic equation of degree k be

$$a_0 z^k + a_1 z^{k-1} + \cdots + a_{k-1} z + a_k = 0 \tag{7.2}$$

where $a_0, a_1, \ldots, a_k \in \mathbb{Z}$. For each positive integer n, there are only finite number of equations with

$$k + |a_0| + |a_1| + \cdots + |a_k| = n. \tag{7.3}$$

Since each equation with degree k has k roots it follows that the set

$$B_n = \{z \in \mathbb{C} : z \text{ is a root of (7.2) with condition (7.3)}\}$$

is a finite set. Hence the set of all algebraic numbers which is given by $\cup_{n=1}^{\infty} B_n$ is countable. Since $\mathbb{C}$ is uncountable and since the complement of algebraic numbers in $\mathbb{C}$ are transcendental numbers, it also follows that the set of
transcendental numbers is uncountable. □

Example 7.9. *Let $f : \mathbb{C} \longrightarrow \mathbb{C}$ be a nonconstant entire function. Then $f(\mathbb{C})$ is dense in $\mathbb{C}$.*

Solution: Suppose $f(\mathbb{C})$ is not dense in $\mathbb{C}$. That is, $\overline{f(\mathbb{C})} \neq \mathbb{C}$. Then for some $w \in \mathbb{C}$ and $\epsilon > 0$, we have $|f(z) - w| \geq \epsilon$ for all $z \in \mathbb{C}$. The function $\frac{1}{f(z)-w}$ is then entire and bounded in modulus by $\frac{1}{\epsilon}$, so by Liouville's theorem, $\frac{1}{f(z)-w}$ is a constant and so is f.

Example 7.10. *An entire function f with $Ref(z) \leq 0$ for $z \in \mathbb{C}$ is a constant.*

Solution: Notice that since f is entire, e^f is entire and $|e^{f(z)}| = |e^{\mathrm{Re}f(z)}| \leq e^0 = 1$ in $\mathbb{C}$. Hence by Liouville's theorem, $\phi(z) = e^{f(z)}$ is a constant function. Thus $e^{f(z)}f'(z) = \phi'(z) = 0$ in $\mathbb{C}$. Thus $f'(z) = 0$ since $e^{f(z)} \neq 0$. Thus f is constant by Theorem 5.8.

Example 7.11. *If f is entire and $Imf(z) \leq M$ for some constant $M > 0$ and for all $z \in \mathbb{C}$ then f is a constant.*

Solution: Since f is entire, e^{-if} is entire and $|e^{-if(z)}| = |e^{\mathrm{Im}f(z)}| \leq e^M$. Thus $\phi(z) = e^{-if(z)}$ is bounded. Thus, by Liouville's theorem, $\phi(z) = e^{-if(z)}$ is a constant function and hence f is a constant function.

Remark 7.7. In the above context one may note that the condition $|f(z)| \leq M$ implies

$$-M \leq \mathrm{Re}f(z) \leq M; -M \leq \mathrm{Im}f(z) \leq M.$$

Applying the above with $f, -f, if$ and $-if$ we can show that any one of the four inequalities suffices to prove that f is constant. Further suppose f is an entire function satisfying any one of the following conditions for all $z \in \mathbb{C}$:

(a) $|f(z)| \geq 1$

(b) $\mathrm{Re}f(z)$ or $\mathrm{Im}f(z)$ has no zeros,

then f is a constant. Notice that if $f = u + iv$ is entire and $\mathrm{Re}f(z) = u$ has no zeros then since $\mathbb{C}$ is connected, $u(\mathbb{C})$ is connected and $0 \notin u(\mathbb{C})$. Thus either $u(z) > 0$ for all $z \in \mathbb{C}$ or $u(z) < 0$ for all $z \in \mathbb{C}$. If $u(z) > 0$ for all $z \in \mathbb{C}$ then $|e^{-f}| = e^{-u} \leq 1$. By Liouville's theorem e^f is a constant and hence f is constant.

7.4 Morera's theorem

The Cauchy's theorem has a useful converse which is known as Morera's theorem.

Theorem 7.14. *(Morera's theorem): Let G be an open set and let $f : G \to \mathbb{C}$ be a continuous function. Suppose $\int_T f(z)dz = 0$ where T is the boundary of a triangle $\triangle \subset G$. Then f is holomorphic in G.*

Proof. Let $a \in G$. Since G is open, there exists $r > 0$ such that $B(a, r) \subset G$. Now by Theorem 7.2, the function f has a primitive F in $B(a, r)$. Since F is holomorphic in $B(a, r)$, so is its derivative f. Now since $a \in G$ is arbitrarily chosen, hence f is differentiable on all of G. □

Corollary 7.10. *Let G be an open set and let $f : G \to \mathbb{C}$ be continuous. Suppose that for any $\epsilon > 0$ there exists a holomorphic function g on G such that $|f(z) - g(z)| < \epsilon$. Then f is holomorphic.*

Proof. To see this, fix ϵ and let g be the corresponding holomorphic function. We write $f = (f - g) + g$. Then for any triangle $\triangle \subset G$ with $\partial\triangle = T$, we have $\int_T f(z)dz = \int_T (f(z) - g(z))dz + \int_T g(z)dz$. Since g is holomorphic, the second term on the right is zero. Now,

$$\begin{aligned} \left|\int_T f(z)dz\right| &= \left|\int_T (f(z) - g(z))dz\right| \\ &\le \int_T |f(z) - g(z)|\, dz < L(T)\epsilon. \end{aligned}$$

Since ϵ is arbitrary, $\int_T f(z)dz = 0$. By Morera's theorem, f is holomorphic. □

We now present some typical applications of Morera's theorem.

Example 7.12. *Let g be a continuous function on $[0, 1]$ and define $F(z) = \int_0^1 g(t)\sin(tz)dt$. Then F is holomorphic in $\mathbb{C}$.*

Solution: It is not so difficult to see that F is continuous. We now use Fubini's theorem for continuous functions on a compact set to see

$$\begin{aligned} \int_T F(z)dz &= \int_T \left[\int_0^1 g(t)\sin(tz)dt\right] dz \\ &= \int_0^1 g(t)\left[\int_T \sin(tz)dz\right] dt = 0. \end{aligned}$$

Example 7.13. *Let f be continuous on $\mathbb{C}$ and analytic on $\{z \in \mathbb{C} : Imz \neq 0\}$. We shall use Morera's theorem to show that f must be analytic on $\mathbb{C}$.*

Solution: By Morera's theorem, it suffices to show that $\int_\gamma f(z)dz = 0$ for all rectangles γ in $\mathbb{C}$. Since f is analytic on $\{z \in \mathbb{C} : \text{Im}z \neq 0\}$, which is simply connected (its complement with respect to the extended complex plane is connected), it is enough to consider rectangles which contain part of the real axis in their interiors.

Let γ be such a rectangle (see Fig.7.3) and l be the segment of $\mathbb{R}$ in its interior. For $\epsilon > 0$ small enough, draw line segments l_1 and l_2 parallel to the

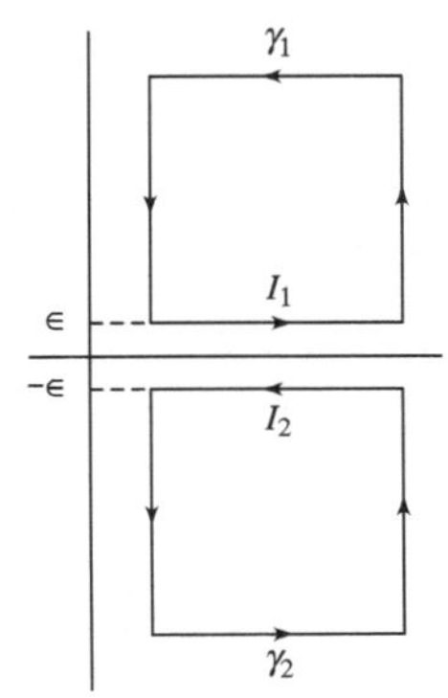

Fig. 7.3: The rectangle γ

real axis at distance ϵ above and below it, forming contours γ_1 and γ_2. Since f is continuous, its integral depends continuously on the path. So as ϵ tends to 0,

$$\int_{\gamma_1} f(z)dz + \int_{\gamma_2} f(z)dz = \int_{\gamma_1+\gamma_2} f(z)dz \to \int_{\gamma} f(z)dz, \qquad (7.4)$$

since the integrals along l_1 and l_2 have opposite orientation, in the limit, they cancel each other. By Cauchy's theorem, the left side of (7.4) is always 0, so $\int_\gamma f(z)dz = 0$.

Example 7.14. *Let f be a continuous complex valued function on $[0,1]$, and define the function g by $g(z) = \int_0^1 f(t)e^{tz}dt, z \in \mathbb{C}$. Then g is holomorphic in the entire complex plane.*

Solution: The integrand in the integral defining g is a continuous function of the pair of variables $(t,z) \in [0,1] \times \mathbb{C}$, implying that g is continuous. If $R \subset \mathbb{C}$ is a rectangle, then by Fubini's theorem and the analyticity of e^{tz} with respect to z, we have $\int_R g(z)dz = \int_R \int_0^1 f(t)e^{tz}dtdz = \int_0^1 \int_R f(t)e^{tz}dzdt = 0$. By Morera's theorem, g is analytic.

7.5 Zeros of a holomorphic function

Definition 7.7. *If $f : G \to \mathbb{C}$ is holomorphic then $a \in G$ is called a zero of f if $f(a) = 0$. Let $\mathcal{Z}(f) = \{z \in G : f(z) = 0\} = f^{-1}(\{0\})$. The set $\mathcal{Z}(f)$ is called the zero set of f. We shall say $a \in G$ is a zero of f of multiplicity (or order) $m \geq 1$, if $f^{(n)}(a) = 0$ for $n = 0, 1, 2, \cdots, m-1$, while $f^{(m)}(a) \neq 0$.*

In such a case the Taylor series for f about a (in an open ball $B(a,r) \subset G$) has the form $\sum_{n=m}^{\infty} a_n(z-a)^n$, where $a_m = \frac{f^{(m)}(a)}{m!} \neq 0$.

Now we can write $f(z) = (z-a)^m g(z)$, where the function g is given by

$$g(z) = \begin{cases} (z-a)^{-m} f(z) & \text{if } \ z \in G \setminus \{a\}; \\ a_m & \text{if } \ z = a. \end{cases}$$

Note that g is holomorphic in G. The function g is nonzero at a, i.e., $g(a) = a_m \neq 0$. Hence there exists a neighborhood $B(a,\delta), \delta > 0$ of a such that $g(z) \neq 0$ for all $z \in B(a,\delta)$ by Lemma 3.3. Thus $f(z) \neq 0$ for all $z \in B(a,\delta) \setminus \{a\}$. Therefore, it follows that zeros of finite multiplicity of holomorphic functions are isolated.

Theorem 7.15. *Let G be a region in $\mathbb{C}$ and let $f : G \to \mathbb{C}$ be a holomorphic function. Then $f \equiv 0$ if and only if there is a point $a \in G$ such that $f^{(n)}(a) = 0$ for each nonnegative integer n.*

Proof. If $f \equiv 0$ then $f^{(n)}(a) = 0$ for each $n \in \mathbb{Z}_+$, the set of nonnegative integers. For the converse, let $A = \{z \in G : f^{(n)}(z) = 0 \text{ for all } n \geq 0\}$. Note that $A \neq \emptyset$ as $a \in A$. We shall show that A is both open and closed in G. Since G is connected it will follow that $A = G$ and hence $f \equiv 0$ in G.

To see that A is closed let (z_k) be a sequence in A such that $z = \lim\limits_{k\to\infty} z_k$. Since $f^{(n)}$ is continuous for each $n \geq 0$, it follows that $f^{(n)}(z) = \lim\limits_{k\to\infty} f^{(n)}(z_k) = 0$. So $z \in A$ and A is closed.

To see that A is open, let $a \in A$ and let $R > 0$ be such that $B(a, R) \subset G$. Then $f(z) = \sum_{n=0}^{\infty} a_n(z-a)^n$ for $z \in B(a, R)$ where $a_n = \frac{f^{(n)}(a)}{n!} = 0$ for each $n \geq 0$ (as $a \in A$). Hence $f(z) = 0$ for all z in $B(a, R)$. Hence $B(a, R) \subset A$. Thus A is open and this completes the proof of the theorem. □

So we have proved that a holomorphic function f can have a zero of infinite order only if it is the zero function. That is if f is a holomorphic function and not the zero function, in the region $G \subseteq \mathbb{C}$, then each zero of f is of finite order and the zeros are isolated.

In contrast to the behavior of holomorphic functions, it is possible for a real-valued C^∞ function on $\mathbb{R}$ to have a zero of infinite order and yet not vanish identically. A standard example is the function that equals 0 at the origin and equals $e^{-\frac{1}{x^2}}$ elsewhere. One can show that the origin is a zero of infinite order of this function.

Theorem 7.16. *Let G be a region in $\mathbb{C}$ and let $f : G \to \mathbb{C}$ be a holomorphic function. Then the zero set of f, $\mathcal{Z}(f) = \{z \in G : f(z) = 0\}$ has a limit point in G if and only if $f \equiv 0$ on G.*

Proof. If $f \equiv 0$ in G then $\mathcal{Z}(f) = G$ which contains a limit point of itself. For the converse suppose that $a \in G$ is a limit point of $\mathcal{Z}(f)$. Hence there exists a sequence (z_k) in $\mathcal{Z}(f)$ such that $\lim\limits_{k\to\infty} z_k = a$. As f is continuous, $f(a) = f(\lim\limits_{k\to\infty} z_k) = \lim\limits_{k\to\infty} f(z_k) = 0$. Since G is open, there exists $R > 0$ such that $B(a, R) \subset G$. Suppose there is an integer $n \geq 1$ such that $f(a) = f'(a) = \cdots = f^{(n-1)}(a) = 0$ and $f^{(n)}(a) \neq 0$. Expanding f in power series in $B(a, R)$ we have $f(z) = \sum_{k=n}^{\infty} a_k(z-a)^k$. Let $g(z) = \sum_{k=n}^{\infty} a_k(z-a)^{k-n}$. Then g is holomorphic in $B(a, R)$, $f(z) = (z-a)^n g(z)$ and $g(a) = a_n \neq 0$. Since g is continuous we can find an r, $0 < r < R$, such that $g(z) \neq 0$ for all $z \in B(a, r)$. But since a is a limit point of $\mathcal{Z}(f)$, the ball $B(a, r)$ contains a point b of $\mathcal{Z}(f)$ such that $b \neq a$. Thus $f(b) = 0, |b-a| < r$ and $b \neq a$. This gives $0 = f(b) = (b-a)^n g(b)$. So $g(b) = 0$ as $b \neq a$ which is a contradiction. Hence we cannot find an integer $n \geq 1$ such that $f(a) = f'(a) = \cdots = f^{(n-1)}(a) = 0$ but $f^{(n)}(a) \neq 0$. Thus $f^{(m)}(a) = 0$ for all $m \in \mathbb{Z}_+$. By Theorem 7.15, $f \equiv 0$. □

Example 7.15. *The theorem fails if we drop the assumption that G is connected: If $G = G_1 \cup G_2$, and G_1 and G_2 are disjoint open sets, put $f = 0$ in G_1 and $f = 1$ in G_2.*

From the above theorem we conclude that if f is a holomorphic function and not the zero function in the region G of $\mathbb{C}$ then the zero set of f, $\mathcal{Z}(f)$ has no limit points in G. Since an uncountable subset of $\mathbb{C}$ always contains limit points of itself (Theorem 2.15), we obtain that the zero set $\mathcal{Z}(f) = f^{-1}(0)$ of a nonzero holomorphic function f is at most countable.

Corollary 7.11. *Let f be a nonconstant holomorphic function on the region $G \subset \mathbb{C}$. Then the set $f^{-1}(w)$ is countable for every complex number w.*

Proof. Suppose $f^{-1}(w)$ is an uncountable subset of G. Let $g(z) = f(z) - w$. Then, g is holomorphic and g has uncountable number of zeros in G. Thus the zero set $\mathcal{Z}(g) = g^{-1}(\{0\})$ of g has a limit point in G. Hence by Theorem 7.16, $g \equiv 0$. Thus f is the constant function w which contradicts the hypothesis that f is not a constant function. □

Theorem 7.17. *(The Identity Theorem): Let f and g be holomorphic functions defined on the region $G \subseteq \mathbb{C}$. Then the set $\{z \in G : f(z) = g(z)\}$ has a limit point in G if and only if $f \equiv g$ in G.*

Proof. If $f \equiv g$ in G then the set $\{z \in G : f(z) = g(z)\} = G$ contains a limit point of itself. For the converse, suppose the set $\{z \in G : f(z) = g(z)\}$ has a limit point in G. Then the zero set of $f - g$ has a limit point in G. Thus $f - g \equiv 0$ in G. □

The above theorem is also sometimes referred to as uniqueness theorem.

Example 7.16. *Let f be analytic in a region R with zeros at a sequence of points z_i which tend to a limit $z^* \in R$. Then f is identically zero in R.*

Solution: As f is continuous, $f(z^*) = \lim_{i\to\infty} f(z_i) = 0$. If f is not identically zero then we can expand f as a power series about z^*,

$$f(z) = \sum_{n=1}^{\infty} a_n (z - z^*)^n,$$

in which not all the coefficients a_n are zero. If a_m is the first non-zero coefficient then

$$f(z) = \sum_{n=m}^{\infty} a_n (z - z^*)^n = (z - z^*)^m g(z),$$

where g is analytic in R and $g(z^*) = a_m \neq 0$. Suppose that $|g(z^*)| = 2\epsilon$. Then as g is continuous there exists $\delta > 0$ such that if $|z - z^*| < \delta$,

$$||g(z)| - 2\epsilon| = ||g(z)| - |g(z^*)|| \leq |g(z) - g(z^*)| < \epsilon,$$

and therefore $|g(z)| > \epsilon$ in this disk. Thus $f(z) \neq 0$ for all $z \in B(z^*, \delta) \setminus \{z^*\}$, which contradicts the hypothesis that z^* is a limit point of zeros of f. Therefore f is identically zero in R.

Let f and g be holomorphic in a region $G \subseteq \mathbb{C}$ and let $U \subset G$ is such that $U' \cap G \neq \emptyset$ where U' is the derived set of U. If $f \equiv g$ on U, from the identity theorem (Theorem 7.17), it follows that $f \equiv g$ on G. Suppose $f \in H(U)$ and $g \in H(G)$ where $U \subset G$ and $U' \cap G \neq \emptyset$. If $g(z) = f(z)$ for each $z \in U$ then we shall say g is the **analytic continuation** of f into G. For more details on analytic continuation see Chapter 16.

Example 7.17. *Does there exist a nonconstant analytic function f in $|z| < 2$ such that $f(\frac{1}{n}) = 0, n = 1, 2, 3, \ldots$.*

Solution: Suppose f is a nonconstant analytic function in $|z| < 2$ such that $f\left(\frac{1}{n}\right) = 0, n = 1, 2, 3, \ldots$. Then $f(0) = f\left(\lim\limits_{n\to\infty} \dfrac{1}{n}\right) = \lim\limits_{n\to\infty} f\left(\dfrac{1}{n}\right) = 0$. Thus the zero set $Z(f)$ of f contains the set $\{0, 1, \frac{1}{2}, \frac{1}{3}, \cdots\}$. That is, $Z(f)$ contains a limit point of itself. Hence by Theorem 7.16, $f \equiv 0$. That is, f is a constant function. But by our hypothesis, f is nonconstant.

Example 7.18. *Let f be an analytic function in the region $G \subset \mathbb{C}$. Assume that for each point z in G, there is a positive integer n such that the nth derivative of f vanishes at z. Then f is a polynomial.*

Solution: Let D be an open disk such that $\overline{D} \subset G$. It will suffice to show that there is an n such that $f^{(n)}$ has infinitely many zeros in D. For then, the zeros of $f^{(n)}$ will have a limit point in G (by Theorem 2.45), forcing $f^{(n)}$ to vanish identically in G (Theorem 7.16), and it follows that f is a polynomial of degree, at most, $n - 1$. By hypothesis, D is the union of the sets $Z_n = \{z \in D : f^{(n)}(z) = 0\}$ for $n = 1, 2, \ldots$. Since D is uncountable, at least one Z_n is, in fact, uncountable.

Example 7.19. *Suppose f is an entire function, and that in every power series*

$$f(z) = \sum_{n=0}^{\infty} a_n (z - a)^n$$

at least one coefficient is zero. Then f is a polynomial.

Solution: The hypothesis say for each $a \in \mathbb{C}$, there exists a $m \in \mathbb{N}$ such that $\frac{f^{(m)}(a)}{m!} = a_m = 0$. By Example 7.18, f is a polynomial.

Example 7.20. *Prove that there is no holomorphic function f from the open unit disk $\mathbb{D}$ into itself such that $f(\frac{1}{n}) = 2^{-n}$, for $n = 2, 3, 4, \ldots$.*

Solution: We prove this by the method of contradiction. So suppose that $f : \mathbb{D} \to \mathbb{D}$ is a holomorphic function such that $f(\frac{1}{n}) = 2^{-n}$, for $n = 2, 3, 4, \ldots$. Since f is holomorphic it is continuous and hence

$$f(0) = f\left(\lim_{n\to\infty} \frac{1}{n}\right) = \lim_{n\to\infty} f\left(\frac{1}{n}\right) = \lim_{n\to\infty} 2^{-n} = 0.$$

We shall now prove that $f^{(k)}(0) = 0, k = 0, 1, 2, 3, \ldots$ We prove this by induction over k. We have already shown that $f^{(k)}(0) = 0$, for $k = 0$. Now suppose that

$$f(0) = f'(0) = f''(0) = \cdots = f^{(m)}(0) = 0, m \geq 0.$$

Then $z = 0$ is a zero of f of order $m + 1 \geq 1$. So there exists a function g analytic in $\mathbb{D}$ such that

$$f(z) = z^{m+1}g(z), \text{ for all } z \in \mathbb{D}. \tag{7.5}$$

Differentiating both the sides $(m + 1)$ times we get (by Leibnitz rule)

$$f^{(m+1)}(z) = \sum_{j=0}^{m+1} \binom{m+1}{j}^2 (m - j + 1)! z^j g^{(j)}(z),$$

for all $z \in \mathbb{D}$.

In particular, for $z = 0$, we get

$$f^{(m+1)}(0) = \binom{m+1}{0}^2 (m + 1)! g(0).$$

Hence

$$f^{(m+1)}(0) = (m + 1)! g(0). \tag{7.6}$$

From (7.5) we get $g(z) = \frac{f(z)}{z^{m+1}}$ for all $z \in \mathbb{D}, z \neq 0$. So

$$g(0) = \lim_{z \to 0} g(z) = \lim_{z \to 0} \frac{f(z)}{z^{m+1}} = \lim_{n \to \infty} \frac{f(\frac{1}{n})}{(\frac{1}{n})^{m+1}} = \lim_{n \to \infty} \frac{n^{m+1}}{2^n} = 0.$$

So from (7.6) we get $f^{(m+1)}(0) = (m+1)! \cdot 0 = 0$. Hence $f(0) = f'(0) = f''(0) = \cdots = f^{(m)}(0) = 0$ implies $f^{(m+1)}(0) = 0$. Therefore, by induction principle

$$f^{(k)}(0) = 0, \text{ for all } \ k = 0, 1, 2, \ldots$$

Since f is holomorphic in $\mathbb{D}$,

$$f(z) = \sum_{k=0}^{\infty} a_k z^k, \text{ for all } z \in \mathbb{D}, \tag{7.7}$$

where $a_k = \frac{f^{(k)}(0)}{k!}, k = 0, 1, 2, \ldots$. Thus we get $a_k = 0, k = 0, 1, 2, \ldots$. Hence from (7.7), we have $f(z) = 0$ for all $z \in \mathbb{D}$. This is a contradiction, because $f(\frac{1}{n}) = 2^{-n}$, for $n = 2, 3, 4, \ldots$. Hence there is no holomorphic function f from the open unit disk $\mathbb{D}$ into itself such that $f(\frac{1}{n}) = 2^{-n}$, for $n = 2, 3, 4, \ldots$.

Example 7.21. *Does there exists an analytic function f in G such that $f(\frac{1}{n}) = \frac{1}{n+1}, n = 1, 2, 3, \ldots$ where (i) $G = \{z \in \mathbb{C} : |z| < 1\}$, (ii) $G = \mathbb{C} \setminus \{-1\}$, (iii) $G = \mathbb{C}$*

Solution: Let $g(z) = \frac{z}{z+1}$. Then $g(\frac{1}{n}) = \frac{1}{n+1}$ and g is analytic in $|z| < 1$. Note the function g is analytic in $\mathbb{C} \setminus \{-1\}$ and g is not analytic in $\mathbb{C}$. Further

$$g(0) = g\left(\lim_{n\to\infty} \frac{1}{n}\right) = \lim_{n\to\infty} g\left(\frac{1}{n}\right) = \lim_{n\to\infty} \frac{1}{n+1} = 0.$$

If there exist another analytic function f such that $f(\frac{1}{n}) = \frac{1}{n+1}, n = 1, 2, 3, \ldots$ then $f(0) = 0$. Let $h = f - g$. Then h is analytic in $|z| < 1$. Let $A = \{z \in \mathbb{C} : h(z) = 0\}$. Then $A \supset \{0, 1, \frac{1}{2}, \frac{1}{3}, \ldots\}$. By Theorem 7.16, $h \equiv 0$ in $|z| < 1$. Hence $g \equiv f$ in $|z| < 1$. Thus such a function f exists in $|z| < 1, \mathbb{C} \setminus \{-1\}$ but not in $\mathbb{C}$.

Example 7.22. *Do there exists functions $f(z)$ and $g(z)$ that are analytic at $z = 0$ and that satisfy (a) $f(\frac{1}{n}) = f(-\frac{1}{n}) = \frac{1}{n^2}, n = 1, 2, 3, \ldots$ (b) $g(\frac{1}{n}) = g(-\frac{1}{n}) = \frac{1}{n^3}, n = 1, 2, 3, \ldots$?*

Solution: (a) The function $f(z) = z^2$ is entire and satisfies $f(\frac{1}{n}) = f(-\frac{1}{n}) = \frac{1}{n^2}, n = 1, 2, 3, \ldots$.
(b) By the identity theorem, in a disk centered at the origin, g would have to be z^3 and $-z^3$, which is not possible; therefore, no such function g can exist.

Example 7.23. *Let f be an entire function such that $f(x) = e^x$ for all $x \in \mathbb{R}$. Then $f(z) = e^z$.*

Solution: Let $g(z) = e^z$. Then g is an entire function such that $g(x) = e^x$ for all $x \in \mathbb{R}$. Hence $f - g$ is an entire function and $(f - g)(x) = 0$ for all $x \in \mathbb{R}$. Therefore the set $\{z \in \mathbb{C} : (f - g)(z) = 0\}$ contains a limit point of itself. By Theorem 7.16, $f - g \equiv 0$. Thus $f(z) = e^z$ for all $z \in \mathbb{C}$.

Let G be a region in $\mathbb{C}$ and $H(G)$ be the set of holomorphic functions on G. We shall show below that $H(G)$ is an integral domain with usual operations of addition and multiplication.

Example 7.24. *Let G be a region and let f and g be holomorphic functions on G such that $f(z)g(z) = 0$ for all z in G. Show that either $f \equiv 0$ or $g \equiv 0$.*

Solution: Let $f(z) \neq 0$ for all $z \in G$. Then $g(z) = 0$ for all $z \in G$ and $g \equiv 0$ as $\mathbb{C}$ is an integral domain. Now suppose there exists $a \in G$ such that $f(a) = 0$. This implies there exists $\delta > 0$ such that if $w \in B(a, \delta), w \neq a$ then $f(w) \neq 0$. This is so as zeros of analytic functions are isolated. Since $B(a, \delta) \subset G$, $f(z)g(z) = 0$ for all $z \in B(a, \delta) \setminus \{a\}$, hence $g(z) = 0$ for all $z \in B(a, \delta) \setminus \{a\}$. Thus the set of zeros of g in G has a limit point in G and therefore $g \equiv 0$ in G.

Example 7.25. *If $Q(z) = a_0 + a_1 z + \cdots + a_{n-1} z^{n-1} + z^n$ $(n \geq 1)$ then there exists $R > 0$ such that for all z, $|z| \geq R$,*

$$\frac{|z|^n}{2} \leq |Q(z)| \leq 2|z|^n.$$

Solution: Note if $P(z) = a_0 + a_1 z + a_2 z^2 + \cdots + a_n z^n$ then

$$\begin{aligned}|P(z)| &\leq |a_0| + |a_1||z| + \cdots + |a_n||z|^n \\ &\leq |a_0| + |a_1| r + \cdots + |a_n| r^n\end{aligned}$$

if $|z| \leq r$. That is,

$$|P(z)| \leq |a_n| r^n \left(1 + \frac{|a_{n-1}|}{|a_n|}\frac{1}{r} + \cdots + \frac{|a_0|}{|a_n|}\frac{1}{r^n}\right)$$

if $|z| \leq r$. Letting $r \to \infty$,

$$\frac{|a_{n-1}|}{|a_n|}\frac{1}{r} + \cdots + \frac{|a_0|}{|a_n|}\frac{1}{r^n} \to 0.$$

Hence given $\epsilon > 0$, there exists $r_0 > 0$ such that for $r > r_0$,

$$\frac{|a_{n-1}|}{|a_n|}\frac{1}{r} + \cdots + \frac{|a_0|}{|a_n|}\frac{1}{r^n} < \epsilon.$$

Thus for $r > r_0, |z| \leq r$,

$$|P(z)| \leq |a_n| r^n (1 + \epsilon) \tag{7.8}$$

Similarly, as in Theorem 7.12 one can show that

$$|P(z)| \geq |a_n| r^n (1 - \epsilon) \tag{7.9}$$

if $r > r_1$. Let $R = \max\{r_0, r_1\}$. Thus if $r > R$, we have from (7.8) and (7.9),

$$|a_n| r^n (1 - \epsilon) \leq |P(z)| \leq |a_n| r^n (1 + \epsilon).$$

Taking $a_n = 1, \epsilon = \frac{1}{2}$, the result follows for $Q(z)$.

Example 7.26. *Let f be entire. Suppose that $|f(z)| \to \infty$ as $|z| \to \infty$. Then show that f is a polynomial.*

Solution: Since $f(z) \to \infty$ as $z \to \infty$, given $K > 0$ there is an R such that $|f(z)| > K$ for $|z| \geq R$. Note that f can have only finitely many zeros in $\triangle_R = \{z \in \mathbb{C} : |z| < R\}$ since otherwise the set of all zeros would have a limit point in $\triangle_R$ and by Theorem 7.16, $f \equiv 0$ which is a contradiction. Let the zeros of f be $\alpha_1, \alpha_2, \ldots, \alpha_m \in \mathbb{C}$, of orders $k_1, k_2, \ldots, k_m$ respectively. Then g defined by $g(z) = \frac{f(z)}{P(z)}$, where $P(z) = \prod_{j=1}^{m} (z - \alpha_j)^{k_j}$ is an entire function and $g(z) \neq 0$ in $\mathbb{C}$. Further $\frac{1}{g(z)}$ is entire and $\frac{1}{g(z)} \neq 0$ in $\mathbb{C}$. Thus we see that for sufficiently large $|z|$,

$$\left|\frac{1}{g(z)}\right| = \left|\frac{P(z)}{f(z)}\right| \leq \frac{2}{K}|z|^N$$

where $N = k_1 + k_2 + \cdots + k_m$. From Example 7.5, it follows that $\frac{1}{g(z)}$ is a polynomial. But $\frac{1}{g(z)} \neq 0$ in $\mathbb{C}$, therefore $\frac{1}{g(z)}$ is a nonzero constant. Hence there exist $c \in \mathbb{C}$ such that $f(z) = cP(z)$ is a polynomial.

Example 7.27. *A nonzero entire function cannot have infinitely many zeros in a disk of finite radius.*

Solution: Let f be an entire function and suppose f has infinitely many zeros, say $z_1, z_2, z_3, \dots$ in the disk $B(a, r)$ for some $r > 0, a \in \mathbb{C}$. This implies $A = \{z_1, z_2, z_3, \dots\}$ is an infinite bounded set and hence it has a limit point by Theorem 2.45.Thus by Theorem 7.16, $f \equiv 0$ in $B(a, r)$ and hence $f(z) = 0$ for all $z \in \mathbb{C}$. But by hypothesis f is a nonzero entire function.

Example 7.28. *If f is a nonconstant entire function then show that for each $\lambda \in \mathbb{C}$, the set $\{z \in \mathbb{C} : f(z) = \lambda\}$ is countable and $f(\mathbb{C})$ is uncountable.*

Solution: Let $g(z) = f(z) - \lambda$. Then g is an entire function and if $\{z \in \mathbb{C} : g(z) = 0\} = \{z \in \mathbb{C} : f(z) = \lambda\}$ is an uncountable set then we obtain $g \equiv 0$. Hence $f \equiv \lambda$. But by our hypothesis f is nonconstant. Now suppose $f(\mathbb{C})$ is countable and $f(\mathbb{C}) = \{\lambda_1, \lambda_2, \dots\}$. Then $\mathbb{C} = \cup_{i=1}^{\infty} f^{-1}(\lambda_i)$. Since from the first part it follows that each $f^{-1}(\lambda_i), i = 1, 2, \dots$ is countable, and countable union of countable sets is countable, hence $\mathbb{C}$ is countable. But this is a contradiction.

7.6 Maximum modulus principle

Theorem 7.18. *(Maximum Modulus Theorem): Let G be a region contained in $\mathbb{C}$ and $f : G \to \mathbb{C}$ be a holomorphic function. Suppose there is a point $a \in G$ such that $|f(a)| \geq |f(z)|$ for all z in G. Then f is a constant in G.*

Proof. Since G is open, there exists $r > 0$ such that $\overline{B(a, r)} \subset G$. Let $\gamma(t) = a + re^{it}$ for $0 \leq t \leq 2\pi$. By Mean value property,

$$f(a) = \frac{1}{2\pi} \int_0^{2\pi} f(a + re^{it}) dt.$$

Hence $|f(a)| \leq \frac{1}{2\pi} \int_0^{2\pi} |f(a + re^{it})| dt \leq |f(a)|$ since $|f(a + re^{it})| \leq |f(a)|$ for all t. This gives that

$$0 = \int_0^{2\pi} [|f(a)| - |f(a + re^{it})|] dt;$$

since the integrand is non-negative it follows that $|f(a)| = |f(a + re^{it})|$ for all t. Since r was arbitrary, we have that f maps any disk $B(a, R) \subset G$ into circle $|z| = |\alpha|$ where $\alpha = f(a)$.

But this implies that f is constant on $B(a, R)$ by Example 5.22. In particular $f(z) = \alpha$ for $z \in B(a, R)$. Since the ball $B(a, R)$ contains a limit point of itself hence by Theorem 7.17, $f \equiv \alpha$ in G. □

Corollary 7.12. *Let G be a bounded open set in $\mathbb{C}$ and suppose f is a continuous function from $\overline{G}$ into $\mathbb{C}$ such that f is holomorphic in G. Then either f is a constant or $|f|$ attains its maximum on the boundary of G.*

Proof. Since G is bounded in $\mathbb{C}$ therefore $\overline{G}$ is compact. Hence there is a point $a \in \overline{G}$ such that $|f(a)| \geq |f(z)|$ for all $z \in \overline{G}$. If f is not a constant function then $|f|$ attains its maximum on the boundary of G because of the maximum modulus theorem. $\square$

Can you have a minimum principle in an analogous manner? If we look at $f(z) = z$ on $B(0,1)$, we realize that we need to formulate it carefully.

Theorem 7.19. *(Minimum Modulus Principle): If f is a nonconstant holomorphic function on a region $G \subseteq \mathbb{C}$ then $z \in G$ cannot be a relative local minimum of $|f|$ unless $f(z) = 0$.*

The following theorem is an important application of Cauchy's estimate which plays an important role in analytic continuation.

Theorem 7.20. *For each $n, n = 1, 2, \ldots$, let f_n be a power series $f_n(z) = \sum_{k=0}^{\infty} a_k^{(n)}(z - z_0)^k$ convergent on $\Omega = \{z \in \mathbb{C} : |z - z_0| < R\}$. In addition suppose there exists a constant $M < \infty$ such that $|f_n(z)| \leq M$ for all $z \in \Omega$ and n. If $\lim\limits_{n\to\infty} a_k^{(n)} = a_k$ exists for each k, then $\{f_n\}$ converges uniformly on every compact subsets of Ω to $f(z) = \sum_{k=0}^{\infty} a_k(z - z_0)^k$.*

Proof. From Cauchy's estimate, it follows that for each k and n,

$$\left|a_k^{(n)}\right| \leq \frac{M}{R^k}.$$

Letting $n \to \infty$, we obtain $|a_k| \leq \frac{M}{R^k}$. Therefore $f(z) = \sum_{k=0}^{\infty} a_k(z - z_0)^k$ converges on Ω. To see that $\{f_n\}$ converges uniformly on every compact subsets of Ω to f, it suffices to produce for each $r < R$ and $\epsilon > 0$ an integer N such that

$$\max_{|z-z_0|\leq r} |f_n(z) - f(z)| < \epsilon$$

if $n \geq N$. To this end fix r and ϵ, and let l be an integer large enough that

$$\frac{MR}{R - r}\left(\frac{r}{R}\right)^l < \frac{\epsilon}{4}.$$

This is possible because $\frac{r}{R} < 1$. Let N be large enough so that if $n \geq N$ and if $|z - z_0| \leq r$, then

$$\left|\sum_{k=0}^{l-1} \left(a_k{}^{(n)} - a_k\right)(z - z_0)^k\right| < \frac{\epsilon}{2}.$$

For the same values of n and z the above inequalities together with triangle inequality imply that

$$
\begin{aligned}
&|f_n(z) - f(z)| \\
&= \left|\sum_{k=0}^{\infty}(a_k^{(n)} - a_k)(z-z_0)^k\right| \\
&\leq \left|\sum_{k=0}^{l-1}(a_k^{(n)} - a_k)(z-z_0)^k\right| + \left|\sum_{k=l}^{\infty}(a_k^{(n)} - a_k)(z-z_0)^k\right| \\
&\leq \frac{\epsilon}{2} + \sum_{k=l}^{\infty} |a_k^{(n)} - a_k| r^k \\
&\leq \frac{\epsilon}{2} + 2M \sum_{k=l}^{\infty} \frac{r^k}{R^k} \\
&= \tfrac{\epsilon}{2} + \tfrac{2MR}{R-r}\left(\tfrac{r}{R}\right)^l < \epsilon.
\end{aligned}
$$

Thus, $\{f_n\}$ converges uniformly on every compact subsets of Ω and the theorem is proved. □

7.7 Exercises

1. Let $f(z) = u(z) + iv(z)$ be holomorphic in $|z| < 1$, u and v real. Show that $\int_0^{2\pi} u(re^{i\theta})^2 d\theta = \int_0^{2\pi} v(re^{i\theta})^2 d\theta$ for $0 < r < 1$ if $u(0)^2 = v(0)^2$.

2. Prove that there is no holomorphic function f in the open unit disk such that $f(\frac{1}{n}) = \frac{(-1)^n}{n^2}$, for $n = 2, 3, 4, \ldots$.

3. Let f be an entire function taking a constant value on the interval $\{x+iy : 0 \leq x \leq 1, y = 0\}$. Show that f is a constant function.

4. Can there be an entire function f satisfying $f(\frac{1}{n}) = \frac{1}{1+n^2}$? Justify your answer.

5. The function $\sin\frac{1}{1-z}$ has an infinite sequence of zeros convergent at the point $z = 1$, yet it is not a constant. Doesn't this contradict the identity (uniqueness) theorem?

6. Let f and g be entire functions such that for all $z \in \mathbb{C}$, $\text{Re}f(z) \leq k\text{Re}g(z)$ for some real constant k (independent of z). Show that there are constants $a, b \in \mathbb{C}$ such that $f(z) = ag(z) + b$.

7. Show that if f and g are analytic functions on a region G such that $\bar{f}g$ is analytic then either f is constant or $g \equiv 0$.

8. Evaluate the following integrals:

   (i) $\int_{|z|=3} \frac{e^{3z}}{(z+1)^6} dz$ (ii) $\int_{|z|=2} \frac{dz}{z^2-1}$ (iii) $\int_{|z|=1} \frac{e^z}{6z} dz$

   (iv) $\int_{|z|=2} \frac{dz}{z^2+1}$ (v) $\int_{|z-1|=3} \frac{z^3+e^{z^2}}{z+i} dz$

9. Let f be a holomorphic function in the open subset G of $\mathbb{C}$. Let the point z_0 of G be a zero of f of order m. Prove that there is a branch of $f^{\frac{1}{m}}$ in some open disk centered at z_0.

10. Let $f : [a, b] \to \mathbb{C}$ be continuous. Then show that $F(z) = \int_a^b f(t)e^{-tz}dt$ is holomorphic on $\mathbb{C}$.

11. Let f be holomorphic in the region $G, B(z_0, R) \subset G$. Let $\gamma_r(t) = z_0 + re^{it}$, for any $r \in (0, R), 0 \leq t \leq 2\pi$. Show that

$$\frac{1}{2\pi}\int_0^{2\pi} |f(z_0 + re^{it})|^2 dt = \sum_{n=0}^{\infty} |a_n|^2 r^{2n}.$$

12. Let $f(z)$ be analytic on the open unit disk $\mathbb{D} = \{z \in \mathbb{C} : |z| < 1\}$. Prove that there is a sequence (z_n) in $\mathbb{D}$ such that $|z_n| \to 1$ and $(f(z_n))$ is bounded.

13. Let f be an analytic function in the unit disk $\mathbb{D}$.

    (i) Prove that there is a sequence (z_n) in the unit disk with $\lim\limits_{n\to\infty} |z_n| = 1$ and $\lim\limits_{n\to\infty} f(z_n)$ exists (and is finite).

    (ii) Assume f is nonconstant. Prove that there are two sequences (z_n) and (w_n) in the disk such that $\lim\limits_{n\to\infty} |z_n| = \lim\limits_{n\to\infty} |w_n| = 1$, and such that both limits $\lim\limits_{n\to\infty} f(z_n)$ and $\lim\limits_{n\to\infty} f(w_n)$ exist (finite) and are not equal.

14. Determine the regions in which the following functions are defined and holomorphic:

$$(i)\ \ f(z) = \int_0^1 \frac{dt}{1 + tz}; \qquad (ii)\ \ g(z) = \int_0^1 \frac{e^{tz}}{1 + t^2}dt;$$

$(iii)\ \ h(z) = \int_{-1}^1 \frac{e^{tz}}{1+t^2}dt.$

Hint: Combine Morera's theorem and Fubini's theorem.

15. Show that an odd entire function has only odd terms (odd powers of z) in its power series (Taylor's series) expansion.

16. Let $f(z) = \frac{z^2}{z+2}$. Find the maximum of $|f|$ on $\overline{B(0, 1)}$.

17. Find the maximum and minimum moduli of $|f|$ where $f(z) = e^{z^2}$ on $\overline{B(0, 1)}$.

18. Let $U = \{z \in \mathbb{C} : |\text{Re}z| \leq 1, |\text{Im}z| \leq 1\}$. Let $f \in C(\overline{U}) \cap H(U)$ be such that $f(z) = 0$ if $\text{Re}z = 1$. Show that $f = 0$ on $\overline{U}$. *Hint:* Consider $g(z) = f(z)f(-z)f(iz)f(-iz)$.

19. Let f be continuous on $\overline{B(0,1)}$ and holomorphic on $B(0,1)$. Assume $f(z) = 0$ for z such that $\text{Im}z \geq 0$ and $|z| = 1$. Then $f = 0$ on $\overline{B(0,1)}$.

20. If $f \in H(B(0,R))$ is nonconstant, then $M(r) = \max\{|f(z)| : |z| = r\}$ is a strictly increasing function of r on $(0,R)$.

21. Let U be bounded and open. Let $f \in H(U) \cap C(\overline{U})$ be nonconstant. Assume that $|f(z)| = 1$ for $z \in \partial U$. Then show that f vanishes at some point of U.

22. Let U be bounded and open. Let $f \in C(\overline{U})$ be nonconstant, holomorphic on U. If $|f|$ is constant on ∂U, then show that $f(a) = 0$ for some $a \in U$.

23. Let f and g be continuous on $\overline{U}$ and holomorphic on U. Assume that U is connected and that $f = g$ on ∂U. Then $f = g$ on U.

24. Let G be a region in $\mathbb{C}$ and $f \in H(G)$ be nonconstant. Suppose $K \subset G$ is compact. Then show that for any $\alpha \in \mathbb{C}$, the set $\{z \in K : f(z) = \alpha\}$ is finite.

25. Show that $f(z) = \int_0^1 \frac{dt}{1-tz}$ is holomorphic in $B(0,1)$. Find a power series expansion for f.

26. Let Ω be a region in $\mathbb{C}$ and suppose $f : \Omega \to \mathbb{C}$ is holomorphic and it has no zeros. Show that $|f(z)|$ cannot have a local minimum.

27. Let f be an entire function with Taylor series expansion $f(z) = \sum_{n=0}^{\infty} a_n z^n$ about 0.

    (a) Let $g(z) = e^z f(z)$ for $z \in \mathbb{C}$. Show that the Taylor series expansion of g is $\sum_{n=0}^{\infty}(\sum_{k=0}^{n} \frac{a_k}{(n-k)!})z^n$.

    (b) Show that if $f(z) = \exp(\exp(z))$ then the sequence (a_n) satisfies the recurrence relation $a_n = \frac{1}{n}\sum_{k=0}^{n-1} \frac{a_k}{(n-k-1)!}$ for $n \geq 1$.

28. Let f be an entire function. Suppose that f satisfies $f(z+m+in) = f(z)$ for all $z \in \mathbb{C}, m, n \in \mathbb{Z}$. Show that f is constant.

29. If f is a nonconstant holomorphic function on a bounded open set and continuous on $\overline{U}$, then either f has a zero in U or $|f|$ assumes its minimum value on ∂U.

30. Use Cauchy's integral formula to evaluate the following:

    (i) $\int_{\Gamma_1} \frac{z^4}{z-1} dz$, where $\Gamma_1(t) = 2e^{it}, 0 \leq t \leq 2\pi$;

    (ii) $\int_{\Gamma_2} \frac{z^4}{z-1} dz$, where $\Gamma_2(t) = 2e^{-it}, 0 \leq t \leq 4\pi$;

    (iii) $\int_{\Gamma_3} \frac{\cos z}{z^2-2z} dz$, where Γ_3 is the square with vertices at $\pm 1 \pm i$, traversed once anticlockwise.

31. To which of the following integrals does Cauchy's theorem apply? There is no need to evaluate them. Here the contours $\gamma_1, \gamma_2, \gamma_3$ are the same as in exercise 3 of chapter 6.

 (i) $\int_{\gamma_1} |z|^2 dz$;

 (ii) $\int_{\gamma_1} \frac{z^2}{z-2} dz$;

 (iii) $\int_{\gamma_2} z dz$;

 (iv) $\int_{\gamma_2+\gamma_3} \sin(\frac{1}{z-1}) dz$.

32. Let $\gamma(t) = e^{it} (0 \leq t \leq 2\pi)$. Evaluate the following integrals:

 (i) $\int_\gamma |z+1|^2 dz$;

 (ii) $\int_\gamma \frac{e^{\frac{1}{z}}}{z^2} dz$;

 (iii) $\int_\gamma \frac{e^{\sin z}}{z(z-2)^2} dz$.

33. Let $P(z) = \sum_{k=0}^n a_k z^k$ be a polynomial of degree n such that $|P(z)| \leq M$ for $|z| = 1$. Prove that $|a_0| + |a_n| \leq M$.

Chapter 8

Isolated Singularities and Laurent Series

8.1 Isolated singularities

Recall that a complex-valued function f is said to be holomorphic at z_0 if and only if $f'(z)$ exists for each z in some neighborhood of z_0.

Definition 8.1. *A point z_0 is called a singular point of a complex-valued function f if f is not holomorphic at z_0, but is holomorphic at some point in each neighborhood of z_0.*

Most singularities we encounter are singularities of functions such as given below.

Example 8.1. *(i) the function $f(z) = \frac{1}{z-a}, z \neq a$ has a singularity at $z = a$.*

$$(ii) f(z) = \begin{cases} 1 & \text{if} \quad z \neq 0 \\ 0 & \text{if} \quad z = 0 \end{cases}$$

has a singularity at the origin.

Definition 8.2. *A function f has an isolated singularity at z_0 if there exists a number $R > 0$ such that f is holomorphic in the punctured disk*

$$\{z \in \mathbb{C} : 0 < |z - z_0| < R\}$$

but not in $B(z_0, R)$.

Example 8.2. *Consider the following functions:*

(i) $f(z) = \frac{\sin z}{z}, z \neq 0;$

(ii) $f(z) = \frac{1}{z}, z \neq 0;$

(iii) $f(z) = \frac{z}{e^z - 1}, z \neq 0;$

(iv) $f(z) = \frac{\cos z}{z}, z \neq 0;$

(v) $f(z) = e^{\frac{1}{z}}, z \neq 0.$

The functions in (i)-(v) have isolated singularities at $z = 0$.

(i) If we examine $z \mapsto \frac{\sin z}{z}$ more closely, we find that there is nothing wrong with it at $z = 0$. As $z \to 0$, $\frac{\sin z}{z} \to 1$. So the function

$$f(z) = \begin{cases} \frac{\sin z}{z} & \text{if} \quad z \neq 0; \\ 1 & \text{if} \quad z = 0 \end{cases}$$

is continuous at $z = 0$. By a sharper version of Cauchy's theorem, we know that this later function is holomorphic. Thus by defining a suitable value for the function at 0, we have been able to remove the singularity. Accordingly, we call 0 as a removable singularity of the function $\frac{\sin z}{z}$ defined on $\mathbb{C} \setminus \{0\}$.

(ii) The function $f(z) = \frac{1}{z}(z \neq 0)$ has an isolated singularity at 0 because the function f is holomorphic in a deleted neighborhood of 0 and it is not holomorphic in any disk containing the origin.

(iii) Consider the function $f(z) = \frac{z}{e^z - 1}, z \neq 0$. The function $e^z - 1$ that appears in the denominator is entire and has a simple zero at the origin, so it can be written as $e^z - 1 = zg(z)$, where g is entire and $g(0) \neq 0$. In fact $g(0) = 1$. The function $\frac{1}{g}$ is thus holomorphic in a neighborhood of the origin, and it coincides with f in a punctured disk centered at the origin. It follows therefore that the singularity of f at the origin can be removed by defining $f(0) = 1$. Thus we can say 0 is a removable singularity of the function $f(z) = \frac{z}{e^z - 1}$.

(iv) Now consider the function $f(z) = \frac{\cos z}{z}$. It has an isolated singularity at the origin. Since $\cos 0 = 1$, hence by continuity there exists a neighbourhood of 0 on which $|\cos z| > \frac{1}{2}$. If we take z sufficiently small, we can make $\left|\frac{\cos z}{z}\right|$ greater than any positive number. This means that $\frac{\cos z}{z}$ cannot approach any limit as $z \to 0$.

(v) On the other hand, $e^{\frac{1}{z}}$ does not have any limit as $z \to 0$. For $z = x > 0$, $x \to 0, e^{\frac{1}{z}} = e^{\frac{1}{x}} \to \infty$ and $z = -x > 0, x \to 0, e^{\frac{1}{z}} = e^{-\frac{1}{x}} \to 0$. For $z = iy, y \to 0, y \in \mathbb{R}, e^{\frac{1}{z}} = e^{\frac{1}{iy}}$ does not have a limit.

Example 8.3. *Now we shall discuss about the singularities of the function* $f(z) = \frac{1}{\sin(\frac{1}{z})}$. *Note that* f *has singularities at the point where* $sin\frac{1}{z} = 0$. *Therefore the singularities of* f *are the points* $z = \frac{1}{n\pi}$, $n = 0, \pm 1, \pm 2, \pm 3, \cdots$. *Since every neighborhood* N *of the origin, contains singular points of* f, *distinct from* $z = 0$, *the point* $z = 0$ *cannot be an isolated singularity of* f. *We shall say* $z = 0$ *is a* ***non-isolated singularity*** *of* f. *Since for each of the points* $z = \frac{1}{n\pi}, n \in \mathbb{Z}$, *we can obtain neighborhoods (open balls with sufficiently small radius* $r > 0$*) which contain no other singularity of* f, *hence these points are isolated singularities of* f.

Thus any neighborhood of a non-isolated singular point of f contains other singularities and hence a non-isolated singular point of f is a limit point of the singular points. Observe also that f must be discontinuous at an isolated singular point.

Example 8.4. *(i) Consider* $\tan\frac{1}{z}$, *for* $z \neq 0$. *The point* $z = 0$ *is a non-isolated singularity of the function.*

(ii) The function f *given by* $f(z) = \frac{1}{\sin\frac{\pi}{z}}$ *has infinitely many isolated singularities in the interval* $(0, 1]$. *The origin is a singular point but not an isolated singular point. Hence* $z = 0$ *is a non-isolated singularity of the function.*

There are broadly three kinds of isolated singularities for a complex-valued function: (1) removable singularity, (2) pole and (3) essential singularity.

Definition 8.3. *An isolated singularity* z_0 *of* f, $f \in H(B(z_0, R) \setminus \{z_0\})$; *is called a* ***removable singularity*** *at* z_0 *if there is a holomorphic function* $g : B(z_0, R) \to \mathbb{C}$ *such that* $g(z) = f(z)$ *for* $0 < |z - z_0| < R$.

Example 8.5. *Consider the functions*

(i) $f(z) = \frac{\sin z}{z}, z \neq 0$;

(ii) $g(z) = \frac{z^2-1}{z-1}, z \neq 1$;

(iii) $h(z) = \frac{2(1-\cos z)}{z^2}, z \neq 0$.

These functions have removable singularities at $z = 0$, $z = 1$ *and at* $z = 0$ *respectively. Now if we define* $f(0) = 1$, $g(1) = 2$ *and* $h(0) = 1$ *then the singularities of the functions can be removed.*

Theorem 8.1. *(Riemann's theorem on removable singularity): Suppose* f *has an isolated singularity at* z_0. *Then the following are equivalent.*

(1) The point $z = z_0$ *is a removable singularity of* f.

(2) f *is bounded in a deleted neighborhood of* z_0 *(i.e.,* $\limsup\limits_{z\to z_0} |f(z)| < \infty$*).*

(3) $\lim\limits_{z\to z_0} f(z)$ *exists.*

(4) $\lim\limits_{z\to z_0} (z - z_0)f(z) = 0$.

Proof. For proving (4) $\Rightarrow$ (1), we define

$$h(z) = \begin{cases} (z - z_0)f(z) & \text{if } z \neq z_0; \\ 0 & \text{if } z = z_0. \end{cases}$$

By hypothesis, h is continuous at z_0. Since h is analytic in a deleted neighborhood $0 < |z - z_0| < \delta$ (as f is holomorphic in the deleted neighborhood), it follows that h is holomorphic in $B(z_0, \delta)$ by Theorem 7.5 and Theorem 7.14. Let g be defined by $g(z) = \frac{h(z)-h(z_0)}{z-z_0} (= \frac{h(z)}{z-z_0}), z \in B(z_0, \delta) \setminus \{z_0\}$. Then g is holomorphic on $B(z_0, \delta)\setminus\{z_0\}$. As $\lim\limits_{z\to z_0} g(z)$ exists and equals $h'(z_0)$, hence g has a removable singularity at z_0, which can be removed by letting $g(z_0) = h'(z_0)$.

Note $g = f$ for $z \neq z_0$. So if we define $f(z_0) = h'(z_0)$ then f is holomorphic in $B(z_0, \delta)$. Thus the singularity of f at $z = z_0$ is a removable singularity.

To prove (3) $\Rightarrow$ (4), we assume $\lim_{z \to z_0} f(z)$ exists. Now $\lim_{z \to z_0} (z - z_0) f(z) = \lim_{z \to z_0} (z - z_0) \lim_{z \to z_0} f(z) = 0$.

Now we shall prove (2) $\Rightarrow$ (3). Suppose f is holomorphic and bounded in $B(z_0, \delta) \setminus \{z_0\}$. Consider the function g defined by

$$g(z) = \begin{cases} (z - z_0)^2 f(z) & \text{if } z \neq z_0; \\ 0 & \text{if } z = z_0. \end{cases}$$

Then g is holomorphic at z_0 because $g'(z)$ exists for $z \neq z_0$ and $g'(z_0) = \lim_{z \to z_0} \dfrac{g(z) - g(z_0)}{z - z_0} = \lim_{z \to z_0} (z - z_0) f(z) = 0$ (as f is bounded).

We expand g as a Taylor series about z_0. Since $g(z_0) = g'(z_0) = 0$, the first two terms of the Taylor series are zero. So

$$g(z) = \sum_{n=2}^{\infty} \frac{g^{(n)}(z_0)}{n!} (z - z_0)^n = (z - z_0)^2 h(z)$$

where h is holomorphic at z_0. Hence $(z - z_0)^2 h(z) = (z - z_0)^2 f(z)$ for all $z \neq z_0$. Thus $h(z) = f(z)$ in $B(z_0, \delta) \setminus \{z_0\}$. Moreover, $\lim_{z \to z_0} h(z)$ exists. Hence $\lim_{z \to z_0} f(z)$ exists. (If we redefine $f(z_0)$ as this limit, then we have $f(z) = h(z)$ and so f is holomorphic at $z = z_0$.)

To prove (1) $\Rightarrow$ (2), suppose f has a removable singularity at z_0. Then there is a holomorphic function $g : B(z_0, R) \to \mathbb{C}$, such that $g(z) = f(z)$ for $0 < |z - z_0| < R$. Choose r such that $0 < r < R$ and $\overline{B(z_0, r)} \subset B(z_0, R)$. The function g is bounded on $\overline{B(z_0, r)}$ as $\overline{B(z_0, r)}$ is compact and g is continuous. But as $g = f$ in $0 < |z - z_0| < r$, hence f is bounded in $0 < |z - z_0| < r$. □

Remark 8.1. (i) From Theorem 8.1 it follows that if f is holomorphic and bounded in some punctured disk around z_0, then $\lim_{z \to z_0} f(z)$ exists. If $f(z_0)$ denotes this limit, then f becomes holomorphic at z_0.

(ii) Note that the boundedness of $|f(z)|$ can be replaced by $|f(z)| = o(|z - z_0|^{-c}), 0 \leq c < 1$ (i.e., $\lim_{z \to z_0} |z - z_0|^c |f(z)| = 0, 0 \leq c < 1$). As it holds for $0 \leq c < 1$ and hence for all $c \geq 1$. Thus $\lim_{z \to z_0} (z - z_0) f(z) = 0$, so z_0 is a removable singularity of f.

Corollary 8.1. *Let the function f be analytic in the entire complex plane, and suppose that $\frac{f(z)}{z} \to 0$ as $|z| \to \infty$. Then f is a constant function.*

Proof. Let $g(z) = f(z) - f(0)$. Then $g(0) = 0$, so $\frac{g(z)}{z}$ has a removable singularity at 0 and extends to an entire function. Further, $\frac{g(z)}{z}$ tends to 0 as $|z| \to \infty$ since $\frac{f(z)}{z}$ does. Let $\epsilon > 0$. There is an $R > 0$ such that $|\frac{g(z)}{z}| < \epsilon$ for $|z| \geq R$.

By the Maximum modulus principle, $|\frac{g(z)}{z}| < \epsilon$ for all z. Since ϵ is arbitrary, $\frac{g(z)}{z}$ is identically 0. Hence, $g(z) = 0$ for all z and f is constant. □

Definition 8.4. *If f has an isolated singularity at z_0 and $\lim_{z \to z_0} |f(z)| = \infty$ then z_0 is called a* ***pole*** *of f.*

We have already seen that if $\limsup_{z \to z_0} |f(z)| < \infty$, then z_0 is a removable singularity of f. Thus if z_0 is not a removable singularity of f then f is not bounded in a neighborhood of z_0. In this case it may be possible that $(z-z_0)^n f$ is bounded near z_0. If $(z - z_0)^n f$ is bounded near z_0 then we say that z_0 is a pole of f and the smallest positive integer n such that $(z - z_0)^n f$ is bounded near z_0 is called the order of the pole of f at z_0. A pole of order one is called a simple pole of f.

(i) The function $f(z) = \frac{z^8}{(z-1)^2}$ has a double pole at $z = 1$.

(ii) The functions $\frac{\cos z}{z}$ and $\frac{1}{z}$ have poles of order 1 (simple poles) at $z = 0$.

(iii) The function $\frac{e^{2z}}{z^4}$ has a pole of order 4 at $z = 0$.

(iv) Let $f(z) = \frac{e^z}{z^2-1}$. The poles of f are $+1$ and -1 (simple poles).

(v) The rational function $\frac{3z+2}{(z^2+4)(z+1)^3}$ has poles at $\pm 2i$ of order 1 and a pole at -1 of order 3.

Theorem 8.2. *If f has an isolated singularity at z_0 then z_0 is a pole of order n of f if and only if*

(i) $\lim_{z \to z_0} (z - z_0)^{n+1} f(z) = 0$ *(i.e., $(z - z_0)^n f$ has a removable singularity at z_0).*

(ii) $\lim_{z \to z_0} (z - z_0)^n f(z) \neq 0$.

Proof. The proof follows from the observation that the $\lim_{z \to z_0} (z - z_0)^n f(z) \neq 0$ (i.e., the limit exists and is nonzero) if and only if n is the smallest positive integer such that $(z - z_0)^n f$ has a removable singularity at z_0. □

Remark 8.2. Let f be holomorphic in $B(z_0, R), R > 0$. Suppose f has a zero of multiplicity (order) n at z_0. Then $f^{(k)}(z_0) = 0, 0 \leq k \leq n-1$ but $f^{(n)}(z_0) \neq 0$. Conversely, if $f^{(k)}(z_0) = 0$ for $0 \leq k \leq n-1$ and $f^{(n)}(z_0) \neq 0$ then f has a zero of order n at z_0. Thus from the Taylor's expansion in $B(z_0, R)$ it follows that f has a zero of order n if and only if $f(z) = (z - z_0)^n g(z)$ where g is holomorphic in $B(z_0, R)$, and $g(z_0) = \frac{f^{(n)}(z_0)}{n!} \neq 0$. Thus we obtain that f has a zero of order n at z_0 if and only if $\frac{1}{f}$ has a pole of order n at z_0.

Now let us have a closer look at functions with a pole. Let z_0 be an isolated singularity of f. If $\lim_{z\to z_0} |f(z)| = \infty$ then given $M > 0$, there exist $\delta > 0$ such that $|z - z_0| < \delta$ implies $|f(z)| > M > 0$. Let $g(z) = \frac{1}{f(z)}$ in $B(z_0, \delta)$. In $0 < |z - z_0| < \delta$, g is holomorphic, bounded and $|g(z)| \leq \frac{1}{M}$. So g has a removable singularity at z_0. This singularity can be removed by defining $g(z_0) = 0$. Consider the Taylor series expansion of f about z_0 in $B(z_0, \delta)$. In this open ball

$$g(z) = \sum_{k=0}^{\infty} c_k(z - z_0)^k = \sum_{k=m}^{\infty} c_k(z - z_0)^k = (z - a)^m \sum_{k=m}^{\infty} c_k(z - z_0)^{k-m},$$

where m is the first positive integer such that $c_k \neq 0$. We can therefore write $g(z) = (z - z_0)^m h(z)$ where $h(z)$ is holomorphic in $B(z_0, \delta)$ and $h(z_0) \neq 0$. Hence $f(z) = (z - z_0)^{-m}\frac{1}{h(z)} = (z - z_0)^{-m}\phi(z)$ (that is, $\frac{1}{h(z)} = \phi(z)$, say). Note ϕ is holomorphic in a neighborhood of z_0 and f has a pole of order n at z_0.

Thus the order of the pole of f is the order of the zero of g or the smallest integer n such that $(z - z_0)^n f(z)$ is bounded in a neighborhood of z_0.

We summarize the above in a theorem as follows.

Theorem 8.3. *Let G be an open set in $\mathbb{C}$ and $z_0 \in G$. Let $f : G \setminus \{z_0\} \to \mathbb{C}$ be holomorphic. Then f has a pole at $z = z_0$ if and only if there exists a holomorphic function $\phi \in H(G)$ and a positive integer m such that $\phi(z_0) \neq 0$ and $f(z) = \frac{\phi(z)}{(z-z_0)^m}$. The integer m is called the order of the pole of f at $z = z_0$.*

Lemma 8.1. *Let f be holomorphic in $0 < |z - z_0| < R$. Assume that f has a pole of order m at z_0. Let $g(z) = (z - z_0)^m f(z)$ in $0 < |z - z_0| < R$. Then*

$$f(z) = \frac{a_{-m}}{(z - z_0)^m} + \frac{a_{-m+1}}{(z - z_0)^{m-1}} + \cdots + \frac{a_{-1}}{(z - z_0)} + h(z),$$

$0 < |z - z_0| < R$ where h is holomorphic in $B(z_0, R)$ and $a_k = \frac{g^{(m+k)}(z_0)}{(m+k)!} = \frac{1}{2\pi i}\int_\gamma \frac{f(w)}{(w-z_0)^{k+1}}dw$ for $\gamma(t) = z_0 + re^{it}$, $0 \leq t \leq 2\pi$, $0 < r < R$.

Proof. Note g has a removable singularity at z_0. We assume that $g(z_0)$ is suitably defined so that g is holomorphic in $B(z_0, R)$. Using the Taylor expansion for g in $B(z_0, R)$ about z_0, we get

$$g(z) = \sum_{k=0}^{m-1} \frac{g^{(k)}(z_0)}{k!}(z - z_0)^k + (z - z_0)^m h(z). \tag{8.1}$$

Dividing (8.1) by $(z - z_0)^m$, we obtain the result. □

Theorem 8.4. *Let f be holomorphic in $0 < |z - z_0| < R$, and f have an isolated singularity at z_0. Then f has a pole of order m at z_0 if and only if there exists constants $c_1 > 0$ and $c_2 > 0$ such that*

$$c_1|z - z_0|^{-m} \leq |f(z)| \leq c_2|z - z_0|^{-m}, \tag{8.2}$$

in $0 < |z - z_0| < r$ where $r < R$.

Proof. Let f have a pole of order m at z_0. We then write $f(z) = \frac{g(z)}{(z-z_0)^m}$ in $0 < |z-z_0| < R$ with $g \in H(B(z_0, R))$ and $g(z_0) \neq 0$. Hence there exist $r > 0, r < R$ such that $g(z) \neq 0$ for all $z \in \overline{B(z_0, r)}$. We define $c_1 = \inf\{|g(z)| : z \in \overline{B(z_0, r)}\}$ and $c_2 = \sup\{|g(z)| : z \in \overline{B(z_0, r)}\}$. Then $c_1 \leq |g(z)| = |z-a|^m |f(z)| \leq c_2$ in $0 < |z - z_0| < r$.

Conversely, suppose that the inequalities hold. Then right part of the above inequalities shows that $(z - z_0)^m f$ is bounded near z_0 so that

$$\lim_{z \to z_0} (z - z_0)^{m+1} f(z) = 0.$$

The left inequality implies

$$|(z - z_0)^m f(z)| \geq c_1$$

as $z \to z_0$. Hence $\lim_{z \to z_0} (z - z_0)^m f(z) \neq 0$. From Theorem 8.2, f has a pole of order m at z_0. □

Remark 8.3. There is no holomorphic function which approaches ∞ like a fractional power of $\frac{1}{z-z_0}$ in the neighborhood of an isolated singularity. For example, if f were holomorphic in a deleted neighborhood of 0 and satisfies $|f(z)| \leq \frac{1}{\sqrt{|z|}}$, then f would be bounded since the singularity will be removable.

Similarly, given that $|f(z)| < \frac{1}{|z|^{\frac{5}{2}}}$, we conclude that $z^2 f(z)$ has a removable singularity at 0. Hence f has a pole of order at most 2 and hence $|f(z)| \leq \frac{c}{|z|^2}$ for some constant c.

Note that the Theorem 8.4 gives a characterization of poles of f through the behavior of the values of f near z_0. Suppose that f has a pole of order n at z_0. Then

$$f(z) = \frac{g(z)}{(z - z_0)^n}$$

where g is holomorphic and is nonzero at z_0. So, by continuity of g at z_0 there is a neighborhood N of z_0 on which f is defined (except at z_0) with $|g(z)| \geq \frac{|g(z_0)|}{2}$ for $z \in N$. Hence

$$|f(z)| = \left|\frac{g(z)}{(z - z_0)^n}\right| \geq \frac{1}{|z - z_0|^n} \frac{|g(z_0)|}{2}$$

for $z \in N \setminus \{z_0\}$.

Thus we can also make the following definition for poles.

Definition 8.5. *Let $G \subset \mathbb{C}$ be open, $p \in G, f : G \setminus \{p\} \to \mathbb{C}$ is analytic. We say that p is a pole of f of order N if for some $\delta > 0$, f has an expansion*

$$f(p + h) = \sum_{n=-N}^{\infty} a_n h^n, \ (0 < |h| < \delta)$$

with $a_{-N} \neq 0$. Equivalently, $h^N f(p+h)$ tends to a finite nonzero limit as $h \to 0$.

Definition 8.6. *A function holomorphic in a region except at (isolated singularities which are) poles is said to be meromorphic in that region.*

Recall that a function f of the form $f(z) = \frac{P(z)}{Q(z)}$ where P and Q are polynomials is called a rational function.

Example 8.6. *Any rational function is meromorphic in $\mathbb{C}$. The functions $\tan z, \cot z, \frac{\tan z}{z^2+1}$ are meromorphic in $\mathbb{C}$ but $e^{\frac{1}{z}}$ is not meromorphic in $\mathbb{C}$.*

Definition 8.7. *Suppose f has an isolated singularity at z_0. The point z_0 is called an essential singularity of f if it is neither a removable singularity nor a pole of f. That is, the point z_0 is an essential singularity of f if $|f(z)|$ is unbounded but $\lim\limits_{z\to a} |f(z)| \neq \infty$.*

Example 8.7. *Consider the function $f(z) = e^{\frac{1}{z-z_0}}$. It has an isolated singularity at $z = z_0$. Let $z = x + iy$ and $z_0 = x_0 + iy_0$. Let $y = y_0$. Then $x - x_0 = z - z_0$. Thus $z \to z_0$ if and only if $x \to x_0$. Now $e^{\frac{1}{z-z_0}} = e^{\frac{1}{x-x_0}}$. Notice that $\lim\limits_{x-x_0\to 0^+} e^{\frac{1}{x-x_0}} = \infty$ and $\lim\limits_{x-x_0\to 0^-} e^{\frac{1}{x-x_0}} = 0$. Thus z_0 is an essential singularity of f.*

Example 8.8. *The functions $\sin\frac{1}{z}, \cos\frac{1}{z}, ze^{\frac{1}{z^2}}$ have essential singularities at $z = 0$.*

Theorem 8.5. *(Casorati-Weierstrass theorem) Let z_0 be an essential singularity of f defined in an open set $G \subseteq \mathbb{C}$. Then $f(B'(z_0, r))$ is dense in $\mathbb{C}$, for any $r > 0$ with $B(z_0, r) \subset G$ and where $B'(z_0, r) = \{z \in \mathbb{C} : 0 < |z - z_0| < r\}$.*

Proof. If $\overline{f(B'(z_0, r))} \neq \mathbb{C}$, then this implies there exist $\alpha \in \mathbb{C}$ such that $\alpha \notin \overline{f(B'(z_0, r))}$. Hence there exist a neighborhood $B(\alpha, \epsilon), \epsilon > 0$ of α such that $B(\alpha, \epsilon) \cap f(B'(z_0, r)) = \emptyset$. That is, if $z \in B'(z_0, r)$ then $f(z) \notin B(\alpha, \epsilon)$. Thus $|f(z) - \alpha| \geq \epsilon$ whenever $0 < |z - z_0| < r$. Hence $|f(z) - \alpha| \geq \epsilon$ for all $z \in B(z_0, r)$ (as z_0 is an essential singularity of f implies $|f(z)|$ is unbounded). Therefore, $\frac{1}{f(z)-\alpha}$ is bounded in the deleted neighborhood of z_0. Hence $\frac{1}{f(z)-\alpha}$ has a removable singularity at z_0. Thus $\lim\limits_{z\to z_0} \dfrac{1}{f(z)-\alpha}$ exists. Let $\lim\limits_{z\to z_0} \dfrac{1}{f(z)-\alpha} = \lambda$. Now there are two possibilities. If $\lambda = 0$, then $|f(z) - \alpha| \to \infty$ as $z \to z_0$. Hence $|f(z)| \to \infty$ as $z \to z_0$. Thus f has a pole at z_0, a contradiction. If $\lambda \neq 0$, then $f(z) - \alpha \to \frac{1}{\lambda}$, so f has a removable singularity at z_0, again a contradiction. □

Corollary 8.2. *Suppose the function f have an essential singularity at the point z_0. Then, for any complex number w, there is a sequence $(z_n)_{n=1}^{\infty}$ in the* *region where f is defined such that $\lim_{n\to\infty} z_n = z_0$ and $\lim_{n\to\infty} f(z_n) = w$ (i.e., f takes values arbitrarily close to every complex number in every neighbourhood of an essential singularity.)*

Proof. By hypothesis, f is holomorphic in a deleted neighborhood of z_0. Suppose there exists a complex number w_0 and a sequence (z_n) with $z_n \to z_0$ such that $f(z_n) \not\to w_0$. In other words suppose, there exists a w_0, an $\epsilon > 0$ and $\delta > 0$ such that $|f(z) - w_0| \geq \epsilon$ for $0 < |z - z_0| < \delta$. From the proof of Casorati-Weierstrass theorem, note that this leads to a contradiction. □

A deep theorem of E. Picard reveals more. It says in any punctured disk centered at an essential singularity, a holomorphic function actually assumes every complex value infinitely often, with at most a single exception. In other words, except possibly for one value of w, the equality $\lim_{n\to\infty} f(z_n) = w$ in the corollary of the Casorati-Weierstrass theorem can be replaced by $f(z_n) = w, n = 1, 2, \ldots$.

Definition 8.8. *(Lacunary values): A complex number a is said to be a lacunary value of a function f if $f(z) \neq a$ in the region where f is defined.*

For instance, 0 is a lacunary value of e^z in the whole plane. A weaker version of Picard's theorem can also be stated as follows which is known as "Little Picard theorem".

Theorem 8.6. *An entire function with more than one finite lacunary value reduces to a constant.*

Theorem 8.7. *(Picard's theorem): A holomorphic function takes every value with at most a single exception in the neighborhood of an essential singularity.*

That is if f has an isolated singularity at a point z_0 and if f omits two values in some neighborhood of z_0, then z_0 is a removable singularity or a pole of f. The so-called "big Picard theorem" (Theorem 8.7) is a remarkable strengthening of the Casorati-Weierstrass theorem (Theorem 8.5) which merely asserts that the image of every neighborhood of z_0 is dense in $\mathbb{C}$ if f has an essential singularity at z_0.

The proof of the theorem is difficult and we refer the reader to some more advanced books on complex analysis.

Picard's theorem can be illustrated by the function e^z. The function $f(z) = e^z$ on $\mathbb{C}$ assumes every value other than 0 infinitely often in any punctured disk centered at ∞.

Remark 8.4. If a function f does not satisfy either equation (8.2) or the property of Theorem 8.5 near z_0, then we must conclude that f is not holomorphic in a punctured disk around z_0.

For instance, consider the problem of defining z^α for nonintegral real α. If w is any logarithm of z, then $|z^\alpha| = e^{\alpha w}$. There is no integer n for which $|z|^{\alpha - n}$

is bounded away from 0 on any annulus $\{z \in \mathbb{C} : 0 < |z| < r\}$ nor does $|z|^\alpha$ come arbitrarily close to all the real values on any annulus $\{z \in \mathbb{C} : 0 < |z| < r\}$. Therefore, there is no way of choosing logarithm so as to make $z \to z^\alpha$ differentiable on any annulus $\{z \in \mathbb{C} : 0 < |z| < r\}$.

Example 8.9. *An entire function with positive real part is a constant.*

Solution: Let f be an entire function. Let u and v be the real and imaginary parts of f respectively. Then $f(z) = u(z) + iv(z)$ for all $z \in \mathbb{C}$ and $u(z) > 0$ for all $z \in \mathbb{C}$(by hypothesis). By Little Picard's theorem, if g is an entire function that omits two values of $\mathbb{C}$ then g is a constant. In this case since $u(z) > 0$ for all $z \in \mathbb{C}$; the given entire function f omits all values of the left half plane. Hence according to the Little Picard's theorem (Theorem 8.6), the function f is constant.

Example 8.10. *There cannot be a nonconstant entire function whose real part is bounded.*

Solution: The argument is similar to Example 8.9.

Corollary 8.3. *Let f be a nonconstant entire function. Then f is not a polynomial if and only if ∞ is an essential singularity of f (i.e., $g(z) = f(\frac{1}{z}), z \in \mathbb{C}^*$ has an essential singularity at $z = 0$).*

Proof. Let $f(z) = \sum_{k=0}^{\infty} a_k z^k$. If 0 is not an essential singularity for $g(z) = f(\frac{1}{z})$, then for large $n, n \in \mathbb{N}$, $z^n g(z)$ must have a removable singularity at 0 by Theorem 8.2. For these values of n, by Cauchy's theorem, for $\gamma(t) = e^{it}$, $0 \le t \le 2\pi$,

$$0 = \int_\gamma w^n g(w)dw = \sum_k a_k \int_\gamma w^{n-k} dw = 2\pi i a_{n+1}.$$

Thus f is a polynomial of degree at most n. Converse is easy to verify. □

8.2 Laurent series

Definition 8.9. *A series $\sum_{n=-\infty}^{\infty} a_n$ of complex numbers is convergent if there exists $s \in \mathbb{C}$ such that for a given $\epsilon > 0$ there exists $N \in \mathbb{N}$ such that $\left|\sum_{k=-m}^{n} a_k - s\right| < \epsilon$, for $m, n > N$. The number s is called the sum of the series $\displaystyle\sum_{n=-\infty}^{\infty} a_n$.*

Thus the series $\sum_{n=-\infty}^{\infty} a_n$ is convergent if and only if $\sum_{n=0}^{\infty} a_n$ and $\sum_{n=1}^{\infty} a_{-n}$ are convergent. In this case $\sum_{n=-\infty}^{\infty} a_n = \sum_{n=0}^{\infty} a_n + \sum_{n=1}^{\infty} a_{-n}$.

A series $\sum_{n=-\infty}^{\infty} f_n(z)$ of functions defined on a set G is said to be uniformly convergent on G if there exists a function f on G such that for a given $\epsilon > 0$

there exists $N \in \mathbb{N}$ with the property that

$$\left| \sum_{k=-m}^{n} f_k(z) - f(z) \right| < \epsilon,$$

for $m, n > N$ and for all $z \in G$.

Lemma 8.2. *Let* $f(z) = \sum_{n=0}^{\infty} a_n(z-z_0)^n$ *for* $|z-z_0| < R$. *Then* $\sum_{n=0}^{\infty} a_n(z-z_0)^{-n}$ *is convergent for* $|z-z_0| > \frac{1}{R}$. *In fact the convergence is uniform on* $\{z : |z-z_0| \geq r\}$ *where* $r > \frac{1}{R}$. *If* $g(z) = \sum_{n=0}^{\infty} a_n(z-z_0)^{-n}$ *for* $|z-z_0| > \frac{1}{R}$, *then* $g'(z) = -\sum_{n=1}^{\infty} na_n(z-z_0)^{-n-1}$, *for* $|z-z_0| > \frac{1}{R}$.

Proof. Note that $g(z) = f(\frac{1}{z-z_0})$, for $|z-z_0| > \frac{1}{R}$. Given $\epsilon > 0$, by the uniform convergence of $\sum a_n(z-z_0)^n$ in $B(z_0, \rho)$ for $0 < \rho < R$, there exists $N \in \mathbb{N}$ such that $|f(z) - \sum_{k=1}^{n} a_k(z-z_0)^k| < \epsilon$ for $n \geq N$ and $z \in B(z_0, \rho)$. Hence $|g(z) - \sum_{k=1}^{n} a_k(z-z_0)^{-k}| < \epsilon$ whenever $n \geq N$ and $|z-z_0| > \frac{1}{\rho}$. Since $f'(z) = \sum_{n=1}^{\infty} na_n(z-z_0)^{n-1}$, we have by the chain rule,

$$g'(z) = f'\left(\frac{1}{z-z_0}\right)\left(-\frac{1}{(z-z_0)^2}\right) = -\sum_{n=1}^{\infty} na_n(z-z_0)^{-n-1}$$

for $|z-z_0| > \frac{1}{R}$. □

A power series of the form $\sum_{n=-\infty}^{\infty} a_n(z-z_0)^n$ is called a Laurent series centered at the point z_0 of $\mathbb{C}$. Such series arise in particular, when one examines the behavior of holomorphic functions near an isolated singularity. Here the coefficients a_n are complex constants. We already know that the series converges at the point z if both of the series $\sum_{n=0}^{\infty} a_n(z-z_0)^n$ and $\sum_{n=1}^{\infty} a_{-n}(z-z_0)^{-n}$ converge. In that case the sum of the series is equal to $\lim_{N\to\infty} \sum_{n=-N}^{N} a_n(z-z_0)^n$.

Note that the series $\sum_{n=1}^{\infty} a_{-n}(z-z_0)^{-n}$ converges (absolutely and locally uniformly) in the region $|z-z_0| > R_1$ where $R_1 = \limsup_{n\to\infty} |a_{-n}|^{\frac{1}{n}}$ and the series $\sum_{n=0}^{\infty} a_n(z-z_0)^n$ converges (absolutely and locally uniformly) in the disk $|z-z_0| < R_2$ where $R_2 = \dfrac{1}{\limsup_{n\to\infty} |a_n|^{\frac{1}{n}}}$.

Hence, if $R_1 < R_2$, then the Laurent series $\sum_{n=-\infty}^{\infty} a_n(z-z_0)^n$ converges absolutely and locally uniformly in the annulus $R_1 < |z-z_0| < R_2$ (called the annulus of convergence of the Laurent series).

Definition 8.10. *Let* $0 < r < R$. *The set* $A(z_0, r, R) = \{z \in \mathbb{C} : r < |z-z_0| < R\}$ *is called the annulus centered at the point* z_0 *with inner radius* r *and outer radius* R.

Let f be holomorphic in $B'(z_0, R) = B(z_0, R) \setminus \{z_0\}$. The aim in this section is to represent f as a Laurent series in an annulus centered at z_0.

Theorem 8.8. *Let f be holomorphic in the annulus $A(z_0, R_1, R_2)$. Let $C_r = \{z \in \mathbb{C} : |z - z_0| = r\}$, the circle with center z_0 and radius r (oriented counterclockwise) and let $R_1 < r < R_2$. Then the map $F : (R_1, R_2) \to \mathbb{C}$ defined by $F(r) = \int_{C_r} f(z)dz$ is a constant. (Here $(R_1, R_2) = \{x \in \mathbb{R} : R_1 < x < R_2\}$.) Thus, if $r, s \in (R_1, R_2)$ then $\int_{C_r} f(z)dz = \int_{C_s} f(z)dz$. (That is, $\int_{C_r} f(z)dz$ is independent of $r, R_1 < r < R_2$.)*

Proof. We need to show $F'(r) = 0$ where

$$F(r) = \int_{C_r} f(z)dz = \int_0^{2\pi} f(z_0 + re^{it})ire^{it}dt.$$

The integrand in the preceding integral as a function of the two real variables r and t, has continuous partial derivatives. By Leibniz theorem (Theorem 6.2),

$$F'(r) = \int_0^{2\pi} \frac{\partial}{\partial r}[f(z_0 + re^{it})ire^{it}]dt.$$

But

$$\begin{aligned} \tfrac{\partial}{\partial r}[f(z_0 + re^{it})ire^{it}] &= f'(z_0 + re^{it})ire^{2it} + f(z_0 + re^{it})ie^{it} \\ &= \tfrac{\partial}{\partial t}[f(z_0 + re^{it})e^{it}]. \end{aligned}$$

Hence

$$F'(r) = \int_0^{2\pi} \frac{\partial}{\partial t}[f(z_0 + re^{it})e^{it}]dt = f(z_0 + re^{it})e^{it}\Big|_{t=0}^{t=2\pi} = 0.$$

□

Lemma 8.3. *(Cauchy's formula for an annulus): Let f be holomorphic in the annulus $A(z_0, R_1, R_2)$. For $R_1 < r < R_2$, let C_r denote the circle $|z - z_0| = r$, oriented counterclockwise. If $R_1 < r_1 < |z - z_0| < r_2 < R_2$, then*

$$f(z) = \frac{1}{2\pi i}\int_{C_{r_2}} \frac{f(w)}{w - z}dw - \frac{1}{2\pi i}\int_{C_{r_1}} \frac{f(w)}{w - z}dw.$$

Proof. Consider the function g defined in the annulus $A(z_0, R_1, R_2)$ as follows:

$$g(w) = \begin{cases} \frac{f(w)-f(z)}{w-z} & \text{if } w \neq z; \\ f'(z) & \text{if } w = z. \end{cases}$$

By Riemann's theorem on removable singularity, the function g is holomorphic in $A(z_0, R_1, R_2)$. Applying Theorem 8.8 to $\frac{1}{2\pi i}g$, we get

$$\frac{1}{2\pi i}\int_{C_{r_2}} g(w)dw = \frac{1}{2\pi i}\int_{C_{r_1}} g(w)dw.$$

Since $z \notin C_{r_i}, i = 1, 2$, we obtain that

$$0 = \frac{1}{2\pi i}\int_{C_{r_2}} g(w)dw - \frac{1}{2\pi i}\int_{C_{r_1}} g(w)dw$$

$$= \frac{1}{2\pi i}\int_{C_{r_2}} \frac{f(w)-f(z)}{w-z}dw - \frac{1}{2\pi i}\int_{C_{r_1}} \frac{f(w)-f(z)}{w-z}dw$$

$$= \left[\frac{1}{2\pi i}\int_{C_{r_2}} \frac{f(w)}{w-z}dw - \frac{1}{2\pi i}\int_{C_{r_1}} \frac{f(w)}{w-z}dw\right]$$

$$- \left[\frac{1}{2\pi i}\int_{C_{r_2}} \frac{f(z)}{w-z}dw - \frac{1}{2\pi i}\int_{C_{r_1}} \frac{f(z)}{w-z}dw\right].$$

Note that

$$\frac{1}{2\pi i}\int_{C_{r_2}} \frac{dw}{w-z} = 1 \text{ and } \frac{1}{2\pi i}\int_{C_{r_1}} \frac{dw}{w-z} = 0.$$

Using this observation, we get

$$f(z) = \frac{1}{2\pi i}\int_{C_{r_2}} \frac{f(w)}{w-z}dw - \frac{1}{2\pi i}\int_{C_{r_1}} \frac{f(w)}{w-z}dw.$$

□

Theorem 8.9. *(Laurent series representation): Let the function f be holomorphic in the annulus $A(z_0, R_1, R_2), 0 < R_1 < R_2 \leq \infty$. Then the function f is given by $f(z) = \sum_{n=-\infty}^{\infty} a_n(z-z_0)^n, z \in A(z_0, R_1, R_2)$, where*

$$a_n = \frac{1}{2\pi i}\int_{C_r} \frac{f(w)}{(w-z_0)^{n+1}}dw$$

and $C_r = \{z \in \mathbb{C} : |z - z_0| = r\}$, for any $r > 0$ such that $R_1 < r < R_2$.

The series converges absolutely and locally uniformly (uniformly on compact subsets of $A(z_0, R_1, R_2)$) in $A(z_0, R_1, R_2)$.

Proof. Choose positive real numbers r_1 and r_2 satisfying $R_1 < r_1 < r_2 < R_2$. Then

$$f(z) = \frac{1}{2\pi i}\int_{C_{r_2}} \frac{f(w)}{w-z}dw - \frac{1}{2\pi i}\int_{C_{r_1}} \frac{f(w)}{w-z}dw, r_1 < |z-z_0| < r_2.$$

In other words, in the annulus $A(z_0, r_1, r_2)$, we have $f = f_2 + f_1$, where

$$f_2(z) = \frac{1}{2\pi i}\int_{C_{r_2}} \frac{f(w)}{w-z}dw, |z-z_0| < r_2$$

and

$$f_1(z) = -\frac{1}{2\pi i}\int_{C_{r_1}} \frac{f(w)}{w-z}dw, |z-z_0| > r_1.$$

To find the series expansion for f_j's, we argue as in the power series expansion from Cauchy's integral formula:

$$\frac{1}{w-z} = \frac{1}{w-z_0+z_0-z}$$

$$= \frac{1}{(w-z_0)-(z-z_0)} = \begin{cases} -\frac{1}{(z-z_0)\left(1-\frac{w-z_0}{z-z_0}\right)} & \text{if } w \in C_{r_1}; \\ \frac{1}{(w-z_0)\left(1-\frac{z-z_0}{w-z_0}\right)} & \text{if } w \in C_{r_2}. \end{cases}$$

□

In each of the cases, we have a geometric series, uniformly convergent on the corresponding circles $C_{r_j}, j = 1, 2$. We can therefore interchange the order of integration and summation to get

$$f(z) = \sum_{n=0}^{\infty} \left(\frac{1}{2\pi i} \int_{C_{r_2}} \frac{f(w)}{(w-z_0)^{n+1}} dw \right) (z-z_0)^n$$
$$+ \sum_{k=0}^{\infty} \left(\frac{1}{2\pi i} \int_{C_{r_1}} f(w)(w-z_0)^k dw \right) (z-z_0)^{-k-1}.$$

In the second summation let $n = -k-1$. Then the second series is

$$\sum_{n=-\infty}^{-1} \left(\frac{1}{2\pi i} \int_{C_{r_1}} \frac{f(w)}{(w-z_0)^{n+1}} dw \right) (z-z_0)^n.$$

In view of Theorem 8.8, we can write the coefficients a_n of $(z-z_0)^n$ as

$$a_n = \frac{1}{2\pi i} \int_{C_r} \frac{f(w)}{(w-z_0)^{n+1}} dw$$

for any r such that $R_1 < r < R_2$.

The series of the theorem is called the Laurent series (or expansion) of f in the annulus $A(z_0, R_1, R_2)$. There is a uniqueness assertion about the Laurent expansion.

Proposition 8.1. *Let f be holomorphic in an annulus $A(z_0, R_1, R_2)$. Let a series of the form $\sum_{n=-\infty}^{\infty} a_n(z-z_0)^n$ converge uniformly on compact subsets of the annulus to f. Then a_n are given by*

$$a_n = \frac{1}{2\pi i} \int_{C_r} \frac{f(w)}{(w-z_0)^{n+1}} dw$$

for any $R_1 < r < R_2$.

Proof. This can be proved in a way similar to that of Proposition 6.11. □

Remark 8.5. The advantage of this proposition is that if we can find (by some algebraic means or other) a series as in the proposition, then it is necessarily the Laurent series of f. However, the reader should realize that the same function may have different Laurent expansion in different annuli.

Example 8.11. *Let $f(z) = \frac{1}{(z-1)(z-2)}$. The function f has singularity at $z = 1$ and $z = 2$. We shall find the Laurent series expansion of f in $1 < |z| < 2$. Since $1 < |z| < 2$ hence $|\frac{z}{2}| < 1$ and $\frac{1}{|z|} < 1$. Therefore $\frac{1}{(z-1)(z-2)} = \frac{1}{z-2} - \frac{1}{z-1} = \frac{1}{2(\frac{z}{2}-1)} - \frac{1}{z(1-\frac{1}{z})} = -\frac{1}{2}\sum_{n=0}^{\infty}(\frac{z}{2})^n - \frac{1}{z}\sum_{n=0}^{\infty}\frac{1}{z^n} = \sum_{n=0}^{\infty}(-1)z^{-(n+1)} + \sum_{n=0}^{\infty}\frac{(-1)}{2^{n+1}}z^n$.*

Example 8.12. *Consider $f(z) = \frac{1}{z(z-1)(z-2)}$. By the method of partial fractions we obtain $f(z) = \frac{1}{2z} - \frac{1}{z-1} + \frac{1}{2(z-2)}$. In $0 < |z| < 1$,*

$f(z) = \frac{1}{2z} + \sum_{n=0}^{\infty} z^n + \frac{1}{2}\frac{1}{2(\frac{z}{2}-1)} = \frac{1}{2z} + \sum_{n=0}^{\infty} z^n - \frac{1}{4}\frac{1}{1-\frac{z}{2}} = \frac{1}{2z} + \sum_{n=0}^{\infty} z^n - \frac{1}{4}\sum_{n=0}^{\infty}(\frac{z}{2})^n = \frac{1}{2z} + \sum_{n=0}^{\infty} z^n - \sum_{n=0}^{\infty}\frac{z^n}{2^{n+2}} = \frac{1}{2z} + \sum_{n=0}^{\infty}(1 - \frac{1}{2^{n+2}})z^n$. In $1 < |z| < 2$, $f(z) = \frac{1}{2z} - \frac{1}{z}\frac{1}{(1-\frac{1}{z})} + \frac{1}{2}\frac{1}{2(\frac{z}{2}-1)} = \frac{1}{2z} - \frac{1}{z}\sum_{n=0}^{\infty}\frac{1}{z^n} - \frac{1}{4}\sum_{n=0}^{\infty}(\frac{z}{2})^n = \frac{1}{2z} - \sum_{n=0}^{\infty}\frac{1}{z^{n+1}} - \sum_{n=0}^{\infty}\frac{z^n}{2^{n+2}} = -\frac{1}{2z} - \sum_{n=2}^{\infty}\frac{1}{z^n} - \frac{1}{4}\sum_{n=0}^{\infty}\frac{z^n}{2^n}$.

Example 8.13. *Let $f(z) = \frac{1-2z}{z(1-z)}, z \in \mathbb{C}^*$. We shall find the Laurent series expansion of f about $z = 0$ in $0 < |z| < 1$.*

Solution: We can write, since $|z| < 1$, $f(z) = \frac{1}{z} - \frac{1}{1-z} = \frac{1}{z} - \sum_{n=0}^{\infty} z^n$. This is the Laurent series expansion of f in the annulus $0 < |z| < 1$. The coefficients a_n can be computed as follows: Let $\gamma(t) = \frac{e^{it}}{2}, 0 \le t \le 2\pi$. Then by Cauchy integral formula for derivatives it follows that

$$a_n = \frac{1}{2\pi i}\int_\gamma \frac{1-2z}{z^{n+2}(1-z)}dz = \frac{1}{(n+1)!}\frac{d^{n+1}}{dz^{n+1}}\left(\frac{1-2z}{1-z}\right)\bigg|_{z=0}$$

$$= \begin{cases} -1 & \text{if} \quad n \ge 0; \\ 1 & \text{if} \quad n = -1. \end{cases}$$

Example 8.14. *Consider $f(z) = \frac{1}{z^2(1-z)}$ in $0 < |z| < 1$ and in $0 < |z-1| < 1$. Notice that in $0 < |z| < 1$, $\frac{1}{z^2(1-z)} = \frac{1}{z^2}(1 + z + z^2 + \cdots) = \frac{1}{z^2} + \frac{1}{z} + 1 + z + \cdots$.*

In $0 < |z-1| < 1$, $\frac{1}{z^2(1-z)} = \frac{-1}{z^2(z-1)} = \frac{-1}{[1+(z-1)]^2(z-1)} = \frac{-1}{z-1} + 2 - 3(z-1) + 4(z-1)^2 + \cdots$.

Example 8.15. *Let $f(z) = \frac{e^z}{z}, z \in \mathbb{C} \setminus \{0\}$. We shall find the Laurent series expansion of f around origin. We need to compute $a_n = \frac{1}{2\pi i}\int_\gamma \frac{e^z}{z^{n+2}}dz$ where $\gamma(t) = e^{it}, 0 \le t \le 2\pi$. By Cauchy integral formula,*

$$a_n = \begin{cases} \frac{1}{(n+1)!} & \textit{if} \quad n \ge -1; \\ 0 & \textit{if} \quad n \le -2. \end{cases}$$

Hence $f(z) = \sum_{n=-1}^{\infty}\frac{z^n}{(n+1)!}$. We could have derived the series by looking at the series expansion of $e^z = \sum_{n=0}^{\infty}\frac{z^n}{n!}$.

Example 8.16. *Let $f(z) = e^{z+\frac{1}{z}}, z \in \mathbb{C}\setminus\{0\}$. We shall find the Laurent series expansion of f around origin. Notice that*

$$f(z) = e^{z+\frac{1}{z}} = e^z e^{\frac{1}{z}} = \left(1 + \sum_{k=1}^{\infty}\frac{z^k}{k!}\right)\left(1 + \sum_{k=1}^{\infty}\frac{1}{k!z^k}\right).$$

To get the expansion of f we take the Cauchy product of the two series and get $f(z) = \sum_{n=-\infty}^{\infty} c_n z^n = a_0 + \sum_{n=1}^{\infty} a_n(z^n + z^{-n})$ and $a_n = \sum_{k=0}^{\infty} \frac{1}{(n+k)!k!}$ since $c_{-n} = \overline{c_n} = c_n$. This can be verified as follows. Let $\gamma(t) = e^{it}, 0 \le t \le 2\pi$. Then

$$c_n = \frac{1}{2\pi i}\int_\gamma \frac{f(w)}{w^{n+1}}dw = \frac{1}{2\pi i}\int_0^{2\pi} e^{e^{i\theta}+e^{-i\theta}} e^{-in\theta}\cdot i\cdot d\theta = \frac{1}{2\pi}\int_0^{2\pi} e^{2\cos\theta}e^{-in\theta}d\theta$$

and therefore $\overline{c_n} = \frac{1}{2\pi}\int_0^{2\pi} e^{2\cos\theta}e^{in\theta}d\theta = \frac{1}{2\pi}\int_0^{2\pi} e^{2\cos(2\pi-\theta)}e^{-in(2\pi-\theta)}d\theta$ $= \frac{1}{2\pi}\int_0^{2\pi} e^{2\cos\theta}e^{-in\theta}d\theta = c_n$. One may note here that as a by product we obtain that $\frac{1}{\pi}\int_0^{\pi} e^{2\cos\theta}\cos n\theta d\theta = \sum_{j=0}^{\infty}\frac{1}{(n+j)!j!}$. So the Laurent series expansion also helps us in evaluating certain integrals.

Definition 8.11. *If $f(z) = \sum_{k=-\infty}^{\infty} a_k(z-z_0)^k$ is the Laurent expansion of f about an isolated singularity z_0, then $\sum_{k=-\infty}^{-1} a_k(z-z_0)^k$ is called the principal part or singular part of f at z_0 and $\sum_{k=0}^{\infty} a_k(z-z_0)^k$ is called the analytic part.*

8.3 Characterization of singularities

The intimate connection between the nature of singularity of f at z_0 and the Laurent expansion of f at z_0 is given by the following.

Theorem 8.10. *Let the function f have an isolated singularity at $z = z_0$ and let $f(z)$ be holomorphic in $0 < |z - z_0| < R$. Let $\sum_{n=-\infty}^{\infty} a_n(z-z_0)^n$ be the Laurent expansion of f in the annulus $A(z_0, 0, R)$. Then*

(1) f has a removable singularity at z_0 if and only if $a_n = 0$ for all $n < 0$.

(2) f has a pole at z_0 if and only if there exists a positive integer m such that $a_n = 0$ for $n < -m$ but $a_{-m} \neq 0$.

(3) f has an essential singularity at z_0 if and only if there are infinitely many negative integers n such that $a_n \neq 0$.

Proof. (1) If $a_j = 0$ for all $j < 0$, then $f(z) = \sum_{n=0}^{\infty} a_n(z-z_0)^n$ on $A(z_0, 0, R)$. If we define $f(z_0) = a_0$, then f becomes holomorphic in $B(z_0, R)$. The converse is equally as easy.

(2) Let $f(z) = \sum_{n=0}^{\infty} a_n(z-z_0)^n + \sum_{k=1}^{m}\frac{a_{-k}}{(z-z_0)^k}$ with $a_{-m} \neq 0$. Then $(z-z_0)^m f(z)$ is bounded in a deleted neighborhood of z_0 but $(z-z_0)^{m-1}f(z)$ is not. Hence f has a pole of order m at z_0.

To see the converse, assume that f has a pole of order m at z_0. Let $g(z) = (z-z_0)^m f(z)$ in $B(z_0, R) \setminus \{z_0\}$. Then g has a removable singularity at z_0. We assume that $g(z_0)$ is suitably defined so that g is holomorphic around z_0. Using the power series expansion for g at z_0, we get

$$g(z) = \sum_{k=0}^{m-1}\frac{g^{(k)}(z_0)}{k!}(z-z_0)^k + (z-z_0)^m h(z), \tag{8.3}$$

where h is holomorphic on $B(z_0, R)$. Dividing equation (8.3) by $(z - z_0)^m$ we obtain

$$f(z) = \frac{a_{-m}}{(z-z_0)^m} + \frac{a_{-m+1}}{(z-z_0)^{m-1}} + \cdots + \frac{a_{-1}}{(z-z_0)} + h(z),$$

for $0 < |z - z_0| < R$. This proves (2).

(3) Since f has an essential singularity at $z = z_0$ when it has neither a removable singularity nor a pole, (3) follows from (1) and (2). □

8.4 Exercises

1. Let f be holomorphic in $B(0, R)$ and have a zero of order m at $z = 0$. If g is holomorphic in $0 < |z| < R$ and has a pole of order l and if $l \leq m$ at $z = 0$, then fg has a removable singularity at $z = 0$.

2. Let f and g be holomorphic in $B(a, r)$. Assume that both have a zero of order m at a. Then $\frac{f}{g}$ has a removable singularity at a and we have
$$\lim_{z\to a} \frac{f(z)}{g(z)} = \lim_{z\to a} \frac{f^{(m)}(z)}{g^{(m)}(z)}.$$

3. Let f be entire and not a polynomial. Then $f(\mathbb{C} \setminus B(0, R))$ is dense in $\mathbb{C}$ for any $R > 0$.

4. Let a be an isolated singularity of a nonzero f in a region U. Assume further that a is a cluster point of zeros of f in U. Show that a is an isolated singularity of f.

5. Classify the singularities of the following functions at the indicated points:

(i)	$\frac{\cos z}{1-\sin z}$ at $z = \frac{\pi}{2}$	(ii)	$\frac{e^z}{z}$ at $z = 0$
(iii)	$\frac{\sin z}{(z-\pi)^2}$ at $z = \pi$	(iv)	$\frac{1-\cos z}{z}$ at $z = 0$
(v)	$\frac{(e^z-1)^2}{z^2}$ at $z = 0$	(vi)	$\frac{z}{e^z-1}$ at $z = 0$
(vii)	$\frac{\cos z}{z^2}$ at $z = 0$	(viii)	$\frac{e^{z^2}-1}{z^3}$ at $z = 0$
(ix)	$\tan^2 z$ at $z = \frac{\pi}{2}$	(x)	$\frac{e^z-1}{z^2}$ at $z = 0$.

6. Locate and classify the singularities of the following functions:

(i)	$\frac{1-\cos z}{z}$	(ii)	$\frac{1-\cos z}{z^2}$
(iii)	$\frac{1}{z(e^z-1)}$	(iv)	$\tan z$
(v)	$\frac{\sin z}{z^2}$	(vi)	$\frac{\cos z}{z}$
(vii)	$\frac{\log(z+1)}{z^2}$	(viii)	$\frac{\cos(z^{-1})}{z^{-1}}$
(ix)	$z \sin \frac{1}{z}$	(x)	$\frac{z^2+1}{z(z-1)}$
(xi)	$z^n \sin \frac{1}{z}$.		

7. Let f be holomorphic in $U = \mathbb{C} \setminus \overline{B(0, R)}$. Assume that $|f(z)| \leq M$ for $z \in U$. Let c_n be the coefficients of the Laurent series of f in $R < |z| < \infty$. Show that $c_n = 0$ for $n > 0$.

8. Find the Laurent series expansion of the following function in the annulus indicated:

(i)	$\frac{(z+1)^2}{z}$ in $\lvert z\rvert > 0$	(ii)	$\frac{1}{z(1-z)}$ in $\lvert z\rvert > 1$
(iii)	$\frac{1}{(z-a)(z-b)}$, $0 < \lvert a\rvert < \lvert z\rvert < \lvert b\rvert$	(iv)	$\frac{2z-2}{(z+1)(z-3)}$, $1 < \lvert z\rvert < 3$
(v)	$\frac{1}{(z-1)(z-2)}$, $0 < \lvert z\rvert < 1$	(vi)	$e^{\frac{1}{z}}$, $0 < \lvert z\rvert < \infty$
(vii)	$\frac{z^2-2z+5}{(z-2)(z^2+1)}$, $1 < \lvert z\rvert < 2$	(viii)	$e^{z+\frac{1}{z}}$, $0 < \lvert z\rvert < \infty$
(ix)	$\frac{1}{(z+1)(z+3)}$, $1 < \lvert z\rvert < 3$	(x)	$\frac{1}{(z+1)(z+3)}$, $0 < \lvert z+1\rvert < 2$.

9. Let G be a region in $\mathbb{C}$. If $f : G \to \mathbb{C}$ is analytic except for poles, show that the poles of f cannot have a limit point in G.

10. Suppose R is a rational function such that $|R(z)| = 1$ for $|z| = 1$. How are the poles and zeros of R located?

11. Suppose R is a rational function such that $|R(z)| = 1$ if $|z| = 1$. Prove that

$$R(z) = cz^m \prod_{n=1}^{k} \frac{z-\alpha_n}{1-\bar{\alpha}_n z}$$

where c is a constant, m is an integer, and $\alpha_1, \cdots, \alpha_k$ are complex numbers such that $\alpha_n \neq 0$ and $|\alpha_n| \neq 1$. Note that each of the above factors has absolute value 1 if $|z| = 1$.

12. Locate and classify the singularities of the following functions:

(i)	$\cot \pi z$	(ii)	$\frac{1}{(z^2-1)^4(z^3+1)^2}$	(iii)	$\frac{\cos \pi z}{1+z^2}$
(iv)	$\sin \frac{1}{z}$	(v)	$\frac{e^{az}}{\sin \pi z}$	(vi)	$\frac{e^z}{e^z-1}$
(vii)	$\frac{e^{e^z}}{e^z+e^{-z}}$	(viii)	$\tan z$	(ix)	$e^{\frac{1}{z}}$
(x)	$\frac{z^3+1}{z^4+1}$	(xi)	$\frac{1}{(z^2+1)^3(z+1)}$	(xii)	$\frac{\sin \pi z}{z^4+1}$
(xiii)	$\cos \frac{1}{z}$	(xiv)	$\frac{e^{az}}{1+z}$	(xv)	$\frac{e^{-z^3}}{1+z^3}$
(xvi)	$\frac{1}{\sinh z}$	(xvii)	$\frac{\sin z}{z}$	(xviii)	$\sin \frac{\pi}{z}$
(xix)	$\cot z$	(xx)	$\frac{1}{e^z-1}$	(xxi)	$e^{\frac{1}{z-3}}$
(xxii)	$\frac{z^5}{1+z+z^2+z^3+z^4}$	(xxiii)	$\frac{1}{\sin^2 z}$.		

13. Suppose f is a nonconstant entire function such that $f(\mathbb{R}) \subset \mathbb{R}$. Show that f has no nonreal lacunary values.

14. Let f be an analytic function in the region $\text{Re}z \geq 0$, except possibly for poles on the imaginary axis. Furthermore, let $f(\bar{z}) = \overline{f(z)}$ and $\text{Re}f(z) \geq 0$ in $\text{Re}z \geq 0$. Prove that for $\text{Re}z \geq 0$, $|f'(z)| \leq \frac{f(z)+f(\bar{z})}{z+\bar{z}}$.

Chapter 9

The Winding Number of a Closed Curve

This chapter concerns closed curves. It will be devoted to the definition and discussion of an integer called the "winding number" of a closed curve about a point. Roughly speaking, the winding number $n(\gamma, a)$ of a closed curve γ about a point $a \notin [\gamma]$ is the net number of times that γ winds about a, anticlockwise counted positive. That is, it indicates how many times a closed curve winds around a fixed point not on the curve.

9.1 Winding number

Definition 9.1. *Let γ be a closed path in $\mathbb{C}$ and let $a \in \mathbb{C}$ be a point not on $[\gamma]$. Then the winding number of γ about a(also called the index of γ with respect to the point a) is defined as*

$$n(\gamma, a) = \frac{1}{2\pi i} \int_\gamma \frac{dz}{z-a}.$$

Example 9.1. *Let γ be the unit circle parametrized by $\gamma(t) = e^{2\pi i t}, 0 \leq t \leq 1$, and let $a = 0$. Then $n(\gamma, 0) = 1$.*

Example 9.2. *Let $\gamma(t) = a + re^{int}, 0 \leq t \leq 2\pi, r > 0$. Then*

$$n(\gamma, a) = \frac{1}{2\pi i} \int_0^{2\pi} \frac{rine^{int}}{re^{int}} dt = \frac{1}{2\pi i} \int_0^{2\pi} indt = n.$$

Theorem 9.1. *Let $\gamma : [a, b] \to \mathbb{C}$ be a closed path. Let $z_0 \notin [\gamma]$. Then $n(\gamma, z_0)$ is an integer.*

Proof. Suppose γ is a smooth path. Define $g : [a, b] \to \mathbb{C}$ by

$$g(t) = \int_a^t \frac{\gamma'(s)}{\gamma(s) - z_0} ds. \tag{9.1}$$

Then $g(a) = 0$ and $g(b) = 2\pi i n(\gamma, z_0)$. Further, $g'(t) = \frac{\gamma'(t)}{\gamma(t)-z_0}$, for $a \le t \le b$. But this gives

$$\begin{aligned}\frac{d}{dt}[e^{-g(t)}(\gamma(t)-z_0)] &= e^{-g(t)}\gamma'(t) - g'(t)e^{-g(t)}(\gamma(t)-z_0) \\ &= e^{-g(t)}[\gamma'(t) - \gamma'(t)(\gamma(t)-z_0)^{-1}(\gamma(t)-z_0)] \\ &= 0. \qquad (9.2)\end{aligned}$$

So $e^{-g(t)}(\gamma(t)-z_0)$ is a constant function on $[a,b]$. So we get $e^{-g(a)}(\gamma(a)-z_0) = e^{-g(b)}(\gamma(b)-z_0)$. But $\gamma(a) = \gamma(b)$ so that $e^{-g(a)} = e^{-g(b)}$ as $z_0 \notin [\gamma]$. But $g(a) = 0$. So $e^{-g(b)} = 1$. Hence $g(b) = 2\pi i n$ for $n \in \mathbb{Z}$.

If γ is piecewise smooth, proceed in the same way. Observe that in this case $g'(t) = \frac{\gamma'(t)}{\gamma(t)-z_0}$ at points where the integrand of (9.1) is continuous and (9.2) implies $e^{-g(t)}(\gamma(t)-z_0)$ is piecewise constant on $[a,b]$. But it is continuous on $[a,b]$ and hence must be a constant. □

Proposition 9.1. *If γ and σ are closed paths in $\mathbb{C}$ having the same initial points then*

(a) $n(\gamma, a) = -n(-\gamma, a)$ for every $a \notin [\gamma]$;

(b) $n(\gamma+\sigma, a) = n(\gamma, a) + n(\sigma, a)$ for every $a \notin [\gamma] \cup [\sigma]$.

Theorem 9.2. *Let γ be a rectifiable curve in $\mathbb{C}$. Then the function defined by $\alpha \mapsto \int_\gamma \frac{1}{z-\alpha}dz$ for $\alpha \notin [\gamma]$ is a continuous function of α.*

Proof. Let $\alpha_0 \notin [\gamma]$ and since γ is rectifiable, there exist a constant $M > 0$ such that $L(\gamma) \le M$. The map $t \mapsto |\gamma(t) - \alpha_0|$ is continuous, not 0 and hence (see Theorem 2.57) $\delta = d(\alpha_0, [\gamma]) > 0$. For $\alpha \in B(\alpha_0, \frac{\delta}{2})$, we have,

$$\begin{aligned}|\alpha - \gamma(t)| &\ge |\alpha_0 - \gamma(t)| - |\alpha - \alpha_0| \\ &\ge \delta - \frac{\delta}{2} \\ &= \frac{\delta}{2}\end{aligned}$$

for all t. Therefore, for $z \in [\gamma]$,

$$\left|\frac{1}{z-\alpha} - \frac{1}{z-\alpha_0}\right| = \left|\frac{\alpha-\alpha_0}{(z-\alpha)(z-\alpha_0)}\right| \le \frac{2}{\delta^2}|\alpha-\alpha_0|.$$

Hence

$$\begin{aligned}\left|\int_\gamma \left(\frac{1}{z-\alpha} - \frac{1}{z-\alpha_0}\right) dz\right| &\le \frac{2}{\delta^2}|\alpha-\alpha_0| L(\gamma) \\ &\le \frac{2M}{\delta^2}|\alpha-\alpha_0|.\end{aligned}$$

□

Let γ be a closed path in $\mathbb{C}$ and let $\Omega = \mathbb{C} - [\gamma]$. Since $[\gamma]$ is compact (closed and bounded), its complement Ω is open in $\mathbb{C}$. Hence Ω can be represented as a union of disjoint regions, the components of the complement. We shall say that γ determines these regions. If the complementary regions are considered in the extended complex plane, there is exactly one which contains the point at infinity. Consequently, γ determines one and only one unbounded region. In other words, $\{z : |z| > R\} \subset \Omega$ for some sufficiently large R. Hence Ω has one, and only one, unbounded component.

Theorem 9.3. *Let γ be a closed path in $\mathbb{C}$. Then*

(1) $n(\gamma, .) : \mathbb{C} \setminus [\gamma] \longrightarrow \mathbb{Z}$ is locally constant (i.e., $n(\gamma, a)$ is constant for a belonging to a component of $\Omega = \mathbb{C} - [\gamma]$).

(2) If S is the unbounded component of $\mathbb{C} \setminus [\gamma]$, then $n(\gamma, z) = 0$ for all $z \in S$.

Proof. (1) Define $f : \Omega \longrightarrow \mathbb{Z}$ by $f(a) = n(\gamma, a)$. Let D be a (bounded or unbounded) component of Ω. Since f is continuous, the set $f(D)$ is a connected subset of $\mathbb{Z}$ by Theorem 2.55. We know the only connected subset of $\mathbb{Z}$ are singleton subsets of $\mathbb{Z}$. Hence f is a constant on D.

(2) Suppose S is the unbounded component of $\Omega = \mathbb{C} \setminus [\gamma]$. Hence there is an $R > 0$ such that $\{z : |z| > R\} \subset S$. If $\epsilon > 0$ choose a with $|a| > R$ and $|z - a| > (2\pi\epsilon)^{-1} V(\gamma)$ uniformly for z on $[\gamma]$ where $V(\gamma)$ is the total variation of γ. (Notice that γ is piecewise smooth implies γ is a function of bounded variation and hence γ is rectifiable and $L(\gamma) = V(\gamma)$.) Then

$$\begin{aligned} |n(\gamma, a)| &= \left| \frac{1}{2\pi i} \int_\gamma \frac{dz}{z - a} \right| \\ &\leq \frac{1}{2\pi} \int_\gamma \frac{|dz|}{|z - a|} \\ &< \frac{1}{2\pi} \frac{2\pi\epsilon}{V(\gamma)} V(\gamma) \\ &= \epsilon. \end{aligned}$$

Hence $n(\gamma, a) \rightarrow 0$ as $a \rightarrow \infty$. Since $n(\gamma, a)$ is constant on S, it must be zero. □

We have seen before that if $\gamma(t) = a + e^{2\pi int}$ for $0 \leq t \leq 1$, then $n(\gamma, a) = n$. In fact if $|b - a| < 1$ then $n(\gamma, b) = n$ and if $|b - a| > 1$, then $n(\gamma, b) = 0$. This can be seen by invoking Theorem 9.3.

The rest of this chapter is to define index for any curve in a more geometric way and deduce the integral representation of the index in the case of a closed path from it again.

9.2 Index of a closed path

We discussed about continuous logarithm in Chapter 4. We formalize them here and develop a theory which will enable us to obtain an equivalent definition of winding number.

Definition 9.2. *Let X be a topological space and $f : X \to \mathbb{C}^* = \mathbb{C} - \{0\}$ be continuous. A function $g : X \to \mathbb{C}$ is a continuous logarithm of f if (1) g is continuous and (2) $f(x) = e^{g(x)}$ for each $x \in X$. We say $\theta : X \to \mathbb{R}$ is a continuous argument if $f(x) = |f(x)|e^{i\theta(x)}$ for $x \in X$.*

Theorem 9.4. *(1) Let $f : X \to \mathbb{C}^*$ be continuous. Then f has a continuous logarithm if and only if it has a continuous argument. More precisely if g is a continuous logarithm of f, then Img is a continuous argument of f and if θ is a continuous argument of f, then $log|f| + i\theta$ is a continuous logarithm of f.*

(2) Let X be connected. If g_1 and g_2 are continuous logarithms of f, then $g_1 - g_2 = 2\pi i n$ for some integer n. If θ_1 and θ_2 are continuous arguments of f, then $\theta_1 - \theta_2 = 2\pi m$ for some integer m.

(3) If X is connected and $x, y \in X$, then

$$g(x) - g(y) = log|f(x)| - log|f(y)| + i[\theta(x) - \theta(y)]$$

for any continuous logarithm g and any continuous argument θ of f.

Remark 9.1. Let $S^1 = \{z \in \mathbb{C} : |z| = 1\}$. Suppose $g : S^1 \to [0, 2\pi)$ is defined by $g(e^{it}) = t$. Then g is not continuous on S^1. The set $V = [0, \frac{\pi}{2})$ is open in $[0, 2\pi)$ but $g^{-1}(V)$ is not open in S^1. Hence g is not continuous at the point $(1, 0)$.

We shall now prove that the map $f : S^1 \to \mathbb{C}^*$ defined by $f(z) = z$, does not have a continuous argument.

Lemma 9.1. *If $S^1 = \{z \in \mathbb{C} : |z| = 1\}$ and $f : S^1 \to \mathbb{C}^*$ is given by $f(z) = z$, then f does not have a continuous argument.*

Proof. To prove this assume that $\theta : S^1 \to \mathbb{R}$ is a continuous argument of f. Let $\gamma(t) = e^{it}$ for $0 \le t \le 2\pi$. Then we have $e^{it} = e^{i\theta(e^{it})}, 0 \le t \le 2\pi$. Hence $t \mapsto t$ and $t \mapsto \theta(e^{it})$ are continuous arguments of γ. Hence $\theta(e^{it}) = t + 2k\pi$ for some $k \in \mathbb{Z}$. Letting $t \to 0$ and $t \to 2\pi$ we get $2k\pi = \theta(1) = 2(k+1)\pi$. □

If $z \neq 0$, arg$z$, the argument of z, is defined to be any real number t such that $e^{it} = \frac{z}{|z|}$. Note that arg$z$ = arg$\frac{z}{|z|}$. Again for any real number v_0 there is a unique value of arg$z$ in the interval $[v_0 - \pi, v_0 + \pi)$.

Remark 9.2. If v_0 is real, consider the function $\gamma(t) = e^{it}$ on the open interval $(v_0 - \pi, v_0 + \pi)$. The image of this interval contains all points of the unit circle save one, $e^{i(v_0 - \pi)} = -e^{iv_0}$. It is not so difficult to verify that the mapping $e^{it} \mapsto t$ is continuous on the set $\{e^{it} : v_0 - \pi < t < v_0 + \pi\}$. The function is not continuous on the full unit circle.

Lemma 9.2. *Let U be an open disk which does not contain zero. If $z_0 \in U$, and if v_0 is a prescribed value of argz_0, the function h defined by*

$$h(z) = \arg z, v_0 - \pi \le \arg z < v_0 + \pi$$

is continuous on U.

Proof. Let $U = B(z_0, r), r > 0$ and $0 \notin U$. Suppose $z_0 = r_0 e^{iv_0}, r_0 > 0$. Define a map $g : U \to S^1 = \{z \in \mathbb{C} : |z| = 1\}$ by $g(z) = \frac{z}{|z|}$. The map is well-defined as $0 \notin U$. We claim that $e^{i(v_0+\pi)}$ is not in the image of g. If it were, then there exists $z \in U$ such that $g(z) = \frac{z}{|z|} = e^{i(v_0+\pi)}$. That is, $z = |z|e^{i(v_0+\pi)} = -|z|e^{iv_0} = -\left(\frac{|z|}{|z_0|}\right) z_0$. Thus z is a positive multiple of $-z_0$. The set U is an open disk and hence is a convex set. Therefore the line segment joining z_0 to z belongs to U. But 0 lies on this segment contradicting the assumption that $0 \notin U$. Thus $-e^{iv_0} \notin g(U)$. Now the map $\phi : (v_0 - \pi, v_0 + \pi) \to S^1 \setminus \{-e^{iv_0}\}$ defined by $\phi(t) = e^{it}$ is one-one, onto and continuous. It is a homeomorphism on each of the compact subintervals so that it has a continuous inverse, $\psi : S^1 \setminus \{-e^{iv_0}\} \to (v_0 - \pi, v_0 + \pi)$. Note $h(z) = \psi(g(z))$. Since g and ψ both are continuous on U, hence the function h is continuous on U. □

We can restate the above lemma in the following way:

Let $|z_0| = r_0 > 0$. Let $U = B(z_0, r_0)$. Then there exists a continuous argument θ for the identity function on U. In particular, there exists a continuous (and hence holomorphic) logarithm of the identity function on U.

Theorem 9.5. *Let $\gamma : [a, b] \to \mathbb{C}^*$ be continuous. Then γ has a continuous argument (and hence a continuous logarithm).*

Proof. Let $m = \min_{a \le t \le b} |\gamma(t)| > 0$. Since $[a, b]$ is compact and γ is continuous on $[a, b]$, the function γ is uniformly continuous. By uniform continuity of γ we can find a partition $P = \{a = t_0 < t_1 < \cdots < t_n = b\}$ such that $\|P\| = \max_{1 \le i \le n} |t_i - t_{i-1}| < \delta$ and $|\gamma(t) - \gamma(t_j)| < m$ for $t_j \le t \le t_{j+1}, 0 \le j \le n-1$.

Hence $\gamma(t) \in B(\gamma(t_j), m) = B_j(\text{say})$ for $t \in [t_j, t_{j+1}], 0 \le j \le n-1$. By Lemma 9.2, we have a continuous argument, say h_j of the identity function on B_j. In particular, $\gamma(t) = |\gamma(t)|e^{ih_j(\gamma(t))}$ for $t \in [t_j, t_{j+1}]$.

Now $h_0(t_1) = h_1(t_1) + 2\pi k_1$ for a fixed $k_1 \in \mathbb{Z}$. We may replace h_1 by $h_1 + 2\pi k_1$ to get a continuous argument on $[t_1, t_2]$. The resulting function is continuous on $[t_0, t_2]$. Proceeding similarly, we get a continuous argument of γ on $[a, b]$. □

Recall that a curve γ in $\mathbb{C}$ is a continuous map from some interval $[a, b]$ into $\mathbb{C}$.

Lemma 9.3. *Let γ be a closed curve and assume that $0 \notin [\gamma]$. Let θ be a continuous argument of γ. Then $\frac{[\theta(b)-\theta(a)]}{2\pi}$ is an integer and it is independent of the choice of the continuous argument.*

Proof. γ is a closed curve implies $\gamma(b) = \gamma(a)$. Hence

$$|\gamma(b)|e^{ih(\gamma(b))} = |\gamma(a)|e^{ih(\gamma(a))}$$

where h is a continuous argument of γ on $[a, b]$. Hence

$$e^{i[h(\gamma(b))-h(\gamma(a))]} = 1$$

which implies $e^{i(\theta(b)-\theta(a))} = 1$. Thus $\theta(b) - \theta(a) = 2\pi n$ for some integer n. □

Remark 9.3. Compare and contrast Lemma 9.1 and Lemma 9.3.

Definition 9.3. *Let γ be a curve in $\mathbb{C}$ and $z_0 \notin [\gamma]$. We then consider the curve $\gamma - z_0 : t \mapsto \gamma(t) - z_0$. Let θ be a continuous argument of this curve. We let*

$$Ind_\gamma(z_0) = \frac{[\theta(b) - \theta(a)]}{2\pi}$$

where $Ind_\gamma(z_0)$ is called the index of the curve γ at z_0. Note that the orientation of the curve γ is also taken into account.

Example 9.3. *Let $\gamma(t) = e^{it}, \sigma(t) = e^{-it}$ and $\eta(t) = e^{2it}$ for $0 \leq t \leq 2\pi$. Then $Ind_\gamma(0) = 1, Ind_\sigma(0) = -1$ and $Ind_\eta(0) = 2$.*

Lemma 9.4. *Let γ be a curve in an open set $U \subset \mathbb{C}$. Then there is a partition $a = t_0 < t_1 < \cdots < t_n = b$ and open disk $B_j \subset U$ such that $\gamma(t) \in B_j$ for $t \in [t_{j-1}, t_j], 1 \leq j \leq n$.*

Proof. Let $\epsilon = d([\gamma], \mathbb{C} \setminus U) > 0$. For this ϵ, by uniform continuity of γ, there exists a $\delta > 0$. Use this δ to get a partition. □

Theorem 9.6. *Let $\gamma : [a, b] \to U \subset \mathbb{C}$ be a closed path. If $z_0 \notin [\gamma]$, we have*

$$Ind_\gamma(z_0) = \frac{1}{2\pi i} \int_\gamma \frac{dz}{z - z_0}.$$

Proof. By Lemma 9.4, we can find a partition $a = t_0 < t_1 < \cdots < t_n = b$ of $[a, b]$ and disks B_j in such a way that $z_0 \notin B_j$. Hence, by Lemma 9.2 there exists a continuous logarithm log_j on B_j of $\gamma(t) - z_0$ for $t \in [t_{j-1}, t_j]$. The function $\log_j$ is holomorphic on B_j with $\log_j'(z - z_0) = \frac{1}{z - z_0}$ (see also the inverse function theorem). Using this fact, we see

$$\int_{\gamma|_{[t_{j-1}, t_j]}} \frac{dz}{z - z_0} = \log_j(\gamma(t_j) - z_0) - \log_j(\gamma(t_{j-1}) - z_0). \tag{9.3}$$

Let θ be a continuous argument of $\gamma(t) - z_0$ on $[a, b]$. From Equation (9.3) and Theorem 9.4 we get,

$$\frac{1}{2\pi i}\int_\gamma \frac{dz}{z-z_0}$$

$$= \frac{1}{2\pi i}\sum_{j=1}^{n}[\log_j(\gamma(t_j)) - \log_j(\gamma(t_{j-1}))]$$

$$= \frac{1}{2\pi i}\sum_{j=1}^{n}[\log|\gamma(t_j) - z_0| - \log|\gamma(t_{j-1}) - z_0| + i\theta(t_j) - i\theta(t_{j-1})]$$

$$= \frac{1}{2\pi i}[\log|\gamma(b) - z_0| - \log|\gamma(a) - z_0| + i\theta(b) - i\theta(a)]$$

$$= \frac{1}{2\pi}[\theta(b) - \theta(a)] = \mathrm{Ind}_\gamma(z_0)$$

as $\gamma(b) = \gamma(a)$. □

Thus we have proved that if γ is a closed path in $\mathbb{C}$ and $z_0 \notin [\gamma]$ then

$$n(\gamma, z_0) = \mathrm{Ind}_\gamma(z_0).$$

Hence the two definitions of index coincide.

Definition 9.4. *Let γ be a closed curve in $\mathbb{C}$. We say that $z \in \mathbb{C} \setminus [\gamma]$ is*

- *inside γ if $n(\gamma, z) \neq 0$,*
- *outside γ if $n(\gamma, z) = 0$.*

Whenever we say z is inside γ or outside γ, it is with reference to this definition. This coincides with our intuitive notions in familiar cases.

Example 9.4. *Let $\gamma(t) = r(t)e^{it}, 0 \leq t \leq 2\pi$ be a closed path. Intuitively, $z = \rho e^{it} \notin [\gamma]$ is inside γ if $\rho < r(t)$ where $t \in [0, 2\pi]$ and is outside γ if $\rho > r(t)$. We show that these agree with our definition. If $\rho > r(t)$, then $\frac{1}{w-z}$ is holomorphic in the star-shaped open set $\mathbb{C}$ minus the closed half line $\{Re^{it} : R \geq \rho\}$ and hence has a primitive over there. It follows that $n(\gamma, z) = 0$ in this case. Since the set $A = \{\rho e^{it} : 0 \leq \rho < r(t) : t \in [0, 2\pi]\}$ is star-shaped at 0, it is connected. Hence $n(\gamma, z) = n(\gamma, 0)$ for $z \in A$ by Theorem 9.3.*

It is easy to check that $n(\gamma, 0) = 1$ by a trivial computation.

Notice that if σ is a reparametrization of γ in a region U and $z_0 \notin [\gamma]$, then $[\sigma] = [\gamma]$ and hence $n(\gamma, z_0) = n(\sigma, z_0)$. Further if γ is a closed path such that $[\gamma] \subset B(0, r)$ then $n(\gamma, a) = 0$ if $|a| > r$.

Definition 9.5. *Let $\gamma_0, \gamma_1 : [a, b] \to U$ be two curves. Assume that they have the same endpoints : $\gamma_0(a) = \gamma_1(a)$ and $\gamma_0(b) = \gamma_1(b)$. γ_0 is said to be path homotopic to γ_1 in U if there exists a continuous map $H : [a, b] \times [0, 1] \to U$ such that* *(i)* $H(s,0) = \gamma_0(s)$ *and* $H(s,1) = \gamma_1(s)$ *for all* $s \in [a,b]$ *and*

(ii) $H(0,t) = \gamma_0(a)$ *and* $H(1,t) = \gamma_0(b)$ *for all* $t \in [0,1]$.

If we let $\gamma_t(s) = H(s,t)$, *then each* γ_t *is a curve in* U *and* H *deforms* γ_0 *continuously into* γ_1 *within* U *via the intermediate curves* γ_t. H *is called a path homotopy from* γ_0 *to* γ_1.

Lemma 9.5. *Let* $R = \{z \in \mathbb{C} : a \leq Rez \leq b, c \leq Imz \leq d\}$. *Let* $f : R \to \mathbb{C}^*$ *be continuous. Then* f *has a continuous argument.*

Proof. Note that R is compact and f is continuous. Let $\epsilon = \min\{|f(z)| : z \in R\}$, by uniform continuity of $|f|$ on R, there exists a $\delta > 0$. We divide R into small squares whose diagonals are less than δ. If S is one such squares and $z_0 \in S$, then $S \subset B(z_0, \delta)$ and $f(z) \in B(f(z_0), \epsilon)$ for $z \in S$. By Lemma 9.2, we have a continuous argument of the identity function and hence a continuous argument of f. If S and S' are adjacent squares having a vertical side common, then we can adjust the arguments on S and S' so that we end up with a continuous argument of the identity function on their union and hence a continuous argument of f on this union of squares. Starting from the bottom most squares we can move upwards and get a continuous argument of the identity function on all of R and hence a continuous argument of f on all of R. □

Lemma 9.6. *Let* $\gamma_0, \gamma_1 : [a,b] \to U$ *be closed curves with the same endpoint. If they are path homotopic in* U *and* $z_0 \notin U$ *then* $n(\gamma_0, z_0) = n(\gamma_1, z_0)$.

Proof. Let H be a path homotopy from γ_0 to γ_1. Let θ be a continuous argument for $H - z_0$ on the domain of H. In particular, if θ_t is a continuous argument of γ_t then since $\gamma_t(a) = \gamma_t(b)$, we have

$$\begin{aligned} e^{i\theta_t(b)} &= \frac{\gamma_t(b) - z_0}{|\gamma_t(b) - z_0|} \\ &= \frac{\gamma_t(a) - z_0}{|\gamma_t(a) - z_0|} \\ &= e^{i\theta_t(a)}. \end{aligned}$$

The map $t \mapsto \frac{\theta_t(b) - \theta_t(a)}{2\pi}$ is an integer valued continuous function on $[a,b]$ and therefore is a constant. □

9.3 Holomorphic logarithms

The reader might have observed that in the last two sections we had to deal with integrals of the form $\int_\gamma \frac{f'}{f}$. Taking inspiration from calculus one is tempted to write

$$\int_{\gamma}[\log(f(z))]'dz = \log f(\gamma(b)) - \log f(\gamma(a)).$$

We investigate the conditions under which this can be made rigorous. This will be put to use when we come to argument principle. We discussed about holomorphic logarithm in chapter 5. Recall the following definition.

Definition 9.6. *Let $f : U \subset \mathbb{C} \to \mathbb{C}^*$ be holomorphic in the region U. We say that $g : U \to \mathbb{C}$ is a holomorphic logarithm of f in U if $exp(g) = f$ on U.*

Example 9.5. *If $U = \mathbb{C}^*$ and f is the identity, then we have seen there exists no holomorphic logarithm of f in U. But, however, there exists holomorphic logarithms on $\mathbb{C} \setminus L_\alpha$.*

Theorem 9.7. *Let $f : U \to C^*$ be holomorphic. Then f has a holomorphic logarithm in U if and only if the "logarithmic derivative" $\frac{f'}{f}$ has a primitive in U, that is, if and only if $\int_\gamma \frac{f'}{f} = 0$ for any closed path γ in U.*

Proof. Suppose f has a holomorphic logarithm in U. Then there exists a function g such that $e^g = f$ on U. We differentiate and see that $g' = \frac{f'}{f}$. Conversely, suppose $\frac{f'}{f}$ has a primitive in U. Then there exists $g : U \to \mathbb{C}$ holomorphic such that $g' = \frac{f'}{f}$. This implies $(fe^{-g})' = 0$ on U. Hence fe^{-g} is locally constant. Since f never takes the value 0, these constants are never equal to 0 and hence are of the form e^α. □

Corollary 9.1. *Let U be a convex open set and $f : U \to \mathbb{C}^*$ be holomorphic. Then f has a holomorphic logarithm.*

Proof. Since f never takes the value 0, the function $\frac{f'}{f}$ is holomorphic on U and hence by Cauchy's theorem for convex sets, $\int_\gamma \frac{f'}{f} = 0$ for any closed path γ in U. The corollary follows from Theorem 9.7. □

Corollary 9.2. *If g is a holomorphic logarithm of $f : U \to \mathbb{C}^*$, and $\gamma : [a, b] \to U$ is a path, then $\int_\gamma \frac{f'}{f}dz = g(\gamma(b)) - g(\gamma(a))$.*

Proof. Given that $e^g = f$. Hence $\frac{f'}{f} = g'$. Thus

$$\begin{aligned}
\int_\gamma \frac{f'}{f}dz &= \int_a^b g'(\gamma(t))\gamma'(t)dt \\
&= \int_a^b (g \circ \gamma)'(t)dt \\
&= (g \circ \gamma)(b) - (g \circ \gamma)(a) \\
&= g(\gamma(b)) - g(\gamma(a)).
\end{aligned}$$

The result follows. □

Theorem 9.8. *Let $\gamma : [a,b] \to U \subset \mathbb{C}$ be a closed path. Let $f : U \to \mathbb{C}$ be holomorphic. Assume that $z_0 \notin [f\circ\gamma]$. Then we have $n(f\circ\gamma, z_0) = \frac{1}{2\pi i}\int_\gamma \frac{f'(z)}{f(z)-z_0}dz$.*

Proof. The proof is similar to Theorem 9.6. Without loss of generality, We may assume that $f(z) - z_0 \neq 0$ in U (by replacing U by a smaller open set containing γ if necessary). As in Lemma 9.4 , we can find a partition of $[a,b]$ and disks B_j. By Corollary 9.1, there exists a holomorphic logarithm g_j of $f - z_0$ on B_j. Hence, by Corollary 9.2,

$$\int_{\gamma|_{[t_{j-1},t_j]}} \frac{f'(z)}{f(z)-z_0}dz = g_j(\gamma(t_j)) - g_j(\gamma(t_{j-1})).$$

Let θ be a continuous argument of $f\circ\gamma - z_0$. Since $g_j\circ\gamma$ is a continuous logarithm of $f\circ\gamma - z_0$ on $[t_{j-1}, t_j]$, we use Theorem 9.4 to obtain

$$\begin{aligned}
&\frac{1}{2\pi i}\int_\gamma \frac{f'(z)}{f(z)-z_0}dz \\
&= \frac{1}{2\pi i}\sum_{j=1}^{n}[g_j(\gamma(t_j)) - g_j(\gamma(t_{j-1}))] \\
&= \frac{1}{2\pi i}\sum_{j=1}^{n}[\log|f(\gamma(t_j)) - z_0| - \log|f(\gamma(t_{j-1})) - z_0| + i\theta(t_j) - i\theta(t_{j-1})] \\
&= \frac{1}{2\pi i}[\log|f(\gamma(b)) - z_0| - \log|f(\gamma(a)) - z_0|] + \frac{1}{2\pi}[\theta(b) - \theta(a)].
\end{aligned}$$

As γ is a closed path, we have $\gamma(b) = \gamma(a)$. Hence

$$\frac{1}{2\pi i}\int_\gamma \frac{f'(z)}{f(z)-z_0}dz = \frac{1}{2\pi}[\theta(b) - \theta(a)] = n(f\circ\gamma, z_0).$$

□

9.4 Exercises

1. Show that the index is invariant under translations, magnifications (multiplication by a positive scalar) and rotations (multiplication by a complex number of unit modulus).

2. Let $\gamma(t) = 1 + 2e^{3it}, 0 \leq t \leq 2\pi$. Find $n(\gamma, z)$ when (i) $z = \frac{3}{2}$ and (ii) $z = 4 + i$.

3. Let $\gamma : [a,b] \to \mathbb{C}$ be a closed curve and $z_0 \notin [\gamma]$. Let $\epsilon = d(z_0, [\gamma])$. Let σ be a closed curve defined on $[a,b]$ such that $|\sigma(t) - \gamma(t)| < \epsilon$. Then $n(\gamma, z_0) = n(\sigma, z_0)$.

4. Let $G = \mathbb{C} - \{a,b\}, a \neq b$, and let γ be the curve in Fig. 9.1.

   Show that $n(\gamma, a) = n(\gamma, b) = 0$.

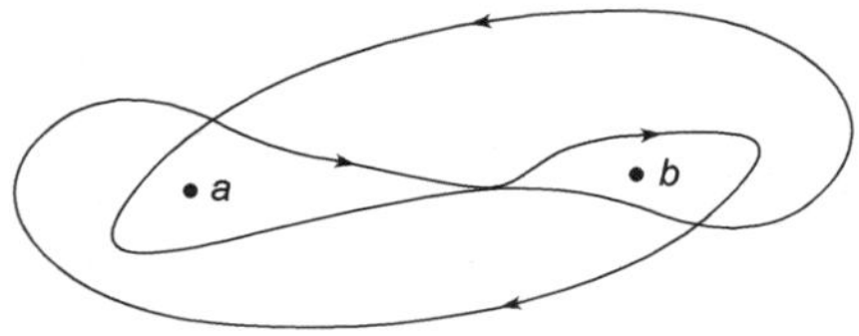

Fig. 9.1: The curve γ

5. Let $P(z)$ be a polynomial of degree n and suppose all zeros of P lie in $B(0, R), R > 0$. Show that $\frac{1}{2\pi i}\int_\gamma \frac{P'(z)}{P(z)}dz = n$ where $\gamma(t) = Re^{it}$, $0 \leq t \leq 2\pi$.

6. Give an example of a closed rectifiable curve γ in $\mathbb{C}$ such that for any integer k there is a point $a \notin [\gamma]$ with $n(\gamma, a) = k$.

7. Fix $z = re^{i\theta} \neq 0$ and let γ be a rectifiable curve in $\mathbb{C} \setminus \{0\}$ from 1 to z. Show that there is an integer k such that $\int_\gamma \frac{1}{w}dw = \log r + i\theta + 2\pi i k$.

Chapter 10

The General Form of Cauchy's Theorem

In our earlier treatment of Cauchy's theorem and the integral formula we considered only the case of a circular region (in fact a star-shaped region) and we studied about the local properties of analytic functions. In this chapter we shall characterize the regions in which Cauchy's theorem is valid for every holomorphic function and for every closed rectifiable path.

10.1 Homology version of Cauchy's theorem

Before we state and prove this theorem, which will remove the restriction to convex regions that was imposed in Corollary 7.1 and Corollary 7.2, we require to prove some preliminary lemmas. We shall no longer restrict ourselves to integrals over single paths but consider integrals over finite "sum" of paths instead.

Definition 10.1. *A formal combination such as $\gamma = \gamma_1 + \cdots + \gamma_n$ is called a cycle in an open set Ω where each γ_i is a closed path in Ω. We let $[\gamma] = \cup_i[\gamma_i]$. We define the index $n(\gamma, z)$ of such a cycle γ by setting $n(\gamma, z) = \sum_{i=1}^{n} n(\gamma_i, z)$ for $z \in \mathbb{C} \setminus [\gamma]$. If $f \in C(\gamma)$, we define*

$$\int_{\gamma} f dz = \sum_{i} \int_{\gamma_i} f dz.$$

We say that a cycle γ is homologous to zero ($\gamma \approx 0$) in Ω if $n(\gamma; z) = 0$ for any $z \notin \Omega$.

Lemma 10.1. *Given a closed path γ in an open set Ω and a point $z \in [\gamma]$, there is a closed path σ in Ω not passing through z and such that $\int_{\gamma} f(z)dz = \int_{\sigma} f(z)dz$ for all $f \in H(\Omega)$.*

Proof. The idea of the proof is quite simple. Assume for the moment that $\gamma : [0, 1] \to \Omega$ passes through z only once. That is, there is a unique $t \in [0, 1]$ such

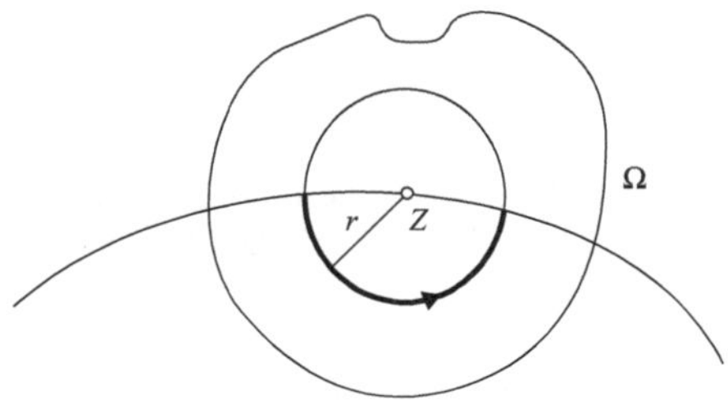

Fig. 10.1: The open set Ω and the circle $S(z,r)$

that $\gamma(t) = z$. Then we can find an $r > 0$ such that $B(z, 2r) \subset \Omega$ and replace the part of γ with the corresponding part of $S(z,r) = \{w \in \mathbb{C} : |w - z| = r\}$. Cauchy's theorem for disks will then be applicable and we get the result. See Fig.10.1.

To prove the lemma, we assume that there exists $w \in [\gamma]$ and $w \neq z$ (otherwise there is nothing to prove). Choose $r > 0$ such that $B(z, 2r) \subset \Omega$ and $w \notin B(z,r)$. We may assume that γ is defined as above with $\gamma(0) = w = \gamma(1)$.

By the uniform continuity of γ, we can find $\delta > 0$ such that if $s, t \in [0,1]$ and $|s - t| < \delta$, then $|\gamma(s) - \gamma(t)| < r$. Let $U = \gamma^{-1}(B(z,r))$. Since $\gamma(0) = w = \gamma(1)$ and $w \notin B(z,r)$, it follows that U is an open subset of $(0,1)$ and that each of the connected component of U is an open interval in $(0,1)$. If (a,b) is one such interval, then by continuity, $\gamma(a)$ and $\gamma(b)$ are points of $S(z,r)$. Moreover, if $t \in (a,b)$ is such that $\gamma(t) = z$, then $b - a \geq 2\delta$, since $|\gamma(b) - \gamma(t)| = r = |\gamma(a) - \gamma(t)|$. Thus there are only finitely many connected components $(a_i, b_i), 1 \leq i \leq N$ of U containing points $t \in (a_i, b_i)$ with $\gamma(t) = z$. We replace the part of the curve γ which is the image of the closure $[a_i, b_i]$ of each such component with the circular arc on $S(z,r)$ going from $\gamma(a_i)$ to $\gamma(b_i)$ in the positive direction of γ.

By Cauchy's theorem for the disk $B(z, 2r)$ the integral of a holomorphic f over this portion of γ is same as that of f over the circular arc. This way we get a new path σ which avoids z and meets our requirement. □

Lemma 10.2. *Let γ be a rectifiable path in $\mathbb{C}$ and suppose ϕ is a function defined and continuous on $[\gamma]$. For each $m \geq 1$, let $F_m(z) = \int_\gamma \frac{\phi(w)}{(w-z)^m} dw$ for $z \notin [\gamma]$. Then each F_m is analytic on $\mathbb{C} - [\gamma]$ and $F'_m(z) = mF_{m+1}(z)$.*

Proof. The lemma is already proved for $m = 1$ in Theorem 6.7. We first claim that each F_m is continuous. Note that $[\gamma]$ is compact, ϕ is continuous on $[\gamma]$. Hence ϕ is bounded on $[\gamma]$. Further,

$$\frac{1}{(w-z)^m} - \frac{1}{(w-a)^m}$$

$$= \left[\frac{1}{w-z} - \frac{1}{w-a}\right] \sum_{k=1}^{m} \frac{1}{(w-z)^{m-k}} \frac{1}{(w-a)^{k-1}}$$

$$= \quad (z-a)\left[\frac{1}{(w-z)^m(w-a)}+\cdots+\frac{1}{(w-z)(w-a)^m}\right]. \qquad (10.1)$$

Proceed as in Proposition 6.13 to show that F_m is continuous on $[\gamma]$. Now fix $a \in \Omega = \mathbb{C} - [\gamma]$ and let $z \in \Omega, z \neq a$. It follows from (10.1) that

$$\frac{F_m(z)-F_m(a)}{z-a} = \int_\gamma \frac{\phi(w)(w-a)^{-1}}{(w-z)^m}dw+\cdots+\int_\gamma \frac{\phi(w)(w-a)^{-m}}{w-z}dw. \quad (10.2)$$

Since $a \notin [\gamma]$, $\frac{\phi(w)}{(w-a)^k}$ is continuous on $[\gamma]$ for each k. By the first part of this proof, each integral on the right hand side of (10.2) defines a continuous function of $z, z \in \Omega$. Hence letting $z \to a$, (10.2) gives that the limit exists and

$$F_m'(a) = \int_\gamma \frac{\phi(w)}{(w-a)^{m+1}}dw+\cdots+\int_\gamma \frac{\phi(w)}{(w-a)^{m+1}}dw = mF_{m+1}(a).$$

□

Lemma 10.3. *Let f be holomorphic in an open set Ω. Let $\phi : \Omega \times \Omega \to \mathbb{C}$ be defined by*

$$\phi(w,z) = \begin{cases} \frac{f(w)-f(z)}{w-z}, & w \neq z \\ f'(w), & w = z. \end{cases}$$

Let γ be a closed path in Ω. Then for each $z \in \Omega$, the function $w \mapsto \phi(w,z)$ is holomorphic in Ω. Moreover the function g defined by $g(z) = \int_\gamma \phi(w,z)dw$ is holomorphic in Ω.

Proof. Given $z \in \Omega$, the function $w \mapsto \phi(w,z)$ is continuous on Ω and holomorphic in $\Omega\backslash\{z\}$. Thus, by Remark 8.1, it follows that $\phi(w,z)$ is holomorphic in Ω as a function of w. If $z \notin [\gamma]$, we have

$$g(z) = \int_\gamma \phi(w,z)dw = \int_\gamma \frac{f(w)}{w-z}dz - f(z)\int_\gamma \frac{dw}{w-z}.$$

From Lemma 10.2 it follows that g is holomorphic in $\Omega\backslash[\gamma]$. We now fix $z \in [\gamma]$. Then by Lemma 10.1, we can replace γ by a closed path σ which misses z and therefore an open ball $B(z,\epsilon)$, without changing the value of g on $B(z,\epsilon)$. It follows therefore that g is holomorphic in $B(z,\epsilon)$. Thus g is holomorphic at each point of Ω. □

The homology version of Cauchy's theorem is as follows:

Theorem 10.1. *Let Ω be a nonempty open subset of $\mathbb{C}$. Let γ be a cycle in Ω homologous to zero in Ω. Then we have*

(1) If $f \in H(\Omega)$, then $\int_\gamma f(z)dz = 0$.

(2) The Cauchy integral formula is true:

$$n(\gamma, z)f(z) = \frac{1}{2\pi i}\int_\gamma \frac{f(w)}{w-z}dw \tag{10.3}$$

for $z \in \Omega \setminus [\gamma]$.

Proof. Let $f \in H(\Omega)$. Let ϕ, g be as in Lemma 10.3. Note $\{0\} = \mathbb{Z} \cap (-1,1)$ is open in $\mathbb{Z}$. Let $V = \{z \in \mathbb{C} : n(\gamma, z) = 0\}$. Then V is open, since $z \mapsto n(\gamma, z)$ is a continuous integer-valued function on $\mathbb{C} \setminus [\gamma]$. Moreover $\mathbb{C} = V \cup \Omega$ and hence $V \cap \Omega \neq \emptyset$, as $\mathbb{C}$ is connected. We define a function h on V by setting $h(z) = \int_\gamma \frac{f(w)}{w-z}dw$, $z \in V$. The function h is holomorphic by Lemma 10.2. We claim that $h = g$ on $\Omega \cap V$: if $z \in \Omega \cap V$,

$$\begin{aligned} g(z) &= \int_\gamma \frac{f(w)}{w-z}dw - f(z)\int_\gamma \frac{dw}{w-z} \\ &= h(z) - f(z)2\pi i\, n(\gamma, z) \\ &= h(z) - f(z)2\pi i \cdot 0 = h(z). \end{aligned}$$

We define a function from $\mathbb{C}$ into itself which is equal to $g(z)$ if $z \in \Omega$ and equal to $h(z)$ if $z \in V$. Note that the function is analytic on $\mathbb{C}$. We thus get an entire function, which we still denote by g, and is such that $g|_V = h$. Clearly, we have $\lim_{z\to\infty} g(z) = 0$.

This is because Theorem 9.3 implies that V contains a neighborhood of ∞ in $\mathbb{C}_\infty$. Again f is bounded on the compact set $[\gamma]$ and $\lim_{z\to\infty} \dfrac{1}{w-z} = 0$ uniformly for w in $[\gamma]$ and $\lim_{z\to\infty} g(z) = \lim_{z\to\infty} h(z) = \lim_{z\to\infty} \displaystyle\int_\gamma \frac{f(w)}{w-z}dw = 0$. In particular this implies there is an $R > 0$ such that $|g(z)| \le 1$ for $|z| \ge R$. Since g is bounded on $\overline{B(0,R)}$ it follows that g is a bounded entire function. Hence g is constant by Liouville's theorem. This constant is obviously 0 since $g(z) \to 0$ as $z \to \infty$. Thus we get if $z \in \Omega \setminus [\gamma]$ then $0 = \int_\gamma \frac{f(w)-f(z)}{w-z}dw = \int_\gamma \frac{f(w)}{w-z}dw - f(z)\int_\gamma \frac{dw}{w-z} = \int_\gamma \frac{f(w)}{w-z}dw - 2\pi i n(\gamma, z)f(z)$. Thus $n(\gamma, z)f(z) = \frac{1}{2\pi i}\int_\gamma \frac{f(w)}{w-z}dw$. To prove Cauchy's theorem, we fix $a \in \Omega \setminus [\gamma]$ and observe

$$\begin{aligned} \int_\gamma f(w)dw &= \int_\gamma \frac{f(w)(w-a)}{w-a}dw \\ &= 2\pi i\, n(\gamma, a)\, [f(w)(w-a)]_{w=a} = 0. \end{aligned}$$

□

Corollary 10.1. *Let Ω be an open subset of the plane and $f : \Omega \to \mathbb{C}$ be an analytic function. Let γ be a cycle in Ω which is homologous to 0 in Ω. Then for $a \in \Omega - [\gamma]$ and $k \ge 1$,*

$$f^{(k)}(a)\, n(\gamma, a) = \frac{k!}{2\pi i}\int_\gamma \frac{f(z)}{(z-a)^{k+1}}dz.$$

Proof. This follows immediately by differentiating both sides of the formula in (10.3) and applying Lemma 10.2. □

10.2 Homotopy version of Cauchy's theorem

There are several forms of Cauchy's theorem. They all assert that if γ is a closed path or cycle in an open set G, and if γ and G satisfy certain topological conditions, then the integral of every $f \in H(G)$ over γ is 0. In Chapter 7, we derived a simple version of this theorem which is quite sufficient for many applications. A more general global form will be established here.

We shall first make a brief discussion on a topological concept that is relevant to prove the homotopy version of Cauchy's theorem.

Definition 10.2. *Let G be a region and let $\gamma_0, \gamma_1 : [0,1] \to G$ be two rectifiable paths in G such that $\gamma_i(0) \neq \gamma_i(1), i = 0, 1$. We say that γ_0 is homotopic to γ_1 in G if there is a continuous function $\Gamma : [0,1] \times [0,1] \longrightarrow G$ such that*

$$\Gamma(s,0) = \gamma_0(s) \text{ and } \Gamma(s,1) = \gamma_1(s) \ (0 \le s \le 1) \tag{10.4}$$

The map Γ is called a homotopy between γ_0 and γ_1. If γ_0 is homotopic to γ_1, we write $\gamma_0 \simeq \gamma_1$. We think of a homotopy as a continuous one-parameter family of paths from $[0,1]$ into G. If we imagine the parameter t as representing time, then the homotopy Γ represents a continuous "deforming" of the path γ_0 to the path γ_1, as t goes from 0 to 1.

Definition 10.3. *If γ_0, $\gamma_1 : [0,1] \to G$ are two closed rectifiable paths in G then γ_0 is homotopic to γ_1 in G if there is a continuous function $\Gamma : [0,1] \times [0,1] \longrightarrow G$ such that*

$$\Gamma(s,0) = \gamma_0(s) \text{ and } \Gamma(s,1) = \gamma_1(s) \ (0 \le s \le 1)$$

and

$$\Gamma(0,t) = \Gamma(1,t) \ (0 \le t \le 1).$$

That is, if we define $\gamma_t : [0,1] \longrightarrow G$ by $\gamma_t(s) = \Gamma(s,t)$ then each γ_t is a closed path and they form a continuous family of paths which start at γ_0 and go to γ_1. In this case also we shall write $\gamma_0 \simeq \gamma_1$.

Notice however that there is no requirement that each γ_t be rectifiable. In practice when γ_0 and γ_1 are rectifiable (or smooth) each of the γ_t will also be rectifiable (or smooth).

Lemma 10.4. *The relation $\simeq$ is an equivalence relation.*

Proof. Given a (closed) path $\gamma : [0,1] \to G$ it is trivial that $\gamma \simeq \gamma$; the map $\Gamma(s,t) = \gamma(s)$ is the required homotopy. Given two (closed) paths $\gamma, \gamma' : [0,1] \to G$ and $\gamma \simeq \gamma'$, we show that $\gamma' \simeq \gamma$. Let Γ be a homotopy between γ and γ'. Then $\Lambda(s,t) = \Gamma(s, 1-t)$ is a homotopy between γ' and γ. Suppose that $\gamma, \gamma', \gamma''$ are closed paths from $[0,1]$ into G, $\gamma \simeq \gamma', \gamma' \simeq \gamma''$. We show that $\gamma \simeq \gamma''$. Let Γ be a homotopy between γ and γ', and let Γ' be a homotopy between γ' and γ''. Define $\Lambda : [0,1] \times [0,1] \to G$ by the equation

$$\Lambda(s,t) = \begin{cases} \Gamma(s,2t) & \text{if } t \in [0,\frac{1}{2}]; \\ \Gamma'(s,2t-1) & \text{if } t \in [\frac{1}{2},1]. \end{cases}$$

The map Λ is well-defined, since if $t = \frac{1}{2}$, $\Gamma(s,2t) = \Gamma(s,1) = \gamma'(s) = \Gamma'(s,0) = \Gamma'(s,2t-1)$. Because Λ is continuous on the two closed subsets $[0,1]\times[0,\frac{1}{2}]$ and $[0,1]\times[\frac{1}{2},1]$ of $[0,1]\times[0,1]$, it is continuous on all of $[0,1]\times[0,1]$, by the Pasting lemma. Thus Λ is the required homotopy between γ and γ''. □

The equivalence classes are called homotopy classes.

Example 10.1. *Let $G = B(a;R)$ and let $\gamma : [0,1] \to G$ be a closed path. If $0 \le t \le 1$ and $0 \le s \le 1$, and we put $z = ta + (1-t)\gamma(s)$; then z lies on the straight line segment from a to $\gamma(s)$. Hence z must lie in G(as G is convex). Let $\gamma_t(s) = ta + (1-t)\gamma(s)$ for $0 \le s \le 1$ and $0 \le t \le 1$. So $\gamma_0 = \gamma$ and γ_1 is the path constantly equal to a; the paths γ_t are somewhere in between. We were able to draw γ down to a because there were no holes in G and hence $\gamma_0 \simeq \gamma_1$. But if a point inside γ were missing from G, then as γ shrinks it would get caught on the hole and could not be continuously deformed to the constant path.*

Definition 10.4. *(Linear Homotopy): Let γ_0 and γ_1 be any two paths from $[0,1]$ into $\mathbb{C}$. It is easy to see that $\gamma_0 \simeq \gamma_1$. The map $\Gamma(s,t) = (1-t)\gamma_0(s)+t\gamma_1(s)$ is a homotopy between them. It is called the straight line homotopy or linear homotopy because it moves the point $\gamma_0(s)$ to the point $\gamma_1(s)$ along the straight line segment joining them.*

Proposition 10.1. *Let G be an open set which is star-shaped at $a \in G$. If γ_0 is the path which is constantly equal to a then every closed rectifiable path in G is homotopic to γ_0.*

Proof. Let γ_1 be a closed rectifiable curve in G and put $\Gamma(s,t) = t\gamma_1(s) + (1-t)a$. Because the set G is star-shaped at $a \in G$, hence $\Gamma(s,t) \in G$ for $0 \le s,t \le 1$. It is easy to see that Γ is a homotopy between γ_0 and γ_1. □

Definition 10.5. *If γ is a closed rectifiable path in G then γ is homotopic to zero ($\gamma \simeq 0$) if γ is homotopic to a constant path.*

Theorem 10.2. *If γ_0 and γ_1 are two closed rectifiable paths in a region G such that $\gamma_0 \simeq \gamma_1$ and if f is holomorphic on G then $\int_{\gamma_0} f = \int_{\gamma_1} f$.*

Proof. First we consider the case in which γ_0 and γ_1 are linearly homotopic in G. Let Γ be a linear homotopy between γ_0 and γ_1. That is,

$$\Gamma(s,t) = (1-t)\gamma_0(s) + t\gamma_1(s).$$

Hence Γ has continuous second partial derivatives and

$$\frac{\partial^2\Gamma}{\partial s\partial t} = \frac{\partial^2\Gamma}{\partial t\partial s}$$

for all $(s,t) \in [0,1] \times [0,1]$.

Define

$$g(t) = \int_0^1 f(\Gamma(s,t))\frac{\partial\Gamma}{\partial s}(s,t)ds;$$

then $g(0) = \int_{\gamma_0} f$ and $g(1) = \int_{\gamma_1} f$. By Leibniz's rule g has a continuous derivative,

$$g'(t) = \int_0^1 \left[f'(\Gamma(s,t))\frac{\partial\Gamma}{\partial s}\frac{\partial\Gamma}{\partial t} + f(\Gamma(s,t))\frac{\partial^2\Gamma}{\partial t\partial s}\right] ds.$$

But

$$\frac{\partial}{\partial s}\left[(f\circ\Gamma)\frac{\partial\Gamma}{\partial t}\right] = (f'\circ\Gamma)\frac{\partial\Gamma}{\partial s}\frac{\partial\Gamma}{\partial t} + (f\circ\Gamma)\frac{\partial^2\Gamma}{\partial s\partial t}.$$

Hence

$$\begin{aligned} g'(t) &= \int_0^1 \frac{\partial}{\partial s}\left[(f\circ\Gamma)\frac{\partial\Gamma}{\partial t}\right] ds \\ &= f(\Gamma(1,t))\frac{\partial\Gamma}{\partial t}(1,t) - f(\Gamma(0,t))\frac{\partial\Gamma}{\partial t}(0,t). \end{aligned}$$

Since $\Gamma(1,t) = \Gamma(0,t)$ for all t we get $g'(t) = 0$ for all t. So g is a constant; in particular $g(0) = \int_{\gamma_0} f = \int_{\gamma_1} f = g(1)$. So the theorem follows in the special case when Γ has continuous second order partial derivatives. We shall now prove the theorem for the general case. Let $I = [0,1]$ and $\Gamma : I \times I \to G$ be a homotopy between γ_0 and γ_1. Since Γ is continuous and $[0,1]\times[0,1]$ is compact, Γ is uniformly continuous and $\Gamma([0,1]\times[0,1])$ is a compact subset of G. Let $r = d(\Gamma(I^2), \mathbb{C} - G)$ where $I^2 = [0,1]\times[0,1]$. Then $r > 0$ by Theorem 2.57. There is also an integer N such that if $(s-s')^2 + (t-t')^2 < \frac{4}{N^2}$ then $|\Gamma(s,t) - \Gamma(s',t')| < r$.

Let $Z_{jk} = \Gamma(\frac{j}{n}, \frac{k}{n}), 0 \le j,k \le n$ and put $J_{jk} = \left[\frac{j}{n}, \frac{j+1}{n}\right] \times \left[\frac{k}{n}, \frac{k+1}{n}\right]$ for $0 \le j,k \le n-1$. Since the diameter of the square (length of the diagonal) J_{jk} is $\frac{\sqrt{2}}{n}$, it follows that $\Gamma(J_{jk}) \subset B(Z_{jk}, r)$. So if we let P_{jk} be the closed polygon $[Z_{jk}, Z_{j+1,k}, Z_{j+1,k+1}, Z_{j,k+1}, Z_{jk}]$; then, because disks are convex, $P_{jk} \subset B(Z_{jk}, r)$. But from Corollary 7.1 it is known that

$$\int_{P_{jk}} f = 0 \tag{10.5}$$

for any function f holomorphic in G. It can now be shown that $\int_{\gamma_0} f = \int_{\gamma_1} f$ by going up the ladder we have constructed, one rung at a time. That is, let Θ_k be the closed polygon $[Z_{0,k}, Z_{1,k}, \ldots, Z_{n,k}]$. We will show that $\int_{\gamma_0} f = \int_{\Theta_0} f = \int_{\Theta_1} f = \cdots = \int_{\Theta_n} f = \int_{\gamma_1} f$. To see that $\int_{\gamma_0} f = \int_{\Theta_0} f$, observe that if

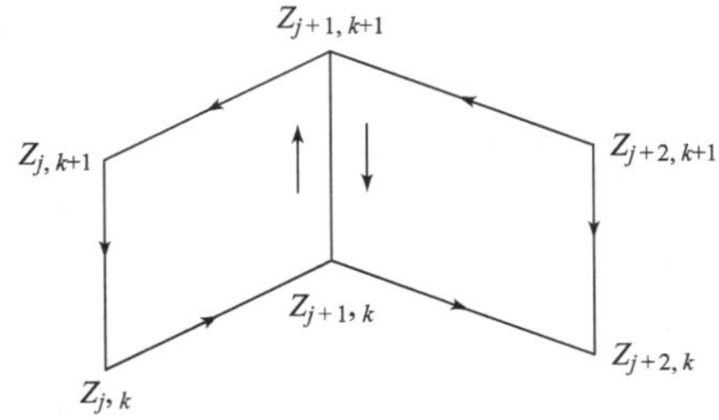

Fig. 10.2: The closed polygons P_{jk} and $P_{j+1,k}$

$\sigma_j(t) = \gamma_0(t)$ for $\frac{j}{n} \le t \le \frac{j+1}{n}$ then $\sigma_j + [Z_{j+1,0}, Z_{j,0}]$ (the plus sign '+' indicates that σ_j is to be followed by the polygon) is a closed rectifiable curve in the disk $B(Z_{j,0}, r) \subset G$. Hence $\int_{\sigma_j} f = -\int_{[Z_{j+1,0},Z_{j,0}]} f = \int_{[Z_{j,0},Z_{j+1,0}]} f$. Adding both sides of this equation for $0 \le j \le n$ yields $\int_{\gamma_0} f = \int_{\Theta_0} f$. Similarly $\int_{\gamma_1} f = \int_{\Theta_1} f$. To see that $\int_{\Theta_k} f = \int_{\Theta_{k+1}} f$ use equation (10.5); this gives

$$\sum_{j=0}^{n-1} \int_{P_{jk}} f = 0. \tag{10.6}$$

However, notice that the integral $\int_{P_{jk}} f$ includes the integral over $[Z_{j+1,k}, Z_{j+1,k+1}]$, which is negative of the integral over $[Z_{j+1,k+1}, Z_{j+1,k}]$, which is part of the integral $\int_{P_{j+1,k}} f$. Also,

$$Z_{0,k} = \Gamma\left(0, \frac{k}{n}\right) = \Gamma\left(1, \frac{k}{n}\right) = Z_{1,k}$$

so that $[Z_{0,k+1}, Z_{0,k}] = -[Z_{1,k}, Z_{1,k+1}]$. Hence, taking these cancellations into consideration, equation (10.6) becomes $0 = \int_{\Theta_k} f - \int_{\Theta_{k+1}} f$. This completes the proof. $\square$

Definition 10.6. *Two paths $\gamma_0, \gamma_1 : [0,1] \to G$ are said to be path homotopic (we write $\gamma_0 \simeq_p \gamma_1$) if they have the same initial point x_0 (that is, $\gamma_0(0) = \gamma_1(0) = x_0$) and the same final point x_1 (that is $\gamma_0(1) = \gamma_1(1) = x_1$) and if there is a continuous map $\Gamma : I \times I \to G$ such that $\Gamma(s,0) = \gamma_0(s)$ and $\Gamma(s,1) = \gamma_1(s)$, $\Gamma(0,t) = x_0$ and $\Gamma(1,t) = x_1$, for each $s \in I$ and each $t \in I$. (Here $I = [0,1]$.) We call Γ a path homotopy between γ_0 and γ_1.*

The first condition says that Γ is a homotopy between γ and γ', that is, Γ represents a continuous way of deforming the path γ_0 to the path γ_1. The second condition says that the end points of the path remain fixed during the deformation. That is, for each t, the path $s \mapsto \Gamma(s,t)$ is a path from x_0 to x_1. We can show similarly as in Lemma 10.4 that path homotopy is an equivalence relation between paths having fixed end points. The equivalence classes are called path homotopy classes.

Corollary 10.2. *If γ is a closed rectifiable path in a region G such that $\gamma \simeq_p 0$ then $n(\gamma, w) = 0$ for all w in $\mathbb{C} - G$.*

Proof. The proof follows from Lemma 9.6. □

Note that there exists closed path in a region G such that $n(\gamma, w) = 0$ for all w in $\mathbb{C} - G$ but γ is not path homotopic to a constant curve.

Example 10.2. *Let $G = \mathbb{C} - \{a, b\}, a \neq b$, and let γ be the curve in the figure 9.1. Convince yourself that γ is not homotopic to zero. Notice that this example shows that it is possible to have a closed curve γ in a region such that $n(\gamma, z) = 0$ for all z not in G without γ being homotopic to zero. That is the converse to Corollary 10.2 is false.*

If G is an open set and γ_0 and γ_1 are two closed rectifiable paths in G such that $\gamma_0 \simeq_p \gamma_1$ then it follows from Lemma 9.6 that $n(\gamma_0, w) = n(\gamma_1, w)$ for each $w \in \mathbb{C} - G$. Let $\gamma_0(t) = e^{2\pi i t}$ and $\gamma_1(t) = e^{-2\pi i t}, 0 \leq t \leq 1$, then $n(\gamma_0, 0) = 1$ and $n(\gamma_1, 0) = -1$. Hence the two paths γ_0 and γ_1 are not homotopic in $\mathbb{C} - \{0\}$.

Thus what we have shown above is that between any two homotopic paths, we can interpolate a finite number of polygonal paths, each of which is linearly homotopic to its successor.

An immediate consequence of the above result is the homotopic version of Cauchy's theorem.

Theorem 10.3. *Let G be a region in $\mathbb{C}$ and γ be a closed path in G such that $\gamma \simeq 0$ in G. Suppose $f : G \to \mathbb{C}$ is holomorphic. Then $\int_\gamma f = 0$.*

Proof. Note that if $\gamma \simeq 0$ then $n(\gamma, w) = \frac{1}{2\pi i} \int_\gamma \frac{dz}{z-w} = 0$ for all $w \in \mathbb{C} - G$ by Lemma 9.6. The result follows from Theorem 10.1. □

If γ_0 and γ_1 are two rectifiable paths from x_0 to x_1 then $\gamma_0 - \gamma_1$ is a closed rectifiable path. If Γ is a path homotopy between γ_0 and γ_1, then define $\gamma : [0, 1] \to G$ by

$$\gamma(s) = \begin{cases} \gamma_0(3s) & \text{if } 0 \leq s \leq \frac{1}{3}; \\ x_1 & \text{if } \frac{1}{3} \leq s \leq \frac{2}{3}; \\ \gamma_1(3 - 3s) & \text{if } \frac{2}{3} \leq s \leq 1. \end{cases}$$

Now define $\Lambda : I^2 \to G$ by

$$\Lambda(s, t) = \begin{cases} \Gamma(3s(1-t), t) & \text{if } 0 \leq s \leq \frac{1}{3}; \\ \Gamma(1 - t, 3s - 1 + 2t - 3st) & \text{if } \frac{1}{3} \leq s \leq \frac{2}{3}; \\ \gamma_1((3 - 3s)(1 - t)) & \text{if } \frac{2}{3} \leq s \leq 1. \end{cases}$$

It is not so difficult to show that Λ is a path homotopy between γ and the constant path γ_2 given by $\gamma_2(s) = x_0$ for $0 \leq s \leq 1$. Thus $\gamma \simeq 0$. Hence if f is holomorphic on G, we have $0 = \int_\gamma f = \int_{\gamma_0} f - \int_{\gamma_1} f$.

Theorem 10.4. *If γ_0 and γ_1 are two rectifiable paths in G from x_0 to x_1 and if γ_0 is path homotopic to γ_1 and f is holomorphic on G then $\int_{\gamma_0} f = \int_{\gamma_1} f$.*

Definition 10.7. *An open set $G \subseteq \mathbb{C}$ is simply connected if G is connected and every closed path in G is homotopic to a constant path in G.*

Example 10.3. *(i) Every star shaped region is simply connected and convex sets are simply connected.*

(ii) The set $\Omega = \mathbb{C} - \{\theta e^{i\theta} : 0 \leq \theta < \infty\}$ is simply connected.

There are many equivalent definitions for simple connectivity in the complex plane $\mathbb{C}$.

(1) A region $G \subseteq \mathbb{C}$ is simply connected if its complement with respect to the extended complex plane $\mathbb{C}_\infty = \mathbb{C} \cup \{\infty\}$ is connected.

(2) A region $G \subseteq \mathbb{C}$ is simply connected if and only if $n(\gamma, a) = 0$ for all cycles (sum of closed curves) γ in G and all points a which do not belong to G.

Indeed, a region without holes is obviously one whose complement consists of a single piece and hence the region is simply connected. We can also say a region G is simply connected if any closed curve in G can be contracted to a point. Equivalently, a closed curve in a simply connected region cannot wind around any point which does not belong to the region. It seems quite evident that this condition is not fulfilled in the case of a region with a hole. Notice that a disk, a half plane, and a parallel strip are simply connected.

Corollary 10.3. *If G is simply connected and γ is a closed path in G, then $\int_\gamma f = 0$ for all holomorphic functions f on G.*

We now list some important consequences of this general form of Cauchy's theorem.

Proposition 10.2. *If G is simply connected and $f : G \to \mathbb{C}$ is holomorphic in G then f has a primitive in G.*

Proof. Fix a point a in G and let γ_1, γ_2 be any two rectifiable paths in G from a to a point z in G. Since G is open and connected there is always a path from a to any other point of G. Hence by Corollary 10.3, $\int_{\gamma_1} f - \int_{\gamma_2} f = \int_{\gamma_1 - \gamma_2} f = 0$. Here $\gamma_1 - \gamma_2$ is the closed path from a to z along γ_1 and then back to a along $-\gamma_2$. Hence we get a well defined function $F : G \to \mathbb{C}$ by setting $F(z) = \int_\gamma f$ where γ is any rectifiable path from a to z. We claim that F is a primitive of f. If $z_0 \in G$ and $r > 0$ is such that $B(z_0, r) \subset G$, then let γ be a path from a to z_0. For z in $B(z_0, r)$ let $\gamma_z = \gamma + [z_0, z]$; that is, γ_z is the path γ followed by the straight line segment from z_0 to z. Hence

$$\begin{aligned}\frac{F(z)-F(z_0)}{z-z_0} &= \frac{1}{z-z_0}\left[\int_{[a,z]} f(z)dz - \int_{[a,z_0]} f(z)dz\right]\\ &= \frac{1}{z-z_0}\int_{[z_0,z]} f(z)dz.\end{aligned}$$

□

Now proceed as in the proof of Morera's theorem (Theorem 7.14) to show that $F'(z_0) = f(z_0)$.

One of the direct consequence of the Proposition 10.2 is the fact that a branch of $\log f(z)$ where f is analytic and never vanishes, can be defined on a simply connected region.

Corollary 10.4. *Let G be a simply connected region and let $f : G \to \mathbb{C}$ be a holomorphic function such that $f(z) \neq 0$ for any z in G. Then there exists a holomorphic logarithm of f in G. That is, there exists a holomorphic function $g : G \to \mathbb{C}$ such that $f(z) = e^{g(z)}$ on G. If $z_0 \in G$ and $e^{w_0} = f(z_0)$, we may choose g such that $g(z_0) = w_0$. Any two such holomorphic logarithms differ by $2\pi in$ for some $n \in \mathbb{Z}$.*

Proof. Since f never vanishes, $\frac{f'}{f}$ is holomorphic on G; so by Proposition 10.2 it must have a primitive ϕ in G. If $h(z) = e^{\phi(z)}$ then h is holomorphic on G and never vanishes. So, $\frac{f}{h} = fe^{-\phi}$ is holomorphic and its derivative is $\frac{h(z)f'(z)-h'(z)f(z)}{(h(z))^2} = \frac{h(z)f'(z)-\phi'(z)h(z)f(z)}{(h(z))^2} = \frac{h(z)f'(z)-f'(z)h(z)}{(h(z))^2} = 0.$

Hence $\frac{f}{h} = fe^{-\phi}$ is a constant C for all z in G. Note that $C \neq 0$(otherwise $f = 0$). Thus we can write $C = e^{\alpha}$ for some $\alpha \in \mathbb{C}$. We then have $f = Ce^{\phi} = e^{\phi+\alpha}$. By letting $g(z) = \phi(z)+\alpha+2\pi ik$ for an appropriate k, such that $g(z_0) = w_0$, the result is proved. Further, if $e^{g_1} = f = e^{g_2}$ with $g_i \in H(G), i = 1, 2$, we let $g = g_1 - g_2$. Hence $e^g = 1$ so that $g'e^g = 0$. Thus g is a constant with $e^g = 1$. But then $g = 2\pi in$ for some $n \in \mathbb{Z}$. □

Corollary 10.5. *Let the hypotheses on G and f be as in the last proposition. Then there exists $h \in H(G)$ such that $h^n = f$ for any $n \neq 0$.*

In fact, Riemann mapping theorem for connected open sets in the plane says that all the above three statements are equivalent to the topological condition of simply connectedness of the domain.

Theorem 10.5. *(Riemann mapping theorem): Given any simply connected region Ω which is not the whole plane, and a point $z_0 \in \Omega$, there exists a unique analytic function $f(z)$ in Ω, normalized by the conditions $f(z_0) = 0, f'(z_0) > 0$, such that f defines one-to-one mapping of Ω onto the disk $|w| < 1$.*

For proof of Theorem 10.5, see Chapter 15.

From the above results it follows that there is a branch of $\log(z-a)$ in any simply connected domain G, not containing $z = a$; take $f(z) = z - a$ in

Corollary 10.4. Further one can also show that there is a branch of $(z-a)^{\frac{1}{2}}$ in any simply connected domain G, not containing $z = a$. This last statement can be verified as follows: Let g be a branch in G of $\log(z-a)$. Then g and $\frac{1}{2}g$ are analytic in G. Hence $h(z) = e^{\frac{g}{2}}$ is analytic in G since e^z is an entire function. But $[h(z)]^2 = [e^{\frac{g}{2}}]^2 = e^g = z - a$ on G since $g(z) = \log(z-a)$ on G. Thus $h(z)$ is one of the values of $(z-a)^{\frac{1}{2}}$ for each z in G. Hence h is a branch in G of $(z-a)^{\frac{1}{2}}$.

Lemma 10.5. *Let U be $\mathbb{C}$ minus any ray starting from the origin. Then U is simply connected.*

Proof. For simplicity, let us assume that the ray is the negative real axis including the origin. Let γ be any closed path in G. Let us write $\gamma(t) = r(t)e^{i\theta(t)}$, for $a \le t \le b$ and $0 < \theta < \pi$. We define the homotopy by $H(s,t) = r(sa + (1-s)t)e^{i(1-s)\theta(t)}$. Geometrically, we are folding back the angle towards zero and we are contracting the distance from $r(t)$ to $r(a)$. □

Definition 10.8. *We say f is a holomorphic branch of logz in a domain Ω if*

(1) f is holomorphic in Ω.

(2) f is an inverse of the exponential function, that is, $e^{f(z)} = z$ for all $z \in \Omega$.

Note if $e^{f(z)} = z$ for all $z \in \Omega$ then $e^{f(z)+2\pi ki} = z$ for all $z \in \Omega$. Let $z = Re^{i\theta}, R \neq 0$. If we set $f(z) = \log z = u(z) + iv(z)$, then $e^{f(z)} = e^{u(z)}e^{iv(z)} = Re^{i\theta}$, which is possible if and only if $e^{u(z)} = |z| = R$ and $v(z) = \text{Arg}z = \theta + 2k\pi$. Hence a function satisfying (2) can always be found by setting $f(z) = u(z) + iv(z) = \log|z| + i\text{Arg}z$. Since, Arg is not even continuous and hence f is not holomorphic. However, if Ω is simply connected not containing zero, we can always define a holomorphic branch of log in Ω.

Theorem 10.6. *Let Ω be simply connected, $0 \notin \Omega$. Choose $z_0 \in \Omega$, and fix a value of $\log z_0$. Set*

$$f(z) = \int_{z_0}^{z} \frac{dw}{w} + \log z_0.$$

Then f is a holomorphic logarithm.

Proof. The map f is well defined since $\frac{1}{w}$ is holomorphic in Ω and the integral along any two paths from z_0 to z yield the same value by homotopy version of Cauchy's theorem (Theorem 10.3). Consider $g(z) = ze^{-f(z)}$. Then $f(z) = \log \frac{z}{g(z)}$ is a holomorphic logarithm. □

10.3 Exercises

1. Let f and g be any two continuous maps from a metric space X into $\mathbb{R}^2$. Show that f and g are homotopic.

2. Let X denote the punctured plane $\mathbb{R}^2 - \{0\}$. Define the paths f, g, h from $[0, 1]$ into X as follows:

$$
\begin{aligned}
f(s) &= (\cos \pi s, \sin \pi s) \\
g(s) &= (\cos \pi s, 2 \sin \pi s) \\
h(s) &= (\cos \pi s, - \sin \pi s).
\end{aligned}
$$

   Prove that f and g are path homotopic but f and h are not path homotopic.

3. Let X, Y, Z be three metric spaces. Show that if $h, h' : X \to Y$ are homotopic and $k, k' : Y \to Z$ are homotopic, then $k \circ h$ and $k' \circ h'$ are homotopic.

4. Show that $\mathbb{R}^n$ and $\mathbb{R}^2$ are not homeomorphic for $n > 2$.

5. Let G be any simply connected domain in $\mathbb{C}$. Let $a, b \in \mathbb{C} \setminus G$. Prove that there is a branch of $[(z - a)(z - b)]^{\frac{1}{2}}$.

Chapter 11

The Residue Theorem and its Applications

11.1 Residues

In a previous chapter, we have already seen that if G is an open, star-shaped subset of $\mathbb{C}$ and $f \in C(G)$ and $f \in H(G - \{p\})$ then for every closed path γ in G, $\int_\gamma f(z)dz = 0$. We shall in this chapter answer the following question: If f has an isolated singularity at $z = z_0$ what are the possible values for $\int_\gamma f(z)dz$ when $n(\gamma, w) = 0$ for all $w \in \mathbb{C} \setminus G$ and for $z_0 \notin [\gamma]$?

Definition 11.1. *(Residues): Let $U \subset \mathbb{C}$ be open and $a \in U$. Let f be holomorphic on $U - \{a\}$. Let $B(a, R) \subset U$ and $0 < r < R$. The value of the integral $\frac{1}{2\pi i}\int_\gamma f(z)dz$, where $\gamma(t) = a + re^{it}, 0 \le t \le 2\pi$ is independent of r and is called the residue of f at a. It is denoted by Res(f, a). Note that Res(f, a) is the coefficient of $\frac{1}{z-a}$ in the Laurent expansion of f at a.*

Example 11.1. *Consider the function $f(z) = \frac{\sin z}{z}$ on $\mathbb{C} \setminus \{0\}$. The function has an isolated singularity at 0 and $\frac{\sin z}{z} = \frac{1}{z}(z - \frac{z^3}{3!} + \frac{z^5}{5!} - \cdots) = 1 - \frac{z^2}{3!} + \frac{z^4}{5!} - \cdots$. So Res$(f, 0) = 0$.*

Example 11.2. *If f is holomorphic at a, then Res$(f, a) = 0$. The converse is not true.*

The following observations can be made:

(i) Let $U \subset \mathbb{C}$ be open and $a \in U$. Suppose f is holomorphic on $U - \{a\}$. If the point $z = a$ is a removable singularity of f then by Theorem 8.10, $\text{Res}(f, a) = 0$. See Example 11.1. Further suppose $\gamma(t) = a + re^{it}, 0 \le t \le 2\pi, 0 < r < R$ and $B(a, R) \subset U$. Then by Theorem 8.9, $\int_\gamma f(z)dz = 0$.

(ii) If f has an isolated singularity at $z = a$ and if $f(z - a) = f(-(z - a))$ then $\text{Res}(f, a) = 0$. Because in this case the Laurent series expansion around a cannot have odd powers of $z - a$. For examples, verify that (a) $\text{Res}\left(\frac{1}{(\sin z)^2}, k\pi\right) = 0, k \in \mathbb{Z}$, (b) $\text{Res}\left(e^{\frac{1}{z^2}}, 0\right) = 0$, (c) $\text{Res}\left(\frac{1}{z^2(z^2-4)}, 0\right) = 0$.

(iii) Let f have a simple pole at a. Then $f(z) = \frac{c_{-1}}{z-a} + g(z)$ where g is the holomorphic part of the Laurent expansion of f at a. Then we have

$$\text{Res}(f, a) = \lim_{z \to a}(z-a)f(z) = c_{-1}$$

if a is a simple pole of f.

Example 11.3. *We shall compute the residue of* $f(z) = \frac{e^{-iz}}{z^2+1}$ *at* $z = i$. *By the formula above,*

$$Res(f, i) = \lim_{z \to i}(z-i)\frac{e^{-iz}}{(z+i)(z-i)} = \frac{e}{2i}.$$

(iv) Let $f(z) = \frac{g(z)}{h(z)}$, g, h holomorphic around a. Assume that h has a simple zero at a and $g(a) \neq 0$. Then $\text{Res}(f, a) = \frac{g(a)}{h'(a)}$, a is a simple zero of h and $g(a) \neq 0$. This is because

$$\begin{aligned}\text{Res}(f, a) &= \lim_{z \to a}(z-a)\frac{g(z)}{h(z)} \\ &= \lim_{z \to a}\frac{g(z)}{\frac{h(z)-h(a)}{z-a}} \\ &= \frac{g(a)}{h'(a)}.\end{aligned}$$

Example 11.4. *Consider* $f(z) = \frac{1}{z^6+1} = \frac{g(z)}{h(z)}$. *Then* f *has six poles. They are the sixth roots of* -1 : $\zeta_1 = e^{\frac{i\pi}{6}}, \zeta_2 = e^{\frac{i\pi}{2}}, \zeta_3 = e^{\frac{5i\pi}{6}}, \zeta_4 = \overline{\zeta_1}, \zeta_5 = \overline{\zeta_2}$ *and* $\zeta_6 = \overline{\zeta_3}$. *Hence* $Res(f, \zeta_1) = \frac{g(\zeta_1)}{h'(\zeta_1)} = \frac{1}{6z^5}|_{z=\zeta_1} = \frac{1}{6}e^{-\frac{5i\pi}{6}}$.

One can also compute the residue by observing that $z^6 + 1 = \prod_{k=1}^{6}(z - \zeta_k)$. *Hence*

$$Res(f, \zeta_1) = \lim_{z \to \zeta_1}(z - \zeta_1)\frac{1}{z^6+1} = \prod_{k=2}^{6}(\zeta_1 - \zeta_k)^{-1}.$$

Remark 11.1. Let f have a pole of order m at $z = a$ and its Laurent expansion be

$$f(z) = \frac{c_{-m}}{(z-a)^m} + \cdots + \frac{c_{-1}}{z-a} + g(z),$$

g being the holomorphic part. Then

$$(z-a)^m f(z) = c_{-m} + \cdots + c_{-1}(z-a)^{m-1} + (z-a)^m g(z).$$

It follows that

$$\text{Res}(f, a) = c_{-1} = \frac{1}{(m-1)!}\frac{d^{m-1}}{dz^{m-1}}\left[(z-a)^m f(z)\right]_{z=a}.$$

Example 11.5. *We use the above formula to find the residue of* $f(z) = \frac{e^{3z}}{(z-2)^3}$ *at* $a = 2$*. In this case*

$$Res(f,a) = \frac{1}{2!}\frac{d^2}{dz^2}\left[(z-2)^3\frac{e^{3z}}{(z-2)^3}\right]_{z=2}$$

$$= \frac{1}{2!}\frac{d^2}{dz^2}(e^{3z})\bigg|_{z=2} = \frac{9}{2}e^6.$$

Example 11.6. *Let* $f(z) = e^{\frac{1}{z}}$ *and* $a = 0$*. The Laurent series expansion of* $e^{\frac{1}{z}}$ *about* $a = 0$ *is* $\sum_{n=0}^{\infty}\frac{z^{-n}}{n!}$*, giving* $Res(f,0) = 1$*.*

Example 11.7. *Let* $f(z) = \frac{z^3}{z-1}$ *and* $a = 1$*. The Taylor expansion of* z^3 *about the point 1 is* $z^3 = 1 + 3(z-1) + 3(z-1)^2 + (z-1)^3$*. Hence, the Laurent expansion of* $\frac{z^3}{z-1}$ *about the point 1 is* $\frac{z^3}{z-1} = \frac{1}{z-1} + 3 + 3(z-1) + (z-1)^2$*, giving* $Res(f,1) = 1$*.*

Example 11.8. *Let* $f(z) = \frac{e^z}{z^2}, z \neq 0$*. Then* $f(h) = \frac{1}{h^2}(1 + h + \frac{h^2}{2!} + \cdots) = \frac{1}{h^2} + \frac{1}{h} + \frac{1}{2!} + \cdots$*. So* $Res(f,0) = 1 = \frac{1}{1!}\frac{d}{dz}\left[z^2\frac{e^z}{z^2}\right]_{z=0}$*.*

Example 11.9. *Let* $\cot\pi z = \sum_{n=-\infty}^{\infty} a_n z^n$ *be the Laurent series expansion for* $\cot\pi z$ *on the annulus* $1 < |z| < 2$*. We shall compute the* a_n *for* $n < 0$*. Recall that* $\cot\pi z$ *has simple poles at all integers* z*, with residues* $\frac{1}{\pi}$*, and no other singularities. We have* $a_{-1} = Res(\cot\pi z, -1) + Res(\cot\pi z, 0) + Res(\cot\pi z, 1) = \frac{3}{\pi}$*. For* $n < -1$*, the coefficients are given by*

$$\begin{aligned} a_n &= \frac{1}{2\pi}\int_{|z|=\frac{3}{2}}\frac{\cot\pi z}{z^{n+1}}dz \\ &= Res\left(\frac{\cot\pi z}{z^{n+1}}, -1\right) + Res\left(\frac{\cot\pi z}{z^{n+1}}, 1\right) \\ &= \lim_{z\to -1}(z+1)\frac{\cot\pi z}{z^{n+1}} + \lim_{z\to 1}(z-1)\frac{\cot\pi z}{z^{n+1}} \\ &= ((-1)^{-n-1} + 1)\frac{1}{\pi}. \end{aligned}$$

Proposition 11.1. *Suppose that* f, g *are analytic on an open set* U *containing* p*, and that* $g(p) = 0, g'(p) \neq 0$*. Then if* $f(p) = 0$*, the function* $\frac{f}{g}$ *has a removable singularity at* p*, and if* $f(p) \neq 0$*, the function* $\frac{f}{g}$ *has a simple pole at* p *with* $Res(\frac{f}{g}, p) = \frac{f(p)}{g'(p)}$*.*

Proof. By (Taylor's expansion),

$$\begin{aligned}\frac{f(p+h)}{g(p+h)} &= \frac{f(p)+f'(p)h+\cdots}{g(p)+g'(p)h+\cdots}\\ &= \frac{f'(p)+\frac{h}{2}f''(p)+\cdots}{g'(p)+\dfrac{h}{2}g''(p)+\cdots}\end{aligned}$$

if $f(p) = 0, g(p) = 0$ and $g'(p) \neq 0$; this tends to $\frac{f'(p)}{g'(p)}$ as $h \to 0$, and the singularity is removable, since the denominator is nonzero at $h = 0$ and hence by continuity remains nonzero for $|h| < \delta$, for some $\delta > 0$. Thus the quotient $\frac{f(p+h)}{g(p+h)}$ is a differentiable function in $|h| < \delta$ and hence the singularity at p is removable. If $f(p) \neq 0$; then

$$\frac{f(p+h)}{g(p+h)} = \frac{1}{h}\left[\frac{f(p)+f'(p)h+\cdots}{g'(p)+\frac{h}{2}g''(p)+\cdots}\right]$$

and again the quotient inside the bracket is analytic on some $|h| < \delta$, equals $\sum_{n=0}^{\infty} b_n h^n$ say, with $b_0 = \frac{f(p)}{g'(p)}$. So $\frac{f(p+h)}{g(p+h)} = \frac{1}{h}\sum_{n=0}^{\infty} b_n h^n = \frac{f(p)}{g'(p)}\frac{1}{h} + b_1 + b_2 h + \cdots$ for $0 < |h| < \delta$. Hence $\text{Res}(\frac{f}{g}, p) = \frac{f(p)}{g'(p)}$. □

Example 11.10. *Find Res*$(\frac{\pi\cot\pi z}{z^2}, 0)$.

Solution: For $0 < |h| < 1$,

$$\begin{aligned}\frac{\pi\cot\pi h}{h^2} &= \frac{\pi\cos\pi h}{h^2\sin\pi h}\\ &= \frac{\pi(1-\frac{(\pi h)^2}{2!}+O(h^4))}{h^2(\pi h-\frac{(\pi h)^3}{3!}+O(h^5))}\\ &= \frac{1}{h^3}(1-\frac{\pi^2h^2}{2!}+O(h^4))(1-\frac{\pi^2}{6}h^2+O(h^4))^{-1}.\end{aligned}$$

Now write $(1-\frac{\pi^2}{6}h^2+O(h^4))^{-1} = b_0+b_1h+b_2h^2+\cdots$. We only need terms up to h^2, since we are looking for coefficient of $\frac{1}{h}$. Since $(1-\frac{\pi^2}{6}h^2+\cdots)(b_0+b_1h+\cdots) = 1$, equating coefficients we obtain $b_0 = 1, b_1 = 0, b_2-\frac{\pi^2}{6} = 0, b_3 = 0, \cdots$. Hence $\frac{\pi\cot\pi h}{h^2} = \frac{1}{h^3}(1-\frac{\pi^2h^2}{2}+O(h^4))(1+\frac{\pi^2}{6}h^2+O(h^4)) = \frac{1}{h^3}-\frac{\pi^2}{3h}+O(h)$. Thus $\text{Res}(\frac{\pi\cot\pi z}{z^2}, 0) = -\frac{\pi^2}{3}$.

The residues at isolated singularities $a_1, \cdots, a_n$ of a function f holomorphic in $\mathbb{C}\setminus\{a_1,\cdots,a_n\}$ can be used in the evaluation of integrals of the function. This is the content of the residue theorem.

Definition 11.2. *If G is an open set then we shall say a closed path γ in G is homologous to zero, in symbols $\gamma \approx 0$, if $n(\gamma, w) = 0$ for all w in $\mathbb{C}\setminus G$.*

By a contour we shall mean a formal sum $\gamma = \sum_{j=1}^{p} n_j\gamma_j$, where $\gamma_1, \cdots, \gamma_p$ are piecewise C^1 closed curves in $\mathbb{C}$ and $n_1, n_2, \cdots, n_p$ are integers. The contour $\gamma = \sum_{j=1}^{n} n_j\gamma_j$ is homologous to zero if and only if $\sum_{j=1}^{n} n_j n(\gamma_j, w) = 0$ for all $w \in \mathbb{C} \setminus G$. If f is a continuous complex valued function defined on each γ_j then we shall say f is defined on γ and define the complex integral of f over γ by

$$\int_\gamma f(z)dz = \sum_{j=1}^{p} n_j \int_{\gamma_j} f(z)dz.$$

Two piecewise C^1 closed curves will be identified if one of them is a reparametrization of the other. If $\gamma = \sum_{j=1}^{p} n_j\gamma_j$ and $\gamma' = \sum_{j=1}^{p} n'_j\gamma_j$ are two contours, we define their sum by

$$\gamma + \gamma' = \sum_{j=1}^{p}(n_j + n'_j)\gamma_j.$$

Then, if f is a continuous function defined on every γ_j, we have

$$\int_{\gamma+\gamma'} f(z)dz = \int_\gamma f(z)dz + \int_{\gamma'} f(z)dz.$$

Let $\gamma = \sum_{j=1}^{p} n_j\gamma_j$ be a contour, and assume that each n_j is nonzero. By the complement of γ, written $\mathbb{C} \setminus [\gamma]$, we shall mean the set of points of $\mathbb{C}$ that are in the complement of each $[\gamma_j]$. If z_0 is such a point, we define the winding number of γ about z_0, by

$$n(\gamma, z_0) = \sum_{j=1}^{p} n_j n(\gamma_j, z_0).$$

Recall that in the case of a single closed curve γ, we have the integral formula $n(\gamma, z_0) = \frac{1}{2\pi i}\int_\gamma \frac{1}{z-z_0}dz$. We have already shown that for a contour γ, $n(\gamma, z_0)$ is constant on each connected component of $\mathbb{C} \setminus [\gamma]$ and equals 0 on the unbounded component.

Let γ_1 be the circle $|z| = 1$ and γ_2 the circle $|z| = 2$, both oriented counter-clockwise. For the contour $\gamma = \gamma_2 - \gamma_1$ we have

$$n(\gamma, z_0) = \begin{cases} 1 & \text{for} \quad 1 < |z_0| < 2 \\ 0 & \text{for} \quad |z_0| < 1 \text{ and for } |z_0| > 2. \end{cases}$$

We define the interior of the contour γ, denoted intγ, to be the set of points z_0 in $\mathbb{C} \setminus [\gamma]$ such that $n(\gamma, z_0) \neq 0$ and the exterior, denoted extγ, to be the set of points z_0 in $\mathbb{C} \setminus [\gamma]$ such that $n(\gamma, z_0) = 0$. Each of these sets is a union of certain of the connected components of $\mathbb{C} \setminus [\gamma]$. As those components are open sets, the sets intγ and extγ are open. Because the unbounded component of $\mathbb{C} \setminus [\gamma]$ is contained in extγ, the set intγ is bounded. All boundary points of intγ and of extγ are contained in γ.

A contour Γ will be called simple if $n(\Gamma, z_0)$ is either 0 or 1 for each point z_0 in $\mathbb{C} \setminus [\Gamma]$.

Theorem 11.1. *(The residue theorem): Let G be a star-shaped, open set and γ a closed path in G such that $\gamma \approx 0$ in G. Let the function f be holomorphic in G except for isolated singularities at the points $a_1, \cdots, a_n$ (counted according to multiplicity); none of which lies on γ. Then*

$$\frac{1}{2\pi i}\int_\gamma f(z)dz = \sum_{k=1}^{n} n(\gamma, a_k)Res(f, a_k).$$

Proof. To prove this, we choose positive numbers $r_1, r_2, \cdots, r_n$ such that no two disks $\overline{B(a_k, r_k)}$ intersect, none of them intersects $[\gamma]$, and each disk is contained in G. This can be done for example, if we choose $\delta > 0$ such that $\delta < \min\{d(a_j, [\gamma]),\ d(a_j, a_k) : 1 \le j < k \le n\}$ and such that $\overline{B(a_j, \frac{\delta}{2})} \subset G$. Note that $B(a_j, \frac{\delta}{2}) \cap [\gamma] = \emptyset$ and $B(a_j, \frac{\delta}{2}) \cap B(a_k, \frac{\delta}{2}) = \emptyset$ for $k \ne j$. Let $m_k = n(\gamma, a_k)$ for $1 \le k \le m$. Let $\gamma_k(t) = a_k + r_k e^{-2\pi i m_k t}$ for $0 \le t \le 1$ and $1 \le k \le n$. Then for $1 \le j \le n$,

$$n(\gamma, a_j) + \sum_{k=1}^{n} n(\gamma_k, a_j) = 0 \tag{11.1}$$

because, $n(\gamma, a_j) = m_j$ and $n(\gamma_k, a_j) = 0$ if $k \ne j$ and $n(\gamma_k, a_j) = -m_j$ if $k = j$.

Now $\gamma \approx 0$ in G and $\overline{B(a_k, r_k)} \subset G$,

$$n(\gamma, a) + \sum_{k=1}^{n} n(\gamma_k, a) = 0 \tag{11.2}$$

for all $a \in \mathbb{C} \setminus G$. From (11.1) and (11.2), we have

$$n(\gamma, a) + \sum_{k=1}^{n} n(\gamma_k, a) = 0$$

for all $a \in \mathbb{C} \setminus G - \{a_1, a_2, \ldots, a_n\}$. Since f is holomorphic in $G - \{a_1, \cdots, a_n\}$ by Theorem 10.1, we have

$$\int_\gamma f(z)dz + \sum_{k=1}^{n} \int_{\gamma_k} f(z)dz = 0. \tag{11.3}$$

If the Laurent series expansion of $f(z)$ about $z = a_k$ is equal to $f(z) = \sum_{n=-\infty}^{\infty} c_n (z - a_k)^n$ then the Laurent series converges uniformly on $\overline{B(a_k, r_k)}$. Hence

$$\begin{aligned}
\frac{1}{2\pi i}\int_{\gamma_k} f(z)dz &= \sum_{n=-\infty}^{\infty} c_n \left[\frac{1}{2\pi i}\int_{\gamma_k} (z - a_k)^n dz\right] \\
&= c_{-1} n(\gamma_k, a_k) = c_{-1}(-m_k) \\
&= (-1)c_{-1} n(\gamma, a_k)
\end{aligned}$$

because $\int_{\gamma_k}(z-a_k)^n dz = 0$ if $n \neq -1$ (in this case the integrand has a primitive). Thus

$$\frac{1}{2\pi i}\int_{\gamma_k} f(z)dz = (-1)n(\gamma, a_k)\text{Res}(f, a_k).$$

Hence from (11.3),

$$\frac{1}{2\pi i}\int_{\gamma} f(z)dz = \sum_{k=1}^{n} n(\gamma, a_k)\text{Res}(f, a_k).$$

□

We shall illustrate below some of the basic methods for evaluating certain integrals using calculus of residue.

Example 11.11. *We shall evaluate* $\int_{|z|=1}\frac{e^z}{z^2}dz$ *using calculus of residues. By Residue theorem* $\int_{|z|=1}\frac{e^z}{z^2}dz = 2\pi i Res(\frac{e^z}{z^2}, 0) = 2\pi i$.

Remark 11.2. If we parametrize the circle as $\gamma(t) = e^{it}(0 \leq t \leq 2\pi)$ we obtain $\int_{|z|=1}\frac{e^z}{z^2}dz = \int_0^{2\pi}\frac{e^{e^{it}}}{e^{2\pi i t}}ie^{it}dt = i\int_0^{2\pi}e^{e^{it}-it}dt = i\int_0^{2\pi}e^{\cos t}e^{i(\sin t - t)}dt$. Taking real and imaginary parts we get

$$\int_0^{2\pi} e^{\cos t}\cos(\sin t - t)dt = 2\pi$$

and

$$\int_0^{2\pi} e^{\cos t}\sin(\sin t - t)dt = 0.$$

Example 11.12. *Find* $\int_{|z|=2}\frac{ze^z}{(z-1)^3}dz$.

Solution: The function $f(z) = \frac{ze^z}{(z-1)^3}$ has a pole of order 3 inside the circle at $z = 1$.

$$\begin{aligned} f(1+h) &= \frac{(1+h)e^{1+h}}{h^3} = \frac{1+h}{h^3}\cdot e \cdot e^h \\ &= \frac{e}{h^3}(1+h)\left(1 + h + \frac{h^2}{2!} + \frac{h^3}{3!} + \cdots\right) \\ &= \frac{e}{h^2} + \frac{2e}{h^2} + \frac{3e}{2h} + \frac{2e}{3} + \cdots. \end{aligned}$$

Hence $\text{Res}(f, 1) = \frac{3e}{2}$ and $\int_{|z|=2}\frac{ze^z}{(z-1)^3}dz = 2\pi i \cdot \frac{3e}{2} = 3\pi e i$.

Example 11.13. *Find* $\int_{\gamma}\frac{2z^2}{1+z^4}dz$ *where* γ *is the closed path consisting of the line segment* $[-R, R], R > 1$ *followed by* $\gamma_R(t) = Re^{it}, 0 \leq t \leq \pi$.

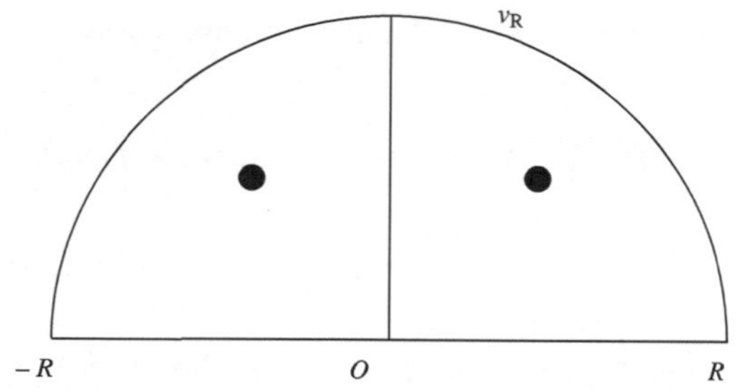

Fig. 11.1: The closed path $\gamma = [-\mathbf{R}, \mathbf{R}] + \gamma_{\mathbf{R}}$

Solution: The function $f(z) = \frac{2z^2}{1+z^4}$ has singularities at $e^{\frac{i\pi}{4}}, e^{\frac{3i\pi}{4}}, e^{\frac{-i\pi}{4}}$ and $e^{\frac{-3i\pi}{4}}$. The poles $e^{\frac{i\pi}{4}}$ and $e^{\frac{3i\pi}{4}}$ are inside γ and the poles $e^{\frac{-i\pi}{4}}$ and $e^{\frac{-3i\pi}{4}}$ are outside γ. Again

$$\text{Res}\left(\frac{2z^2}{1+z^4}, e^{\frac{i\pi}{4}}\right) = \left.\frac{2z^2}{4z^3}\right|_{z=e^{\frac{i\pi}{4}}} = \frac{e^{\frac{-i\pi}{4}}}{2}$$

and

$$\text{Res}\left(\frac{2z^2}{1+z^4}, e^{\frac{3i\pi}{4}}\right) = \left.\frac{2z^2}{4z^3}\right|_{z=e^{\frac{3i\pi}{4}}} = \frac{e^{\frac{-3i\pi}{4}}}{2}.$$

Hence

$$\begin{aligned}\int_\gamma \frac{2z^2}{1+z^4}dz &= 2\pi i\left(\frac{1}{2}e^{\frac{-i\pi}{4}} + \frac{1}{2}e^{\frac{-3i\pi}{4}}\right)\\ &= \pi i\left(\cos\frac{\pi}{4} - i\sin\frac{\pi}{4} + \cos\frac{3\pi}{4} - i\sin\frac{3\pi}{4}\right)\\ &= i\pi\left(\frac{-2i}{\sqrt{2}}\right) = \sqrt{2}\pi.\end{aligned}$$

Example 11.14. *Evaluate* $\int_{|z|=1} \frac{e^{e^z}}{z^2}dz$.

Solution: The function $f(z) = \frac{e^{e^z}}{z^2}$ has a pole of order 2 at $z = 0$ and by Remark 11.1,

$$\begin{aligned}\text{Res}\left(\frac{e^{e^z}}{z^2}, 0\right) &= \lim_{z\to 0}\frac{d}{dz}\left(z^2 \cdot \frac{e^{e^z}}{z^2}\right)\\ &= \lim_{z\to 0} e^{e^z} \cdot e^z = e.\end{aligned}$$

So $\int_{|z|=1} \frac{e^{e^z}}{z^2}dz = 2\pi i \cdot e$.

Remark 11.3. A path $\gamma : [a, b] \to \mathbb{C}$ is said to be simple positively oriented closed path if it is closed, γ is 1-1 on (a, b) and $n(\gamma, z)$ is either 0 or 1 for all $z \notin [\gamma]$. Any point z with $n(\gamma, z) = 1$ is said to be inside of γ. If $n(\gamma, z) = 0$ then z is said to be outside of γ. Examples of simple, positively oriented closed paths are $\gamma(t) = a + re^{it}, a \in \mathbb{C}, 0 \leq t \leq 2\pi$. More generally, if $\gamma(t) = r(t)e^{it}, 0 \leq t \leq$

2π and $r(0) = r(2\pi)$, then γ is simple, positively oriented and closed. In the residue theorem, if we assume that γ is simple, positively oriented closed path, then we get $\frac{1}{2\pi i}\int_\gamma f(z)dz =$ sum of the residues at isolated singularities inside γ of the function f.

Lemma 11.1. *Let f be differentiable on $\mathbb{C}$ except for a finite number of simple poles $a_k, 1 \le k \le n$ and that there exist constants M and R such that $|f(z)| \le M$ for $|z| \ge R$. Then for $z \ne a_k, 1 \le k \le n$, we have*

$$f(z) = c + \sum_{j=1}^{n} \frac{Res(f, a_j)}{z - a_j},$$

where $c = \lim_{z\to\infty} f(z)$.

Proof. Let $g(z) = \sum_{j=1}^{n} \frac{\text{Res}(f,a_j)}{z-a_j}$. Then $f - g$ is holomorphic on $\mathbb{C}$. Also $\frac{f(z)-g(z)}{z} \to 0$ as $z \to \infty$. Hence, $f - g$ is a constant. □

Remark 11.4. Suppose f is holomorphic and has a zero of order m at $z = a$. Then $f(z) = (z-a)^m g(z)$ where g is holomorphic and in $B'(a, r) = B(a, r)\setminus\{a\}$ for some $r > 0$, $g(a) \ne 0$. Hence $\frac{f'(z)}{f(z)} = \frac{m}{z-a} + \frac{g'(z)}{g(z)}$. Let γ be any circle contained in $B'(a, r)$ with center at a. Hence $\text{Res}(\frac{f'}{f}, a) = \frac{1}{2\pi i}\int_\gamma \frac{f'(z)}{f(z)}dz = m$. Note that the function $\frac{g'}{g}$ is holomorphic near $z = a$ since $g(a) \ne 0$.

Now suppose that f has a pole of order n at $z = p$; that is $f(z) = (z - p)^{-n} g(z)$ where g is holomorphic in $B'(p, r') = B(p, r') \setminus \{p\}$ for some $r' > 0$ and $g(p) \ne 0$. This gives

$$\frac{f'(z)}{f(z)} = \frac{-n}{z - p} + \frac{g'(z)}{g(z)}$$

and again $\frac{g'}{g}$ is holomorphic near $z = p$. Hence $\text{Res}(\frac{f'}{f}, p) = -n$. Thus we conclude that if f is a meromorphic function in G with a pole of order n at $z = p$ and a zero of order m at $z = a$ then $\frac{f'}{f}$ has simple poles at p and a and $\text{Res}(\frac{f'}{f}, p) = -n$ and $\text{Res}(\frac{f'}{f}, a) = m$.

11.2 Argument principle and Rouché's theorem

Theorem 11.2. *(Argument Principle): Let G be a star-shaped, open set in $\mathbb{C}$ and let γ be a closed path in G such that $\gamma \approx 0$ in G. Let f be meromorphic in G with poles $p_1, p_2, \cdots, p_m$ and zeros $a_1, a_2, \cdots, a_n$, counted according to multiplicity. Further assume that none of these poles and zeros of f lie on $[\gamma]$. Then*

$$\frac{1}{2\pi i}\int_\gamma \frac{f'(z)}{f(z)}dz = \sum_{j=1}^{n} n(\gamma, a_j) - \sum_{k=1}^{m} n(\gamma, p_k).$$

Proof. Note that $\frac{f'}{f}$ is holomorphic except at the zeros and poles of f. Hence by residue theorem,

$$\frac{1}{2\pi i}\int_\gamma \frac{f'(z)}{f(z)}dz = \sum_{j=1}^n n(\gamma, a_j)\text{Res}\left(\tfrac{f'}{f}, a_j\right) + \sum_{k=1}^m n(\gamma, p_k)\text{Res}\left(\tfrac{f'}{f}, p_k\right).$$

Since all the zeros $a_1, \cdots, a_n$ and poles $p_1, \cdots, p_m$ are counted according to their multiplicities hence we can treat each of them as simple zeros and poles of f. Hence $\text{Res}\left(\frac{f'}{f}, a_j\right) = 1, 1 \leq j \leq n$ and $\text{Res}\left(\frac{f'}{f}, p_k\right) = -1, 1 \leq k \leq m$. Thus

$$\frac{1}{2\pi i}\int_\gamma \frac{f'(z)}{f(z)}dz = \sum_{j=1}^{n} n(\gamma, a_j) - \sum_{k=1}^{m} n(\gamma, p_k).$$

□

We explain why this is called the argument principle. If γ is given by $\gamma(t), 0 \leq t \leq 1$, we choose a logarithm of f along γ (possible by Theorem 9.5). We compute:

$$\begin{aligned}\frac{1}{2\pi i}\int_\gamma \frac{f'(z)}{f(z)}dz &= \frac{1}{2\pi i}[\log f(\gamma(1)) - \log f(\gamma(0))] \\ &= \frac{1}{2\pi} \triangle \operatorname{Arg} f(z),\end{aligned}$$

$\triangle$Arg being the difference in the arguments of $f(\gamma(1))$ and $f(\gamma(0))$ as $\gamma(t)$ travels around γ starting from $\gamma(0)$ to $\gamma(1)$.

Definition 11.3. *A function f is said to be meromorphic in an open set $\Omega \subseteq \mathbb{C}$ if there is a set $A \subset \Omega$ such that (a) A has no limit point in Ω; (b) $f \in H(\Omega - A)$; (c) f has a pole at each point of A.*

Note that the possibility $A = \emptyset$ is not excluded. Thus every $f \in H(\Omega)$ is meromorphic in Ω. Note also that (a) implies that no compact subset of Ω contains infinitely many points of A, and that A is therefore at most countable.

Theorem 11.3. *Every function which is meromorphic on the whole plane is the quotient of two entire functions.*

Proof. If $F(z)$ is meromorphic in the whole plane, we can find an entire function $g(z)$ with the poles of $F(z)$ for zeros. The product $F(z)g(z)$ is then an entire function $f(z)$, and we obtain $F(z) = \frac{f(z)}{g(z)}$. □

The proposition that follows is the most important special case of argument principle.

Proposition 11.2. *Let f be meromorphic on the star-shaped open set U and let γ be simple positively oriented closed path in U. If neither of the zeros and poles of f lie on $[\gamma]$, then*

$$\frac{1}{2\pi i}\int_\gamma \frac{f'(z)}{f(z)}dz = N - P,$$

where N (respectively P) is the number of zeros (respectively poles) of f inside γ, the number being computed with multiplicity. That is, if a_j is a zero of order r_j for $1 \leq j \leq n$ of f inside γ and p_k is a pole of order s_k for $1 \leq k \leq m$ of f inside γ then $N = \sum_{j=1}^{n} r_j$ and $P = \sum_{k=1}^{m} s_k$.

In the following example we illustrate the use of the argument principle.

Example 11.15. *Let $f(z) = \frac{z}{z^2+1}$ for $z \neq \pm i$ and $\gamma(t) = 2e^{it}, 0 \leq t \leq 2\pi$. We can compute the integral $\int_\gamma f(z)dz$ in two ways: one by residue theorem and the other by argument principle.*

Solution: If we take $g(z) = z^2 + 1$ then $f = \frac{1}{2}\frac{g'}{g}$. Hence

$$\begin{aligned}\frac{1}{2\pi i}\int_\gamma f(z)dz &= \tfrac{1}{2}\left[\frac{1}{2\pi i}\int_\gamma \frac{g'(z)}{g(z)}dz\right] \\ &= \frac{1}{2} \times \text{the number of zeros of } g \\ &= \frac{1}{2} \times 2 = 1.\end{aligned}$$

Thus $\int_\gamma f(z)dz = 2\pi i$. Compute by using residue theorem.

Theorem 11.4. *(Rouché's theorem): Let G be a star-shaped open set in $\mathbb{C}$. Let γ be a simple, positively oriented closed path in G. Let f and g be holomorphic on G and $|f(z) - g(z)| < |f(z)|, z \in [\gamma]$. Then f and g have the same number of zeros inside γ. Equivalently if we assume that $|g(z)| < |f(z)|$ on $[\gamma]$, then f and $f + g$ have the same number of zeros inside γ.*

Proof. Note that the assumption implies that f, g have no zeros on $[\gamma]$. We have

$$\left|\frac{g(z)}{f(z)} - 1\right| < 1 \text{ on } [\gamma].$$

Thus the values of $\frac{g}{f}$ are contained in $B(1,1)$. Let $F = \frac{g}{f}$. Thus $F \circ \gamma$ is a closed path contained in $B(1,1)$. As $0 \notin B(1,1)$, $n(F \circ \gamma, 0) = 0$.

But then

$$\begin{aligned}0 = n(F\circ\gamma, 0) &= \frac{1}{2\pi i}\int_{F\circ\gamma}\frac{dz}{z} \\ &= \frac{1}{2\pi i}\int_a^b \frac{F'(\gamma(t))\gamma'(t)}{F(\gamma(t))}dt \\ &= \frac{1}{2\pi i}\int_\gamma \frac{F'(z)}{F(z)}dz = \frac{1}{2\pi i}\left[\int_\gamma \frac{g'(z)}{g(z)}dz - \int_\gamma \frac{f'(z)}{f(z)}dz\right] \\ &= \text{No. of zeros of } g - \text{No. of zeros of } f.\end{aligned}$$

□

Roughly speaking the theorem states that the number of zeros a holomorphic function has in a compact set is stable under small perturbations of the function, provided zeros on the boundary of the set are absent.

Remark 11.5. (Dog on a leash)

There is a geometric interpretation of the result. If we assume that a man $f(\gamma(t))$ walks a dog which is on a leash $g(\gamma(t))$ whose length is less than the distance $|f(z)|$ of the man from the hydrant 0, then the man and the dog will go around the hydrant the same number of times. We give some typical applications of Rouché's theorem below.

Example 11.16. *All the roots of* $z^7 - 5z^3 + 12 = 0$ *lie between the circle* $|z| = 1$ *and* $|z| = 2$.

Solution: Consider the circle $C_1 = \{z \in \mathbb{C} : |z| = 1\}$. Let $f(z) = 12$, $g(z) = z^7 - 5z^3$. On C_1 we have

$$|g(z)| = \left|z^7 - 5z^3\right| \leq |z|^7 + \left|5z^3\right| \leq 6 < 12 = |f(z)|.$$

Hence by Rouché's theorem, $f(z) + g(z) = z^7 - 5z^3 + 12$ has the same number of zeros inside $|z| = 1$ as $f(z) = 12$, i.e., there are no zeros inside C_1. Consider the circle $C_2 = \{z \in \mathbb{C} : |z| = 2\}$. Let $f(z) = z^7, g(z) = 12 - 5z^3$. On C_2 we have

$$|g(z)| = \left|12 - 5z^3\right| \leq 12 + \left|5z^3\right| \leq 60 < 2^7 = |f(z)|.$$

Hence by Rouché's theorem $f(z) + g(z) = z^7 - 5z^3 + 12$ has the same number of zeros inside $|z| = 2$ as $f(z) = z^7$, i.e., all the zeros are inside C_2. Hence all the roots lie inside $|z| = 2$ but outside $|z| = 1$, as required.

Example 11.17. *Find the number of roots of* $z^7 - 4z^3 - 11 = 0$ *which lie between the two circles* $|z| = 1$ *and* $|z| = 2$.

Solution: Let $P(z) = z^7 - 4z^3 - 11$. For $|z| = 1$, we have

$$\left|z^7 - 4z^3\right| \leq 5 < 11$$

so, by Rouché's theorem, the given polynomial has no zeros in the unit disk. For $|z| = 2$,

$$\left|-4z^3 - 11\right| \leq 43 < 128 = \left|z^7\right|$$

so there are seven zeros of the given polynomial inside the disk $\{z : |z| < 2\}$ and they are all between the circles $|z| = 1$ and $|z| = 2$.

Example 11.18. *Let* $P(z) = z^5 + z^3 + 5z^2 + 2$. *How many zeros (counting multiplicities) does p have in the annular region* $1 < |z| < 2$.

Solution: If $|z| = 1$,

$$\left|5z^2\right| = 5 > 4 \geq \left|z^5 + z^3 + 2\right|.$$

Hence $P(z)$ has two zeros in $|z| < 1$. For $|z| = 2$,

$$|z|^5 = 32 > 30 \geq \left|z^3 + 5z^2 + 2\right|.$$

So $P(z)$ has 5 zeros in $|z| < 2$. Thus $P(z)$ has three zeros in $1 < |z| < 2$.

Note that the residue theorem, the argument principle and Rouche's theorem remain true if we replace the condition that G is star-shaped by the requiring that G is simply connected. Observe that we invoked the hypothesis that G is star-shaped only to conclude that $\int_\gamma f = 0$ for any closed path γ in G and $f \in H(G)$.

11.3 Mapping properties of holomorphic functions

As we have already seen certain statements about individual holomorphic functions enjoy the property of being "preserved" under local uniform limits. Let P denote a statement whose truth can be tested for a holomorphic function. If (f_n) converges locally uniformly to f on a region, and if P is true of f_n for each n, then P is true of f, at least if f is not constant. Such statements as "-is nonzero" or "-is one-to-one" will be cases in point. Theorem 11.6 which correspond to "-is nonzero," is due to A. Hurwitz. We have used the result of Hurwitz as a basis for the proof of some other results in this section. The following theorem gives a clearer picture of local uniform convergence.

Theorem 11.5. *Let (f_n) be a sequence of holomorphic functions on a region Ω, and suppose (f_n) converges uniformly on every compact subset of Ω to a function f. If f is not constant, there exists for each $z \in \Omega$, an integer $n_1 = n_1(z)$ and a sequence $(z_n), n \geq n_1$, such that*

$$f_n(z_n) = f(z), n \geq n_1 \tag{11.4}$$

$$\lim_{n\to\infty} z_n = z. \tag{11.5}$$

Proof. We shall assume f is not constant since otherwise there is nothing to prove. Fix $z \in \Omega$, and set $w = f(z)$. By Theorem 7.10, f is holomorphic. Because f is not constant and zeros of a holomorphic function are isolated, there exists $\delta > 0$ such that $\overline{B(z, \delta)} \subset \Omega$ and if $0 < |\zeta - z| \leq \delta$, then $f(\zeta) \neq w$.

For each integer $k > 0$ define γ_k to be the circle of radius $\frac{\delta}{k}$ about z, and let $\delta_k = \min\limits_{\zeta\in\gamma_k} |f(\zeta) - w|$. The continuity of f, the compactness of γ_k, and the definition of of δ combine to imply $\delta_k > 0, k = 1, 2, \cdots$. Using the uniform convergence of (f_n) on γ_k, choose n_k so large that if $n \geq n_k$, then

$$\max_{\zeta\in\gamma_k} |f_n(\zeta) - f(\zeta)| < \frac{\delta_k}{2}.$$

Further, choose the sequence (n_k) so that $n_1 < n_2 < n_3 < \cdots$.

For each k, if $n \geq n_k$, and if $\zeta \in \gamma_k$, the inequalities

$$|(f_n(\zeta) - w) - (f(\zeta) - w)| = |f_n(\zeta) - f(\zeta)| < \frac{\delta_k}{2} < \delta_k \leq |f(\zeta) - w|$$

are true. Therefore, by Rouché's theorem, the functions $(f_n(\zeta) - w)$ and $(f(\zeta) - w)$ have the same number of zeros inside γ_k. By assumption $(f(\zeta) - w)$ has a zero of order at least 1 at z, and therefore there exists at least one point z_n inside γ_k with $f_n(z_n) = w$. Define a sequence $(z_n), n \geq n_1$, as follows: if $n_k \leq n < n_{k+1}$, select z_n inside γ_k with $f_n(z_n) = w$. By construction, $\lim_{n\to\infty} z_n = z$, and the theorem is proved. □

Theorem 11.6. *(Hurwitz theorem): Let G be an open set. Let $f_n \in H(G)$ be nonconstant. Assume that f_n converges to f uniformly on compact subsets of G and that f_n have no zeros in G. Then either $f \equiv 0$ on G or f has no zeros in G. More generally, if $f_n(z) \neq w$ for all $z \in G$ and n, then either $f \equiv w$ or else $f(z) \neq w$ for all $z \in G$.*

Proof. Note that f is holomorphic in G. Let $f(a) = 0$ for some $a \in G$. If f is not identically zero on G, we can choose $r > 0$ such that $f(z) \neq 0$ for $z \in \overline{B(a,r)} \setminus \{a\} \subset G$. Let $\epsilon = \min\{|f(z)| : z \in S(a,r)\}$ where $S(a,r) = \{z \in \mathbb{C} : |z-a| = r\}$. Note that $\epsilon > 0$(as continuous function on a compact set attains its maximum and minimum and $|f(z)| > 0$ for all $z \in S(a,r)$.) Choose $N \in \mathbb{N}$ such that $|f(z) - f_n(z)| < \epsilon \leq |f(z)|$ for $z \in S(a,r)$ and for $n \geq N$. By Rouché's theorem, f_n has a zero in $B(a,r)$ for $n \geq N$, contradicting the hypothesis. □

Example 11.19. *Let $f_n(z) = \frac{e^z}{n}$ on $\mathbb{C}$. Then f_n has no zeros in $\mathbb{C}$ but f_n converges uniformly to the identically zero function.*

Proposition 11.3. *Let G be an open subset of $\mathbb{C}$. Assume that f_n is a sequence of one-to-one, holomorphic functions on G and that f_n converges to f uniformly on compact subsets of G. Then either f is a constant or f is one-to-one.*

Proof. Suppose $f(z_1) = f(z_2)$ but $z_1 \neq z_2$. This implies $d(z_1, z_2) > 0$. The function $f(z) - f(z_1)$ has a zero at z_1. Hence there exists $r_1 > 0$ such that $f(z) - f(z_1) \neq 0$ for all $z \in \overline{B(z_1, r_1)} \setminus \{z_1\}$. Similarly the function $f(z) - f(z_2)$ has a zero at z_2 and therefore there exists $r_2 > 0$ such that $f(z) - f(z_2) \neq 0$ for all $z \in \overline{B(z_2, r_2)} \setminus \{z_2\}$. Let

$$r' = \min\{r_1, r_2, d(z_1, z_2)\}.$$

Let r be such that $0 < r < \frac{r'}{2}$.

Then $B(z_1, r) \cap B(z_2, r) = \emptyset$. Let $\epsilon = \min\{|f(z) - f(z_1)|, |f(z) - f(z_2)| : z \in S(z_1, r) \cup S(z_2, r)\}$. Since f_n converges to f uniformly on compact subsets of G, hence there exists N such that for $n \geq N, j = 1, 2$,

$$|f_n(z) - f(z)| < \epsilon \leq |f(z) - f(z_j)|$$

for $z \in S(z_j, r) = \{z \in \mathbb{C} : |z - z_j| = r\}$. Hence by Rouché's theorem, for $n \geq N$, the function $f_n(z) - f(z) + f(z) - f(z_1)$ and $f(z) - f(z_1)$ have the same number of zeros inside $S(z_1, r)$. Since $f(z) - f(z_1)$ has a zero at $z = z_1$ hence $f_N(z) - f(z_1)$ has at least one zero w_1 in $B(z_1, r)$. Again applying Rouché's theorem, we can say that for $n \geq N$, the function $f_n(z) - f(z) + f(z) - f(z_2)$ and $f(z) - f(z_2)$ have the same number of zeros in $S(z_2, r)$. Since $f(z) - f(z_2)$ has a zero at z_2 hence $f_N(z) - f(z_2)$ has a zero w_2 in $B(z_2, r)$. Note that $w_1 \neq w_2$ because $B(z_1, r) \cap B(z_2, r) = \emptyset$. Thus we have, $f_N(w_1) - f(z_1) = 0 = f_N(w_2) - f(z_2)$. Hence $f_N(w_1) - f_N(w_2) = f(z_1) - f(z_2) = 0$. Thus $f_N(w_1) = f_N(w_2)$ but $w_1 \neq w_2$. Hence f_N is not one-one. This is a contradiction. Hence f is one-one. □

Example 11.20. *Let f_n be one to one and analytic in a region Ω for each n. If $z_0 \in \Omega$ is such that $f_n'(z_0) = 1, n = 1, 2, \cdots$ and if (f_n) converges uniformly on every compact subset of Ω to a function f, then f is also one one.*

Solution: By Weierstrass theorem, f_n' converges uniformly to f' on every compact subset of Ω and $f'(z_0) = 1$. By Hurwitz's theorem, $f'(z) \neq 0$ for all $z \in \Omega$. We shall show that if $z \neq z_1$ then $f(z) \neq f(z_1)$. So let $z \in \Omega \setminus \{z_1\}$. Then $(f_n(z) - f_n(z_1))$ converges to $f(z) - f(z_1)$ uniformly on every compact subset of Ω. Since f_n's are one-one, $f_n(z) - f_n(z_1) \neq 0$ for all $z \in \Omega \setminus \{z_1\}$. By Hurwitz's theorem, either $f(z) - f(z_1) \equiv 0$ or $f(z) - f(z_1) \neq 0$ for all $z \in \Omega \setminus \{z_1\}$. If $f(z) - f(z_1) \equiv 0$ then $f(z) = f(z_1)$ for all $z \in \Omega \setminus \{z_1\}$. Hence f is a constant function and therefore $f'(z) = 0$ which is a contradiction. If $f(z) \neq f(z_1)$ for all $z \in \Omega \setminus \{z_1\}$ then f is one one.

11.4 Local mapping theorem and its consequences

Theorem 11.7. *(The local mapping theorem): Let f be a nonconstant holomorphic function defined in the region $G \subset \mathbb{C}$. Let $z_0 \in G$ and $w_0 = f(z_0)$. Let m be the multiplicity with which f assumes the value w_0 at z_0 (i.e., m is the order of the zero of $f - w_0$ at z_0). For every sufficiently small positive number δ, there is a positive number ϵ such that every value w satisfying $0 < |w - w_0| < \epsilon$ is assumed by f at exactly m distinct points in the punctured disk $0 < |z - z_0| < \delta$, with multiplicity 1 at each of those points.*

Proof. Let $\delta > 0$ be as described and consider the closed disk $\overline{B(z_0, \delta)}$ with boundary $S(z_0, \delta) = \{z \in G : |z - z_0| = \delta\}$. The set $S(z_0, \delta)$ is a compact subset of $\mathbb{C}$. Let $\epsilon = \min\{|f(z) - w_0| : z \in S(z_0, \delta)\}$. Note $\epsilon > 0$. To prove the theorem we choose a positive number δ_0 such that the closed disk $\overline{B(z_0, \delta_0)}$ is contained in G, and such that neither $f(z) - w_0$ nor f' vanishes for $0 < |z - z_0| \leq \delta_0$; (If $m \geq 2$ then $f'(a) = 0$, hence there exists deleted neighborhood of a at which $f'(z) \neq 0$.) such a choice is possible because the zeros of nonconstant holomorphic function are isolated. We shall show that the desired conclusion holds for every positive number δ not exceeding δ_0.

Note $\epsilon > 0$ because f does not assume the value w_0 in $\overline{B(z_0, \delta)} \setminus \{z_0\}$. Fix w such that $|w - w_0| < \epsilon$. Then for $z \in S(z_0, \delta)$, we have

$$|(f(z) - w_0) - (f(z) - w)| = |w - w_0| < \epsilon \leq |f(z) - w_0| .$$

It therefore follows by Rouché's theorem that the functions $f(z) - w_0$ and $f(z) - w$ have the same number of zeros in $B(z_0, \delta)$. The function $f(z) - w_0$ has a zero of order m at z_0 and no other zeros in $\overline{B(z_0, \delta)}$. Hence, the function $f(z) - w$ has a total of m zeros in $B(z_0, \delta)$. Each of its zeros in $B(z_0, \delta)$ is of order 1, because f' is nonzero in $\overline{B(z_0, \delta)} \setminus \{z_0\}$. Thus, f assumes the value w at exactly m distinct points in $B(z_0, \delta)$, with multiplicity 1 at each of those points. Hence the theorem is proved. □

Note that the local mapping theorem says that $B(w_0, \epsilon) \subset f(B(z_0, \delta))$.

Remark 11.6. For functions of a real variable, the situation is less satisfactory; for example, the function $f : \mathbb{R} \to \mathbb{R}$ given by $f(x) = x^3$ has a triple root at $x = 0$, but is locally one-to-one; the function $f(x) = x^2$ has a double root at $x = 0$, but the equations $f(x) = c'$ have two solutions for $c' > 0$, and none for $c' < 0$.

We have already seen that there exists bijective, continuous maps that are not open map. Define $f(x) = x^2, x \in \mathbb{R}$. Then f is not an open map. We shall show below that if f is holomorphic and nonconstant then f is an open map.

Theorem 11.8. *(The open mapping theorem): If f is a nonconstant holomorphic function in a region $G \subseteq \mathbb{C}$, then f maps open subsets of G onto open sets (i.e., f is an open map).*

Proof. Let U be open in G. We claim that $f(U)$ is an open subset of $\mathbb{C}$. Let $w_0 \in f(U)$. This implies there exists $z_0 \in U$ such that $f(z_0) = w_0$. Since U is open, there exists $\delta > 0$ such that $B(z_0, \delta) \subset U$. By local mapping theorem, there exists $\epsilon > 0$ such that $B(w_0, \epsilon) \subset f(B(z_0, \delta)) \subset f(U)$. Hence $f(U)$ is open in $\mathbb{C}$. □

Thus if Ω is a region and $f \in H(\Omega)$ then $f(\Omega)$ is either a region or a point.

Theorem 11.9. *If the holomorphic function f has a nonvanishing derivative at the point z_0, then there is a disk centered at z_0 in which f is one-one.*

Proof. Suppose $f'(z_0) \neq 0$ and let $w_0 = f(z_0)$. Then $f(z) - w_0$ has a simple zero at $z = z_0$. Let $f(z) - w_0 \neq 0$ for all $z \in \overline{B(z_0, \delta_0)} \setminus \{z_0\}$. Choose δ such that $0 < \delta \leq \delta_0$. Then by local mapping theorem, there exist $\epsilon > 0$ such that for each $w \in \overline{B(w_0, \epsilon)} \setminus \{w_0\}$, the equation $f(z) = w$ has exactly one solution in $0 < |z - z_0| < \delta$. Thus f is one-one in $B(z_0, \delta)$. □

This criterion for one-oneness is strictly local. The function e^z, for example, has a nowhere vanishing derivative, yet it is not globally one-one.

Theorem 11.10. *A one-one, holomorphic function has a nowhere vanishing derivative.*

Proof. Let G be a region in $\mathbb{C}$ and assume that $f : G \to \mathbb{C}$ is holomorphic and one-one in G. Suppose $a \in G$ and $f'(a) = 0$. Let $b = f(a)$. Consider the function $g(z) = f(z) - b$. The function g is holomorphic in G and $g(a) = 0$ and $g'(a) = f'(a) = 0$. Thus $g(z) = f(z) - b$ has a zero of order $m \geq 2$ at $z = a$. By local mapping theorem, for every sufficiently small positive number δ, there is a positive number ϵ such that every value w satisfying $0 < |w - w_0| < \epsilon$ is assumed by f at at least two distinct points in the punctured disk $0 < |z - z_0| < \delta$. Thus f is not one-one which is a contradiction to the hypothesis. □

Thus if the derivative of a holomorphic function vanishes at some point, then, in every neighborhood of that point, the function fails to be univalent (i.e., the function is not one-one).

There is a partial analogue of these results in real analysis. If $f : (a, b) \to \mathbb{R}$ is continuously differentiable and if $f'(c) \neq 0$, then f is one-one around a. If $f'(c) > 0$, then $f'(x) > 0$ around c so that f is strictly increasing around c and hence is one-one around c. However the converse is not true. For instance, consider $f(x) = x^3$. Then f is one-one around 0 but $f'(0) = 0$.

Theorem 11.11. *(Inverse function theorem): Let G be a region in $\mathbb{C}$. Suppose $f : G \to \mathbb{C}$ is holomorphic, one-one, and $f(G) = \Omega$. Then $f^{-1} : \Omega \to \mathbb{C}$ is holomorphic and $(f^{-1})'(w) = \frac{1}{f'(z)}$ where $w = f(z)$.*

Proof. By the open mapping theorem the set $\Omega = f(G)$ is open in $\mathbb{C}$ and f^{-1} is continuous (as f is an open map). Because f is one-one, $f'(z) \neq 0$ for all $z \in G$. Since $z = f^{-1}(f(z))$ for each $z \in G$, the result follows from Theorem 5.9. □

The following is an application of Cauchy's theorem. It is shown how to count the number of zeros inside a closed path. Let G be a region and suppose $f : G \to \mathbb{C}$ is holomorphic and $a \in G$ is a zero of f of multiplicity m. Then we can write $f(z) = (z - a)^m g(z)$ where g is holomorphic in G and $g(a) \neq 0$. Suppose G is a region and let f be holomorphic in G with zeros at $a_1, a_2, \cdots, a_n$(where some of the a_k may be repeated according to the multiplicity of the zero.) So we write $f(z) = (z - a_1)(z - a_2) \cdots (z - a_n) g(z)$ where g is holomorphic on G and $g(z) \neq 0$ for any z in G. This gives

$$\frac{f'(z)}{f(z)} = \frac{1}{z - a_1} + \frac{1}{z - a_2} + \cdots + \frac{1}{z - a_n} + \frac{g'(z)}{g(z)}$$

for $z \neq a_1, \cdots, a_n$.

Theorem 11.12. *Let G be a star-shaped open set contained in $\mathbb{C}$. Let f be a holomorphic function on G with zeros $a_1, \cdots, a_n$(repeated according to multiplicity). Suppose γ is a closed path in G such that $\gamma \approx 0$ in G and no zero of* *f lie on* $[\gamma]$. *Then*

$$\frac{1}{2\pi i}\int_\gamma \frac{f'(z)}{f(z)}dz = \sum_{k=1}^{n} n(\gamma, a_k).$$

Proof. From the discussion prior to the theorem it follows that $\frac{g'(z)}{g(z)}$ is holomorphic in G (as $g(z) \neq 0$ for all $z \in G$). Since $\gamma \approx 0$, by Cauchy's theorem we have $\int_\gamma \frac{g'(z)}{g(z)}dz = 0$. The result follows from the definition of winding number $n(\gamma, a_k), 1 \leq k \leq n$. □

Corollary 11.1. *Let* f, G *and* γ *be as in the preceding theorem except that* $a_1, \cdots, a_n$ *are the points in* G *that satisfy the equation* $f(z) = \alpha$ *then*

$$\frac{1}{2\pi i}\int_\gamma \frac{f'(z)}{f(z)-\alpha}dz = \sum_{k=1}^{m} n(\gamma, a_k).$$

Example 11.21. *As an illustration, let us calculate* $\int_\gamma \frac{2z}{z^2+1}dz$ *where* γ *is the circle* $|z| = 2$.

Since the denominator of the integrand factors into $(z - i)(z + i)$, *hence* $\int_\gamma \frac{2z}{z^2+1}dz = 2\pi i \cdot 2 = 4\pi i$.

Corollary 11.2. *Let* U *be open in* $\mathbb{C}, f : U \to \mathbb{C}$ *be analytic. Let* $\gamma : [a, b] \to \mathbb{C}$ *be a closed path such that* γ *and its interior lying in* U. *Suppose that* $f(z)-w$ *has zeros at* $z_1, \cdots, z_m$ *inside* γ *with multiplicity* $M_1, \cdots, M_m$. *That is,* $f(z_i + h) = w + \sum_{k=M_i}^{\infty} a_k h^k$ *with* $a_{M_i} \neq 0$ *and* $f(z) \neq w$ *on* γ. *Then* $n(f \circ \gamma, w) = \sum_{i=1}^{m} M_i n(\gamma, z_i)$. *In particular, if* γ *only winds round points in its interior once then* $n(f \circ \gamma, w)$ *is the number of solutions of* $f(z) = w$ *inside* γ *counting multiplicities.*

Proof. Notice that $n(f \circ \gamma, w) = \frac{1}{2\pi i}\int_a^b \frac{(f\circ\gamma)'(t)}{(f\circ\gamma)(t)-w}dt = \frac{1}{2\pi i}\int_a^b \frac{f'(\gamma(t))\gamma'(t)}{f(\gamma(t))-w}dt = \frac{1}{2\pi i}\int_\gamma \frac{f'(z)}{f(z)-w}dz = \sum_{i=1}^{m} M_i n(\gamma, z_i)$. The last equality follows from argument principle. One can also show it directly using residue theorem as follows: $\frac{f'}{f-w}$ has poles only if $f(z) = w$. That is, $\frac{f'}{f-w}$ has poles (simple poles) at $z_1, \cdots, z_m$ and

$$\left(\frac{f'}{f-w}\right)(z_i + h) = \frac{\sum_{k=M_i}^{\infty} k a_k h^{k-1}}{\sum_{k=M_i}^{\infty} a_k h^k}.$$

Thus $h\left(\frac{f'}{f-w}\right)(z_i + h) = \frac{\sum_{k=M_i}^{\infty} k a_k h^k}{\sum_{k=M_i}^{\infty} a_k h^k} \to M_i$ as $h \to 0$. Hence $n(f \circ \gamma, w) = \sum_{i=1}^{m} M_i n(\gamma, z_i)$ as required. □

Example 11.22. *Let* $\gamma = \{z \in \mathbb{C} : |z| = 2\}, U = \mathbb{C}$, *and* $f(z) = z^2 - z$. *Thus if* $w \in \{z \in \mathbb{C} : 2 < |z| < 6\}$, *then there exist one solution of* $z^2 - z - w = 0$ *inside* γ. *For example, if we take* $w = 3\frac{3}{4}$, *then* $z^2 - z - \frac{15}{4} = 0$ *implies* $z = \frac{1\pm\sqrt{1+15}}{2}$. *That is* $z = 2\frac{1}{2}, -1\frac{1}{2}$. *So there is only one root inside the circle*

$\{z \in \mathbb{C} : |z| = 2\}$. Verify also that if $w = 0$, then $n(f \circ \gamma, w) = 2$, and if $w = \frac{1}{4}$ then $n(f \circ \gamma, w) = 2$.

Theorem 11.13. *(Generalized argument principle): Let f be meromorphic in the region G with zeros $a_1, a_2, \cdots, a_n$ and poles $p_1, p_2, \cdots, p_m$ counted according to multiplicity. Suppose g is holomorphic in G and γ is a closed path in G with $\gamma \approx 0$ and further assume that none of the zeros and poles of f lie on $[\gamma]$. Then*

$$\frac{1}{2\pi i}\int_\gamma g(z)\frac{f'(z)}{f(z)}dz = \sum_{j=1}^{n} g(a_j)n(\gamma, a_j) - \sum_{k=1}^{m} g(p_k)n(\gamma, p_k).$$

Proof. First calculate that if a is a zero of f of order r then $g(z)\frac{f'(z)}{f(z)}$ has a pole with residue $rg(a)n(\gamma, a)$ at the point a and if p is a pole of f of order s then $g(z)\frac{f'(z)}{f(z)}$ has a pole with residue $-sg(p)n(\gamma, p)$ at the point p. Thus

$$\frac{1}{2\pi i}\int_\gamma g(z)\frac{f'(z)}{f(z)}dz = \sum_{j=1}^{n} g(a_j)n(\gamma, a_j) - \sum_{k=1}^{m} g(p_k)n(\gamma, p_k).$$

□

By the inverse function theorem, every one-one holomorphic function has a holomorphic inverse. The generalized argument principle gives a formula for calculating this inverse.

Proposition 11.4. *Let f be holomorphic on an open set containing $\overline{B(a, R)}$ and suppose that f is one-one on $B(a, R)$. If $\Omega = f[B(a, R)]$ and γ is the circle $|z - a| = R$ then $f^{-1}(\zeta)$ is defined for each ζ in Ω by the formula*

$$f^{-1}(\zeta) = \frac{1}{2\pi i}\int_\gamma \frac{zf'(z)}{f(z) - \zeta}dz.$$

Proof. Let $\zeta \in \Omega = f[B(a, R)]$. Then there exists $w \in B(a, R)$ such that $\zeta = f(w)$. Then since f is one-one, $f(z) - \zeta$ has one and only one zero in $B(a, R)$. If we choose $g(z) = z$, in Theorem 11.13 then we obtain

$$\begin{aligned}\frac{1}{2\pi i}\int_\gamma z\frac{f'(z)}{f(z) - \zeta}dz &= \frac{1}{2\pi i}\int_\gamma g(z)\frac{f'(z)}{f(z) - \zeta}dz \\ &= g(w) = w = f^{-1}(\zeta).\end{aligned}$$

□

Theorem 11.14. *Suppose f and g are meromorphic in a neighborhood of $\overline{B(a, R)}$ with no zeros or poles on the circle $\gamma = \{z : |z - a| = R\}$. If $N_f, N_g (P_f, P_g)$ are the number of zeros(poles) of f and g inside γ counted according to their multiplicities and if*

$$|f(z) + g(z)| < |f(z)| + |g(z)|$$

on $[\gamma]$ then $Z_f - P_f = Z_g - P_g$.

Proof. From the hypothesis

$$\left|\frac{f(z)}{g(z)}+1\right| < \left|\frac{f(z)}{g(z)}\right|+1$$

on $[\gamma]$. If $\lambda = \frac{f(z)}{g(z)}$ and if λ is a positive real number then this inequality becomes $\lambda + 1 < \lambda + 1$, a contradiction. Hence $\lambda \notin [0, \infty)$. Thus the meromorphic function $\frac{f}{g}$ maps γ onto $\Omega = \mathbb{C} - [0, \infty)$. If l is a branch of the logarithm on Ω then $l\left(\frac{f(z)}{g(z)}\right)$ is a well defined primitive for $\left(\frac{f}{g}\right)'\left(\frac{f}{g}\right)^{-1}$ in a neighborhood of $[\gamma]$. Thus

$$\begin{aligned} 0 = \frac{1}{2\pi i}\int_\gamma \left(\frac{f}{g}\right)'\left(\frac{f}{g}\right)^{-1} dz &= \frac{1}{2\pi i}\int_\gamma \frac{f'(z)}{f(z)}dz - \frac{1}{2\pi i}\int_\gamma \frac{g'(z)}{g(z)}dz \\ &= (Z_f - P_f) - (Z_g - P_g). \end{aligned}$$

□

11.5 Exercises

1. Consider $f_n(x) = x^2 + \frac{1}{n}$ for $x \in (-1, 1)$. What is $\lim f_n$? What is the relevance of this sequence to Hurwitz theorem?

2. Show that for each positive integer $m, p_m(z) = z^m$ defines an open mapping.

3. If E is any straight line in $\mathbb{C}$, and if a holomorphic function f on an open set G takes values in E, then prove that f is a constant.

4. Suppose Ω_1 and Ω_2 are regions in $\mathbb{C}$, f and g are nonconstant complex functions defined in Ω_1 and Ω_2 respectively and $f(\Omega_1) \subset \Omega_2$. Put $h = g \circ f$. If f and g are holomorphic, we know that h is holomorphic. Suppose we know that f and h are holomorphic. Can we conclude anything about g? What if we know that g and h are holomorphic?

5. Suppose (f_n) is a sequence of holomorphic functions defined on a region Ω converging locally uniformly to a function f. If $f_n'(z) \neq 0$ for all $z \in \Omega$ and n then show that either f is constant or else $f'(z) \neq 0$ for all $z \in \Omega$.

6. Let (f_n) be a sequence of holomorphic functions defined on a region Ω converging locally uniformly to a function f. Suppose Ω_0 is a region such that $f_n(\Omega) \subseteq \Omega_0$ for each n. Show that either f is constant or else $f(\Omega) \subseteq \Omega_0$.

7. Suppose (f_n) is a sequence of holomorphic functions defined on a region Ω converging locally uniformly to a function f. Let f_n be one-to-one on

Ω for each n, and suppose f is not constant. If D is an open set which belongs to both $f(\Omega)$ and $f_n(\Omega)$ for all n, then show that on D,

$$\lim_{n\to\infty} f_n^{-1}(w) = f^{-1}(w)$$

holds.

8. How many zeros does the function $f(z) = 3z^{100} - e^z$ have inside the unit circle (counting multiplicities)?

9. How many zeros does the complex polynomial $3z^9 + 8z^6 + z^5 + 2z^3 + 1$ have in the annulus $1 < |z| < 2$?

10. Let $f(z) = \frac{z^4}{z^2+1}$ and $\gamma(t) = 2e^{it}, 0 \le t \le 2\pi$. Evaluate $\int_\gamma f(z)dz$.

11. Evaluate $\int_0^{2\pi} \frac{d\theta}{a+\cos\theta}$, $a > 1$ using the residue theorem.

12. Prove that $\lim\limits_{n\to\infty} \left(1 + \dfrac{z}{n}\right)^n = e^z$ uniformly on all compact sets.

13. Locate and classify the singularities of the following functions and find the residue at the singularities:

(i)	$\tan z$	(ii)	$\dfrac{z^3+1}{z^4+1}$
(iii)	$e^{-\frac{1}{z}}$	(iv)	$\dfrac{z^p}{1-z^q}$ (p, q positive integers)
(v)	$\dfrac{1}{\sin z}$	(vi)	$\dfrac{z^5}{(z^2-1)^2}$
(vii)	$\dfrac{z}{(z-1)(z+1)^2}$	(viii)	$\dfrac{\cos z}{1+z+z^2}$
(ix)	$\dfrac{z^2-2z}{(z+1)^2(z^2+4)}$	(x)	$e^z \csc^2 z$.

14. Calculate the residues of the following functions at the points stated:

(i)	$\dfrac{\cos 2z}{z^3}$ at 0	(ii)	$\frac{e^{iz}}{z^2+4}$ at $2i$
(iii)	$\dfrac{1+\cos z}{(z-\pi)^3}$ at π	(iv)	$\dfrac{e^{\sin z}}{z(z-2)^2}$ at 0
(v)	$\dfrac{1}{e^z-1}$ at 0	(vi)	$\dfrac{1}{(e^z-1)^2}$ at 0
(vii)	$\dfrac{\cot z \coth z}{z^3}$ at $z = 0$.		

15. Use residue theorem to calculate the following integrals (all contours traced anticlockwise):

    (i) $\int_{\gamma_1} \frac{\cos 2z}{z^3} dz$, where γ_1 is the square with vertices $\pm 1 \pm i$.

    (ii) $\int_{\gamma_2} \frac{e^{iz}}{z^2+4} dz$, where $\gamma_2 = \{z \in \mathbb{C} : |z| \leq 10, \text{Im } z \geq 0\}$.

    (iii) $\int_{\gamma_3} \frac{e^{\sin z}}{z(z-2)^2} dz$, where γ_3 is the unit circle.

    (iv) $\int_{\gamma_4} \frac{dz}{e^z - 1}$, where γ_4 is the parallelogram with vertices $1, 10i, -1, -10i$.

Chapter 12

Evaluation of Definite Integrals

The residue theorem is useful not only in evaluating the integrals of functions of a complex variable, but also in evaluating various definite integrals of functions of a real variable. Very often one is able to obtain an answer in cases where the use of other methods of analysis proves complicated. In this chapter we shall consider a number of typical cases.

12.1 Preliminaries on improper integrals

Definition 12.1. *Let $f : (a, \infty) \to \mathbb{R}$ be given. We say that $\int_a^\infty f(x)dx$ exists if $\int_a^R f(x)dx$ exists for all $R \geq a$ and if $\lim_{R\to\infty} \int_a^R f(x)dx$ exists. If the latter limit is s, we then say $\int_a^\infty f(x)dx$ converges to s and write $\int_a^\infty f(x)dx = s$. We similarly assign a meaning to $\int_{-\infty}^a g(x)dx$ for $g : (-\infty, a) \to \mathbb{R}$. Finally, we say that $\int_{-\infty}^\infty f(x)dx$ exists and is s if $\int_{-\infty}^0 f(x)dx = s_1$ and $\int_0^\infty f(x)dx = s_2$ with $s = s_1 + s_2$.*

It thus follows from Definition 12.1 that if f is a real valued function defined on $\mathbb{R}$ then $\int_{-\infty}^\infty f(x)dx = s$ if and only if for any $\epsilon > 0$, there exists $R > 0$ such that $\left|\int_{-a}^b f(x)dx - s\right| < \epsilon$ for $a \geq R$ and $b \geq R$.

There is another way of assigning a meaning to $\int_{-\infty}^\infty f(x)dx$.

Definition 12.2. *We say that the Cauchy principal value of the integral is L if $\int_{-R}^R f(x)dx$ exists for all R and $\lim_{R\to\infty} \int_{-R}^R f(x)dx = L$. In such a case, we write P.V. $\int_{-\infty}^\infty f(x)dx = L$.*

Let $f : \mathbb{R} \to \mathbb{R}$ be any odd continuous function. Then *P.V.* $\int_{-\infty}^\infty f(x)dx = 0$.

Example 12.1. *The integral $\displaystyle\int_{-\infty}^\infty x^3dx$ is not convergent but its Cauchy principal value exists. In fact P.V. $\displaystyle\int_{-\infty}^\infty x^3dx = 0$.*

We shall clarify the relation between these two concepts. Suppose $\int_{-\infty}^{\infty} f(x)dx$ exists. Then

$$P.V. \int_{-\infty}^{\infty} f(x)dx = \int_{-\infty}^{\infty} f(x)dx.$$

Most often, we may first of all establish the convergence of $\int_{-\infty}^{\infty} f(x)dx$ and then compute $P.V. \int_{-\infty}^{\infty} f(x)dx$ which is equal to the value of the improper integral $\int_{-\infty}^{\infty} f(x)dx$. Note that $\int_{-\infty}^{\infty} f(x)dx = \lim_{R,S\to\infty} \int_{-R}^{S} f(x)dx$ where R and S have to be allowed to run to ∞ independently of each other. Thus the existence of $\lim_{R\to\infty} \int_{-R}^{R} f(x)dx$ does not imply the existence of $\int_{-\infty}^{\infty} f(x)dx$ as pointed out above.

Now suppose f is continuous on the closed interval $[a,b]$ except at a point $x_0 \in (a,b)$, where f has a singularity. Then one of the most natural way to define $\int_a^b f(x)dx$ is as follows: If δ_1 and δ_2 are positive we define

$$\int_a^b f(x)dx = \lim_{\substack{\delta_1\to 0\\ \delta_2\to 0}} \left[\int_a^{x_0-\delta_1} f(x)dx + \int_{x_0+\delta_2}^{b} f(x)dx\right]. \tag{12.1}$$

We may get different values if we evaluate the limit

$$\lim_{\delta\to 0} \left[\int_a^{x_0-\delta} f(x)dx + \int_{x_0+\delta}^{b} f(x)dx\right]. \tag{12.2}$$

However, if $\int_a^b f(x)dx$ exists then these two limits in (12.1) and (12.2) are equal.

Example 12.2. *Consider the function*

$$f(x) = \frac{1}{x^3}, x \neq 0, x \in [-1,1].$$

The point $x_0 = 0$ is a singularity of f and

$$\left[\int_{-1}^{-\delta_1} \frac{dx}{x^3} + \int_{\delta_2}^{1} \frac{dx}{x^3}\right] = \frac{1}{2}\left[\frac{1}{\delta_2^2} - \frac{1}{\delta_1^2}\right].$$

The limit of $\frac{1}{2}\left[\frac{1}{\delta_2^2} - \frac{1}{\delta_1^2}\right]$ does not exist as $\delta_1 \to 0$ and $\delta_2 \to 0$. Thus the integrals $\int_{-1}^{0} \frac{dx}{x^3}$ and $\int_0^1 \frac{dx}{x^3}$ are not convergent. However, observe that

$$\int_{-1}^{-\delta} \frac{dx}{x^3} + \int_{\delta}^{1} \frac{dx}{x^3} = 0$$

for all δ such that $0 < \delta < 1$.

Recall the following criterion of convergence of improper integrals. It tests the convergence of the improper integral even if it is not possible to evaluate it explicitly.

Let $f : (c, \infty) \to \mathbb{R}$ be such that $\int_c^d f(t)dt$ exist for each $d > c$. Then $\int_c^\infty f(t)dt$ exists if and only if for each $\epsilon > 0$, there is a number $R > c$ such that $|\int_a^b f(t)dt| < \epsilon$ if $b > a > R$. The proof is immediate from the following fact: if $f : \mathbb{R} \to \mathbb{R}$ is such that for any $\epsilon > 0$, there exists $M > 0$ such that $|f(x) - f(y)| < \epsilon$ whenever $x > M$ and $y > M$ then $\lim_{x\to\infty} f(x)$ exists.

The following results on improper integrals will be very useful in the sequel. We list them here for ready reference.

(1) If f is a continuous real valued function on (a, ∞) and if $\int_a^\infty |f(x)|dx$ exists, then $\int_a^\infty f(x)dx$ exists.

(2) Suppose that $\int_a^\infty g(t)dt$ exists and $g(x) > 0$ for $x \geq a$. If $f : (a, \infty) \to \mathbb{R}$ is such that $\int_a^x f(t)dt$ exists for $x > a$, and $\lim_{x\to\infty} \frac{f(x)}{g(x)} = 1$ then $\int_a^\infty f(t)dt$ exists.

(3) If $\int_a^\infty |f(x)|dx$ converges, $\int_a^\infty g(t)dt$ exists and $|g(x)| \leq |f(x)|$ for $x \in [a, \infty)$, then $\int_a^\infty |g(x)|dx$ converges.

(4) Let $f(x, y)$ be a real valued function defined on the semi-infinite strip $S = \{(x, y) : a \leq x \leq b \text{ and } c \leq y < +\infty\}$. Let $f, f_x \in C(S)$. Suppose $\int_c^\infty f(x, y)dy = g(x)$ on $[a, b]$ and $\int_c^\infty f_x(x, y)dy$ converges uniformly on $[a, b]$ to the function $h(x)$. Then $g'(x) = h(x)$ on $[a, b]$.

(5) Let $\int_c^r f(x, t)dt$ exist for each $r \geq c$ and for each $x \in [a, b]$. If $|f(x, t)| \leq M(t)$ for each $x \in [a, b]$ and for each t in $[c, +\infty)$ and if $\int_c^\infty M(t)dt$ exists, then $\int_c^r f(x, t)dt$ converges uniformly on $[a, b]$ to $g(x)$ where $g(x) = \int_c^\infty f(x, t)dt$.

(6) Let $f : (a, \infty) \to \mathbb{R}$ be such that $\int_a^s f(t)dt$ exists for all $s > a > 0$. Suppose there exists a $\mu > 1$ such that $x^\mu f(x)$ is bounded for $x > a$. Then $\int_a^\infty f(x)dx$ converges. In particular the following holds:

    If $f : \mathbb{R} \to \mathbb{R}$ is continuous and there exist $K > 0$ and M such that $|f(x)| \leq \frac{M}{|x|^2}$ for all x with $|x| \geq K$ then $\int_{-\infty}^\infty f(x)dx$ exists.

12.2 Evaluation of real integrals

Integrals of type $\int_\alpha^{2\pi+\alpha} R(\cos\theta, \sin\theta)d\theta$

We consider the integral $I = \int_\alpha^{2\pi+\alpha} R(\cos\theta, \sin\theta)d\theta$ where R is a rational function of its arguments. Such integrals can easily be reduced to integrals of an analytic function of a complex variable over a closed contour.

Example 12.3. *Find* $\int_{-\pi}^{\pi} \frac{\sin^2\theta}{5+4\cos\theta}d\theta.$

Solution: Let γ be the unit circle $\gamma(\theta)=e^{i\theta}, -\pi\le\theta\le\pi$. Let $z=e^{i\theta}$ and therefore $dz=ie^{i\theta}d\theta$. Now

$$\begin{aligned}\int_{-\pi}^{\pi}\frac{\sin^2\theta}{5+4\cos\theta}d\theta &= \int_\gamma \frac{\left(\frac{z-\frac{1}{z}}{2i}\right)^2}{5+4\left(\frac{z+\frac{1}{z}}{2}\right)}\cdot\frac{dz}{iz}\\ &= \int_\gamma \frac{z^2-2+\frac{1}{z^2}}{(-4)\left(5+2z+\frac{2}{z}\right)}\frac{dz}{iz}\\ &= \frac{i}{4}\int_\gamma\frac{z^4-2z^2+1}{z^2(2z^2+5z+2)}dz\\ &= \frac{i}{4}\int_\gamma\frac{z^4-2z^2+1}{z^2(z+2)(2z+1)}dz.\end{aligned}$$

Let $f(z)=\frac{z^4-2z^2+1}{z^2(z+2)(2z+1)}$. The singularities of the function f in the interior of γ are the points $z_1=0$ and $z_2=-\frac{1}{2}$.
Now

$$\begin{aligned}\text{Res}(f,0) &= \lim_{z\to0}\frac{d}{dz}\frac{z^4-2z^2+1}{2z^2+5z+2}\\ &= \left.\frac{(2z^2+5z+2)(4z^3-4z)-(z^4-2z^2+1)(4z+5)}{(2z^2+5z+2)^2}\right|_{z=0}=-\frac{5}{4}.\end{aligned}$$

Further, $\text{Res}\left(f,-\frac{1}{2}\right)=\left.\frac{z^4-2z^2+1}{2z^2(z+2)}\right|_{z=-\frac{1}{2}}=\frac{3}{4}$. Thus

$$\begin{aligned}\int_{-\pi}^{\pi}\frac{\sin^2\theta}{5+4\cos\theta}d\theta &= \tfrac{i}{4}\int_\gamma\frac{z^4-2z^2+1}{z^2(z+2)(2z+1)}dz\\ &= \tfrac{i}{4}\left[2\pi i\{\text{Res}(f,0)+\text{Res}\left(f,-\frac{1}{2}\right)\}\right]\\ &= \tfrac{i}{4}2\pi i\left(-\tfrac{5}{4}+\tfrac{3}{4}\right)=\tfrac{\pi}{4}.\end{aligned}$$

Example 12.4. *Use the residue theorem to evaluate the integral*

$$\int_0^{2\pi}\frac{d\theta}{1-2r\cos\theta+r^2}\quad \textit{where } r^2\ne1.$$

Solution: Let f be the function defined by $f(z)=\frac{i}{(z-r)(rz-1)}$. We have $\int_{|z|=1}f(z)dz=\int_0^{2\pi}\frac{d\theta}{1-2r\cos\theta+r^2}$. For $|r|<1$, $\text{Res}(f(z),r)=\frac{i}{r^2-1}$ and

$\int_0^{2\pi} \frac{d\theta}{1-2r\cos\theta+r^2} = \frac{2\pi}{1-r^2}$. For $|r|>1$, we get $\text{Res}\left(f(z), \frac{1}{r}\right) = \frac{i}{1-r^2}$ and $\int_0^{2\pi} \frac{d\theta}{1-2r\cos\theta+r^2} = \frac{2\pi}{r^2-1}$.

Integrals of the form $\int_{-\infty}^{\infty} f(x)dx$

We now consider applying the calculus of residues to evaluate improper integrals of the form $\int_{-\infty}^{\infty} f(x)dx$ where $|f(x)| \le \frac{C}{|x|^2}$ for $|x|$ very large. We illustrate some of the basic methods.

Example 12.5. *Evaluate* $\int_{-\infty}^{\infty} \frac{1}{1+x^2}dx$.

Solution: By the facts listed above on improper integrals, $\int_{-\infty}^{\infty} f(x)dx$ exists. Let $f(z) = \frac{1}{1+z^2}$. The poles of f are at $z = \pm i$. Let C_R be the closed path;

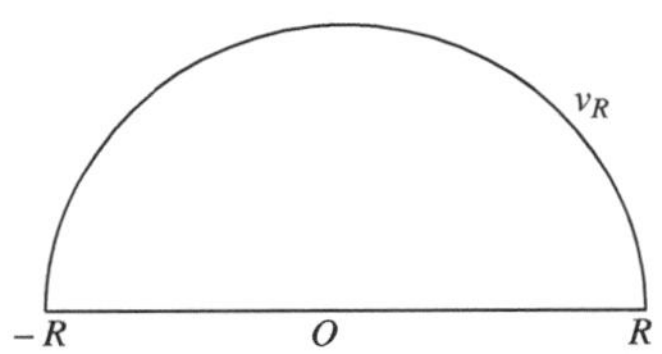

Fig. 12.1: The closed path $\mathbf{C_R} = [-\mathbf{R}, \mathbf{R}] + \gamma_\mathbf{R}$

the line segment $[-R, R]$ followed by $\gamma_R(t) = Re^{it}, 0 \le t \le \pi$. If $R > 1$, then the pole $z = i$ is inside C_R. Hence by the residue theorem we have

$$\int_{C_R} f(z)dz = 2\pi i \text{Res}(f, i) = 2\pi i \cdot \frac{1}{2i} = \pi.$$

Now, $\int_{C_R} f(z)dz = \int_{-R}^{R} \frac{1}{1+x^2}dx + \int_{\gamma_R} f(z)dz$. We claim that the latter integral goes to 0, as $R \to \infty$. Note $\lim_{R\to\infty} \int_{-R}^{R} \frac{1}{1+x^2}dx$ gives us the principal value of the integral $\int_{-\infty}^{\infty} f(x)dx$. But since $\int_{-\infty}^{\infty} f(x)dx$ exists the principal value of the integral coincides with the value of the integral $\int_{-\infty}^{\infty} f(x)dx$. Now, the claim follows from the estimate below:

$$\begin{aligned} \left|\int_{\gamma_R} f(z)dz\right| &= \left|\int_0^{\pi} \frac{iRe^{i\theta}}{1+R^2e^{2i\theta}}d\theta\right| \\ &\le \int_0^{\pi} \frac{R}{R^2-1}d\theta \\ &= \frac{R\pi}{R^2-1}. \end{aligned}$$

Hence $\lim_{R\to\infty}\int_{\gamma_R} f(z)dz = 0$. Thus $\int_{-\infty}^{\infty}\frac{1}{1+x^2}dx = \pi$.

Example 12.6. *Evaluate* $\int_{-\infty}^{\infty}\frac{x^2}{x^2+a^2}dx$, *where* $a > 0$.

Solution: Let $f(z) = \frac{z^2}{z^2+a^2}$. Then f has poles at $z = \pm ia$. Now $\text{Res}(f, ia) = \frac{1}{4ia}$. If C_R is as in Example 12.5, we have $\int_{C_R} f(z)dz = 2\pi i\text{Res}(f, ia) = \frac{1}{4ia}$. It is easy to show that $\int_{\gamma_R} f(z)dz \to 0$ as $R \to \infty$ and we conclude that $\int_{-\infty}^{\infty}\frac{x^2}{x^2+a^2}dx = \frac{\pi}{2a}$.

These examples suggest the following result.

Lemma 12.1. *Let the function* f *be holomorphic in the upper half plane* $Imz > 0$, *with the exception of a finite number of isolated singular points and let there exist positive numbers* R_0, M *and* δ *such that for all points* z *of the upper half plane which satisfy the condition* $|z| > R_0$, *we have* $|f(z)| < \frac{M}{|z|^{1+\delta}}$. *Then*

$$\lim_{R\to\infty}\int_{\gamma_R} f(z)dz = 0 \tag{12.3}$$

where the contour of integration γ_R *is a semicircle* $|z| = R, Imz > 0$ *in the upper half of the complex plane.*

Proof. The lemma follows because

$$\begin{aligned}\left|\int_{\gamma_R} f(z)dz\right| &\leq \int_{\gamma_R} |f(z)||dz| \\ &< \frac{M}{R^{1+\delta}}\int_0^{\pi} Rd\theta \\ &= \frac{M\pi R}{R^{1+\delta}} \\ &= \frac{\pi M}{R^{\delta}} \to 0\end{aligned}$$

as $R \to \infty$. □

Remark 12.1. If the conditions of the lemma are fulfilled in some sector $\phi_1 < \arg z < \phi_2$ of the complex plane, then formula (12.3) is valid in integration along the arc γ_R of a circle lying in the given sector.

Remark 12.2. Lemma 12.1 finds broad application in computing a number of improper integrals of the form $\int_{-\infty}^{\infty} f(x)dx$. The conditions of the lemma will obviously be fulfilled if the function $f(z)$ is holomorphic in the neighborhood of the point at infinity and the point $z = \infty$ is a zero of order at least two of the function f. Indeed, in this case, a Laurent series expansion of f in the neighborhood of $z = \infty$ is of the form

$$f(z) = \frac{c_{-2}}{z^2} + \frac{c_{-3}}{z^3} + \cdots = \frac{\Psi(z)}{z^2}$$

and $|\Psi(z)| < M$. Thus $|f(z)| \leq \frac{M}{|z|^2}$. So it satisfies the condition of the lemma.

Theorem 12.1. *Let f be holomorphic on $\mathbb{C}$ except at a finite number of isolated singular points. Suppose f does not have singularities on the real axis and that those in the upper half plane $\{z \in \mathbb{C} : Imz > 0\}$ are $z_1, z_2, \cdots, z_n$. Assume that there exist positive numbers R_0 and M such that $|f(z)| \leq \frac{M}{|z|^\alpha}$ for $|z| > R_0$ and for some $\alpha > 1$. Then*

$$\int_{-\infty}^{\infty} f(x)dx = 2\pi i \sum_{k=1}^{n} Res(f, z_k).$$

Proof. By hypothesis, the function f in the upper half plane has a finite number of singularities $z_1, z_2, \cdots, z_n$ and they all satisfy the condition $|z_k| < R_0, k = 1, \cdots, n$. Consider in the upper half plane a closed path C_R consisting of the line segment $[-R, R], R > R_0$, followed by $\gamma_R(t) = Re^{it}, 0 \leq t \leq \pi$ (the semicircular arc of radius R) as shown in Fig.12.1. By Residue theorem

$$\begin{aligned} \int_{C_R} f(z)dz &= \int_{-R}^{R} f(x)dx + \int_{\gamma_R} f(z)dz \\ &= 2\pi i \sum_{k=1}^{n} \text{Res}(f, z_k). \end{aligned}$$

Further, it follows from Lemma 12.1 that $\int_{\gamma_R} f(z)dz \to 0$ as $R \to \infty$. A direct proof is as follows:

$$\begin{aligned} \left|\int_{\gamma_R} f(z)dz\right| &= \left|\int_0^{\pi} f(Re^{i\theta})iRe^{i\theta}d\theta\right| \\ &\leq \frac{M}{R^\alpha}R\pi \\ &= \frac{M\pi}{R^{\alpha-1}} \end{aligned}$$

which approaches to zero as $R \to \infty$, since $\alpha > 1$. Thus

$$\lim_{R\to\infty} \int_{\gamma_R} f(z)dz = 0.$$

Since $|f(z)| \le \frac{M}{|z|^\alpha}$ for $|z| > R_0$ and for some $\alpha > 1$, the limit $\int_{-\infty}^{\infty} f(x)dx$ exists. Hence

$$\int_{-\infty}^{\infty} f(x)dx = \lim_{R\to\infty}\int_{-R}^{\mathbb{R}} f(x)dx = 2\pi i\sum_{k=1}^{n}\text{Res}(f, z_k), \tag{12.4}$$

since the right side of (12.4) is independent of R for $R > R_0$. □

We can make similar statements for the lower half plane. We state the result for future reference.

Theorem 12.2. *Let f be holomorphic on $\mathbb{C}$ except at a finite number of points, none of which are real and that those in the lower half plane $\{z \in \mathbb{C} : Imz < 0\}$ are $z_1, z_2, \cdots, z_n$. Assume that there exist positive constants R_0 and M such that $|f(z)| \le \frac{M}{|z|^\alpha}$ for $|z| > R_0$ and for some $\alpha > 1$. Then*

$$\int_{-\infty}^{\infty} f(x)dx = -2\pi i\sum_{k=1}^{n} Res(f, z_k). \tag{12.5}$$

Observe that the minus sign in (12.5) appears because of the counter clockwise orientation.

Corollary 12.1. *Suppose P and Q are polynomials and $f(z) = \frac{P(z)}{Q(z)}$ is a rational function with no poles on the real axis. If $degQ(z) \ge degP(z) + 2$ then*

$$\int_{-\infty}^{\infty} f(x)dx = \begin{cases} 2\pi i\sum_{k=1}^{n} Res(f, z_k); \\ -2\pi i\sum_{k=1}^{m} Res(f, w_k) \end{cases}$$

where $z_1, \cdots, z_n$ are poles of f in the upper half plane and $w_1, \cdots, w_m$ are poles of f in the lower half plane.

Proof. Consider $f(z) = \frac{P(z)}{Q(z)}$. Suppose $\deg P(z) = k$ and $\deg Q(z) = l + k$ with $l \ge 2$. This implies (see Example 7.25) there exist constants C, L_1, L_2 such that

$$C|z|^k \le |P(z)| \le L_1|z|^k$$

for $|z| \ge R$ and

$$|Q(z)| \ge L_2|z|^{k+l}$$

for $|z| \ge R$ $(R \ge 1)$. Then $\left|\frac{P(z)}{Q(z)}\right| \le \frac{L_1}{L_2|z|^l} \le \frac{L_1}{L_2}\cdot\frac{1}{|z|^2} = \frac{M}{|z|^2}$, say for $|z| \ge R$. Thus the result follows from Theorem 12.1 and Theorem 12.2. □

Corollary 12.2. *Let p and q be polynomials such that $degq > degp + 1$. Let C be a circle whose interior contains all of the roots of q. Then $\int_C \frac{p(z)}{q(z)}dz = 0$.*

That is, if p and q are polynomials such that $\deg q > \deg p + 1$. then the sum of the residues of the rational function $\frac{p}{q}$, taken over all of its poles in $\mathbb{C}$, is 0. The proof of this fact follows from the following theorem.

Theorem 12.3. *Let p and q be polynomials with the degree of q at least two more than the degree of p. Then there is an $r > 0$ such that if C is a closed curve outside $|z| = r$, then $\int_C \frac{p(z)}{q(z)}dz = 0$.*

Proof. Let the disc centered at the origin with radius r contain all the zeros of q. Let C_R be a circle centered at the origin with radius $R > r$. Then $\int_C \frac{p(z)}{q(z)}dz = \int_{C_R} \frac{p(z)}{q(z)}dz$ where C is any closed curve outside $|z| = r$. As

$$\frac{p(z)}{q(z)} = O(|z|^{-2}) \ (|z| \to \infty)$$

we have

$$\int_{C_R} \frac{p(z)}{q(z)}dz = O(|z|^{-2})2\pi R = o(1) \ (R \to \infty)$$

and the result follows. □

Example 12.7. *Evaluate the integral* $\displaystyle\int_{-\infty}^{\infty} \frac{\cos x}{x^2 + a^2}dx$.

Solution: The integral $\int_{-\infty}^{\infty} \frac{\cos x}{x^2+a^2}dx$ exists, for $a > 0$. The obvious choice $f(z) = \frac{\cos z}{z^2+a^2}$ is bad, as will be seen below. We consider $f(z) = \frac{e^{iz}}{z^2+a^2}$. Let C_R be as earlier. Then the only pole inside C_R is at $z = ia$, since $a > 0$. The residue $\text{Res}(f, ia) = \frac{e^{-a}}{2ia}$. Proceeding as usual, we see that the real integral $\int_{-\infty}^{\infty} \frac{e^{ix}}{x^2+a^2}dx$ converges to $\frac{\pi e^{-a}}{a}$, provided that we show that $\int_{\gamma_R} f(z)dz \to 0$ as $R \to \infty$. On γ_R, we note that $|e^{iz}| = e^{-y} \le 1$ so that

$$\left| \int_{\gamma_R} \frac{e^{iz}}{z^2 + a^2}dz \right| \le \frac{\pi R}{R^2 - a^2},$$

for $R > a$ which is what we wanted. Note that had we chosen $\cos z$ in place of e^{iz}, we would be in trouble, since $|\cos z|$ becomes large when $\text{Im}z$ is large. Thus $\int_{-\infty}^{\infty} \frac{\cos x}{x^2+a^2}dx + i\int_{-\infty}^{\infty} \frac{\sin x}{x^2+a^2}dx = \frac{\pi e^{-a}}{a}$. Equating real and imaginary parts we obtain $\int_{-\infty}^{\infty} \frac{\cos x}{x^2+a^2}dx = \frac{\pi e^{-a}}{a}$ and $\int_{-\infty}^{\infty} \frac{\sin x}{x^2+a^2}dx = 0$.

Integrals of the form $\int_{-\infty}^{\infty} f(x)e^{iax}dx$, $\int_{-\infty}^{\infty} f(x)\cos ax dx$ etc.

Evaluation of the following important class of improper integrals by means of the calculus of residues is based on the use of the of the Jordan's lemma which says that the integral converges if $f(z) \to 0$ as $|z| \to \infty$ in the upper(lower) half plane. Let us first prove the Jordan's inequality which we shall use in proving the lemma.

Theorem 12.4. *(Jordan's inequality): $\frac{2}{\pi} \le \frac{\sin t}{t} \le 1$, for $0 < t \le \frac{\pi}{2}$.*

Proof. We shall first draw figures of the functions involved. Let $\Phi(t) = \frac{\sin t}{t}$. Then $\Phi'(t) = \frac{\Psi(t)}{t^2}$, with $\Psi(t) = t\cos t - \sin t$. Since $\Psi(0) = 0$ and since $\Psi'(t) =$

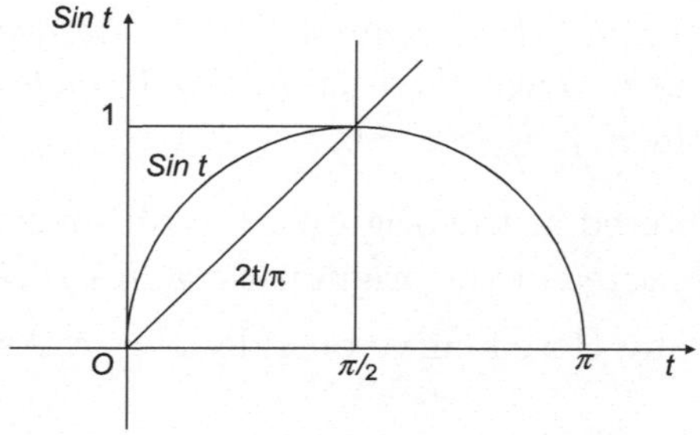

Fig. 12.2: Illustration of Jordan's inequality

$-t\sin t \leq 0$ for $0 < t \leq \frac{\pi}{2}$, we see that Ψ decreases as t increases. This means that $\Psi(t) \leq \Psi(0) = 0$ for $0 < t \leq \frac{\pi}{2}$ from which we get that $\Phi'(t) \leq 0$ for $0 < t \leq \frac{\pi}{2}$. Thus Φ decreases as t increases. Hence, for $0 \leq t \leq \frac{\pi}{2}$, we have $\frac{2}{\pi} \leq \frac{\sin t}{t} \leq 1$. □

Lemma 12.2. *(Jordan's Lemma): Let $\gamma_R(t) = Re^{it}, 0 \leq t \leq \pi$. Assume that (1) f is continuous on γ_R for $R \geq R_0$, (2) $|f(z)| \leq M_R$ on γ_R, where $M_R \to 0$ as $R \to \infty$ and (3) $a > 0$. Then*

$$\left|\int_{\gamma_R} f(z)e^{iaz}dz\right| \leq \frac{\pi}{a}M_R$$

and hence $\int_{\gamma_R} f(z)e^{iaz}dz \to 0$ as $R \to \infty$.

Proof. For the proof observe that

$$\begin{aligned}
\left|\int_{\gamma_R} f(z)e^{iaz}dz\right| &= \left|\int_0^{\pi} f(Re^{it})e^{iaRe^{it}}iRe^{it}dt\right| \\
&\leq \int_0^{\pi} |f(Re^{it})||e^{iaRe^{it}}||iRe^{it}|dt \\
&\leq RM_R\int_0^{\pi} e^{-aR\sin t}dt \\
&= RM_R\left[\int_0^{\frac{\pi}{2}} e^{-aR\sin t}dt + \int_{\frac{\pi}{2}}^{\pi} e^{-aR\sin t}dt\right] \\
&= RM_R\left[\int_0^{\frac{\pi}{2}} e^{-aR\sin t}dt + \int_0^{\frac{\pi}{2}} e^{-aR\sin\theta}d\theta\right] \\
&= 2RM_R\int_0^{\frac{\pi}{2}} e^{-aR\sin t}dt.
\end{aligned}$$

So using Jordan's inequality

$$\begin{aligned}\left|\int_{\gamma_R} f(z)e^{iaz}dz\right| &\leq 2RM_R\int_0^{\frac{\pi}{2}} e^{-aR(\frac{2t}{\pi})}dt \\ &= \frac{-\pi M_R}{a}\left[e^{-aR\frac{2t}{\pi}}\right]_0^{\frac{\pi}{2}} \\ &= \frac{\pi M_R}{a}(1-e^{-aR}) \\ &\leq \frac{\pi M_R}{a}.\end{aligned}$$

□

Remark 12.3. Note estimate theorem gives the bound $M_R \cdot \pi R$ which is not useful in this context.

Most often when we use this version of Jordan's lemma to calculate the value of an improper integral what we get is the Cauchy Principal value of $\int_{-\infty}^{\infty} f(x)dx$. So one needs some standard test of convergence of improper integral to assert the existence of the integral under consideration.

Remark 12.4. If $a < 0$ and the function $f(z)$ satisfies the conditions of Jordan's lemma in the lower half plane Im$z \leq 0$, then also

$$\int_{\gamma_R'} f(z)e^{iaz}dz \to 0$$

as $R \to \infty$ where γ_R' is the semicircular arc $|z| = R$ in the lower half of the z-plane. Similar assertions hold for $a = \pm i\alpha(\alpha > 0)$ as well when integrating, respectively, in the right half plane (Re$z \geq 0$) or in the left half plane (Re$z \leq 0$). The proofs of these statements are carried out in an exactly similar manner, so we leave them to the reader.

Remark 12.5. (i) The following form of Jordan's lemma, which refers to integration in the right half plane is worth noting:

$$\lim_{R\to\infty}\int_{\gamma_R} e^{-az}f(z)dz = 0, a > 0$$

where γ_R is the semicircular arc $|z| = R$ in the right half plane Re$z \geq 0$.

(ii) Jordan's lemma holds true for the case when the function $f(z)$ satisfies the above formulated conditions in the half plane Im$z \geq y_0$ where y_0 is a fixed number which may be either positive or negative, and the integration is carried out along the semicircular arc $|z - iy_0| = R$ in the half plane Im$z \geq y_0$. The proof is similar to the previous proof. The Jordan's lemma also holds true under considerably weaker restrictions imposed on the function f than in the way it is stated above.

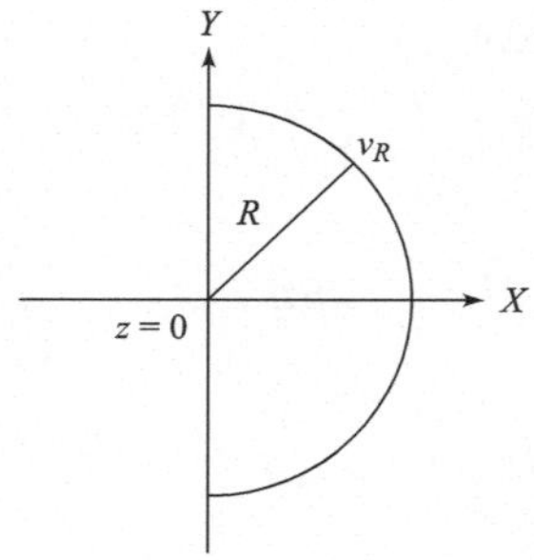

Fig. 12.3: The semicircular arc $|z| = R$ in the right half plane

We proved Lemma 12.1 and Lemma 12.2 on the assumption that the function f has only a finite number of singularities in the upper half plane. However, as will readily be seen, a slight change in the statements of these lemmas makes them hold true for the case of an infinite number of isolated singularities of the function f. Let us require that there be a sequence of numbers $R_n \to \infty$ as $n \to \infty$, such that on the semicircular arcs γ_{R_n} in the upper half plane the conditions of the lemmas are fulfilled (i.e., $|f(z)| \leq M_{R_n}$ on $|z| = R_n$ where $M_{R_n} \to 0$ as $R_n \to \infty$). Then the assertions of Lemma 12.1 and Lemma 12.2 will hold true provided that the limit process in the integrals under consideration occurs with respect to the sequence of arcs γ_{R_n} as $n \to \infty$. It is also clear that if the appropriate integrals exist we can extend the integration methods at hand to the case of functions with an infinite number of isolated singularities.

Theorem 12.5. *Let f be holomorphic on $\mathbb{C}$ except at a finite number of poles none of which are real. If there exist positive numbers R_0 and M such that $|f(z)| \leq \frac{M}{|z|^\alpha}$ for $|z| \geq R_0$ and $\alpha > 0$, then*

$$\int_{-\infty}^{\infty} f(x)e^{iax}dx = \begin{cases} 2\pi i \sum_{k=1}^{n} Res(g, z_k) & if \;\; a > 0; \\ -2\pi i \sum_{j=1}^{m} Res(g, b_j) & if \;\; a < 0 \end{cases}$$

where $g(z) = f(z)e^{iaz}$ and $z_1, z_2, \cdots, z_n$ are poles of g in the upper half plane and $b_1, b_2, \cdots, b_m$ are the poles of g in the lower half plane.

Proof. Suppose $a > 0$. By hypothesis, the singular points $z_1, z_2, \cdots, z_n$ of the function f satisfy the condition $|z_k| < R_0, k = 1, 2, \cdots, n$ in the upper half plane. Consider a closed path $C_R = [-R, R] \cup \gamma_R$ where $\gamma_R = \{Re^{it} : 0 \leq t \leq \pi\}, R > R_0$ in the upper half plane with direction chosen in the positive sense (counter-clockwise orientation). By the Residue theorem

$$2\pi i \sum_{k=1}^{n} \text{Res}(g, z_k) = \int_{C_R} g(z)dz = \int_{-R}^{R} e^{iax} f(x)dx + \int_{\gamma_R} g(z)dz.$$

By Jordan's lemma, $\int_{\gamma_R} g(z)dz \to 0$ as $R \to \infty$. We can make a direct proof as follows:

$$\begin{aligned}
\left|\int_{\gamma_R} g(z)dz\right| &= \left|\int_{\gamma_R} e^{iaz} f(z)dz\right| \\
&= \left|\int_0^{\pi} e^{iaRe^{i\theta}} f(Re^{i\theta}) iRe^{i\theta} d\theta\right| \\
&\leq \frac{M}{R^{\alpha-1}} \int_0^{\pi} e^{-aR\sin\theta} d\theta \\
&= \frac{2M}{R^{\alpha-1}} \int_0^{\frac{\pi}{2}} e^{-aR\sin\theta} d\theta \\
&\leq \frac{2M}{R^{\alpha-1}} \int_0^{\frac{\pi}{2}} e^{-aR\frac{2\theta}{\pi}} d\theta \\
&= \frac{\pi M}{aR^{\alpha}}(1 - e^{-aR}).
\end{aligned}$$

Since $a, \alpha > 0$, $\left|\int_{\gamma_R} g(z)dz\right| \to 0$ as $R \to \infty$ and the result follows. For the case $a < 0$, we consider the curve in the lower half plane and proceed similarly. The minus sign appears because of the counter-clockwise orientation. □

Corollary 12.3. *Let P and Q be polynomials such that $degQ \geq 1 + degP$ and let $f(z) = \frac{P(z)}{Q(z)}$. Suppose Q has no zeros on the real axis and $f(\mathbb{R}) \subset \mathbb{R}$. If $a > 0$, then*

$$\int_{-\infty}^{\infty} \frac{P(x)}{Q(x)} e^{iax} dx = 2\pi i \sum_{j=1}^{k} Res(g, z_j)$$

where $g(z) = f(z)e^{iaz}$ and $z_1, z_2, \cdots, z_k$ are poles of f in the upper half plane.

Proof. Note deg$Q \geq 1 +$ deg$P$ if and only if there exist positive constants R_0 and M such that $|f(z)| \leq \frac{M}{|z|}$ for $|z| \geq R_0$. The proof then follows from the theorem. □

Alternative proof of Theorem 12.5:

Proof. Suppose $a > 0$. Consider the rectangle C with vertices $-R, S, S + iT$ and $-R + iT$. So $C = \gamma_1 + \gamma_2 + \gamma_3 + \gamma_4$. Choose R and S large enough so

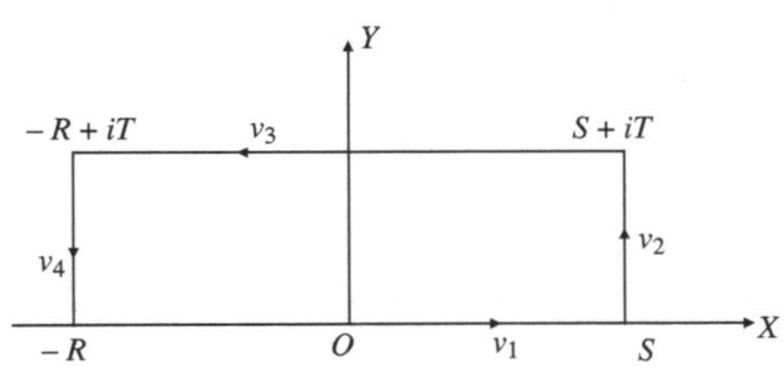

Fig. 12.4: The rectangle C with vertices $-R, S, S + iT$ and $-R + iT$

that all the singularities of f in the upper half plane lie in the interior of the rectangle C. By residue theorem,

$$\int_C g(z)dz = 2\pi i \sum_{k=1}^{n} \text{Res}(g, z_k). \tag{12.6}$$

But

$$\int_C g(z)dz = \sum_{k=1}^{4} \int_{\gamma_k} f(z)e^{iaz}dz. \tag{12.7}$$

Let $I_k = \int_{\gamma_k} f(z)e^{iaz}dz, k = 1, 2, 3, 4$. Now,

$$\begin{aligned}
|I_2| &= \left|\int_{\gamma_2} e^{iaz}g(z)dz\right| \\
&= \left|\int_0^T e^{ia(S+iy)}f(S+iy)dy\right| \\
&\leq \int_0^T e^{-ay}\,|f(S+iy)|\,dy \\
&\leq \max_{z\in\gamma_2}|f(z)| \int_0^T e^{-ay}dy \\
&= \max_{z\in\gamma_2}|f(z)| \left(\frac{1-e^{-aT}}{a}\right) \\
&\leq \tfrac{1}{a}\left\{\max_{z\in\gamma_2}|f(z)|\right\}.
\end{aligned}$$

Similarly,

$$\begin{aligned}
|I_3| &= \left|\int_S^{-R} e^{ia(x+iT)}f(x+iT)dx\right| \\
&\leq \max_{z\in\gamma_3}|f(z)| \int_S^{-R} e^{-aT}dx \\
&= -\max_{z\in\gamma_3}|f(z)|\,e^{-aT}(R+S).
\end{aligned}$$

Further it is easy to show that

$$|I_4| \leq \frac{1}{a}\left\{\max_{z\in\gamma_4}|f(z)|\right\}.$$

By hypothesis, $\lim_{z\to\infty} f(z) = 0$, that is, $\lim_{R,S\to\infty} \max_{z\in\gamma_j}|f(z)| = 0$ for $j = 2, 3, 4$. Thus $\int_{-\infty}^{\infty} f(x)e^{iax}dx = 2\pi i \sum_{k=1}^{n} \text{Res}(g, z_k)$. For the case $a < 0$, consider a rectangular curve in the lower half plane with counter clockwise orientation. □

Note that in the alternative method here we actually get the value of the integral but in the previous method we get only the principal value of the integral.

Example 12.8. *Evaluate* $\displaystyle\int_{-\infty}^{\infty} \frac{cosx}{x^2-2x+2}dx$ *using calculus of residue.*

Solution: Note that the integral converges. So we shall find the principal value of the integral. Let us consider $I = \int \frac{e^{iz}}{z^2-2z+2}dz$, taken round the contour $C_R = [-R, R] \cup \gamma_R$ where $\gamma_R = \{Re^{it} : 0 \le t \le \pi\}$. The function $f(z) = \frac{e^{iz}}{z^2-2z+2}$ has two poles namely $z = 1+i$ and $z = 1-i$. Take $R > \sqrt{2}$, so the pole $z = 1+i$ of f lie in the interior of C_R. Hence $\int_{C_R} \frac{e^{iz}}{z^2-2z+2}dz = 2\pi i \text{Res}\left(\frac{e^{iz}}{z^2-2z+2}, 1+i\right) = 2\pi i \frac{e^{i(1+i)}}{2(1+i)-2} = \frac{\pi}{e}(\cos 1 + i\sin 1)$. On γ_R, $|e^{iz}| = |e^{ix-y}| = e^{-y} \le 1$ since $y \ge 0$. Further, $|z^2 - 2z + 2| = |z^2 - (2z-2)| \ge |z|^2 - |2z-2| \ge |z|^2 - (|2z| + 2) = R^2 - 2R - 2$. So $|\int_{\gamma_R} \frac{e^{iz}}{z^2-2z+2}dz| \le \int_{\gamma_R} \frac{|e^{iz}|}{|z^2-2z+2|}|dz| \le \int_0^\pi \frac{1}{R^2-2R-2}Rd\theta = \frac{\pi R}{R^2-2R-2} \to 0$ as $R \to \infty$. Thus $\frac{\pi}{e}(\cos 1 + i\sin 1) = \lim_{R\to\infty} \int_{C_R} f(z)dz = \lim_{R\to\infty}\left[\int_{-R}^{R} f(x)dx + \int_{\gamma_R} f(z)dz\right] = \lim_{R\to\infty}\int_{-R}^{R} f(x)dx = \int_{-\infty}^{\infty} \frac{\cos x}{x^2-2x+2}dx + i\int_{-\infty}^{\infty} \frac{\sin x}{x^2-2x+2}dx$. Hence

$$\int_{-\infty}^{\infty} \frac{\cos x}{x^2-2x+2}dx = \frac{\pi}{e}\cos 1$$

and

$$\int_{-\infty}^{\infty} \frac{\sin x}{x^2-2x+2}dx = \frac{\pi}{e}\sin 1.$$

Example 12.9. *Find P.V.* $\displaystyle\int_{-\infty}^{\infty} \frac{x^3 sinx}{1+x^4}dx.$

Solution: Let $I = \int_{C_R} \frac{z^3e^{iz}}{1+z^4}dz$ where $C_R = [-R, R] \cup \gamma_R$ and $\gamma_R = \{Re^{it} : 0 \le t \le \pi\}$. The function $f(z) = \frac{z^3e^{iz}}{1+z^4}$ has poles at $z = \frac{\pm1\pm i}{\sqrt{2}}$. But only $\frac{\pm1+i}{\sqrt{2}}$ are inside C_R if $R > 1$. Hence

$$\begin{aligned}\int_{C_R} f(z)dz &= 2\pi i\left[\text{Res}\left(\frac{z^3e^{iz}}{1+z^4}, \frac{1+i}{\sqrt{2}}\right) + \text{Res}\left(\frac{z^3e^{iz}}{1+z^4}, \frac{-1+i}{\sqrt{2}}\right)\right] \\ &= 2\pi i\left[\frac{1}{4}e^{\frac{i(1+i)}{\sqrt{2}}} + \frac{1}{4}e^{\frac{i(-1+i)}{\sqrt{2}}}\right] \\ &= i\pi e^{-\frac{1}{\sqrt{2}}}\cos\left(\frac{1}{\sqrt{2}}\right).\end{aligned}$$

On γ_R, $\left|\frac{z^3}{1+z^4}\right| \le \frac{|z|^3}{|z|^4-1} = \frac{R^3}{R^4-1}$. Thus $\left|\frac{z^3}{1+z^4}\right| = O\left(\frac{1}{R}\right)$. By Jordan's lemma, $\left|\int_{\gamma_R} \frac{z^3e^{iz}}{1+z^4}dz\right| = O\left(\frac{1}{R}\right) \to 0$ as $R \to \infty$. Hence $\int_{C_R} \frac{z^3e^{iz}}{1+z^4}dz = \int_{-R}^{R} \frac{x^3e^{ix}}{1+x^4}dx + \int_{\gamma_R} \frac{z^3e^{iz}}{1+z^4}dz$. Thus

$$\lim_{R\to\infty} \int_{-R}^{R} \frac{x^3e^{ix}}{1+x^4}dx = i\pi e^{-\frac{1}{\sqrt{2}}}\cos\left(\frac{1}{\sqrt{2}}\right).$$

Therefore,

$$P.V. \int_{-\infty}^{\infty} \frac{x^3 \sin x}{1+x^4} dx = \pi e^{-\frac{1}{\sqrt{2}}} \cos\left(\frac{1}{\sqrt{2}}\right).$$

Note that in this case we have only calculated the Cauchy principal value using Jordan's lemma. Since $\left|\frac{z^3}{1+z^4}\right| = O\left(\frac{1}{R}\right)$ hence $\int_{-\infty}^{\infty} \frac{x^3 \sin x}{1+x^4} dx$ exists by Theorem 12.5.

Example 12.10. *We show that* $\displaystyle\int_{-\infty}^{\infty} \frac{x \sin 3x}{x^2+9} dx = \frac{\pi}{2e^9}$.

Solution: Consider $f(z) = \frac{ze^{3iz}}{z^2+9}$. Then f has a simple pole at $z = 3i$ in the upper half plane. Its residue there is $\text{Res}(f, 3i) = \lim_{z\to 3i}(z-3i)\frac{ze^{3iz}}{z^2+9} = \frac{3ie^{-9}}{6i} = \frac{1}{2e^9}$. Hence $\int_{C_R} f(z)dz = 2\pi i \frac{1}{2e^9} = \frac{\pi i}{e^9}$ where C_R is as defined before. Further, by Jordan's lemma, $\int_{\gamma_R} \frac{z}{z^2+9} e^{3iz} dz \to 0$ as $R \to \infty$ since $|\frac{z}{z^2+9}| \leq \frac{|z|}{|z|^2-9} = \frac{R}{R^2-9}$ on γ_R. The result follows from the theorem on separating the real and imaginary parts.

Integrals of the form $\int_{-\infty}^{\infty} f(x)dx$ which have singularities on the real line

Previously, we assumed that the function f does not have singularities on the real axis. However, slight supplementary considerations enable one to use these theorems to evaluate improper integrals even when the function has singularities on the real axis.
To evaluate integrals of this type we shall use the following result.

Theorem 12.6. *Assume that f has a simple pole at p. Let $\gamma_r(t) = p + re^{it}$ $(\theta_1 \leq t \leq \theta_2)$ with counter-clockwise orientation. Then* $\lim_{r\to 0} \int_{\gamma_r} f(z)dz = i(\theta_2 - \theta_1)$ *Res*(f, p).

Proof. Write $f(z) = \frac{a_{-1}}{z} + g(z)$ where g is holomorphic in a ball around p, say and hence bounded in a closed ball $\overline{B(p, \epsilon)}$. Then $|g|$ is bounded on this compact set, say by M. Thus for $0 < r < \epsilon$, we have $\int_{\gamma_r} f(z)dz = \int_{\gamma_r} \frac{a_{-1}}{z-p} dz + \int_{\gamma_r} g(z)dz$. Now $\int_{\gamma_r} \frac{a_{-1}}{z-p} dz = a_{-1} \int_{\theta_1}^{\theta_2} \frac{1}{re^{it}} ire^{it} dt = a_{-1}(\theta_2 - \theta_1)i = i(\theta_2 - \theta_1)\text{Res}(f, p)$. But $|\int_{\gamma_r} g(z)dz| \leq \int_{\gamma_r} |g(z)||dz| = \sup_{z\in\gamma_r}|g(z)| \int_{\gamma_r} |dz| = \sup_{z\in\gamma_r}|g(z)| \int_{\theta_1}^{\theta_2} r d\theta = \sup_{z\in\gamma_r}|g(z)| \cdot (\theta_2 - \theta_1)r \leq M(\theta_2 - \theta_1)r$ which goes to 0 as $r \to 0$. Thus $\lim_{r\to 0} \int_{\gamma_r} f(z)dz = i(\theta_2 - \theta_1)\text{Res}(f, p)$. □

Let us illustrate this by a simple example.

Example 12.11. *Evaluate* $\displaystyle\int_{0}^{\infty} \frac{\sin x}{x} dx$.

Solution: This is an improper integral. It is not obvious that the integral converges. The integrand is continuous and as such it has no singularity anywhere. But although the integrand is well-behaved at the left end point of the interval of integration, there is trouble at the far end, because $\int_0^\infty \left|\frac{\sin x}{x}\right| dx$ diverges. We shall show below that the given integral does converge on the basis of Cauchy's theorem.

We consider the function $f(z) = \frac{e^{iz}}{z}$. The function f has a simple pole at $z = 0$. Consider the rectangular path C as in Theorem 12.5 with vertices $-R, S, S+iT, -R+iT$ (here $R > 0$ and $S > 0$) with a slight detour near the pole $z = 0$. More precisely, we replace the bottom side of the rectangle C

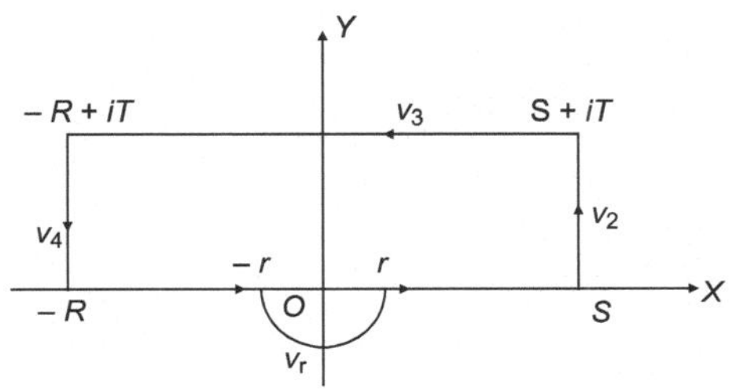

Fig. 12.5: Rectangular path with vertices $-R, S, S+iT, -R+iT$ with a detour near $z = 0$

by the path γ which is made up of the following: the line segment $[-R, -r]$, (for $0 < r < R, 0 < r < S$) followed by the semicircular arc (in the lower half plane)$\gamma_r(t) = re^{it}, \pi \le t \le 2\pi$, the line segment $[r, S]$. Let the new closed path be denoted by σ_R. Proceeding as in Theorem 12.5 we get: $2\pi i = 2\pi i \cdot \text{Res}(f, 0) = \int_{\sigma_R} f(z)dz = \int_\gamma f(z)dz + \int_{\gamma_2} f(z)dz + \int_{\gamma_3} f(z)dz + \int_{\gamma_4} f(z)dz$. By Theorem 12.6,

$$\lim_{r\to 0} \int_{\gamma_r} f(z)dz = \pi i \text{Res}(f, 0) = \pi i.$$

We have already shown in the proof of Theorem 12.5 that as $R, S \to \infty$, $\displaystyle\int_{\gamma_k} f(z)dz \to 0$ for $k = 2, 3, 4$. Let $S = R$. Again $\displaystyle\int_\gamma f(z)dz = \int_{-R}^{-r} f(x)dx + \int_{\gamma_r} f(z)dz + \int_r^R f(x)dx$. Hence

$$\begin{aligned} 2\pi i &= \left[\int_\gamma f(z)dz + \sum_{k=2}^{4} \int_{\gamma_k} f(z)dz\right] \\ &= \int_{-R}^{-r} f(x)dx + \int_{\gamma_r} f(z)dz + \int_r^R f(x)dx + \sum_{k=2}^{4} \int_{\gamma_k} f(z)dz. \end{aligned}$$

Thus $2\pi i - \int_{\gamma_r} f(z)dz = \int_{-R}^{-r} f(x)dx + \int_r^R f(x)dx + \sum_{k=2}^4 \int_{\gamma_k} f(z)dz$. It follows therefore that

$$\pi i = \lim_{\substack{R\to\infty \\ r\to 0}} \left\{ \int_{-R}^{-r} f(x)dx + \int_{r}^{R} f(x)dx \right\}.$$

Note that

$$\begin{aligned} \int_r^R \frac{\sin x}{x} dx &= \tfrac{1}{2i} \int_r^R \frac{e^{ix} - e^{-ix}}{x} dx \\ &= \frac{1}{2i} \int_r^R \frac{e^{ix}}{x} dx + \frac{1}{2i} \int_{-R}^{-r} \frac{e^{ix}}{x} dx. \end{aligned}$$

Hence

$$\int_0^\infty \frac{\sin x}{x} dx = \frac{1}{2i} \pi i = \frac{\pi}{2}.$$

To find the principal value of $\int_0^\infty \frac{\sin x}{x} dx$ we can also consider the following path. For $r > 0$, let γ_r denote the semicircle in the upper half plane with center 0 and radius r, oriented counterclockwise. For $0 < r < R$, let $\Gamma_{r,R}$ denote the closed curve consisting of the interval $[r, R]$ followed by the semicircle γ_R followed by the interval $[-R, -r]$ followed by the semicircle $-\gamma_r$(see Fig.12.6). The

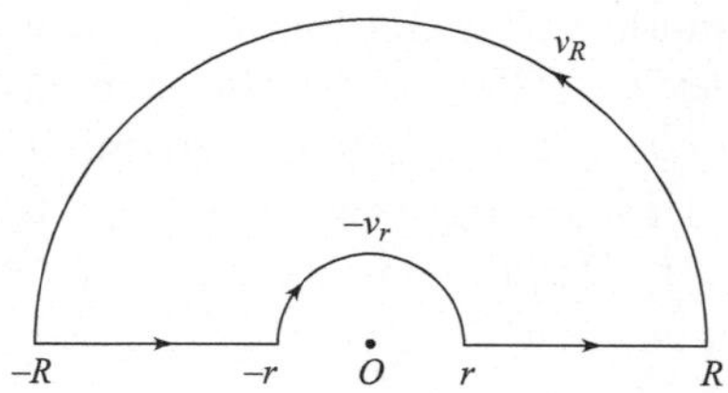

Fig. 12.6: $\Gamma_{r,R} = [r, R] + \gamma_R + [-R, -r] + \{-\gamma_r\}$

function $\frac{e^{iz}}{z}$, whose imaginary part on the real axis equals $\frac{\sin x}{x}$, is holomorphic in $\mathbb{C} \setminus \{0\}$, a domain containing $\Gamma_{r,R}$ and its interior. By Cauchy's theorem, $\displaystyle\int_{\Gamma_{r,R}} \frac{e^{iz}}{z} dz = 0$. That is,

$$\int_{-R}^{-r} f(x)dx + \int_{-\gamma_r} f(z)dz + \int_r^R f(x)dx + \int_{\gamma_R} f(z)dz = 0. \tag{12.8}$$

Since $\text{Res}(f, 0) = \lim_{z\to 0} zf(z) = 1$, we note that

$$\lim_{r\to 0} \int_{-\gamma_r} f(z)dz = i(0 - \pi)\text{Res}(f, 0) = -\pi i. \tag{12.9}$$

Further by Jordan's inequality, we observe that

$$\begin{aligned}\left|\int_{\gamma_R} f(z)dz\right| &\leq \int_0^{\pi} e^{-R\sin\theta} d\theta \\ &= 2\int_0^{\frac{\pi}{2}} e^{-R\sin\theta} d\theta \\ &\leq 2\int_0^{\frac{\pi}{2}} e^{-R\frac{2\theta}{\pi}} d\theta \\ &= \frac{\pi}{R}\left(1-e^{-R}\right).\end{aligned}$$

Hence

$$\lim_{R\to\infty}\int_{\gamma_R} f(z)dz = 0. \tag{12.10}$$

So if we allow $R \to \infty$ and $r \to 0$ in (12.8), by (12.9) and (12.10) we obtain $\lim\limits_{\substack{R\to\infty \\ r\to 0}}\left\{\int_r^R \frac{e^{ix}}{x}dx + \int_{-R}^{-r}\frac{e^{ix}}{x}dx\right\} = \pi i$. Thus $\int_0^{\infty}\frac{\sin x}{x}dx = \frac{\pi}{2}$.

Trigonometric integrals of the form $\int_{\alpha}^{2\pi+\alpha} f(\cos\theta, \sin\theta)d\theta$

We have already seen examples of this kind in Example 12.3 and Example 12.4.

Example 12.12. *Evaluate* $\int_0^{\pi} sin^{2n}t dt.$

Solution: Note

$$\int_0^{\pi}\sin^{2n}t dt = \frac{1}{2}\int_0^{2\pi}\sin^{2n}t dt$$

and

$$\begin{aligned}\int_0^{2\pi}\sin^{2n}t dt &= \int_{|z|=1}\left(\frac{z-\frac{1}{z}}{2i}\right)^{2n}\frac{dz}{iz} \\ &= \frac{(-i)}{(2i)^{2n}}\int_{|z|=1}\frac{1}{z}\left(z-\frac{1}{z}\right)^{2n}dz \\ &= \frac{(-i)}{4^n(-1)^n}\int_{|z|=1} f(z)dz\end{aligned}$$

where $f(z) = \frac{1}{z}\left(z-\frac{1}{z}\right)^{2n} = \frac{(z^2-1)^{2n}}{z^{2n+1}}$. So the function has a pole of order $2n+1$

at $z = 0$. Now, by binomial theorem,

$$\begin{aligned} f(z) &= \frac{1}{z}\sum_{k=0}^{2n}\binom{2n}{k}z^k\left(-\frac{1}{z}\right)^{2n-k} \\ &= \sum_{k=0}^{n}\binom{2n}{k}z^{k-1}(-1)^{2n-k}z^{k-2n} \\ &= \sum_{k=0}^{n}\binom{2n}{k}(-1)^{2n-k}z^{2k-2n-1}. \end{aligned}$$

Hence $\text{Res}(f,0) = \binom{2n}{n}(-1)^n$. Thus

$$\begin{aligned} \int_0^{\pi}\sin^{2n}t dt &= \tfrac{1}{2}\int_0^{2\pi}\sin^{2n}t dt \\ &= \frac{1}{2}\frac{(-i)}{4^n(-1)^n}2\pi i \text{Res}(f,0) \\ &= \frac{\pi}{4^n}\frac{(2n)!}{(n!)^2}. \end{aligned}$$

Example 12.13. *We show that* $\displaystyle\int_0^{\infty}\frac{x^{m-1}}{1+x^n}dx = \frac{\pi}{n}\left(\frac{1}{sin\frac{m\pi}{n}}\right)$ *for integers* $0 < m < n$.

Solution: Note that the integrals exist. We consider $f(z) = \frac{z^{m-1}}{1+z^n}$. The path under consideration is the sector C_R (see Fig.12.7) formed of the line segment $[0,R]$, followed by the arc $\gamma_R(t) = Re^{it}, 0 \le t \le \frac{2\pi}{n}$ and the line segment from $Re^{\frac{2\pi i}{n}}$ to 0. The point $p = e^{\frac{i\pi}{n}}$ is only one pole inside this closed path. We

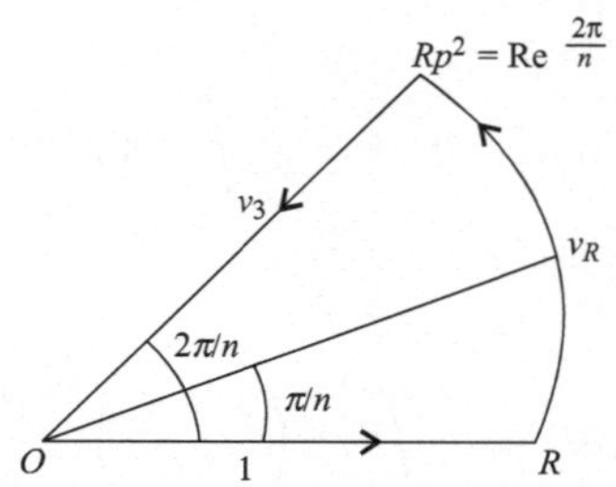

Fig. 12.7: The sector C_R

have $\text{Res}(f,p) = \frac{-p^m}{n}$. It is easy to see that the integral over the arc goes to zero as $R \to \infty$. We compute the integral over the line segment $[Rp^2, 0]$ as follows. Let $\gamma_3(t) = tp^2, 0 \le t \le R$. Then $\int_{[Rp^2,0]} f(z)dz = -\int_{\gamma_3} f(z)dz$ $= -\int_0^R \frac{t^{m-1}p^{2m-2}}{1+t^np^{2n}}p^2dt = -p^{2m}\int_0^R f(x)dx$, that is, a multiple of the original integral. Thus $\int_{C_R} f(z)dz = \int_0^R f(x)dx + \int_{\gamma_R} f(z)dz + \int_{\gamma_3} f(z)dz$. Hence

$2\pi i\left(\frac{-p^m}{n}\right) = \int_0^\infty f(x)dx - p^{2m}\int_0^\infty f(x)dx = (1-p^{2m})\int_0^\infty f(x)dx$. It follows therefore that

$$\begin{aligned}\int_0^\infty f(x)dx &= \tfrac{-2\pi i}{n}\left(\tfrac{p^m}{1-p^{2m}}\right) = \tfrac{2\pi i}{n}\left(\tfrac{p^m}{p^{2m}-1}\right)\\ &= \tfrac{2\pi i}{n}\left(\tfrac{1}{p^m-p^{-m}}\right) = \tfrac{2\pi i}{n}\left(\frac{1}{2i\left(\frac{e^{\frac{im\pi}{n}}-e^{\frac{-im\pi}{n}}}{2i}\right)}\right)\\ &= \tfrac{\pi}{n}\left(\tfrac{1}{\sin\frac{m\pi}{n}}\right).\end{aligned}$$

Example 12.14. *Evaluate* $\displaystyle\int_0^\infty \frac{1}{1+x^a}dx, a>1.$

Solution: By means of the comparison test one can see that the integral converges. Let $G=\{re^{i\theta} : r>0, \left|\theta-\frac{\pi}{a}\right|<\pi\}$. We consider the branch of the function z^a that takes the value 1 at the point $z=1$ in the domain G. We also denote this function simply by z^a. The function $f(z)=\frac{1}{1+z^a}$ is holomorphic in G except for a simple pole at the point $p=e^{\frac{\pi i}{a}}$. For $r>0$, let A_r denote the circular arc $\{re^{i\theta} : 0<\theta<\frac{2\pi}{a}\}$, oriented counterclockwise. For $0<\epsilon<1$, let $\Gamma_{\epsilon,R}$ denote the closed path consisting of the interval $[\epsilon,R]$ followed by the arc A_R followed by the segment $\left[Re^{\frac{2\pi i}{a}}, \epsilon e^{\frac{2\pi i}{a}}\right]$ followed by the arc $-A_\epsilon$ (see Fig. 12.8). By the residue theorem,

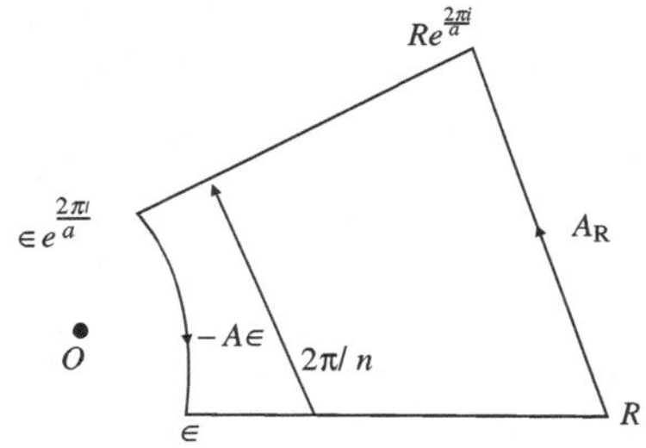

Fig. 12.8: The closed path $\Gamma_{\epsilon,R}$

$$\begin{aligned}\int_{\Gamma_{\epsilon,R}}\frac{1}{1+z^a}dz &= 2\pi i\text{Res}\left(\frac{1}{1+z^a},p\right)\\ &= \frac{2\pi i}{ap^{a-1}}\\ &= \frac{2\pi i e^{-\frac{i\pi(a-1)}{a}}}{a}\\ &= \frac{-2\pi i e^{\frac{i\pi}{a}}}{a}.\end{aligned}$$

The integral of $\frac{1}{1+z^a}$ over the interval $[\epsilon,R]$ equals $\displaystyle\int_\epsilon^R\frac{1}{1+x^a}dx$. By using the parametrization $t \mapsto e^{\frac{2\pi i}{a}}t(\epsilon \le t \le R)$, the integral of the function $f(z)$ along the segment $\left[\epsilon e^{\frac{2\pi i}{a}}, Re^{\frac{2\pi i}{a}}\right]$ equals $e^{\frac{2\pi i}{a}} \int_\epsilon^R \frac{1}{1+x^a}dx$. Further

$$\begin{aligned}\left|\int_{-A_\epsilon} f(z)dz\right| &= \left|\int_0^{\frac{2\pi}{a}} f\left(\epsilon e^{i\theta}\right)\epsilon e^{i\theta} i d\theta\right| \\ &= \left|\int_0^{\frac{2\pi}{a}} \frac{1}{1+\epsilon^a e^{ia\theta}}\epsilon e^{i\theta} i d\theta\right| \\ &\le \int_0^{\frac{2\pi}{a}} \frac{\epsilon}{\epsilon^a - 1} d\theta \\ &= \frac{\epsilon}{\epsilon^a - 1}\frac{2\pi}{a} \\ &< \frac{1}{\epsilon^a - 1}\frac{2\pi}{a} \to 0\end{aligned}$$

as $\epsilon \to 0$. Similarly when we consider the integral of the function f over A_R, the integrand remains bounded in absolute value by $\frac{1}{R^a-1}$, so the integral itself is bounded in absolute value by $\frac{2\pi R}{a(R^a-1)}$, which tends to a as $R \to \infty$. Thus we obtain

$$\lim_{\substack{\epsilon\to 0\\ R\to\infty}} \left[\int_\epsilon^R \frac{1}{1+x^a}dx - e^{\frac{2\pi i}{a}} \int_\epsilon^R \frac{1}{1+x^a}dx + \int_{-A_\epsilon} \frac{1}{1+z^a}dz\right.$$

$$\left. + \int_{A_R} \frac{1}{1+z^a}dz\right] = \frac{-2\pi i e^{\frac{\pi i}{a}}}{a}.$$

Hence

$$\left(1 - e^{\frac{2\pi i}{a}}\right) \int_0^\infty \frac{1}{1+x^a}dx = \frac{-2\pi i e^{\frac{\pi i}{a}}}{a}.$$

Therefore

$$\int_0^\infty \frac{1}{1+x^a}dx = \frac{\pi}{a\sin\frac{\pi}{a}}.$$

In all the preceding examples we dealt with only single-valued holomorphic function. In these cases when the holomorphic function f is multiple-valued in the extended complex plane, the contour of integration must be chosen so that there are no branch points of the function $f(z)$ inside it, and we consider only the single-valued branch $f(z)$ of the holomorphic function f. This reasoning enables us to extend the foregoing methods to a number of improper integrals that are frequently encountered in applications. We consider a few typical cases.

Integrals of the form $\int_0^\infty x^{a-1}\Theta(x)dx, 0 < a < 1$.

Here Θ is a rational function of x with no poles on the real axis. We can evaluate such integrals by contour integration. Since z^{a-1} is a multiple-valued function, it is necessary to choose a suitable domain where a single-valued, analytic branch of the function can be defined. That is, to evaluate such integrals we must consider a branch of the function z^{a-1}. The point $z = 0$ is a branch point of z^{a-1}, and the process used to evaluate this integral is sometimes called integration around a branch point.

Example 12.15. *Show that* $\displaystyle\int_0^\infty \frac{x^{a-1}}{1+x}dx = \frac{\pi}{\sin a\pi}$ *if* $0 < a < 1$.

Solution: Let $G = \{z \in \mathbb{C} : z \neq 0 \text{ and } 0 < \arg z < 2\pi\}$, define a branch of the logarithm on G by putting $l(re^{i\theta}) = \log r + i\theta$ where $0 < \theta < 2\pi$. For z in G, put $f(z) = \exp((a-1)l(z))$; so f is a branch of z^{a-1}. We now select an appropriate curve (see Fig.12.9) γ in G. Let $0 < r < 1 < R$ and let $\delta > 0$. Let L_1 be the line segment $[r + \delta i, R + \delta i]$; γ_R the part of the circle $|z| = R$ from $R + \delta i$ counterclockwise to $R - \delta i$; L_2 the line segment $[R - \delta i, r - \delta i]$; and γ_r the part of the circle $|z| = r$ from $r - \delta i$ clockwise to $r + \delta i$. Put $\gamma = L_1 + \gamma_R + L_2 + \gamma_r$. Now

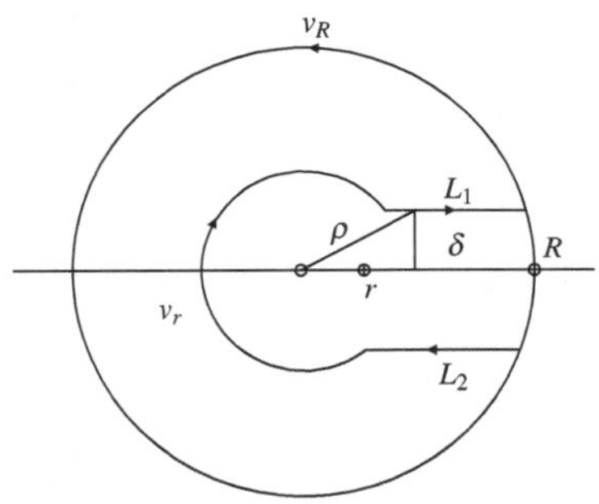

Fig. 12.9: The curve $\gamma = L_1 + \gamma_R + L_2 + \gamma_r$

$$\text{Res}\left(\frac{f(z)}{1+z}; -1\right) = e^{i\pi(a-1)},$$

the Residue theorem gives

$$\int_\gamma \frac{f(z)}{1+z}dz = 2\pi i e^{i\pi(a-1)} = -2\pi i e^{i\pi a}. \tag{12.11}$$

Using the definition of a line integral

$$\int_{L_1} \frac{f(z)}{1+z}dz = \int_r^R \frac{f(x+i\delta)}{1+x+i\delta}dx.$$

Let $g(x, \delta)$ be defined on the compact set $[r, R] \times [0, \frac{\pi}{2}]$ by

$$g(x, \delta) = \left|\frac{f(x+i\delta)}{1+x+i\delta} - \frac{x^{a-1}}{1+x}\right|$$

when $\delta > 0$ and $g(x,0) \equiv 0$. Then g is continuous and hence uniformly continuous. If $\epsilon > 0$, then there is a δ_0 such that if $(x-x')^2+(\delta-\delta')^2 < \delta_0^2$ then $|g(x,\delta)-g(x',\delta')| < \frac{\epsilon}{R}$. In particular, $g(x,\delta) < \frac{\epsilon}{R}$ when $r \le x \le R$ and $\delta < \delta_0$. Thus $\int_r^R g(x,\delta)dx \le \epsilon$ for $\delta < \delta_0$. Hence it follows that

$$\int_r^R \frac{x^{a-1}}{1+x}dx = \lim_{\delta\to 0^+}\int_{L_1}\frac{f(z)}{1+z}dz. \tag{12.12}$$

Similarly, using the fact that $l(\overline{z}) = \overline{l(z)} + 2\pi i$,

$$-e^{2\pi i(a-1)}\int_r^R \frac{x^{a-1}}{1+x}dx = \lim_{\delta\to 0^+}\int_{L_2}\frac{f(z)}{1+z}dz. \tag{12.13}$$

Now the value of the integral (12.11) does not depend on δ. Therefore, letting $\delta \to 0^+$ and using (12.12) and (12.13) gives

$$\begin{aligned}-2\pi i e^{i\pi a} &= -(1-e^{2\pi i a})\int_r^R \frac{x^{a-1}}{1+x}dx \\ &= \lim_{\delta\to 0^+}\left[\int_{\gamma_R}\frac{f(z)}{1+z}dz + \int_{\gamma_r}\frac{f(z)}{1+z}dz\right]. \end{aligned} \tag{12.14}$$

Now if $\rho > 0$ and $\rho \ne 1$ and if γ_ρ is the part of the circle $|z| = \rho$ from $\sqrt{\rho^2-\delta^2}+i\delta$ to $\sqrt{\rho^2-\delta^2}-i\delta$ then

$$\left|\int_{\gamma_\rho}\frac{f(z)}{1+z}dz\right| \le \frac{\rho^{a-1}}{|1-\rho|}2\pi\rho.$$

Since this estimate is independent of δ, (12.14) implies that

$$\begin{aligned}&\left|2\pi i e^{i\pi(a-1)} - (1-e^{2\pi i(a-1)})\int_r^R \frac{x^{a-1}}{1+x}dx\right| \\ &\le \frac{r^{a-1}}{|1-r|}2\pi r + \frac{R^{a-1}}{|1-R|}2\pi R.\end{aligned}$$

But as $r \to 0^+$ and $R \to \infty$ the right hand side of this last inequality converges to zero. Hence $-2\pi i e^{i\pi a} = (1-e^{2\pi i a})\int_0^\infty \frac{x^{a-1}}{1+x}dx$; that is,

$$\begin{aligned}\int_0^\infty \frac{x^{a-1}}{1+x}dx &= \frac{-2\pi i e^{i\pi a}}{1-e^{2\pi i a}} \\ &= \frac{2\pi i}{e^{\pi i a}-e^{-i\pi a}} \\ &= \frac{\pi}{\sin a\pi}.\end{aligned}$$

Example 12.16. *Evaluate*

$$\int_0^\infty \frac{logx}{1+x^2}dx.$$

Solution: We shall first look for a suitable region G in which a single valued analytic branch of $\log z$ can be defined. So, let $G = \{z = re^{i\theta} \in \mathbb{C} : r \neq 0 \text{ and } -\frac{\pi}{2} < \theta < \frac{3\pi}{2}\}$. Now if $z \in G$, let $l(z) = \log|z| + i\theta$. Let $0 < r < R$ and let γ be the path as in Fig.12.10. Hence, $\int_\gamma \frac{l(z)}{1+z^2}dz =$

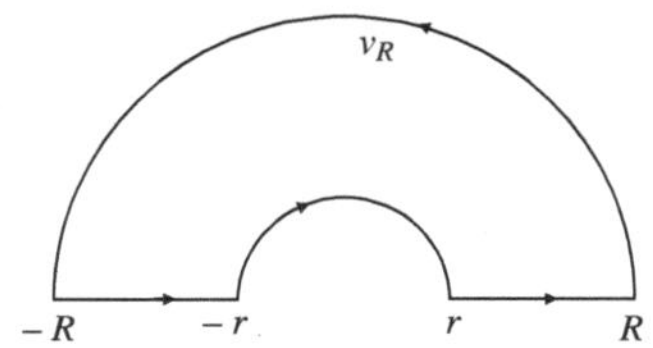

Fig. 12.10: The path γ containing inside the pole $z = i$ of $\frac{l(z)}{1+z^2}$

$\int_r^R \frac{\log x}{1+x^2}dx + iR\int_0^\pi \frac{\log R+i\theta}{1+R^2e^{2i\theta}}e^{i\theta}d\theta + \int_{-R}^{-r}\frac{\log|x|+i\pi}{1+x^2}dx + ir\int_\pi^0 \frac{\log r+i\theta}{1+r^2e^{2i\theta}}e^{i\theta}d\theta$. Note $z = i$ is the only singularity of $\frac{l(z)}{1+z^2}$ inside γ and it is a simple pole and furthermore $\text{Res}\left(\frac{l(z)}{1+z^2}, i\right) = \frac{1}{2i}\left[\log|i| + \frac{\pi}{2}i\right] = \frac{\pi}{4}$. So, by Residue theorem,

$$\int_\gamma \frac{l(z)}{1+z^2}dz = 2\pi i\frac{\pi}{4} = \frac{\pi^2 i}{2}.$$

Also,

$$\int_r^R \frac{\log x}{1+x^2}dx + \int_{-R}^{-r}\frac{\log|x| + i\pi}{1+x^2}dx$$
$$= 2\int_r^R \frac{\log x}{1+x^2}dx + i\pi\int_r^R \frac{dx}{1+x^2}.$$

Letting $r \to 0^+$ and $R \to \infty$, and using the fact (Example 12.5) that

$$\int_0^\infty \frac{dx}{1+x^2} = \frac{\pi}{2}$$

it follows that

$$\frac{\pi^2 i}{2} - 2\int_0^\infty \frac{\log x}{1+x^2}dx - (\pi i)\frac{\pi}{2}$$
$$= -\lim_{r\to 0^+} ir\int_0^\pi \frac{\log r + i\theta}{1+r^2e^{2i\theta}}e^{i\theta}d\theta + \lim_{R\to\infty} iR\int_0^\pi \frac{\log R + i\theta}{1+R^2e^{2i\theta}}e^{i\theta}d\theta.$$

That is, $\displaystyle\int_0^\infty \frac{\log x}{1+x^2}dx$

$$= \frac{1}{2}\left[\lim_{r\to 0^+} ir\int_0^\pi \frac{\log r + i\theta}{1+r^2e^{2i\theta}}e^{i\theta}d\theta - \lim_{R\to\infty} iR\int_0^\pi \frac{\log R + i\theta}{1+R^2e^{2i\theta}}e^{i\theta}d\theta\right].$$

We now show that both of these limits are zero. If $\rho > 0$, then

$$\begin{aligned}\left|i\rho\int_0^\pi \frac{\log\rho + i\theta}{1+\rho^2e^{2i\theta}}e^{i\theta}d\theta\right| &\le \tfrac{\rho|\log\rho|}{|1-\rho^2|}\int_0^\pi d\theta + \frac{\rho}{|1-\rho^2|}\int_0^\pi \theta d\theta \\ &= \frac{\pi\rho|\log\rho|}{|1-\rho^2|} + \frac{\rho\pi^2}{2|1-\rho^2|};\end{aligned}$$

letting $\rho \to 0^+$ or $\rho \to \infty$, the limit of this expression is zero. Thus $\displaystyle\int_0^\infty \frac{\log x}{1+x^2}dx = 0$.

Example 12.17. The normal (Gaussian) probability integral

Evaluate $\displaystyle\int_{-\infty}^{\infty} e^{-\frac{x^2}{2}}dx$.

Solution: Let $f(z) = \frac{e^{i\pi z^2}}{\sin\pi z}$. Let $w = e^{\frac{i\pi}{4}} = \frac{1+i}{\sqrt{2}}$, so $w^2 = i$. We choose $Y > 0$ large and discuss $f(z)$ in the parallelogram Γ, with vertices at $\frac{1}{2}+Yw\sqrt{2}, -\frac{1}{2}+Yw\sqrt{2}, -\frac{1}{2}-Yw\sqrt{2}$, and $\frac{1}{2}-Yw\sqrt{2}$. That is, $\Gamma = \gamma_1 - \gamma_Y - \gamma_2 + \gamma_{-Y}$ as in Fig. 12.11. Thus

$$\int_\Gamma f(z)dz = \int_{\gamma_1} f(z)dz - \int_{\gamma_Y} f(z)dz - \int_{\gamma_2} f(z)dz + \int_{\gamma_{-Y}} f(z)dz. \quad (12.15)$$

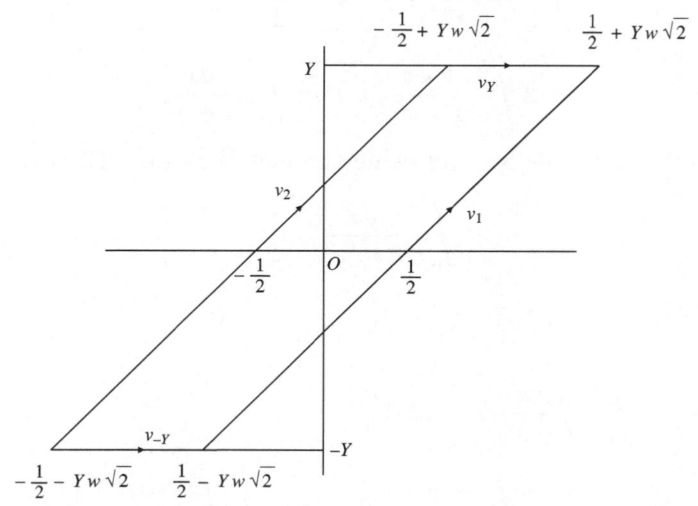

Fig. 12.11: The parallelogram $\Gamma = \gamma_1 - \gamma_Y - \gamma_2 + \gamma_{-Y}$

Since $\sin \pi z$ has zeroes at the integers and only 0 lies inside Γ, so by Residue theorem,

$$\begin{aligned}\int_\Gamma \frac{e^{i\pi z^2}}{\sin\pi z}dz &= 2\pi i\text{Res}\left(\frac{e^{i\pi z^2}}{\sin\pi z}, 0\right)\\ &= 2\pi i\frac{1}{\pi\cos\pi 0} = 2i.\end{aligned}$$

We shall now evaluate $\int_{\gamma_1} f(z)dz - \int_{\gamma_2} f(z)dz$. Consider the parametrization $\gamma_1(t) = wt + \frac{1}{2}, -\frac{1}{\sqrt{2}} \leq t \leq \frac{1}{\sqrt{2}}$ and $\gamma_2(t) = wt - \frac{1}{2}, -\frac{1}{\sqrt{2}} \leq t \leq \frac{1}{\sqrt{2}}$.
Then

$$\begin{aligned}\int_{\gamma_1} \frac{e^{i\pi z^2}}{\sin\pi z}dz - \int_{\gamma_2} \frac{e^{i\pi z^2}}{\sin\pi z}dz &= \int_{-\sqrt{2}Y}^{\sqrt{2}Y} \left\{\frac{e^{i\pi(wt+\frac{1}{2})^2}}{\sin\pi(wt+\frac{1}{2})} - \frac{e^{i\pi(wt-\frac{1}{2})^2}}{\sin\pi(wt-\frac{1}{2})}\right\} wdt\\ &= \int_{-\sqrt{2}Y}^{\sqrt{2}Y} \frac{e^{-\pi t^2}e^{\frac{i\pi}{4}}}{\cos\pi wt}\left\{e^{i\pi wt} + e^{-i\pi wt}\right\} wdt\\ &= 2i\int_{-\sqrt{2}Y}^{\sqrt{2}Y} e^{-\pi t^2}dt.\end{aligned}$$

Now we shall obtain an estimate of $\left|\int_{\gamma_Y} \frac{e^{i\pi z^2}}{\sin\pi z}dz\right|$. Note that if $z = x+iY \in \gamma_Y$ then $-\frac{1}{2} + Y \leq x \leq \frac{1}{2} + Y$. Hence

$$\left|e^{i\pi z^2}\right| = \left|e^{i\pi(x^2+2ixY-Y^2)}\right| = e^{-2\pi xy} \leq e^{-2\pi(Y-\frac{1}{2})Y}$$

and

$$|\sin\pi z| = \left|\frac{1}{2i}(e^{i\pi(x+iY)} + e^{-i\pi(x+iY)})\right| \geq \frac{1}{2}\left(e^{\pi Y} - e^{-\pi Y}\right).$$

Hence $\left|\frac{e^{i\pi z^2}}{\sin\pi z}\right| \leq \frac{e^{-2\pi Y\left(Y-\frac{1}{2}\right)}}{\frac{1}{2}(e^{\pi Y}-e^{-\pi Y})}$. By estimate theorem

$$\begin{aligned}\left|\int_{\gamma_Y} \frac{e^{i\pi z^2}}{\sin\pi z}dz\right| &\leq \frac{e^{-2\pi Y\left(Y-\frac{1}{2}\right)}}{\frac{1}{2}(e^{\pi Y}-e^{-\pi Y})}\int_{-\frac{1}{2}+Y}^{\frac{1}{2}+Y} dx\\ &= \frac{e^{-2\pi Y\left(Y-\frac{1}{2}\right)}}{\frac{1}{2}\left(e^{\pi Y} - e^{-\pi Y}\right)} \cdot 1 \to 0\end{aligned}$$

as $Y \to \infty$. Similarly we can show that $\left|\int_{\gamma_{-Y}} \frac{e^{i\pi z^2}}{\sin\pi z}dz\right| \to 0$ as $Y \to \infty$.

Now let $Y \to \infty$ in (12.15). We get $2i = \lim_{Y\to\infty}\left(2i\int_{-\sqrt{2}Y}^{\sqrt{2}Y} e^{-\pi t^2}dt\right)$. Hence $\int_{-\infty}^{\infty} e^{-\pi t^2}dt = 1$. If we make a change of variable and write $x = \sqrt{2\pi}t$ we obtain

$$\int_{-\infty}^{\infty} e^{-\frac{x^2}{2}}dx = (\sqrt{2\pi})\int_{-\infty}^{\infty} e^{-\pi t^2}dt = \sqrt{2\pi}.$$

Example 12.18. ***Fresnel Integrals***

Evaluate $\int_0^\infty sin(x^2)dx, \int_0^\infty cos(x^2)dx.$

Solution: It is not even obvious that these integral converge, but we will soon see that they do. Consider the function $f(z) = e^{iz^2}$. Choose $R > 0$ large and consider the closed path γ consisting of the line segment $[0, R]$ (or the path $\gamma_1(x) = x, 0 \leq x \leq R$), the path $\gamma_R = \{z = Re^{i\theta} : 0 \leq \theta \leq \frac{\pi}{4}\}$ and the path $\gamma_2(t) = te^{\frac{i\pi}{4}}, 0 \leq t \leq R$ with counterclockwise orientation. That is,

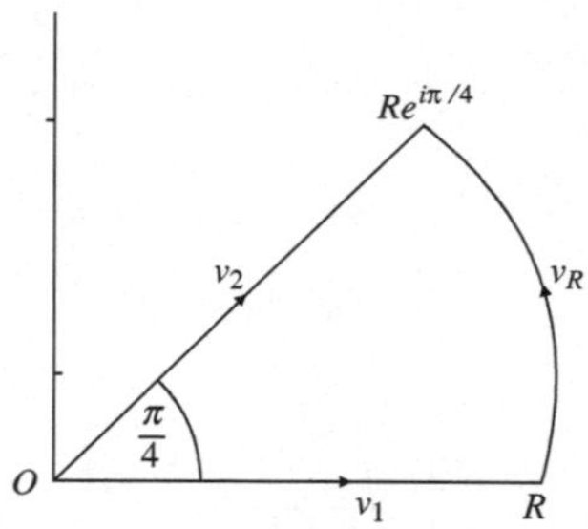

Fig. 12.12: The closed path $\gamma = \gamma_1 + \gamma_R - \gamma_2$

$\gamma = \gamma_1 + \gamma_R - \gamma_2$ and

$$\int_\gamma e^{iz^2} dz = \int_{\gamma_1} e^{iz^2} dz + \int_{\gamma_R} e^{iz^2} dz - \int_{\gamma_2} e^{iz^2} dz. \tag{12.16}$$

Since e^{iz^2} is an entire function, so $\int_\gamma e^{iz^2} dz = 0$. Now $\int_{\gamma_1} e^{iz^2} dz = \int_0^R e^{ix^2} dx = \int_0^R \cos(x^2)dx + i\int_0^R \sin(x^2)dx$. Let $w = e^{\frac{i\pi}{4}}$, then $\gamma_2(t) = wt$ $(0 \leq t \leq R)$. Hence $\int_{\gamma_2} e^{iz^2} dz = \int_0^R e^{i(wt)^2} wdt = w\int_0^R e^{-t^2} dt$ as $w^2 = i$. Further, by using Jordan's inequality we have

$$\begin{aligned}
\left|\int_{\gamma_R} e^{iz^2} dz\right| &= \left|\int_0^{\frac{\pi}{4}} e^{i(Re^{i\theta})^2} iRe^{i\theta} d\theta\right| \\
&\leq R\int_0^{\frac{\pi}{4}} e^{-2R^2 \text{sin}\theta\text{cos}\theta} d\theta \\
&= \tfrac{R}{2}\int_0^{\frac{\pi}{2}} e^{-R^2 \text{sin}\phi} d\phi \quad (\text{ Take } \phi = 2\theta.) \\
&\leq \tfrac{R}{2}\int_0^{\frac{\pi}{2}} e^{-R^2 . \frac{2}{\pi}\phi} d\phi \\
&= \tfrac{R}{2}\left[e^{-R^2 \frac{2}{\pi}\phi}\right]_{\phi=0}^{\frac{\pi}{2}} \left(-\tfrac{\pi}{2R^2}\right) \\
&\leq \tfrac{\pi}{4R} \to 0
\end{aligned}$$

as $R \to \infty$. Letting $R \to \infty$ in (12.16) we obtain

$$0 = \lim_{R\to\infty} \left[\int_0^R \cos(x^2)dx + \int_0^R \sin(x^2)dx - w \int_0^R e^{-t^2} dt \right].$$

Since $w = e^{\frac{i\pi}{4}} = \cos\frac{\pi}{4} + i\sin\frac{\pi}{4} = \frac{1}{\sqrt{2}} + \frac{i}{\sqrt{2}}$ we have

$$\begin{aligned} \int_0^\infty \cos(x^2)dx &= \tfrac{1}{\sqrt{2}} \int_0^\infty e^{-t^2} dt \\ &= \tfrac{1}{4} \int_{-\infty}^\infty e^{-\frac{u^2}{2}} du \quad \left(\text{take } t = \tfrac{u}{\sqrt{2}}\right) \\ &= \sqrt{\tfrac{\pi}{8}}. \end{aligned}$$

So $\int_0^\infty \cos(x^2)dx$ converges and its value is $\sqrt{\frac{\pi}{8}}$ and similarly one can show that $\int_0^\infty \sin(x^2)dx = \sqrt{\frac{\pi}{8}}$.

Example 12.19. *To evaluate* $\int_0^\infty \frac{1}{\sqrt{x}} cos x dx$, *let* $x = u^2$. *Hence* $dx = 2udu$ *and* $\int_0^\infty \frac{1}{\sqrt{x}} cos x dx = 2\int_0^\infty cos(u^2)du = \sqrt{\frac{\pi}{2}}$. *Similarly, the integral* $\int_0^\infty \frac{1}{\sqrt{x}} sin x dx = 2\int_0^\infty sin(u^2)du = \sqrt{\frac{\pi}{2}}$.

Example 12.20. *Show that* $\displaystyle\int_{-\infty}^\infty \frac{e^{ax}}{1+e^x} dx = \frac{\pi}{\sin \pi a}, 0 < a < 1.$

Solution: For the proof, we integrate the function $f(z) = \frac{e^{az}}{1+e^z}$ over the perimeter of a rectangle (see Fig.12.13) with the vertices $X, X + 2\pi i, -X + 2\pi i$, and $-X$ where $X > 0$. That is, $\gamma = \gamma_1 + \gamma_X - \gamma_2 - \gamma_{-X}$, where $\gamma_1(x) = x, -X \le x \le X$ and $\gamma_2(x) = x + 2\pi i (-X \le x \le X)$, $\gamma_X(t) = X + it, 0 \le t \le 2\pi$ and $\gamma_{-X}(t) = -X + it, 0 \le t \le 2\pi$.

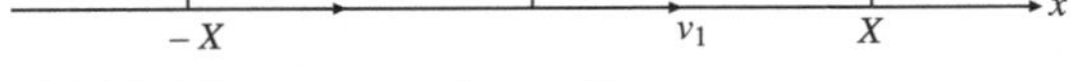

Fig. 12.13: The rectangular path $\gamma = \gamma_1 + \gamma_X - \gamma_2 - \gamma_{-X}$

The poles of f are the roots of $1 + e^z = 0$, i.e., when

$$z = \cdots, -3\pi i, -\pi i, \pi i, 3\pi i, \cdots.$$

The only pole of f inside γ is at $z = \pi i$, hence by residue theorem,

$$\int_\gamma f(z)dz = 2\pi i \text{Res}(f, \pi i) = (-2\pi i)e^{i\pi a}.$$

Now

$$\begin{aligned}\int_{\gamma_1} \frac{e^{az}}{1+e^z}dz - \int_{\gamma_2} \frac{e^{az}}{1+e^z}dz &= \int_{-X}^{X} \frac{e^{ax}}{1+e^x}dx - \int_{-X}^{X} \frac{e^{a(x+2\pi i)}}{1+e^{(x+2\pi i)}}dx \\ &= (1-e^{a.2\pi i}) \int_{-X}^{X} \frac{e^{ax}}{1+e^x}dx.\end{aligned}$$

On $\gamma_X, \left|\frac{e^{az}}{1+e^z}\right| \leq \frac{|e^{az}|}{|e^z|-1} = \frac{e^{aX}}{e^X-1}$. Hence $\left|\int_{\gamma_X} \frac{e^{az}}{1+e^z}dz\right| \leq \frac{e^{aX}}{e^X-1}.2\pi = \frac{2\pi e^{(a-1)X}}{1-e^{-X}}$ $\to 0$ as $X \to \infty$ since $a < 1$. On $\gamma_{-X}, \left|\frac{e^{az}}{1+e^z}\right| \leq \frac{|e^{az}|}{1-|e^z|} = \frac{e^{-aX}}{1-e^{-X}}$. Therefore $\left|\int_{\gamma_{-X}} \frac{e^{az}}{1+e^z}dz\right| \leq \frac{e^{-aX}}{1-e^{-X}} \cdot 2\pi \to 0$ as $X \to \infty$ since $a > 0$.

Thus

$$\lim_{X\to\infty} (1-e^{2\pi ia}) \int_{-X}^{X} \frac{e^{ax}}{1+e^x}dx = -2\pi ie^{i\pi a}.$$

Hence $\int_{-\infty}^{\infty} \frac{e^{ax}}{1+e^x}dx = (-2\pi i)\frac{e^{\pi ai}}{1-e^{2\pi ai}} = \pi\left(\frac{2i}{e^{\pi ai}-e^{-\pi ai}}\right) = \frac{\pi}{\sin \pi a}$.

Other integrals can sometimes be reduced to this integral. For example, for $\alpha > 1$,

$$\int_0^\infty \frac{dt}{1+t^\alpha} = \int_{-\infty}^{\infty} \frac{\frac{1}{\alpha}e^{\frac{x}{\alpha}}}{1+e^x}dx = \frac{\frac{\pi}{\alpha}}{\sin\frac{\pi}{\alpha}}.$$

Now observe that if we write $t = e^x$, then

$$\int_{-\infty}^{\infty} \frac{e^{ax}}{1+e^x}dx = \int_0^\infty \frac{t^{a-1}}{1+t}dt = \frac{\pi}{\sin\pi a} \text{ if } 0 < a < 1 \tag{12.17}$$

(see Example 12.15).

Further if we use the transformation $t = x^n$ with $a = \frac{m}{n}$ in (12.17), it becomes

$$\int_0^\infty \frac{x^{m-1}}{1+x^n}dx = \frac{\pi}{n}\frac{1}{\sin\frac{m\pi}{n}}$$

for $0 < m < n$ (see Example 12.15).

Example 12.21. *Evaluate* $\displaystyle\int_{-\infty}^{\infty} \frac{e^{-ix}}{x^2-2x+4}dx.$

Solution: For large $|x|$, the absolute value of the integrand is of the same order of magnitude as $\frac{1}{|x|^2}$. Hence it follows that the integral converges. Let $f(z) = \dfrac{e^{-iz}}{z^2-2z+4}$. Let $R > 2$. Consider the closed path Γ consisting of the segment $[-R, R]$ and the semicircular arc γ_R in the lower half plane with the same end points as that segment and with clockwise orientation.

The function f has two simple poles at $z = 1 \pm i\sqrt{3}$ but it has only one pole inside Γ which is at the point $z = 1 - i\sqrt{3}$. By Residue theorem,

$$\int_\Gamma f(z)dz = -2\pi i\text{Res}\left(\frac{e^{-iz}}{z^2-2z+4}, 1-i\sqrt{3}\right) = (-2\pi i)\frac{e^{-\sqrt{3}-i}}{-2i\sqrt{3}}$$

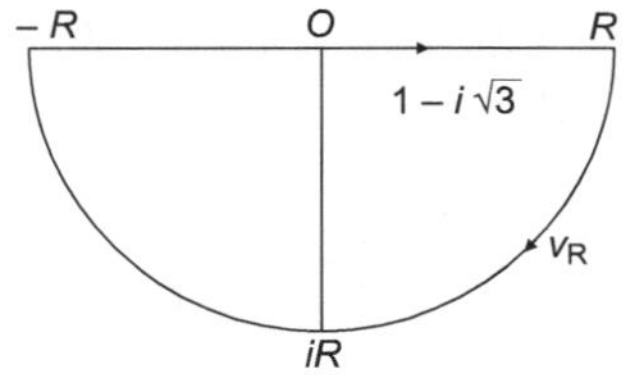

Fig. 12.14: The closed path $\Gamma = [-R, R] + \gamma_R$

(The minus sign appears because of the clockwise orientation.)

$$= \frac{\pi}{\sqrt{3}} e^{-\sqrt{3}-i}.$$

If z lies in the lower half plane, then $z = x + iy, x, y > 0$. Hence $|e^{-iz}| = |e^{-i(x-iy)}| = |e^{-y}| \leq 1$ for z in the lower half plane. Now

$$\left| \int_{\gamma_R} \frac{e^{-iz}}{z^2 - 2z + 4} dz \right| \leq \frac{2\pi R}{R^2 - 2R - 4} \to 0$$

as $R \to \infty$.
Hence,

$$\frac{\pi e^{-\sqrt{3}-i}}{\sqrt{3}} = \int_{-R}^{R} \frac{e^{-ix}}{x^2 - 2x + 4} dx + \int_{\gamma_R} \frac{e^{-iz}}{z^2 - 2z + 4} dz.$$

Letting $R \to \infty$, we obtain

$$\int_{-\infty}^{\infty} \frac{e^{-ix}}{x^2 - 2x + 4} dx = \frac{\pi e^{-\sqrt{3}-i}}{\sqrt{3}}.$$

Example 12.22. *Evaluate* $\int_{-\infty}^{\infty} \frac{dx}{1 + x^{2n}}$ *where* n *is a positive integer.*

Solution: Observing that the function is even, doubling the integral from 0 to ∞ and making the substitution $y = x^{2n}$, we get

$$\begin{aligned} \int_{-\infty}^{\infty} \frac{dx}{1 + x^{2n}} &= 2 \int_{0}^{\infty} \frac{dx}{1 + x^{2n}} \\ &= \frac{1}{n} \int_{0}^{\infty} \frac{y^{\frac{1}{2n}-1}}{y + 1} dy \\ &= \frac{\pi}{n \sin(\frac{\pi}{2n})} \end{aligned}$$

as $n > 0$ and $0 < \frac{1}{2n} < 1$ by Example 12.15.

Example 12.23. *Evaluate the integral* $\int_C \frac{5z - 2}{z(z - 1)} dz$, *where* C *is the circle* $|z| = 2$, *with counterclockwise orientation.*

Solution: Let $f(z) = \frac{5z-2}{z(z-1)}$. The two singularities, $z = 0$ and $z = 1$, of the function f lie inside the closed contour C. Observe that

$$\begin{aligned} \frac{5z-2}{z(z-1)} &= \left(5 - \frac{2}{z}\right)\left(\frac{-1}{1-z}\right) \\ &= \left(-5 + \frac{2}{z}\right)(1 + z + z^2 + \cdots) \\ &= \frac{2}{z} - 3 - 3z - 3z^2 - \cdots \ (0 < |z| < 1). \end{aligned}$$

Hence $\text{Res}(f, 0) = 2$. To find the residue of f at $z = 1$, we may use the Taylor series

$$\frac{1}{z} = 1 - (z-1) + (z-1)^2 - \cdots \ (|z-1| < 1)$$

in order to find the coefficients in the appropriate Laurent expansion of the function f about the point $z = 1$. Thus

$$\frac{5z-2}{z(z-1)} = \left(5 + \frac{3}{z-1}\right)\frac{1}{z} = \left(5 + \frac{3}{z-1}\right)\left[1 - (z-1) + (z-1)^2 - \cdots\right]$$

when $0 < |z-1| < 1$. The coefficient of $\frac{1}{z-1}$ in this product is 3; that is, $\text{Res}(f, 1) = 3$. Therefore, $\int_C \frac{5z-2}{z(z-1)}dz = 2\pi i(2+3) = 10\pi i$. In this example it is simpler to write the integrand as the sum of its partial fractions. Then $\int_C \frac{5z-2}{z(z-1)}dz = \int_C \frac{2}{z}dz + \int_C \frac{3}{z-1}dz = 2\pi i(2+3) = 10\pi i$.

As still another method of evaluating the integral, we can note that the integral divided by $2\pi i$ is the coefficient b_1 of the Laurent expansion of the function f about the origin in the domain outside the circle $|z| = 1$, according to the formula (for the coefficient of Laurent series expansion). That expansion is easily written as follows: If $|z| > 1$, $\frac{5z-2}{z(z-1)} = \frac{5z-2}{z^2}\frac{1}{1-\frac{1}{z}} = (5z-2)\sum_{n=0}^{\infty}\frac{1}{z^{n+2}}$. The coefficient b_1 of $\frac{1}{z}$ in this product is 5. Hence the value of the integral is $10\pi i$.

Example 12.24. $\displaystyle\int_0^{\infty} e^{-x^2} cos(2ax)dx = \frac{1}{2}\sqrt{\pi}e^{-a^2}, a \in \mathbb{R}.$

Solution: If $a = 0$, it is just the **Gaussian probability integral** and we have already shown that $\int_0^{\infty} e^{-x^2}dx = \frac{\sqrt{\pi}}{2}$. Choose $R > 0$ large and suppose $\alpha > 0$. If $a > 0$, consider the function $f(z) = e^{-z^2}e^{2iaz}$ and the closed contour Γ, which is a rectangle with vertices $R, R+\alpha i, -R+\alpha i, -R$ with counterclockwise orientation.

That is, $\Gamma = \gamma_1 + \gamma_2 + \gamma_3 + \gamma_4$ as in Fig.12.15 and

$$\int_{\Gamma} f(z)dz = \sum_{i=1}^{4}\int_{\gamma_i} f(z)dz. \qquad (12.18)$$

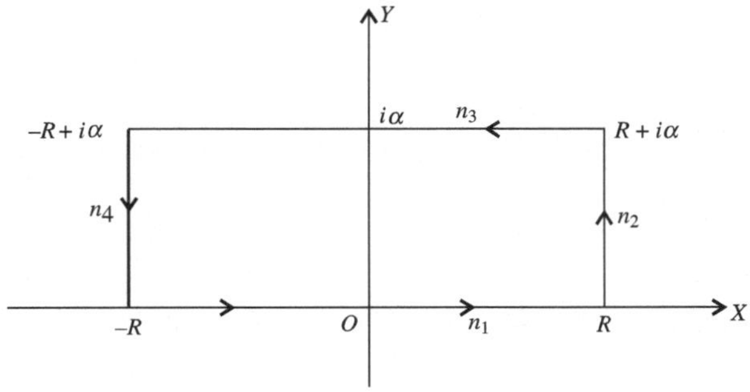

Fig. 12.15: The closed contour $\Gamma = \gamma_1 + \gamma_2 + \gamma_3 + \gamma_4$

Since f is entire, $\int_\Gamma f(z)dz = 0$. (If $a < 0$, then draw the contour below the x-axis.) Now,

$$\int_{\gamma_1} f(z)dz = \int_{-R}^{R} e^{-x^2} e^{i2ax} dx$$

and

$$\int_{\gamma_3} f(z)dz = \int_{R}^{-R} e^{-(x+\alpha i)^2} e^{i2ax} dx = -e^{\alpha^2 - 2a\alpha} \int_{-R}^{R} e^{-x^2} e^{2i(a-\alpha)} dx.$$

Further if $z \in \gamma_2$,

$$|f(z)| = \left| e^{-(R+iy)^2} e^{i2a(R+iy)} \right| = e^{-R^2+y^2} e^{-2ay} \leq e^{-R^2} e^{\alpha^2+2a\alpha}.$$

Similarly if $z \in \gamma_4$,

$$|f(z)| = \left| e^{-(-R+iy)^2 e^{i2a(-R+iy)}} \right| \leq e^{-R^2} e^{\alpha^2+2a\alpha}.$$

Thus

$$\left| \int_{\gamma_2} f(z)dz \right| \leq \alpha e^{-R^2} e^{\alpha^2+2a\alpha}$$

and

$$\left| \int_{\gamma_3} f(z)dz \right| \leq \alpha e^{-R^2} e^{\alpha^2+2a\alpha}.$$

Hence $\left|\int_{\gamma_2} f(z)dz\right|$ and $\left|\int_{\gamma_3} f(z)dz\right|$ tend to 0 as $R \to \infty$. Hence letting $R \to \infty$ in (12.18) we have

$$0 = \int_{-\infty}^{\infty} e^{-x^2} e^{2aix} dx - e^{\alpha^2-2a\alpha} \int_{-\infty}^{\infty} e^{-x^2} e^{2i(a-\alpha)x} dx.$$

Since this holds for all α and a, let $a = \alpha$ and the integral becomes

$$\int_{-\infty}^{\infty} e^{-x^2} e^{2aix} dx = e^{-a^2} \int_{-\infty}^{\infty} e^{-x^2} dx = e^{-a^2} \sqrt{\pi}$$

and the result follows.

Example 12.25. *Evaluate* $\displaystyle\int_0^\infty \frac{(log x)^2}{x^2+1}dx.$

Solution: For $0 < r < 1 < R$, let $\Gamma_{r,R}$ denote the contour drawn below (see Fig.12.16).

Let $G = \{z \in \mathbb{C} : z \neq 0 \text{ and } -\frac{\pi}{2} < \arg z < \frac{3\pi}{2}\}$ and define $\log z$ for $z \in G$ as follows. If $z = |z|e^{i\theta} \neq 0$ with $-\frac{\pi}{2} < \arg z < \frac{3\pi}{2}$, let $l(z) = \log|z| + i\theta$. (That is, $l(z)$ denote the branch of the logarithm function in the plane slit along the negative imaginary axis.)

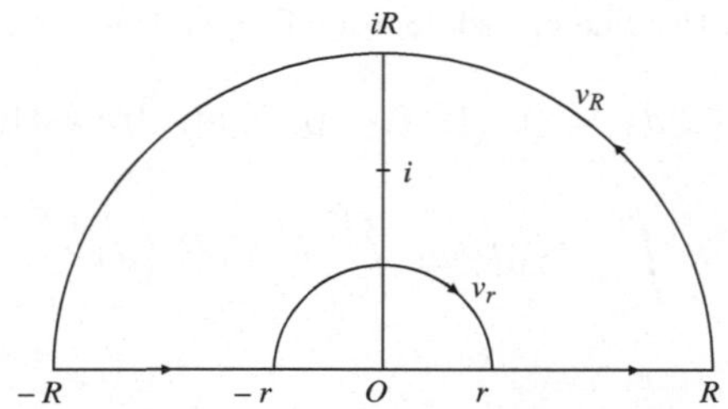

Fig. 12.16: The contour $\Gamma_{r,R}$

Let $f(z) = \frac{(\log z)^2}{z^2+1}$. The integrand has two simple poles at $z = i$ and $z = -i$. Only the point $z = i$ is inside the contour $\Gamma_{r,R}$. By Residue theorem,

$$\int_{\Gamma_{r,R}} \frac{(l(z))^2}{z^2+1}dz = 2\pi i \text{Res}\left(\frac{(l(z))^2}{z^2+1}, i\right) = -\frac{\pi^3}{4}.$$

Notice $l(x) = \log x$ for $x > 0$, and $l(x) = \log|x| + \pi i$ for $x < 0$. Hence if $x < 0$, $(l(x))^2 = (\log|x|)^2 - \pi^2 + 2\pi i \log|x|$. Moreover,

$$\begin{aligned} -\frac{\pi^3}{4} &= \int_{\Gamma_{r,R}} \frac{(l(z))^2}{z^2+1}dz \\ &= \int_r^R \frac{(\log x)^2}{1+x^2}dx + iR\int_0^\pi \frac{[\log R + i\theta]^2}{1+R^2e^{2i\theta}}e^{i\theta}d\theta \\ &+ \int_{-R}^{-r} \frac{[\log|x| + \pi i]^2}{1+x^2}dx + ir\int_\pi^0 \frac{[\log r + i\theta]^2}{1+r^2e^{2i\theta}}e^{i\theta}d\theta. \end{aligned} \tag{12.19}$$

Now

$$\begin{aligned} \left|iR\int_0^\pi \frac{[\log R + i\theta]^2}{1+R^2e^{2i\theta}}e^{i\theta}d\theta\right| &\leq R\int_0^\pi \frac{(\log R)^2 + \theta^2}{R^2-1}d\theta \\ &= \frac{\pi R(\log R)^2}{R^2-1} + \frac{R}{R^2-1}\frac{\pi^3}{3} = o(1) \end{aligned}$$

as $R \to \infty$.

Similarly,

$$\left| ir \int_{\pi}^{0} \frac{[\log r + i\theta]^2}{1 + r^2 e^{2i\theta}} e^{i\theta} d\theta \right| \leq \frac{\pi r (\log r)^2}{r^2 - 1} + \frac{r}{r^2 - 1} \frac{\pi^3}{3} = o(1)$$

as $r \to 0$. Further,

$$\int_{-R}^{-r} \frac{[\log|x| + \pi i]^2}{1 + x^2} dx = \int_{r}^{R} \frac{[\log x + \pi i]^2}{1 + x^2} dx.$$

Taking limit $r \to 0$ and $R \to \infty$ in (12.19), we have

$$\frac{-\pi^3}{4} = 2 \int_0^{\infty} \frac{(\log x)^2}{1 + x^2} dx - \pi^2 \int_0^{\infty} \frac{dx}{1 + x^2} + 2i\pi \int_0^{\infty} \frac{\log x}{1 + x^2} dx.$$

Using the fact that $\int_0^{\infty} \frac{dx}{1 + x^2} = \frac{\pi}{2}$, we have

$$\frac{-\pi^3}{4} + \pi^2 . \frac{\pi}{2} = 2 \int_0^{\infty} \frac{(\log x)^2}{1 + x^2} dx + 2i\pi \int_0^{\infty} \frac{\log x}{1 + x^2} dx.$$

Equating the real parts, we obtain $\int_0^{\infty} \frac{(\log x)^2}{1 + x^2} dx = \frac{\pi^3}{8}$. Equating the imaginary parts, we obtain $\int_0^{\infty} \frac{\log x}{1 + x^2} dx = 0$. This we have already seen in Example 12.16.

Example 12.26. *Prove that* $\int_0^{\infty} \frac{x}{e^x - e^{-x}} dx = \frac{\pi^2}{8}$.

Solution: We have

$$\begin{aligned} I &= \int_0^{\infty} \frac{x}{e^x - e^{-x}} dx \\ &= \tfrac{1}{2} \int_{-\infty}^{\infty} \frac{x}{e^x - e^{-x}} dx \\ &= \tfrac{1}{2} \int_0^{\infty} \frac{\log u}{(u - \frac{1}{u}) u} du \\ &= \tfrac{1}{4} \int_{-\infty}^{\infty} \frac{\log|u|}{u^2 - 1} du \end{aligned}$$

where we used $u = e^x$. Integrate $f(z) = \frac{\log z}{z^2 - 1}$ over the contour $\Gamma = \Gamma_R + \gamma_{-1} + \gamma_0 + \gamma_1$.

By Cauchy's theorem, we have $\int_{\Gamma} f(z) dz = 0$. Also, $\int_{\gamma_{-1}} f(z) dz = -i\pi \text{Res}(f, -1) = \frac{\pi^2}{2}$ and $\int_{\gamma_1} f(z) dz = -i\pi \frac{\log 1}{2} = 0$. Further, $\int_{\Gamma_R} f(z) dz \to 0$

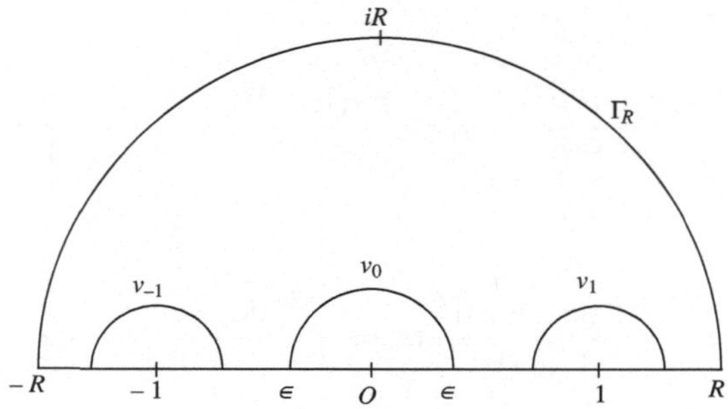

Fig. 12.17: The contour $\Gamma = \Gamma_R + \gamma_{-1} + \gamma_0 + \gamma_1$

as $R \to \infty$ since $\lim\limits_{R\to\infty} \dfrac{R\log R}{R^2-1} = 0$ and $\lim\limits_{\epsilon\to 0} \displaystyle\int_{\gamma_0} f(z)dz = 0$ since $\lim\limits_{\epsilon\to 0} \dfrac{\epsilon\log\epsilon}{\epsilon^2-1} = 0$.
So we get $\displaystyle\int_{-\infty}^{\infty} \frac{\log|u|}{u^2-1}du = \frac{\pi^2}{2}$ and $I = \frac{\pi^2}{8}$.

Example 12.27. *Evaluate* $\displaystyle\int_{-\infty}^{\infty} \frac{sin^2 x}{x^2}dx.$

Solution 1: Assuming the Dirichlet integral $\int_0^\infty \frac{\sin x}{x}dx = \frac{\pi}{2}$ (Example 12.11), we can use integration by parts:

$$\begin{aligned}
\int_0^\infty \frac{\sin^2 x}{x^2}dx &= -\tfrac{1}{x}\sin^2 x\Big|_0^\infty - \int_0^\infty -\frac{1}{x}\frac{\sin 2x}{x}dx \\
&= 0 + \int_0^\infty \frac{\sin y}{y}dy \\
&= \frac{\pi}{2}.
\end{aligned}$$

Solution 2: Consider the function f defined by $f(z) = \frac{e^{2iz}-1}{z^2}$. We shall integrate f along the contour (see Fig. 12.18) C_R given by $C_R = [-R, -\epsilon] \cup \gamma_\epsilon \cup [\epsilon, R] \cup \Gamma_R$.

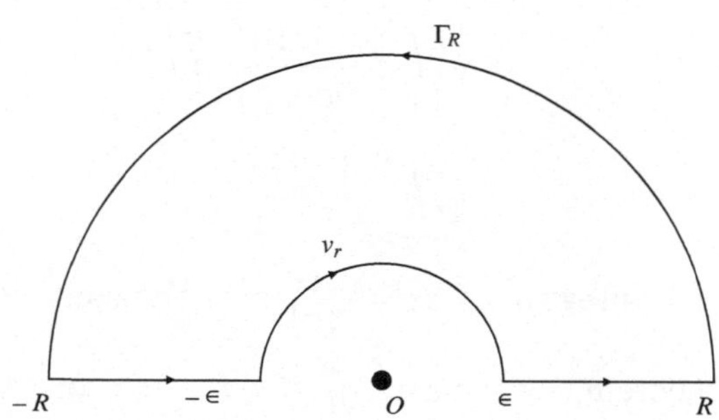

Fig. 12.18: The contour $C_R = [-R, -\epsilon] \cup \gamma_\epsilon \cup [\epsilon, R] \cup \Gamma_R$

Since $\lim_{z\to 0}(z-0)f(z) = -2i$, we have $\lim_{\epsilon\to 0}\int_{\gamma_\epsilon} f(z)dz = i(-2i)(0-\pi) = -2\pi$. Further, we note that

$$\begin{aligned}\left|\int_{\Gamma_R} f(z)dz\right| &\leq \int_0^\pi \frac{|e^{2iz}-1|}{|z|^2}|dz| \\ &\leq \int_0^\pi \frac{|e^{2iz}|+1}{|z|^2}|dz| \\ &= \int_0^\pi \frac{2}{R^2}Rd\theta \leq \frac{2\pi}{R}\end{aligned}$$

and so $\lim_{R\to\infty}\int_{\Gamma_R} f(z)dz = 0$. It is now easy to deduce that

$$\int_{-\infty}^{\infty} \frac{1-e^{2ix}}{x^2}dx = 2\pi.$$

Example 12.28. *Evaluate* $I = \int_{-\infty}^{\infty} \frac{sin^3x}{x^3}dx.$

Solution: The integrand is absolutely integrable, because it is bounded in absolute value by 1 near the origin and by $\frac{1}{|x|^3}$ away from the origin. We have

$$\begin{aligned}\sin^3 x &= -\tfrac{1}{8i}(e^{ix}-e^{-ix})^3 \\ &= -\tfrac{1}{8i}\left(e^{3ix}-e^{-3ix}-3e^{ix}+3e^{-ix}\right) \\ &= \text{Im}\left(\tfrac{3}{4}e^{ix}-\tfrac{1}{4}e^{3ix}\right) \\ &= \tfrac{1}{4}\text{Im}(3e^{ix}-e^{3ix}).\end{aligned}$$

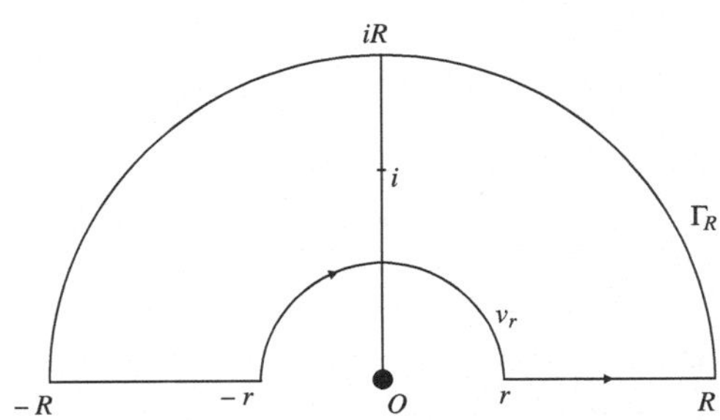

Fig. 12.19: The contour $\Gamma_{r,R} = \Gamma_R \cup \gamma_r \cup [-R,-r] \cup [r,R]$

For $0 < r < R$, consider the contour $\Gamma_{r,R} = \Gamma_R \cup \gamma_r \cup [-R,-r] \cup [r,R]$, by

Cauchy's theorem,

$$\begin{aligned}0 &= \int_{\Gamma_{r,R}} \frac{3e^{iz}-e^{3iz}}{z^3}dz \\ &= \int_{\Gamma_R} \frac{3e^{iz}-e^{3iz}}{z^3}dz + \int_{[-R,-r]} \frac{3e^{iz}-e^{3iz}}{z^3}dz \\ &+ \int_{\gamma_r} \frac{3e^{iz}-e^{3iz}}{z^3}dz + \int_{[r,R]} \frac{3e^{iz}-e^{3iz}}{z^3}dz.\end{aligned}$$

The integrand over Γ_R is bounded in absolute value by $(4R^{-3})(2\pi R)$ since $|e^{iz}|$ and $|e^{3iz}|$ are bounded by 1 in the upper half plane, so it tends to 0 as $R \to \infty$. To estimate the integral over γ_r, we note that

$$\begin{aligned}\frac{3e^{iz}-e^{3iz}}{z^3} &= \frac{1}{z^3}(2+\frac{3(iz)^2}{2}-\frac{(3iz)^2}{2}+O(z^3)) \\ &= \frac{2}{z^3}+\frac{3}{z}+O(1)\end{aligned}$$

as $z \to 0$. Hence,

$$\begin{aligned}\int_{\gamma_r} \frac{3e^{iz}-e^{3iz}}{z^3}dz &= \int_\pi^0 \left(\frac{2}{\epsilon^3 e^{3i\theta}}-\frac{3}{\epsilon e^{i\theta}}+O(1)\right) i\epsilon e^{i\theta}d\theta \\ &= \tfrac{2i}{\epsilon^2}\int_\pi^0 e^{-2i\theta}d\theta + 3i\int_\pi^0 d\theta + O(\epsilon) \\ &= 0-3\pi i+O(\epsilon) \to -3\pi i\end{aligned}$$

as $\epsilon \to 0$. Thus,

$$\lim_{\substack{R\to\infty \\ r\to 0}} \left(\int_{-R}^{-r} \frac{3e^{ix}-e^{3ix}}{x^3}dx + \int_r^R \frac{3e^{ix}-e^{3ix}}{x^3}dx\right) = 3\pi i.$$

Hence $\displaystyle\int_{-\infty}^{\infty} \frac{\sin^3 x}{x^3}dx = \frac{1}{4}\text{Im}\int_{-\infty}^{\infty} \frac{3e^{ix}-e^{3ix}}{x^3}dx = \frac{3\pi}{4}$.

Example 12.29. *Evaluate the following integral for $a \in \mathbb{R}$, and $|a| < 1$.*

$$\int_0^\infty \frac{coshax}{coshx}dx$$

Solution: Consider the function $f(z) = \frac{e^{az}}{\cosh z}$ and integrate it along the rectangle Γ having vertices $-R, R, R+\pi i, -R+\pi i$ (see Fig.12.20).

The poles of f are simple and occur where $\cosh z = 0$, i.e., $z = (n+\frac{1}{2})\pi i$, $n = 0, \pm 1, \pm 2, \cdots$. The only pole that is inside Γ is at $z = \frac{\pi i}{2}$.

Now $\text{Res}\left(f, \frac{\pi i}{2}\right) = \lim_{z\to\frac{\pi i}{2}}(z-\frac{\pi i}{2})\frac{e^{az}}{\cosh z} = \frac{e^{\frac{a\pi i}{2}}}{\sinh\frac{\pi i}{2}} = \frac{e^{\frac{a\pi i}{2}}}{i\sin\frac{\pi}{2}} = (-i)e^{\frac{a\pi i}{2}}$.

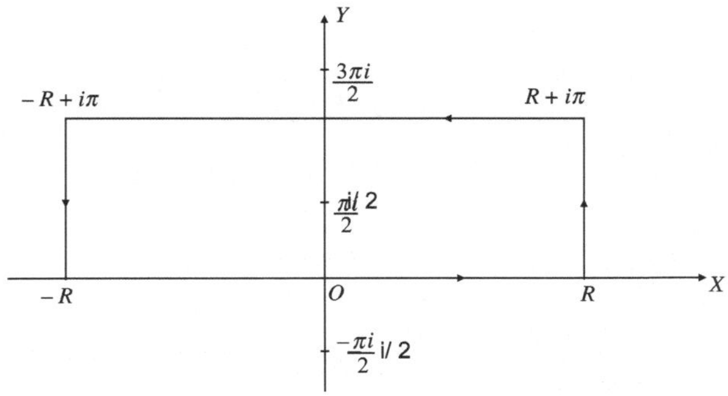

Fig. 12.20: The rectangle Γ having vertices $-R, R, R+\pi i, -R+\pi i$

By Residue theorem,

$$\int_\Gamma f(z)dz = 2\pi i(-ie^{\frac{a\pi i}{2}}) = 2\pi e^{\frac{a\pi i}{2}}.$$

That is,

$$\int_{-R}^{R} \frac{e^{ax}}{\cosh x}dx + \int_0^\pi \frac{e^{a(R+iy)}}{\cosh(R+iy)}idy$$

$$+\int_R^{-R} \frac{e^{a(x+\pi i)}}{\cosh(x+\pi i)}dx + \int_\pi^0 \frac{e^{a(-R+iy)}}{\cosh(-R+iy)}idy = 2\pi e^{\frac{a\pi i}{2}}. \qquad (12.20)$$

As $R \to \infty$, the second and fourth integrals on the left approach zero. To see this let us consider the second integral. Since

$$\begin{aligned} |\cosh(R+iy)| &= \left|\frac{e^{R+iy}+e^{-R-iy}}{2}\right| \\ &\geq \frac{1}{2}\left\{|e^{R+iy}| - |e^{-R-iy}|\right\} \\ &= \frac{1}{2}(e^R - e^{-R}) \geq \frac{1}{4}e^R \end{aligned}$$

we have

$$\begin{aligned} \left|\int_0^\pi \frac{e^{a(R+iy)}}{\cosh(R+iy)}idy\right| &\leq \int_0^\pi \frac{e^{aR}}{\frac{1}{4}e^R}dy \\ &= 4\pi e^{(a-1)R} \end{aligned}$$

and the term on the right tends to zero as $R \to \infty$ since $|a| < 1$.
Similarly, we can show that the fourth integral on the left of (12.20) approaches zero as $R \to \infty$. Since $\cosh(x+\pi i) = -\cosh x$, it follows from (12.20) that

$$\lim_{R\to\infty}\left\{\int_{-R}^{R}\frac{e^{ax}}{\cosh x}dx + e^{a\pi i}\int_{-R}^{R}\frac{e^{ax}}{\cosh x}dx\right\} = 2\pi e^{\frac{a\pi i}{2}}.$$

Thus

$$\int_{-\infty}^{\infty}\frac{e^{ax}}{\cosh x}dx = \frac{2\pi e^{\frac{a\pi i}{2}}}{1+e^{a\pi i}} = \frac{2\pi}{e^{\frac{a\pi i}{2}}+e^{\frac{-a\pi i}{2}}} = \frac{\pi}{\cos(\frac{\pi a}{2})}.$$

Hence we get $\int_{-\infty}^{0}\frac{e^{ax}}{\cosh x}dx + \int_{0}^{\infty}\frac{e^{ax}}{\cosh x}dx = \frac{\pi}{\cos(\frac{\pi a}{2})}$. Now replacing x by $-x$ in the first integral, we have $\int_{0}^{\infty}\frac{e^{-ax}}{\cosh x}dx + \int_{0}^{\infty}\frac{e^{ax}}{\cosh x}dx = 2\int_{0}^{\infty}\frac{\cosh ax}{\cosh x}dx = \frac{\pi}{\cos(\frac{\pi a}{2})}$.

Example 12.30. *Prove that* $\int_{0}^{\infty}\frac{log(x^2+1)}{x^2+1}dx = \pi\, log\ 2.$

Solution: Consider the contour γ consisting of the real axis from $-R$ to R and the semicircle γ_R of radius R (see Fig.12.21). Let $\log z$ denote the branch of the logarithm function in the plane slit along the negative imaginary axis and $f(z) = \frac{\log(z+i)}{z^2+1}$. The only pole of f that is inside γ is the simple pole at $z=i$, and

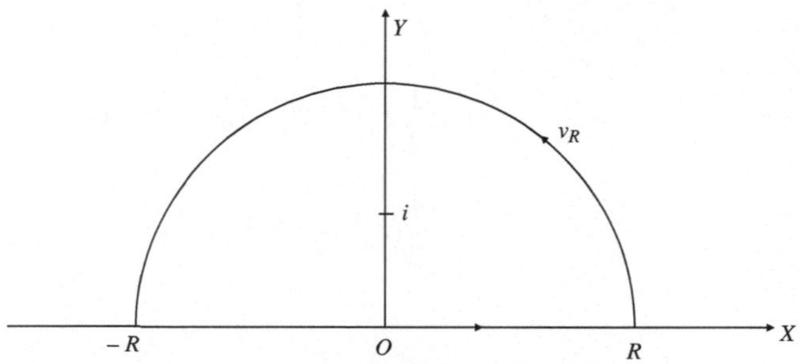

Fig. 12.21: The contour $\gamma = [-R, R] + \gamma_R$

$$\begin{aligned}\text{Res}(f,i) &= \lim_{z\to i}(z-i)\frac{\log(z+i)}{(z-i)(z+i)} \\ &= \frac{\log(2i)}{2i}.\end{aligned}$$

By residue theorem,

$$\begin{aligned}\int_{\gamma} f(z)dz &= 2\pi i\left\{\frac{\log(2i)}{2i}\right\}\\ &= \pi\log(2i)\\ &= \pi(\log2+\log i)\\ &= \pi(\log2+\frac{\pi i}{2}).\end{aligned}$$

Thus, we obtain

$$\int_{-R}^{R}\frac{\log(x+i)}{x^2+1}dx+\int_{\gamma_R}\frac{\log(z+i)}{z^2+1}dz=\pi\log2+\frac{\pi^2 i}{2}.$$

Hence

$$\int_{-R}^{0}\frac{\log(x+i)}{x^2+1}dx+\int_{0}^{R}\frac{\log(x+i)}{x^2+1}dx+\int_{\gamma_R}\frac{\log(z+i)}{z^2+1}dz$$
$$=\pi\log2+\frac{\pi^2 i}{2}.$$

Replacing x by $-x$ in the first integral, this can be written as

$$\int_{0}^{R}\frac{\log(-x+i)}{x^2+1}dx+\int_{0}^{R}\frac{\log(x+i)}{x^2+1}dx+\int_{\gamma_R}\frac{\log(z+i)}{z^2+1}dz$$
$$=\pi\log2+\frac{\pi^2 i}{2}.$$

Since

$$\begin{aligned}\log(x+i)+\log(-x+i) &= \log(i^2-x^2)\\ &= \log(-1)+\log(x^2+1)\\ &= i\pi+\log(x^2+1)\end{aligned}$$

we obtain

$$\int_{0}^{R}\frac{\log(x^2+1)}{x^2+1}dx+\int_{0}^{R}\frac{\pi i}{x^2+1}dx+\int_{\gamma_R}\frac{\log(z+i)}{z^2+1}dz \quad = \quad \pi\log2+\frac{\pi^2 i}{2}.$$

As $R\to\infty$, we can show that the last integral around γ_R tends to zero. Since $\int_0^\infty \frac{dx}{x^2+1}=\frac{\pi}{2}$, we get

$$\int_{0}^{\infty}\frac{\log(x^2+1)}{x^2+1}dx=\pi\log2.$$

Example 12.31. *Prove that* $\int_0^{\frac{\pi}{2}} ln sin x dx=\int_0^{\frac{\pi}{2}} ln cos x dx=-\frac{\pi}{2}ln2.$

Solution: Letting $x = \tan\theta$ in Example 12.30, we find

$$\int_0^{\frac{\pi}{2}} \frac{\ln(\tan^2\theta + 1)}{\tan^2\theta + 1}\sec^2\theta d\theta = -2\int_0^{\frac{\pi}{2}} \ln\cos\theta d\theta = \pi\log 2$$

from which it follows that

$$\int_0^{\frac{\pi}{2}} \ln\cos x \, dx = -\frac{\pi}{2}\ln 2. \tag{12.21}$$

Letting $\theta = \frac{\pi}{2} - \psi$ in (12.21), we obtain

$$\int_0^{\frac{\pi}{2}} \ln\sin\psi \, d\psi = -\frac{\pi}{2}\ln 2.$$

Example 12.32. *Evaluate* $\displaystyle\int_0^{\infty} \frac{\log x}{(x^2+1)(x^2+4)}dx.$

Solution: The integral converges absolutely since, for each $0 < \delta < 1$, we have $\log x = o(x^{\delta})$ as $x \to 0^+$. Consider the function $f(z) = \frac{\log z}{(z^2+1)(z^2+4)}$ and, for $0 < \delta < R$, consider the contour $\Gamma_{\delta,R} = [-R, -\delta] \cup \gamma_\delta \cup [\delta, R] \cup \gamma_R$.

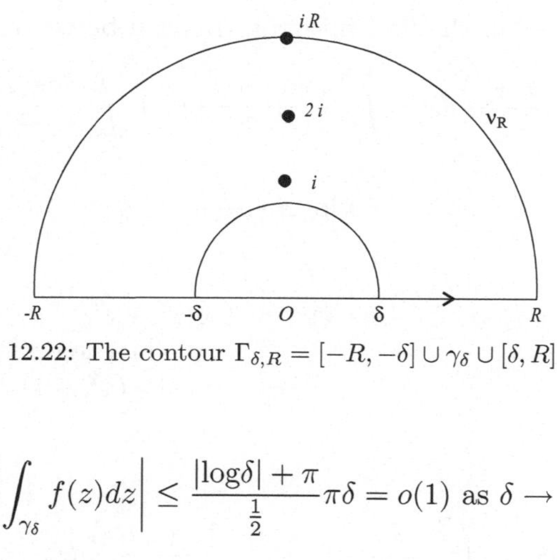

Fig. 12.22: The contour $\Gamma_{\delta,R} = [-R, -\delta] \cup \gamma_\delta \cup [\delta, R] \cup \gamma_R$

Note that,

$$\left|\int_{\gamma_\delta} f(z)dz\right| \le \frac{|\log\delta| + \pi}{\frac{1}{2}}\pi\delta = o(1) \text{ as } \delta \to 0$$

and

$$\left|\int_{\gamma_R} f(z)dz\right| \le \frac{|\log R| + \pi}{(R^2-1)(R^2-4)}\pi R = o(1) \text{ as } R \to \infty.$$

By residue theorem,

$$\int_{-\infty}^{0} f(z)dz + \int_0^{\infty} f(z)dz = 2\pi i(\text{Res}(f, i) + \text{Res}(f, 2i))$$

$$= 2\pi i \left(\frac{\pi}{12} + \frac{i\log 2}{12} - \frac{\pi}{24} \right).$$

We also have

$$\begin{aligned} \int_{-\infty}^{0} f(z)dz &= \int_{-\infty}^{0} \frac{\log(-x) + \pi i}{(x^2+1)(x^2+4)} dx \\ &= \int_{0}^{\infty} \frac{\log(x)}{(x^2+1)(x^2+4)} dx + \int_{0}^{\infty} \frac{\pi i}{(x^2+1)(x^2+4)} dx \end{aligned}$$

and

$$\int_{0}^{\infty} f(z)dz = \int_{0}^{\infty} \frac{\log(x)}{(x^2+1)(x^2+4)} dx.$$

So

$$2\int_{0}^{\infty} \frac{\log(x)}{(x^2+1)(x^2+4)} dx + \frac{\pi^2 i}{12} = 2\pi i \left(\frac{\pi}{12} + \frac{i\log 2}{12} - \frac{\pi}{24} \right)$$

and

$$\int_{0}^{\infty} \frac{\log(x)}{(x^2+1)(x^2+4)} dx = -\frac{\pi \log 2}{12}.$$

12.3 Summation of series

Theorem 12.7. *Suppose f is meromorphic in $\mathbb{C}$, with a finite number of poles at $p_1, p_2, \cdots, p_m$ and suppose there exist constants C, R such that for $|z| > R$, $|f(z)| \le \frac{C}{|z|^\alpha}$ for a fixed $\alpha > 1$. Then*

$$\begin{aligned} (i) \sum_{\substack{n=-\infty \\ n \ne p_j}}^{\infty} f(n) &= -\sum_{j=1}^{m} Res(f(z)\pi \cot \pi z, p_j); \\ (ii) \sum_{\substack{n=-\infty \\ n \ne p_j}}^{\infty} (-1)^n f(n) &= -\sum_{j=1}^{m} Res(f(z)\pi \operatorname{cosec} \pi z, p_j). \end{aligned}$$

Remark 12.6. If $|n| > R$, then $|f(n)| \le \frac{C}{n^\alpha}$ so $\sum |f(n)|$ converges by comparison test and hence $\sum f(n)$ and $\sum (-1)^n f(n)$ converge. Theorem 12.7 tells us what they converge to.

Example 12.33. *Find* $\displaystyle\sum_{n=1}^{\infty} \frac{1}{n^2}$.

Solution: Apply Theorem 12.7(i) with $f(z) = \frac{1}{z^2}$. The function f has a pole at 0. Hence

$$\begin{aligned} \sum_{\substack{n=-\infty \\ n \ne 0}}^{\infty} \frac{1}{n^2} &= -\text{Res}\left(\tfrac{1}{z^2}\pi \cot \pi z, 0\right) \\ &= -\left(-\frac{\pi^2}{3}\right) = \frac{\pi^2}{3}. \end{aligned}$$

So

$$\sum_{\substack{n=-\infty \\ n\neq 0}}^{\infty} \frac{1}{n^2} = \sum_{n=1}^{\infty} \frac{1}{n^2} + \sum_{n=-\infty}^{-1} \frac{1}{n^2}$$
$$= 2\sum_{n=1}^{\infty} \frac{1}{n^2}$$

and $\sum_{n=1}^{\infty} \frac{1}{n^2} = \frac{1}{2}\frac{\pi^2}{3} = \frac{\pi^2}{6}$.

Example 12.34. *Find* $\displaystyle\sum_{n=-\infty}^{\infty} \frac{(-1)^n}{n^2+1}$.

Solution: Apply Theorem 12.7(ii) with $f(z) = \dfrac{1}{z^2+1}$. The function f has poles at $\pm i$. Hence

$$\begin{aligned}\sum_{n=-\infty}^{\infty} \frac{(-1)^n}{n^2+1} &= -\text{Res}\left(\frac{\pi\text{cosec}\pi z}{z^2+1}, i\right) - \text{Res}\left(\frac{\pi\text{cosec}\pi z}{z^2+1}, -i\right) \\ &= \frac{-\pi\text{cosec}\pi i}{2i} - \frac{\pi\text{cosec}\,(-\pi i)}{-2i} \\ &= i\pi\text{cosec}\pi i \\ &= \frac{\pi}{\text{sinh}\pi}.\end{aligned}$$

Lemma 12.3. *Let* Γ_N *be the square with vertices at* $\pm(N+\frac{1}{2})\pm i(N+\frac{1}{2})$, $N \in \mathbb{N}$. *Then for all* N *and for all* $z \in \Gamma_N$,

$$(i)|cot\pi z| \leq \sqrt{2}$$

$$(ii)|cosec\pi z| \leq 1.$$

Proof. Note that if $a, b \in \mathbb{R}$, then $\sin(a+ib) = \text{sin}a \cdot \text{cosh}b + i\text{cos}a \cdot \text{sinh}b$. Hence $|\sin^2(a+ib)| = \sin^2 a \cdot \cosh^2 b + \cos^2 a \cdot \sinh^2 b = \sin^2 a(\sinh^2 b + 1) + \cos^2 a \cdot \sinh^2 b = \sin^2 a + \sinh^2 b$. On the vertical sides $z = \pm(N+\frac{1}{2})| + iy$ of Γ_N, $|\sin^2 \pi z| = \sin^2(N+\frac{1}{2})\pi + \sinh^2 \pi y \geq 1$. On the horizontal sides $z = x \pm i(N+\frac{1}{2})$ of Γ_N, $|\sin^2\pi z| = \sin^2 \pi x + \sinh^2(N+\frac{1}{2})\pi \geq \sinh^2\frac{\pi}{2} \geq 1$ since $\text{sinh}x = x + \frac{x^3}{3!} + \cdots \geq x$ if $x \geq 0$. So $|\sin^2 \pi z| \geq 1$ on Γ_N. Hence $|\text{cosec}^2 \pi z| \leq 1$ and $|\cot^2 \pi z| = |\text{cosec}^2 \pi z - 1| \leq |\text{cosec}^2 \pi z| + 1 \leq 2$. □

Proof. (Proof of the theorem)

(i) Let N be an integer greater than $\max\{|p_1|, |p_2|, \cdots, |p_m|, R\}$ and let Γ_N be the square with vertices at $\pm(N+\frac{1}{2}) \pm i(N+\frac{1}{2})$. Inside Γ_N, the function $f(z)\pi\text{cot}\pi z = f(z)\pi\frac{\text{cos}\pi z}{\text{sin}\pi z}$ has poles at $p_1, p_2, \cdots, p_m$ and at the zeros of $\text{sin}\pi z$ (i.e., at $z = n = -N, -(N-1), \cdots, 0, 1, \cdots, N$).

By the residue theorem

$$\int_{\Gamma_N} f(z)\pi\cot\pi z dz = 2\pi i\left[\sum_{j=1}^{m}\text{Res}(f(z)\pi\cot\pi z, p_j) + \sum_{\substack{n=-N\\ n\neq p_j}}^{N}\text{Res}(f(z)\pi\cot\pi z, n)\right].$$

But, if $n \neq p_j$, then by (10.2)

$$\begin{aligned}\text{Res}(f(z)\pi\cot\pi z, n) &= \text{Res}\left(\frac{f(z)\pi\cos\pi z}{\sin\pi z}, n\right)\\ &= f(n)\frac{\pi\cos\pi n}{\pi\cos\pi n}\\ &= f(n).\end{aligned}$$

So, we obtain

$$\int_{\Gamma_N} f(z)\pi\cot\pi z dz = 2\pi i\left\{\sum_{j=1}^{m}\text{Res}(f(z)\pi\cot\pi z, p_j) + \sum_{\substack{n=-N\\ n\neq p_j}}^{N} f(n)\right\}.$$

Now we shall show that

$$\int_{\Gamma_N} f(z)\pi\cot\pi z dz \to 0 \text{ as } N \to \infty.$$

First note that

$$|\cot\pi z| \leq \sqrt{2} \tag{12.22}$$

for all $z \in \Gamma_N$(see Lemma 12.3). Also, if $z \in \Gamma_N$, then $|z| \geq N + \frac{1}{2}$, so by assumption on f, $|f(z)| \leq \frac{C}{(N+\frac{1}{2})^\alpha}$. Also, $L(\Gamma_N) = 8(N + \frac{1}{2})$. So by the estimation theorem,

$$\left|\int_{\Gamma_N} f(z)\pi\cot\pi z dz\right| \leq \frac{C}{(N+\frac{1}{2})^\alpha}\pi 8\sqrt{2}(N + \frac{1}{2})$$

which tends to zero as $N \to \infty$ since $\alpha > 1$.

Thus

$$\begin{aligned}&\lim_{N\to\infty} 2\pi i\left\{\sum_{j=1}^{m}\text{Res}\left(f(z)\pi\cot\pi z, p_j\right) + \sum_{\substack{n=-N\\ n\neq p_j}}^{N} f(n)\right\}\\ &= \lim_{N\to\infty}\int_{\Gamma_N} f(z)\pi\cot\pi z dz = 0.\end{aligned}$$

Hence it follows that

$$\sum_{\substack{n=-\infty \\ n\neq p_j}}^{\infty} f(n) = -\sum_{j=1}^{m} \text{Res}\left(f(z)\pi\cot\pi z, p_j\right).$$

(ii) same proof with $\pi\cot\pi z$ replaced by $\pi\text{cosec}\pi z$. The main difference is that

$$\begin{aligned}\text{Res}\left(f(z)\pi\text{cosec}\pi z, n\right) &= \text{Res}\left(f(z)\frac{\pi}{\sin\pi z}, n\right) \\ &= \frac{f(n)\pi}{\pi\cos\pi n} = (-1)^n f(n).\end{aligned}$$

Also, we replace inequality (12.22) by

$$|\text{cosec}\pi z| \leq 1 \text{ for all } z \in \Gamma_N. \tag{12.23}$$

For proof see Lemma 12.3. □

Example 12.35. *Take $f(z) = \frac{1}{z^4}$. One can show (proceed as in Example 12.33) that $\sum_{n=1}^{\infty} \frac{1}{n^4} = \frac{\pi^4}{90}$.*

Example 12.36.

$$\sum_{n=-\infty}^{\infty} \frac{1}{n^2+a^2} = \frac{\pi}{a}\coth\pi a \text{ where } a > 0.$$

Solution: Let $f(z) = \frac{1}{z^2+a^2}$ which has simple poles at $z = \pm ai$. Now

$$\begin{aligned}\text{Res}\left(\frac{\pi\cot\pi z}{z^2+a^2}, ai\right) &= \lim_{z\to ai}(z-ai)\frac{\pi\cot\pi z}{(z-ai)(z+ai)} \\ &= \frac{\pi\cot\pi ai}{2ai} \\ &= -\frac{\pi}{2a}\coth\pi a.\end{aligned}$$

Further, $\text{Res}\left(\frac{\pi\cot\pi z}{z^2+a^2}, -ai\right) = -\frac{\pi}{2a}\coth\pi a$. Thus by Theorem 12.7,

$$\begin{aligned}\sum_{n=-\infty}^{\infty} \frac{1}{n^2+a^2} &= -\left(-\frac{\pi}{2a}\coth\pi a - \frac{\pi}{2a}\coth\pi a\right) \\ &= \frac{\pi}{a}\coth\pi a.\end{aligned}$$

Example 12.37. *Find*

$$\sum_{n=1}^{\infty} \frac{(-1)^n}{(2n+1)^3}.$$

Solution: Let $f(z) = \frac{\pi}{z^3 \cos \pi z} = \frac{\pi \sec \pi z}{z^3}$. The poles of f are at $z = 0$ and at the zeros of $\cos\pi z$ (i.e., at $z = n + \frac{1}{2}, n = 0, \pm 1, \pm 2, \cdots$). Since $f(z) = \frac{\pi}{z^3 \cos \pi z} = \frac{\pi}{z^3(1-\frac{\pi^2 z^2}{2!}+\cdots)} = \frac{\pi}{z^3}(1 + \frac{\pi^2 z^2}{2!} + \cdots) = \frac{\pi}{z^3} + \frac{\pi^3}{2z} + \cdots$ hence $\text{Res}(f, 0) = \frac{\pi^3}{2}$.

Further,

$$\begin{aligned} \text{Res}\left(f, n + \frac{1}{2}\right) &= \lim_{z \to n + \frac{1}{2}} \left(z - (n + \frac{1}{2})\right) \frac{\pi}{z^3 \cos \pi z} \\ &= \frac{\pi}{\left(n + \frac{1}{2}\right)^3} \lim_{z \to n + \frac{1}{2}} \frac{z - \left(n + \frac{1}{2}\right)}{\cos \pi z} \\ &= \frac{-(-1)^n}{\left(n + \frac{1}{2}\right)^3} \end{aligned}$$

as $z = n + \frac{1}{2}, n \in \mathbb{Z}$ are simple poles of f.

Now if Γ_N is a square with vertices at $\pm N \pm iN$, then

$$\begin{aligned} \frac{1}{2\pi i} \int_{\Gamma_N} f(z) dz &= -\sum_{\substack{n=-N \\ n \neq 0}}^{N} \frac{(-1)^n}{\left(n + \frac{1}{2}\right)^3} + \frac{\pi^3}{2} \\ &= -8 \sum_{\substack{n=-N \\ n \neq 0}}^{N} \frac{(-1)^n}{(2n+1)^3} + \frac{\pi^3}{2}. \end{aligned}$$

Since $\int_{\Gamma_N} f(z)dz$ approaches zero as $N \to \infty$, we have $-8 \sum_{\substack{n=-\infty \\ n \neq 0}}^{\infty} \frac{(-1)^n}{(2n+1)^3} = -\frac{\pi^3}{2}$. Thus $\sum_{n=1}^{\infty} \frac{(-1)^n}{(2n+1)^3} = \frac{\pi^3}{32}$. Notice $L(\Gamma_N) = 8N$ and on Γ_N, $|z| \geq N$. Further

$$\left| \int_{\Gamma_N} \frac{\pi}{z^3 \cos \pi z} dz \right| = \left| \int_{\Gamma_N} \frac{\pi \sec \pi z}{z^3} dz \right| \leq \frac{\pi 8 \sqrt{2} N}{N^3}$$

which tends to zero as $N \to \infty$. If $a, b \in \mathbb{R}$,

$$|\cos(a + ib)|^2 = \cos^2 a + \sinh^2 b = \frac{1}{2}(\cos 2a + \cosh 2b).$$

On the vertical sides $z = \pm N + iy$ of Γ_N, $|\cos^2 \pi z| = \cos^2 N\pi + \sinh^2 \pi y \geq 1$. On the horizontal sides $z = x \pm iN$ of Γ_N, $|\cos^2 \pi z| = \frac{1}{2}(\cos 2\pi x + \cosh 2\pi N) \geq \frac{1}{2}$ as $\cosh x = \frac{e^x + e^{-x}}{2} = \frac{(1+x+\frac{x^2}{2!}+\frac{x^3}{3!}+\cdots)+(1-x+\frac{x^2}{2!}-\frac{x^3}{3!}+\cdots)}{2} = 1 + \frac{x^2}{2!} + \cdots > 1$ if $x > 0$. Thus $|\cos^2 \pi z| \geq \frac{1}{2}$ for all $z \in \Gamma_N$. Hence $|\sec^2 \pi z| \leq 2$ for all $z \in \Gamma_N$.

Example 12.38.

$$\sum_{n=-\infty}^{\infty} \frac{(-1)^n}{(n+\alpha)^2} = \frac{\pi^2 cos\pi\alpha}{sin^2\pi\alpha}$$

where α is real but $\alpha \notin \mathbb{Z}$.

Solution: Let $f(z) = \dfrac{1}{(z+\alpha)^2}$. The function f has a double pole at $z = -\alpha$ and

$$\begin{aligned}\text{Res}(f, -\alpha) &= \lim_{z\to-\alpha} \frac{d}{dz}\left\{(z+\alpha)^2 \frac{\pi\text{cosec}\pi z}{(z+\alpha)^2}\right\} \\ &= -\pi^2\text{cosec}\pi\alpha \cdot \cot\pi\alpha.\end{aligned}$$

Thus by Theorem 12.7,

$$\begin{aligned}\sum_{n=-\infty}^{\infty} \frac{(-1)^n}{(n+\alpha)^2} &= -\left(-\pi^2\text{cosec}\pi\alpha \cdot \cot\pi\alpha\right) \\ &= \frac{\pi^2\cos\pi\alpha}{\sin^2\pi\alpha}.\end{aligned}$$

12.4 The Laplace transform and its inversion

Given a function $f(t)$ defined for real variable t, and a complex parameter $s = \sigma + i\tau$, the Laplace transform of $f(t)$ is defined by

$$F(s) = \mathcal{L}\{f(t)\} = \int_0^{\infty} e^{-st} f(t)dt \tag{12.24}$$

provided that the integral exists.

The integral (12.24) is an improper integral, dependent on the variable s as a parameter. In general the integral (12.24) does not converge for all values of the parameter s. In fact, if the function $f(t)$ approaches a nonzero limit as $t \to \infty$, and Re $s = \sigma < 0$, then the integral (12.24) diverges. It is therefore natural to pose the problem of region of convergence of the integral (12.24) and hence also that of the domain of definition of the function $F(s)$. Further, the integral (12.24) is not meaningful for every function $f(t)$. Therefore we begin with a class of functions $f(t)$ for which the given transformation can be realized.

Consider the function $f(t)$ defined for $t \in \mathbb{R}$ and satisfying the following conditions:

(i) $f(t) = 0$ for $t < 0$.

(ii) For $t \geq 0$, the function f is piecewise continuous (has at most a finite number of discontinuities) in every finite interval in $\mathbb{R}$.

(iii) The function $f(t)$ has a bounded order of growth at $t \to +\infty$, that is, there exist positive constants M and a, such that for all $t > 0$,

$$|f(t)| \leq Me^{at}. \tag{12.25}$$

The greatest lower bound of those values of a for which inequality (12.25) holds is called the growth exponent (index of the order of growth) of the function $f(t)$ and the function f that satisfies condition (12.25) is said to be of exponential order.

Example 12.39. *The growth exponent of the function $f(t) = t^n$ is zero.*

Theorem 12.8. *The integral (12.24) converges in the region Re $s = \sigma > a$, where a is the index of the order of growth of the function $f(t)$, and in the region Re $s = \sigma \geq \sigma_0 > a$ the integral converges uniformly and absolutely.*

Proof. For any $s = \sigma + i\tau$ for $\sigma > a$, we can specify an $\epsilon > 0$ such that $\sigma > a_1 = a + \epsilon$, and $|f(t)| < Me^{a_1 t}$. Thus we get

$$|F(s)| = \left| \int_0^\infty e^{-st} f(t) dt \right| \leq M \int_0^\infty e^{-\sigma t} e^{a_1 t} dt = \frac{M}{\sigma - a_1}, \sigma > a_1. \tag{12.26}$$

Hence the integral (12.24) converges for $\sigma > a$. If $\sigma \geq \sigma_0 > a$, then

$$|F(s)| \leq \int_0^\infty e^{-(\sigma_0 - a_1)t} dt = \frac{M}{\sigma_0 - a_1}. \tag{12.27}$$

By Weierstrass test, it follows that the integral (12.24) converges uniformly with respect to the parameter s in the region Re $s = \sigma \geq \sigma_0 > a$. □

Further from (12.27) it follows that $|F(s)| \to 0$ as Re $s \to \infty$.

In summary, any piecewise continuous function of exponential order has a Laplace transform. These conditions are sufficient but not necessary. For example, the function $f(t) = \frac{1}{\sqrt{t}}$ has a singularity at $t = 0$ and so it is not continuous, but its Laplace transform exists.

Further note that the proof of Theorem 12.8 depend substantially upon the conditions (ii) and (iii) of the definition of this class of functions $f(t)$ of the real variable t. However, it is possible to extend the class of functions $f(t)$ which admit the Laplace transformation.

Lemma 12.4. *Let the function $f(t)$ of the real variable t be defined for all $t \geq 0$ and let s_0 be a complex number such that the integral*

$$\left| \int_0^\infty e^{-s_0 t} f(t) dt \right| < M \tag{12.28}$$

converges. Then for all s satisfying the condition Re $s >$ Re s_0 the integral

$$\int_0^\infty e^{-st} f(t) dt \tag{12.29}$$

converges.

Proof. Let $\phi(t) = e^{-s_0 t} f(t)$ and $F(t) = -\int_t^\infty \phi(\tau)d\tau$. Then $F'(t) = \phi(t)$. Again from (12.28) it follows that for a given $\epsilon' > 0$, there exists a T_0 such that $|F(t)| < \epsilon'$ for $t \geq T_0$. Let T_1 and T_2 be two real numbers such that $T_2 > T_1$ and consider the integral $\int_{T_1}^{T_2} e^{-st} f(t)dt$.

Notice that $\int_{T_1}^{T_2} e^{-st} f(t)dt = \int_{T_1}^{T_2} e^{-(s-s_0)t}\phi(t)dt = \int_{T_1}^{T_2} e^{-(s-s_0)t}F'(t)dt$. Evaluating the last integral by parts, we obtain

$$\begin{aligned}\int_{T_1}^{T_2} e^{-(s-s_0)t}F'(t)dt &= e^{-(s-s_0)T_2}F(T_2) - e^{-(s-s_0)T_1}F(T_1)\\ &+ (s-s_0)\int_{T_1}^{T_2} e^{-(s-s_0)t}F(t)dt.\end{aligned}$$

From this, for $T_1, T_2 > 0$ and Re $(s - s_0) > 0$, we obtain

$$\begin{aligned}\left|\int_{T_1}^{T_2} e^{-st}f(t)dt\right| &\leq \left(e^{-\text{Re }(s-s_0)T_2} + e^{-\text{Re }(s-s_0)T_1}\right)\epsilon'\\ &+ \epsilon'\frac{|s-s_0|}{\text{Re }(s-s_0)}\left(e^{-\text{Re }(s-s_0)T_1} - e^{-\text{Re }(s-s_0)T_2}\right)\\ &< \epsilon'\left[2 + \frac{|s-s_0|}{\text{Re }(s-s_0)}\right]e^{-\text{Re }(s-s_0)T_0}.\end{aligned}$$

Since it is always possible to choose the value of T_0 such that the expression obtained is less than any prescribed $\epsilon > 0$, hence the integral (12.29) converges. □

Note that in the region Re $s \geq$ Re $s_1 >$ Re s_0, the integral (12.29) converges uniformly in the parameter s. From Lemma 12.4, it follows that we can regard the functions satisfying the condition (12.28) as the basic class of functions $f(t), t \in \mathbb{R}$ for which the Laplace transform (12.24) can be defined.

We shall now show that the Laplace transform is analytic in the region Re $s > a$ in the s-plane.

Theorem 12.9. *The Laplace transform (12.24) of the function $f(t)$ is an analytic function of the complex variable s in the domain Re $s > a$, where a is the index of the order of growth of $f(t)$.*

Proof. By Theorem 12.8, the improper integral converges in the domain Re $s > a$. Now partition the integration interval into subintervals $[t_i, t_{i+1}]$ of arbitrary finite length. Take $t_0 = 0$ and also assume $t_n \to \infty$ as $n \to \infty$. Then the function $F(s)$, for Re $s > a$, is the sum of the convergent series

$$F(s) = \sum_{n=0}^{\infty}\int_{t_n}^{t_{n+1}} e^{-st}f(t)dt = \sum_{n=0}^{\infty} u_n(s). \tag{12.30}$$

Note that since the nth remainder of the series (12.30) is equal to $\int_{t_n}^{t_{n+1}} e^{-st}f(t)dt$, by Theorem 12.8 the series (12.30) converges uniformly in the region Re $s \geq \sigma_0 > a$. Now each of the functions $u_n(s) = \int_{t_n}^{t_{n+1}} e^{-st} f(t)dt$ is defined as an integral, dependent on the parameter s, over a subinterval of finite length in the complex plane. It can easily be checked that the functions $u_n(p)$ are entire functions of p. Thus it follows from Weierstrass test that the series in (12.30) converges uniformly in the region Re $s > a$, and the function $F(s)$ is analytic in the domain Re $s > a$ and its derivative may be computed by differentiating the integrand function in (12.24) with respect to the variable s. □

The Laplace transform of some elementary functions are shown below:

$$\begin{aligned}
\mathcal{L}\{e^{at}\} &= \int_0^\infty e^{-st}e^{at}dt &&= \tfrac{1}{s-a}, \text{Re } s > a.\\
\mathcal{L}\{t^n\} &= \int_0^\infty e^{-st}t^n dt &&= \tfrac{n}{s}\int_0^\infty e^{-st}t^{n-1}dt = \tfrac{n!}{s^{n+1}}, \text{Re } s > 0.\\
\mathcal{L}\{\sin wt\} &= \text{Im}\int_0^\infty e^{-st}e^{iwt}dt &&= \text{Im}\tfrac{1}{s-iw} = \tfrac{w}{s^2+w^2}, \text{Re } s > 0.\\
\mathcal{L}\{\cosh \beta t\} &= \int_0^\infty e^{-st}\tfrac{e^{\beta t}+e^{-\beta t}}{2}dt &&= \tfrac{1}{2}\left(\tfrac{1}{s-\beta}+\tfrac{1}{s+\beta}\right) = \tfrac{s}{s^2-\beta^2}, s > |\beta|.
\end{aligned}$$

Inverse Laplace transforms

In this section we consider methods for determining the original function from a given transform, and we also give certain sufficient conditions under which a given function $F(s)$ of complex variable s is a transform of the function $f(t)$ of the real variable t. We begin with the case when it is known that the given function $F(s)$ of the complex variable s is the transform of a piecewise smooth function $f(t)$ with bounded order of growth $|f(t)| < Me^{at}$; the value of the constant a is given. It is required, from the given function $F(s)$, to construct the desired function $f(t)$. This problem is solved with the aid of the following theorem.

Theorem 12.10. *Suppose in the region Re $s > a$, the function $F(s)$ is the Laplace transform of a piecewise smooth function $f(t)$ of a real variable t and possesses an order of growth a. Then*

$$f(t) = \frac{1}{2\pi i}\int_{\sigma-i\infty}^{\sigma+i\infty} e^{st}F(s)ds, \sigma > a. \tag{12.31}$$

Proof. By hypothesis, the function $f(t)$ exists and we know its order of growth. Let $\phi(t) = e^{-\sigma t}f(t), \sigma > a$. This is a piecewise smooth function which on any bounded interval of the $t-$axis has a finite number of discontinuities and exponentially tends to zero as $t \to \infty$. We apply the Fourier transform inversion formula to the function to obtain

$$\phi(t) = \frac{1}{2\pi}\int_{-\infty}^{\infty} d\xi \int_{-\infty}^{\infty} \phi(\eta)e^{i\xi(t-\eta)}d\eta. \tag{12.32}$$

Putting into (12.32) the expression of the function $\phi(t)$ in terms of the sought-for function $f(t)$, we get

$$\begin{aligned} e^{-st}f(t) &= \frac{1}{2\pi}\int_{-\infty}^{\infty} d\xi \int_{-\infty}^{\infty} e^{-\sigma\eta} f(\eta) e^{i\xi(t-\eta)} d\eta \\ &= \frac{1}{2\pi}\int_{-\infty}^{\infty} e^{i\xi t} d\xi \int_{0}^{\infty} e^{-(\sigma+i\xi)\eta} f(\eta) d\eta \end{aligned} \tag{12.33}$$

since $f(\eta) = 0$ for $\eta < 0$.

Denote $s = \sigma + i\xi$ and note that the inner integral in (12.33) is the given transform $F(s)$ of the sought-for function $f(t)$. Then the expression (12.33) becomes

$$f(t) = \frac{1}{2\pi}\int_{-\infty}^{\infty} e^{(\sigma+i\xi)t} F(s) d\xi = \frac{1}{2\pi i}\int_{\sigma-i\infty}^{\sigma+i\infty} e^{st} F(s) ds.$$

This proves the theorem. □

Formula (12.31) is often called Mellin's formula. In a sense, it is the inverse of the Laplace transformation since it expresses the original function in terms of a given transform. Observe that in formula (12.31) the integration is performed in the complex plane along a straight line parallel to the imaginary axis and passing to the right of the straight line Re $s = a$. The value of the integral (12.31) is independent of σ, provided that the straight line of integration lies to the right of the straight line Re $s = a$. Further note that the integral in (12.31) converges to $f(t)$ only at the points of continuity of this function and the formula (12.31) defines the function $f(t)$ only at its points of continuity. The vertical line Re $s = a$ is called the Bromwich line, and the inversion integral in (12.31) is called the Bromwich integral. Below we will consider certain sufficient conditions under which a given function $F(s), s \in \mathbb{C}$ is the Laplace transform of some function $f(t), t \in \mathbb{R}$, and will show how the latter may be found.

Theorem 12.11. *Let the function $F(s), s = \sigma + i\tau \in \mathbb{C}$ satisfy the following conditions:*

(i) $F(s)$ is an analytic function in the domain Re $s > a$;

(ii) In the region Re $s = \sigma > a$, the function $F(s)$ tends to zero as $|s| \to \infty$ uniformly in arg s;

(iii) For all Re $s = \sigma > a$, the following integral converges:

$$\int_{\sigma-i\infty}^{\sigma+i\infty} |F(s)| d\tau < M, \sigma > a. \tag{12.34}$$

Then $F(s)$, for $Res > a$, is the Laplace transform of the function $f(t), t \in \mathbb{R}$, which is defined by the expression

$$f(t) = \frac{1}{2\pi i} \int_{\sigma-i\infty}^{\sigma+i\infty} e^{st} F(s) ds, \sigma > a. \tag{12.35}$$

The last integral is to be taken in the sense of the principal value, that is,

$$\int_{\sigma-i\infty}^{\sigma+i\infty} e^{st} F(s) ds = \lim_{A\to\infty} \int_{\sigma-iA}^{\sigma+iA} e^{st} F(s) ds.$$

Proof. We have to show that the integral (12.35) is the function whose Laplace transform is the function $F(s)$. First the question arises as to the existence of this improper integral. Clearly

$$\begin{aligned} \left| \frac{1}{2\pi i} \int_{\sigma-i\infty}^{\sigma+i\infty} e^{st} F(s) ds \right| &\leq \frac{1}{2\pi} \int_{\sigma-i\infty}^{\sigma+i\infty} |e^{st} F(s)||ds| \\ = \frac{e^{\sigma t}}{2\pi} \int_{\sigma-i\infty}^{\sigma+i\infty} |F(s)| d\tau &\leq \frac{M}{2\pi} e^{\sigma t}. \end{aligned} \tag{12.36}$$

This proves the convergence of the integral (12.35) for any $\sigma > a$. It also follows from (12.36) that the integral (12.35) converges uniformly with respect to the variable t on any finite interval $0 \leq t \leq T$.

In order to prove that the integral (12.35) represents a function whose Laplace transform is $F(s)$, we have to establish that:

(1) the integral (12.35) is independent of σ and defines the function $f(t)$ and this function f has a bounded order of growth;

(2) for $t < 0$, $f(t) = 0$;

(3) the given function $F(s)$ is the Laplace transform of the function $f(t)$.

For proof of (1) observe the following: In the domain Re $s = \sigma > a$, consider a closed contour Γ consisting of segments of straight lines $[\sigma_1 - iA, \sigma_1 + iA]$ and $[\sigma_2 - iA, \sigma_2 + iA]$ parallel to the imaginary axis, and of the straight lines connecting them $[\sigma_1 - iA, \sigma_2 - iA]$, $[\sigma_1 + iA, \sigma_2 + iA]$, which are parallel to the real axis (see Fig.12.23). In Fig.12.23, we have $p = s, x = \sigma, x_1 = \sigma_1$ and $x_2 = \sigma_2$. Here, $A > 0$ and σ_1, σ_2 are arbitrary numbers greater than a. Since the function $F(s)$ is analytic in the region Re $s = \sigma > a$, then by the Cauchy theorem the integral of the function $e^{st}F(s)$ around the contour Γ is zero. Let A tend to infinity and let σ_1 and σ_2 remain fixed. Since $|F(s)| \to 0$ as $|s| \to \infty$ in the region Re $s = \sigma > a$, the integrals over the horizontal segments of the contour will in the limit yield zero whereas the integrals over the vertical straight lines will pass into the integral (12.35). This yields

$$\int_{\sigma_1-i\infty}^{\sigma_1+i\infty} e^{st} F(s) ds = \int_{\sigma_2-i\infty}^{\sigma_2+i\infty} e^{st} F(s) ds$$

which, since σ_1 and σ_2 are arbitrary, proves (1). Thus the integral (12.35) is a function of the variable t alone.

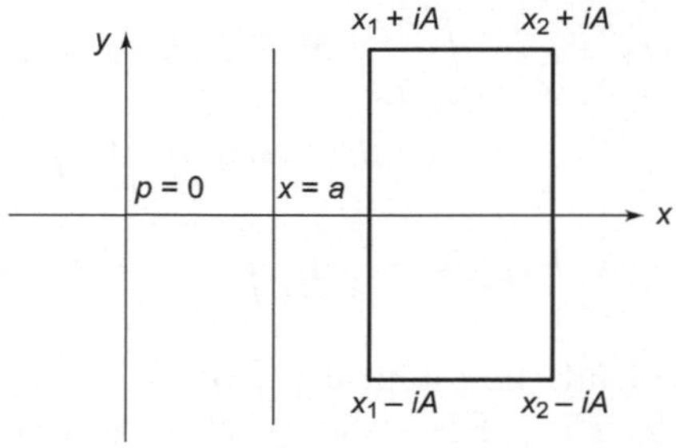

Fig. 12.23: The closed contour Γ

It follows from (12.36) that the integral (12.35) is a function of bounded order growth with respect to t, and the index of the order of growth of this function is a.

For proof of (2), consider the value of the integral (12.35) for $t < 0$. To do this, consider, in the domain Re $s = \sigma > a$, a closed contour C (see Fig.12.24) consisting of the straight-line segment $[\sigma - iR, \sigma + iR], \sigma > a$, and of the arc C'_R of the semicircle $|s - \sigma| = R$ completing it. In Fig.12.24, we have $p = s, x = \sigma$. By the Cauchy theorem, the integral of the function $e^{st}F(s)$ over the given contour is zero. By the remark concerning Jordan's lemma (Lemma 12.2), the integral over the arc C'_R tends to zero for $t < 0$ as $R \to \infty$. Therefore

$$f(t) = \frac{1}{2\pi i} \int_{\sigma - i\infty}^{\sigma + i\infty} e^{st} F(s) ds = 0, t < 0, \text{Re } s > a$$

and (2) is proved.

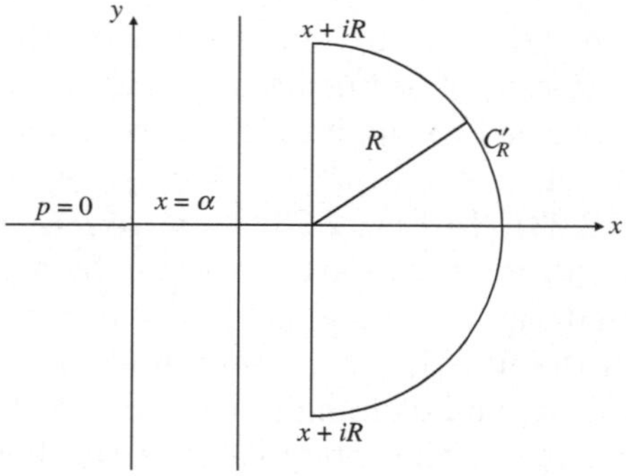

Fig. 12.24: The closed contour C

For proof of (3), construct the Laplace transform of the function (12.35) and consider its value for some arbitrary s_0, Re $s_0 > a$;

$$\int_0^\infty e^{-s_0 t} f(t)dt = \frac{1}{2\pi i}\int_0^\infty e^{-s_0 t}dt \int_{\sigma-i\infty}^{\sigma+i\infty} e^{st}F(s)ds. \tag{12.37}$$

The inner integral in (12.37) is independent of σ. Choose a value of σ that satisfies the condition $a < \sigma < \text{Re } s_0$ and change the order of integration. This is possible by virtue of the uniform convergence of the corresponding integrals. Now

$$\begin{aligned}\int_0^\infty e^{-s_0 t} f(t)dt &= \frac{1}{2\pi i}\int_{\sigma-i\infty}^{\sigma+i\infty} F(s)ds \int_0^\infty e^{-(s_0-s)t}dt \\ &= \frac{1}{2\pi i}\int_{\sigma-i\infty}^{\sigma+i\infty} F(s)\frac{ds}{s_0-s}.\end{aligned} \tag{12.38}$$

The integral (12.38) can be computed with the aid of residues, since by condition (ii) of the theorem, the integrand function tends to zero faster than the function $\frac{1}{p}$ as $|p| \to \infty$. Therefore, taking into account that the only singularity of the integrand function-a simple pole at $s = s_0$ and, upon completion of (12.38) in the right half plane, the integration is performed in the negative direction, we obtain

$$\mathcal{L}(f(t)) = \int_0^\infty e^{-s_0 t} f(t)dt = F(s_0).$$

Since s_0 is an arbitrary point in the region Re$s > a$, the theorem is proved. □

With the aid of earlier methods of computing contour integrals we can also get an expression of the function $f(t)$ whose Laplace transform is given to be $F(s)$. Suppose $F(s)$ is defined on the region Re $s > a$, be analytically extended to the entire $s-$plane. Let its analytic extension satisfy, for Re $s < a$, the conditions of Jordan's lemma. For $t > 0$,

$$\int_{C'_R} e^{st}F(s)ds \to 0, R \to \infty$$

where C'_R is the arc of the semicircle $|s - \sigma| = R$ in the left half plane. In this case, the integral (12.35) may be computed with the help of calculus of residues. Let us consider some examples.

Example 12.40. *Find the function $f(t)$ whose Laplace transform is $F(s) = \frac{w}{s^2+w^2}$, Re $s > 0, w^2 > 0$. Since the condition of Theorem 12.11 are fulfilled,*

$$f(t) = \frac{1}{2\pi i}\int_{\sigma-i\infty}^{\sigma+i\infty} e^{st}\frac{w}{s^2+w^2}ds, \sigma > 0.$$

The analytic continuation of the function $F(s)$ into the left half plane Re $s < 0$, the function $\frac{w}{s^2+w^2}$, satisfies the conditions of the Jordan lemma, and has two singularities-simple poles at $p_1 = +iw, p_2 = -iw$. Therefore, for t ≥ 0,

$$f(t) = \sum_{k=1}^{2} Res \left[e^{st} \frac{w}{s^2 + w^2}, p_k \right] = \frac{we^{iwt}}{2iw} - \frac{we^{-iwt}}{2iw} = \sin wt, t \geq 0.$$

The conditions of Theorem 12.11, in particular (iii), are sufficient conditions for the existence of the function $f(t)$ whose Laplace transform is $F(s)$, analytic in the region Re $s > a$. It is easy to indicate examples which show that if this condition does not hold, then the function $F(s)$ can still be the Laplace transform of some function of a real variable.

Example 12.41. *The function $f(t)$ whose Laplace transform is $F(s) = \frac{1}{s^{\alpha+1}}$, $-1 < \alpha < 0, Re\ s > 0$ is given by $f(t) = \frac{t^\alpha}{\Gamma(1+\alpha)}$.*

12.5 Exercises

1. Evaluate the following integral using calculus of residue:

(a) (i) $\displaystyle\int_0^{2\pi} \frac{d\theta}{a + \cos\theta}, a \in \mathbb{R}, a > 1$ (ii) $\displaystyle\int_0^{\pi} \frac{\cos 4\theta}{1 + \cos^2\theta} d\theta$

(iii) $\displaystyle\int_{-\pi}^{\pi} \frac{\sin n\theta}{\sin\theta} d\theta, n = 1, 2, \cdots$ (iv) $\displaystyle\int_0^{\pi} \frac{d\theta}{(a + \cos\theta)^2}, a > 1$

(v) $\displaystyle\int_0^{\frac{\pi}{2}} \frac{d\theta}{a + \sin^2\theta}, a > 0$ (vi) $\displaystyle\int_{-\pi}^{\pi} \frac{d\theta}{2 - \cos\theta}$

(b)

(i) $\displaystyle\int_0^{\infty} \frac{dx}{(x^2 + a^2)^2}, a > 0$ (ii) $\displaystyle\int_{-\infty}^{\infty} \frac{x^2}{x^4 + 5x^2 + 4} dx$

(iii) $\displaystyle\int_0^{\infty} \frac{x^2 + 1}{x^4 + 1} dx$ (iv) $\displaystyle\int_0^{\infty} \frac{x^2}{x^6 + 1} dx$

(v) $\displaystyle\int_{-\infty}^{\infty} \frac{dx}{(x^2 + 1)^2 (x^2 + 4)}$ (vi) $\displaystyle\int_{-\infty}^{\infty} \frac{x \sin x}{(1 + x^2)^2} dx$

(vii) $\displaystyle\int_0^{\infty} \frac{\cos ax}{(1 + x^2)^2} dx, a > 0$ (viii) $\displaystyle\int_{-\infty}^{\infty} \frac{x \sin x}{x^2 + 1} dx$

(ix) $\displaystyle\int_{-\infty}^{\infty} \frac{\cos x}{(x^2 + a^2)(x^2 + b^2)} dx,$ $0 < a < b$ (x) $\displaystyle\int_{-\infty}^{\infty} \frac{dx}{x^4 + 1}$

(c)

(i) $\displaystyle\int_{-\infty}^{\infty} \frac{x^3 \sin x}{(1+x^2)^2} dx$

(ii) $\displaystyle\int_{0}^{\infty} \frac{x - \sin x}{x^3} dx$

(iii) $\displaystyle\int_{-\infty}^{\infty} \frac{\cos nx}{x^4+1} dx$

(d)

(i) $\displaystyle\int_{0}^{\infty} \frac{1-\cos ax}{x^2} dx$ (ii) $\displaystyle\int_{0}^{\pi} \cos^{2n}\theta d\theta, n \in \mathbb{N}$

(iii) $\displaystyle\int_{0}^{\infty} \frac{(\log x)^3}{1+x^2} dx$ (iv) $\displaystyle\int_{0}^{\infty} \frac{\log x}{(1+x^2)^2} dx$

(v) $\displaystyle\int_{0}^{\infty} \frac{x^{\frac{1}{3}}}{1+x^2} dx$

2. Verify the following using calculus of residue:

(i) $\displaystyle\sum_{n=-\infty}^{\infty} \frac{1}{a^2-n^2} = \frac{\pi \cot \pi a}{2a}, a \notin \mathbb{Z}$

(ii) $\displaystyle\sum_{n=0}^{\infty} \frac{1}{(2n+1)^2} = \frac{\pi^2}{8}$

(iii) $\displaystyle\sum_{n=1}^{\infty} \frac{1}{n^4} = \frac{\pi^4}{90}$

(iv) $\displaystyle\sum_{n=1}^{\infty} \frac{(-1)^n}{n^2} = \frac{-\pi^2}{12}$

(v) $\displaystyle\sum_{n=-\infty}^{\infty} \frac{1}{(2n+1)(3n+1)} = \frac{\pi}{\sqrt{3}}$

(vi) $\displaystyle\sum_{n=-\infty}^{\infty} \frac{(-1)^n}{n^2+a^2} = \frac{\pi}{a} \csc \pi a$

(vii) $\displaystyle\sum_{n=1}^{\infty} \frac{n^2}{n^4+5n^2+4} = \frac{\pi}{6}(2\coth 2\pi - \coth \pi).$

3. Sum the following using calculus of residue:

(i) $\sum\limits_{n=-\infty}^{\infty} \frac{1}{n^2+a^2}$

(ii) $\sum\limits_{n=-\infty}^{\infty} \frac{n^3}{n^4+a^4}$

(iii) $\sum\limits_{n=-\infty}^{\infty} (-1)^n \frac{n \sin na}{a^2-n^2}, a \notin \mathbb{Z}, -\pi < a < \pi.$

4. Show that

(a) $\int_0^\infty \frac{\sin \alpha x}{x(x^2+a^2)} dx = \frac{\pi}{2a^2}(1-e^{-a\alpha})$

(b) $\int_0^\infty \frac{\sin \alpha x}{x(x^2+a^2)^2} dx = \frac{\pi}{2a^4}(1-\frac{a\alpha+2}{2}e^{-a\alpha}).$

(Hints: Apply Theorem 6.17 to the equation in 4(a) to verify the result in 4(b).)

Chapter 13

Extended Complex Plane

As we have seen there are several advantages in using the set $\mathbb{C}$ of complex numbers as the domain of definition of functions. The complex numbers form a field which is algebraically closed, that is, polynomials of degree n have n roots in $\mathbb{C}$, counting multiplicities. Geometrically, $\mathbb{C}$ can be regarded as the Euclidean plane $\mathbb{R}^2$, probably the most familiar geometric structure of all (hence we sometimes call $\mathbb{C}$ the complex plane). As a domain of definition of functions, $\mathbb{C}$ has the following remarkable property: if f is a function of a complex variable and is differentiable on some region $G \subseteq \mathbb{C}$, then f is infinitely differentiable on G, and for each $a \in G$ we can expand f as a convergent power series in some sufficiently small disk containing a. In contrast, there are functions of a real variable which are once but not twice differentiable, or which are infinitely differentiable but cannot be represented by power series. There are, however, some disadvantages in using $\mathbb{C}$. Division by 0 is impossible, and so some standard functions are not defined everywhere; for example, $\frac{1}{z}$ is undefined at $z = 0$. There is a rather less obvious disadvantage in that $\mathbb{C}$ is not compact, so that certain sequences have no convergent subsequences. We can avoid these disadvantages, and retain some of the advantages, by using the extended complex plane $\mathbb{C}_\infty$. Geometrically, $\mathbb{C}_\infty$ is still very well behaved, for as we shall now show, $\mathbb{C}_\infty$ may be regarded as being a sphere. The extended complex plane is often referred to as the Riemann sphere.

13.1 Point at infinity

We add an ideal point ∞ (call it the point at infinity) to $\mathbb{C}$ and denote the new set by $\mathbb{C}_\infty = \mathbb{C} \cup \{\infty\}$. The set $\mathbb{C}_\infty$ is called the extended complex plane. We extend the algebraic operations on $\mathbb{C}$ to $\mathbb{C}_\infty$ in a limited way as follows:

$$\infty \pm z = \infty = z \pm \infty \text{ for all } z \in \mathbb{C};$$

$$\infty \cdot z = \infty = z \cdot \infty \text{ for } z \in \mathbb{C} \setminus \{0\} = \mathbb{C}^*;$$

$$\frac{z}{0} = \infty \text{ for } z \in \mathbb{C}^*;$$

$$\frac{z}{\infty} = 0 \text{ for } z \in \mathbb{C};$$
$$\infty \cdot \infty = \infty.$$

Note that we do not assign any meaning to any of the following expressions: $\infty \pm \infty, \frac{\infty}{\infty}, \infty \cdot 0$ and $0 \cdot \infty$. It is obvious that though $\mathbb{C}$ is a field, $\mathbb{C} \cup \{\infty\}$ is not a field.

We now endow $\mathbb{C}_\infty$ with a topology, called the topology of one point compactification, on $\mathbb{C}_\infty$. A subset $U \subset \mathbb{C}_\infty$ is said to be open either if $\infty \notin U$(and hence $U \subset \mathbb{C}$) and U is open in $\mathbb{C}$ or if $\infty \in U$ and the set $\mathbb{C}_\infty \setminus U \subset \mathbb{C}$ is compact in $\mathbb{C}$. It is easy to show that the class of open sets in $\mathbb{C}_\infty$ defines a topology on $\mathbb{C}_\infty$. (If V is open in $\mathbb{C}_\infty$, then $V \setminus \{\infty\}$ is open in $\mathbb{C}$.) Thus a set $A \subset \mathbb{C}_\infty$ is a neighborhood of ∞ if $\infty \in A$ and there exist $r > 0$ such that $\{z : |z| > r\} \subset A$. The sets $\{\infty\} \cup \mathbb{C} \setminus \overline{B(0,n)}, n \in \mathbb{N}$ are open in $\mathbb{C}_\infty$ and they form a local basis at ∞. We have seen in Theorem 13.2 that $\mathbb{C}_\infty$ is compact.

We shall give a geometrical description of $\mathbb{C}_\infty$ which we obtained by adding an extra point to $\mathbb{C}$ to compactify it. In fact we shall show that $\mathbb{C}_\infty$ is homeomorphic to the 2−sphere

$$S^2 = \{(x_1, x_2, x_3) \in \mathbb{R}^3 : x_1^2 + x_2^2 + x_3^2 = 1\}.$$

Note that the sphere is given the subspace topology.

Theorem 13.1. *The spaces $\mathbb{C}_\infty$ and S^2 are homeomorphic.*

Proof. Since both spaces are compact (see Theorem 13.2 and Theorem 2.43) and Hausdorff, it is enough to find a continuous bijection from one to the other (see Theorem 2.53). The bijection is the stereographic projection defined as follows. Let N denote the north pole $(0,0,1) \in S^2$. Also, identify $\mathbb{C}$ with $\{(x_1, x_2, 0) : x_1, x_2 \in \mathbb{R}\}$ so that $\mathbb{C}$ cuts S^2 along the equator. Now for each point z in $\mathbb{C}$ consider the straight line in $\mathbb{R}^3$ through z and N. This intersects the sphere in exactly one point $Z \neq N$. Now put $z = x+iy$ and let $Z = (x_1, x_2, x_3)$ be the corresponding point on S^2. We shall find equations expressing x_1, x_2 and x_3 in terms of x and y and vice versa. The line in $\mathbb{R}^3$ through z and N is given by $\{tN + (1-t)z : -\infty < t < \infty\}$, or by $\{((1-t)x, (1-t)y, t) : -\infty < t < \infty\}$.

Hence, we can find the coordinates of Z if we can find the value of t at which this intersects S. In analytical terms, we are looking for the values of t so that

$$(1-t)^2 x^2 + (1-t)^2 y^2 + t^2 = 1,$$

i.e.,

$$(1-t)^2 |z|^2 + t^2 = 1.$$

From this we get

$$1 - t^2 = (1-t)^2 |z|^2.$$

Since $t \neq 1$ $(z \neq \infty)$ we arrive at $t = \frac{|z|^2 - 1}{|z|^2 + 1}$.

Thus $x_1 = \frac{2x}{|z|^2+1}$, $x_2 = \frac{2y}{|z|^2+1}$, $x_3 = \frac{|z|^2-1}{|z|^2+1}$. But this gives

$$x_1 = \frac{z+\bar{z}}{|z|^2+1}, x_2 = \frac{-i(z-\bar{z})}{|z|^2+1} \text{ and } x_3 = \frac{|z|^2-1}{|z|^2+1}. \tag{13.1}$$

Now suppose the point Z is given ($Z \neq N$) and we wish to find z. In analytical terms, we are looking for the values of t so that $tZ + (1-t)N$ has the last co-ordinate zero. Hence if t is this value then

$$t(x_1, x_2, x_3) + (1-t)(0,0,1) = (x, y, 0).$$

Thus $t = \frac{1}{1-x_3}$, $x = tx_1 = \frac{x_1}{1-x_3}$, $y = tx_2 = \frac{x_2}{1-x_3}$. Therefore, $z = \frac{x_1+ix_2}{1-x_3}$. □

Remark 13.1. Now observe that if $|z| > 1$ then Z is in the northern hemisphere and if $|z| < 1$ then Z is in the southern hemisphere; also, for $|z| = 1$, $Z = z$. Further Z approaches N as $|z| \to \infty$. Hence we identify N and the point ∞ in $\mathbb{C}_\infty$. Thus $\mathbb{C}_\infty$ is represented as the sphere S^2. Thus in this correspondence the unit circle in $\mathbb{C}$ corresponds to the equator of S^2, the interior of the unit circle corresponds to the southern hemisphere of S^2, and straight lines in $\mathbb{C}$ correspond to circles on S^2 passing through the north pole.

Now define a metric on $\mathbb{C}_\infty$ as follows: For z, z' in $\mathbb{C}_\infty$ define the distance from z to z', $d(z, z')$, to be the distance between the corresponding points Z and Z' in $\mathbb{R}^3$. If $Z = (x_1, x_2, x_3)$ and $Z' = (x_1', x_2', x_3')$ then

$$d(z, z') = [(x_1 - x_1')^2 + (x_2 - x_2')^2 + (x_3 - x_3')^2]^{\frac{1}{2}}. \tag{13.2}$$

Using the fact that Z and Z' are on S^2, (13.2) gives

$$[d(z, z')]^2 = 2 - 2(x_1x_1' + x_2x_2' + x_3x_3').$$

By using equation (13.2) we get for $z, z' \in \mathbb{C}$,

$$d(z, z') = \frac{2|z - z'|}{\sqrt{1+|z|^2}\sqrt{1+|z'|^2}}. \tag{13.3}$$

In a similar manner we get for $z \in \mathbb{C}$,

$$d(z, \infty) = \frac{2}{\sqrt{1+|z|^2}}. \tag{13.4}$$

It is easy to verify that $(\mathbb{C}_\infty, d)$ where d is given by (13.3) and (13.4) is a metric space.

Thus, a sequence $(z_n)_{n=1}^{\infty}$ in $\mathbb{C}$ converges to ∞ in this metric if and only if $|z_n| \to \infty$ (in the standard sense).

This correspondence between points of S and $\mathbb{C}_\infty$ is called the **stereographic projection**.

Theorem 13.2. *The space $\mathbb{C}_\infty$ is compact, Hausdorff and is second countable. Any infinite subset of $\mathbb{C}_\infty$ has a cluster point in $\mathbb{C}_\infty$. The map $f(z) = \frac{1}{z}$ defined on $\mathbb{C}_\infty$ is a homeomorphism.*

Proof. We shall only verify that $\mathbb{C}_\infty$ is compact. The rest follows easily because $\mathbb{C}_\infty$ is a metric space. Let $\{U_\alpha\}_{\alpha\in J}$ be an open cover for $\mathbb{C}_\infty = \mathbb{C}\cup\{\infty\}$. That is, $\mathbb{C}\cup\{\infty\} = \cup_{\alpha\in J}U_\alpha$. Let $\infty \in U_{\alpha_0}$. Therefore there is an $r > 0$ such that $\{z \in \mathbb{C} : |z| > r\} \subseteq U_{\alpha_0}$. Now the set $\{z \in \mathbb{C} : |z| \leq r\} \subset \cup_{\alpha\in J}U_\alpha$. The set $\{z \in \mathbb{C} : |z| \leq r\}$ is closed and bounded and hence compact. Therefore there exists a finite subcover $\{U_1, U_2, \ldots, U_n\}$ of $\{U_\alpha\}_{\alpha\in J}$. Thus $\mathbb{C}\cup\{\infty\} \subseteq U_1 \cup U_2 \cdots \cup U_n \cup U_{\alpha_0}$. □

Now if we define a map $\phi : S^2 \to \mathbb{C}_\infty$ as

$$\phi(x_1, x_2, x_3) = \left(\frac{x_1}{1-x_3}, \frac{x_2}{1-x_3}, 0\right)$$

if $(x_1, x_2, x_3) \neq (0, 0, 1)$ and $\phi(0, 0, 1) = \infty$. Then ϕ is a bijection from S^2 onto $\mathbb{C}_\infty$. The inverse of ϕ is given by

$$\phi^{-1}(z) = \begin{cases} \left(\frac{z+\bar{z}}{|z|^2+1}, \frac{-i(z-\bar{z})}{|z|^2+1}, \frac{|z|^2-1}{|z|^2+1}\right) & \text{if } z \in \mathbb{C} \\ N & \text{if } z = \infty. \end{cases}$$

Further ϕ is continuous from the compact space S^2 to the Hausdorff space $\mathbb{C}_\infty$. To check the continuity of ϕ at N, let $V = \{z \in \mathbb{C} : |z| > R\}$. We show that the set $U = \{(x_1, x_2, x_3) \in S^2 : x_3 > r\}$ is such that $\phi(U) \subset V$ if r is near to 1. Now, for $p \in U$,

$$\|\phi(p)\|^2 = \frac{x_1^2 + x_2^2}{(1-x_3)^2} = \frac{1-x_3^2}{(1-x_3)^2} = \frac{1+x_3}{1-x_3}$$

which is $> R^2$ if $x_3 > 1 - \frac{1}{R^2} = r$. Thus ϕ is a homeomorphism.

Definition 13.1. *Let $X = \mathbb{C}^2 \setminus \{(0,0)\}$, the nonzero vectors in the 2-dimensional complex vector space $\mathbb{C}^2$. We endow X with the subspace topology. (The topology on $\mathbb{C}^2$ is the product topology.) We let the multiplicative group $\mathbb{C}^*$ of nonzero complex numbers "act" on $X : \alpha(z, w) = (\alpha z, \alpha w)$, that is the scalar multiplication. The orbits $\mathbb{C}^*.(z, w) = \{(\alpha z, \alpha w) : \alpha \in \mathbb{C}^*\}$ form a partition of X and hence induce an equivalence relation $\sim$ on $X : (z_1, w_1) \sim (z_2, w_2)$ if and only if there exists a nonzero complex number α such that $(z_1, w_1) = \alpha(z_2, w_2)$. Let $X\backslash \sim$ denote the quotient set of equivalence classes. We introduce a topology on $X\backslash \sim$, called the quotient topology as follows: Let $\pi : X \to X\backslash \sim$ be the quotient map $x \mapsto [x]$. A set $V \subset X\backslash \sim$ is open if and only if $\pi^{-1}(V)$ is open in X (which is the same as requiring that $\pi^{-1}(V)$ is open in $\mathbb{C}^2$). The resulting space is denoted by $P'(\mathbb{C})$. It is called the one dimensional complex projective space. Our aim is to prove that $P'(\mathbb{C})$ is homeomorphic to $\mathbb{C}_\infty$. We need a preliminary observation.*

Lemma 13.1. *Let Y be any topological space. Then a map $f : P'(\mathbb{C}) \to Y$ is continuous if and only if the map $f \circ \pi : X \to Y$ is continuous.*

Proof. If f is continuous then $f \circ \pi$ is continuous as π is continuous. Now suppose $f \circ \pi$ is continuous. Let $V \subset Y$ be open. Since $f \circ \pi$ is continuous, $(f \circ \pi)^{-1}(V) = \pi^{-1}(f^{-1}(V))$ is open in X. Thus, $f^{-1}(V)$ is open in $X\backslash \sim$ by the very definition of the quotient topology. □

Proposition 13.1. *The map $\phi : P'(\mathbb{C}) \to \mathbb{C}_\infty$ given by*

$$\phi([z,w]) = \begin{cases} \frac{z}{w} & \text{if } w \neq 0 \\ \infty & \text{if } w = 0 \end{cases}$$

is a homeomorphism.

Proof. Note that ϕ is well-defined. It is onto: if $\tau \in \mathbb{C}$, then $\phi([\tau, 1]) = \tau$. In view of Lemma 13.1, we need to show that $\phi \circ \pi$ is continuous at any (z, w). So, we need only check the continuity at $[(z, 0)]$. Thus if $(z_n, w_n) \to (z, 0)$ in X, then $z \neq 0$. Therefore, if $\epsilon > 0$ is given there exists $N \in \mathbb{N}$ such that $|z_n| \geq \frac{|z|}{2}$ and $|w_n| < \epsilon$ for $n \geq N$. Hence $|(\phi \circ \pi)(z_n, w_n)| \geq \frac{|z|}{2\epsilon}$ for all $n \geq N$ so that $(\phi \circ \pi)(z_n, w_n)$ lies in a neighborhood of ∞. Note that $P'(\mathbb{C}) = \pi(K)$ where $K = \{(z,w) \in X : |z|^2 + |w|^2 = 1\}$ is compact. Hence ϕ is a continuous bijection from the compact space $P'(\mathbb{C})$ onto a Hausdorff space $\mathbb{C}_\infty$ and hence is a homeomorphism. □

13.2 Fractional linear transformation

Definition 13.2. *A mapping S of the form $S(z) = \frac{az+b}{cz+d}$ is called a fractional linear transformation. If $a, b, c,$ and d also satisfy $ad - bc \neq 0$ then S is called a Möbius transformation.*

If S is a Möbius transformation then S as a map from $\mathbb{C}_\infty$ into $\mathbb{C}_\infty$ is bijective and $S^{-1}(z) = \frac{dz-b}{-cz+a}$. That is, $S(S^{-1}(z)) = S^{-1}(S(z)) = z$. Note that $S\left(-\frac{d}{c}\right) = \infty$ and $S(\infty) = \frac{a}{c}$. If S and T are Möbius transformations then $S \circ T$ is also a Möbius transformation. Hence, the set of Möbius transformations forms a group under composition. Notice that we cannot have $a = 0 = c$ or $d = 0 = c$ since either situation would contradict $ad - bc \neq 0$. If

$$A = \begin{bmatrix} a & b \\ c & d \end{bmatrix},$$

$ad - bc \neq 0$ then A is called the matrix associated with $S = S_A$. Note that A is nonsingular. Let $GL(2, \mathbb{C})$ denote the set of 2×2 nonsingular complex matrices.

Remark 13.2. There are three classes of Möbius transformations.

(i) The translations $z \mapsto z + b$, which translates any set of points in the z-plane through the vector b, that is, through the distance $|b|$ and in the direction $\arg b$;

(ii) the multiplication by a nonzero scalar $z \mapsto az$, which is geometrically magnification by $|a|$ and a rotation by $e^{i\theta}$ if $a = |a|e^{i\theta}$, that is, the image of z is obtained by rotating the vector z through the angle θ and magnifying (or contracting) the length of z by the factor $|a|$;

(iii) the inversion $z \mapsto \frac{1}{z}$.

In fact, any Möbius transformation can be written as a composition of these three. To see this observe that

$$\phi(z) = \frac{az+b}{cz+d} = \frac{a}{c} + \frac{bc-ad}{c^2}\frac{1}{z+\frac{d}{c}}.$$

Thus ϕ is given by

$$z \mapsto w_1 = z + \left(\frac{d}{c}\right);$$

$$w_1 \mapsto w_2 = \frac{1}{w_1} = \frac{1}{z+\frac{d}{c}};$$

$$w_2 \mapsto w_3 = \frac{bc-ad}{c^2}w_2 = \frac{bc-ad}{c^2}\frac{1}{z+\frac{d}{c}};$$

$$w_3 \mapsto w_4 = \frac{a}{c} + w_3 = \frac{a}{c} + \frac{bc-ad}{c^2}\frac{1}{z+\frac{d}{c}}.$$

Proposition 13.2. *The following statements about the set of Möbius transformations are true:*

(1) If $A \in GL(2,\mathbb{C})$ and $\lambda \in \mathbb{C}^*$, then $S_A = S_{\lambda A}$.

(2) If $A, B \in GL(2,\mathbb{C})$, then $S_A \circ S_B = S_{AB}$ where AB denotes the product matrix.

(3) If $A \in GL(2,\mathbb{C})$, the inverse of S_A is $S_{A^{-1}}$. (Note $S_{A^{-1}}(z) = \frac{dz-b}{-cz+a}$.)

Proof. (1) If

$$A = \begin{bmatrix} a & b \\ c & d \end{bmatrix},$$

$ad-bc \neq 0$, then $S_A(z) = \frac{az+b}{cz+d}$. If λ is any nonzero complex number, then $S_{\lambda A}(z) = \frac{(\lambda a)z+(\lambda b)}{(\lambda c)z+(\lambda d)}$. That is, the coefficients a, b, c, d are not unique.

(2) If $A = \begin{bmatrix} a & b \\ c & d \end{bmatrix}$, $ad-bc \neq 0$, and $B = \begin{bmatrix} a' & b' \\ c' & d' \end{bmatrix}$, $a'd'-b'c' \neq 0$, then $S_A(z) = \frac{az+b}{cz+d}$ and $S_B(z) = \frac{a'z+b'}{c'z+d'}$. It is easy to verify that $S_A \circ S_B = S_{AB}$.

(3) If $A = \begin{bmatrix} a & b \\ c & d \end{bmatrix}$, $ad-bc \neq 0$, then $A^{-1} = \begin{bmatrix} d & -b \\ -c & a \end{bmatrix}$, and $S_A(z) = \frac{az+b}{cz+d}$. Hence $S_A^{-1}(z) = \frac{dz-b}{-cz+a} = S_{A^{-1}}(z)$.

□

Theorem 13.3. *A Möbius transformation $S_A(z) = \frac{az+b}{cz+d}$, $ad-bc \neq 0$ can have at most two fixed points unless $S_A(z) = z$ for all z.*

Proof. For this we want to find out what are the points z satisfying $S_A(z) = z$. If z satisfies this condition then $cz^2 + (d-a)z - b = 0$ which has at most two distinct solutions. □

Remark 13.3. Now suppose S is a Möbius transformation and let a, b, c be distinct points in $\mathbb{C}_\infty$ with $\alpha = S(a), \beta = S(b), \gamma = S(c)$. Suppose T is another Möbius transformation with this property. Then $T^{-1} \circ S$ has a, b and c as fixed points and, therefore, $T^{-1} \circ S = I =$ the identity. That is, $T = S$. Hence a Möbius transformation is uniquely determined by its action on any three given points in $\mathbb{C}_\infty$.

Theorem 13.4. *Let z_2, z_3, z_4 be points in $\mathbb{C}_\infty$. Then there exists a unique Möbius transformation $S : \mathbb{C}_\infty \to \mathbb{C}_\infty$ such that $S(z_2) = 1, S(z_3) = 0$, $S(z_4) = \infty$.*

Proof. A map that vanishes at z_3 is given by $z - z_3$. If we also demand that it is infinite at z_4, then we may try $\frac{z-z_3}{z-z_4}$. This takes the value $\frac{z_2-z_3}{z_2-z_4}$ at $z = z_2$. Hence if we define

$$S(z) = \begin{cases} \left(\frac{z-z_3}{z-z_4}\right)\left(\frac{z_2-z_4}{z_2-z_3}\right) & \text{if} \quad z_2, z_3, z_4 \in \mathbb{C}; \\ \frac{z-z_3}{z-z_4} & \text{if} \quad z_2 = \infty \\ \frac{z_2-z_4}{z-z_4} & \text{if} \quad z_3 = \infty \\ \frac{z-z_3}{z_2-z_3} & \text{if} \quad z_4 = \infty \end{cases}$$

then S is a Möbius transformation and S is as required. The map S is the only transformation having this property. This follows from Remark 13.3. □

Definition 13.3. *If $z_1 \in \mathbb{C}_\infty$ then (z_1, z_2, z_3, z_4) (the cross ratio of z_1, z_2, z_3, and z_4) is the image of z_1 under the unique Möbius transformation which takes z_2 to 1, z_3 to 0, and z_4 to ∞.*

Example 13.1. *For example, $(z_3, z_2, z_3, z_4) = 0$ and $(z, 1, 0, \infty) = z$. Also, if M is any Möbius transformation and w_2, w_3, w_4 are the points such that $Mw_2 = 1, Mw_3 = 0, Mw_4 = \infty$ then $Mz = (z, w_2, w_3, w_4)$.*

Proposition 13.3. *If z_2, z_3, z_4 are distinct points and T is any Möbius transformation then*

$$(z_1, z_2, z_3, z_4) = (Tz_1, Tz_2, Tz_3, Tz_4).$$

for any point z_1.

Proof. Let $Sz = (z, z_2, z_3, z_4)$; then S is a Möbius map. If $M = ST^{-1}$ then $M(Tz_2) = 1, M(Tz_3) = 0, M(Tz_4) = \infty$; hence, $ST^{-1}z = (z, Tz_2, Tz_3, Tz_4)$ for all z in $\mathbb{C}_\infty$. In particular, if $z = Tz_1$ the desired result follows. □

Proposition 13.4. *If z_2, z_3, z_4 are distinct points in $\mathbb{C}_\infty$ and w_2, w_3, w_4 are also distinct points of $\mathbb{C}_\infty$, then there is one and only one Möbius transformation S such that $Sz_2 = w_2, Sz_3 = w_3, Sz_4 = w_4$.*

Proof. Let $Tz = (z, z_2, z_3, z_4), Mz = (z, w_2, w_3, w_4)$ and put $S = M^{-1}T$. Clearly S has the desired property. If R is another Möbius transformation with $Rz_j = w_j$ for $j = 2, 3, 4$ then $R^{-1} \circ S$ has three fixed points z_2, z_3 and z_4. Hence $R^{-1} \circ S = I$ or $S = R$. □

At times one may use the invariance of the cross ratio to compute the Möbius transformation.For if $w_k = S(z_k), 2 \le k \le 4$ and $w = S(z)$, then

$$(z, z_2, z_3, z_4) = (Sz, Sz_2, Sz_3, Sz_4).$$

That is,

$$\frac{(z - z_3)(z_2 - z_4)}{(z - z_4)(z_2 - z_3)} = \frac{(w - w_3)(w_2 - w_4)}{(w - w_4)(w_2 - w_3)}, \tag{13.5}$$

from which the expression for w can be found.

Example 13.2. *We shall find the Möbius transformation which takes* 0, $\infty, 1$ *to* $\infty, 0, i$ *respectively. We get from (13.5) that* $\frac{-1}{z-1} = \frac{w}{w-i}$ *which implies* $-w + i = w(z - 1)$. *Hence* $z - 1 = \frac{i-w}{w}$. *Thus* $z = \frac{i-w}{w} + 1 = \frac{i}{w}$ *or* $w = \frac{i}{z}$.

Theorem 13.5. *If $z, z_1, z_2,$ and z_3 are four distinct points in $\mathbb{C}_\infty$, then $z, z_1, z_2,$ and z_3 are on one circle or one line if and only if their cross ratio (z, z_1, z_2, z_3) is a real number.*

Proof. (1) Assume that $z_1, z_2,$ and z_3 are on a circle Γ. Let $\theta, -\pi < \theta \le \pi$, be the argument of $\frac{z-z_1}{z-z_3}$ and let $\psi, -\pi < \psi \le \pi$, be the argument of $\frac{z_2-z_1}{z_2-z_3}$. Since the argument of (z, z_1, z_2, z_3) equals $\theta - \psi, -2\pi < \theta - \psi < 2\pi$, (z, z_1, z_2, z_3) is a real number if and only if $\theta - \psi = 0$ or $\theta - \psi = \pm\pi$ and it is clear from simple geometric considerations that z is on the circle Γ if and only if $\theta - \psi = 0$ or $\theta - \psi = \pm\pi$, hence, if $z_1, z_2,$ and z_3 are on a circle, the points $z, z_1, z_2,$ and z_3 are on a circle if and only if (z, z_1, z_2, z_3) is a real number.

(2) Assume that $z_1, z_2,$ and z_3 are on a line $l \subset \mathbb{C}$ since $\frac{z_2-z_1}{z_2-z_3}$ is a real number, (z, z_1, z_2, z_3) is a real number if and only if $\frac{z-z_1}{z-z_3}$ is a real number, that is, if and only if z is on l or $z = \infty$.

(3) Assume that one of the points $z_1, z_2,$ and z_3 is ∞. Let us assume that $z_1 = \infty$(the other cases are similar). Since

$$(z, \infty, z_2, z_3) = \frac{z_2 - z_3}{z - z_3},$$

(z, ∞, z_2, z_3) is real if and only if $z, z_2,$ and z_3 are on a line l, that is, if and only if $z, z_1 = \infty, z_2$ and z_3 are on a line $l \cup \{\infty\}$. □

It is well known from high school geometry that three points in the plane determine a circle. Recall that a circle in $\mathbb{C}_\infty$ passing through ∞ corresponds to a straight line in $\mathbb{C}$(under the stereographic projection map). From now on a straight line in the plane $\mathbb{C}$ will be called a circle. We agree that every straight line shall pass through the point at infinity. By contrast, no half plane shall contain the ideal point.

Thus with this convention the above theorem can be stated as follows:

If z, z_1, z_2, z_3 are four distinct points in $\mathbb{C}_\infty$ then (z, z_1, z_2, z_3) is a real number if and only if all four points lie on a circle.

Theorem 13.6. *A Möbius transformation maps circles onto circles.*

Proof. Let Γ be any circle in $\mathbb{C}_\infty$ and let ϕ be any Möbius transformation. Let z_1, z_2, z_3 be three distinct points on Γ and put $w_j = \phi(z_j)$ for $j = 1, 2, 3$. Then w_1, w_2, w_3 determine a circle Γ'. We claim that $\phi(\Gamma) = \Gamma'$. In fact, for any z in $\mathbb{C}_\infty$,

$$(z, z_1, z_2, z_3) = (\phi(z), w_1, w_2, w_3) \tag{13.6}$$

by Proposition 13.3. By the preceding theorem (Theorem 13.5) if z is on Γ then both sides of (13.6) are real. But this says that $\phi(z) \in \Gamma'$. $\square$

We indicate an alternative proof of the above fact below.

The transformations $z \mapsto z + b$ and $z \mapsto kz$ clearly map circles and straight lines into circles and straight lines.

Let $S = \{z : |z - \alpha| = r\}, \alpha = x_0 + iy_0$, and $f(z) = \frac{1}{z} = w = u + iv$. The equation for S is

$$(z - \alpha)(\bar{z} - \bar{\alpha}) = r^2$$

or

$$\frac{1}{w\bar{w}} - \frac{\alpha}{\bar{w}} - \frac{\bar{\alpha}}{w} = r^2 - |\alpha|^2.$$

Case 1: If $r = |\alpha|$, that is, when S contains the origin, we get $1 - \alpha w - \overline{\alpha w} = 0$ or $\operatorname{Re}(\alpha w) = \frac{1}{2}$. This is equivalent to $ux_0 - vy_0 = \frac{1}{2}$ which represents a straight line.

Case 2: If $r \neq |\alpha|$, we obtain

$$w\bar{w} - \left(\frac{\bar{\alpha}}{|\alpha|^2 - r^2}\right)\bar{w} - \left(\frac{\alpha}{|\alpha|^2 - r^2}\right)w = \frac{-1}{|\alpha|^2 - r^2}.$$

Letting $\beta = \frac{\bar{\alpha}}{|\alpha|^2 - r^2}$ we get $w\bar{w} - \beta\bar{w} - \bar{\beta}w + |\beta|^2 = \frac{r^2}{(|\alpha|^2 - r^2)^2}$ and

$$|w - \beta|^2 = \left(\frac{r}{|\alpha|^2 - r^2}\right)^2$$

which represents the circle centered at β with radius $\left(\frac{r}{|\alpha|^2 - r^2}\right)$. If S is a straight line, then, for some real constants a, b, and c, we have, for $z = x + iy \in S$, $ax + by = c$.

Letting $\alpha = a - ib$, we get $\text{Re}(\alpha z) = c$ or $\alpha z + \overline{\alpha z} = 2c$ and it follows, as above, that $f(S)$ is a straight line or a circle. Finally, let $f(z) = \frac{az+b}{cz+d}$. If $c = 0$, f is linear, so it is the sum of two functions that map circles and lines into circles and lines, so f itself has that mapping property. If $c \neq 0$, we have $\frac{az+b}{cz+d} = \frac{1}{c}\left(a - \frac{ad-bc}{cz+d}\right)$ so $f(z) = f_3(f_2(f_1(z)))$ where $f_1(z) = cz + d$, $f_2(z) = \frac{1}{z}$, $f_3(z) = \frac{a}{c} - \frac{ad-bc}{c}z$, each of which has the desired property, and so does f.

Now we shall see given two circles Γ and Γ' in $\mathbb{C}_\infty$ how to obtain the Möbius transformation that maps Γ onto Γ'. Let $z_1, z_2, z_3 \in \Gamma$; $w_1, w_2, w_3 \in \Gamma'$. Put $\phi(z) = (z, z_1, z_2, z_3)$, $\psi(z) = (z, w_1, w_2, w_3)$. Then $\theta = \psi^{-1} \circ \phi$ maps Γ onto Γ'. In fact, $\theta(z_j) = w_j$ for $j = 1, 2, 3$ and, $\theta(\Gamma) = \Gamma'$.

Thus given any two circles Γ and Γ' in $\mathbb{C}_\infty$ there is a Möbius transformation T such that $T(\Gamma) = \Gamma'$. Furthermore, we can specify that T take any three points on Γ onto any three points of Γ'. If we do specify Tz_j for $j = 2, 3, 4$ (distinct z_j on Γ) then T is unique.

Definition 13.4. *Let Γ be a circle through points z_2, z_3, z_4. The points $z, z*$ in $\mathbb{C}_\infty$ are said to be symmetric with respect to Γ if $(z*, z_2, z_3, z_4) = \overline{(z, z_2, z_3, z_4)}$.*

Theorem 13.7. *(Symmetry Principle): If a Möbius transformation S takes a circle Γ_1 onto the circle Γ_2 then any pair of points symmetric with respect to Γ_1 are mapped by S onto a pair of points symmetric with respect to Γ_2.*

Proof. Let $z_2, z_3, z_4 \in \Gamma_1$. If z and $z*$ are symmetric with respect to Γ_1 then by Proposition 13.3, $(Sz*, Sz_2, Sz_3, Sz_4) = (z*, z_2, z_3, z_4) = \overline{(z, z_2, z_3, z_4)} = \overline{(Sz, Sz_2, Sz_3, Sz_4)}$. Thus $Sz*$ and Sz are symmetric with respect to Γ_2. □

We shall now define orientation for circles in $\mathbb{C}_\infty$; which will enable us to distinguish between the "inside" and "outside" of a circle in $\mathbb{C}_\infty$.

Definition 13.5. *If Γ is a circle then an orientation for Γ is an ordered triple of points (z_1, z_2, z_3) such that each z_j is in Γ. In fact these three points give a direction to Γ. That is, we move from z_1 to z_2 to z_3. If only two points were given, this would be ambiguous. If (z_1, z_2, z_3) is an orientation of Γ then we define the right side of Γ with respect to (z_1, z_2, z_3) to be $\{z : Im(z, z_1, z_2, z_3) > 0\}$. Similarly, we define the left side of Γ to be $\{z : Im(z, z_1, z_2, z_3) < 0\}$.*

Theorem 13.8. *(Orientation Principle): Let Γ_1 and $\Gamma_2$be two circles in $\mathbb{C}_\infty$ and let S be a Möbius transformation such that $S(\Gamma_1) = \Gamma_2$. Let (z_1, z_2, z_3) be an orientation for Γ_1. Then S takes the right side and the left side of Γ_1 onto the right side and left side of Γ_2 with respect to the orientation (Sz_1, Sz_2, Sz_3).*

Proof. The proof follows from Proposition 13.3. □

Example 13.3. *Let Γ be the unit circle in $\mathbb{C}$. Then the Möbius transformation $f(z) = \frac{1}{z}$ preserves the unit circle. It maps the right side of Γ to the left side of Γ (the point 0 to ∞) and vice-versa with respect to any orientation of Γ.*

13.3 Conformal maps

A holomorphic function can be approximated locally by fractional linear transformations. If the function f is holomorphic in a neighborhood of the point z_0, one would expect it to behave near z_0 approximately like the function $z \mapsto f'(z_0)(z - z_0) + f(z_0)$. If $f'(z_0) = 0$ this will tell us very little, but if $f'(z_0) \neq 0$, it should say something about the "infinitesimal" deformation produced by f near z_0. We shall see below that if $f'(z_0) \neq 0$, the holomorphic function f preserves the angles between curves intersecting at z_0. To make this precise we need some preliminaries about curves in the complex plane.

Let G be a region in $\mathbb{C}$. Suppose $\gamma : [a,b] \to G$ is a smooth path and that for some $t_0 \in (a,b), \gamma'(t_0) \neq 0$. Then γ has a tangent line at the point $z_0 = \gamma(t_0)$. This line goes through the point z_0 in the direction of the vector $\gamma'(t_0)$; or, the slope of the line is $\tan(\arg \gamma'(t_0))$. If γ_1 and γ_2 are two smooth paths with $\gamma_1(t_1) = \gamma_2(t_2) = z_0$ and $\gamma_1'(t_1) \neq 0, \gamma_2'(t_2) \neq 0$, then we define the angle between the paths γ_1 and γ_2 at z_0 to be

$$\arg \gamma_2'(t_2) - \arg \gamma_1'(t_1).$$

Suppose γ is a smooth path in G and $f : G \to \mathbb{C}$ is analytic. Then $\sigma = f \circ \gamma$ is also a smooth path and $\sigma'(t) = f'(\gamma(t))\gamma'(t)$. Let $z_0 = \gamma(t_0)$, and suppose that $\gamma'(t_0) \neq 0$ and $f'(z_0) \neq 0$; then $\sigma'(t_0) \neq 0$ and $\arg \sigma'(t_0) = \arg f'(z_0) + \arg \gamma'(t_0)$. That is,

$$\arg \sigma'(t_0) - \arg \gamma'(t_0) = \arg f'(z_0).$$

Now let γ_1 and γ_2 be smooth paths in G with $\gamma_1(t_1) = \gamma_2(t_2) = z_0$ and $\gamma_1'(t_1) \neq 0 \neq \gamma_2'(t_2)$. Let $\sigma_1 = f \circ \gamma_1$ and $\sigma_2 = f \circ \gamma_2$. Also, suppose that the paths γ_1 and γ_2 are not tangent to each other at z_0; that is, suppose $\gamma_1'(t_1) \neq \gamma_2'(t_2)$. Thus we get

$$\arg \gamma_2'(t_2) - \arg \gamma_1'(t_1) = \arg \sigma_2'(t_2) - \arg \sigma_1'(t_1).$$

This says that given any two paths through z_0; f maps these paths onto two paths through $w_0 = f(z_0)$ and, when $f'(z_0) \neq 0$, the angles between the curves are preserved both in magnitude and direction. Thus if $f : G \to \mathbb{C}$ is holomorphic and $f'(z_0) \neq 0$ for some $z_0 \in G$ then f preserves angles at the point z_0.

Definition 13.6. *A function $f : G \to \mathbb{C}$ which has the angle preserving property and also has $\lim\limits_{z \to a} \dfrac{|f(z) - f(a)|}{|z - a|}$ existing is called a conformal map.*

If f is holomorphic in G and $f'(z) \neq 0$ for any $z \in G$, then f is conformal in G. The converse of this statement is also true.

Example 13.4. *If $f(z) = e^z$, then f is conformal throughout $\mathbb{C}$.*

Example 13.5. *If $f(z) = z^2$, then f is conformal in $\mathbb{C} \setminus \{0\}$.*

Example 13.6. *Every Möbius map $S(z) = \frac{az+b}{cz+d}, ad - bc \neq 0$ is conformal on $\mathbb{C} \setminus \{-\frac{d}{c}\}$, since $S'(z) = \frac{ad-bc}{(cz+d)^2} \neq 0$.*

Example 13.7. *Let $f(z) = \frac{1}{z-1}$. It maps 1 to ∞. So image of the unit circle Γ under f is unbounded. That is, it cannot be a circle. It takes -1 to $-\frac{1}{2}$ and i to $\frac{1}{i-1} = \frac{1}{2}(-1-i)$. So image is the straight line $x = -\frac{1}{2}$. Thus by orientation principle f maps the open unit disk $\mathbb{D}$ onto the half plane $z = x + iy \in \mathbb{C} : x \leq -\frac{1}{2}$.*

Example 13.8. *Let $f(z) = \frac{1}{z-2}$. The image of the unit circle Γ under f is a circle since no point maps to ∞. It passes through $f(1) = -1, f(-1) = -\frac{1}{3}$ and $f(i) = -\frac{2}{5} - \frac{1}{5}i$. The image is the circle with center at $-\frac{2}{3}$ and radius $\frac{1}{3}$.*

It follows from Theorem 11.10 that every one-one holomorphic maps are conformal. But one may note that every conformal map is not one-one. For example, e^z is holomorphic and $\frac{d}{dz}e^z = e^z \neq 0$ for all $z \in \mathbb{C}$. Hence e^z is conformal but $e^z = e^{z+2\pi ik}$ for all $k \in \mathbb{Z}$. Hence e^z is not one-one. However every conformal map is locally one-one.

Example 13.9. *In this example we exhibit a conformal map f from the right half plane $H = \{z \in \mathbb{C} : Rez > 0\}$ onto the open unit disk $\mathbb{D} = \{z \in \mathbb{C} : |z| < 1\}$.*

Solution: We solve this problem by finding a Möbius map which takes the imaginary axis onto the unit circle and, by the orientation principle; takes H onto $\mathbb{D}$. The same map will also take the left half plane $\{z \in \mathbb{C} : \text{Re}z < 0\}$ onto $\{z \in \mathbb{C} : |z| > 1\}$. If we give the imaginary axis the the orientation $(-i, 0, i)$ then $\{z \in \mathbb{C} : \text{Re}z > 0\}$ is on the right of this axis and $(z, -i, 0, i) = \frac{2z}{z-i} = S(z)$. Giving the unit circle Γ the orientation $(-i, -1, i)$ we have that the unit disk $\mathbb{D}$ lies on the right of the unit circle Γ and $(z, -i, -1, i) = \frac{2i}{i-1}\frac{z+1}{z-i} = R(z)$. Now $T = R^{-1} \circ S$ maps H onto $\mathbb{D}$ and $T(z) = R^{-1}(Sz) = R^{-1}\left(\frac{2z}{z-i}\right) = \frac{z-1}{z+1}$. Note that S maps the imaginary axis to $\mathbb{R} \cup \{\infty\} = \mathbb{R}_\infty$ and R maps the unit circle to $\mathbb{R}_\infty$.

The function $g(z) = e^{-\frac{i\pi}{2}}z$ maps the upper half plane to the right half plane. Thus $T \circ g$ maps the upper half plane into the open unit disk $\mathbb{D}$ and $(T \circ g)(z) = T(g(z)) = \frac{z-i}{z+i}$. The map $T \circ g$ is conformal. Similarly we can find conformal maps from the left half plane and lower half plane onto $\mathbb{D}$.

Example 13.10. *An entire function with positive real part is constant.*

Solution: Suppose f is an entire function with positive real part. Then f takes values in the right half plane. Let $g(z) = \frac{z-1}{z+1}$. Then g is a conformal map from the right half plane onto the unit disk $\mathbb{D}$ (see Example 13.9). Now $g \circ f : \mathbb{C} \to \mathbb{D}$ is an entire function and $|(g \circ f)(z)| < 1$ for all $z \in \mathbb{C}$. By Liouville's theorem, $(g \circ f)(z) = c$, a constant. This implies $g(f(z)) = \frac{f(z)-1}{f(z)+1} = c$. Hence $f(z) = \frac{1+c}{1-c}$. Thus f is constant.

Similarly one can show that an entire function f with negative real part or positive imaginary part or negative imaginary part is constant by taking the corresponding conformal maps onto the unit disk $\mathbb{D}$.

Example 13.11. *Exhibit a conformal map from* $\{z \in \mathbb{C} : |z| < 1, Re\ z > 0\}$ *onto* $\mathbb{D} = \{z \in \mathbb{C} : |z| < 1\}$.

Solution: The map $z \mapsto iz$ maps the given region conformally onto $A = \mathbb{D} \cap \{z : \text{Im } z > 0\}$. The map $w \mapsto \frac{1+w}{1-w}$ maps A onto the first quadrant Q. The square function takes Q onto $\{z \in \mathbb{C} : \text{Im} z > 0\}$. Finally, $\zeta \mapsto \frac{\zeta - i}{\zeta + i}$ takes $\{\zeta : \text{Im } \zeta > 0\}$ onto $\mathbb{D}$. Composing these maps, we get the required map as $\phi(z) = \frac{(1+iz)^2 - i(1-iz)^2}{(1+iz)^2 + i(1-iz)^2}$.

Example 13.12. *Find a one-to-one conformal map of the semidisk* $\{z \in \mathbb{C} : Im\ z > 0, |z - \frac{1}{2}| < \frac{1}{2}\}$ *onto the upper half plane.*

Solution: The map $\phi_1(z) = 2z - 1$ maps conformally the semidisk $\{z \in \mathbb{C} : \text{Im} z > 0, |z - \frac{1}{2}| < \frac{1}{2}\}$ onto the upper half of the unit disk. The map $\phi_2(z) = \frac{1+z}{1-z}$ maps the unit disk conformally onto the right half plane. Letting $z = re^{i\theta}$, it becomes

$$\frac{1 + re^{i\theta}}{1 - re^{i\theta}} = \frac{1 - r^2 + 2ir \sin\theta}{|1 + re^{i\theta}|^2}.$$

Since $\sin\theta > 0$ for $0 < \theta < \pi$, ϕ_2 maps the upper half of $\mathbb{D}$ onto the upper right quadrant. The map $\phi_3(z) = z^2$ maps the upper right quadrant conformally onto the upper half plane. The composition of ϕ_1, ϕ_2, and ϕ_3 is the desired map, namely the function $z \mapsto \frac{z^2}{(1-z)^2}$.

13.4 Functions on the extended plane

Let $U \subset \mathbb{C}_\infty$ be open with $\infty \in U$. Let $\widetilde{U} = \{w \in \mathbb{C} : w = \frac{1}{z} \text{ for some } z \in U\}$. Then $\widetilde{U}$ is an open set containing 0. We say that a complex valued function $f : U \to \mathbb{C}$ is differentiable at $z \in U$, if the function $\widetilde{f} : \widetilde{U} \to \mathbb{C}$ defined by $\widetilde{f}(w) = f\left(\frac{1}{w}\right)$ is differentiable at $w = \frac{1}{z}$. In particular, f is differentiable at ∞ if $\widetilde{f}$ is differentiable at 0. Observe that if $z \in U \cap \mathbb{C}^*$ is a complex number, then f is differentiable at z if and only if $\widetilde{f}$ is differentiable at $w = \frac{1}{z}$. However, note that we do not define $f'(\infty)$, even if f is differentiable at ∞.

It is not so difficult to verify that if U is a neighborhood of ∞ in $\mathbb{C}_\infty$ and if $f : U \to \mathbb{C}$ is differentiable, then f is continuous.

One can mimic this approach to define the zeros and singularities of f at ∞. First of all we say that ∞ is a isolated singularity of f if f is holomorphic in a deleted neighborhood of ∞, i.e., f is holomorphic in $\{z \in \mathbb{C} : |z| > R\}$. We say ∞ is a pole of f of order n if $z = 0$ is a pole of $\widetilde{f}$ of order n.

Example 13.13. *(i) A polynomial of degree $n \geq 1$ has a pole of order n at ∞.*

(ii) The function $\cot \pi z$ has poles at $n \in \mathbb{Z}$ so that ∞ is a cluster point of poles of this function. Thus $z = \infty$ is a non-isolated singularity of $\cot \pi z$.

(iii) The function $f(z) = e^z$ has an essential singularity at ∞.

(iv) The function $f(z) = \frac{1}{z}$ defined on $\mathbb{C}_\infty$ has ∞ as a simple zero of f. Further any rational function of the form $f(z) = \frac{P(z)}{Q(z)}$ with $\deg P \leq \deg Q$ is holomorphic at ∞.

Definition 13.7. *Let $U \subset \mathbb{C}_\infty$ be open. Let $S \subset U$ be a closed subset. Let $f : U \setminus S \to \mathbb{C}$ be holomorphic. Assume further that each point of S is an isolated singularity of f which is a pole. Then we say that f is meromorphic in U.*

Theorem 13.9. *(1) Let $f : \mathbb{C}_\infty \to \mathbb{C}$ be holomorphic. Then f is a constant.*
(2) Let $f : \mathbb{C}_\infty \to \mathbb{C}_\infty$ be meromorphic. Then f is a rational function.

Proof. (1) First note that f is continuous on the compact set $\mathbb{C}_\infty$ and hence is bounded. Hence $f(\mathbb{C}) \subset f(\mathbb{C}_\infty)$ is bounded. That is, the restriction of f to $\mathbb{C}$ is a holomorphic function on $\mathbb{C}$ which is bounded. By Liouville's theorem, f is a constant on $\mathbb{C}$. By continuity, f is a constant on $\mathbb{C}_\infty$.

(2) We note first that f cannot have an infinite number of poles. For otherwise, there exists $z \in \mathbb{C}_\infty$ which is a cluster point of the set S of poles (this is because $S \subset \mathbb{C}_\infty$, $\mathbb{C}_\infty$ is compact, and every infinite subset of $\mathbb{C}_\infty$ has a limit point in $\mathbb{C}_\infty$. So S has a limit point in $\mathbb{C}_\infty$.) Now z cannot be one of the poles, since in every neighborhood of z, there exist infinitely many poles and hence z is not an isolated singularity. In fact it is a non-isolated essential singularity of f. But the function f is meromorphic. Therefore, S is a finite set. Let z_k be a pole of order $m_k, 1 \leq k \leq N$. Let f_k be the principal part of f at z_k and let f_0 be the principal part of f at ∞(if ∞ is a pole). Then $g = f - \sum_{k=0}^{N} f_k$ is holomorphic in $\mathbb{C}_\infty$. By the first part of the theorem, g is a constant so that f is a rational function. □

Theorem 13.10. *A function $f : \mathbb{C}_\infty \to \mathbb{C}_\infty$ is rational if and only if it is meromorphic on $\mathbb{C}_\infty$.*

Proof. Let $f(z) = \frac{p(z)}{q(z)}$ where p and q are polynomials with complex coefficients and $q(z)$ is not identically zero. By the fundamental theorem of algebra we can express every polynomial as a product of linear factors. Hence we can write

$$f(z) = c(z - \alpha_1)^{m_1} \cdots (z - \alpha_r)^{m_r} (z - \beta_1)^{-n_1} \cdots (z - \beta_s)^{-n_s},$$

where $c \in \mathbb{C}$, $\alpha_1, \cdots, \alpha_r$ are the zeros of p of orders $m_1, \cdots, m_r$, and $\beta_1, \cdots, \beta_s$ are the zeros of q of orders $n_1, \cdots, n_s$. If we decompose a rational function f as above, then f is differentiable at each $z \neq \infty, \beta_j (1 \leq j \leq s)$, so f is analytic on $\mathbb{C} \setminus \{\beta_1, \ldots, \beta_s\}$. At each β_j, f has a pole of order n_j, while at ∞, f is analytic if $\deg(p) \leq \deg(q)$ and f has a pole of order $\deg(p) - \deg(q)$ if $\deg(p) > \deg(q)$. Thus f is meromorphic on $\mathbb{C}_\infty$. Conversely, suppose that f is meromorphic on $\mathbb{C}_\infty$. Note that f has finitely many poles in $\mathbb{C}$, say $\beta_1, \cdots, \beta_s$ of orders $n_1, \cdots, n_s$. Then the function

$$g(z) = (z - \beta_1)^{n_1}(z - \beta_2)^{n_2} \cdots (z - \beta_s)^{n_s} f(z)$$

is analytic on $\mathbb{C}$, so g has a Taylor expansion $g(z) = a_0 + a_1 z + a_2 z^2 + \cdots$ valid for all $z \in \mathbb{C}$. Now g is meromorphic at ∞ (since f is), so

$$g\left(\frac{1}{z}\right) = a_0 + \frac{a_1}{z} + \frac{a_2}{z^2} + \cdots$$

is meromorphic at 0, and hence $a_j = 0$ for all sufficiently large j. Thus g is a polynomial, so

$$f(z) = g(z)(z - \beta_1)^{-n_1} \cdots (z - \beta_s)^{-n_s}$$

is a rational function. □

Theorem 13.11. *A nonconstant meromorphic function $f : \mathbb{C}_\infty \to \mathbb{C}_\infty$ takes any given value $c \in \mathbb{C}_\infty$ only finitely many times, counting multiplicities (that is, the sum of the multiplicities of the solutions of $f(z) = c$ is finite).*

Proof. If $z \in \mathbb{C}_\infty$ and $f(z) = c$, then there exists a neighborhood N_z of z such that f does not take the value c on $N_z \setminus \{z\}$: for if $c = \infty$ then the poles of f are the zeros of $\frac{1}{f(z)}$, and since $\frac{1}{f(z)}$ is meromorphic and not constant these are isolated, by theorem 7.16; if $c \neq \infty$ we use the fact that the zeros of $f - c$ are isolated. Being compact, $\mathbb{C}_\infty$ is covered by finitely many such neighborhoods $N_{z_1}, \ldots, N_{z_k}$, so $f^{-1}(c) = \{z_1, \ldots, z_k\}$, a finite set. Since f is meromorphic, each solution of $f(z) = c$ has finite multiplicity, so f takes the value c only finitely many times. □

Theorem 13.12. *Let f and g be meromorphic functions on $\mathbb{C}_\infty$ with poles at the same points in $\mathbb{C}_\infty$, and with the same principal parts at these points. Then $f(z) = g(z) + c$ for some constant c. Thus meromorphic functions on $\mathbb{C}_\infty$ are determined, up to additive constants, by their principal parts.*

Proof. The function $h = f - g$ is meromorphic and therefore continuous on $\mathbb{C}_\infty$; since $\mathbb{C}_\infty$ is compact, the image $h(\mathbb{C}_\infty)$ is compact. Since the principal parts of f and g cancel, h has no poles, so $h(\mathbb{C}_\infty)$ is a subset of $\mathbb{C}$, and being compact it is bounded. Liouville's theorem shows that h, being analytic and bounded, must be constant on $\mathbb{C}$, and hence (by continuity) on $\mathbb{C}_\infty$, so that $f = g + c$ for some constant c. □

Theorem 13.13. *Let f and g be meromorphic functions on $\mathbb{C}_\infty$ with zeros and poles of the same orders at the same points of $\mathbb{C}$. Then $f(z) = cg(z)$ for some constant $c \neq 0$.*

Proof. We may assume that neither f nor g is identically zero. Then both $\frac{f}{g}$ and $\frac{g}{f}$ are meromorphic on $\mathbb{C}_\infty$, and neither of them has any poles in $\mathbb{C}$, so both are analytic on $\mathbb{C}$. At least one of them (call it h) is finite at ∞, so h is analytic on $\mathbb{C}_\infty$. As in Theorem 13.12, Liouville's theorem implies that h is constant, so that $f = cg$ for some constant c. We have $c \neq 0$ since f is not identically zero. □

Notice that in the hypothesis of the above theorem we do not need to assume that f and g behave similarly at ∞; this follows from the fact that $f = cg$.

Remark 13.4. Suppose that $f : \mathbb{C}_\infty \to \mathbb{C}_\infty$ is meromorphic and not constant. If $f(a) = c$ with multiplicity k, where $a, c \in \mathbb{C}$, then local mapping theorem shows that f is a k-to-one function near a; more precisely, if U is any sufficiently small open set containing a then there exists an open set V containing c such that for each $c' \in V \setminus \{c\}$ the equation $f(z) = c'$ has exactly k solutions in U, all simple. By considering the functions $f\left(\frac{1}{z}\right), \frac{1}{f(z)}, \frac{1}{f\left(\frac{1}{z}\right)}$, it is easily seen that we have the same behavior near a when either a, c or both are equal to ∞.

We shall show below that non-constant meromorphic functions are open maps.

Theorem 13.14. *A nonconstant meromorphic function $f : \mathbb{C}_\infty \to \mathbb{C}_\infty$ maps open sets onto open sets.*

Proof. Let X be an open subset of $\mathbb{C}_\infty$ and let $c \in f(X)$. Let $c = f(a)$ with multiplicity k for some $a \in X$. Then by local mapping theorem and from the discussion prior to this theorem, it follows that c belongs to an open set $V \subseteq f(U) \subseteq f(X)$, where U is a neighborhood of a in X. Thus $f(X)$ is open, as required. □

Polynomials are characterized among the class of entire functions by the following proposition.

Proposition 13.5. *Let $f : \mathbb{C} \to \mathbb{C}$ be holomorphic and nonconstant. Then f is not a polynomial if and only if ∞ is an essential singularity of f.*

Proof. A polynomial of degree n has a pole of order n at ∞. Now suppose f is not a polynomial. Then we have $f(z) = \sum_{n=0}^{\infty} c_n z^n$ for $z \in \mathbb{C}$ with $c_n \neq 0$ for infinitely many $n \in \mathbb{N}$. Hence for $z \in \mathbb{C}^*$, we have $\widetilde{f}(z) = f\left(\frac{1}{z}\right) = \sum_{n=0}^{\infty} c_n z^{-n}$. By uniqueness of Laurent series, this is the Laurent series of $\widetilde{f}$ at $z = 0$. Thus 0 is an essential singularity of $\widetilde{f}$. Hence ∞ is an essential singularity of f. □

Theorem 13.15. *A rational function has no essential singularities.*

Proof. Suppose $R(z)$ is a rational function. Then $R(z) = \frac{P(z)}{Q(z)}$ for some polynomials $P(z)$ and $Q(z)$. So the zeros of $Q(z)$ are the singularities of $R(z)$. Now it is required to prove that $R(z)$ has no essential singularity. Thus our claim is if $z = a$ is a singularity of $R(z)$ then it is either a removable singularity or a pole of $R(z)$. Let $z = a$ be a singularity of $f(z)$. Then $z = a$ is a zero of $Q(z)$ of multiplicity m_1(say), and $0 < m_1 \leq n_1 = \deg Q(z)$. So there exists a polynomial $g_1(z)$ of degree $(n_1 - m_1)$ such that $g_1(a) \neq 0$ and $Q(z) = (z-a)^{m_1} g_1(z)$. Again, $z = a$ is a zero of $P(z)$ of multiplicity $m_2 \geq 0$ and $m_2 \leq n_2 = \deg P(z)$. So there exists a polynomial $g_2(z)$ such that $g_2(a) \neq 0$ and $P(z) = (z-a)^{m_2} g_2(z)$.

Consider first the case when $m_1 \leq m_2$. In this case

$$\lim_{z \to a} R(z) = \lim_{z \to a} \frac{P(z)}{Q(z)} = \lim_{z \to a} \frac{(z-a)^{m_2} g_2(z)}{(z-a)^{m_1} g_1(z)}$$

$$= \lim_{z \to a} (z-a)^{m_2 - m_1} \frac{g_2(z)}{g_1(z)} = \begin{cases} 0 & \text{if } m_2 > m_1 \\ \frac{g_2(a)}{g_1(a)} & \text{if } m_2 = m_1. \end{cases}$$

Hence in this case $z = a$ is a removable singularity of $R(z)$.

When $m_1 > m_2$, we have $(m_1 - m_2)$ is a positive integer and $\lim_{z \to a} (z-a)^{m_1 - m_2} R(z) = \lim_{z \to a} \frac{g_2(z)}{g_1(z)} = \frac{g_2(a)}{g_1(a)} \neq 0$. Hence $z = a$ is a pole of $R(z)$ of order $m_1 - m_2$. Thus $R(z)$ has no essential singularity. □

Definition 13.8. *Let ∞ be an isolated singularity of f. Then the residue of f at ∞ defined by*

$$Res(f, \infty) = -\frac{1}{2\pi i} \int_{S(0,R)} f(z) dz.$$

The residue $\text{Res}(f, \infty)$ is $-a_{-1}$ where a_{-1} is the coefficient of $\frac{1}{z}$ in the Laurent expansion of f near ∞, i.e., $-\text{Res}(f, \infty)$ is the coefficient of w in the expansion of $f(\frac{1}{w})$ near 0. It is the coefficient of $\frac{1}{w}$ in the expansion of $F(w) = \frac{1}{w^2} f\left(\frac{1}{w}\right)$ near 0. Hence $\text{Res}(f, \infty) = -\text{Res}(F, 0)$. Using the change of variable $w = \frac{1}{z}$, we get

$$\text{Res}(F, 0) = \frac{1}{2\pi i} \int_{|w| = \frac{1}{R}} f\left(\frac{1}{w}\right) \frac{dw}{w^2} = \frac{1}{2\pi i} \int_{|z| = R} f(z) dz = -\text{Res}(f, \infty). \tag{13.7}$$

13.5 Automorphisms of simply connected domains

Definition 13.9. *Let U and V be open sets in $\mathbb{C}$. A bijection $f : U \to V$ is called an analytic isomorphism if f and f^{-1} are both holomorphic. In classical books an analytic isomorphism is known as a conformal map. In other words, we shall say two regions G_1 and G_2 in $\mathbb{C}$ are conformally equivalent if there is a univalent holomorphic function f in G_1 such that $f(G_1) = G_2$. Let U be open in $\mathbb{C}$. An (analytic) automorphism of U is an analytic isomorphism $f : U \to U$. A map $f : \mathbb{C}_\infty \to \mathbb{C}_\infty$ is an (analytic) automorphism if f is a homeomorphism of $\mathbb{C}_\infty$ onto itself such that f is differentiable at all points except at $\alpha = f^{-1}(\infty)$ and f^{-1} is differentiable at all points except at $f(\infty)$.*

Example 13.14. *(i) Any open disk in $\mathbb{C}$ is conformally equivalent to the unit disk via a linear fractional transformation.*

(ii) Any open half plane is conformally equivalent to the unit disk in $\mathbb{C}$.

(iii) The complex plane $\mathbb{C}$ is not conformally equivalent to the unit disk $\mathbb{D}$. This is so, because if there is a conformal map f from $\mathbb{C}$ onto $\mathbb{D}$, then by Liouville's theorem f is a constant. This contradicts the hypothesis that f is conformal. Here one may note that there exist continuous one-one and onto map from $\mathbb{C}$ onto $\mathbb{D}$, for example, $f(z) = \frac{z}{1+|z|}$.

We let $\mathrm{Aut}(E)$ stand for the set of automorphisms of E.

Example 13.15. *Any two open balls $B(z, r)$ and $B(w, R)$ are analytically isomorphic. The map $h(x) = \frac{R}{r}(x - z) + w$ is the required isomorphism.*

Example 13.16. *Let $U = \{z \in \mathbb{C} : Im(z) > 0\}$. Then U is analytically isomorphic to $B(0, 1)$ via the map $f(z) = \frac{z-i}{z+i}$.*

Example 13.17. *$\mathbb{C}$ and $B(0, 1)$ are not analytically isomorphic. We have already seen that if there exists such an isomorphism f then by Liouville's theorem f is a constant and therefore f is not bijective.*

Note that the set of automorphisms $\mathrm{Aut}(U)$ of an open set in $\mathbb{C}$ forms a group under composition of maps. $\mathrm{Aut}(\mathbb{C}_\infty)$ is also a group likewise.

We determine the automorphism group of U, where U is one of the simply connected domains $B(0, 1), \mathbb{C}$ and $\mathbb{C}_\infty$. By a very deep theorem of Riemann, upto analytic isomorphism, the only simply connected open subsets of $\mathbb{C}$ are $B(0, 1)$ and $\mathbb{C}$. A uniformization theorem, says that the only simply connected Riemann surfaces (upto analytic isomorphisms) are the open unit disk, the complex plane and the Riemann sphere $\mathbb{C}_\infty$.

Notice that if U and V are analytically isomorphic, say, via ϕ, there is a group isomorphism of $\mathrm{Aut}(U)$ onto $\mathrm{Aut}(V)$.

Definition 13.10. *Let $f = \frac{p}{q}$ be a rational function, with p and q having no common factors. Then the degree*

$$deg\ f = max\{deg\ p, deg\ q\}.$$

Note that f is a constant if and only if $deg\ f = 0$.

Lemma 13.2. *If $f : \mathbb{C}_\infty \to \mathbb{C}_\infty$ is a rational function of degree $n > 0$, then f takes each value $\lambda \in \mathbb{C}_\infty$ exactly $n$times counted with multiplicity.*

Proof. Let $f = \frac{p}{q}$, with p and q having no common factors.

Case 1: Let $\lambda = \infty$. Then $f(z) = \infty$ if and only if $q(z) = 0$. This accounts for $\deg q$ preimages $f^{-1}(\infty)$. If $\deg p > \deg q$, then $f(\infty) = \infty$ with multiplicity $\deg p - \deg q$.

Case 2: Let $\lambda \neq \infty$. Since f is of degree $n > 0$, f is not a constant. Then $g(z) = \frac{1}{f(z)-\lambda} = \frac{q(z)}{p(z)-\lambda q(z)}$ is a rational function. The solutions of $f(z) = \lambda$ are the poles of g.

By the previous argument, there are deg g such poles. Now, q and $p - \lambda q$ have no common factors. So

$$\deg g = \max\{\deg q, \deg (p - \lambda q)\} = \max\{\deg q, \deg p\} = \deg f.$$

□

Theorem 13.16. *Any automorphism ϕ of $\mathbb{C}_\infty$ is of the form $\frac{az+b}{cz+d}$ where $ad - bc \neq 0$.*

Proof. Using the notation of the Definition 13.9, let $\alpha = \phi^{-1}(\infty)$. Then it can be checked that α is a pole for $\phi : \mathbb{C}_\infty \to \mathbb{C}$. It is clearly an isolated singularity. Since ϕ is continuous, $\lim_{z\to\alpha} \phi(z) = \infty$ so that α is a pole. Similarly, $\phi(\infty)$ is a pole of ϕ^{-1}. Since ϕ is meromorphic in $\mathbb{C}_\infty$ it is a rational function, say $\frac{p}{q}$ by Theorem 13.9. By Lemma 13.2, the pole $z = \alpha$ is simple. If $\alpha \in \mathbb{C}$, then $\phi(z) = \frac{\lambda}{z-\alpha} + \beta$. If $\alpha = \infty$, then $\phi(z) = az + b$. □

An immediate consequence of this is the determination of the automorphism of $\mathbb{C}$.

Theorem 13.17. *Let f be an automorphism of $\mathbb{C}$. Then $f(z) = az + b$ for $a, b \in \mathbb{C}$ with $a \neq 0$.*

Proof. Let f be an automorphism of $\mathbb{C}$. Then f is an entire function. Let $f(z) = \sum_{k=0}^{\infty} a_k z^k$, the Taylor series expansion of f around origin (Remark 7.1). Notice that the Taylor series converges absolutely for all $z \in \mathbb{C}$. There are three possibilities.

(i) $f(z) = a_0$. (That is, f is a constant function.)

(ii) $f(z) = \sum_{k=0}^{n} a_k z^k, a_n \neq 0, n \geq 1$. (That is, f is a polynomial of degree n.)

(iii) $f(z) = \sum_{k=0}^{\infty} a_k z^k, a_k \neq 0$ for infinitely many k.

Since f is one-one, the case (i) cannot occur. In case (ii), f is a polynomial of degree $n \geq 1$. But by Fundamental theorem of algebra, f has n zeros which lie inside some circle $|z| = R$, R large enough. Therefore, f is one-one only when n is not greater than 1. Thus $n = 1$ and $f(z) = a_0 + a_1 z$, where $a_1 \neq 0$. In case (iii), f has an essential singularity at ∞. Then, by Casorati-Weierstrass theorem (Theorem 8.5) for given any complex number w, there exists a sequence (z_k) such that $z_k \to \infty$ and $f(z_k) \to w$. In particular, for $w = 0$, we have $\lim_{k\to\infty} z_k = \lim_{k\to\infty} f^{-1}(f(z_k)) = f^{-1}(0) \neq \infty$. This contradiction shows that f is a polynomial which by case (ii) yields this result. □

Notice also that an automorphism $\phi : \mathbb{C}_\infty \to \mathbb{C}_\infty$ induces an automorphism of $\mathbb{C}$ if and only if $\phi(\infty) = \infty$. Thus $\phi(z) = az + b$.

Remark 13.5. Any one-one holomorphic function f on $\mathbb{C}$ is of the form $f(z) = az + b$ for $a, b \in \mathbb{C}$ with $a \neq 0$.

Example 13.18. *If a region G_1 is simply connected and G_1 is conformally equivalent to the region G_2 then G_2 is simply connected.*

Solution: Suppose f is a conformal, one-one map from G_1 onto G_2. By inverse function theorem the inverse of f is also a conformal one-one map from G_2 onto G_1. Let $g = f^{-1}$. Then $f \circ g = i$ and $g \circ f = j$ where i is the identity map from G_2 onto itself and j is the identity map from G_1 onto itself. Let $\gamma : [0, 1] \to G_2$ be a closed curve in G_2. Then $g \circ \gamma$ is homotopic to a point $z_0 \in G_1$ as G_1 is simply connected. Let $H : [0, 1] \times [0, 1] \to G_1$ be a continuous function such that $H(s, 0) = (g \circ \gamma)(s)$, $H(s, 1) = z_0$, and $H(0, t) = H(1, t)$ for $0 \leq s \leq 1$ and $0 \leq t \leq 1$. Define $H_1 : [0, 1] \times [0, 1] \to G_2$ as $H_1(s, t) = (f \circ H)(s, t)$. Then H_1 is continuous and $H_1(s, 0) = \gamma(s)$ and $H_1(s, 1) = f(z_0)$. Further $H_1(0, t) = H_1(1, t)$. Thus γ is homotopic to the point $f(z_0)$ and H_1 is the corresponding homotopy. Hence G_2 is simply connected.

13.6 Schwarz's lemma

Lemma 13.3. *(Schwarz's lemma) Let f be analytic and bounded by 1 for $|z| < 1$. If $f(0) = 0$, then either*

$$|f(z)| < |z|, z \neq 0, \tag{13.8}$$

$$|f'(0)| < 1$$

or else $f(z) = cz$ for some constant c of absolute value 1.

Proof. Define h by

$$h(z) = \begin{cases} \frac{f(z)}{z} & \text{if } z \neq 0 \\ f'(0) & \text{if } z = 0. \end{cases}$$

If $|z| < r < 1$, the maximum principle applies to the analytic function h to give

$$|h(z)| \leq \max_{|w|=r} |h(w)| .$$

The assumption on f guarantees that $\max_{|w|=r} |h(w)| \leq \frac{1}{r}$. Letting $r \to 1$ gives $|h(z)| \leq 1, z \in \mathbb{D}$. If $|h(z)| = 1$ for some $z \in \mathbb{D}$, then by the maximum principle h is constant. In this case $h(z) \equiv c$, where $|c| = 1$, and a fortiori $f(z) = cz$. On the other hand, if $|h(z)| < 1$ for each z, (13.8) is true. The lemma is proved. □

As a first application of Schwarz's lemma we prove the following.

Corollary 13.1. *Let f be a one-to-one analytic mapping of the unit disk $\mathbb{D}$, $\mathbb{D} = \{z||z| < 1\}$, onto itself. If $f(0) = 0$, then $f(z) = cz$ for a certain constant c of modulus 1.*

Proof. Both f and f^{-1} satisfy the hypotheses of Schwarz's lemma. Therefore, since $z = f^{-1}(f(z))$, we have

$$|z| = \left|f^{-1}f(z)\right| \leq |f(z)|.$$

Applying Schwarz's lemma $f(z) = cz$ as claimed. □

Corollary 13.2. *If f is a holomorphic function from the open unit disk $\mathbb{D}$ into itself then*

$$\left|\frac{f(z) - f(0)}{1 - \overline{f(0)}f(z)}\right| \leq |z|$$

for all $z \in \mathbb{D}$.

Proof. We define $g : \mathbb{D} \longrightarrow \mathbb{C}$ by $g(z) = \frac{f(z)-f(0)}{1-\overline{f(0)}f(z)}$. Since $f : \mathbb{D} \to \mathbb{D}$ is holomorphic and $|f(0)| < 1$, so g is a holomorphic function from $\mathbb{D}$ into $\mathbb{D}$. Now $g(0) = 0$. Therefore by Schwarz's lemma, $|g(z)| \leq |z|$ for all $z \in \mathbb{D}$. Hence the result follows. □

Lemma 13.4. *The linear fractional transformation*

$$Tz = e^{i\theta}\frac{z-a}{1-\overline{a}z}, |a| < 1, \theta \text{ real},$$

maps $|z| \leq 1$ onto itself.

Proof. If $|z| = 1$, then $z\overline{z} = 1$ and

$$|Tz| = \left|\frac{z-a}{1-\overline{a}z}\frac{\overline{z}}{\overline{z}}\right| = \left|\frac{z-a}{\overline{z}-\overline{a}}\overline{z}\right| = 1.$$

Thus, T sends $|z| = 1$ onto itself. If $|z| < 1$, Lemma 1.1 shows that $|Tz| < 1$. From Theorem 13.8 (Orientation principle) imply that $|Tz| > 1$ if $|z| > 1$, and the lemma is proved. □

It is now possible to prove that the transformation T gives the most general one-to-one analytic mapping of $\mathbb{D}$ onto itself where $\mathbb{D} = \{z \in \mathbb{C} : |z| < 1\}$.

Theorem 13.18. *Let f be a one-to-one analytic mapping of $\mathbb{D}$ onto itself. Then for some θ real and a complex number a with$|a| < 1$*

$$f(z) = e^{i\theta}\frac{z-a}{1-\overline{a}z}.$$

Proof. Given that (i) $f : \mathbb{D} \longrightarrow \mathbb{D}$ is analytic (ii) f is bijective. It is required to prove that

$$f(z) = e^{i\theta}\frac{z-a}{1-\overline{a}z}, \text{ for all } z \in \mathbb{D}$$

where θ is a real number and a is a complex number with $|a| < 1$. Let $f(0) = b$. Then $|b| < 1$. We define $\phi : \mathbb{D} \longrightarrow \mathbb{D}$ by

$$\phi(z) = \frac{z-b}{1-\bar{b}z}, \text{ for all } z \in \mathbb{D}.$$

Then ϕ is a one-one analytic map of $\mathbb{D}$ onto itself and $\phi(b) = 0$. So inverse of ϕ exists and

$$\phi^{-1}(z) = \frac{z+b}{1+\bar{b}z}, \text{ for all } z \in \mathbb{D}.$$

Let $g = \phi \circ f$ be the composition of the bijective analytic functions ϕ and f. So g is a one-one, analytic map of $\mathbb{D}$ onto itself. Let $h = g^{-1}$. Then h is one-one, analytic map of $\mathbb{D}$ onto itself and $(h \circ g)(z) = z$, for all $z \in \mathbb{D}$. Hence $(h \circ g)(0) = h(g(0)) = h(0) = 0$ as $g(0) = \phi(f(0)) = \phi(b) = 0$. Also $|h(z)| = |g^{-1}(z)| < 1$, for all $z \in \mathbb{D}$. Hence by Schwarz's lemma

$$|h(z)| \leq |z|, \text{ for all } z \in \mathbb{D}. \tag{13.9}$$

Again since $|g(z)| < 1$, by Schwarz's lemma

$$|g(z)| \leq |z|, \text{ for all } z \in \mathbb{D}. \tag{13.10}$$

Now from (13.9) and (13.10) it follows that for all $z \in \mathbb{D}$, $|g(z)| \leq |z| = |(h \circ g)(z)| = |h(g(z))| \leq |g(z)|$. Thus $|g(z)| = |z|$ for all $z \in \mathbb{D}$. By Schwarz's lemma, $g(z) = cz$, for some constant c with $|c| = 1$. Thus $(\phi \circ f)(z) = cz$ for all $z \in \mathbb{D}$. Hence

$$f(z) = \phi^{-1}(cz) = \frac{cz+b}{1+\bar{b}cz} = c\frac{z+b\bar{c}}{1+\bar{b}cz}. \tag{13.11}$$

Let $a = -b\bar{c}$. Then $|a| = |b\bar{c}| = |b||c| = |b| < 1$. Since $|c| = 1$, there exists a real θ such that $c = e^{i\theta}$. Thus from (13.11), we get

$$f(z) = e^{i\theta}\frac{z-a}{1-\bar{a}z}$$

for all $z \in \mathbb{D}$ where θ is real and a is a complex number with $|a| < 1$. □

Theorem 13.19. *If f is a holomorphic map from the open unit disk $\mathbb{D}$ into itself then*

$$\frac{|f'(z)|}{1-|f(z)|^2} \leq \frac{1}{1-|z|^2}.$$

Proof. Let $z_0 \in \mathbb{D}$ be fixed and let $f(z_0) = w_0$. Then $w_0 \in \mathbb{D}$. We define $T : \mathbb{D} \longrightarrow \mathbb{D}$ and $S : \mathbb{D} \longrightarrow \mathbb{D}$ by $Tz = \frac{z-z_0}{1-\overline{z_0}z}$ and $Sz = \frac{z-w_0}{1-\overline{w_0}z}$. Then both T and S are one-one, analytic maps of $\mathbb{D}$ onto itself. Also $Tz_0 = Sw_0 = 0$. Further the map $T^{-1} : \mathbb{D} \longrightarrow \mathbb{D}$ is analytic. Now consider the map $g : \mathbb{D} \longrightarrow \mathbb{C}$ defined by $g(z) = (S \circ f \circ T^{-1})(z)$. Since g is composition of the analytic functions S, f and T^{-1}, so it is also analytic. Thus because $g(0) = 0$ by

Schwarz's lemma, $|g(z)| \leq |z|$ for all $z \in \mathbb{D}$. Hence $|(S \circ f)(T^{-1}(z))| \leq |z|$. Therefore $|S(f(z))| \leq |Tz|$ for all $z \in \mathbb{D}$. This implies

$$\left|\frac{f(z) - w_0}{1 - \overline{w_0} f(z)}\right| \leq \left|\frac{z - z_0}{1 - \overline{z_0} z}\right|$$

for all $z \in \mathbb{D}$. Hence

$$\left|\frac{f(z) - f(z_0)}{z - z_0}\right| \leq \left|\frac{1 - \overline{f(z_0)} f(z)}{1 - \overline{z_0} z}\right|$$

for all $z \in \mathbb{D}$. Taking limit $z \to z_0$ both the sides of the inequality we get

$$|f'(z_0)| \leq \frac{1 - |f(z_0)|^2}{1 - |z_0|^2}.$$

But $z_0 \in \mathbb{D}$ is arbitrary. Hence the proof follows. □

Corollary 13.3. *If f is a holomorphic function from the open unit disk $\mathbb{D}$ into itself then*

$$|f'(z)| \leq \frac{1}{1 - |z|}$$

for all $z \in \mathbb{D}$.

Proof. From Theorem 13.19, it follows that for all $z \in \mathbb{D}$,

$$|f'(z)| \leq \frac{1 - |f(z)|^2}{1 - |z|^2}.$$

The right hand side of the above inequality is less than or equal to $\frac{1}{1-|z|^2} = \frac{1}{1-|z|}\frac{1}{1+|z|}$ as $0 < 1 - |f(z)|^2 < 1$. Since $\frac{1}{1+|z|} \leq 1$, the result follows. □

Theorem 13.20. *Let f be holomorphic from the half-plane $Rez > 0$ into the open unit disk $\mathbb{D}$. Then*

$$|f'(z)| \leq \frac{1 - |f(z)|^2}{2Rez}$$

for all z in the half plane $Rez > 0$.

Proof. Given that $f : H \longrightarrow \mathbb{D}$ is a holomorphic function where $H = \{z \in \mathbb{C} : \text{Re}z > 0\}$ is the right half plane. It is required to prove that

$$|f'(z)| \leq \frac{1 - |f(z)|^2}{2\text{Re}z}$$

for all $z \in H$. Let $z_0 \in H$ be arbitrary and let $f(z_0) = w_0$. Then $|w_0| < 1$. We define $S : \mathbb{D} \longrightarrow \mathbb{D}$ by $Sz = \frac{z - w_0}{1 - \overline{w_0} z}$. Then S is a one-one analytic map of $\mathbb{D}$ onto itself and $S(w_0) = 0$. We define another map $T : H \longrightarrow \mathbb{D}$ by $Tz = \frac{1-z}{1+z}$. Then T is a one-one analytic map of H onto $\mathbb{D}$. Again we define

$U : \mathbb{D} \longrightarrow \mathbb{D}$ by $Uz = \frac{z - Tz_0}{1 - \overline{Tz_0}\, z}$. Then U is a one-one analytic map of $\mathbb{D}$ onto itself as $|Tz_0| < 1$ and $U(Tz_0) = 0$. Now consider the map $M : \mathbb{D} \longrightarrow \mathbb{D}$ defined by $Mz = (S \circ f \circ T^{-1} \circ U^{-1})(z)$ for all $z \in \mathbb{D}$. It is clear that M is an analytic function from $\mathbb{D}$ into itself. Note that $M(0) = 0$ and $|Mz| < 1$. By Schwarz's lemma, $|Mz| \leq |z|$ for all $z \in \mathbb{D}$. Hence $|(S \circ f \circ T^{-1} \circ U^{-1})(z)| \leq |z|$ for all $z \in \mathbb{D}$. Therefore by putting $U^{-1}z = w$, we obtain $|(S \circ f \circ T^{-1})(w)| \leq |Uw|$ for all $w \in \mathbb{D}$. This is possible because the map U is bijective. Now since $T : H \longrightarrow \mathbb{D}$ is bijective hence $|(S \circ f)(z)| \leq |U(T(z))|$ for all $z \in H$. This last inequality implies

$$\left|\frac{f(z) - w_0}{1 - \overline{w_0} f(z)}\right| \leq \left|\frac{1 + \overline{z_0}}{1 + z_0} \cdot \frac{z_0 - z}{z + \overline{z_0}}\right|$$

for all $z \in H$. Thus

$$\left|\frac{f(z) - f(z_0)}{1 - \overline{f(z_0)} f(z)}\right| \leq \left|\frac{z_0 - z}{z + \overline{z_0}}\right|$$

for all $z \in H$ because $\left|\frac{1 + \overline{z_0}}{1 + z_0}\right| = 1$. Hence it follows that

$$\left|\frac{f(z) - f(z_0)}{z - z_0}\right| \leq \left|\frac{1 - \overline{f(z_0)} f(z)}{z + \overline{z_0}}\right|$$

for all $z \in H$. Taking limit $z \longrightarrow z_0$, we get

$$|f'(z_0)| \leq \left|\frac{1 - |f(z_0)|^2}{z_0 + \overline{z_0}}\right|.$$

Thus

$$|f'(z_0)| \leq \frac{1 - |f(z_0)|^2}{2\text{Re} z_0}.$$

Since $z_0 \in H$ is arbitrary, we have

$$|f'(z)| \leq \frac{1 - |f(z)|^2}{2\text{Re} z}$$

for all $z \in H$. This completes the proof. □

Theorem 13.21. *Except for the identity function, a holomorphic map from the open unit disk* $\mathbb{D}$ *into itself has at most one fixed point.*

Proof. Suppose that (i) $f : \mathbb{D} \longrightarrow \mathbb{D}$ is a holomorphic function (ii) f is not the identity map. It is required to prove that the function f has no more than one fixed point in $\mathbb{D}$. We prove this by the method of contradiction. So suppose that f has more than one fixed point in $\mathbb{D}$. Then there exists two complex numbers $a, b \in \mathbb{D}$ such that $a \neq b$, $f(a) = a$ and $f(b) = b$. We define $T : \mathbb{D} \longrightarrow \mathbb{D}$ by

$$Tz = \frac{z - a}{1 - \overline{a} z}.$$

Since $|a| < 1$, T is a one-one analytic map of $\mathbb{D}$ onto itself. So T^{-1} exists and is a one-one analytic map of $\mathbb{D}$ onto itself. Now consider the map $M = T \circ f \circ T^{-1}$, defined by

$$M(z) = (T \circ f \circ T^{-1})(z)$$

for all $z \in \mathbb{D}$. It is clear that M is a holomorphic function from $\mathbb{D}$ into $\mathbb{D}$ and $M(0) = (T \circ f \circ T^{-1})(0) = T(f(a)) = T(a) = 0$. Let $z_0 = T(b)$. Then $z_0 \in \mathbb{D}, z_0 \neq 0$(as T is one-one, $T(a) = 0, a \neq b$) and $M(z_0) = M(T(b)) = (T \circ f \circ T^{-1})(T(b)) = (T \circ f)(b) = T(f(b)) = T(b) = z_0$. So by Schwarz's lemma, there exists a complex number c with $|c| = 1$ such that

$$M(z) = cz, \tag{13.12}$$

for all $z \in \mathbb{D}$. Now $M(z_0) = z_0$ implies $c = 1$ as $z_0 \neq 0$. Hence from (13.12) we get $M(z) = z$ for all $z \in \mathbb{D}$. Thus $(T \circ f \circ T^{-1})(z) = z$ for all $z \in \mathbb{D}$. Therefore $(T \circ f)(w) = T(w)$ for all $w \in \mathbb{D}$. It follows by putting $T^{-1}(z) = w$ and because T is bijective. Hence $T(f(w)) = T(w)$ for all $w \in \mathbb{D}$. Since T is one-one, it follows that $f(w) = w$ for all $w \in \mathbb{D}$. This implies f is an identity map. So this contradicts to the given condition that f is not an identity map. Hence every holomorphic function $f : \mathbb{D} \longrightarrow \mathbb{D}$ has at most one fixed point in $\mathbb{D}$. □

Theorem 13.18 enables us to determine the the most general mapping of one disk $D_1 = \{z \in \mathbb{C} : |z-a| \leq R_1\}$ onto another disk $D_2 = \{z \in \mathbb{C} : |z-b| \leq R_2\}$. Let U, V be linear transformations which map $\mathbb{D} = \{z \in \mathbb{C} : |z| < 1\}$ onto D_1, D_2 respectively. If T is a one-to-one analytic transformation of D_1 onto D_2, then $Sz = V^{-1}TUz$ is a one-to-one analytic transformation of $\mathbb{D}$ onto $\mathbb{D}$. Therefore, Sz is given by a linear fractional transformation as in Theorem 13.18. Thus T can be computed as the composition of three linear transformations

$$Tw = VSU^{-1}w.$$

In particular T itself is linear. In case D_1, D_2 are both the disk of radius R centered at 0, U and V can be taken to be $Uz = Vz = R \cdot z$. If $Sz = e^{i\theta}\frac{z-a}{1-\bar{a}z}$ is as in Theorem 13.18, then

$$Tw = VSU^{-1}w = Re^{i\theta}\frac{(\frac{w}{R}) - a}{1 - \bar{a}(\frac{w}{R})} = Re^{i\theta}\frac{w - aR}{R - \bar{a}w}.$$

Thus, letting $b = aR$,

$$Tw = R^2 e^{i\theta}\frac{w-b}{R^2 - \bar{b}w}. \tag{13.13}$$

If $D_1 = \{w : |w| \leq R\}$ and $D_2 = \{w : |w| \leq S\}$, similar calculations give

$$Tw = Se^{i\theta}\frac{w - aR}{R - \bar{a}w},$$

or letting $b = aR$,

$$Tw = RSe^{i\theta}\frac{w-b}{R^2 - \bar{b}w}, \tag{13.14}$$

as the most general one-to-one analytic mapping between D_1 and D_2.

We shall now give the generalization due to Pick of Schwarz's lemma.

Let f be analytic on $D_1 = \{z \in \mathbb{C} : |z| < R\}$, and suppose $|f(z)| < S$ if $z \in D_1$. That is, f takes its values in the disk $D_2 = \{z \in \mathbb{C} : |z| < S\}$. Fixing $z_0 \in D_1$ with $f(z_0) = w_0 \in D_2$, let T_1, T_2 be transformations taking D_1, D_2 onto $\mathbb{D}$ with z_0, w_0 going into 0. By (13.14) with first $R = R, S = 1, b = z_0, \theta = 0$ and then $R = S, S = 1, b = w_0, \theta = 0$, we have

$$T_1 z = R\frac{z - z_0}{R^2 - \bar{z}_0 z},$$

$$T_2 w = S\frac{w - w_0}{S^2 - \bar{w}_0 w}.$$

Define an analytic function ϕ on $\mathbb{D}$ by $\phi(z) = T_2 f(T_1^{-1} z)$. Then $\phi(0) = 0$ and $|\phi(z)| \leq 1$ if $|z| \leq 1$. By Schwarz's lemma either

$$|\phi(z)| < |z|, z \neq 0$$

or else

$$\phi(z) = cz$$

for some constant c of modulus 1.

In the latter case let $z = T_1\xi, \xi \in D_1$. We have $f(\xi) = T_2^{-1}\phi(T_1\xi) = T_2^{-1}c(T_1\xi)$, and f is a linear transformation of D_1 onto D_2. In the former case we find

$$|T_2 f(\xi)| < |T_1 \xi|$$

for all $\xi \in D_1, \xi \neq z_0$. Therefore, substituting the values of $T_1\xi$ and $T_2 f(\xi)$, we obtain

$$\left| S\frac{f(\xi) - f(z_0)}{S^2 - \overline{f(z_0)}f(\xi)} \right| < \left| R\frac{\xi - z_0}{R^2 - \bar{z}_0 \xi} \right|.$$

Theorem 13.22. *(Pick) If f is analytic on D_1 with values in D_2, then either f is linear, or else for each $z_0 \in D_1$ the inequalities*

$$\left| S\frac{f(z) - f(z_0)}{S^2 - \overline{f(z_0)}f(z)} \right| < \left| R\frac{z - z_0}{R^2 - \bar{z}_0 z} \right|, \quad z \neq z_0 \tag{13.15}$$

and

$$|f'(z_0)| < \frac{R}{S}\frac{S^2 - |f(z_0)|^2}{R^2 - |z_0|^2} \tag{13.16}$$

will hold.

Proof. If f is linear, (13.15) holds. We need only to show that if f is not linear, (13.16) holds. For fixed z_0 define g by

$$g(z) = \begin{cases} S\frac{f(z)-f(z_0)}{S^2-\overline{f(z_0)}f(z)}\frac{1}{R}\frac{R^2-\bar{z}_0 z}{z-z_0} & \text{if} \quad z \neq z_0; \\ \frac{S}{R}f'(z_0)\frac{R^2-|z_0|^2}{S^2-|f(z_0)|^2} & \text{if} \quad z = z_0. \end{cases}$$

The function g is analytic, and by (13.15), $|g(z)| < 1$ for $z \neq z_0$. From the maximum principle we conclude that $|g(z_0)| < 1$, which is equivalent with (13.16). □

Example 13.19. *All conformal mappings which map the upper half plane H^+ onto the open unit disk $\mathbb{D} = \{z \in \mathbb{C} : |z| < 1\}$ and the point $\alpha \in H^+$ onto 0 are given by $Tz = e^{i\theta}\frac{z-\alpha}{z-\bar{\alpha}}, \theta \in \mathbb{R}$.*

Solution: Let $\alpha = a + ib \in \mathbb{C}, b > 0$ and consider the linear fractional transformation $Tz = \frac{z-\alpha}{z-\bar{\alpha}}$. Let $H^+ = \{z \in \mathbb{C} : \text{Im}z > 0\}$ be the upper half plane and $H^- = \{z \in \mathbb{C} : \text{Im}z < 0\}$ be the lower half plane. Notice that $a, a+b$, and ∞ are three points in $\mathbb{R} \cup \{\infty\}$ and $Ta = -1, T(a+b) = -i, T(\infty) = 1$. Hence $T(\mathbb{R} \cup \{\infty\})$ is the circle Γ passing through $-1, -i$ and 1, that is, the unit circle. Consider the orientations $(a, a+b, \infty)$ of $\mathbb{R} \cup \{\infty\}$ and $(-1, -i, 1)$ of Γ. With respect to these orientations $T(H^+) = \{z \in \mathbb{C} : |z| < 1\}$ and $T(H^-) = \{z \in \mathbb{C} : |z| > 1\}$. Also note that $T(\alpha) = 0, T(\bar{\alpha}) = \infty$. Thus the linear fractional transformation T maps the upper half plane H^+ onto the open unit disk $\mathbb{D} = \{z \in \mathbb{C} : |z| < 1\}$ and the point α onto 0. Thus it follows from Theorem 13.18 that all conformal mappings which map the upper half plane H^+ onto the open unit disk $\mathbb{D} = \{z \in \mathbb{C} : |z| < 1\}$ and the point $\alpha \in H^+$ onto 0 are given by $Tz = e^{i\theta}\frac{z-\alpha}{z-\bar{\alpha}}, \theta \in \mathbb{R}$.

Example 13.20. *Let γ_1 and γ_2 be two arcs (see Fig.13.1) in the z-plane with initial point p and the terminal point q. Suppose the angle between γ_1 and γ_2 at p equal $\frac{\pi}{m}$, m a natural number, $m \geq 2$. We want to determine a conformal mapping which maps the area G enclosed between γ_1 and γ_2 onto the open unit disk $\mathbb{D}$.*

Solution:

Let L be the line segment connecting p and q, ψ be the angle between γ_1 and L at p and put $f(z) = -e^{i\psi}\frac{z-p}{z-q}$.

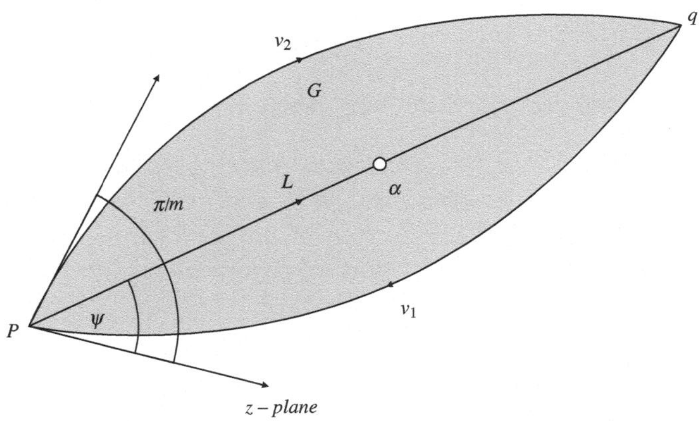

Fig. 13.1: The area G enclosed between the arcs γ_1 and γ_2

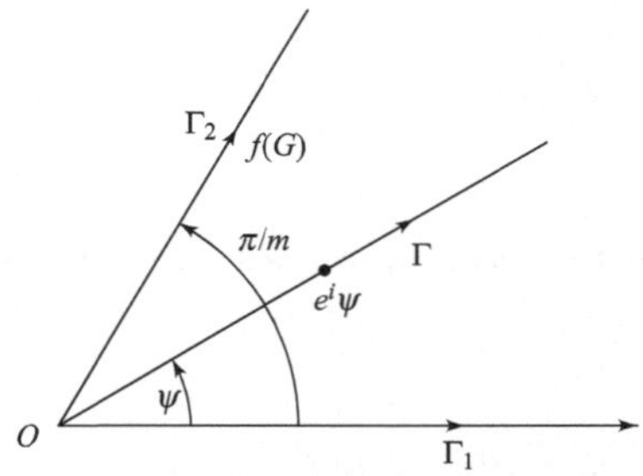

Fig. 13.2: The region $f(G)$

Since $f(p) = 0$ and $f(q) = \infty$, f maps γ_1, L, and γ_2 onto rays $\Gamma_1 = f(\gamma_1), \Gamma = f(L)$, and $\Gamma_2 = f(\gamma_2)$ with initial points 0 in the extended ζ-plane.

Let $\alpha = \frac{p+q}{2}$ be the mid point of L. Since $f(\alpha) = e^{i\psi}$, the ray Γ passes through $e^{i\psi}$, hence the angle between the positive real axis of the ζ-plane and Γ equals ψ.

Since f is a conformal mapping on the z-plane minus the point q, the angle between Γ_1, Γ_2 at 0 equals $\frac{\pi}{m}$ and the angle between Γ_1 and Γ equals ψ. Hence Γ_1 coincides with the nonnegative part of the real axis of the ζ-plane. Since G is to the left of γ_1 and to the right of γ_2, $f(G)$ is to the left (see Fig.13.2) of Γ_1 and to the right of Γ_2, that is, $f(G) = \{\zeta : \zeta = \rho e^{i\theta}, 0 < \rho < +\infty, 0 < \theta < \frac{\pi}{m}\}$. The map $w = \zeta^m = \rho^m e^{im\theta}$ is a holomorphic function of $\zeta = \rho e^{i\theta}$, which is univalent on $f(G)$ and $\{w : w = \zeta^m, \zeta \in f(G)\}$ is the upper half plane H^+ of the w-plane. Therefore the map $\zeta \mapsto w = \zeta^m$ is a conformal mapping which maps $f(G)$ onto H^+. Thus $g(z) = (f(z))^m = e^{im\psi}\left[\frac{z-p}{q-z}\right]^m$, $z \in G$, is a conformal map from G onto H^+. From Example 13.19 we know that $Sz = \dfrac{z - g(\alpha)}{z - \overline{g(\alpha)}}$ is a conformal mappings which maps the upper half plane H^+ onto the open unit disk $\mathbb{D} = \{z \in \mathbb{C} : |z| < 1\}$ and the point $g(\alpha) = (f(\alpha))^m = e^{im\psi} \in H^+$ onto 0. Thus $h(z) = S(g(z)) = \frac{g(z)-g(\alpha)}{g(z)-\overline{g(\alpha)}}$ is a conformal mappings which maps the region G onto the open unit disk $\mathbb{D} = \{z \in \mathbb{C} : |z| < 1\}$. Since $h(\alpha) = 0$ and $g(\alpha) = e^{im\psi}$, we have $h(z) = \frac{(z-p)^m-(q-z)^m}{(z-p)^m-e^{-2im\psi}(q-z)^m}$. If G is symmetric with respect to the segment L, i.e., if $2\psi = \frac{\pi}{m}$, then $h(z) = \frac{(z-p)^m-(q-z)^m}{(z-p)^m+(q-z)^m}$.

Example 13.21. *Let Γ_1 be the circle with center $\frac{i}{2}$ passing through* 0, *let Γ_2 be the circle with center $\frac{i}{4}$ passing through* 0, *and let G be the region enclosed by Γ_1 and Γ_2, that is, G is the collection of all points that are inside Γ_1 and outside Γ_2. See Fig.13.3. We shall find a conformal mapping which maps G onto the open unit disk $\mathbb{D} = \{z \in \mathbb{C} : |z| < 1\}$.*

Solution: First, let $\zeta = T(z) = \dfrac{1}{z}$. The points $i, z_1 = \frac{i-1}{2}$, and 0 are points of Γ_1 and $T(i) = -i, T(z_1) = -i-1$, and $T(0) = \infty$, hence $T(\Gamma_1) = l_1 \cup \{\infty\}$, where l_1 is the line through $-i$ and $-i-1$ in the ζ-plane. Similarly, $T(\Gamma_2) = l_2 \cup \{\infty\}$,

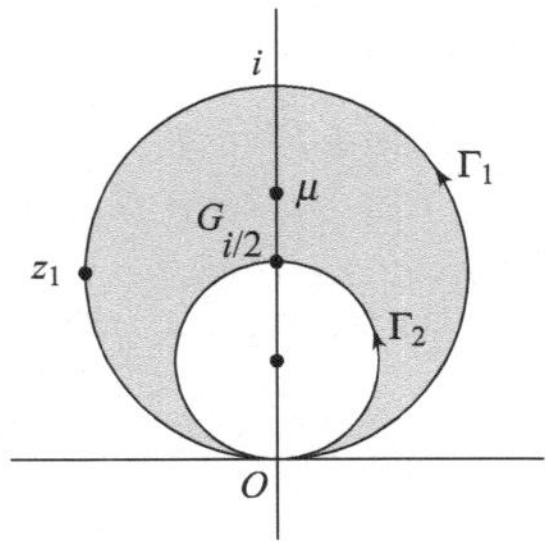

Fig. 13.3: The region G enclosed by Γ_1 and Γ_2

where l_2 is the line through $-2i$ and $-2i-2$ in the ζ-plane. Consider the orientations $(-i, -i-1, \infty)$ in $l_1 \cup \{\infty\}$ and $(-2i, -2i-2, \infty)$ in $l_2 \cup \{\infty\}$. Since T maps the interior of Γ_1 onto the region to the left of l_1 and the exterior of Γ_2 onto the region to the right of l_2, $T(G)$ is the region of the ζ-plane between l_1 and l_2. See Fig. 13.4.

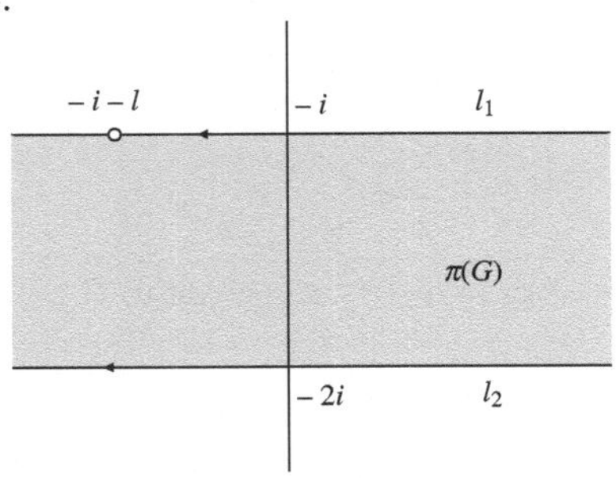

Fig. 13.4: The region $T(G)$ between l_1 and l_2

The map $w = e^{\pi\zeta}$ is a holomorphic function of ζ, which is univalent on $T(G)$ and $\{w : w = e^{\pi\zeta}, \zeta \in T(G)\}$ is the upper half plane H^+ of the w-plane. Thus the map $\zeta \mapsto w = e^{\pi\zeta}$ is a conformal mapping that maps $T(G)$ onto H^+ and therefore $z \mapsto w = e^{\pi T(z)} = e^{\frac{\pi}{z}}$ is a conformal mapping that maps G onto H^+. Hence $h(z) = \dfrac{e^{\frac{\pi}{z}} - i}{e^{\frac{\pi}{z}} + i}$, $z \in G$, is a conformal map that maps the region G onto the unit disk $\mathbb{D}$. Note that $h(\frac{2i}{3}) = 0$.

Example 13.22. *(Zhukovsky's function): The function $w = f(z) = \frac{1}{2}(z + \frac{1}{z})$, $z \in \mathbb{C}$ is called the Zhukovsky's function. This function was widely used by N.E. Zhukovsky in the solution of numerous problems in hydrodynamics and aerodynamics. The function f is analytic in $\mathbb{C} \setminus \{0\}$ and $z = 0$ is a simple pole of the function f. Computing the derivative of the function, we get $f'(z) = \frac{1}{2}(1 - \frac{1}{z^2})$. Thus $f'(z) \neq 0$ for all $z \in \mathbb{C} \setminus \{-1, 1\}$. Thus f is conformal in $\mathbb{C} \setminus \{-1, 1\}$. Now suppose $z_1 \neq z_2$ but $f(z_1) = f(z_2)$. Then $z_1 + \frac{1}{z_1} = z_2 + \frac{1}{z_2}$. That is, $z_1 - z_2 = \frac{z_1 - z_2}{z_1 z_2}$. Since $z_1 \neq z_2$, it follows that $z_1 z_2 = 1$. Thus the regions of univalence of the function f are, the open unit disk $\mathbb{D} = \{z \in \mathbb{C} : |z| < 1\}$ and the region $\{z \in \mathbb{C} : |z| > 1\}$. It can be verified that both of these regions* *are mapped conformally by f onto one and the same region of the w-plane.*

Definition 13.11. *The points P and P' are called symmetric with respect to a circle C if they lie on a common ray passing through the center O of the circle C, and the product of their distances from the center is equal to the square of the radius of the circle. That is, $OP \cdot OP' = R^2$.*

Proposition 13.6. *Every circle C' passing through the symmetric points P and P' with respect to the circle C is orthogonal to the circle C.*

Proof. Drawing a ray OP' and a radius OA to the point of intersection of the circles C and C', we get $OP \cdot OP' = R^2 = (OA)^2$.

This implies that OA is a tangent to the circle C' drawn from the point O. Thus C' is orthogonal to C. See Fig. 13.5. □

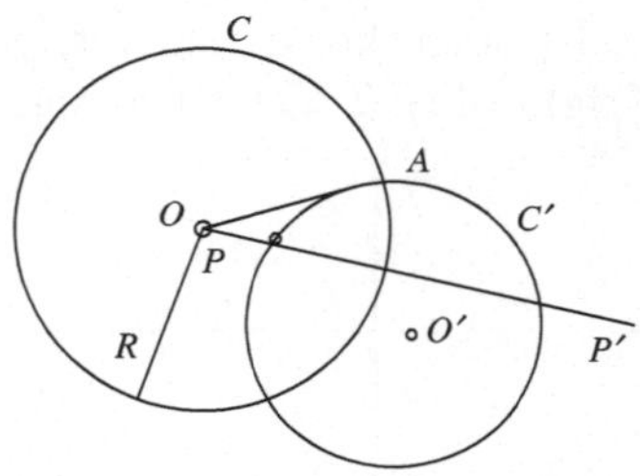

Fig. 13.5: C' orthogonal to C

Proposition 13.7. *Suppose two mutually intersecting circles C' and C'' orthogonal to one and the same circle C intersect at the points P and P'. Then the points P and P' are symmetric with respect to the circle C.*

Proof. Through the point P of intersection of the circles C' and C'', which lies inside C, draw a ray OP. See Fig. 13.6.

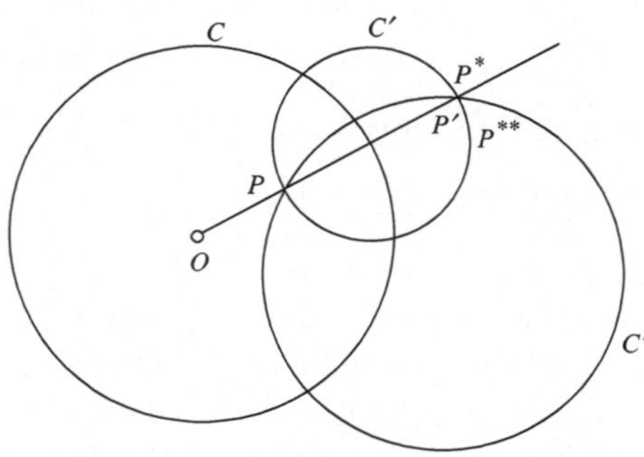

Fig. 13.6: Points P and P' are symmetric with respect to C

Suppose that OP intersects C' and C'' at different points, P^* and P^{**}, respectively. Since the circles C' and C'' are orthogonal to C, we obtain

$$OP \cdot OP^* = R^2,$$

$$OP \cdot OP^{**} = R^2.$$

Since the points P^* and P^{**} lie on one ray, hence the points P^* and P^{**} coincide. Thus $P^* = P^{**} = P'$. □

This actually shows that points symmetric with respect to any circle are mapped by a linear fractional transformation into points that are symmetric with respect to the image of the circle.

Suppose the points P and P' are symmetric with respect to the circle C. Through these points draw two circles C' and C''. By Proposition 13.6 the circles C' and C'' are orthogonal to C. The circles C, C' and C'' are mapped by a conformal map f to some circles K, K' and K'' and the circles K' and K'' will be orthogonal to K. The points P and P' of intersection of the circles C' and C'' will be transformed into the points Q and Q' of intersection of their images the circles K' and K''. By Proposition 13.7, the points Q and Q' must be symmetric with respect to K.

It follows from Definition 13.4 that given a circle γ, there is for each point z, a unique point z^* symmetric to z with respect to γ. The mapping ϕ which sends z into z^* is called **reflection** in γ. Notice that reflection is not an analytic mapping. To see this let z_2, z_3, z_4 be distinct points on γ. Let $Tz = \overline{(z, z_2, z_3, z_4)}$. Then it follows that $\overline{Tz} = Tz^*$. Thus $\overline{Tz} = T\phi(z)$. Hence $\phi(z) = T^{-1}(\overline{Tz})$ which is not an analytic function.

If γ is the circle $|w| = 1$, reflection in γ is given by $w^* = \frac{1}{\overline{w}}$. The proof goes as follows: The mapping $Tz = \frac{z-i}{z+i}$ sends the extended real line onto $|w| = 1$ and clearly satisfies $T(-z) = \frac{1}{Tz}$. The inverse of T is $T^{-1}w = \frac{iw+i}{-w+1} = -i\frac{w+1}{w-1}$, and $\overline{T^{-1}w} = i\frac{\overline{w}+1}{\overline{w}-1} = -T^{-1}\overline{w}$. It follows that $w^* = T\overline{T^{-1}w} = T(-T^{-1}\overline{w}) = \frac{1}{T(T^{-1}\overline{w})} = \frac{1}{\overline{w}}$ as claimed.

Theorem 13.23. *If γ is the circle of radius R about a point a, then*

$$z^* = \frac{R^2}{\overline{z} - \overline{a}} + a.$$

Proof. Let $Tz = \frac{z-a}{R}$ map γ onto $|w| = 1$. Then T is a linear fractional transformation. By Theorem 13.7, $(Tz)^* = Tz^*$, the first reflection is with respect to $|w| = 1$, the second with respect to γ. From our above discussion it follows that

$$Tz^* = \frac{R}{\overline{z} - \overline{a}}$$

and applying T^{-1} to both sides

$$z^* = \frac{R^2}{\overline{z} - \overline{a}} + a.$$

□

Thus if γ is the circle of radius R about a point a, and if z, z^* are symmetric with respect to γ then $z^* - a = \frac{R^2}{\bar{z}-\bar{a}}$. That is, $|z^* - a||z - a| = R^2$. Hence the Definition 13.4 of symmetric points implies the Definition 13.11.

We demonstrate below another method for obtaining a conformal map from the open unit disk $\mathbb{D}$ onto itself.

Example 13.23. *Find a conformal map f from the open unit disk $\mathbb{D}$ onto itself such that $f(z_0) = 0$ where z_0 is a point of $\mathbb{D}$.*

Solution: The point z_0 and the point z_1, symmetric to it with respect to the circle $|z| = 1$, will be transformed by a linear fractional transformation into points symmetric with respect to the circle $|w| = 1$.

Since a point that is symmetric to the center of a circle is the point at infinity and the point z_0 must be transformed into the point $w = 0$, it follows that the point z_1 will have to transform into the point $w = \infty$. Hence the linear fractional transformation has the form $w = f(z) = \lambda\frac{z-z_0}{z-z_1}$. Since $z_1 = \frac{1}{\bar{z}_0}$, hence $f(z) = \lambda\bar{z}_0\frac{z-z_0}{z\bar{z}_0-1}$. Since f maps the circle $|z| = 1$ into the circle $|w| = 1$, the following must hold:

$$|\lambda\bar{z}_0|\left|\frac{e^{i\theta} - z_0}{e^{i\theta}\bar{z}_0 - 1}\right| = |\lambda\bar{z}_0|\left|\frac{e^{i\theta} - z_0}{\bar{z}_0 - e^{-i\theta}}\right| = |\lambda\bar{z}_0| = 1.$$

This implies $\lambda\bar{z}_0 = e^{i\phi}$, where $\phi \in \mathbb{R}$ and $f(z) = e^{i\phi}\frac{z-z_0}{z\bar{z}_0-1}$. Note that we have obtained a solution defined to within one arbitrary parameter ϕ, which obviously determines the rotation of the circle $|w| = 1$ about the center. Specification of the value of the argument of the derivative of the function w at the point $z = z_0$ completely defines the function w.

Example 13.24. *We shall now construct a conformal map from a region bounded by two circles with noncoinciding centers onto some concentric annulus.*

Solution: For convenience, let us suppose that the center of the larger circle C lies at the point $z = 0$, its radius is R, and the center of the smaller circle C', of radius r, lies at the point $z = a$ on the real axis.

We find the points P_1 and P_2, which are simultaneously symmetric with respect to both circles C and C'. These points clearly lie on the real axis. Let $P_1 = x_1 + i \cdot 0$ and $P_2 = x_2 + i \cdot 0$. Thus we get $(x_1 - a)(x_2 - a) = r^2$ and $x_1x_2 = R^2$. Hence x_1 and x_2 are roots of the quadratic equation

$$ax^2 - (R^2 - r^2 + a^2)x + aR^2 = 0. \tag{13.17}$$

The discriminant of this equation $(R^2 - r^2 + a^2)^2 - 4a^2R^2$ is positive since the obvious relation $R - r > a$ holds. We construct the linear-fractional transformation

$$w = \lambda\frac{z - x_1}{z - x_2} \tag{13.18}$$

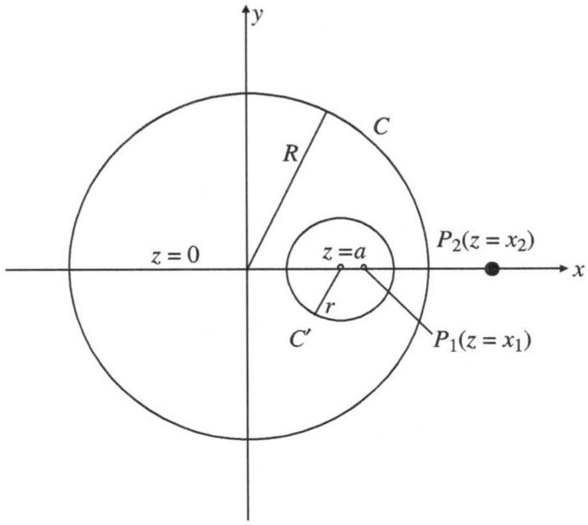

Fig. 13.7: The region bounded by circles C and C'

where x_1 and x_2 are the abscissae of the points P_1 and P_2 obtained from equation (13.17). The linear fractional transformation (13.18) will map the circles C and C' onto some circles K and K' of the w-plane and it will map the point P_2, which is exterior to the circles C and C', into the point $w = \infty$. The point P_1, which is symmetric to the point P_2 with respect to the circles C and C', must be mapped into a point that is symmetric to the point $w = \infty$ with respect to the circles K and K'. But a point that is symmetric to the point $w = \infty$ is the center of the circle. Hence, the linear fractional transformation (13.18) maps the point P_1 to the common center of the circles K and K'. Thus the desired linear fractional transformation has been obtained. In (13.18), the parameter λ was arbitrary; however, any variation in the parameter results only in a magnification and a rotation. But this cannot change the ratio of the radii of the circles of the concentric annulus thus obtained.

A lune (two-sided polygon) is a plane figure formed by the intersection of the arcs of two circles of different radii. It is clear that the angles at the vertices of the lune are equal.

Example 13.25. *Let there be given a lune with vertices at the points $A(z_1)$ and $B(z_2)$ and angle α at the vertex, and let it be required to construct a conformal mapping from the interior region of the given lune (see Fig. 13.8) onto the upper half plane $Im w > 0$.*

Solution: Consider the linear fractional transformation

$$\zeta = \frac{z_1 - z}{z_2 - z} \quad (\zeta = \xi + i\eta).$$

It maps the extended z-plane conformally onto the extended ζ-plane; the point $z = z_1$ is transformed to the point $\zeta = 0$, and the point $z = z_2$ into the point $\zeta = \infty$. By virtue of the circular property of a linear fractional transformation the circles forming the lune are mapped by the transformation onto circles. But the circle passing through the points $\zeta = 0$ and $\zeta = \infty$ has

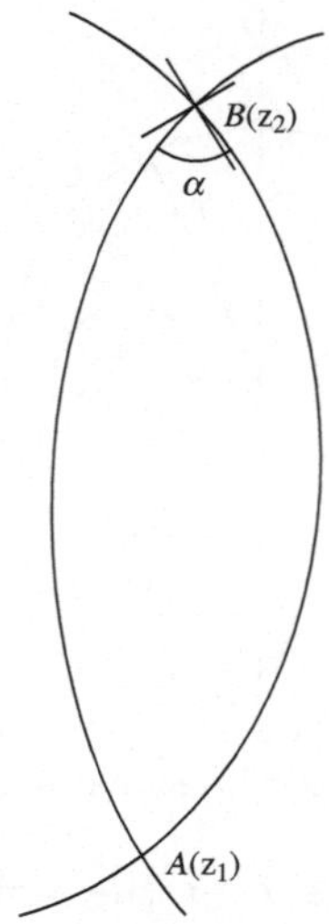

Fig. 13.8: The lune with vertices at $A(z_1)$ and $B(z_2)$ and angle α

an infinitely large radius. This means that the linear fractional transformation under consideration maps the sides of the lune into rays (I and II) emanating from the point $\zeta = 0$, and the angle between these rays will be equal to the angle α at the vertex of the lune.

Thus it is a conformal map from the given lune in the z-plane onto a sector with central angle α in the ζ-plane (see Fig. 13.9), and the ray I forms with the positive ξ-axis an angle α_0 whose value is determined by the position of the vertices A and B of the lune.

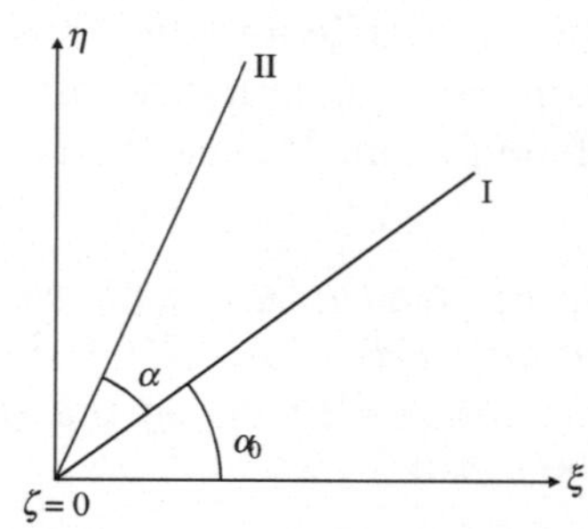

Fig. 13.9: The sector $\alpha_0 < \arg\zeta < \alpha_0 + \alpha$

The map $w = \zeta^{\frac{\pi}{\alpha}}$ is a direct analytic continuation of the real function $x^{\frac{\pi}{\alpha}}, x > 0$, maps the domain inside the sector $\alpha_0 < \arg\zeta < \alpha_0 + \alpha$ conformally onto the half-plane $\frac{\alpha_0}{\alpha}\pi < \arg w < \frac{\alpha_0}{\alpha}\pi + \pi$. It now remains to transform the half-plane thus obtained to the half-plane $\mathrm{Im}w > 0$. To do this, it is sufficient to rotate the entire plane as a whole through the angle $-\frac{\alpha_0}{\alpha}\pi$. This may be done by multiplying the map $w = \zeta^{\frac{\pi}{\alpha}}$ by the complex number $e^{-i\frac{\alpha_0}{\alpha}\pi}$. Thus the desired linear fractional transformation that will map conformally the lune AB onto the upper half plane Im$w > 0$ takes the form

$$w = e^{-i\frac{\alpha_0}{\alpha}\pi}\left(\frac{z_1 - z}{z_2 - z}\right)^{\frac{\pi}{\alpha}}.$$

Note that the conformality of the mapping is violated at the points z_1 and z_2.

Example 13.26. *Construct a conformal map of the upper half of the circle* $|z| < 1, Imz > 0$, *onto the upper half plane* $Imw > 0$.

Solution: Clearly, the given domain is a lune with vertices at the points $z_1 = -1$ and $z_2 = 1$, and angle $\alpha = \frac{\pi}{2}$ at the vertex. The function $\zeta = \frac{1+z}{1-z}$ maps this lune conformally onto the first quadrant of the $\zeta-$plane and the function $w = (\frac{1+z}{1-z})^2$ yields the desired mapping.

Example 13.27. *Construct a function that maps the strip* $0 < Rez < a$ *conformally onto the upper half plane* $Imw > 0$.

Solution: The function $z_1 = \frac{\pi}{a}z$ maps the original strip onto the strip $0 <$ Re$z_1 < \pi$. The function $z_2 = iz_1$ transforms the resulting strip into the strip $0 <$ Im$z_2 < \pi$. Finally, the function $w = e^{z_2}$ maps the given strip conformally onto the upper half plane Im$w > 0$. Therefore, the desired conformal map is given by $w = e^{i\frac{\pi}{a}z}$.

Example 13.28. *Construct a linear fractional transformation that map conformally the disk* $|z - 1 - i| \leq 2$ *onto the unit disk* $|w| \leq 1$.

Solution: It is easy to see that the desired function is of the form $w = a(z-1-i)$ where $|a| = \frac{1}{2}$, and the argument a of the complex number can have any value and determines the rotation of the $w-$plane about the point $w = 0$.

Example 13.29. *Consider the function* $f(z) = z^n$ *where* n *is an integer and* $n > 1$. *We shall discuss about the mapping properties of these functions.*

Solution: The function carries out a one-to-one mapping of the domain of its univalence–the sector $\Psi_0 < \arg z < \Psi_0 + \frac{2\pi}{n}$–onto the extended w-plane cut along the ray $\arg w = n\Psi_0$. Its derivative $f'(z) = nz^{n-1}$ is nonzero and is bounded everywhere within the given sector and at points of its boundary, with the exception of $z = 0$ and $z = \infty$. The function maps conformally the domain (see Fig. 13.10) within the indicated sector onto the cut w-plane.

Any infinitely small plane figures lying inside the given sector is transformed into a similar infinitely small figures in the w-plane; for instance the parallelogram $ABCD$, whose sides are the coordinate lines of the polar system of coordinates, will be transformed into a similar infinitely small parallelogram $A'B'C'D'$ whose sides are also the coordinate lines of the polar system of coordinates in the w-plane. However, at the boundary point $z = 0$ the conformality of the mapping is violated.

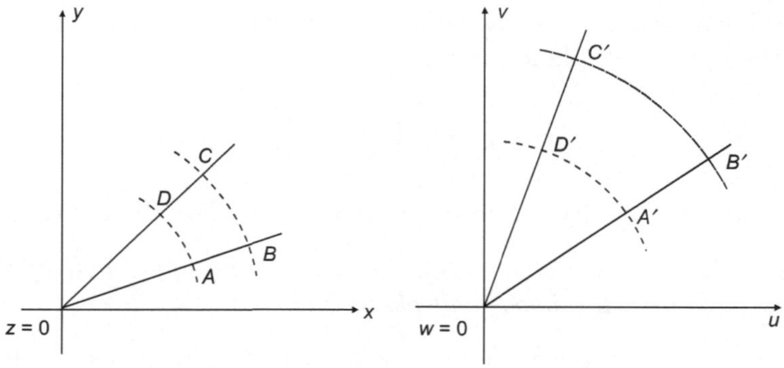

Fig. 13.10: The mapping properties of $f(z) = z^n$

Consider the curves γ_1 and γ_2 lying inside the given sector and intersecting at the point $z = 0$ at an angle ϕ_0. Clearly the function $w = z^n$ transforms these curves into the curves Γ_1 and Γ_2 which intersect at the point $w = 0$ at an angle $\Phi_0 = n\phi_0 \neq \phi_0$. Thus, the given function will map an infinitely small triangle with vertex at the point $z = 0$ onto a triangle which is no longer similar to the original one. Note that at the point $z = 0$ where conformality of the mapping is violated, the derivative of the function $f(z) = z^n$ is zero.

Example 13.30. *Construct a function that maps the first quadrant of the z-plane ($Rez > 0, Imz > 0$) conformally onto the upper half of the w-plane ($Imw > 0$).*

Solution: It is easy to see that the function $w = az^2 + b$ where $a > 0$ and b are arbitrary real constants, maps the first quadrant conformally onto the upper half plane. At the points $z = 0$ and $z = \infty$ conformal property of the function is violated.

Example 13.31. *Consider the exponential function $w = f(z) = e^z$. Find a strip in which the function is one-one and conformal.*

Solution: The function is one-one on the strip $y_0 < \text{Im}z < y_0 + 2\pi$ and maps the strip onto the extended w-plane cut along the ray $\arg w = y_0$. Further since $f'(z) = e^z$, is nonzero everywhere inside the given strip, the mapping is conformal. It is easy to see that in this mapping an orthogonal grid of cartesian coordinates $x = C_1, y = C_2$ inside the given strip is transformed into an orthogonal grid of polar coordinates $|w| = e^{C_1}, \arg w = C_2$ in the w-plane.

13.7 Exercises

1. Show that $\{z \in \mathbb{C} : \text{Im}z > 0\} \cup \{\infty\}$ is not a neighborhood of ∞.
2. Find the linear fractional transformation which carries $0, i, -i$ into $1, -1, 0$.

3. Exhibit the linear fractional transformation that maps $0, 1, \infty$ to $1, \infty, -i$ respectively.

4. Find a conformal map from the upper half plane into the open unit disk.

5. Prove that an entire function with negative real part is constant.

6. Show that the functions $e^z, \sin z, \cos z$ have essential singularity at ∞.

7. Show that any function which is meromorphic in the extended plane is rational.

8. Show that a function which is analytic in the whole plane and has a nonessential singularity at ∞ reduces to a polynomial.

9. Classify the singularity at $z = \infty$ of (i)$\frac{1}{z-z^3}$ (ii) $\frac{z^5}{(1-z)^2}$ (iii) $\frac{e^z}{1+z^2}$.

10. Show that the function $f(z) = z + \frac{1}{z}$ is holomorphic on $\mathbb{C}_\infty$ except at 0 and ∞.

11. Let $f(z) = \sin(\frac{z}{1+z})$. Compute $\text{Res}(f, \infty)$.

12. Let P be a polynomial of degree $n \geq 1$. Let $f(z) = \frac{P'(z)}{P(z)}$. Show that $\text{Res}(f, \infty) = -n$.

13. If the singularity of f in $\mathbb{C}_\infty$ are finite in number and are isolated, then show that the sum of residues (including the one at ∞)is zero.

14. If $f(\infty) = 0$, show that $\text{Res}(f, \infty) = -\lim_{z\to\infty} zf(z)$. Hint: F has a simple pole at $w = 0$ and $\text{Res}(F, 0) = \lim_{w\to 0} wF(w)$.

15. Prove that an entire function has a removable singularity at ∞ if and only if it is a constant.

16. Prove that an entire function has a pole at infinity of order m if and only if it is a polynomial of degree m.

17. Let G be a region and f be meromorphic on G; show that neither the poles nor the zeros of f have a limit point in G.

18. Investigate the nature of the following functions at ∞ :

    (i) $f(z) = 3z + 4$
    (ii) $g(z) = \exp(\frac{z}{z+1})$
    (iii) $h(z) = \sin^2(\frac{1}{z+2})$
    (iv) $k(z) = z \sin \frac{1}{z}$

19. Give an example of a meromorphic function $f : \mathbb{C} \to \mathbb{C}_\infty$ with a simple pole at each $z \in \mathbb{Z}$; does it extend to a meromorphic function from $\mathbb{C}_\infty$ into $\mathbb{C}_\infty$?

20. Let f be a nonconstant meromorphic function. Show that there are, at most, a finite number of periods of f in any bounded region of the complex plane.

21. Map the quarter plane $\{z \in \mathbb{C} : \text{Re}z > 0, \text{Im}z > 0\}$ conformally into the unit disk.

22. Find the image of the wedge $\{z = re^{i\theta} : r > 0, 0 < \theta < \frac{\pi}{4}\}$ under the mapping $w = z^4$.

23. Let $f : \mathbb{C}_\infty \setminus F \to \mathbb{C}$ be a holomorphic function where F is a finite subset of $\mathbb{C}_\infty$. If f has no essential singularities then f is a rational function.

Chapter 14

Harmonic Functions and Their Properties

Harmonic functions—the solutions of Laplace's equation—play a crucial role in many areas of mathematics, physics and engineering. The word "harmonic" is commonly used to describe a quality of sound. Physicists label the movement of a point on a vibrating string as "harmonic motion". Such motion may be described using sine and cosine functions and in this context the sine and cosine functions are sometimes called harmonics.

14.1 Harmonic functions

In this chapter throughout we shall assume that Ω is a region in $\mathbb{C}$.

Definition 14.1. *A real-valued function $u : \Omega \to \mathbb{R}$ is said to be harmonic in Ω if u has continuous second order partial derivatives (i.e., u is in C^2) and satisfies the Laplace's equation*

$$\frac{\partial^2 u}{\partial x^2} + \frac{\partial^2 u}{\partial y^2} = 0.$$

Note that the following holds.

1. If u and v are harmonic then $\alpha u + \beta v$ is harmonic for all $\alpha, \beta \in \mathbb{R}$.
2. The functions $u(x, y) = ax + by, a, b \in \mathbb{R}$ and $v(x, y) = x^2 - y^2$ are harmonic.
3. The function $u(z) = \arg z$ is harmonic whenever the argument θ is uniquely defined.

Theorem 14.1. *If $f : \Omega \to \mathbb{C}$ is holomorphic and $f = u + iv$ then u and v are harmonic.*

Proof. By Cauchy Riemann equation,
$u_x = v_y$ and $u_y = -v_x$, so that $u_{xx} = v_{xy} = -u_{yy}$, or $u_{xx} + u_{yy} = 0$.
Apply the same argument to $-if$ to conclude that v is harmonic. □

Theorem 14.2. *If u is harmonic in Ω, then $u_x - iu_y$ is holomorphic in Ω.*

Proof. Let $f = u_x - iu_y$. Since $u \in C^2$, f is in C^1. Further $f_y = u_{xy} - iu_{yy} = u_{yx} + iu_{xx} = if_x$. Thus f satisfies the Cauchy-Riemann equations and hence f is holomorphic in Ω. The theorem is thus proved. □

Theorem 14.3. *Suppose that u is a real-valued harmonic function in an open disk D, and suppose that u^2 is also harmonic. Then u is a constant.*

Proof. Let $v = u^2$. Since u is harmonic, $\frac{\partial^2 u}{\partial x^2} + \frac{\partial^2 u}{\partial y^2} = 0$. If v is harmonic then $\frac{\partial^2 v}{\partial x^2} + \frac{\partial^2 v}{\partial y^2} = 0$. That is, $2\left[\left(\frac{\partial u}{\partial x}\right)^2 + \left(\frac{\partial u}{\partial y}\right)^2 + u\left(\frac{\partial^2 u}{\partial x^2} + \frac{\partial^2 u}{\partial y^2}\right)\right] = 0$. Hence $2\left[\left(\frac{\partial u}{\partial x}\right)^2 + \left(\frac{\partial u}{\partial y}\right)^2\right] = 0$. Thus $\frac{\partial u}{\partial x} = 0$ and $\frac{\partial u}{\partial y} = 0$ and hence u is a constant. □

Theorem 14.4. *A harmonic function satisfies the formal differential equation*

$$\frac{\partial^2 u}{\partial \bar{z} \partial z} = 0.$$

Proof. Let $z = x + iy$. We shall prove this by showing that

$$\frac{\partial^2}{\partial \bar{z} \partial z} = \frac{1}{4}\left(\frac{\partial^2}{\partial x^2} + \frac{\partial^2}{\partial y^2}\right).$$

Since u is harmonic, the second order partial derivatives exist and are continuous; hence second order mixed partial derivatives are equal. Notice that $x = \frac{z+\bar{z}}{2}$, $y = \frac{z-\bar{z}}{2i}$. Hence $\frac{\partial x}{\partial z} = \frac{1}{2}$, $\frac{\partial y}{\partial z} = \frac{1}{2i}$. Now

$$\frac{\partial u}{\partial z} = \frac{\partial u}{\partial x}\frac{\partial x}{\partial z} + \frac{\partial u}{\partial y}\frac{\partial y}{\partial z} = \frac{\partial u}{\partial x}\frac{1}{2} + \frac{\partial u}{\partial y}\frac{1}{2i}.$$

Hence

$$\begin{aligned}
\frac{\partial}{\partial \bar{z}}\left(\frac{\partial u}{\partial z}\right) &= \frac{\partial}{\partial x}\left(\frac{\partial u}{\partial z}\right)\frac{\partial x}{\partial \bar{z}} + \frac{\partial}{\partial y}\left(\frac{\partial u}{\partial z}\right)\frac{\partial y}{\partial \bar{z}} \\
&= \frac{\partial}{\partial x}\left[\frac{1}{2}\frac{\partial u}{\partial x} + \frac{1}{2i}\frac{\partial u}{\partial y}\right]\frac{1}{2} + \frac{\partial}{\partial y}\left[\frac{1}{2}\frac{\partial u}{\partial x} + \frac{1}{2i}\frac{\partial u}{\partial y}\right]\left(-\frac{1}{2i}\right) \\
&= \frac{1}{4}\left[\frac{\partial^2 u}{\partial x^2} - i\frac{\partial^2 u}{\partial x \partial y} + i\frac{\partial^2 u}{\partial y \partial x} + \frac{\partial^2 u}{\partial y^2}\right] = 0.
\end{aligned}$$

□

Theorem 14.5. *The real and imaginary parts of a holomorphic function $f(z) = u(z)+iv(z)$ of a complex variable when expressed in polar form satisfy the equation (Laplace's equation in polar form)*

$$\frac{\partial^2 \Psi}{\partial r^2} + \frac{1}{r}\frac{\partial \Psi}{\partial r} + \frac{1}{r^2}\frac{\partial^2 \Psi}{\partial \theta^2} = 0.$$

Proof. In polar form the Cauchy-Riemann equations can be written as

$$\frac{\partial v}{\partial \theta} = r\frac{\partial u}{\partial r}, \tag{14.1}$$

$$\frac{\partial v}{\partial r} = -\frac{1}{r}\frac{\partial u}{\partial \theta}. \tag{14.2}$$

To eliminate v differentiate (14.1) partially with respect to r and (14.2) with respect to θ. Then

$$\frac{\partial^2 v}{\partial r \partial \theta} = \frac{\partial}{\partial r}\left(\frac{\partial v}{\partial \theta}\right) = \frac{\partial}{\partial r}\left(r\frac{\partial u}{\partial r}\right) = r\frac{\partial^2 u}{\partial r^2} + \frac{\partial u}{\partial r}. \tag{14.3}$$

$$\frac{\partial^2 v}{\partial \theta \partial r} = \frac{\partial}{\partial \theta}\left(\frac{\partial v}{\partial r}\right) = \frac{\partial}{\partial \theta}\left(-\frac{1}{r}\frac{\partial u}{\partial \theta}\right) = -\frac{1}{r}\frac{\partial^2 u}{\partial \theta^2}. \tag{14.4}$$

But

$$\frac{\partial^2 v}{\partial r \partial \theta} = \frac{\partial^2 v}{\partial \theta \partial r}$$

assuming the second partial derivatives are continuous. Hence from (14.3) and (14.4),

$$r\frac{\partial^2 u}{\partial r^2} + \frac{\partial u}{\partial r} = -\frac{1}{r}\frac{\partial^2 u}{\partial \theta^2}$$

or

$$\frac{\partial^2 u}{\partial r^2} + \frac{1}{r}\frac{\partial u}{\partial r} + \frac{1}{r^2}\frac{\partial^2 u}{\partial \theta^2} = 0.$$

Similarly by elimination of u we find

$$\frac{\partial^2 v}{\partial r^2} + \frac{1}{r}\frac{\partial v}{\partial r} + \frac{1}{r^2}\frac{\partial^2 v}{\partial \theta^2} = 0$$

so that the required result is proved. □

Theorem 14.6. *If Ω is simply connected and u is harmonic in Ω then u is the real part of a holomorphic function in Ω.*

Proof. If Ω is simply connected, there is a primitive F for $f = u_x - iu_y$. Let $F = U + iV$. Then $F'(z) = U_x + iV_x = U_x - iU_y$. But $F'(z) = f(z) = u_x - iu_y$. Comparing the two expressions of F' we get $U_x = u_x$ and $U_y = u_y$. Hence $U(x, y) = u(x, y) + C$. Thus u is the real part of the holomorphic function $F - C$ in Ω. This proves the theorem. □

Remark 14.1. In fact one can show that a region Ω is simply connected if and only if for each harmonic function u on Ω there is a harmonic function v on Ω such that $f = u + iv$ is holomorphic on Ω.

Corollary 14.1. *Let Ω' be a region in $\mathbb{C}$. If u is harmonic in Ω and $h : \Omega' \to \Omega$ is holomorphic, then $u \circ h : \Omega' \to \mathbb{R}$ is harmonic.*

Proof. Fix $h(z) = z' \in \Omega$. Let $u = \text{Re}\phi$ in a neighborhood (which is simply connected) of z', where ϕ is holomorphic. Then $\phi \circ h$ is holomorphic in a neighborhood of z and $u \circ h = \text{Re}(\phi \circ h)$ in the same neighborhood. Hence $u \circ h$ is harmonic. □

Definition 14.2. *If Ω is a region and $f : \Omega \to \mathbb{C}$ is a holomorphic function then $u = Ref$ and $v = Imf$ are called harmonic conjugates.*

Example 14.1. *Let $u(x, y) = x^3 - 3xy^2$. We shall show that u is harmonic in $\mathbb{C}$ and we shall find its harmonic conjugate v.*

Solution: Notice that $\frac{\partial^2 u}{\partial x^2} = 6x = -\frac{\partial^2 u}{\partial y^2}$. So $\frac{\partial^2 u}{\partial x^2} + \frac{\partial^2 u}{\partial y^2} = \triangle u = 0$. If v is the harmonic conjugate of u then $\frac{\partial u}{\partial x} = \frac{\partial v}{\partial y}$ and $\frac{\partial u}{\partial y} = -\frac{\partial v}{\partial x}$. Thus $\frac{\partial v}{\partial y} = 3x^2 - 3y^2$ and $\frac{\partial v}{\partial x} = -(-6xy) = 6xy$. Thus $v(x, y) = 6y\frac{x^2}{2} + \phi(y) = 3x^2y + \phi(y)$. Now $3x^2 - 3y^2 = \frac{\partial v}{\partial y} = 3x^2 + \phi'(y)$. Hence $\phi'(y) = -3y^2$. That is, $\phi(y) = -3\frac{y^3}{3} + c = -y^3 + c$ where c is an arbitrary constant.

Note that if v_1 and v_2 are both harmonic conjugates of u then $i(v_1 - v_2) = (u + iv_1) - (u + iv_2)$ is an analytic function whose range is contained in the imaginary axis; hence $v_1 = v_2 + c$, for some constant c. Thus any two harmonic conjugates of u differ by a constant.

Thus from Remark 14.1 it follows that every harmonic function on a simply connected region has a harmonic conjugate. We shall give below an example of a harmonic function u in a region Ω which has no harmonic conjugate in the region.

Example 14.2. *Let $u(x, y) = log(x^2 + y^2)^{\frac{1}{2}}$ defined on the region $\Omega = \{z = x + iy : R_1 < |z| < R_2\}$ where $R_2 > R_1 \geq 0$. We shall show that u is harmonic in Ω but u has no harmonic conjugate in Ω.*

Solution: The function u is harmonic in Ω as it satisfies Laplace's equation and has continuous second partial derivatives. Let v be a harmonic conjugate of u in Ω. Then $f = u + iv$ is holomorphic in Ω. Hence f' is also holomorphic in Ω. Now

$$u_x(x, y) = \frac{x}{x^2 + y^2} \text{ and } u_y(x, y) = \frac{y}{x^2 + y^2}.$$

Hence $f'(z) = u_x + iv_x = u_x - iu_y = \frac{x - iy}{x^2 + y^2} = \frac{\bar{z}}{|z|^2} = \frac{1}{z}$.

Let $z_0 = R$ and $z = Re^{i\theta}, R_1 < R < R_2, 0 \leq \theta \leq 2\pi$. Then $z_0, z \in \Omega$. By Theorem 6.8, the value of the integral $\int_{z_0}^{z} f'(z)dz$ is independent of the path joining z_0 and z in Ω. Thus we have

$$f(z) - f(z_0) = \int_{z_0}^{z} f'(z)dz = \int_{z_0}^{z} \frac{1}{z}dz. \tag{14.5}$$

Thus in particular, if we choose $z_0 = R$ and $z = Re^{2\pi i}$, from (14.5) it follows that

$$0 = f(Re^{2\pi i}) - f(R) = \int_0^{2\pi} \frac{iRe^{i\theta}}{Re^{i\theta}} d\theta = 2\pi i.$$

This contradiction shows that no such function v can exist in Ω.

Note that in the Example 14.2 the region Ω is not simply connected. If the region Ω is simply connected, then every harmonic function defined on Ω will have a harmonic conjugate. The proof of the above is beyond the scope of this book. However there are some simply connected regions Ω (for example, any disk or the whole plane) in which the proof of the above fact is not so difficult.

Theorem 14.7. *Let Ω be either the whole complex plane $\mathbb{C}$ or some open disk. If $u : \Omega \to \mathbb{R}$ is a harmonic function then u has a harmonic conjugate.*

Proof. Without loss of generality, let $\Omega = B(0, R), 0 < R \leq \infty$ and let $u : \Omega \to \mathbb{R}$ be a harmonic function. If v is a harmonic conjugate of u in Ω then $v(x, y) = \int_0^y u_x(x, t)dt + g(x)$ and the functions u and v must satisfy the Cauchy-Riemann equation. Thus differentiating both sides of this equation with respect to x (apply Leibniz's rule for differentiation, Theorem 6.2), we get

$$\begin{aligned} v_x(x, y) &= \int_0^y u_{xx}(x, t)dt + g'(x) \\ &= -\int_0^y u_{yy}(x, t)dt + g'(x) \\ &= -u_y(x, y) + u_y(x, 0) + g'(x). \end{aligned}$$

Thus $\phi'(x) = -u_y(x, 0)$. By fundamental theorem of calculus,

$$v(x, y) = \int_0^y u_x(x, t)dt - \int_0^x u_y(s, 0)ds.$$

Hence, we obtain the harmonic conjugate v of u in Ω. □

Thus we have shown that if u is a harmonic function on Ω and D is a disk that is contained in Ω then there is a harmonic function v on D such that $u + iv$ is holomorphic on D. In other words, each harmonic function has a harmonic conjugate locally.

Example 14.3. *If the harmonic function v is a harmonic conjugate of the harmonic function u, then the functions uv and $u^2 - v^2$ are both harmonic.*

Solution: By hypothesis, $u + iv$ is holomorphic and $u_x = v_y$ and $u_y = -v_x$. Now $\frac{\partial^2(uv)}{\partial x^2} + \frac{\partial^2(uv)}{\partial y^2} = v(\frac{\partial^2 u}{\partial x^2} + \frac{\partial^2 u}{\partial y^2}) + u(\frac{\partial^2 v}{\partial x^2} + \frac{\partial^2 v}{\partial y^2}) = 0$. Hence uv is harmonic. Similarly, we can show that $u^2 - v^2$ is harmonic.

Example 14.4. *If u is a real-valued harmonic function then the function $\frac{\partial u}{\partial z}$ is holomorphic.*

Solution: Let $z = x + iy$. Then $x = \frac{z+\bar{z}}{2}$ and $y = \frac{z-\bar{z}}{2i}$. Thus $\frac{\partial u}{\partial z} = \frac{\partial u}{\partial x}\frac{\partial x}{\partial z} + \frac{\partial u}{\partial y}\frac{\partial y}{\partial z} = \frac{\partial u}{\partial x}\frac{1}{2} + \frac{\partial u}{\partial y}\frac{1}{2i} = \frac{1}{2}\frac{\partial u}{\partial x} - \frac{i}{2}\frac{\partial u}{\partial y}$. Let $U = \frac{1}{2}\frac{\partial u}{\partial x}$ and $V = -\frac{1}{2}\frac{\partial u}{\partial y}$. Note U and V have continuous first order partial derivatives and they satisfy the Cauchy-Riemann equations: $\frac{\partial U}{\partial x} = \frac{1}{2}\frac{\partial^2 u}{\partial x^2} = -\frac{1}{2}\frac{\partial^2 u}{\partial y^2} = \frac{\partial V}{\partial y}$ as u is harmonic. Further, $\frac{\partial U}{\partial y} = \frac{1}{2}\frac{\partial^2 u}{\partial y \partial x} = \frac{1}{2}\frac{\partial^2 u}{\partial x \partial y} = -\frac{\partial V}{\partial x}$ since mixed derivatives are equal.

Theorem 14.8. *(Mean value property for harmonic functions): Let $u : \Omega \to \mathbb{R}$ be a harmonic function and let $\overline{B(a,r)} \subset \Omega$. If $\gamma = \{z \in \mathbb{C} : |z - a| = r\}$ then*

$$u(a) = \frac{1}{2\pi}\int_0^{2\pi} u(a + re^{i\theta})d\theta.$$

Proof. Let D be a disk such that $\overline{B(a,r)} \subset D \subset \Omega$ (the existence of D follows from Corollary 2.7 as $\overline{B(a,r)}$ is compact and Ω is open.) Let v be the harmonic conjugate (by Theorem 14.7, the harmonic conjugate exists) of u in D. Let $f = u + iv$. Then f is holomorphic in D. Thus by Cauchy's integral formula

$$\begin{aligned} f(a) &= \frac{1}{2\pi i}\int_\gamma \frac{f(z)}{z-a}dz \\ &= \frac{1}{2\pi i}\int_0^{2\pi} \frac{f(a+re^{i\theta})}{a+re^{i\theta}-a} re^{i\theta} i d\theta \\ &= \frac{1}{2\pi}\int_0^{2\pi} f(a+re^{i\theta})d\theta. \end{aligned}$$

By taking the real part of each side of this equation we complete the proof. □

14.2 Maximum principle for harmonic functions

Theorem 14.9. *(Maximum principle for harmonic functions): If u is a nonconstant harmonic function in a region Ω, then u has no maximum or minimum in Ω.*

Proof. Assume that u is nonconstant, harmonic in Ω and that it attains its maximum M at a point $a \in \Omega$. Let $E = \{z \in \Omega : u(z) = u(a)\}$. Note that $E \neq \emptyset$ as $a \in E$. Since u is continuous, the set E is closed in Ω. For let (z_n) be a sequence in E such that $z_n \to z, z \in \Omega$. Since u is continuous, $u(z_n) \to u(z)$. Now, $u(z) = \lim_{n\to\infty} u(z_n) = \lim_{n\to\infty} u(a) = u(a)$. Hence $z \in E$. (One can also note that $E = u^{-1}(\{b\})$ if $u(a) = b$ and therefore E is closed.)

Now we shall show that E is open. Let $z_0 \in E$ and let $r > 0$ be chosen such that $\overline{B(z_0,r)} \subset \Omega$. Suppose there is a point w in $B(z_0,r)$ such that $u(w) \neq u(a)$; then $u(w) < u(a)$. By continuity, $u(z) < u(a) = u(z_0)$ for all z in a neighborhood of w. (Let $\epsilon = u(a) - u(w) > 0$. Since u is continuous, there exists $\delta > 0$ such that whenever $|z - w| < \delta$, we have $|u(z) - u(w)| < \epsilon =$

$u(a) - u(w)$. Thus $-\epsilon < u(z) - u(w) < u(a) - u(w)$. This implies $u(z) < u(a)$ if $z \in B(w, \delta)$.)

In particular, if $\rho = |z_0 - w|$ and $w = z_0 + \rho e^{i\alpha}, 0 \leq \alpha < 2\pi$ then there is a proper interval I of $[0, 2\pi]$ such that $\alpha \in I$ and $u(z_0 + \rho e^{i\theta}) < u(z_0)$ for all $\theta \in I$. Hence by the mean value property (Theorem 14.8),

$$u(z_0) = \frac{1}{2\pi} \int_0^{2\pi} u(z_0 + re^{i\theta}) d\theta < u(z_0).$$

This is a contradiction. So $B(z_0, r) \subset E$ and E is open. By the connectedness of Ω, $E = \Omega$. Hence u is a constant function. This is again a contradiction. So u has no maximum in Ω. Applying the above argument to $-u$ proves the assertion concerning the minimum. □

Corollary 14.2. *Let G be a bounded region in $\mathbb{C}$ and suppose that $u : \overline{G} \to \mathbb{R}$ is a continuous function that satisfies the mean value property on G. If $u(z) = 0$ for all z in ∂G(boundary of G) then $u(z) = 0$ for all z in G.*

Example 14.5. *Let f be a complex valued function which is analytic on an open set containing the disk $|z| \leq 1$, and which is real valued on the unit circle. We shall prove that f is constant using maximum principle for harmonic functions.*

Solution: Let $f = u + iv$. Then v is identically 0 on the unit circle. By the maximum modulus principle for harmonic functions, v is identically zero on the open unit disk $\mathbb{D}$. By the Cauchy-Riemann equations, the partial derivatives of u vanish; hence, u is constant also and so is f.

Corollary 14.3. *Let Ω be a region such that $\overline{\Omega}$ is bounded. Let $w : \overline{\Omega} \to \mathbb{R}$ be a continuous function such that w is harmonic in Ω. Then w attains its maximum M and minimum m on the boundary $\partial\Omega$ of Ω. Further if u and v are two continuous real valued functions on $\overline{\Omega}$, both harmonic in Ω, and if $u = v$ on the boundary of Ω, then $u = v$ on $\overline{\Omega}$.*

Proof. If w is constant then clearly w attains its maximum and minimum on the boundary of Ω. Assume therefore that w is nonconstant. Since w is continuous on the compact set $\overline{\Omega}$ (as it is closed and bounded), w attains its maximum M and minimum m in $\overline{\Omega}$. since w is nonconstant and harmonic in Ω, the maximum and minimum of w are not attained in Ω. Hence they are attained in $\partial\Omega$. This proves the first part of the corollary. To prove the second part put $w = u - v$, which is continuous in $\overline{\Omega}$, harmonic in Ω, and vanishes on $\partial\Omega$. Clearly by the first part of the corollary the maximum and the minimum of w are both zero in Ω, hence $u = v$ in Ω. □

Remark 14.2. The hypothesis of compactness of $\partial\Omega$ in the above corollary cannot be dropped. One can consider a continuous function on the closed upper half plane which is harmonic in the open upper half plane and which does not attain its maximum or the minimum on the closed upper half plane.

Let $z = x + iy$, $u(z) = 0, v(z) = y$ agree on the boundary of the upper half plane(real axis). But they are not the same function.

Remark 14.3. If the domain Ω does not have a compact closure the solution may not be unique. consider for example, $\overline{\Omega} = \{z : \text{Im}z \geq 0\}$ and the function $f(z) = 0$ for real z. Then $u_1(z) = \text{Im}e^{\frac{1}{z}}$ on $\Omega = \{z : \text{Im}z > 0\}$, $u_1(z) = 0$ for real z, and, $u_2(z) = 0$ for all $z \in \overline{\Omega}$, are two distinct functions, both harmonic in the open upper half plane, continuous on the closed upper half plane and which agree with f on the real axis, the boundary of Ω. Indeed a simpler function $v(z) = v(x+iy) = y$ also serves as an illustrative example. It is also a nonconstant (but unbounded) harmonic function on the upper half plane and vanishes on the boundary.

Let $u(z) = \text{Log}|z|$ on $\overline{G}$ and let $v(z) = 0$ on $\overline{G}$ where $G = \{z \in \mathbb{C} : |z| > 1\}$. Now u, v are harmonic on G. Also u and v are continuous on $\overline{G}$ and $u = v$ on $\partial(G)$. But $u \neq v$ on G. Thus boundedness is essential in Corollary 14.3.

Theorem 14.10. *(Identity theorem for harmonic functions) If G is a region in $\mathbb{C}$, and if u and v are harmonic functions in G that agree on a nonempty open subset of G then $u = v$ throughout G.*

Proof. It is sufficient to prove this for the case where u is real-valued and v is the zero function. Suppose then that u is a real-valued harmonic function in G that vanishes on a nonempty open subset of G. Let H be the interior of the set $u^{-1}(\{0\})$. Then H is nonempty and open. Suppose H is not all of G. Then, because G is connected, some point z_0 of G lies in the boundary of H. Choose $r > 0$ such that the disk $|z - z_0| < r$ is contained in G. By Theorem 14.7, there is a holomorphic function g in this disk such that $u = \text{Re}g$ and $g(z_0) = u(z_0) = 0$. Since z_0 is in the boundary of H, the intersection of H with the disk $|z - z_0| < r$ is nonempty. Thus $\text{Re}g$ vanishes on a nonempty subdisk of the preceding disk. On the smaller disk we have $g' = 0$ by the Cauchy-Riemann equations. Hence $g' = 0$ in the disk $|z - z_0| < r$, by the identity theorem for holomorphic functions. Therefore g is constant in the disk $|z - z_0| < r$. Since $g(z_0) = 0$, the constant is 0. We have thus shown that u vanishes in the disk $|z - z_0| < r$, which means z_0 belongs to the set H (the interior of $u^{-1}(0)$). This is a contradiction, because z_0 was chosen to be in the boundary of H. We can conclude that $H = G$, in other words, u is the zero function. □

Theorem 14.11. *If $u : \Omega \to \mathbb{C}$ is harmonic then u is infinitely differentiable.*

Proof. Fix $z_0 = x_0 + iy_0$ in Ω and let δ be chosen such that $B(z_0, \delta) \subset \Omega$. Then u has a harmonic conjugate v on $B(z_0, \delta)$. That is, $f = u + iv$ is analytic and hence infinitely differentiable on $B(z_0, \delta)$. It now follows that u is infinitely differentiable. □

Theorem 14.12. *Let u be a real-valued, harmonic function on Ω, and suppose $\limsup\limits_{k\to\infty} u(a_k) \leq M$ for every sequence (a_k) in Ω converging either to a point in $\partial\Omega$ or to ∞. Then $u \leq M$ on Ω.*

Remark 14.4. In the corollary above, M may be infinite. To say that (a_k) converges to ∞ means that $|a_k| \to \infty$. The corollary is valid if "lim sup" is replaced by "lim inf" and the inequalities are reversed. (Note that the connectivity of Ω is not needed in the proof below. The region Ω may even be unbounded and u is not continuous on $\overline{\Omega}$.)

Proof. Let $M' = \sup\{u(x) : x \in \Omega\}$, and choose a sequence (b_k) in Ω such that $u(b_k) \to M'$. If (b_k) has subsequence converging to some point $b \in \Omega$, then $u(b) = M'$, which implies u is constant on Ω by the maximum principle. □

We shall now prove a property of harmonic function which we have already observed in Corollary 14.1. The proof of Theorem 14.13 below uses only Laplace's equation and it holds only for conformal maps.

Theorem 14.13. *A harmonic function $\phi(x, y)$ remains harmonic under the transformation $w = f(z)$ where f is holomorphic and $f'(z) \neq 0$.*

Proof. The function $\phi(x, y)$ is transformed into a function $\phi[x(u, v), y(u, v)]$ by the transformation. By differentiation we have

$$\frac{\partial \phi}{\partial x} = \frac{\partial \phi}{\partial u}\frac{\partial u}{\partial x} + \frac{\partial \phi}{\partial v}\frac{\partial v}{\partial x},$$

$$\frac{\partial \phi}{\partial y} = \frac{\partial \phi}{\partial u}\frac{\partial u}{\partial y} + \frac{\partial \phi}{\partial v}\frac{\partial v}{\partial y}.$$

Hence

$$\begin{aligned}\frac{\partial^2 \phi}{\partial x^2} &= \frac{\partial \phi}{\partial u}\frac{\partial^2 u}{\partial x^2} + \frac{\partial u}{\partial x}\frac{\partial}{\partial x}\left[\frac{\partial \phi}{\partial u}\right] + \frac{\partial \phi}{\partial v}\frac{\partial^2 v}{\partial x^2} + \frac{\partial v}{\partial x}\frac{\partial}{\partial x}\left[\frac{\partial \phi}{\partial v}\right] \\ &= \frac{\partial \phi}{\partial u}\frac{\partial^2 u}{\partial x^2} + \frac{\partial u}{\partial x}\left[\frac{\partial}{\partial u}\left(\frac{\partial \phi}{\partial u}\right)\frac{\partial u}{\partial x} + \frac{\partial}{\partial v}\left(\frac{\partial \phi}{\partial u}\right)\frac{\partial v}{\partial x}\right] \\ &\quad + \frac{\partial \phi}{\partial v}\frac{\partial^2 v}{\partial x^2} + \frac{\partial v}{\partial x}\left[\frac{\partial}{\partial u}\left(\frac{\partial \phi}{\partial v}\right)\frac{\partial u}{\partial x} + \frac{\partial}{\partial v}\left(\frac{\partial \phi}{\partial v}\right)\frac{\partial v}{\partial x}\right] \\ &= \frac{\partial \phi}{\partial u}\frac{\partial^2 u}{\partial x^2} + \frac{\partial u}{\partial x}\left[\frac{\partial^2 \phi}{\partial u^2}\frac{\partial u}{\partial x} + \frac{\partial^2 \phi}{\partial v \partial u}\frac{\partial v}{\partial x}\right] \\ &\quad + \frac{\partial \phi}{\partial v}\frac{\partial^2 v}{\partial x^2} + \frac{\partial v}{\partial x}\left[\frac{\partial^2 \phi}{\partial v^2}\frac{\partial v}{\partial x} + \frac{\partial^2 \phi}{\partial u \partial v}\frac{\partial u}{\partial x}\right].\end{aligned}$$

Similarly, $\frac{\partial^2 \phi}{\partial y^2} = \frac{\partial \phi}{\partial u}\frac{\partial^2 u}{\partial y^2} + \frac{\partial u}{\partial y}\left[\frac{\partial^2 \phi}{\partial u^2}\frac{\partial u}{\partial y} + \frac{\partial^2 \phi}{\partial v \partial u}\frac{\partial v}{\partial y}\right] + \frac{\partial \phi}{\partial v}\frac{\partial^2 v}{\partial y^2} + \frac{\partial v}{\partial y}\left[\frac{\partial^2 \phi}{\partial v^2}\frac{\partial v}{\partial y} + \frac{\partial^2 \phi}{\partial u \partial v}\frac{\partial u}{\partial y}\right].$
Adding,

$$\frac{\partial^2 \phi}{\partial x^2} + \frac{\partial^2 \phi}{\partial y^2} = \frac{\partial \phi}{\partial u}\left(\frac{\partial^2 u}{\partial x^2} + \frac{\partial^2 u}{\partial y^2}\right) + \frac{\partial \phi}{\partial v}\left(\frac{\partial^2 v}{\partial x^2} + \frac{\partial^2 v}{\partial y^2}\right) + \frac{\partial^2 \phi}{\partial u^2}\left[\left(\frac{\partial u}{\partial x}\right)^2 + \left(\frac{\partial u}{\partial y}\right)^2\right]$$

$$+2\frac{\partial^2\phi}{\partial u\partial v}\left[\frac{\partial u}{\partial x}\frac{\partial v}{\partial x}+\frac{\partial u}{\partial y}\frac{\partial v}{\partial y}\right]+\frac{\partial^2\phi}{\partial v^2}\left[\left(\frac{\partial v}{\partial x}\right)^2+\left(\frac{\partial v}{\partial y}\right)^2\right]. \tag{14.6}$$

Since u and v are harmonic, $\frac{\partial^2 u}{\partial x^2}+\frac{\partial^2 u}{\partial y^2}=0$, $\frac{\partial^2 v}{\partial x^2}+\frac{\partial^2 v}{\partial y^2}=0$. Also, by the Cauchy-Riemann equations, $\frac{\partial u}{\partial x}=\frac{\partial v}{\partial y}$, $\frac{\partial u}{\partial y}=-\frac{\partial v}{\partial x}$. Then $\left(\frac{\partial u}{\partial x}\right)^2+\left(\frac{\partial u}{\partial y}\right)^2=\left(\frac{\partial v}{\partial x}\right)^2+\left(\frac{\partial v}{\partial y}\right)^2=\left(\frac{\partial u}{\partial x}\right)^2+\left(\frac{\partial v}{\partial x}\right)^2=\left|\frac{\partial u}{\partial x}+i\frac{\partial v}{\partial x}\right|^2=|f'(z)|^2$ and $\frac{\partial u}{\partial x}\frac{\partial v}{\partial x}+\frac{\partial u}{\partial y}\frac{\partial v}{\partial y}=0$. Hence (14.6) becomes $\frac{\partial^2\phi}{\partial x^2}+\frac{\partial^2\phi}{\partial y^2}=|f'(z)|^2\left(\frac{\partial^2\phi}{\partial u^2}+\frac{\partial^2\phi}{\partial v^2}\right)$. So if $\frac{\partial^2\phi}{\partial x^2}+\frac{\partial^2\phi}{\partial y^2}=0$ and $f'(z)\neq 0$ then $\frac{\partial^2\phi}{\partial u^2}+\frac{\partial^2\phi}{\partial v^2}=0$. □

Proposition 14.1. *Let f be analytic in a region G and suppose f' is never zero in G. Let g be a real-valued function defined on $E=f(G)$ and let h be defined on G by $h(z)=g(f(z))$ for each $z\in G$. Then h is harmonic on G if and only if g is harmonic on E.*

Proof. Let $f=u+iv$. Then $h(x,y)=g(u,v)$ where $u=u(x,y)$ and $v=v(x,y)$. It is not difficult to check that $h_{xx}=g_u u_{xx}+g_{uu}(u_x)^2+g_{vu}u_xv_x+g_vv_{xx}+g_{uv}u_xv_x+g_{vv}(v_x)^2$ and $h_{yy}=g_uu_{yy}+g_{uu}(u_y)^2+g_{vu}u_yv_y+g_vv_{yy}+g_{uv}u_yv_y+g_{vv}(v_y)^2$.

As in Theorem 14.13, one can also verify that $h_{xx}+h_{yy}=|f'(z)|^2(g_{uu}+g_{vv})$. Suppose now that g is harmonic on E. Then $g_{uu}+g_{vv}=0$ on E. Since $f'(z)\neq 0$ on G, from the above identity it follows that $h_{xx}+h_{yy}=0$. Conversely, suppose h is harmonic on G. Let $w_0\in E=f(G)$ and let $z_0\in G$ such that $f(z_0)=w_0$. By Theorem 11.9, there exists $\rho>0$ such that f is one-to-one on $B(z_0,\rho)$. Let $W=f(B(z_0,\rho))$. From Theorem 11.8 and Theorem 2.55, it follows that W is a region and $w_0\in W$. Let $f_\rho=f|_{B(z_0,\rho)}$, the restriction of f to $B(z_0,\rho)$. Thus the map $f_\rho: B(z_0,\rho)\to W$ is one one and onto. Let $F=f_\rho^{-1}$ on W. This implies $g(w)=h(F(w))$ for each $w\in W=f(B(z_0,\rho))$. By Theorem 11.11, the function F is holomorphic and $F'\neq 0$ in W. Further,

$$(g_{uu}+g_{vv})=|F'(w)|^2(h_{xx}+h_{yy})$$

on W. Thus h is harmonic on G implies $g_{uu}+g_{vv}=0$ on W and hence g is harmonic on W. Since w_0 is an arbitrary point of E, it follows that g is harmonic on E. □

Definition 14.3. *If f is a complex valued function defined in open subset of $\mathbb{C}$, such that $Ref=u$, $Imf=v$ and if u and v have first partial derivatives then we define (see Chapter 5, section 5.3) $\frac{\partial f}{\partial x}=\frac{\partial u}{\partial x}+i\frac{\partial v}{\partial x}$ and $\frac{\partial f}{\partial y}=\frac{\partial u}{\partial y}+i\frac{\partial v}{\partial y}$. Higher order derivatives are defined analogously. The function f is said to be of class C^k, where k is a positive integer, if it has continuous partial derivatives up to order k. It is said to be of class C^∞ if it is of class C^k for every positive integer k. We recall the theorem from calculus (which extends immediately to complex-valued functions) stating that if f is of class C^2 then its second-order mixed partial derivatives are equal: $\frac{\partial^2 f}{\partial x\partial y}=\frac{\partial^2 f}{\partial y\partial x}$.*

The complex-valued function f is said to be harmonic if it is of class C^2 and satisfies Laplace's equation:

$$\frac{\partial^2 f}{\partial x^2} + \frac{\partial^2 f}{\partial y^2} = 0.$$

It is not so difficult to prove that the complex-valued function f is harmonic if and only if its real and imaginary part are harmonic.

The theorem and corollary above apply only to real valued functions. The next corollary is a version of the maximum principle for complex valued functions.

Corollary 14.4. *Let Ω be connected, and let u be harmonic on Ω. If $|u|$ has a maximum in Ω, then u is a constant.*

Proof. Suppose $|u|$ attains a maximum value of M at some point $a \in \Omega$. Choose $\lambda \in \mathbb{C}$ such that $|\lambda| = 1$and $\lambda u(a) = M$. Then the real-valued harmonic function $\text{Re}\lambda u$ attains its maximum value M at a; thus by the maximum principle for harmonic functions, $\text{Re}\lambda u \equiv M$ on Ω. Because $|\lambda u| = |u| \leq M$, we have $\text{Im}\lambda u \equiv 0$ on Ω. Thus λu, and hence u, is constant on Ω. □

No minimum principle holds for $|u|$ (consider $u(z) = x$ on $\mathbb{D}$, the open unit disk in $\mathbb{C}$.) The Corollary 14.4 is the analogue of Theorem 14.9 (Maximum principle for harmonic function) for complex-valued harmonic functions; the corresponding analogues of Corollary 14.2 and Corollary 14.3 are also valid. We will be able to prove a local version of the maximum principle after we prove that harmonic functions are real analytic.

Corollary 14.5. *The zeros of a real valued harmonic function are never isolated.*

Proof. Assume u is harmonic and real valued on $\Omega, a \in \Omega$, and $u(a) = 0$. Suppose $\overline{B(a,r)} \subset \Omega$. If $u \equiv 0$ on $B(a,r)$, then we are done. Otherwise u takes on positive and negative values on $\partial B(a,r)$ by the maximum principle. This implies that the connected set $u(\partial B(a,r)) \subset \mathbb{R}$ contains 0. Thus u has a zero on $\partial B(a,r)$ for every small $r > 0$, proving that a is not an isolated zero. □

Theorem 14.14. *Suppose Ω is connected, f is real analytic in Ω, and $f = 0$ on a non-empty open subset of Ω. Then $f \equiv 0$ in Ω.*

Proof. Let Ω^0 denote the interior of $\{x \in \Omega : f(x) = 0\}$. Then Ω^0 is an open subset of Ω. The vanishing of all derivatives of f on Ω^0 shows that Ω^0 is also closed in Ω. If $a \in \Omega$ is a limit point of Ω^0, then all derivatives of f vanish at a by continuity, implying that the power series of f at a is identically zero; hence $a \in \Omega^0$. Since Ω^0 is nonempty by hypothesis, we must have $\Omega^0 = \Omega$ by connectivity, giving $f \equiv 0$ in Ω. □

Theorem 14.15. *If u is harmonic in Ω, then u is real analytic in Ω.*

Proof. It suffices to show that if u is harmonic on $\overline{\mathbb{D}}$, where $\mathbb{D}$ is the open unit disk in $\mathbb{C}$ then u has a power series expansion converging to u in a neighborhood of 0. □

As mentioned earlier, the real analyticity of harmonic functions allows us to prove a local maximum principle.

Theorem 14.16. *(The local maximum principle): Suppose Ω is connected, u is real valued and harmonic in Ω, and u has a local maximum in Ω. Then u is constant in Ω.*

Proof. If u has a local maximum at $a \in \Omega$, then there exists a ball $B(a,r) \subset \Omega$ such that $u \leq u(a)$ in $B(a,r)$. By Theorem 14.9 (maximum principle) u is constant in $B(a,r)$. Because u is real analytic in Ω, $u \equiv u(a)$ in Ω by Theorem 14.14. □

Theorem 14.17. *A bounded harmonic function on $\mathbb{C}$ is a constant.*

Proof. Suppose u is a harmonic function on $\mathbb{C}$ and $|u(z)| \leq M$ for $z \in \mathbb{C}$. Let v be the harmonic conjugate of u on $\mathbb{C}$. Then $f = u + iv$ is entire with $|\text{Re} f| = |u| \leq M$ on $\mathbb{C}$. Since an entire function with bounded real part is constant, hence f is constant and hence $u = \frac{1}{2}(f + \bar{f})$ is a constant function. □

Theorem 14.18. *A positive harmonic function on $\mathbb{C}$ is a constant.*

Proof. Let $u : \mathbb{C} \to \mathbb{R}$ be a positive harmonic function. Then $u(z) > 0$ for all $z \in \mathbb{C}$. It is required to prove that u is a constant function. By Theorem 14.7, the harmonic function u has a harmonic conjugate v in $\mathbb{C}$. Thus $f = u + iv$ is an entire function. Also the real part u of f is positive. So by Example 13.10, f is constant. That is, $f(z) = c$, where $c \in \mathbb{C}$. Hence $u + iv = c = x + iy, x, y \in \mathbb{R}$. This implies $u(z) = x$ for all $z \in \mathbb{C}$. □

Corollary 14.6. *A positive harmonic function on $\mathbb{C} \setminus \{0\}$ is constant.*

Proof. If u is positive and harmonic on $\mathbb{C} \setminus \{0\}$, then the function $z \mapsto u(e^z)$ is positive and harmonic on $\mathbb{C}$ and hence (by Theorem 14.18) is constant. This proves that u is constant. □

Example 14.6. *Let u be a nonconstant real valued harmonic function defined in a region Ω. Then $u(\Omega)$ is open in $\mathbb{R}$.*

Solution: Since Ω is connected, u is continuous, the set $u(\Omega)$ is a connected subset of $\mathbb{R}$. But we know connected subsets of $\mathbb{R}$ are intervals. Thus $u(\Omega)$ is an interval in $\mathbb{R}$. The harmonic function u does not attain its minimum and maximum at any point in the open set Ω. Thus $u(\Omega)$ must be an open interval (a, b) in $\mathbb{R}$ which is an open set in $\mathbb{R}$.

Theorem 14.19. *Let u be a nonconstant real valued harmonic function in $\mathbb{C}$. Then the set $u^{-1}(c)$ is unbounded for every real number c.*

Proof. Let $c \in \mathbb{R}$ and $v(z) = u(z) - c$ for all $z \in \mathbb{C}$. The function v is harmonic on $\mathbb{C}$. Then $u^{-1}(c) = \{z \in \mathbb{C} : u(z) = c\} = \{z \in \mathbb{C} : v(z) = 0\}$. We first verify that $u^{-1}(c) \neq \emptyset$. Suppose $u^{-1}(c) = \emptyset$. The function u is harmonic in $\mathbb{C}$ which is simply connected. So u has a harmonic conjugate w in $\mathbb{C}$. Thus $f = u + iw$ is an entire function and $f(z) \neq c+ik$ for all $k \in \mathbb{R}$ and for all $z \in \mathbb{C}$. By Picard's theorem f is a constant function. But u is nonconstant. Thus $u^{-1}(c) \neq \emptyset$. It therefore follows that there exists $a \in u^{-1}(c)$ such that $v(a) = 0$. Hence a is a zero of the harmonic function v. From the proof of Corollary 14.5, it follows that for any $R > 0$, there exists a $p \in \mathbb{C}$ such that $|p - a| = R$ and p is a zero of $v(z)$ (that is, $p \in u^{-1}(c)$). Thus $u^{-1}(c)$ is unbounded. $\square$

14.3 The Dirichlet problem on the unit disk

We now discuss about a famous problem in harmonic function theory: Let $\mathbb{D} = B(0,1) = \{z \in \mathbb{C} : |z| < 1\}$ and $\partial\mathbb{D} = \{z \in \mathbb{C} : |z| = 1\} = S^1$. Given a real-valued continuous function f on the unit circle S^1, does there exist a continuous function u on $\overline{\mathbb{D}}$, with u harmonic on $\mathbb{D}$, such that $u = f$ on S^1? If so, how do we find u ? This is the Dirichlet problem for the open unit disk. Recall that by the maximum principle, if a solution exists, then it is unique.

To motivate the construction of such a u, assume that such a u exists. Then by the mean value property we have

$$u(0) = \frac{1}{2\pi}\int_0^{2\pi} f(e^{i\theta})d\theta.$$

Now we want to know the value of $u(z)$ for other values of $z \in \mathbb{D}$. Consider the automorphism of the unit disk which takes 0 to z, i.e., $Tw = \frac{w+z}{1+\overline{z}w}$. Then T maps $\mathbb{D}$ onto itself biholomorphically and T is a homeomorphism of S^1. Further $(u \circ T)(w)$ is continuous on $\overline{\mathbb{D}}$ and harmonic in $\mathbb{D}$. Hence by the mean value property of harmonic functions,

$$\begin{aligned}
u(re^{i\theta}) &= u(z) \\
&= u(T(0)) \\
&= (u \circ T)(0) \\
&= \tfrac{1}{2\pi}\int_0^{2\pi} (u \circ T)(e^{is})ds \\
&= \tfrac{1}{2\pi}\int_0^{2\pi} u(T(e^{is}))ds \\
&= \tfrac{1}{2\pi}\int_0^{2\pi} f(T(e^{is}))ds \\
&= \tfrac{1}{2\pi}\int_0^{2\pi} f\left(\frac{e^{is}+z}{1+\overline{z}e^{is}}\right) ds.
\end{aligned}$$

We make a change of variable $e^{it} = T(e^{is})$ or $e^{is} = T^{-1}(e^{it})$.

A calculation shows that

$$ds = |(T^{-1})'(e^{it})|dt = \frac{1-|z|^2}{|1-\bar{z}e^{it}|^2}dt.$$

Thus we obtain

$$\begin{aligned} u(z) &= \frac{1}{2\pi}\int_0^{2\pi} f\left(\frac{e^{is}+z}{1+\bar{z}e^{is}}\right) ds \\ &= \frac{1}{2\pi}\int_0^{2\pi} f(e^{it})\frac{1-|z|^2}{|1-\bar{z}e^{it}|^2}dt \\ &= \frac{1}{2\pi}\int_0^{2\pi} f(e^{it})\frac{1-|z|^2}{|e^{it}-z|^2}dt. \end{aligned}$$

Hence

$$u(z) = u(re^{i\theta}) = \frac{1}{2\pi}\int_0^{2\pi} f(e^{it})P(re^{i\theta}, e^{it})dt \tag{14.7}$$

where

$$\begin{aligned} P(re^{i\theta}, e^{it}) &= P(z, e^{it}) \\ &= \frac{1-|z|^2}{|e^{it}-z|^2} \\ &= \frac{1-|z|^2}{|e^{it}-re^{i\theta}|^2} \\ &= \text{Re}\left[\frac{e^{it}+re^{i\theta}}{e^{it}-re^{i\theta}}\right]. \end{aligned}$$

The Equation (14.7) is called the Poisson formula and P is called the Poisson kernel. The function $P(z, e^{it})$ is harmonic in z because it is the real part of the function $\frac{e^{it}+z}{e^{it}-z}$ which is holomorphic in z.

Now the right hand side of Equation (14.7) makes sense whether u in question exists or not. We shall now show that u defined in (14.7) is harmonic in $\mathbb{D}$.

Theorem 14.20. *Suppose $f : S^1 \to \mathbb{R}$ is a continuous function and define for $z \in \mathbb{D}$, $u(z) = \frac{1}{2\pi}\int_0^{2\pi} f\left(\frac{e^{is}+z}{1+\bar{z}e^{is}}\right) ds$. Then u is harmonic in $\mathbb{D}$ and the function u is a continuous extension of f into the open unit disk $\mathbb{D}$ (i.e., u is continuous on $\overline{\mathbb{D}}$ and $u(z) = f(z)$ for $z \in S^1$).*

Proof. Note f is real-valued. Now, as seen above,

$$\begin{aligned} u(z) &= \tfrac{1}{2\pi}\int_0^{2\pi} f(e^{it})P(re^{i\theta}, e^{it})dt \\ &= \tfrac{1}{2\pi}\int_0^{2\pi} f(e^{it})\mathrm{Re}\left[\frac{1+re^{i(\theta-t)}}{1-re^{i(\theta-t)}}\right]dt \\ &= \mathrm{Re}\left\{\tfrac{1}{2\pi}\int_0^{2\pi} f(e^{it})\left[\frac{1+re^{i(\theta-t)}}{1-re^{i(\theta-t)}}\right]dt\right\} \\ &= \mathrm{Re}\left\{\tfrac{1}{2\pi}\int_0^{2\pi} f(e^{it})\left[\frac{e^{it}+re^{i\theta}}{e^{it}-re^{i\theta}}\right]dt\right\}. \end{aligned}$$

Thus the function u is the real part of the function g defined on $\mathbb{D}$ by

$$g(z) = \frac{1}{2\pi}\int_0^{2\pi} f(e^{it})\frac{e^{it}+z}{e^{it}-z}dt.$$

Now g is holomorphic in the open unit disk as we see by Morera's theorem or by differentiating under the integral sign. Hence u, being the real part of g, is harmonic in the unit disk. To see that u is a continuous extension of f inside the unit disk, we note that $\lim\limits_{z\to e^{i\theta}} \dfrac{e^{is}+z}{1+\bar{z}e^{is}} = e^{i\theta}$ except when $e^{i\theta} = -e^{is}$. Since f is continuous on the compact set S^1, it is bounded on it. By bounded convergence theorem,

$$\begin{aligned} \lim_{z\to e^{i\theta}} u(z) &= \frac{1}{2\pi}\int_0^{2\pi} \lim_{z\to e^{i\theta}} f\left(\frac{e^{is}+z}{1+\bar{z}e^{is}}\right)ds \\ &= \frac{1}{2\pi}\int_0^{2\pi} f(e^{i\theta})ds = f(e^{i\theta}) \end{aligned}$$

which shows that u is a continuous extension of f into the open unit disk. □

The above proof of the solution of the Dirichlet problem for the unit disk seems simpler than the usual proof which uses a further property of the Poisson kernel i.e., its being an approximate identity, a property which we explain below. The function

$$\begin{aligned} P(re^{i\theta}, e^{it}) &= P(z, e^{it}) \\ &= \tfrac{1-|z|^2}{|e^{it}-z|^2} = \tfrac{1-r^2}{|1-re^{i(\theta-t)}|^2} \\ &= \tfrac{1-r^2}{1-2r\cos(\theta-t)+r^2}, 0 \le r < 1, 0 \le \theta \le 2\pi \end{aligned}$$

is called the Poisson kernel. In particular,

$$\begin{aligned} P(re^{i\theta},1) &= \frac{1-r^2}{|1-re^{i\theta}|^2} \\ &= \mathrm{Re}\left[\frac{1+re^{i\theta}}{1-re^{i\theta}}\right] \\ &= \mathrm{Re}\left(1+2\sum_{n=1}^{\infty} r^n e^{in\theta}\right) \\ &= 1+2\sum_{n=1}^{\infty} r^n \cos n\theta \\ &= 1+\sum_{n=1}^{\infty} r^n \left(e^{in\theta}+e^{-in\theta}\right) \\ &= \sum_{n=-\infty}^{\infty} r^{|n|} e^{in\theta}. \end{aligned}$$

Poisson kernel as an Approximate identity:
The Poisson kernel has the following properties which makes it an approximate identity.

(1) For $0 \le r < 1, P(re^{i\theta}, e^{it}) > 0$ since $P(re^{i\theta}, e^{it}) = \frac{1-r^2}{|e^{it}-re^{i\theta}|^2}$.

(2) We also have $\frac{1}{2\pi}\int_0^{2\pi} P(re^{i\theta}, e^{it})dt = 1$. This follows from the residue theorem since the left hand side of the above equation is equal to

$$\mathrm{Re}\left[\frac{1}{2\pi}\int_{|w|=1} \frac{w+z}{w-z}\frac{dw}{iw}\right] = 1$$

where $z = re^{i\theta}$. This may also be seen using the Fourier expansion

$$P(re^{i\theta}, e^{it}) = P(re^{i(\theta-t)}, 1) = \sum_{n=-\infty}^{\infty} r^{|n|} e^{in(\theta-t)}.$$

(3) For $0 < \delta < \frac{\pi}{2}$, $\sup\limits_{|t-\theta|\ge\delta} P(re^{i\theta}, e^{it}) \to 0$ as $r \to 1^-$.

This is because

$$\begin{aligned} P(re^{i\theta}, e^{it}) &= \frac{1-r^2}{1-2r\cos(\theta-t)+r^2} \\ &\leq \frac{1-r^2}{1-2\cos(\theta-t)+r^2} \\ &= \frac{1-r^2}{(1-r)^2+2\,(1-\cos(\theta-t))} \\ &= \frac{1-r^2}{(1-r)^2+4\sin^2\dfrac{\theta-t}{2}} \\ &\leq \frac{1-r^2}{\sin^2\frac{\delta}{2}} \end{aligned}$$

which tends to 0 as $r \to 1^-$.

The Theorem 14.20 can also be proved using the harmonicity and the approximate identity properties of the Poisson kernel. But first note the following: The mean-value property shows that if u is harmonic on $\overline{\mathbb{D}}$, then

$$u(0) = \frac{1}{2\pi}\int_0^{2\pi} u(e^{is})ds.$$

We now show that for any $z \in \mathbb{D}$, $u(z)$ is a weighted average of u over the unit circle S^1. More precisely, we will show that there exists a function P on $\mathbb{D} \times S^1$ such that

$$u(z) = u(re^{i\theta}) = \frac{1}{2\pi}\int_0^{2\pi} P(re^{i\theta}, e^{it})u(e^{it})dt$$

for any $z \in \mathbb{D}$ and every u harmonic on $\overline{\mathbb{D}}$.

Now we shall try to find out what P might be. Suppose u is a real-valued harmonic function on $\overline{\mathbb{D}}$, the closed unit disk in $\mathbb{C}$. Then $u = \mathrm{Re} f$ for some function f holomorphic on a neighborhood of the closed disk (see Theorem 14.7). Because $u = \frac{f+\bar{f}}{2}$, the Taylor series expansion of f implies that u has the form $u(re^{it}) = \sum_{n=-\infty}^{\infty} a_n r^{|n|} e^{int}$ where $0 \leq r < 1$. In this formula, take $r = 1$, multiply both sides by e^{-ikt}, then integrate over the unit circle to obtain $a_k = \frac{1}{2\pi}\int_0^{2\pi} u(e^{it})e^{-ikt}dt$. Now let z be a point in the open unit disk $\mathbb{D}$, and write $z = re^{i\theta}, 0 \leq r < 1, 0 \leq \theta \leq 2\pi$. Then

$$
\begin{aligned}
u(z) &= u(re^{i\theta}) \\
&= \sum_{n=-\infty}^{\infty} \left(\tfrac{1}{2\pi} \int_0^{2\pi} u(e^{it}) e^{-int} dt \right) r^{|n|} e^{in\theta} \\
&= \tfrac{1}{2\pi} \int_0^{2\pi} \left(\sum_{n=-\infty}^{\infty} r^{|n|} e^{in(\theta - t)} \right) u(e^{it}) dt \\
&= \tfrac{1}{2\pi} \int_0^{2\pi} \frac{1 - r^2}{|e^{it} - re^{i\theta}|^2} u(e^{it}) dt.
\end{aligned}
$$

Thus, letting $P(re^{i\theta}, e^{it}) = \frac{1-r^2}{|e^{it}-re^{i\theta}|^2}$, we obtain the desired formula, $u(z) = \frac{1}{2\pi}\int_0^{2\pi} P(re^{i\theta}, e^{it}) u(e^{it}) dt$. To prove the Theorem 14.20, we take our clue from the above discussion.

If f happens to be the restriction to S^1 of a function u harmonic on $\overline{\mathbb{D}}$, then define

$$
u(z) = \frac{1}{2\pi} \int_0^{2\pi} P(re^{i\theta}, e^{it}) f(e^{it}) dt \tag{14.8}
$$

for $z = re^{i\theta} \in \mathbb{D}$. We claim the function u has the desired properties.

To show u is harmonic on $\mathbb{D}$ we can proceed as before or else the Laplacian of u can be computed by differentiating under the integral sign in (14.8); and since $P(re^{i\theta}, e^{it})$ is harmonic, so the function u is harmonic in $\mathbb{D}$. The only part that remains to be proved is that u is continuous on S^1. Note u is continuous on $\mathbb{D}$ as u is harmonic in $\mathbb{D}$.

Definition 14.4. *Let S^1 be the unit circle in the complex plane. If f is any function on S^1 and if F is defined on $\mathbb{R}$ by*

$$
F(t) = f(e^{it}), \tag{14.9}
$$

then F is a periodic function of period 2π. This means that $F(t+2\pi) = F(t)$ for all real t. Conversely, if F is a function on $\mathbb{R}$, with period 2π, then there is a function f on S^1 such that (14.9) holds. Thus we may identify functions on S^1 with 2π-periodic functions on $\mathbb{R}$; and, for simplicity of notation, we shall sometimes write $f(t)$ rather than $f(e^{it})$, even if we think of f as being defined on S^1.

With these conventions in mind, we shall proceed to prove u is continuous.

Proof. Since $f(\theta)$ is continuous on the compact set $[0, 2\pi]$ hence $f(\theta)$ is uniformly continuous on $[0, 2\pi]$. Hence for given $\epsilon > 0$, there exists $\delta > 0$ (depending only on ϵ) such that $|\theta - t| < \delta$ implies $|f(\theta) - f(t)| < \epsilon$. If $|\theta - t| \geq \delta$ then by the property of the Poisson kernel we have $\lim\limits_{r \to 1^-} P(re^{i\theta}, e^{it}) = 0$.

In other words there exists $r_0, 0 < r_0 < 1$, such that if $|\theta - t| \geq \delta$ then $P(re^{i\theta}, e^{it}) < \epsilon$ for $0 \leq r_0 \leq r < 1$. Hence

$$\begin{aligned}
\left|u(re^{i\theta}) - f(\theta)\right| &= \left|\tfrac{1}{2\pi}\int_0^{2\pi} P(re^{i\theta}, e^{it})f(t)dt - \frac{1}{2\pi}\int_0^{2\pi} P(re^{i\theta}, e^{it})f(\theta)dt\right| \\
&= \left|\tfrac{1}{2\pi}\int_0^{2\pi} P(re^{i\theta}, e^{it})(f(t) - f(\theta))dt\right| \\
&\leq \tfrac{1}{2\pi}\int_0^{2\pi} P(re^{i\theta}, e^{it})\left|f(\theta) - f(t)\right| dt \\
&= \tfrac{1}{2\pi}\int_{|\theta - t| < \delta} P(re^{i\theta}, e^{it})\left|f(\theta) - f(t)\right| dt \\
&+ \tfrac{1}{2\pi}\int_{|\theta - t| \geq \delta} P(re^{i\theta}, e^{it})|f(\theta) - f(t)|dt \\
&\leq \epsilon\tfrac{1}{2\pi}\int_{|\theta - t| < \delta} P(re^{i\theta}, e^{it})dt \\
&+ \tfrac{1}{2\pi}\left(2\max_{0\leq\theta\leq 2\pi}|f(\theta)|\right)\int_{|\theta - t| \geq \delta} P(re^{i\theta}, e^{it})dt \\
&\leq \epsilon\tfrac{1}{2\pi}\int_0^{2\pi} P(re^{i\theta}, e^{it})dt + \frac{2}{2\pi}\max_{0\leq\theta\leq 2\pi}|f(\theta)|\epsilon\left(\int_0^{2\pi} dt\right) \\
&= \epsilon + \tfrac{2\pi\epsilon}{2\pi}2\max_{0\leq\theta\leq 2\pi}|f(\theta)| \\
&= \epsilon\left[1 + 2\max_{0\leq\theta\leq 2\pi}|f(\theta)|\right].
\end{aligned}$$

Thus $\lim_{r\to 1^-} u(re^{i\theta}) = f(\theta)$. Hence u is continuous at $e^{i\theta} \in S^1$. Since $e^{i\theta} \in S^1$ is arbitrarily chosen, hence u is continuous on S^1. □

Corollary 14.7. *Let $a \in \mathbb{C}, r > 0$, and suppose g is a continuous real-valued function on $\{z \in \mathbb{C} : |z - a| = r\}$; then there is a unique continuous function $v : \overline{B(a, r)} \to \mathbb{R}$ such that v is harmonic on $B(a, r) = \{z \in \mathbb{C} : |z - a| < r\}$ and $v(z) = g(z)$ for $|z - a| = r$.*

Proof. Consider $f(e^{i\theta}) = g(a + re^{i\theta})$, then f is continuous on S^1. Let $u : \overline{\mathbb{D}} \to \mathbb{R}$ be the continuous function which is harmonic in $\mathbb{D}$ and $u(e^{i\theta}) = f(e^{i\theta})$, $0 \leq \theta \leq 2\pi$ (by Theorem 14.20, such a function u exists and it is unique). Let $v(z) = u(\frac{z-a}{r})$. Then v maps $\overline{B(a, r)}$ into $\mathbb{R}$ and v is harmonic in $B(a, r)$

(as u is harmonic in $\mathbb{D}$) and $v(a+re^{i\theta}) = u(e^{i\theta}) = f(e^{i\theta}) = g(a+re^{i\theta})$ for $0 \le \theta < 2\pi$. Hence v is the desired function in $\overline{B(a,r)}$. □

Thus we obtained an integral representation of harmonic functions which is closely related to the Cauchy formula for holomorphic functions. Further, we proved among other things, that every real harmonic function is locally the real part of a holomorphic function, and it also yields information about the boundary behavior of certain classes of holomorphic functions in open disks. Consequently, every harmonic function has continuous partial derivatives of all orders.

We have already shown that every harmonic function satisfies the mean value property. We shall now prove the converse of the mean value theorem.

Theorem 14.21. *Let G be a region in $\mathbb{C}$. If $u : G \to \mathbb{R}$ is a continuous function which has the mean value property (i.e., $u(a) = \frac{1}{2\pi}\int_0^{2\pi} u(a+re^{i\theta})d\theta$ whenever $\overline{B(a,r)} \subset G, r > 0$) then u is harmonic.*

Proof. Let $a \in G$ and choose $r > 0$ such that $\overline{B(a,r)} \subset G$; it is sufficient to show that u is harmonic on $B(a,r)$. By Corollary 14.7, there is a continuous function $v : \overline{B(a,r)} \to \mathbb{R}$ which is harmonic in $B(a,r)$ and $v(a+re^{i\theta}) = u(a+re^{i\theta})$ for $0 \le \theta \le 2\pi$. Since v is harmonic it satisfies the mean value property on $\overline{B(a,r)}$ and u satisfies the mean value property(given) hence $u-v$ satisfies the mean value property and $(u-v)(z) = 0$ for $|z-a| = r$, it follows from Corollary 14.2 that $u \equiv v$ in $B(a,r)$; hence u is harmonic on $B(a,r)$. □

The Poisson integral also yields information about sequences of harmonic functions. The following theorem should remind the reader of the behavior of a uniformly convergent sequence of holomorphic functions.

Corollary 14.8. *Let Ω be a region in $\mathbb{C}$. Suppose (u_m) is a sequence of harmonic functions on Ω such that u_m converges uniformly to a function u on each compact subset of Ω. Then u is harmonic on Ω.*

Proof. Let $a \in \Omega$ and choose $R > 0$ such that $\overline{B(a,R)} \subset \Omega$. We need only show that u is harmonic on $B(a,R)$.

$$u_m(a) = \frac{1}{2\pi}\int_0^{2\pi} u_m(a+Re^{it})dt$$

for every m. Taking the limit of both sides, we obtain

$$u(z) = \frac{1}{2\pi}\int_0^{2\pi} u(a+Re^{it})dt$$

This is possible as u_m converges uniformly to u on each compact subset of Ω. By Theorem 14.21, u is harmonic on $B(a,R)$. □

Corollary 14.9. *If u is a real-valued continuous function on $\overline{\mathbb{D}}$ that is harmonic on $\mathbb{D}$, then*

$$u(re^{i\theta}) = \frac{1}{2\pi}\int_0^{2\pi} P(re^{i\theta}, e^{it})u(e^{it})dt$$

for $0 \leq r < 1$ and $\theta \in [0, 2\pi]$.

Proof. The function

$$g(re^{i\theta}) = u(re^{i\theta}) - \frac{1}{2\pi}\int_0^{2\pi} P(re^{i\theta}, e^{it})u(e^{it})dt$$

is harmonic on $\mathbb{D}$ and extends continuously to be 0 on S^1. The maximum principle (Corollary 14.2) now implies that $g = 0$ on $\mathbb{D}$. □

Remark 14.5. It is possible to derive a formula for the Poisson kernel of an arbitrary disk as follows. If $R > 0$ and $0 \leq r < R$ then $0 \leq \frac{r}{R} < 1$. Define the Poisson kernel $P(re^{i\theta}, Re^{it}) = P(\frac{r}{R}e^{i\theta}, e^{it}) = \frac{1-(\frac{r}{R})^2}{1-2\frac{r}{R}\cos(\theta-t)+(\frac{r}{R})^2} = \frac{R^2-r^2}{R^2-2Rr\cos(\theta-t)+r^2}$ for $0 \leq \theta \leq 2\pi$ and $0 \leq t \leq 2\pi$. So if v is continuous on $\overline{B(a,R)}$ and harmonic in $B(a,R)$ then suppose $0 \leq r < R$. Let $f(e^{it}) = v(a + Re^{it})$. Then f is continuous on S^1. Let $u : \overline{\mathbb{D}} \to \mathbb{R}$ be defined by $u(\frac{r}{R}e^{i\theta}) = \frac{1}{2\pi}\int_0^{2\pi} \frac{1-(\frac{r}{R})^2}{1-2\frac{r}{R}\cos(\theta-t)+(\frac{r}{R})^2} f(e^{it})dt$ if $0 \leq r < R$ and $u(e^{i\theta}) = f(e^{i\theta}), 0 \leq \theta \leq 2\pi$. The function u is harmonic in $\mathbb{D}$. Let $w(z) = u\left(\frac{z-a}{R}\right)$ for $z \in \overline{B(a,R)}$. This implies $w(a+Re^{i\theta}) = v(a+Re^{i\theta})$. By Corollary 14.3 , $w \equiv v$ on $\overline{B(a,R)}$. Hence

$$\begin{aligned} v(a+re^{i\theta}) &= u\left(\frac{a+re^{i\theta}-a}{R}\right) \\ &= \frac{1}{2\pi}\int_0^{2\pi} \frac{R^2-r^2}{R^2-2Rr\cos(\theta-t)+r^2} v(a+Re^{it})dt. \end{aligned}$$

Thus

$$\begin{aligned} v(a+re^{i\theta}) &= \frac{1}{2\pi}\int_0^{2\pi} P(re^{i\theta}, Re^{it})v(a+Re^{it})dt \\ &= \frac{1}{2\pi}\int_0^{2\pi} \frac{R^2-r^2}{|Re^{it}-re^{i\theta}|^2} v(a+Re^{it})dt \end{aligned}$$

and $P(re^{i\theta}, Re^{it})$ is the Poisson kernel. Since $R - r \leq |Re^{it} - re^{i\theta}| \leq R + r$, therefore

$$\begin{aligned} \frac{R-r}{R+r} &\leq \frac{R^2-r^2}{|Re^{it}-re^{i\theta}|^2} \\ &= \frac{R^2-r^2}{R^2-2Rr\cos(\theta-t)+r^2} \\ &\leq \frac{R+r}{R-r}. \end{aligned}$$

If $v \geq 0$ is harmonic in $B(a,R)$ and continuous in $\overline{B(a,R)}$ then

$$\begin{aligned}\tfrac{R-r}{R+r}v(a) &= \tfrac{1}{2\pi}\int_0^{2\pi}\frac{R-r}{R+r}v(a+Re^{it})dt \\ &\leq \tfrac{1}{2\pi}\int_0^{2\pi}P(re^{i\theta},Re^{it})v(a+Re^{it})dt \\ &\leq \tfrac{1}{2\pi}\int_0^{2\pi}\frac{R+r}{R-r}v(a+Re^{it})dt = \frac{R+r}{R-r}v(a).\end{aligned}$$

Thus we obtain

$$\frac{R-r}{R+r}v(a) \leq v(a+re^{i\theta}) \leq \frac{R+r}{R-r}v(a). \tag{14.10}$$

Equation (14.10) is often referred to as Harnack's inequality.

It says that positive harmonic functions cannot oscillate too much on a compact set $K \subset \Omega$ if Ω is connected. For future reference we shall state Harnack's inequality.

Theorem 14.22. *(Harnack's inequality): If $v : \overline{B(a,R)} \to \mathbb{R}$ is continuous, harmonic in $B(a,R)$,and $v \geq 0$ then for $0 \leq r < R$ and for $\theta \in [0,2\pi]$,*

$$\frac{R-r}{R+r}v(a) \leq v(a+re^{i\theta}) \leq \frac{R+r}{R-r}v(a).$$

14.4 Harnack's principle

Harnack's inequality leads to an important convergence theorem for harmonic functions known as Harnack's principle. Consider for a moment a monotone sequence of continuous functions on Ω. The pointwise limit of such a sequence need not behave well-it could be infinite at some points and finite at other points. Even if it is finite everywhere, there is no reason to expect that our sequence converges uniformly on every compact subset of Ω. Harnack's principle shows that none of this bad behavior can occur for a monotone sequence of harmonic functions. Harmonic functions can be defined on open subsets of $\mathbb{C}$. The domain need not be connected. For simplicity we had assumed that the domain of a harmonic function is connected.

Remark 14.6. If G is open in $\mathbb{C}$ then there is a sequence (K_n) of compact subsets of G such that $G = \cup_{n=1}^{\infty} K_n$. Moreover, the sets K_n can be chosen to satisfy the following conditions:

(a) $K_n \subset \text{int}K_{n+1}$;

(b)$K \subset G$ and K compact implies $K \subset K_n$ for some n;

(c) Every component of $\mathbb{C}_\infty - K_n$ contains a component of $\mathbb{C}_\infty - G$.

Let $C(G, \mathbb{R})$ be the space of all continuous functions from G into $\mathbb{R}$. Now define $\rho_n(f,g) = \sup\{d(f(z), g(z)) : z \in K_n\}$ for all functions f and g in $C(G, \mathbb{R})$.

Also define

$$\rho(f,g) = \sum_{n=1}^{\infty} \left(\frac{1}{2}\right)^n \frac{\rho_n(f,g)}{1+\rho_n(f,g)}.$$

It is not so difficult to verify that ρ is a metric on $C(G, \mathbb{R})$ and $C(G, \mathbb{R})$ is complete (see the exercise at the end of this chapter) with respect to ρ.

Definition 14.5. *Let G be an open subset of $\mathbb{C}$. Let $C(G, \mathbb{R})$ be the space of all continuous function from G into $\mathbb{R}$ and Har(G) is the space of all harmonic functions on G. Then Har(G) $\subset C(G, \mathbb{R})$ and Har(G) is a metric space with the metric it inherits as a subset of $C(G, \mathbb{R})$.*

Theorem 14.23. *Let G be a region. The set Har(G) is a complete metric space with respect to the metric it inherits from $C(G, \mathbb{R})$.*

Proof. To show that Har(G) is complete, it is sufficient to show that it is a closed subspace of $C(G, \mathbb{R})$. So let (u_n) be a sequence in Har(G) such that $u_n \to u$ in $C(G, \mathbb{R})$. By Theorem 2.58 it follows that u is continuous as (u_n) converges uniformly to u uniformly on every compact subset of G. Then by Corollary 14.8, u satisfies the mean value property. Hence by Theorem 14.21, u is harmonic in G. □

For $0 \leq t < 1$, define $\alpha(t) = \frac{1-t}{1+t}$ and $\beta(t) = \frac{1+t}{1-t}$. After a translation and a dilation, the Harnack's inequality tells us that if u is positive and harmonic on $B(a, R)$, and $|z-a| \leq r < R$, then

$$\alpha\left(\frac{r}{R}\right) u(a) \leq u(z) \leq \beta\left(\frac{r}{R}\right) u(a). \tag{14.11}$$

Theorem 14.24. *Let Ω be a region in $\mathbb{C}$ and K be a compact subset of Ω. Then there is a constant $c \in (1, \infty)$ such that*

$$\frac{1}{c} \leq \frac{u(w)}{u(z)} \leq c$$

for all points z and w in K and all positive harmonic functions u on Ω.

Proof. We will prove that there is a constant $c \in (1, \infty)$ such that $\frac{u(w)}{u(z)} \leq c$ for all $z, w \in K$ and all positive harmonic functions u on Ω. Because z and w play symmetric roles, the other inequality will also hold. For $(z,w) \in \Omega \times \Omega$, define $s(z,w) = \sup\{\frac{u(w)}{u(z)} : u \text{ is positive and harmonic on } \Omega\}$.

We first show that $s < \infty$ on $\Omega \times \Omega$. Fix $z \in \Omega$, and define

$$E = \{w \in \Omega : s(z,w) < \infty\}.$$

Because $z \in E$, E is not empty. If $w \in E$, we may choose $r > 0$ such that $B(w, 2r) \subset \Omega$. By (14.11), $u \leq \beta(\frac{1}{2})u(w)$ on $B(w, r)$ for all positive harmonic u on Ω. We then have $B(w, r) \subset E$, proving that E is open. If $p \in \Omega$ is a limit point of E, there exists an $r > 0$ and a $w \in E$ such that $p \in B(w, r) \subset B(w, 2r) \subset \Omega$. By (14.11), $u(p) \leq \beta(\frac{1}{2})u(w)$ for all positive harmonic u on Ω. We then have $p \in E$, proving that E is closed. The connectivity of Ω therefore shows that $E = \Omega$. We now know that s is finite at every point of $\Omega \times \Omega$. Let $K \subset \Omega$ be compact, and let $(a, b) \in K \times K$. Then by (14.11),

$$\frac{u(w)}{u(z)} \leq \frac{\beta(\frac{1}{2})u(b)}{\alpha(\frac{1}{2})u(a)} \leq \frac{\beta(\frac{1}{2})}{\alpha(\frac{1}{2})}s(a, b)$$

for all (z, w) in a neighborhood of (a, b), and for all positive harmonic functions u on Ω. Because $K \times K$ is covered by finitely many such neighborhoods, s is bounded above on $K \times K$, as desired. □

Note that the constant c in Theorem 14.24 may depend upon Ω and K, but that c is independent of z, w and u.

An intuitive way to remember Theorem 14.24 is shown in Fig. 14.1. Here we have covered K with a finite chain of overlapping balls. This is possible since Ω is connected and K is compact. To compare the values of a positive harmonic function at any two points in K, we can think of a finite chain of inequalities of the kind expressed in (14.11).

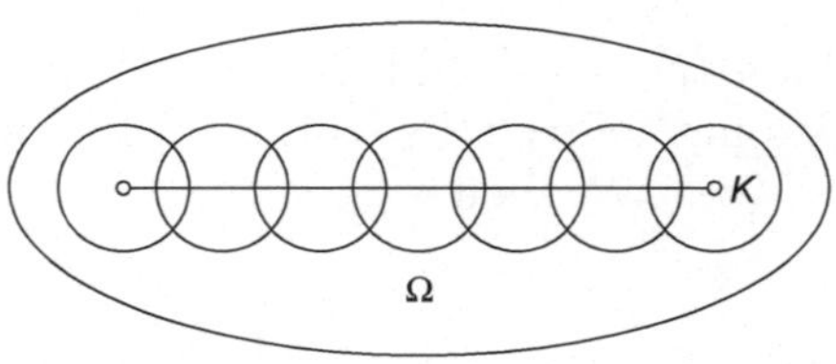

Fig. 14.1: A finite chain of overlapping balls covered set K

Theorem 14.25. *(Harnack's principle): Suppose Ω is a region in $\mathbb{C}$ and (u_m) is a pointwise increasing sequence of harmonic functions on Ω. There are only two possibilities: Either $u_m(z) \to \infty$ for every $z \in \Omega$, or (u_m) converges uniformly on compact subsets of Ω to a function harmonic on Ω.*

Proof. Replacing u_m by $u_m - u_1 + 1$, we can assume that each u_m is positive on Ω. Set $u(z) = \lim_{m\to\infty} u_m(z)$ for each $z \in \Omega$.

Suppose $u(z) = \infty$ for some $z \in \Omega$. Let $w \in \Omega$. Then Theorem 14.24, applied to the compact set $K = \{z, w\}$, shows that there is a constant $c \in (1, \infty)$ such that $u_m(z) \leq cu_m(w)$ for every m. Because $u_m(z) \to \infty$, we also have $u_m(w) \to \infty$, and so $u(w) = \infty$. This implies u is identically ∞ on Ω.

Now suppose u is finite everywhere on Ω. Let K be a compact subset of Ω. Fix $z \in K$. Theorem 14.24 shows there is a constant $c \in (1, \infty)$ such that

$u_m(w)-u_k(w) \leq c(u_m(z)-u_k(z))$ for all $w \in K$, whenever $m > k$. This implies (u_m) is uniformly Cauchy on K, and thus $u_m \to u$ uniformly on K, as desired. Corollary 14.8 shows that the limit function u is harmonic on Ω. □

It is possible to give alternate proofs of Harnack's principle. One involves applying Dini's theorem. Another involves using the Monotone convergence theorem from measure theory to obtain that u has the mean value property. The following theorem gives an alternate proof of Harnack's principle using Harnack's inequality and the connectedness of the region on which the sequence of harmonic functions are defined.

Proof. (Alternate proof of Harnack's principle): We may assume that $u_1 \geq 0$ (if not, consider the sequence $(u_n - u_1)$). Let $u(z) = \sup\{u_n(z) : n \geq 1\}$ for each z in Ω. So for each z in Ω one of two possibilities occurs: $u(z) = \infty$ or $u(z) \in \mathbb{R}$ and $u_n(z) \to u(z)$. Define

$$A = \{z \in \Omega : u(z) = \infty\},$$

$$B = \{z \in \Omega : u(z) < \infty\};$$

then $\Omega = A \cup B$ and $A \cap B = \emptyset$. We will show that both A and B are open.

If $a \in \Omega$, let R be chosen such that $\overline{B(a, R)} \subset G$. By Harnack's inequality

$$\frac{R - |z-a|}{R + |z-a|} u_n(a) \leq u_n(z) \leq \frac{R + |z-a|}{R - |z-a|} u_n(a) \tag{14.12}$$

for all z in $B(a, R)$ and all $n \geq 1$. If $a \in A$ then $u_n(a) \to \infty$ so that the left half of (14.12) gives that $u_n(z) \to \infty$ for all z in $B(a, R)$. That is, $B(a, R) \subset A$ and so A is open. Similarly, if $a \in B$ then the right half of (14.12) gives that $u(z) < \infty$ for $|z - a| < R$. That is B is open. Since Ω is connected, either $A = \Omega$ or $B = \Omega$. Suppose $A = \Omega$; that is $u \equiv \infty$. Again if $\overline{B(a, R)} \subset \Omega$ and $0 < \rho < R$ then $M = \frac{R-\rho}{R+\rho} > 0$ and (14.12) gives that $Mu_n(a) \leq u_n(z)$ for $|z - a| \leq \rho$. Hence $u_n(z) \to \infty$ uniformly for z in $\overline{B(a, \rho)}$. In other words, we have shown that for each a in Ω there is a $\rho > 0$ such that $u_n(z) \to \infty$ uniformly for $|z - a| \leq \rho$. From this it is easy to deduce that $u_n(z) \to \infty$ uniformly for z in any compact set.

Now suppose $B = \Omega$, or that $u(z) < \infty$ for all z in Ω. If $\rho < R$ then, as above there is a constant N, which depends only on a and ρ such that $Mu_n(a) \leq u_n(z) \leq Nu_n(a)$ for $|z - a| \leq \rho$ and all n. So if $m \leq n$,

$$0 \leq u_n(z) - u_m(z) \leq Nu_n(a) - Mu_m(a) \leq C[u_n(a) - u_m(a)]$$

for some constant C. Thus, $(u_n(z))$ is a uniformly Cauchy sequence on $\overline{B(a, \rho)}$. It follows that (u_n) is a Cauchy sequence in $\mathrm{Har}(\Omega)$ and so, by Theorem 14.23, must converge to a harmonic function. Since $u_n(z) \to u(z)$, the function u is thus a harmonic function. □

14.5 Subharmonic functions

An important problem in harmonic function theory is the Dirichlet problem. Poisson's formula solves the problem for a disk, but the case of an arbitrary region is much more difficult. In order to solve the Dirichlet problem for bounded regions in $\mathbb{C}$, generalizations of harmonic functions are introduced. The method of O.Perron is based on the use of subharmonic functions. This method constructs a solution to the Dirichlet problem as the supremum of a family of subharmonic functions. In this section we shall only discuss about some of the properties of subharmonic and superharmonic functions.

Definition 14.6. *Let Ω be a region in $\mathbb{C}$ and let $\phi : \Omega \to \mathbb{R}$ be a continuous function. The function ϕ is subharmonic if whenever $\overline{B(a,r)} \subset \Omega$,*

$$\phi(a) \leq \frac{1}{2\pi} \int_0^{2\pi} \phi(a + re^{i\theta}) d\theta. \tag{14.13}$$

This property of ϕ is often referred to as sub-mean value property of ϕ. The function ϕ is superharmonic if whenever $\overline{B(a,r)} \subset \Omega$,

$$\phi(a) \geq \frac{1}{2\pi} \int_0^{2\pi} \phi(a + re^{i\theta}) d\theta. \tag{14.14}$$

The following are easy to verify:

Theorem 14.26. *Let Ω be a region in $\mathbb{C}$. The following hold:*

(1) $\phi : \Omega \to \mathbb{R}$ is subharmonic if and only if $-\phi$ is superharmonic.

(2) The function $u : \Omega \to \mathbb{R}$ is harmonic if and only if u is both subharmonic and superharmonic.

(3) If ϕ_1 and ϕ_2 are subharmonic functions on Ω and $a, b \geq 0$ then $a\phi_1 + b\phi_2$ is subharmonic.

Example 14.7. *If f is holomorphic in a region Ω and $p > 0$, then $g(z) = |f(z)|^p$ is subharmonic in Ω. To see this we need to verify only (14.13)(the sub-mean value property of g).*

Suppose $\overline{B(a,r)} \subset \Omega$. Note (14.13) is obvious if $f(a) = 0$ as $g(a) = |f(a)|^p = 0 \leq \frac{1}{2\pi} \int_0^{2\pi} g(a + re^{i\theta}) d\theta = \frac{1}{2\pi} \int_0^{2\pi} |f(a + re^{i\theta})|^p d\theta$. If $f(a) \neq 0$, then some single-valued branch of $[f(z)]^p$ is holomorphic in a neighborhood of a, so that

$$[f(a)]^p = \frac{1}{2\pi} \int_0^{2\pi} [f(a + \rho e^{i\theta})]^p d\theta$$

for sufficiently small ρ.

Thus

$$|f(a)|^p \leq \frac{1}{2\pi} \int_0^{2\pi} |f(a + \rho e^{i\theta})|^p d\theta.$$

Example 14.8. *If u is harmonic in a region Ω and $p \geq 1$, then $|u(z)|^p$ is subharmonic in Ω.*

To see this we need to verify only (14.13). If $p = 1$, there is nothing to prove because by mean value property of u whenever $\overline{B(a,r)} \subset \Omega$

$$u(a) = \frac{1}{2\pi} \int_0^{2\pi} u(a + re^{i\theta}) d\theta \tag{14.15}$$

and hence

$$|u(a)| \leq \frac{1}{2\pi} \int_0^{2\pi} |u(a + re^{i\theta})| d\theta. \tag{14.16}$$

If $p > 1$, we apply Holder's inequality to (14.16) and obtain

$$\begin{aligned} |u(a)| &\leq \tfrac{1}{2\pi} \left[\int_0^{2\pi} |u(a + re^{i\theta})|^p d\theta \right]^{\frac{1}{p}} \left[\textstyle\int_0^{2\pi} 1^q d\theta \right]^{\frac{1}{q}} \\ &= \tfrac{1}{2\pi} \left(\int_0^{2\pi} |u(a + re^{i\theta})|^p d\theta \right)^{\frac{1}{p}} (2\pi)^{\frac{1}{q}} \\ &= \left(\tfrac{1}{2\pi} \int_0^{2\pi} |u(a + re^{i\theta})|^p d\theta \right)^{\frac{1}{p}} \end{aligned}$$

where $\frac{1}{p} + \frac{1}{q} = 1$.

This implies the sub-mean value property

$$|u(a)|^p \leq \frac{1}{2\pi} \int_0^{2\pi} |u(a + re^{i\theta})|^p d\theta.$$

Example 14.9. *Let*

$$log^+ x = \begin{cases} logx & if \quad x \geq 1; \\ 0 & if \quad 0 \leq x < 1. \end{cases}$$

Then $g(z) = log^+|f(z)|$ is subharmonic in Ω if f is holomorphic in Ω.

If $|f(a)| \leq 1$, then there is nothing to prove. If $|f(a)| > 1$, then notice that this inequality persists in some neighborhood $|z - a| \leq \rho$. Thus $log^+|f(z)|$ coincides with the harmonic function $\log|f(z)|$ in $|z - a| \leq \rho$.

Example 14.10. *Consider the function $\phi(x, y) = x^2 + y^2$. We shall show that ϕ is subharmonic in a region $\Omega \subset \mathbb{C}$ by verifying the sub-mean value property of ϕ in Ω.*

Suppose $a = c + id, r > 0$ and $\overline{B(a,r)} \subset \Omega$. Then $\phi(a) = c^2 + d^2$. Further

$$\begin{aligned}
\frac{1}{2\pi}\int_0^{2\pi} \phi(a + re^{i\theta})d\theta &= \frac{1}{2\pi}\int_0^{2\pi} \phi\left(a + r\text{cos}\theta + ir\text{sin}\theta\right)d\theta \\
&= \frac{1}{2\pi}\int_0^{2\pi} \phi\left(c + id + r\text{cos}\theta + ir\text{sin}\theta\right)d\theta \\
&= \frac{1}{2\pi}\int_0^{2\pi} \phi([c + r\text{cos}\theta] + i[d + r\text{sin}\theta])\ d\theta \\
&= \frac{1}{2\pi}\int_0^{2\pi} \left[(c + r\text{cos}\theta)^2 + (d + r\text{sin}\theta)^2\right] d\theta \\
&= \frac{1}{2\pi}\int_0^{2\pi} \left(c^2 + d^2 + r^2 + 2cr\text{cos}\theta + 2dr\text{sin}\theta\right) d\theta \\
&= c^2 + d^2 + r^2 + 2cr\frac{1}{2\pi}\int_0^{2\pi} \cos\theta\ d\theta \\
&+ 2dr\frac{1}{2\pi}\int_0^{2\pi} \sin\theta\ d\theta \\
&= c^2 + d^2 + r^2 + 2cr\frac{1}{2\pi}\int_0^{2\pi} \frac{e^{i\theta} + e^{-i\theta}}{2}d\theta \\
&+ 2dr\frac{1}{2\pi}\int_0^{2\pi} \frac{e^{i\theta} - e^{-i\theta}}{2i}d\theta = c^2 + d^2 + r^2.
\end{aligned}$$

Thus $\phi(a) = c^2 + d^2 \leq c^2 + d^2 + r^2 = \frac{1}{2\pi}\int_0^{2\pi} \phi(a + re^{i\theta})\ d\theta$. Thus ϕ is subharmonic.

Theorem 14.27. *(Maximum principle for subharmonic functions): If ϕ is a nonconstant real valued subharmonic function in a region Ω, then ϕ has no maximum in Ω.*

Proof. The proof is the same as the proof of Theorem 14.9 and will not be repeated here. □

Note that from Theorem 14.27 we can also prove that the minimum principle holds for superharmonic functions.

Remark 14.7. A subharmonic function ϕ may assume a minimum value in a region Ω. For example, we have already checked $\phi(x, y) = x^2 + y^2$ is a subharmonic function and it assumes a minimum value at the origin in $\mathbb{C}$. This failure of the minimum principle is due to the fact that if u has the mean value property then so does $-u$; however, if ϕ is subharmonic then $-\phi$ is never subharmonic unless it is harmonic.

Theorem 14.28. *Let Ω be a region in $\mathbb{C}$ and let $\phi : \Omega \to \mathbb{R}$ be continuous. Then ϕ is subharmonic if and only if for every region $\Omega_1 \subset \Omega$ and every harmonic function u_1 on Ω_1, $\phi - u_1$ satisfies the maximum principle on Ω_1.*

Proof. Suppose ϕ is subharmonic on Ω, and u_1 is harmonic on $\Omega_1 \subset \Omega$. Hence $-u_1$ is harmonic on Ω_1 and thus $\phi - u_1$ is subharmonic on Ω_1. It follows therefore by Theorem 14.27 that $\phi - u_1$ satisfy the maximum principle.

Now suppose ϕ is continuous and suppose $\phi - u_1$ satisfy maximum principle whenever u_1 is harmonic on a region $\Omega_1 \subset \Omega$. Let $\overline{B(a,r)} \subset \Omega$. By Corollary 14.7, there is a continuous function $u : \overline{B(a,r)} \to \mathbb{R}$ which is harmonic in $B(a,r)$ and $u(z) = \phi(z)$ for $|z-a| = r$. By hypothesis, $\phi - u$ satisfies the maximum principle. But $(\phi - u)(z) \equiv 0$ for $|z-a| = r$. So $\phi \le u$ in $B(a,r)$ and

$$\phi(a) \le u(a) = \frac{1}{2\pi}\int_0^{2\pi} u(a + re^{i\theta})d\theta = \frac{1}{2\pi}\int_0^{2\pi} \phi(a + re^{i\theta})d\theta.$$

Thus ϕ is subharmonic. □

Corollary 14.10. *Let Ω be a region in $\mathbb{C}$ and $\phi : \Omega \to \mathbb{R}$ be a continuous function; then ϕ is subharmonic if and only if for every bounded region Ω_1 such that $\overline{\Omega_1} \subset \Omega$ and for every continuous function $g : \overline{\Omega_1} \to \mathbb{R}$ that is harmonic in Ω_1 and satisfies $\phi(z) \le g(z)$ for z on $\partial\Omega_1, \phi(z) \le g(z)$ for z in Ω_1.*

Corollary 14.11. *Let Ω be a region in $\mathbb{C}$ and ϕ_1 and ϕ_2 be two subharmonic functions on Ω; if $\phi(z) = max\{\phi_1(z), \phi_2(z)\}$ for each z in Ω then ϕ is a subharmonic function.*

Proof. Let Ω_1 be a region such that $\overline{\Omega_1} \subset \Omega$ and let g be a continuous function on $\overline{\Omega_1}$ which is harmonic on Ω_1 with $\phi(z) \le g(z)$ for all z in $\partial\Omega_1$. Then $\phi_1(z) \le g(z)$ and $\phi_2(z) \le g(z)$ for all $z \in \partial\Omega_1$. From Corollary 14.10, we get that $\phi_1(z) \le g(z)$ and $\phi_2(z) \le g(z)$ for all z in Ω_1. So $\phi(z) \le g(z)$ for $z \in \Omega_1$, and again by Corollary 14.10, ϕ is subharmonic. □

Theorem 14.29. *Let Ω and Ω_1 be two regions such that $\phi : \Omega \to \mathbb{R}$ is subharmonic and $f : \Omega_1 \to \Omega$ is conformal. Then $\phi \circ f : \Omega_1 \to \mathbb{R}$ is subharmonic.*

Proof. Let D be any bounded domain such that $\overline{D} \subset \Omega_1$ and f is one one in D (since f is conformal, it is locally one one). Let u be a real valued continuous function on $\overline{D}$ such that u is harmonic in D and $(\phi \circ f)(z) = u(z)$ for all $z \in \partial D$. By Corollary 14.10, we need only to show that $(\phi \circ f)(z) \le u(z)$ for all $z \in D$. Let $f(D) = D_0$. Then $f(\overline{D}) = \overline{D_0}$. Now f is a analytic, one one map from D onto D_0. By inverse function theorem, f^{-1} is also analytic and one one and therefore $f^{-1} : D_0 \to D$ is conformal. Therefore by Theorem 14.13, $u \circ f^{-1}$ is harmonic in D_0. Further $\phi(z) = (u \circ f^{-1})(z)$ for all $z \in \partial D_0$. Since ϕ is subharmonic, $\phi(z) \le (u \circ f^{-1})(z), z \in D_0$. Thus $(\phi \circ f)(z) \le u(z)$ for all $z \in D$. □

Theorem 14.30. *Let Ω be a region in $\mathbb{C}$ and let $\phi : \Omega \to \mathbb{R}$ be continuous. Then ϕ is subharmonic on Ω if and only if ϕ is the limit of a decreasing sequence of subharmonic functions.*

Proof. Suppose $\phi : \Omega \to \mathbb{R}$ is subharmonic. Let $\phi_j = \phi + \frac{1}{j}, j = 1, 2, 3, \ldots$. Then ϕ_j is subharmonic for each $j = 1, 2, 3, \ldots$ and ϕ is the limit of the decreasing sequence $\{\phi_j\}$ of subharmonic functions. Now for the converse, let $\phi : \Omega \to \mathbb{R}$ be continuous and ϕ is the limit of a decreasing sequence $\{\phi_j\}$ of subharmonic functions. Let $\overline{B(a,r)} \subset \Omega$ and $h : \overline{B(a,r)} \to \mathbb{R}$ be continuous such that h is harmonic in $B(a,r)$ and $\phi \leq h$ on $\partial B(a,r)$. Let $\epsilon > 0$. Let $K_j = \{z \in \partial B(a,r) : \phi_j(z) \geq h(z) + \epsilon\}$. Then each K_j is compact, $K_1 \supseteq K_2 \supseteq \cdots$, and $\cap K_j = \emptyset$. Thus $K_j = \emptyset$ for j large. So $\phi \leq \phi_j \leq h + \epsilon$ on $\partial B(a,r)$ for j large. Since each ϕ_j is subharmonic, $\phi \leq h + \epsilon$ on $\overline{B(a,r)}$. Since $\epsilon > 0$ is arbitrary the theorem is proved. □

Theorem 14.31. *If $f : \Omega \to \mathbb{R}$ is subharmonic and $\psi : \mathbb{R} \to \mathbb{R}$ is convex and monotonically increasing then $\psi \circ f$ is subharmonic.*

Proof. Set $\psi(-\infty) = \lim_{t \to -\infty} \psi(t)$. Let $a \in \Omega$, choose $r > 0$ such that $\overline{B(a,r)} \subset \Omega$. Since f is subharmonic,

$$\begin{aligned} (\psi \circ f)(a) &\leq \psi\left(\tfrac{1}{2\pi} \int_0^{2\pi} f\left(a + re^{i\theta}\right) d\theta\right) \\ &\leq \tfrac{1}{2\pi} \int_0^{2\pi} (\psi \circ f)(a + re^{i\theta}) d\theta \end{aligned}$$

by Jensen's inequality. Hence $\psi \circ f$ is subharmonic on Ω. □

14.6 Exercises

1. If G is a region then show that $C(G, \mathbb{R})$ is complete with respect to the metric ρ.

2. Find a "Poisson integral formula" for the solution of the Dirichlet problem in the upper half plane $\{z \in \mathbb{C} : \text{Im} z > 0\}$ with a bounded continuous boundary data. Explain why we cannot conclude the uniqueness of the solutions in the upper half plane even though the upper half plane is biholomorphic to the unit disk.

3. Let U be bounded. Assume f and g are continuous on $\overline{U}$ and holomorphic on U. Assume further Ref = Reg on ∂U. Show that there exists a real number c such that $f - g = ic$ on U.

4. Suppose f is a complex function in a region Ω, and both f and f^2 are harmonic in Ω. Prove that either f or $\bar{f}$ is holomorphic in Ω.

5. Find necessary and sufficient conditions which the complex numbers a, b, c, and d have to satisfy so that the linear fractional transformation $z \mapsto \frac{az+b}{cz+d}$ maps the upper half plane onto itself.

6. Let f be holomorphic and nonconstant on $B(z_0, R)$. Assume that $f(z_0) = 0$. Show that on every circle $S(z_0, r) = \{z_0 \in \mathbb{C} : |z - z_0| = r\}$, $0 < r < R$, $\text{Re}f$ assumes both positive and negative values.

7. Let Ω be a region in $\mathbb{C}$. If each $x \in \Omega$ has a neighborhood $U_x \subseteq \Omega$ so that $f|_{U_x}$ is subharmonic on U_x, then show that f is subharmonic on Ω.

8. If f_α are subharmonic on Ω and $f(z) = \sup\limits_\alpha f_\alpha(z)$ is upper semi continuous, then show that f is subharmonic on Ω.

9. Show that every translation, dilation and rotation of a subharmonic function is subharmonic.

10. Show that if a sequence of functions subharmonic on Ω converges uniformly on compact subsets of Ω, then the limit function is subharmonic on Ω.

11. Suppose $u \in C^2(\Omega)$ is real-valued. Use Taylor's theorem to show that u is subharmonic on Ω if and only if $\Delta u \geq 0$ on Ω. (Hint: To show $\Delta u \geq 0$ implies u is subharmonic, first assume $\Delta u > 0$. Then consider the functions $u(x) + \epsilon|x|^2$.)

12. Which of the following functions are subharmonic and superharmonic:

(i)	$\|x\|^\alpha, \alpha > 1$	(ii)	$\|z\|^3$
(iii)	$x - y^2$	(iv)	$x + y^2$
(v)	$x^2 + y^2 + 2xy$	(vi)	$\log^+\|1 + z\|$.

Chapter 15

Normal Families and the Riemann Mapping Theorem

15.1 Normal families

Normal families were first introduced by P. Montel. The notion of a normal family is an extension of the notion of compactness defined by Bolzano-Weierstrass to a space whose points are functions.

Let G be a region in $\mathbb{C}$ and let (S, d) be a metric space. Let $\mathcal{F}$ denote a family of functions f from G into S. We are interested in the convergence of sequences $\{f_n\}$ formed by functions in $\mathcal{F}$. There exist sequences which do not possess a single convergent subsequence.

Definition 15.1. *The functions in a family $\mathcal{F}$ are said to be equicontinuous on a set $E \subset G$ if and only if, for each $\epsilon > 0$, there exists a $\delta > 0$ such that $d(f(z), f(z_0)) < \epsilon$ whenever $|z - z_0| < \delta$ and $z, z_0 \in E$, simultaneously for all functions $f \in \mathcal{F}$.*

Thus each f in an equicontinuous family is uniformly continuous on E.

Definition 15.2. *A family $\mathcal{F}$ is said to be normal in G if every sequence $\{f_n\}$ of functions $f_n \in \mathcal{F}$ contains a subsequence which converges uniformly on every compact subset of G.*

This definition does not require the limit functions of convergent subsequences to be members of the family $\mathcal{F}$.

Example 15.1. *If $E_k = \{z \in \mathbb{C} : |z| < k\}$, then z^n converges uniformly to 0 on E_k provided $k < 1$. This is so because $|z^n| < \epsilon$ for all $z \in E_k$ if $n > \frac{\ln \epsilon}{\ln k}$. However, if $k = 1$, then although z^n still converges to 0 on E_1, convergence is no longer uniform, since for each fixed n we have $\lim_{z \to 1} z^n = 1$ and hence there is no n satisfying $|z^n| < \epsilon = \frac{1}{2}$ for all $z \in E_1$. Nevertheless, each compact subset $K \subseteq E_1$ is contained in some E_k with $k < 1$, so z^n converges uniformly on all compact subsets of E_1.*

We shall now define a metric on the space of functions on G with values in S, and convergence with respect to this metric should mean precisely the same as uniform convergence on compact sets.

Proposition 15.1. *If U is open in $\mathbb{C}$ then there is a sequence $\{K_n\}$ of compact subsets of U such that $U = \cup_{n=1}^{\infty} K_n$. Moreover, the sets K_n can be chosen to satisfy the following conditions:*

(a) $K_n \subset \text{int}K_{n+1}$;

(b) $K \subset U$ and K compact implies $K \subset K_n$ for some n;

(c) Every component of $\mathbb{C}_\infty - K_n$ contains a component of $\mathbb{C}_\infty - U$.

Proof. For $n \in \mathbb{Z}_+$ (the set of positive integers), let

$$K_n = \{z \in \mathbb{C} : |z| \leq n\} \cap \left\{z \in \mathbb{C} : d(z, \mathbb{C} - U) \geq \frac{1}{n}\right\}.$$

Note that K_n is bounded. Since K_n is the intersection of two closed subsets of $\mathbb{C}$ hence K_n is closed and therefore K_n is compact. The set

$$\{z \in \mathbb{C} : |z| < n+1\} \cap \left\{z \in \mathbb{C} : d(z, \mathbb{C} - U) > \frac{1}{n+1}\right\}$$

is open, contains K_n, and is contained in K_{n+1}. Thus $K_n \subset \text{int}K_{n+1}$. It follows easily that $U = \cup_{n=1}^{\infty} K_n = \cup_{n=1}^{\infty} \text{int}K_n$. So if K is a compact subset of U the sets $\{\text{int}K_n\}_{n=1}^{\infty}$ form an open cover of K. Hence (b) follows.

To prove (c) note that the unbounded component of $\mathbb{C}_\infty - K_n$ must contain ∞ and $\mathbb{C}_\infty - U \subset \mathbb{C}_\infty - K_n$. Thus $\mathbb{C}_\infty - K_n$ contain the component of $\mathbb{C}_\infty - U$ which contains ∞ for all n. Also the unbounded component contains $\{z \in \mathbb{C} : |z| > n\}$. So if D is a bounded component of $\mathbb{C}_\infty - K_n$ it contains a point z with $d(z, \mathbb{C} - U) < \frac{1}{n}$. Thus there exists a $w \in \mathbb{C} - U$ with $|w - z| < \frac{1}{n}$. But then $z \in B(w, \frac{1}{n}) \subset \mathbb{C}_\infty - K_n$; since disks are connected and z is in the component D of $\mathbb{C}_\infty - K_n$, $B(w, \frac{1}{n}) \subset D$. If D_1 is the component of $\mathbb{C}_\infty - U$ that contains w it follows that $D_1 \subset D$. □

Using Proposition 15.1 to the region G in $\mathbb{C}$ we can conclude that $G = \cup_{n=1}^{\infty} K_n$, where each K_n is compact and $K_n \subset \text{int}K_{n+1}$. Let f and g be any two functions on G with values in the metric space (S, d). We shall define a distance $\rho(f, g)$ between these functions. This should not be confused with the distance $d(f(z), g(z))$ between their values.

Let

$$\delta(f(z), g(z)) = \frac{d(f(z), g(z))}{1 + d(f(z), g(z))}.$$

Note that δ satisfies the triangle inequality and it is bounded. Suppose $\delta_k(f, g) = \sup_{z \in E_k} \delta(f(z), g(z))$ which may be described as the distance between f and g on E_k. Define

$$\rho(f, g) = \sum_{k=1}^{\infty} \frac{1}{2^k} \delta_k(f, g).$$

It is not so difficult to verify that ρ is a metric on G^S, the set of all functions from the region G into S.

Theorem 15.1. *Let $\{f_n\}$ be a sequence in G^S. Then $\{f_n\}$ converges uniformly to f on every compact subset of G if and only if $\rho(f_n, f) \to 0$ as $n \to \infty$.*

Proof. Suppose that f_n converges uniformly to f on every compact set. Then $\delta_k(f_n, f) \to 0$ for every k. Now the series $\sum_{k=1}^{\infty} \frac{1}{2^k}\delta_k(f_n, f)$ has a convergent majorant series $\sum_{k=1}^{\infty} \frac{1}{2^k}$ (with terms independent of n). Thus it follows from Weierstrass's $M-$test that $\rho(f_n, f) \to 0$.

Conversely suppose $\rho(f_n, f) \to 0$ as $n \to \infty$. Hence given $\epsilon > 0$, there exists N such that $\rho(f_n, f) < \epsilon$ for all $n \geq N$. Thus $\delta_k(f_n, f) < 2^k\epsilon$ for all k and $n \geq N$. But this implies that $f_n \to f$ uniformly on E_k, first with respect to the $\delta-$metric, but hence also with respect to the $d-$metric. Since every compact set E is contained in an E_k it follows that the convergence is uniform on E. □

Suppose now that S is complete with respect to the metric d. It is not so difficult to verify that if S is complete then G^S is complete with respect to the metric ρ. Further, it may be noted that the open set G is independent of the choices involved in the construction. That is, if $G = \cup_{n=1}^{\infty} K'_n$ where each K'_n is compact and $K'_n \subset \text{int}K'_{n+1}$ and if μ is the metric defined by the sets $\{K'_n\}$ then a set is open in (G^S, ρ) if and only if it is open in (G^S, μ).

We know that a metric space is compact if and only if every infinite sequence has a convergent subsequence. If the result is applied to the set $\mathcal{F}$, equipped with the distance ρ, and we conclude that $\mathcal{F}$ is compact if and only if $\mathcal{F}$ is normal, and if the limit functions are themselves in $\mathcal{F}$. On the other hand, if $\mathcal{F}$ is normal, so is its closure $\overline{\mathcal{F}}$. The reason is as follows: Let $\{f_n\}_{n=1}^{\infty}$ be a sequence in $\overline{\mathcal{F}}$. Then $B_\rho(f_n, \frac{1}{n}) \cap \mathcal{F} \neq \emptyset$ for all $n \in \mathbb{N}$. Let $h_n \in B_\rho(f_n, \frac{1}{n}) \cap \mathcal{F}$ for all $n \in \mathbb{N}$. Thus $\{h_n\}_{n\in\mathbb{N}}$ is a sequence in $\mathcal{F}$ and $\mathcal{F}$ is normal. Therefore $\{h_n\}$ has a subsequence $\{h_{n_k}\}$ which converges uniformly on every compact subset of G. Now consider the sequence $\{f_{n_k}\}$, let K be an arbitrary compact subset of G and suppose $\{h_{n_k}\}$ converges to h uniformly on K. Then

$$\rho(f_{n_k}, h) \leq \rho(f_{n_k}, h_{n_k}) + \rho(h_{n_k}, h) \leq \frac{1}{n_k} + \rho(h_{n_k}, h).$$

Hence $\rho(f_{n_k}, h) \to 0$ as $n_k \to \infty$ uniformly on K. It follows therefore that $\overline{\mathcal{F}}$ is normal.

Thus we obtain the following characterization of normal families.

Theorem 15.2. *A family $\mathcal{F}$ is normal if and only if its closure $\overline{\mathcal{F}}$ with respect to the metric ρ is compact.*

Theorem 15.3. *If (S, d) is complete then $\mathcal{F}$ is normal if and only if $\mathcal{F}$ is totally bounded.*

Proof. Let S be complete with respect to the metric d. Suppose $\mathcal{F}$ is normal. Hence by Theorem 15.2, $\overline{\mathcal{F}}$ is compact. Therefore, $\overline{\mathcal{F}}$ is totally bounded. Thus $\mathcal{F}$ is totally bounded.

Conversely, suppose $\mathcal{F}$ is totally bounded. Then $\overline{\mathcal{F}}$ is totally bounded. Since S is complete, hence G^S is complete and $\overline{\mathcal{F}} \subset G^S$. Thus $\overline{\mathcal{F}}$ is also complete. It follows therefore that $\overline{\mathcal{F}}$ is compact. By Theorem 15.2, $\mathcal{F}$ is normal. □

The family $\mathcal{F}$ is totally bounded if to every $\epsilon > 0$ there exists a finite number of functions $f_1, \cdots, f_n \in \mathcal{F}$ such that every $f \in \mathcal{F}$ satisfies $\rho(f, f_j) < \epsilon$ for some f_j.

The following theorem states the condition of total boundedness in terms of the original metric d on S rather than in terms of the auxiliary metric ρ.

Theorem 15.4. *The family $\mathcal{F}$ is totally bounded if and only if to every compact set $E \subset \Omega$ and every $\epsilon > 0$ it is possible to find $f_1, \cdots, f_n \in \mathcal{F}$ such that every $f \in \mathcal{F}$ satisfies $d(f, f_j) < \epsilon$ on E for some f_j.*

Proof. Suppose $\mathcal{F}$ is totally bounded. Then there exist $f_1, \cdots, f_n$ such that, for any $f \in \mathcal{F}$, $\rho(f, f_j) < \epsilon$ for some f_j. This implies $\delta_k(f, f_j) < 2^k\epsilon$ or $\delta(f, f_j) < 2^k\epsilon$ on E_k. If we fix k beforehand, we can make $\delta(f, f_j)$ arbitrarily small on E_k and the same is true of $d(f, f_j)$. This proves the condition is necessary.

To prove the sufficiency, we choose k_0 such that $2^{-k_0} < \frac{\epsilon}{2}$. By assumption we can find $f_1, \cdots, f_n$ such that any $f \in \mathcal{F}$ satisfies one of the inequalities $\delta(f, f_j) \leq d(f, f_j) < \frac{\epsilon}{2k_0}$ on E_{k_0}. It follows that $\delta_k(f, f_j) < \frac{\epsilon}{2k_0}$ for $k \leq k_0$. But $\delta_k(f, f_j) < 1$ for $k > k_0$. Thus we obtain $\rho(f, f_j) = \sum_{k=1}^{\infty} \frac{1}{2^k}\delta_k(f, f_j) < k_0(\frac{\epsilon}{2k_0}) + \frac{1}{2^{k_0+1}} + \frac{1}{2^{k_0+2}} + \cdots = \frac{\epsilon}{2} + \frac{\frac{1}{2^{k_0+1}}}{1-\frac{1}{2}} = \frac{\epsilon}{2} + \frac{1}{2^{k_0}} < \epsilon$. The theorem is proved. □

Theorem 15.5. *(Arzela-Ascoli theorem): A family $\mathcal{F}$ of continuous functions with values in a metric space S is normal in the region $G \subseteq \mathbb{C}$ if and only if*

(i) $\mathcal{F}$ is equicontinuous on every compact set $E \subset G$;

(ii) For any $z \in G$, the set $\overline{\{f(z) : f \in \mathcal{F}\}}$ is a compact subset of S.

Proof. Suppose $\mathcal{F}$ is normal. Then $\overline{\mathcal{F}}$ is compact. Hence $\overline{\mathcal{F}}$ is totally bounded. Therefore, $\mathcal{F}$ is totally bounded. Let E be a compact subset of G and $\epsilon > 0$. By Theorem 15.4, there exist $f_1, \cdots, f_n \in \mathcal{F}$ such that every $f \in \mathcal{F}$ satisfies $d(f, f_j) < \frac{\epsilon}{3}$ on E for some f_j.

Because $f_1, \cdots, f_n$ are continuous on E and E is compact, hence $f_1, \cdots, f_n$ are uniformly continuous on E. Thus we can find a $\delta > 0$ such that

$$d(f_j(z), f_j(z_0)) < \frac{\epsilon}{3}$$

for $z, z_0 \in E$, $|z - z_0| < \delta, j = 1, \cdots, n$. For any given $f \in \mathcal{F}$ and corresponding f_j we obtain

$$d(f(z), f(z_0)) \leq d(f(z), f_j(z)) + d(f_j(z), f_j(z_0)) + d(f_j(z_0), f(z_0)) < 3 \cdot \frac{\epsilon}{3} = \epsilon$$

and (i) is proved.

To prove (ii), let $z \in G$ and $\{w_n\}$ be a sequence in $\overline{\{f(z) : f \in \mathcal{F}\}}$. To each w_n we determine $f_n \in \mathcal{F}$ such that $f_n(z) \in B(w_n, \frac{1}{n})$. That is, $d(f_n(z), w_n) < \frac{1}{n}$. Since $\mathcal{F}$ is normal and $\{f_n\}$ is a sequence in $\mathcal{F}$, there exists a subsequence $\{f_{n_k}\}$ of $\{f_n\}$ which converges to a function f uniformly on every compact subset of G.

Now

$$d(f(z), w_{n_k}) \leq d(f(z), f_{n_k}(z)) + d(f_{n_k}(z), w_{n_k}) < d(f(z), f_{n_k}(z)) + \frac{1}{n_k}$$

which tends to zero as $n_k \to \infty$.

Thus $\{w_{n_k}\}$ converges to $f(z)$ and (ii) is proved.

We shall prove the sufficiency, by Cantor's diagonal process. Let $\{f_n\}$ be a sequence in $\mathcal{F}$. Note that there exists an everywhere dense sequence of points ξ_k in G, for instance the points with rational coordinates. Now $\xi_1 \in G$ and by (ii) the set $\overline{\{f(\xi_1) : f \in \mathcal{F}\}}$ is compact. Thus the sequence $\{f_n(\xi_1)\}$ has a convergent subsequence $\{f_{n_{1j}}(\xi_1)\}$. Since $\overline{\{f(\xi_2) : f \in \mathcal{F}\}}$ is compact, the sequence $\{f_{n_{1j}}(\xi_2)\}$ has a convergent subsequence $\{f_{n_{2j}}(\xi_2)\}$. Continuing this process we find an infinite set of sequences

$$\begin{array}{cccc} f_{n_{11}} & f_{n_{12}} & \cdots f_{n_{1j}} & \cdots \\ f_{n_{21}} & f_{n_{22}} & \cdots f_{n_{2j}} & \cdots \\ \vdots & \vdots & \vdots & \\ f_{n_{k1}} & f_{n_{k2}} & \cdots f_{n_{kj}} & \cdots \end{array}$$

each of which is a subsequence of its predecessors and such that $\lim_{j \to \infty} f_{n_{kj}}(\xi_k)$ exists for all k. Thus it follows that $\lim_{j \to \infty} f_{n_{jj}}(\xi_k)$ exists for all k. For convenience we replace the notation n_{jj} by n_j.

Consider now a compact set $E \subset G$ and assume that f is equicontinuous on E. We shall prove that $\{f_{n_j}\}$ converges uniformly on E. Since the family $\mathcal{F}$ is equicontinuous, given $\epsilon > 0$, there exists a $\delta > 0$ such that, for $z, z' \in E$ and $f \in \mathcal{F}, |z - z'| < \delta$ implies $d(f(z), f(z')) < \frac{\epsilon}{3}$. Since E is compact hence, E is totally bounded. Thus it can be covered by a finite number of $\frac{\delta}{2}$-neighborhoods. We select a point ξ_k from each of these neighborhoods. Now as $\lim_{j \to \infty} f_{n_j}(\xi_k)$ exists, hence there exists an i_0 such that $i, j > i_0$ implies $d(f_{n_i}(\xi_k), f_{n_j}(\xi_k)) < \frac{\epsilon}{3}$ for all these ξ_k. For each $z \in E$ one of the ξ_k is within distance δ from z; hence

$$d(f_{n_i}(z), f_{n_i}(\xi_k)) < \frac{\epsilon}{3}, d(f_{n_j}(z), f_{n_j}(\xi_k)) < \frac{\epsilon}{3}.$$

The last three inequalities yield $d(f_{n_i}(z), f_{n_j}(z)) < \epsilon$. Because $\overline{\{f(z) : f \in \mathcal{F}\}}$ is compact and consequently a complete subset of S, it follows that $\{f_{n_i}\}$ is uniformly convergent on E. □

We shall prove below Montel's theorem which gives a condition for a family of analytic functions to be normal.

Theorem 15.6. *(Montel's theorem): A family $\mathcal{F}$ of analytic functions defined from a region G into $\mathbb{C}$ is normal with respect to $\mathbb{C}$ if and only if the functions in $\mathcal{F}$ are uniformly bounded in every compact set.*

Proof. Suppose $\mathcal{F}$ is normal. By Arzela-Ascoli's theorem, for any $z \in G$, the set $\{f(z) : f \in \mathcal{F}\}$ is bounded with a bound that may depend on z. Now for a given $z_0 \in G$, find ρ such that $\{z \in \mathbb{C} : |z - z_0| \leq \rho\} \subset G$. Let $M > 0$ be such that $|f(z_0)| \leq M$ for all $f \in \mathcal{F}$. The family $\mathcal{F}$ by Arzela-Ascoli's theorem is equicontinuous on the closed disk $\{z \in \mathbb{C} : |z - z_0| \leq \rho\}$ which is compact. Let $\epsilon > 0$. By definition of equicontinuity there exists $\delta > 0$ ($\delta < \rho$) such that $|f(z) - f(z_0)| < \epsilon$ whenever $|z - z_0| < \delta, z \in G$ and for all $f \in \mathcal{F}$. Thus $|f(z)| \leq |f(z_0)| + \epsilon \leq M + \epsilon$ in $|z - z_0| < \delta$. Because any compact set can be covered by a finite number of neighborhoods with this property, it follows that the functions are uniformly bounded on every compact set, the bound depending on the set.

To prove the sufficiency, let C be the boundary of a closed disk in G of radius r. If z, z_0 are inside C we obtain by Cauchy's integral formula that

$$\begin{aligned} f(z) - f(z_0) &= \tfrac{1}{2\pi i} \int_C \left(\frac{1}{\xi - z} - \frac{1}{\xi - z_0} \right) f(\xi) d\xi \\ &= \tfrac{z - z_0}{2\pi i} \int_C \frac{f(\xi)}{(\xi - z)(\xi - z_0)} d\xi. \end{aligned}$$

If $|f| \leq M$ on C, and if we restrict z and z_0 to the smaller concentric disk of radius $\frac{r}{2}$, it follows that

$$|f(z) - f(z_0)| \leq \frac{4M\,|z - z_0|}{r}. \tag{15.1}$$

Thus $\mathcal{F}$ is equicontinuous on the smaller disk. Let E be a compact set in G. Each point of E is the center of a disk of radius r, as above. The open disks of radius $\frac{r}{4}$ form an open covering of E. We select a finite subcovering and denote the corresponding centers, radii, and bounds by ξ_k, r_k, M_k. Let $r = \min\limits_k r_k$ and $M = \max\limits_k M_k$. Let $\epsilon > 0$ and $\delta = \min\{\frac{r}{4}, \frac{\epsilon r}{4M}\}$. If $|z - z_0| < \delta$ and $|z_0 - \xi_k| < \frac{r_k}{4}$ it follows that $|z - \xi_k| \leq |z - z_0| + |z_0 - \xi_k| < \delta + \frac{r_k}{4} \leq \frac{r_k}{2}$. Hence from (15.1) it follows that

$$|f(z) - f(z_0)| \leq \frac{4M_k\delta}{r_k} \leq \frac{4M\delta}{r} \leq \epsilon.$$

The theorem is proved. □

Definition 15.3. *If a family $\mathcal{F}$ of analytic functions defined from a region G into $\mathbb{C}$ is locally bounded if the functions in $\mathcal{F}$ are uniformly bounded on every compact set.*

Theorem 15.7. *A locally bounded family of analytic functions defined on a region G into $\mathbb{C}$ has locally bounded derivatives.*

Proof. Let $\mathcal{F}$ be a locally bounded family of analytic functions. Suppose $f \in \mathcal{F}$ and let C be the boundary of a closed disk in G of radius r. Suppose $|f| \leq M$ on $\mathbb{C}$ and for $z \in B(0, \frac{r}{2})$, we have

$$f'(z) = \frac{1}{2\pi i} \int_C \frac{f(\xi)}{(\xi - z)^2} d\xi.$$

Thus

$$|f'(z)| \leq \frac{1}{2\pi} \frac{4M}{r^2} 2\pi r = \frac{4M}{r}.$$

Thus the family $\{f' : f \in \mathcal{F}\}$ is locally bounded. □

If a sequence tends to ∞ there is no great scattering of values and for the purposes of normal families such a sequence should be regarded as convergent. So we shall restyle our definition of normal family as follows.

Definition 15.4. *Let $\mathcal{F}$ be a family of analytic functions on a region Ω. The family $\mathcal{F}$ is normal if every sequence $\{f_n\}$ of elements of $\mathcal{F}$ contains a subsequence $\{f_{n_k}\}$ which converges locally uniformly on Ω. The limit is to be everywhere finite or else identically infinite.*

Example 15.2. *If $\mathcal{F}$ is a normal family of analytic functions, define $\mathcal{F}'$ to be the collection of functions consisting of the derivatives of functions in $\mathcal{F}$. There exists normal family $\mathcal{F}$ of analytic functions such that $\mathcal{F}'$ is not normal. For example, consider the family formed by the functions $f_n(z) = n(z^2 - n)$ in $\mathbb{C}$. This family is normal, for it is clear that $f_n \to \infty$ uniformly on every compact set. But the derivatives $f_n'(z) = 2nz$ do not form a normal family, for $f_n'(z)$ tends to ∞ for $z \neq 0$ and to 0 for $z = 0$. Similarly consider the family formed by the functions $f_n(z) = n^n + z^n$ on $\mathbb{C}$. This family is normal but the derivatives $f_n'(z)$ do not form a normal family.*

The following lemma exhibits a line of reasoning which is frequently encountered in the theory of normal families.

Lemma 15.1. *Let $\mathcal{F}$ be a normal family consisting of functions defined on a region Ω. Let $\{f_n\}$ be a sequence of functions from $\mathcal{F}$. If $\{f_n\}$ does not converge locally uniformly, there exist two subsequences of $\{f_n\}$ which converge locally uniformly to different limits.*

Proof. To say that $\{f_n\}$ does not converge locally uniformly on Ω is to say that there exists a closed disk $U \subset \Omega$ upon which $\{f_n\}$ does not converge uniformly, either to ∞ or to a finite limit. To say that $\{f_n\}$ does not converge uniformly on U is to say that there exists an $\epsilon > 0$ and for each n an $m_n > n$ such that

$$\max_{z \in U} |f_n(z) - f_{m_n}(z)| \geq \epsilon. \tag{15.2}$$

Since $\{f_n\}$ does not converge uniformly to ∞ on U, there exists an $M < \infty$ and a subsequence $\{f_{n_k}\}$ such that for each k, $\inf_{z\in U} |f_{n_k}(z)| \leq M$. For convenience, set $g_k = f_{n_k}$ and $h_k = f_{m_{n_k}}$ for each k. By (15.2),

$$\max_{z\in U} |g_k(z) - h_k(z)| \geq \epsilon.$$

The family $\mathcal{F}$ is normal, and so there exists a subsequence $\{g_{k_l}\}$ of $\{g_k\}$ converging locally uniformly on Ω to a limit g. Since $\inf_{z\in U} |g(z)| \leq M$, g is not identically infinite. Also $\max_{z\in U} |g(z) - h_{k_l}(z)| \geq \frac{\epsilon}{2}$ for large l. If h is the limit of a locally uniformly convergent subsequence of $\{h_{k_l}\}$, then whether h is finite or not, $\max_{z\in U} |g(z) - h(z)| \geq \frac{\epsilon}{2}$. Thus $g \neq h$, and the lemma is proved. □

As an application of Lemma 15.1, suppose P is a property which is held by at most one function on a region Ω. If $\{f_n\}$ is a sequence from a normal family $\mathcal{F}$, and if it is somehow known that the limit of any subsequence of $\{f_n\}$ has property P, then $\{f_n\}$ must converge locally uniformly. For otherwise by Lemma 15.1 there would exist at least two functions having property P. This has been illustrated further in Vitali's theorem.

Theorem 15.8. *(Vitali' theorem): Let $\{f_n\}$ be a sequence from a normal family $\mathcal{F}$ of analytic functions defined on a region Ω. If $\lim_{n\to\infty} f_n(z)$ exists for each z in a set E with accumulation point in Ω, then $\{f_n\}$ converges locally uniformly on Ω.*

Proof. If $\{f_n\}$ does not converge locally uniformly, there exist convergent subsequences of $\{f_n\}$ with distinct limits f and g. However, since $f(z) = g(z)$ for $z \in E$, the identity theorem implies that $f \equiv g$. The theorem follows by contradiction. □

Definition 15.5. *Let $\mathcal{F}$ be a family of functions analytic on a region Ω. The family $\mathcal{F}$ is locally normal if for each open disk D in Ω, the set $\mathcal{F}_D$, consisting of the functions of $\mathcal{F}$ restricted to D, is normal.*

Lemma 15.2. *Let $\mathcal{F}$ be a locally normal family of analytic functions on a region Ω. If $\{f_n\}$ is a sequence of elements from $\mathcal{F}$ which converges locally uniformly on a given disk $D \subseteq \Omega$, then $\{f_n\}$ converges locally uniformly on Ω.*

Proof. Let Δ be an open disk in Ω. If $\{f_n\}$ does not converge locally uniformly on Δ, then by Vitali's theorem and the fact that $\mathcal{F}_\Delta$ is normal, $\{f_n\}$ converges locally uniformly on no open subset of Δ. Denote by G_1 the set of points $z \in \Omega$ such that $\{f_n\}$ converges locally uniformly on a disk containing z. It follows that G_1 and its complement in Ω, G_2, are both open sets. Since Ω is connected, and since by assumption $G_1 \neq \emptyset$, we have $G_1 = \Omega$. Thus $\{f_n\}$ converges locally uniformly on each open disk in Ω. Since an arbitrary closed disk in Ω is contained in some open disk in Ω, $\{f_n\}$ converges uniformly on every closed disk in Ω. The lemma is proved. □

Theorem 15.9. *A locally normal family is normal.*

Proof. Suppose $\mathcal{F}$ is locally normal, and let $\{f_n\}$ be a sequence from $\mathcal{F}$. Then fix a disk Δ in Ω. Because $\mathcal{F}_\Delta$ is normal there exists a subsequence $\{f_{n_k}\}$ which converges locally uniformly on Δ. By Lemma 15.2, $\{f_{n_k}\}$ converges locally uniformly on Ω, and the theorem is proved.

□

One can also discuss about convergence in the spaces of meromorphic functions. We can obtain extensions of Weierstrass's theorem and Hurwitz's theorem to meromorphic functions as follows.

If a sequence of meromorphic functions converges in the sense of spherical distance, uniformly on every compact set, then the limit function is meromorphic or identically equal to ∞. Further if a sequence of analytic functions converges in the same sense, then the limit function is either analytic or identically equal to ∞.

Remark 15.1. By Arzela-Ascoli's theorem, a family of meromorphic functions is normal if and only if it is equicontinuous on compact sets. Because for any $z \in G$, $\overline{\{f(z) : f \in \mathcal{F}\}}$ being a closed subset of the compact set $\mathbb{C}_\infty$ is compact. So the second condition of Arzela-Ascoli's theorem is trivially fulfilled. Further the equicontinuity condition can be replaced by a boundedness condition. In fact, a family of analytic or meromorphic functions f is normal in the classical sense if and only if the expressions $\rho(f) = \frac{2|f'(z)|}{1+|f(z)|^2}$ are locally bounded.

Definition 15.6. *An analytic function defined in a region Ω is said to be univalent if $g(z_1) = g(z_2)$ only for $z_1 = z_2$, in other words, if the mapping g is one-one.*

We shall prove that any simply connected region in the plane, other than the plane itself can be mapped conformally onto the unit disk. Thus from this it will follow that any two simply connected regions that are not the whole plane can be mapped conformally onto each other, for we can use the unit disk as an intermediate step.

15.2 Riemann mapping theorem

Theorem 15.10. *(Riemann mapping theorem)*

Given any simply connected region Ω which is not the whole plane and the point $z_0 \in \Omega$, there exists a unique analytic function $f(z)$ in Ω, normalized by the conditions $f(z_0) = 0, f'(z_0) > 0$, such that f defines a one-to-one mapping of Ω onto the disk $|w| < 1$.

Proof. Suppose $\Omega \neq \mathbb{C}$ is a simply connected region. Let $z_0 \in \Omega$. We have to show that there exists an analytic function f on Ω normalized by the condition

$f(z_0) = 0, f'(z_0) > 0$ such that $f(z)$ defines a one-to-one mapping of Ω onto the disk $|w| < 1$. We shall first show that if such a function f exists then it is unique.

Suppose f_1 and f_2 are two such functions on Ω. Then $S = f_1 \circ f_2^{-1}$ is one-one, analytic and maps the open unit disk $\mathbb{D}$ onto itself.Further, $(f_1 \circ f_2^{-1})(0) = f_1(z_0) = 0$. From Theorem 13.18, it follows that

$$S(z) = (f_1 \circ f_2^{-1})(z) = e^{i\theta} \frac{z - a}{1 - \bar{a}z}, \theta \in \mathbb{R}, |a| < 1.$$

Now $S(0) = 0$ implies $a = 0$. Thus $S(z) = e^{i\theta} z$ and therefore $S'(z) = e^{i\theta}$ and hence $S'(0) = e^{i\theta}$. Now $S'(0) = \frac{f_1'(z_0)}{f_2'(z_0)} > 0$. Thus $\theta = \arg S'(0) = 0$. Hence $S(z) = z$ and therefore $f_1(z) = f_2(z)$ for all $z \in \Omega$.

Consider the family $\mathcal{F}$ formed by all functions g with the following properties:

(i) g is analytic and univalent in Ω,

(ii) $|g(z)| \leq 1$ in Ω,

(iii) $g(z_0) = 0$ and $g'(z_0) > 0$.

We shall prove that the required function f is the function in $\mathcal{F}$ for which the derivative $f'(z_0)$ is maximum. The proof will consist of three parts: (1) $\mathcal{F}$ is not empty (2) there exists a function $f \in \mathcal{F}$ which has a maximal derivative at z_0 (3) the function f has the the desired properties. Since $\Omega \neq \mathbb{C}$, there exists a point $a \in \mathbb{C}$ (hence $a \neq \infty$) such that $a \notin \Omega$. As Ω is simply connected, it is possible to define a single-valued analytic branch of $\sqrt{z - a}$ in Ω; denote it by $h(z)$. If $h(z_1) = h(z_2)$, then $h^2(z_1) = h^2(z_2)$, hence $z_1 - a = z_2 - a$. Therefore $z_1 = z_2$; thus h is one-to-one. The same argument shows that there are no two points z_1 and z_2 in Ω such that $h(z_1) = -h(z_2)$. Thus the function h does not take the same value twice, nor does it take opposite values. Since h is an open-mapping, the set $h(\Omega)$ is open. Now as $h(z_0) \in h(\Omega)$ and $-h(z_0) \notin h(\Omega)$, hence there exists $\rho > 0$ such that $B(h(z_0), \rho) \subset h(\Omega)$ and $B(-h(z_0), \rho) \not\subset h(\Omega)$. This is possible by choosing ρ small enough. Thus $|h(z) + h(z_0)| \geq \rho$ for $z \in \Omega$, and in particular $2|h(z_0)| \geq \rho$. Let

$$g_0(z) = \frac{\rho}{4} \frac{|h'(z_0)|}{|h(z_0)|^2} \frac{h(z_0)}{h'(z_0)} \frac{h(z) - h(z_0)}{h(z) + h(z_0)}.$$

The function g_0 is univalent as it is obtained from the univalent function h by means of a linear transformation. That is, $g_0 = \phi \circ h$ where ϕ is a linear transformation. Moreover, $g_0(z_0) = 0, g_0'(z_0) = \frac{\rho}{8}\frac{|h'(z_0)|}{|h(z_0)|^2} > 0$. Further,

$$\begin{aligned}\left|\frac{h(z)-h(z_0)}{h(z)+h(z_0)}\right| &= |h(z_0)|\left|\frac{1}{h(z_0)} - \frac{2}{h(z)+h(z_0)}\right| \\ &\leq |h(z_0)|\left(\frac{1}{|h(z_0)|} + \frac{2}{|h(z)+h(z_0)|}\right) \\ &\leq |h(z_0)|\left(\frac{2}{\rho}+\frac{2}{\rho}\right) \\ &= \frac{4}{\rho}|h(z_0)|.\end{aligned}$$

Thus $|g_0(z)| \leq 1$ in Ω. Hence $g_0 \in \mathcal{F}$ and $\mathcal{F}$ is not empty. Let $B = \text{lub}\{g'(z_0) : g \in \mathcal{F}\}$. We shall prove that $B < +\infty$ and there exists $f \in \mathcal{F}$ such that $f'(z_0) = B$. By definition of least upper bound there is a sequence of functions $g_n \in \mathcal{F}$ such that $\lim\limits_{n\to\infty} g_n'(z_0) = B$. By Montel's theorem, the family $\mathcal{F}$ is normal. Hence there exists a subsequence $\{g_{n_k}\}$ which converges uniformly on every compact subsets of Ω to an analytic function f. It is clear that $|f(z)| \leq 1$ in $\Omega, f(z_0) = 0$ and $f'(z_0) = B$. Thus $B < +\infty$. Since $f'(z_0) = B > 0$, hence f is not a constant. Let $z_1 \in \Omega$ and consider the functions $g_1(z) = g(z) - g(z_1), g \in \mathcal{F}$. Note that $g_1(z) \neq 0$ for all $g \in \mathcal{F}$ and $z \in \Omega \setminus \{z_1\}$. This is so as all functions in $\mathcal{F}$ are univalent. By Hurwitz's theorem, every limit function is either identically zero or never zero. But $f(z) - f(z_1)$ is a limit function and it is not identically zero. Hence $f(z) \neq f(z_1)$ for $z \neq z_1$, and since z_1 was arbitrarily chosen, we have proved that f is univalent. Now we shall show that $f(\Omega) = \mathbb{D} = \{z \in \mathbb{C} : |z| < 1\}$. Suppose to the contrary, there exists a w_0 with $|w_0| < 1$ such that $f(z) \neq w_0$ for all $z \in \Omega$. Since Ω is simply connected, it is possible to define a single-valued analytic branch of

$$F(z) = \sqrt{\frac{f(z) - w_0}{1 - \overline{w_0}f(z)}}.$$

(If Ω is simply connected then if $\phi(z) \neq 0$ in Ω we can define $\log \phi(z)$ by integration of $\frac{\phi'(z)}{\phi(z)}$ and $\sqrt{\phi(z)} = \exp\left(\frac{1}{2}\log\phi(z)\right)$, see Corollary 10.5 and Corollary 10.4).

The function F is univalent and $|F(z)| \leq 1$ for all $z \in \Omega$. Let

$$G(z) = \frac{|F'(z_0)|}{F'(z_0)} \cdot \frac{F(z) - F(z_0)}{1 - \overline{F(z_0)}F(z)}.$$

Now $G(z_0) = 0$, and $G'(z_0) = \frac{|F'(z_0)|}{1-|F(z_0)|^2} = \frac{1+|w_0|}{2\sqrt{|w_0|}}B > B > 0$.

This is a contradiction, and we conclude that $f(\Omega) = \mathbb{D}$. This completes the proof of the theorem. □

Suppose G is a simply connected domain, other than $\mathbb{C}$, and suppose there exists a univalent holomorphic function f mapping G onto the unit disk. Such maps are commonly referred to as a Riemann map of G onto the unit disk.

Example 15.3. *Let f and g be univalent holomorphic functions in the open unit disk $\mathbb{D}$ such that $f(\mathbb{D}) = g(\mathbb{D}), f(0) = g(0)$ and $\arg f'(0) = \arg g'(0)$. Then $f = g$.*

Solution: Let $\Omega = f(\mathbb{D}) = g(\mathbb{D})$ and $h = g^{-1} \circ f$. The map g^{-1} and f are conformal. Therefore, the map h is conformal. It is not difficult to verify that $h(0) = 0$ and $\arg f'(0) = \arg g'(0) + \arg h'(0)$. Thus $\arg h'(0) = 0$. The map h is a one-one, analytic map from $\mathbb{D}$ onto itself. Thus by Theorem 13.18, $h(z) = e^{i\theta} \frac{z-a}{1-\bar{a}z}$ for some $\theta \in \mathbb{R}, a \in \mathbb{D}$. Now $h(0) = 0$ implies $-ae^{i\theta} = 0$, hence $a = 0$. Therefore it follows that $h(z) = e^{i\theta}z$. Thus $h'(z) = e^{i\theta}$. Hence $h'(0) = e^{i\theta}$ and $\arg h'(0) = \theta$. But we have already shown that $\arg h'(0) = 0$. Hence $\theta = 0$. Thus $h(z) = z$ for all $z \in \mathbb{D}$. Therefore $f(z) = g(z)$ for all $z \in \mathbb{D}$.

Example 15.4. *Let G be a simply connected domain in $\mathbb{C}$, different from $\mathbb{C}$, let z_0 be a point of G, and let θ_0 be a real number. Then there is a univalent holomorphic function f mapping the open unit disk $\mathbb{D}$ onto G such that $f(0) = z_0$ and $\arg f'(0) = \theta_0$.*

Solution: Since G is simply connected and $G \neq \mathbb{C}$, by the Riemann mapping theorem there exist $g : G \to \mathbb{D}$ such that $g(z_0) = 0, g'(z_0) > 0$ (hence $\arg g'(z_0) = 0$). Further, g is one-one, onto and analytic. Let S be a map from $\mathbb{D}$ into $\mathbb{D}$ such that $S(z) = e^{i\theta_0}z$. Note that S is one-one, onto and analytic. Let $f(z) = (g^{-1} \circ S)(z)$. Then $f(0) = z_0$ and $\arg S'(0) = \arg g'(z_0) + \arg f'(0) = \arg f'(0)$. Now $\arg S'(0) = \theta_0$ as $S'(z) = e^{i\theta_0}$. Thus $\arg f'(0) = \theta_0$.

Example 15.5. *Let f be an univalent holomorphic function from the region Ω onto $\mathbb{D}$. Suppose $f(z_0) = 0$ where z_0 is a point in Ω. Let g be a holomorphic function from Ω into $\mathbb{D}$ such that $g(z_0) = 0$. Then $|g'(z_0)| \leq |f'(z_0)|$ with equality if and only if $g = \lambda f$ where λ is a unimodular constant.*

Solution: Notice that $g \circ f^{-1}$ is an analytic function from the open unit disk $\mathbb{D}$ onto itself and $(g \circ f^{-1})(0) = g(f^{-1}(0)) = g(z_0) = 0$. By Schwarz's lemma, $|((g \circ f^{-1})'(0))| \leq 1$. Let $h = g \circ f^{-1}$. Then $|h'(0)| \leq 1$ and $h \circ f = g$. Hence $h'(f(z_0))f'(z_0) = g'(z_0)$. Thus $h'(0) = \frac{g'(z_0)}{f'(z_0)}$ as $f(z_0) = 0$. The result follows since $|h'(0)| \leq 1$. Suppose $|g'(z_0)| = |f'(z_0)|$. Then $|h'(0)| = 1$. Now by Schwarz's lemma $|h'(0)| = 1$ implies $h(z) = \lambda z$ for some constant λ of absolute value 1. That is, $g(z) = \lambda f(z)$.

Thus Riemann maps from the region G onto the open unit disk $\mathbb{D}$ vanishing at z_0 maximize the absolute value of the derivative at z_0 among the family of all holomorphic maps of G into the unit disk $\mathbb{D}$ vanishing at z_0.

The proof of Riemann mapping theorem given here is undeniably elegant but it is highly nonconstructive. It does not address the question of how to find usable approximations to Riemann maps in concrete situations. That question is an active topic in numerical analysis.

15.3 Schwartz-Christoffel integral

In the complex plane let there be given a polygon (see Fig.15.1) with n vertices at the points $A_1, A_2, \cdots, A_n$ and with interior angles $\alpha_1\pi, \alpha_2\pi, \cdots, \alpha_n\pi$ at these vertices respectively. Then $\sum_{i=1}^{n} \alpha_i = n-2, n > 2$. We shall construct a conformal map from the upper half plane onto the interior of such a polygon. This can be done using Schwartz-Christoffel integral. Consider a function f defined on the upper half plane as

$$w = f(z) = C\int_{z_0}^{z} (\zeta - a_1)^{\alpha_1 - 1} \cdots (\zeta - a_n)^{\alpha_n - 1}\, d\zeta + C_1. \tag{15.3}$$

Here z_0, C, C_1 are given complex constants; $a_1, \cdots, a_n$ are real numbers such that $a_1 < a_2 < \cdots < a_n$. Further, $\alpha_1, \cdots, \alpha_n$ are positive constants, $\sum_{i=1}^{n} \alpha_i = n-2$ and $0 < \alpha_i < 2$. In the integrand of (15.3) we choose those branches of the functions $(\zeta - a_i)^{\alpha_i - 1}$ which are a direct analytic continuation (see Chapter 16) into the upper half plane of the real functions $(x - a_i)^{\alpha_i - 1}$ of the real variable $x > a_i$. In that case the function $f(z)$ is a single-valued analytic function in the upper half plane $\text{Im} z > 0$. The points a_i lying on the real axis are singularities of this function. The integral in (15.3) is the Schwartz-Christoffel integral.

For an appropriate choice of points a_i, the function $f(z)$ defines a conformal map from the upper half plane onto the region inside some polygon having n-vertices. Suppose all the numbers a_i are bounded. Since $0 < \alpha_i < 2$, the integral (15.3) remains bounded for all z in the neighborhood of the singularities a_i. When $z \to \infty$, write the integrand function as follows:

$$\begin{aligned}\Psi(\zeta) &= \zeta^{\alpha_1+\alpha_2+\cdots+\alpha_n - n}\left(1 - \frac{a_1}{\zeta}\right)^{\alpha_1 - 1} \cdots \left(1 - \frac{a_n}{\zeta}\right)^{\alpha_n - 1} \\ &= \frac{1}{\zeta^2}\left(1 - \frac{a_1}{\zeta}\right)^{\alpha_1 - 1} \cdots \left(1 - \frac{a_n}{\zeta}\right)^{\alpha_n - 1}.\end{aligned}$$

This is possible as $\sum_{i=1}^{n} \alpha_i = n-2$. From this expression it follows that the integral (15.3) is convergent as $z \to \infty$. Thus the function f maps the upper half plane onto some bounded region in $\mathbb{C}$. Now we shall investigate what is the image of $\mathbb{R}$ under f. Let $\mathcal{H} = \{z \in \mathbb{C} : \text{Im} z \geq 0\}$. Now

$$f'(z) = C(z - a_1)^{\alpha_1 - 1} \cdots (z - a_n)^{\alpha_n - 1} \neq 0$$

for all $z \in \mathcal{H} \setminus \{a_1, \cdots, a_n\}$ and $f(a_i)$ is either equal to 0 or ∞.

As z varies on each of the intervals $a_k < x < a_{k+1}$ $(k = 1, \cdots, n-1)$ of the real axis, the argument of the derivative does not change. Indeed on the given intervals of the real axis, the argument of the functions $(z - a_i)^{\alpha_i - 1}$ take the values

$$\arg(x - a_i)^{\alpha_i - 1} = \begin{cases} \pi(\alpha_i - 1), & \text{if } x < a_i \\ 0 & \text{if } x > a_i \end{cases} \tag{15.4}$$

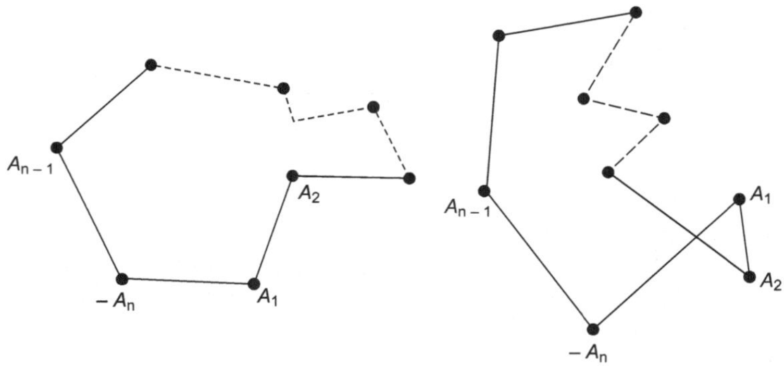

Fig. 15.1: Closed polygons $A_1A_2\cdots A_n$

(with the indicated choice of branches). Thus the segments $a_k < x < a_{k+1}$ of the real axis are mapped by the function $f(z)$ onto rectilinear segments; the points a_k of the real axis are mapped by the function $f(z)$ into points A_k. Therefore the function $f(z)$ maps the segment $[a_k, a_{k+1}]$ of the real axis onto the straight line segments A_kA_{k+1}. Thus, the function $f(z)$ is continuous and single-valued in $\mathbb{R}$ and maps $\mathbb{R}$ onto some closed polygon $A_1A_2\cdots A_n$, the elements of which are the straight-line segments A_kA_{k+1}. Note that, the polygon $A_1A_2\cdots A_n$ can have points of self intersection (as shown in Fig. 15.1). Now let us determine the size of the angles between adjacent segments of the polygon (see Fig. 15.2). Consider the variation of the arguments of the derivative $f'(z)$ as z passes through the point a_i. From (15.4) it follows that as the point z moves along the real axis in the positive direction, the argument of the derivative changes its value by $-\pi(\alpha_i - 1)$.

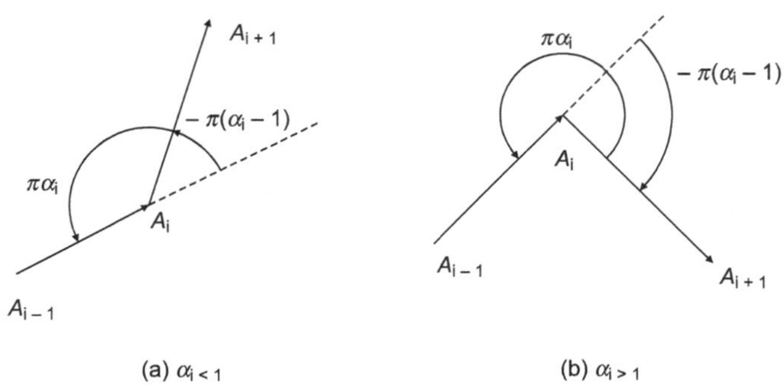

Fig. 15.2: The size of the angles between adjacent segments of the polygon

That is, the angle between the directions of the vectors $\overrightarrow{A_{i-1}A_i}$ and $\overrightarrow{A_iA_{i+1}}$ is equal to $-\pi(\alpha_i - 1)$. For $\alpha_i < 1$, the transition from the direction of the vectors $\overrightarrow{A_{i-1}A_i}$ to the direction of the vector $\overrightarrow{A_iA_{i+1}}$ occurs in the positive sense and for $\alpha_i > 1$ in the negative sense. In both cases, the size of the angle in the positive direction from the direction of the vector $\overrightarrow{A_iA_{i+1}}$ to the direction of the vector $\overrightarrow{A_iA_{i-1}}$ is $\pi\alpha_i$.

Thus if the polygonal line $A_1A_2\cdots A_n$, onto which the function $f(z)$ maps the real axis of the z-plane, does not have points of self-intersection and the direction of traversal is preserved, then the function $f(z)$ maps the upper half plane $\mathrm{Im}z > 0$ conformally onto the interior of the polygon having n vertices bounded by the polygonal line $A_1A_2\cdots A_n$.

If an arbitrary polygon having n vertices is given in the w-plane (the position of its vertices $A_1, A_2, \cdots, A_n$ and the angles at these vertices are known), then it is always possible to specify the values of the constants C, C_1 and the points $a_1, \cdots, a_n$ of the real axis so that a properly constructed function $f(z)$ maps the upper half plane $\mathrm{Im}z > 0$ conformally onto the interior of the given polygon. We will not prove this here but instead we will confine ourselves to a number of remarks and examples.

Remark 15.2. The formula (15.3) involves a number of constants. However, when constructing a conformal map from the upper half plane onto a given polygon $A_1, \cdots, A_n$ of the w-plane, it is only possible to specify three points a_i, a_j, a_k of the real axis that go into any three chosen vertices of the polygon A_i, A_j, A_k. The remaining constants are determined uniquely. In fact, the function $f(z)$ is connected with the function

$$\widehat{f}(z) = \int_{z_0}^{z} (\zeta - a_1)^{\alpha_1 - 1} \cdots (\zeta - a_n)^{\alpha_n - 1}\, d\zeta$$

by a linear fractional transformation (that is, a dilation followed by a translation). Hence, if f maps the upper half plane $\mathrm{Im}z > 0$ onto a given polygon of the w-plane, then the function $\widehat{f}(z)$ maps this half plane onto a polygon that is similar to the given one. For the given values of α_i, in order that an $n-$segment closed polygonal line into which the real axis is mapped by the function $\widehat{f}(z)$ be a polygon similar to the given one, it is sufficient that $n-2$ segments of this polygonal line be proportional to the corresponding sides of the polygon. The two extreme segments are fully defined by specifying their directions. We thus have $n-3$ equations in n constants a_i. If three of these constants are arbitrarily specified, the remaining corresponding equations are defined uniquely.

Remark 15.3. We assumed that all the numbers α_i in formula (15.3) are positive. Then the integral (15.3) converges for all values of z such that $\mathrm{Im}z \geq 0$. If some number α_k is negative, then as $z \to a_k$ the integral (15.3) diverges. This means that the corresponding vertex A_k of the polygon $A_1\cdots A_n$ lies at the point at infinity $w = \infty$. We then take the size of the angle at the vertex A_k to be equal to the size of the angle (with the minus sign) between the prolongation of the segments A_kA_{k-1} and A_kA_{k+1} at the finite point of their intersection. Now the angle at the vertex A_k is equal to $\alpha_k\pi(\alpha_k < 0)$ and by virtue of the condition (15.3) the sum of the interior angles of the resulting polygon with vertex A_k at the point at infinity is equal to $(n-2)\pi$. This also holds true if several numbers α_k are negative.

Remark 15.4. When we investigated formula (15.3) we assumed that all the points a_i are finite. It is easy to get rid of this condition. Let $t \in \mathbb{C}$ be such that

$$z = a_n - \frac{1}{t} \tag{15.5}$$

Then the point $z = a_n$ will pass into the point $t = \infty$. This transformation means that in a mapping of the upper half plane $\operatorname{Im} t > 0$ onto the interior of the polygon A_1, A_2, ..., A_n of the $w-$plane, the point at infinity $t = \infty$ is mapped into the vertex A_n. In the complex t-plane the function (15.3) has the form

$$w = c\int_{t_0}^{t}\left(a_n - a_1 - \frac{1}{\tau}\right)^{\alpha_1 - 1}\cdots\left(a_n - a_{n-1} - \frac{1}{\tau}\right)^{\alpha_{n-1}-1}\left(-\frac{1}{\tau}\right)^{\alpha_n - 1}\frac{d\tau}{\tau^2} + C_1$$

$$= A\int_{t_0}^{t}(\tau - a_1')^{\alpha_1 - 1}\cdots\left(\tau - a_{n-1}'\right)^{\alpha_{n-1}-1}d\tau + C_1 \tag{15.6}$$

where $a_i' = \frac{1}{a_n - a_i}$, $t_0 = \frac{1}{a_n - z_0}$ and

$$A = C\,(a_n - a_1)^{\alpha_1 - 1}\cdots(a_n - a_{n-1})^{\alpha_{n-1}-1}\,(-1)^{\alpha_n - 1}.$$

The relation (15.6) means that when in a conformal mapping of the upper half plane onto the interior of the polygon A_1, A_2, $\cdots$, A_n the point at infinity $t = \infty$ passes into one of the vertices A_n, this mapping is effected by the Schwartz-Christoffel integral (15.6), in the integrand function of which the factor corresponding to the given vertex A_n has been omitted.

Example 15.6. *Find a function that conformally maps the upper half plane* $Imz > 0$ *onto the sector* $0 < \arg w < \alpha\pi$, $0 < \alpha < 2$.

Solution: Since the given sector is a polygon with vertices A_1 $(w = 0)$ and A_2 $(w = \infty)$, the Schwartz-Christoffel integral may be employed to solve the problem. We establish the following correspondence of points of the real axis (in the z-plane) to the vertices of the polygon (in the w-plane):

$$a_1(z = 0) \longrightarrow A_1(w = 0)$$

$$a_2(z = \infty) \longrightarrow A_2(w = \infty). \tag{15.7}$$

Then by (15.6) the mapping function takes the form

$$w = f(z) = C\int_{z_0}^{z}\zeta^{\alpha - 1}d\zeta + C_1.$$

Putting $z_0 = 0$, it follows from (15.7) that $C_1 = 0$. Hence

$$w = C\int_0^z \zeta^{\alpha-1}d\zeta = \frac{C}{\alpha}z^\alpha.$$

Thus the function is defined to within the constant factor defining the similarity transformation. This arbitrariness is due to the fact that the condition (15.7) contain the demand that only two boundary points correspond, but, as we have seen, the function which effects a conformal mapping is defined uniquely by specification of a correspondence of three boundary points. For example, in addition to (15.7) if we impose the condition that $z = 1$ will correspond to $w = 1$ then we get $C = \alpha$. So finally, the function $w = z^\alpha$ defines a conformal mapping of the upper half plane Im$z > 0$ onto the given sector of the w-plane. By virtue of the earlier indicated choice of branches in the integrand function of the Schwartz-Christoffel integral (15.3), we shall consider that branch of the multiple-valued function z^α which is a direct analytic continuation of the real function x^α of the real positive variable x.

Example 15.7. *Find a function which conformally maps the upper half plane Imz > 0 onto the rectangle* $A_1A_2A_3A_4$.

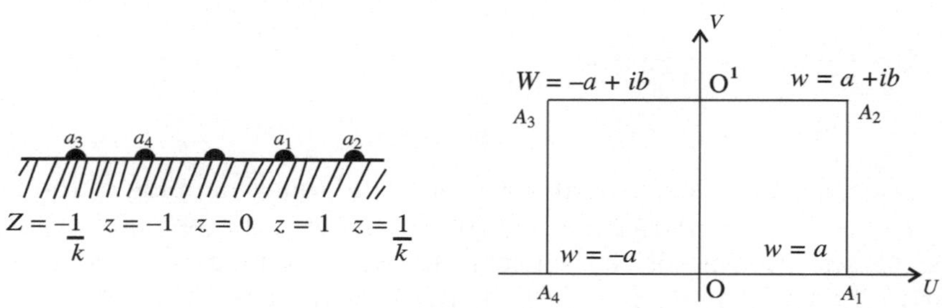

Fig. 15.3: Conformal map from the upper half plane onto rectangle $A_1A_2A_3A_4$

Solution: Let the vertices of the rectangle in the $w-$plane be located at the points $A_1(w = a)$, $A_2(w = a + ib)$, $A_3(w = -a + ib)$, A_4 $(w = -a)$. Note that if we can find a conformal map $f_1(z)$ from the first quadrant of the $z-$plane (Re$z > 0$, Im$z > 0$) onto the right half OA_1A_2O' of the rectangle, (that is, if the right side of the imaginary axis of the z-plane goes into the right side of the segment OO') then by symmetry principle the function which is the analytic continuation of $f_1(z)$ into the domain (Re$z < 0$, Im$z > 0$) maps conformally the given domain onto the left side of the original rectangle. Here, the symmetric points of the real axis (in the z-plane) go into the vertices A_1 and A_4. The same occurs relative to the vertices A_2 and A_3. Therefore, we can establish the following correspondence of points:

$$a_1(z = 1) \longrightarrow A_1(w = a)$$

$$a_4(z = -1) \longrightarrow A_4(w = -a). \tag{15.8}$$

Besides, there must also be the correspondence

$$z = 0 \longrightarrow w = 0. \tag{15.9}$$

We thus establish a correspondence between three boundary points. It is therefore impossible to specify arbitrarily on the real axis (in the z-plane) the point a_2 that goes into the vertex A_2 of the rectangle. Let us suppose that the point a_2 of the real axis (in the z-plane) with coordinate $\frac{1}{k}$ (the value of which will be defined later on) goes into the vertex A_2. Clearly, $k > 0$ and $\frac{1}{k} > 1$. Thus, the function which defines a conformal mapping of the upper half plane onto a given rectangle may be represented in the form

$$\begin{aligned} w &= f(z) \\ &= C' \int_{z_0}^{z} (\zeta - 1)^{\frac{1}{2}-1} \left(\zeta - \tfrac{1}{k}\right)^{\frac{1}{2}-1} \left(\zeta + \tfrac{1}{k}\right)^{\frac{1}{2}-1} (\zeta + 1)^{\frac{1}{2}-1} d\zeta + C_1 \\ &= C \int_{z_0}^{z} \frac{d\zeta}{\sqrt{(1-\zeta^2)(1-k^2\zeta^2)}} + C_1. \end{aligned}$$

Putting $z_0 = 0$ and using (15.9), we get $C_1 = 0$. Then

$$w = C \int_0^z \frac{d\zeta}{\sqrt{(1-\zeta^2)(1-k^2\zeta^2)}}. \tag{15.10}$$

It remains to determine the constants C and k from the correspondence of the points a_1 and a_2 of the real axis (in the z-plane) to the vertices A_1 and A_2. Observe that the integral (15.10) is not expressible in terms of elementary functions and it is called a elliptic integral of the first kind. Since the point $z = 1$ correspond to the point $w = a$, we have

$$a = C \int_0^1 \frac{d\zeta}{\sqrt{(1-\zeta^2)(1-k^2\zeta^2)}}.$$

The integral

$$K(k) = \int_0^1 \frac{d\zeta}{\sqrt{(1-\zeta^2)(1-k^2\zeta^2)}}$$

is called the complete elliptic integral of the first kind. The correspondence of points

$$a_2(z = \frac{1}{k}) \longleftrightarrow A_2(w = a + ib)$$

permits us to write

$$a + ib = C \left[\int_0^1 \frac{d\zeta}{\sqrt{(1-\zeta^2)(1-k^2\zeta^2)}} + \int_1^{\frac{1}{k}} \frac{d\zeta}{\sqrt{(1-\zeta^2)(1-k^2\zeta^2)}} \right].$$

Thus we get

$$b = C \int_1^{\frac{1}{k}} \frac{d\zeta}{\sqrt{(\zeta^2-1)(1-k^2\zeta^2)}}.$$

15.4 Exercises

1. Let $f_n(z) = e^{nz}, n = 1, 2, \cdots$. Show that the sequence $\{f_n\}$ is not normal on $\mathbb{C}$.

2. If $\mathcal{F}$ is locally bounded then show that $\mathcal{F}'$ is normal.

3. Show that the functions z^n, n a nonnegative integer, form a normal family in $|z| < 1$, also in $|z| > 1$, but not in any region that contains a point on the unit circle.

4. Prove that in any region Ω the family of analytic functions with positive real part is normal.

5. If the family $\mathcal{F}$ of analytic (or meromorphic) functions is not normal in Ω, show that there exists a point z_0 such that $\mathcal{F}$ is not normal in any neighborhood of z_0.

6. Let $(G_n)_{n=1}^{\infty}$ be an increasing sequence of simply connected domains whose union, G, is not the whole plane. Let z_0 be a point of G_1. For each n, let f_n be the Riemann map of the open unit disk $\mathbb{D}$ onto G_n satisfying $f_n(0) = z_0$ and $\arg f_n'(0) = 0$. Prove that the sequence $(f_n)_{n=1}^{\infty}$ converges locally uniformly in $\mathbb{D}$ to the Riemann map f of $\mathbb{D}$ onto G satisfying $f(0) = z_0$ and $\arg f'(0) = 0$.

7. Show that if f is analytic on an open set containing the disk $\overline{B(a,R)}$ then
$$|f(a)|^2 \le \frac{1}{\pi R^2} \int_0^{2\pi} \int_0^R |f(a + re^{i\theta})|^2 r\,dr\,d\theta.$$

8. Let G be a region and let M be a fixed positive constant. Let $\mathcal{F}$ be the family of all functions f in $H(G)$ such that $\int\int_G |f(z)|^2 dxdy \le M$. Show that $\mathcal{F}$ is normal.

9. Let $\mathbb{D}$ be the open unit disk in the complex plane $\mathbb{C}$ and for $0 < r < 1$ let $\gamma_r(t) = e^{2\pi irt}, 0 \le t \le 1$. Show that a sequence $\{f_n\}$ in $H(\mathbb{D})$ converges to f if and only if $\int_{\gamma_r} |f(z) - f_n(z)|\,|dz| \to 0$ as $n \to \infty$ for each $r, 0 < r < 1$.

Chapter 16

The Schwarz Reflection Principle and Meromorphic Continuation

16.1 The Schwarz reflection principle

Let Ω be a region in $\mathbb{C}$ and let γ be a circle in $\mathbb{C}$. Denote by Ω^* the set of points symmetric to points of Ω with respect to γ. That is, $\Omega^* = \{z : z^* \in \Omega\}$. If $\gamma = \mathbb{R}$, we associate with an analytic function f on Ω, a function f^* on Ω^*, where f^* is defined on Ω^* by $f^*(z) = \overline{f(\bar{z})}$.

If $z_0 \in \Omega^*$, then about the point $\bar{z}_0 \in \Omega$, f has a Taylor series expansion

$$f(\bar{z}) = \sum_{n=0}^{\infty} a_n(\bar{z} - \bar{z}_0)^n$$

and hence

$$f^*(z) = \overline{f(\bar{z})} = \sum_{n=0}^{\infty} \bar{a}_n(z - z_0)^n.$$

Thus we obtain an expansion of f^* about z_0. Therefore, f^* is analytic on Ω^*.

First version of Schwarz reflection principle

If f is analytic in a region containing a rectangle γ and its inside then by Cauchy integral theorem $\int_\gamma f(z)dz = 0$. Further if f is analytic inside a rectangle and continuous on γ then also it can be deduced from Cauchy integral theorem that $\int_\gamma f(z)dz = 0$. This is the content of the next lemma.

Lemma 16.1. *Let f be analytic inside a rectangle γ and continuous on γ. Then $\int_\gamma f(z)dz = 0$.*

Proof. Let z_0 be the center of γ with $z(t) = z_0 + \zeta(t)$, a parametrization of γ. If $0 < \alpha < 1$, $z_\alpha(t) = z_0 + \alpha\zeta(t)$ is a parametrization of a closed curve γ_α contained inside γ. By the Cauchy integral theorem $\int_{\gamma_\alpha} f(\zeta)d\zeta = 0$.

Let $L = \int |\zeta'(t)|dt$ be the length of γ, and set $M = \max_{z\in\gamma} |f(z)|$. Because f is continuous on the compact set γ, hence $M < \infty$, and if z lies inside γ, $|f(z)| \leq M$ by the maximum principle. Since f is continuous inside and on γ, by Theorem 2.51, the function f is uniformly continuous inside and on γ. Hence given $\epsilon > 0$, let $\delta > 0$ be chosen so that $\delta < \frac{\epsilon(1+L)}{ML}$. Now if α is sufficiently close to 1 (i.e., if $1 - \alpha < \delta$), $|f(z_0 + \zeta(t)) - f(z_0 + \alpha\zeta(t))| < \epsilon$. It follows therefore that

$$\begin{aligned}
\left|\int_\gamma f(z)dz\right| &= \left|\int_\gamma f(z)dz - \int_{\gamma_\alpha} f(z)dz\right| \\
&= \left|\int f(z_0+\zeta(t))\zeta'(t)dt - \int f(z_0+\alpha\zeta(t))\alpha\zeta'(t)dt\right| \\
&= \left|\int [f(z_0+\zeta(t)) - f(z_0+\alpha\zeta(t))]\,\zeta'(t)dt \right.\\
&\quad \left. + \int [f(z_0+\alpha\zeta(t))(1-\alpha)\zeta'(t)]\,dt\right| \\
&\leq \int |f(z_0+\zeta(t)) - f(z_0+\alpha\zeta(t))|\ |\zeta'(t)|dt \\
&\quad + \int |f(z_0+\alpha\zeta(t))|\,(1-\alpha)\ |\zeta'(t)|dt \\
&< \epsilon L + ML(1-\alpha) < \epsilon L + ML\frac{\epsilon(1+L)}{ML} \\
&= \epsilon(1+2L).
\end{aligned}$$

Letting $\epsilon \to 0$, the result is proved. □

From Morera's theorem (Theorem 7.14) we know that if f is continuous on an open disk U, and if $\int_\gamma f(z)dz = 0$ for each closed piecewise differentiable arc γ in U, then f is analytic.

Remark 16.1. Recall that Morera's theorem is proved by fixing a point $z_0 \in U$ and setting up the integral $F(z) = \int_{\gamma_z} f(\zeta)d\zeta$,where for each z, γ_z is a piecewise differentiable arc in U from z_0 to z. Making use of different paths from z_0 to z, one proves $F'(z) = f(z)$. Thus F is analytic, and f, the derivative of an analytic function, is also analytic.

Because f is assumed to be continuous, it is only necessary, in order to prove F is analytic, to demonstrate that $\frac{\partial F}{\partial x} = f$ and $\frac{1}{i}\frac{\partial F}{\partial y} = f$. To compute $\frac{\partial F}{\partial x}$ and $\frac{1}{i}\frac{\partial F}{\partial y}$ at a point z, one can take for γ_z a rectangular arc from z_0 to z. When we compute $\frac{\partial F}{\partial x}$, the last segment of the arc should be horizontal; for $\frac{1}{i}\frac{\partial F}{\partial y}$, the last segment of the arc should be vertical. Thus Morera's theorem needs only to assume $\int_\gamma f(\zeta)d\zeta = 0$ for rectangles having z_0 for one vertex and having sides parallel to the coordinate axes.

Lemma 16.2. *Let f be continuous on a disk U which intersects the real axis in an arc γ. If f is analytic for those $z \in U$ such that $Imz > 0$ or $Imz < 0$, then f is analytic on U.*

Proof. Fix $z_0 \in \gamma$. If Γ is a rectangle in U with vertex z_0 and sides parallel to the axes, it follows from Lemma 16.1 that $\int_\Gamma f(\zeta)d\zeta = 0$. Thus by Morera's theorem and the remark preceding this lemma it follows that, f is analytic. □

We now prove a theorem which we call as first version of Schwarz reflection principle:

We assume Ω is a region in the upper(or lower) half-plane which is bounded in part by an arc γ of the real axis. Let $\widetilde{\Omega}$ be the interior of the set $\Omega \cup \gamma \cup \Omega^*$. (Here certain points of γ may not be interior to $\Omega \cup \gamma \cup \Omega^*$. In fact, $\widetilde{\Omega}$ may not be connected.)

Theorem 16.1. *If f is analytic on Ω, continuous on $\Omega \cup \gamma$, and real on γ, then the function $\widetilde{f}(z)$ defined on $\widetilde{\Omega}$ by*

$$\widetilde{f}(z) = \begin{cases} f(z) & if \;\; z \in \Omega \cup \gamma; \\ f^*(z) & if \;\; z \in \Omega^* \end{cases}$$

is analytic on $\widetilde{\Omega}$. (Note $f^*(z) = \overline{f(\overline{z})}$ is defined at the beginning of the present section.)

Proof. From our earlier remarks we know that $\widetilde{f}$ is analytic on $\Omega \cup \Omega^*$. It will follow from Lemma 16.2 that $\widetilde{f}$ is analytic on $\widetilde{\Omega}$ if we show $\widetilde{f}$ is continuous at points $z \in \gamma$. To this end let z be a point of γ which belongs to $\widetilde{\Omega}$. Given $\epsilon > 0$ there is a $\delta > 0$ such that if $\zeta \in \Omega$ and $|\zeta - z| < \delta$, then $|f(z) - f(\zeta)| < \epsilon$ (since f is continuous on γ).

If $\zeta \in \Omega^*$ and $|\zeta - z| < \delta$, then $|\bar{\zeta} - z| \underline{< \delta}$ and $|f(z) - f(\bar{\zeta})| < \epsilon$. Because $f(z)$ is real on γ, $|f(z) - f(\bar{\zeta})| = |f(z) - \overline{f(\bar{\zeta})}|$ and therefore, $|\widetilde{f}(z) - \tilde{f}(\zeta)| = |f(z) - \overline{f(\bar{\zeta})}| < \epsilon$.

Since ϵ is arbitrary, the function $\widetilde{f}$ is continuous, and the theorem is proved. □

We shall now prove Theorem 16.1 in a more general and natural setting.

Second version of Schwarz reflection principle

Let Ω be a region contained "on one side of" a disk U_1; that is, Ω lies either inside or outside U_1. If γ_1 denotes the boundary circle of U_1, there will be given an arc γ of γ_1 which forms part of the boundary of Ω. If $*$ denotes reflection in γ_1 and $\Omega^* = \{z^* : z \in \Omega\}$, then $\widetilde{\Omega}$ will be the interior of the set $\Omega \cup \gamma \cup \Omega^*$. It is possible that $\widetilde{\Omega}$ contains the point ∞. Suppose now that γ_2 is another circle(or straight line) and that f is analytic on Ω, continuous on $\Omega \cup \gamma$, with values $f(z) \in \gamma_2$ if $z \in \gamma$. On this setting we state the second version of Schwarz reflection principle.

Theorem 16.2. *On $\widetilde{\Omega}$ define*

$$\widetilde{f}(z) = \begin{cases} f(z) & \text{if } z \in \Omega \cup \gamma; \\ f^*(z^*) & \text{if } z \in \Omega^*, \end{cases}$$

where f^* denotes reflection in γ_2. Then $\widetilde{f}(z)$ is meromorphic in $\widetilde{\Omega}$.

Proof. Let S and T be linear fractional transformations which map the real line onto γ_1 and γ_2, respectively. Then

$$h(z) = T^{-1}f(Sz),$$

which is defined on $\Omega_0 = S^{-1}\Omega$, satisfies the hypothesis of Theorem 16.1 if Ω, γ there be replaced by Ω_0, $\gamma_0 = S^{-1}\gamma$. Define $\widetilde{h}$ on $\widetilde{\Omega}_0$ as follows:

$$\widetilde{h}(z) = \begin{cases} h(z) & \text{if } z \in \Omega_0 \cup \gamma_0; \\ h^*(z) & \text{if } z \in \Omega_0^*. \end{cases}$$

We claim

$$\widetilde{f}(w) = T\widetilde{h}(S^{-1}w) \tag{16.1}$$

from which the theorem will follow immediately. If $w \in \Omega \cup \gamma$, (16.1) is immediate. If $w \in \Omega^*$, then $T\widetilde{h}(S^{-1}w) = \overline{Th(\overline{S^{-1}w})} = (T\widetilde{h}(S^{-1}w^*))^* = \widetilde{f}(w^*)^* = \widetilde{f}(w)$, where the second equality follows from symmetry. □

Remark 16.2. If for some $z \in \Omega, f(z)$ is the center of γ_2, then $\widetilde{f}(z^*) = f^*(z) = \infty$, and at z^*, $\widetilde{f}$ will have a pole.

Conformal mapping of an annulus

It is a consequence of Riemann mapping theorem that any two simply connected proper subregions of the plane are conformally equivalent, since each of them is conformally equivalent to the unit disk. This is a very special property of simply connected regions. One may ask whether it extends to the next simple situation, i.e., whether any two annuli are conformally equivalent. The answer is negative.

The following is an application of Schwarz reflection principle which shows the nonexistence of analytic mappings from one region onto another in a certain situation. In fact we shall show that the annuli $\text{ann}(a_1, r_1, R_1)$ and $\text{ann}(a_2, r_2, R_2)$ are conformally equivalent if and only if $\frac{r_1}{R_1} = \frac{r_2}{R_2}$.

We shall first set the background for proving the result. Suppose V is an annulus in the complex plane with inner boundary the unit circle $|z| = 1$ and outer boundary $|z| = R, R > 1$. That is $V = \text{ann}(0, 1, R)$. Let $V^{(1)}$ be the region

$V \cup \gamma \cup V^*$, where $\gamma = \{|z| = 1\}$ is the inner circle and V^* denotes reflection in γ. Having defined annuli $V^{(1)}, V^{(2)}, \cdots, V^{(n)}$ with inner circles $\gamma^{(1)}, \gamma^{(2)}, \cdots, \gamma^{(n)}$ and common outer circle $|z| = R$, let $V^{(n+1)} = V^{(n)} \cup \gamma^{(n)} \cup V^{(n)*}$, where reflection is in $\gamma^{(n)}$. If r_n is the radius of $\gamma^{(n)}$, then it follows from Theorem 13.23 that

$$r_{n+1} = \frac{r_n^2}{R},$$

and so $\lim_{n\to\infty} r_n = 0$.

Now let $V_1 = \text{ann}(0, 1, R_1)$ and $V_2 = \text{ann}(0, 1, R_2)$. Suppose there is a one-to-one holomorphic function f mapping V_1 onto V_2 with the inner circle mapping continuously onto itself and the outer circle $|z| = R_1$ mapping continuously onto $|z| = R_2$. If $V_1^{(n)}, V_2^{(n)}$ are as in the previous paragraph, the function f can be extended by Theorem 16.2 to a mapping of $V_1^{(1)}$ onto $V_2^{(1)}$. Now f is successively extended by Schwarz reflection from a mapping of $V_1^{(n)}$ onto $V_2^{(n)}$ to a mapping of $V_1^{(n+1)}$ onto $V_2^{(n+1)}$. Thus, f becomes defined in the region $0 < |z| < R_1$ mapping onto $0 < |w| < R_2$. By Riemann's removable singularity theorem (Theorem 8.1) f can be defined at 0. Of course, it must be that $f(0) = 0$, and so f becomes a mapping of $|z| < R_1$ onto $|z| < R_2$ which takes 0 into 0. By a similar construction f^{-1} is defined for $|z| < R_2$ with range $|z| < R_1$. since $f(0) = 0 = f^{-1}(0)$, we have from Schwarz's lemma (Lemma 13.3)

$$|f(z)| \leq \frac{R_2}{R_1}|z|$$

and

$$|f^{-1}(z)| \leq \frac{R_1}{R_2}|z|.$$

Taking $|z| = 1$, in which case $|f(z)| = 1 = |f^{-1}(z)|$, gives $\frac{R_1}{R_2} \geq 1$ and $\frac{R_2}{R_1} \geq 1$. Thus $R_1 = R_2$.

Remark 16.3. The mapping $Tz = \frac{R_1}{z}$ puts V_1 in one-to-one correspondence with itself interchanging the boundary circles. If the function $f(z)$ of the preceding paragraph interchanges the inner and outer boundaries of V_1 and V_2, a preliminary application of T reduces us to the situation just covered.

Theorem 16.3. *Let $V_1 = ann(a_1, r_1, R_1)$ and $V_2 = ann(a_2, r_2, R_2)$. Then there is a one-to-one analytic mapping of V_1 onto V_2 if and only if $\frac{r_1}{R_1} = \frac{r_2}{R_2}$.*

Proof. Let $T_1 z = \frac{z-a_1}{r_1}$ and $T_2 z = \frac{z-a_2}{r_2}$. Then T_1 and T_2 map V_1 and V_2 onto the annuli W_1 and W_2 respectively where $W_1 = \text{ann}(0, 1, \frac{R_1}{r_1})$ and $W_2 = \text{ann}(0, 1, \frac{R_2}{r_2})$. For V_1 and V_2 to be analytically equivalent it is necessary and sufficient that W_1 and W_2 be analytically equivalent, and the latter holds by the remarks preceding if, and only if, $\frac{R_1}{r_1} = \frac{R_2}{r_2}$. □

By using elementary techniques one can also prove that there exists no one-to-one conformal maps between certain annular regions.

Example 16.1. *There is no one-to-one conformal map of the punctured disk* $G = \{z \in \mathbb{C} : 0 < |z| < 1\}$ *onto the annulus* $A = \{z \in \mathbb{C} : 1 < |z| < 2\}$.

Solution: Suppose f is such a map. The function f is bounded, so the singularity at the origin is removable, $p = \lim_{z \to 0} f(z)$. Since f is continuous, p is in the closure of A.

Suppose that p is on the boundary of A. Then $f(G) = A \cup \{p\}$, which is not an open set, contradicting the open mapping theorem. Let $p \in A$ and $a \in G$ be such that $f(a) = p$. Take disjoint open neighborhoods U of 0 and V of a. By the open mapping theorem, $f(U)$ and $f(V)$ are open sets containing p. Then $f(U) \cap f(V)$ is a nonempty open set. Take $x \in f(U) \cap f(V), x \neq p$. Then $x = f(z)$ for some nonzero $z \in U$ and $x = f(w)$ for some $w \in V$. Then z and w are distinct elements of G with $f(z) = f(w)$, contradicting the injectivity of f.

Example 16.2. *Does there exist an analytic function mapping the annulus* $A = \{z \in \mathbb{C} : 1 \leq |z| \leq 4\}$ *onto the annulus* $B = \{z \in \mathbb{C} : 1 \leq |z| \leq 2\}$ *and taking* $C_1 \to C_1$, $C_4 \to C_2$ *where* C_r *is the circle of radius* r *?*

Solution: Suppose f is such a function. Let $g : A \to B$ be defined by $g(z) = \frac{f(z)^2}{z}$. Then, as on $C_1 \cup C_4$, the absolute value of g is 1, then g is a constant, c, say. Therefore, $f(z) = \sqrt{cz}$ which is not continuous on A. We conclude that no such function can exist.

16.2 Meromorphic and analytic continuation

In this section we shall discuss about the following two problems:

(i) If f is meromorphic (or analytic) on some region $G \subseteq \mathbb{C}_\infty$, can we extend f to a function which is meromorphic (or analytic) on some larger region $\Omega \supset G$?

(ii) If f is a many-valued function (such as $\sqrt{z}$ or $\log z$), can we represent f by a single-valued function on some suitable domain ?

To solve problem (i) we introduce the concepts of meromorphic and analytic continuation which actually leads to the construction of Riemann surfaces, the 'suitable domains' in problem (ii). We shall discuss in this section about meromorphic continuation along a path and prove the Monodromy theorem. We have discussed about problem (ii) earlier. There will be some repetitions here as we try to explain in detail about meromorphic and analytic continuation.

We define a function element to be a pair (G, f) where G is a region and $f : G \to \mathbb{C}_\infty$ is a single-valued meromorphic function on G. If f is analytic on the region $G \subseteq \mathbb{C}$ then we shall call (G, f) an analytic function element.

Example 16.3. *If f is any rational function then $(\mathbb{C}_\infty, f)$ is a function element.*

Example 16.4. *Let $\mathbb{D} = \{z \in \mathbb{C} : |z| < 1\}$ and $f(z) = \sum_{n=0}^{\infty} z^n$. Then $(\mathbb{D}, f)$ is an analytic function element.*

Notice that in both examples the domain G was chosen to be the largest on which f is meromorphic or analytic, but this need not always be the case-it is often convenient to restrict attention to some smaller region such as $|z| < \frac{1}{2}$ for $\sum_{n=0}^{\infty} z^n$ in Example 16.4 for instance.

Lemma 16.3. *Let (G, f) and (G, g) be function elements defined on the same region G. If $f \equiv g$ on some nonempty open subset U of G, then $f \equiv g$ on G.*

Proof. The function $h = f - g$ is meromorphic on G, and its poles form an isolated set $h^{-1}(\{\infty\})$ and since h is continuous, $h^{-1}(\{\infty\})$ is closed. This implies $G' = G \setminus h^{-1}(\{\infty\})$ is nonempty, open and path connected (hence connected). That is, G' is a region on which h is analytic. We have $h \equiv 0$ on the nonempty open subset $U' = U \cap G'$ of G'. By Theorem 7.16, we obtain that $h \equiv 0$ on G'. Each pole of h is a limit of points in G', so by continuity (since h is meromorphic) we have $h \equiv 0$ on G, and hence $f \equiv g$ on G. □

Corollary 16.1. *If (G_1, f_1) is a function element and G_2 is a region with $G_1 \cap G_2 \neq \emptyset$, then there is at most one meromorphic function f_2 on G_2 such that $f_1 \equiv f_2$ on $G_1 \cap G_2$.*

Proof. Suppose that f_2 and g_2 are two meromorphic functions on G_2 and are identically equal to f_1 on $U = G_1 \cap G_2$. The set U is open since G_1 and G_2 are open and $f_2 \equiv g_2$ on U. By Lemma 16.3, $f_2 \equiv g_2$ on G_2. □

When such a function f_2 exists, we call the function element (G_2, f_2) a direct meromorphic continuation of (G_1, f_1) or a direct analytic continuation if f_2 is analytic; in either case, we write

$$(G_1, f_1) \sim (G_2, f_2),$$

meaning that $G_1 \cap G_2 \neq \emptyset$ and $f_1 \equiv f_2$ on $G_1 \cap G_2$. In such a situation, we have a function element (G, f) where $G = G_1 \cup G_2$ and $f(z) = f_j(z)$ for $z \in G_j, j = 1, 2$. The function f is well defined as $f_1(z) = f_2(z)$ if $z \in G_1 \cap G_2$.

Example 16.5. *Let $G_1 = \mathbb{D} = \{z \in \mathbb{C} : |z| < 1\}, f_1(z) = \sum_{n=0}^{\infty} z^n$ and $f_2(z) = \frac{1}{1-z}$. The function f_2 is meromorphic on $\mathbb{C}_\infty = G_2$, and $f_1 \equiv f_2$ on $G_1 \cap G_2 = \mathbb{D}$. Hence $(G_1, f_1) \sim (G_2, f_2)$, giving a direct meromorphic continuation onto $\mathbb{C}_\infty$. By Corollary 16.1 this continuation is unique.*

Notice that the relation $\sim$ between function elements is reflexive, symmetric but not transitive. For example if $G_1 \cap G_2 \neq \emptyset \neq G_2 \cap G_3$, then it does not follow that $G_1 \cap G_3 \neq \emptyset$, and even if $G_1 \cap G_3 \neq \emptyset$, then it does not follow from $(G_1, f_1) \sim (G_2, f_2) \sim (G_3, f_3)$ that $f_1 \equiv f_3$ on $G_1 \cap G_3$. In Example

16.7 below we shall show that if f_1 is a single-valued analytic branch of the many-valued function $\log(z)$ on G_1, then we have direct analytic continuations $(G_1, f_1) \sim (G_2, f_2) \sim (G_3, f_3)$ but we find that $f_3 \equiv f_1 + 2\pi i$ on $G_1 \cap G_3$, so that (G_3, f_3) is not a direct analytic continuation of (G_1, f_1). Thus one function element (such as (G_1, f_1)) can sometimes give rise to two distinct functions (here f_1 and f_3) on the same region; this illustrates how repeated continuation of a single valued function element can produce a so-called many-valued function. Nevertheless, it is often possible to construct a sequence of direct meromorphic continuations

$$(G_1, f_1) \sim (G_2, f_2) \sim (G_3, f_3) \sim \cdots$$

in such a way that they give a single-valued meromorphic function on $\cup_n G_n$; we call this process meromorphic continuation (or analytic continuation if each f_n is analytic).

It may be worth noting here that "analytic relations" between holomorphic functions are preserved by analytic continuation. For example, if $f_1(z), f_2(z), f_3(z), \cdots$ are analytic continuations of $f_0(z)$, then $f_1'(z), f_2'(z), f_3'(z), \cdots$ are analytic continuations of $f_0'(z)$. Further, if $\phi(w)$ is a holomorphic function defined on a region Ω of the w-plane and if the ranges of $f_0(z)$ and its analytic continuations $f_1(z), f_2(z), f_3(z), \cdots$ are all included in Ω, then $\phi(f_1(z)), \phi(f_2(z)), \phi(f_3(z)), \cdots$ are analytic continuations of $\phi(f_0(z))$. Therefore, if $f_0(z)$ and $g_0(z)$ are holomorphic functions defined on the region G_0, if $f_1(z), f_2(z), \cdots, f_k(z), \cdots$ and $g_1(z), g_2(z), \cdots, g_k(z), \cdots$ are analytic continuations along the regions $G_1, G_2, \cdots, G_k, \cdots$ of $f_0(z)$ and $g_0(z)$ respectively, and if $g_0(z) = f_0'(z)$, then $g_k(z) = f_k'(z)$; if $g_0(z) = f_0''(z)$, then $g_k(z) = f_k''(z); \cdots$; and if $g_0(z) = \phi(f_0(z))$, then $g_k(z) = \phi(f_k(z))$.

Example 16.6. *Consider the improper integral*

$$f_1(z) = \int_0^\infty t^{z-1}e^{-t}dt = \lim_{a\to 0^+}\int_a^1 t^{z-1}e^{-t}dt + \lim_{b\to+\infty}\int_1^b t^{z-1}e^{-t}dt,$$

where $z \in \mathbb{C}$ and the path of integration is along the positive part of the real line.

This integral converges as $b \to +\infty$ for all $z \in \mathbb{C}$, and it converges as $a \to 0^+$ provided $\text{Re}z > 0$ (by comparison with $\int t^{x-1}dx$, for example, where $x = \text{Re}z$). Thus $f_1(z)$ is defined on the region $G_1 = \{z \in \mathbb{C} : \text{Re}z > 0\}$. One can show that $\int_0^\infty (\frac{\partial}{\partial z})(t^{z-1}e^{-t})dt$ is uniformly convergent with respect to z, so we may differentiate $f_1(z)$ with respect to z under the integral sign. This shows that f_1 is analytic on G_1, so (G_1, f_1) is an analytic function element. Integrating by parts, we have

$$f_1(z+1) = \int_0^\infty t^z e^{-t}dt = \left[-t^z e^{-t}\right]_0^\infty + z\int_0^\infty t^{z-1}e^{-t}dt = zf_1(z)$$

so that $f_1(z) = \frac{f_1(z+1)}{z}$ for all $z \in G_1$. Now the function $f_2(z) = \frac{f_1(z+1)}{z}$ is meromorphic on the larger region $G_2 = \{z \in \mathbb{C} : \text{Re}z > -1\}$, since $f_1(z+1)$ and z are analytic on G_2. By direct integration, $f_1(1) = 1$, so f_2 has a simple pole at $z = 0$. We have just seen that $f_2 \equiv f_1$ on the region $G_1 \subset G_2$, so f_2 is a direct meromorphic continuation of f_1 onto G_2. We now iterate this process. For each integer $n \geq 1$, let $G_n = \{z \in \mathbb{C} : \text{Re}z > 1 - n\}$, and suppose that for some $n \geq 1$ there is a function element (G_n, f_n) satisfying $f_n(z) = \frac{f_n(z+1)}{z}$for all $z \in G_n$. We have seen, this is so for $n = 1$. Then the function $f_{n+1}(z) = \frac{f_n(z+1)}{z}$ is meromorphic on G_{n+1} (since $f_n(z+1)$ and z are), and $f_{n+1} \equiv f_n$ on the region $G_n \subset G_{n+1}$, so f_{n+1} is a direct meromorphic continuation of f_n onto G_{n+1}, satisfying $f_{n+1}(z) = \frac{f_{n+1}(z+1)}{z}$ for all $z \in G_{n+1}$ (both sides of this equation being equal to $\frac{f_n(z+1)}{z}$). We may therefore proceed by induction on n to obtain a sequence of direct meromorphic continuations $(G_1, f_1) \sim (G_2, f_2) \sim \cdots$, with each f_n satisfying $f_n(z) = \frac{f_n(z+1)}{z}$ for all $z \in G_n$. We define the Gamma-function $\Gamma(z) = f_n(z)$ for $z \in G_n$. The function Γ is well-defined because if $z \in G_m \cap G_n$ then $f_m(z) = f_n(z)$. Thus $\Gamma(z)$ is meromorphic on $\mathbb{C} = \cup_{n=1}^{\infty} G_n$,and satisfies $\Gamma(z) = \frac{\Gamma(z+1)}{z}$for all $z \in \mathbb{C}$. Since f_1 has a simple pole at $z = 0$, so does Γ, and hence it follows from the above functional equation that Γ has simple poles at $z = -1, -2, \cdots$. Since these poles converge to ∞, Γ is not meromorphic on $\mathbb{C}_\infty$.

Example 16.7. *For each $z \in \mathbb{C} \setminus \{0\}$, the values w of $\log z$ are the solutions of $e^w = z$. Putting $z = re^{i\theta}$, with $r, \theta \in \mathbb{R}$ and $r > 0$, we see that $w = \ln r + i\theta$, where $\ln r = \int_1^r \frac{1}{t} dt$ is the real-valued logarithm of r obtained by integration along $\mathbb{R}$ from 1 to r; there are infinitely many choices for θ, all differing by multiples of 2π, and we must arrange our function elements so that θ is single-valued on the chosen regions.*

The difficulty is that if the region $G \subseteq \mathbb{C} \setminus \{0\}$ encircles the origin, then we cannot make a single-valued continuous choice of $\theta = \arg z$ on G. For example, suppose that G contains the unit circle $\mathbb{T} = \{z \in \mathbb{C} : |z| = 1\}$, and that $\theta(z)$ is a continuous choice of $\arg z$ on $\mathbb{T}$; then the function $\phi(z) = \theta(z) + \theta(\bar{z})$ is continuous on $\mathbb{T}$, and since $\theta(\bar{z}) = -\theta(z) + 2n\pi$ $(n \in \mathbb{Z})$ we see that $\phi(\mathbb{T}) \subset 2\pi\mathbb{Z}$. Now ϕ is continuous and $\mathbb{T}$ is connected, so $\phi(\mathbb{T})$ is a connected subset of $2\pi\mathbb{Z}$ and is therefore a single point $2\pi N$, that is, ϕ is constant on $\mathbb{T}$. Then $\theta(1) = \frac{1}{2}\phi(1) = \pi N = \frac{1}{2}\phi(-1) = \theta(-1)$, which is clearly false since $\theta(1) \in 2\pi\mathbb{Z}$ and $\theta(-1) \in 2\pi\mathbb{Z} + \pi$. To avoid this difficulty, we will use a particular class of regions which do not encircle 0. Let J be any open interval $J = (\alpha, \beta) = \{\theta \in \mathbb{R} : \alpha < \theta < \beta\}$, where $\alpha < \beta \leq \alpha + 2\pi$, let G_J be the region $G_J = \{z = re^{i\theta} : r > 0 \text{ and } \theta \in J\}$, and for each such $z \in G_J$, let $f_J(z) = \ln r + i\theta$.

Since J has length at most 2π, each $z \in G_J$ determines a unique $\theta = \arg z \in J$, so f_J is single valued on G_J. Writing $z = x + iy$ one easily verifies the Cauchy-Riemann equations for f_J :

$$\frac{\partial}{\partial x}(\ln r) = \frac{\partial \theta}{\partial y} \text{ and } \frac{\partial}{\partial y}(\ln r) = -\frac{\partial \theta}{\partial x}.$$

Thus f_J is analytic on G_J, so we have an analytic function element $L_J = (G_J, f_J)$, representing a branch of $\log z$ on G_J. For example, making J as large as possible we could take $J = (-\pi, \pi)$ so that $G_J = \mathbb{C} \setminus \{z \in \mathbb{R} : z \leq 0\}$. However, this is inconvenient for our purposes since there is no direct analytic continuation on L_J beyond G_J : For instance, if (G, f) is a direct analytic continuation with $-1 \in G$ then every neighborhood U of -1 in G contains elements z', z'' with $\text{Im}(z') > 0 > \text{Im}(z'')$ and we have $f(z') = f_J(z') \to i\pi$ as $z' \to -1$, and $f(z'') = f_J(z'') \to -i\pi$ as $z'' \to -1$, so f is not analytic (or even meromorphic) at -1. Similar problems arise at all points $z < 0$. Nevertheless, we can continue f_J outside G_J by first restricting f_J to a smaller subregion of G_J. Let K be the interval $(0, \pi)$, so that $K \subset J$ and hence $G_K \subset G_J$ and $f_K \equiv f_J$ on G_K, giving a direct analytic continuation $L_J \sim L_K$. Since G_K is the upper half-plane, there are no points $z'' \in G_K$ close to the line $z \leq 0$ satisfying $\text{Im}(z'') < 0$, and there is no difficulty in continuing L_K across the negative real line. This suggests that it is wise to use intervals J of length strictly less than 2π. Let G_A, G_B and G_C be the open half-planes corresponding to the intervals $A = (-\frac{\pi}{2}, \frac{\pi}{2}), B = (\frac{\pi}{6}, \frac{7\pi}{6}), C = (\frac{5\pi}{6}, \frac{11\pi}{6})$, and let L_A, L_B and L_C be the corresponding branches of $\log z$. Now $G_A \cap G_B \neq \emptyset$, and for $z = re^{i\theta} \in G_A \cap G_B$ we have $\theta \in A \cap B = (\frac{\pi}{6}, \frac{\pi}{2})$, so $f_A(z) = \ln r + i\theta = f_B(z)$ and hence $L_A \sim L_B$. A similar argument shows that $L_B \sim L_C$. However, although $G_A \cap G_C \neq \emptyset$ we do not have $L_A \sim L_C$: the elements of $G_A \cap G_C$ are of the form $z = re^{i\theta}$ with $\theta \in (-\frac{\pi}{2}, -\frac{\pi}{6}) \subset A$, giving $f_A(z) = \ln r + i\theta$, and also of the form $z = e^{i(\theta+2\pi)}$ with $\theta + 2\pi \in (\frac{3\pi}{2}, \frac{11\pi}{6}) \subset C$, so $f_C(z) = \ln r + i(\theta + 2\pi) = f_A(z) + 2\pi i$. Thus analytic continuation of L_C onto G_A produces the function element $(G_A, f_A + 2\pi i) = L_{A+2\pi}$, where $A + 2\pi$ is the interval $\{\theta + 2\pi : \theta \in A\} = (\frac{3\pi}{2}, \frac{5\pi}{2})$. If we iterate this process n times, using the sequence of intervals $A, B, C, A+2\pi, B+2\pi, \cdots, C+2(n-1)\pi, A+2n\pi$, then we obtain the function element $L_{(A+2n\pi)}$, giving the value $f_{(A+2n\pi)}(z) = f_A(z) + 2n\pi i$; this is done by continuing around the origin n times in the positive direction, and if we continue n times in the negative direction, using the sequence of intervals $A, C - 2\pi, B - 2\pi, A - 2\pi, C - 4\pi, \cdots, B - 2n\pi, A - 2n\pi$, then we obtain the function element $L_{(A-2n\pi)}$, giving the value $f_{(A-2n\pi)}(z) = f_A(z) - 2n\pi i$. In this way, starting at any $z \neq 0$ and continuing around the origin an appropriate number of times, in the positive or negative direction, we can return to z and obtain any of the infinitely many values of $\log z$.

The above example illustrates several of the difficulties associated with meromorphic continuation:

(1) If an unsuitable region G is chosen (such as $G = G_J$ with $J = (-\pi, \pi)$) then a function element (G, f) may admit no further direct meromorphic continuation beyond G.

(2) The relation $\sim$ of direct meromorphic continuation is not transitive since we have seen that $L_A \sim L_B \sim L_C$ does not imply $L_A \sim L_C$, even though $G_A \cap G_C \neq \emptyset$.

(3) If meromorphic continuation starts from, and eventually returns to a particular region G, then the final function defined on G may be quite different from the original function; for example f_A may be continued meromorphically to give $f_A + 2n\pi i$ on $G = G_A$, for any $n \in \mathbb{Z}$;

(4) If (Ω, g) is a meromorphic continuation of (G, f), then the values of the function g on Ω may depend on the chosen sequence of direct meromorphic continuations from G to Ω; for example, taking $(G, f) = L_A$ with $A = (-\frac{\pi}{2}, \frac{\pi}{2})$ as above, and taking Ω to be a small disk containing -1, then continuation from G to Ω via G_B gives the value $f_B(-1) = i\pi$ for $\log(-1)$, whereas continuation via G_C gives $f_C(-1) = -i\pi$, so we obtain two distinct meromorphic continuations g on Ω.

(5) If (G, f) and (Ω, g) are function elements and $f \equiv g$ on a nonempty open subset U of $G \cap \Omega$, it does not follow that $f \equiv g$ on $G \cap \Omega$, so we need not have $(G, f) \sim (\Omega, g)$; the point is that $G \cap \Omega$, although open, need not be path-connected and hence need not be a region, so Lemma 16.3 does not apply. For example, take $(G, f) = L_{A\cup B}$ and $(\Omega, g) = L_C$ as above, where $A \cup B = (-\frac{\pi}{2}, \frac{7\pi}{6})$ and $C = (\frac{5\pi}{6}, \frac{11\pi}{6})$, so $G \cap \Omega$ is the disjoint union of the two regions $G_B \cap G_C$ and $G_C \cap G_A$. On $G_B \cap G_C$ we have $g \equiv f_C \equiv f_B \equiv f_{(A\cup B)} \equiv f$, whereas on $G_C \cap G_A$ we have $g \equiv f_C \equiv f_A + 2\pi i \equiv f_{(A\cup B)} + 2\pi i \equiv f + 2\pi i$.

In the above example the set G is not convex. We can avoid this last difficulty using only convex regions, such as disks: the intersection of two convex sets is convex and therefore path-connected.

16.3 Analytic continuation using power series

An effective method of constructing analytic continuations is to use the fact that a function f is analytic at some point $a \in \mathbb{C}$ if and only if it is represented by a power series $\sum_{n=0}^{\infty} a_n(z-a)^n$ near a. (We can also deal with $a = \infty$ by considering the behavior of $f(\frac{1}{z})$ near 0.) Suppose that (G, f) is an analytic function element with $G \subseteq \mathbb{C}$, and that $a \in G$. Let G_1 be the largest open disk centered at a and contained in G. (We will regard $\mathbb{C}$ as an open disk centered at a, allow for the case $G = \mathbb{C}$.)

Then f is analytic on G_1, so if we define $a_n = \frac{f^{(n)}(a)}{n!}$ for each integer $n \geq 0$, then by Taylor's theorem the power series $f_1(z) = \sum_{n=0}^{\infty} a_n(z-a)^n$ converges absolutely to $f(z)$ on G_1, giving a direct analytic continuation $(G, f) \sim (G_1, f_1)$. It is important that $G_1 \subseteq G$: if we take G_1 to be the largest disk, centered at a, on which $\sum_{n=0}^{\infty} a_n(z-a)^n$ converges, then $G \cap G_1$ need not be a region and we need not have $(G, f) \sim (G_1, f_1)$, with f a branch of $\log z$.

Now let $b \in G$, and let U be the largest open disk centered at b and contained in G_1. Since f_1 is analytic on U, Taylor's theorem gives

$$f_1(z) = \sum_{n=0}^{\infty} b_n(z-b)^n \text{ for all } z \in U, \text{ where } b_n = \frac{1}{n!} f_1^{(n)}(b)$$

$= \dfrac{1}{n!}\dfrac{d^n}{dz^n}\left[\sum_{m=0}^{\infty} a_m(z-a)^m\right]_{z=b} = \sum_{m=n}^{\infty}\binom{m}{n} a_m(b-a)^{m-n}$, using absolute convergence to differentiate term by term. Now the largest region on which this power series $\sum_{n=0}^{\infty} b_n(z-b)^n$converges is an open disk G_2, centered at b and containing U; on G_2 the series represents an analytic function f_2, so we have an analytic function element (G_2, f_2). Since G_1 and G_2 are disks, $G_1 \cap G_2$ is a region, and since $f_1 \equiv f_2$ on the nonempty open set $U \subseteq G_1 \cap G_2$, Lemma 16.3 implies that $f_1 \equiv f_2$ on $G_1 \cap G_2$ and hence $(G_1, f_1) \sim (G_2, f_2)$. If, as is often the case, G_2 strictly contains U, then $G_2 \not\subseteq G_1$ and so we have a non-trivial analytic continuation of (G_1, f_1).

We can now iterate this process, expanding $f_2(z)$ as a power series $f_3(z) = \sum_{n=0}^{\infty} c_n(z-c)^n$ convergent on an open disk G_3, centered at some point $c \in G_2$; since $G_2 \cap G_3$ is a region, we have $(G_2, f_2) \sim (G_3, f_3)$, and so on. Since we are using disks, we can (and generally will) take each G_j to be as large as possible, subject only to f_j being analytic on G_j. In this way, it may eventually be possible to construct an analytic continuation of f outside G.

Example 16.8. *Let $a = 0$ and $f(z) = f_1(z) = \sum_{n=0}^{\infty} z^n$, with region of convergence $G = G_1 = \mathbb{D} = \{z \in \mathbb{C} : |z| < 1\}$.*

If $b \in G_1$, then since $a_m = 1$ for all $m \geq 0$ we have $b_n = \sum_{m=n}^{\infty}\binom{m}{n} b^{m-n}$, and this converges to $(1-b)^{-1-n}$ since $|b| < 1$. This can be seen either by differentiating n times the series $(1-b)^{-1} = \sum_{m=0}^{\infty} b^m$ with respect to b, or by applying the binomial theorem directly to $(1-b)^{-1-n}$ and comparing coefficients. We therefore have an analytic function $f_2(z) = \sum_{n=0}^{\infty} b_n(z-b)^n = \sum_{n=0}^{\infty}(1-b)^{-1-n}(z-b)^n$ on some open disk G_2 centered at b. The radius of convergence of this power series is $\lim_{n\to\infty}(|1-b|^{1+n})^{\frac{1}{n}} = |1-b|$, so G_2 is $\{z \in \mathbb{C} : |z-b| < |1-b|\}$, the largest open disk centered at b and avoiding 1 (where f_1 has a pole). We therefore have $G_2 \not\subseteq G_1$ except when $b \in \mathbb{R}$ and $0 \leq b < 1$,so by avoiding such a choice of b we can continue f_1 analytically outside G_1.

In the examples we have considered so far, there have been certain points $c \in \mathbb{C}_\infty$ at which meromorphic continuation is impossible: for instance $c = \infty$ for $\Gamma(z)$ (which cannot be continued to a meromorphic function at ∞ since it has a sequence of poles converging to ∞), and $c = 0$ or ∞ for $\log z$ (which has no single-valued meromorphic branch near either of these points). If (G, f) is a function element then a point c on the boundary ∂G of G is called a regular point for (G, f) if there is a direct meromorphic continuation $(\Omega, g) \sim (G, f)$

with $c \in \Omega$; there is no loss of generality in taking Ω to be an open disk centered at c. If there is no such direct meromorphic continuation then c is called a singular point for (G, f); thus singular points represent obstacles to meromorphic continuation. If all points $c \in \partial G$ are singular, then ∂G is called the natural boundary of (G, f). (It may be noted that in some books, where the emphasis is on analytic rather than meromorphic functions, c is defined to be a singular point if there is no direct analytic continuation at c : under this convention, though not under ours, $c = 1$ would be a singular point for $f(z) = \sum_{n=0}^{\infty} z^n$ on the open unit disk $\mathbb{D}$.)

Lemma 16.4. *Let (D, f) be a function element with $c \in \partial D$. If c is a regular point for (D, f) then as $z \to c$ with $z \in D$, the function $f(z)$ approaches a limit $l \in \mathbb{C}_\infty$.*

Proof. By hypothesis, there is a direct meromorphic continuation $(E, g) \sim (D, f)$ with $c \in E$. So for $z \in D \cap E$, we have $f(z) = g(z) \to g(c) \in \mathbb{C}_\infty$ as $z \to c$. □

Thus it follows that if $f(z)$ has no limit as $z \to c$ then c is singular. A similar proof shows that if there is a direct analytic continuation at c, then $f(z)$ has a finite limit as $z \to c$.

Example 16.9. *For each $k \in \mathbb{Z}, k \geq 1$, the power series $f_k(z) = \sum_{n=0}^{\infty} z^{kn}$ converges to $\frac{1}{1-z^k}$ on the open unit disk $\mathbb{D}$. So if c is a kth root of unity then there is no direct analytic continuation of $(\mathbb{D}, f_k)$ at c. But meromorphic continuation is possible and so c is regular.*

In the above series $f_k(z)$, the 'gap' between successive nonzero coefficients is equal to k, the larger the gap is, the more obstacles there are to analytic continuation. Hadamard's gap theorem states that if a power series $f(z) = \sum_{n=0}^{\infty} a_n z^n$ has radius of convergence 1, and if the gaps between the suffixes n of successive nonzero coefficients a_n increase sufficiently rapidly as $n \to \infty$, then every point c on the unit circle $\mathbb{T}$ is singular, that is, $\mathbb{T}$ is the natural boundary of $(\mathbb{D}, f)$. In other words, the theorem states that analytic continuation across $\mathbb{T}$ is impossible, but the extension to meromorphic continuation is trivial.

Theorem 16.4. *The power series $\sum_{n=1}^{\infty} z^{n!}$ has the unit circle $\mathbb{T} = \{z \in \mathbb{C} : |z| = 1\}$ as its natural boundary.*

Proof. Let $f(z) = \sum_{n=1}^{\infty} z^{n!}$. From root test, it follows that the series $\sum_{n=1}^{\infty} z^{n!}$ has radius of convergence 1. So f is analytic in the open unit disk $\mathbb{D}$. We shall show that every point c on the unit circle $\mathbb{T} = \partial\mathbb{D}$ (see Fig.14.1) is singular. If c were regular, then there would be a direct meromorphic continuation $(E, g) \sim (\mathbb{D}, f)$ with $c \in E$; since g is meromorphic we could take E sufficiently small so that g is analytic on $E \setminus \{c\}$, and for each point $d \neq c$ in $\mathbb{T} \cap E$ we would have $f(z) = g(z) \to g(d)$ as $z \to d, z \in \mathbb{D}$. However, we shall show that no matter how small E is, there exists some $d \neq c$ in $\mathbb{T} \cap E$ such that $f(z) \to \infty$ as $z \to d$ along a radius from 0 to d.

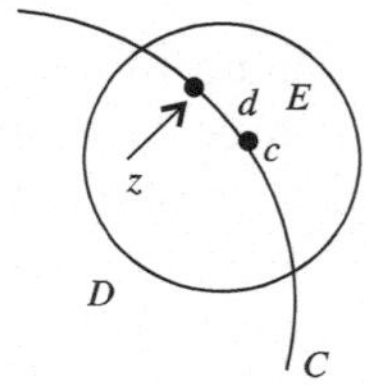

Fig. 16.1: Every point c on $\mathbb{T}$ is singular

We first show that if $r \in \mathbb{R}, 0 \le r < 1$, then $f(r) \to \infty$ as $r \to 1$. Given any $M \in \mathbb{R}$, choose a positive integer $m > M$. Since $0 \le r < 1$, we have

$$f(r) = \sum_{n=1}^{\infty} r^{n!} \ge \sum_{n=1}^{m} r^{n!} \ge mr^{m!}$$

and as $r \to 1$ we have $mr^{m!} \to m > M$. Thus we have $mr^{m!} \ge M$ provided r is sufficiently close to 1 and hence $f(r) \ge M$, as required.

Now take any point $c \in \mathbb{T}$. Then $c = e^{2\pi is}$ for some $s \in \mathbb{R}$. If E is any open neighborhood of c in $\mathbb{C}_\infty$ then we can choose a rational number $t \ne s$ sufficiently close to s that the point $d = e^{2\pi it}$ lies in $(\mathbb{T} \cap E) \setminus \{c\}$. Since $t \in Q$ (the set of rational numbers) we have $t = \frac{p}{q}$ with $p, q \in \mathbb{Z}$ and $q > 0$. Let $z = rd$ where $r \in \mathbb{R}, 0 \le r < 1$, so that $z \in \mathbb{D}$ and

$$z^{n!} = r^{n!}d^{n!} = r^{n!}e^{\frac{2\pi ipn!}{q}}.$$

If $n \ge q$ then q divides $n!$, so $z^{n!} = r^{n!}$ and hence $f(z) = \sum_{n=1}^{q-1} z^{n!} + \sum_{n=q}^{\infty} r^{n!} = P(z) + f(r) - P(r)$, where $P(z)$ is the polynomial $\sum_{n=1}^{q-1} z^{n!}$. As $r \to 1$ we have $z \to d$, so $P(z)$ and $P(r)$ have finite limits $P(d)$ and $P(1)$, whereas we have seen that $f(r) \to \infty$. Thus $f(z) \to \infty$, which is a contradiction. □

There exist power series with radius of convergence 1 such that every point $c \in \mathbb{T}$ (unit circle) is regular. For example, the power series $\sum_{n=0}^{\infty} z^n$ has this property. Notice that at the point $c = 1$ analytic continuation is impossible but meromorphic continuation is possible. The following result shows that this is typical.

Theorem 16.5. *If a power series $f(z) = \sum_{n=0}^{\infty} a_n(z-a)^n$ has radius of convergence $R \ne 0, \infty$, then there is at least one point c on the circle $|z-a| = R$ at which direct analytic continuation is impossible.*

Proof. Let $\zeta = \frac{z-a}{R}$. Then $f(z) = \sum_{n=0}^{\infty} R^n a_n \zeta^n$. Without loss of generality, assume that, $a = 0$ and $R = 1$. So f is analytic on the open unit disk $\mathbb{D} = \{z \in \mathbb{C} : |z| < 1\}$. Let $\mathbb{T} = \partial\mathbb{D} = \{z \in \mathbb{C} : |z| = 1\}$. Suppose that for each point $c \in \mathbb{T}$ there is a direct analytic continuation $(E_c, g_c) \sim (\mathbb{D}, f)$ with $c \in E_c$. Putting $c = e^{is}, s \in \mathbb{R}$, we can take the region E_c to be of the form

$$E_c = \{re^{i\theta} : 1 - r_c < r < 1 + r_c, s - \alpha_c < \theta < s + \alpha_c\},$$

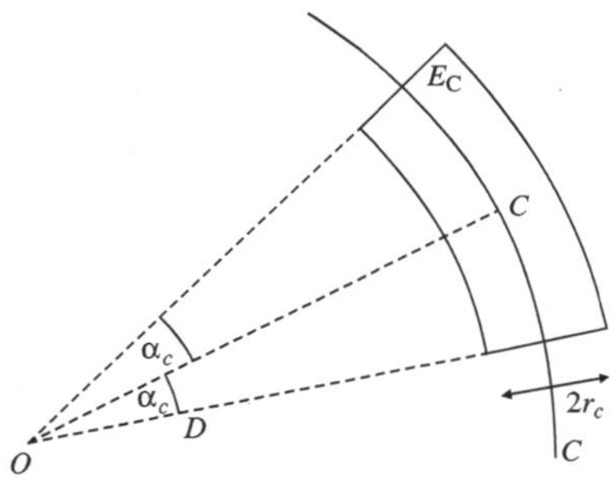

Fig. 16.2: The set E_c

where $0 < r_c < 1$ and $0 < \alpha_c < \pi$. The condition on α_c guarantees that for any $c, d \in \mathbb{T}$, if $E_c \cap E_d \neq \emptyset$, then it is a region being path connected (see Fig. 16.3).

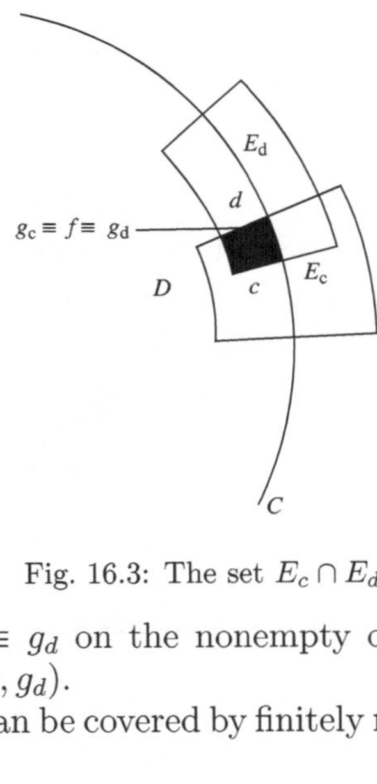

Fig. 16.3: The set $E_c \cap E_d$

In this case, $g_c \equiv f \equiv g_d$ on the nonempty open subset $\mathbb{T} \cap E_c \cap E_d$ of $E_c \cap E_d$, so $(E_c, g_c) \sim (E_d, g_d)$.

Since $\mathbb{T}$ is compact, it can be covered by finitely many such regions $E_{c_1}, \cdots, E_{c_k}$. Let

$$R' = 1 + \min_{1 \leq i \leq k} r_{c_i},$$

so that $R' > 1$, and let $E = \{z \in \mathbb{C} : |z| < R'\}$, so that

$$E \subseteq \mathbb{D} \cup E_{c_1} \cup \cdots E_{c_k}.$$

We define a function g on E by

$$g(z) = \begin{cases} f(z) & \text{if } z \in \mathbb{D}; \\ g_{c_i}(z) & \text{if } z \in E_{c_i}. \end{cases}$$

This definition is consistent since if $z \in \mathbb{D} \cap E_{c_i}$ or if $z \in E_{c_i} \cap E_{c_j}$ then $f(z) = g_{c_i}(z)$ or $g_{c_i}(z) = g_{c_j}(z)$ respectively. Near any point in E, g is identically equal to f and g_{c_i} for some i, so g is analytic on E. By Taylor's theorem the series $\sum_{n=0}^{\infty} \frac{g^{(n)}(0)}{n!} z^n$ must converge to $g(z)$ on the largest open disk centered at 0 and contained in E, and clearly this disk is E itself. Since $g \equiv f$ near 0, we have

$$\frac{g^{(n)}(0)}{n!} = \frac{f^{(n)}(0)}{n!} = a_n$$

for all $n \geq 0$, so we have shown that $\sum_{n=0}^{\infty} a_n z^n$ converges on E and hence has radius of convergence at least $R' > 1$, against our assumption. This contradiction shows that there must be some point $c \in \mathbb{T}$ at which direct analytic continuation of $(\mathbb{D}, f)$ is impossible. □

Thus, for a power series with radius of convergence $\rho \neq 0, \infty$, the 'best possible' behavior is that exhibited by $\sum_{n=0}^{\infty} z^n$, where there is just one point $(c = 1)$ where we need meromorphic rather than analytic continuation.

It is important to note that whether or not a point c is regular for (D, f) depends on D as well as on f: for example, even though direct analytic or meromorphic continuation may be impossible at c, we may nevertheless be able to continue f across ∂D (boundary of D) at c indirectly, using a sequence of direct continuations, as we saw in the example in connection with $\log(z)$. If we take $J = (-\pi, \pi)$ then the function element has domain $D = D_J = \{z = re^{i\theta} : r > 0 \text{ and } -\pi < \theta < \pi\}$, with boundary ∂D consisting of the negative real line together with its end-points 0 and ∞; each $c \in \partial D$ is a singular point for L_J, but nevertheless by restricting to the smaller region D_K with $K = (0, \pi)$ it is possible to continue $\log(z)$ across ∂D at any point $c \neq 0, \infty$, using a sequence of two direct continuations. This suggests that the concept of the natural boundary of a function (rather than a function element) is rather hard to define.

16.4 Meromorphic continuations along a path

To deal with the problems of non-uniqueness of meromorphic continuation, such as those we met in Example 16.7 in connection with $\log(z)$, we introduce the concept of meromorphic continuation along a path. We shall show that meromorphic continuation along a given path is unique, and then we shall consider the relationship between meromorphic continuations along different paths between two given points. Henceforth in this chapter, we shall define a path γ as a continuous function from I into $\mathbb{C}_\infty$ where $I = [0, 1]$. We often refer to $\gamma(I)$ as 'the path γ'. If $a = \gamma(0)$ and $b = \gamma(1)$ then we say that γ is a path 'from a to b'; γ is a closed path if $a = b$, and γ is simple if $\gamma(s) = \gamma(s')$ implies either $s = s'$ or else $s = 0$ and $s' = 1$ (that is, γ has no self-intersections except possibly where $a = b$).

Let (D, f) be a function element, let $a \in D$ and let γ be a path in $\mathbb{C}_\infty$ from a to some point $b \in \mathbb{C}_\infty$. Then a meromorphic continuation of (D, f) along γ

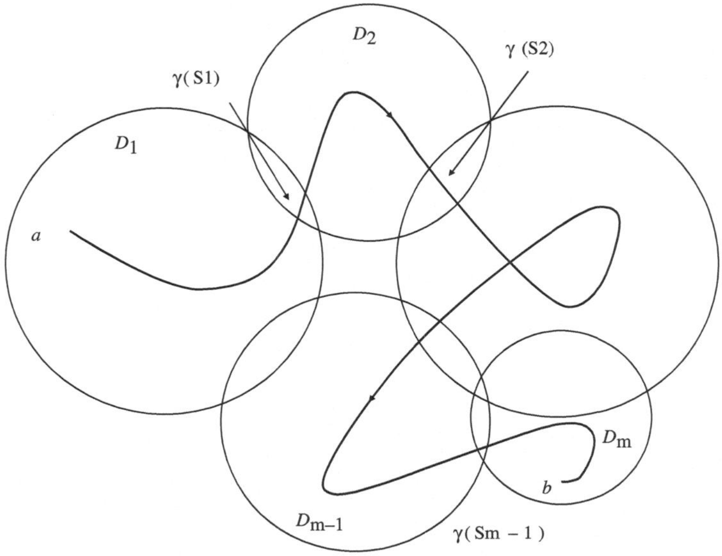

Fig. 16.4: Direct meromorphic continuations

is a finite sequence of direct meromorphic continuations

$$(D, f) \sim (D_1, f_1) \sim (D_2, f_2) \sim \cdots (D_m, f_m)$$

such that

(i) each region D_i is an open disk in $\mathbb{C}_\infty$ with $a \in D_1 \subseteq D$;

(ii) there is a subdivision $0 = s_0 < s_1 < \cdots < s_m = 1$ of I such that $\gamma([s_{i-1}, s_i]) \subseteq D_i$ for $i = 1, 2, \cdots, m$ (and hence $b \in D_m$).

If all the function elements (D_i, f_i) are analytic, we call this an analytic continuation along γ.

Notice that we require the regions D_i to be disks, and why we allow only finitely many of them. There is no loss of generality in using disks, since any region in $\mathbb{C}_\infty$ is a union of open disks, and indeed, if we construct the continuations by means of power series, then we will be forced to use disks. Moreover, it is easily seen that if two open disks on the sphere S^2 have non-empty intersection, then that intersection is path-connected and is therefore a region; using the homeomorphism $\pi : S^2 \to \mathbb{C}_\infty$ we see that disks in $\mathbb{C}_\infty$ also have this useful property. Since $\gamma(I)$ is compact, we can reduce any covering of γ by regions to a finite subcovering, so there is no loss of generality in restricting attention to finite sequences of direct continuations. Given $a \in \mathbb{C}_\infty$, let $\mathcal{F}_a$ be the set of all functions which are meromorphic in some open neighborhood of a. If $f, g \in \mathcal{F}_a$ then we write $f \sim_a g$ if $f \equiv g$ in some neighborhood of a; clearly this is an equivalence relation on $\mathcal{F}_a$, and the equivalence class of f, denoted by $[f]_a$, is called the germ of f at a. We call this germ analytic if f is analytic at a, that is, $f(a) \neq \infty$. We can think of germs $[f]_a (a \in \mathbb{C}_\infty)$ as representing single-valued meromorphic functions defined near a, since by definition we have $[f]_a = [g]_b$ if and only if $a = b$ and $f \equiv g$ near a (in which case, by Lemma 16.3, $f \equiv g$ on any region on which both f and g are meromorphic). The set $\mathcal{M}$ of all germs $[f]_a$, for all $a \in \mathbb{C}_\infty$, is called the sheaf of germs of meromorphic functions.

Example 16.10. *Let γ be the unit circle parametrised by $\gamma(s) = e^{2\pi i s}$, $s \in I$. Thus $a = \gamma(0) = 1$ and $b = \gamma(1) = 1$.*

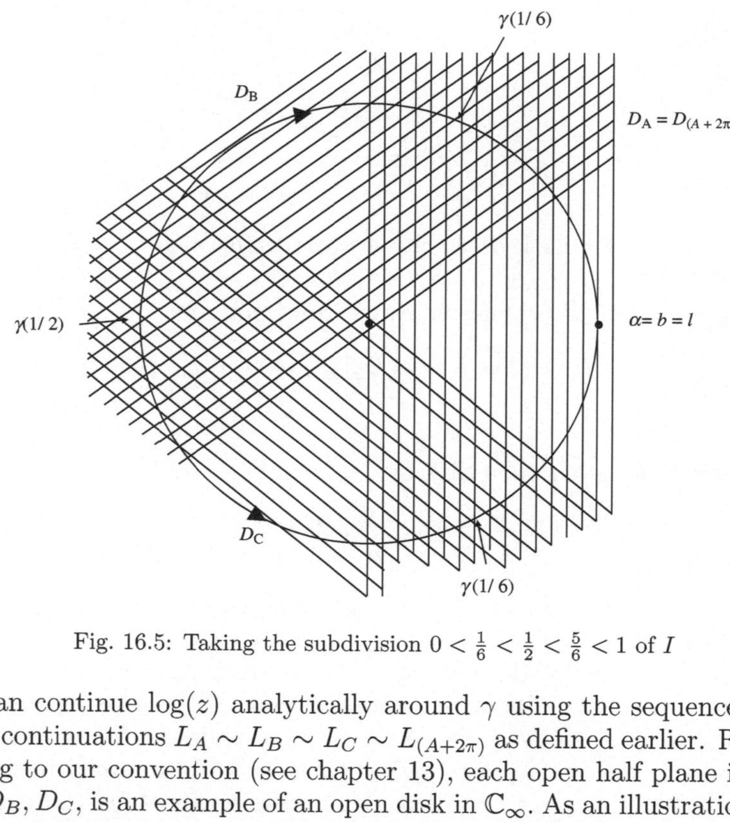

Fig. 16.5: Taking the subdivision $0 < \frac{1}{6} < \frac{1}{2} < \frac{5}{6} < 1$ of I

We can continue $\log(z)$ analytically around γ using the sequence of direct analytic continuations $L_A \sim L_B \sim L_C \sim L_{(A+2\pi)}$ as defined earlier. Recall that according to our convention (see chapter 13), each open half plane in $\mathbb{C}$, such as D_A, D_B, D_C, is an example of an open disk in $\mathbb{C}_\infty$. As an illustration, we can take the subdivision of I in condition (ii) to be $0 < \frac{1}{6} < \frac{1}{2} < \frac{5}{6} < 1$. Thus we start with $f_1(a) = f_A(1) = 0$ at $a = 1$, and end with $f_4(b) = f_{(A+2\pi)}(1) = 2\pi i$ at $b = 1$. Similarly, if $n \in \mathbb{Z}$ and we take $\gamma^n(s) = e^{2\pi n i s}$, then the path γ^n winds n times around 0, and we continue along γ^n from $f_A(1) = 0$ to $f_{(A+2n\pi)}(1) = 2n\pi i$.

The aim of meromorphic continuation along a path γ is to start with a meromorphic function $f = f_1$ at $a = \gamma(0)$ and to allow this function to vary meromorphically along γ until we reach a meromorphic function f_m at $b = \gamma(1)$. The following result shows that the resulting value $f_m(b)$ depends only on the initial function f and the path γ from a to b, and not on the particular choice of disks D_i covering γ, so we may denote this value by $f_\gamma(b)$.

Theorem 16.6. *Let*

$$(D, f) \sim (D_1, f_1) \sim \cdots \sim (D_m, f_m)$$

and

$$(D, f) \sim (E_1, g_1) \sim \cdots \sim (E_n, g_n)$$

be meromorphic continuations of (D, f) *along a path* γ *from* a *to* b, *and let* $0 = s_0 < s_1 < \cdots < s_m = 1$ *and* $0 = t_0 < t_1 < \cdots < t_n = 1$ *be the corresponding subdivisions of the interval* $I = [0,1]$. *Then* $(D_i, f_i) \sim (E_j, g_j)$ *whenever* $[s_{i-1}, s_i] \cap [t_{j-1}, t_j] \neq \emptyset$, *and in particular* $f_m(b) = g_n(b)$. *Further,* $[f_m]_b = [g_n]_b$.

Proof. Suppose that $[s_{i-1}, s_i] \cap [t_{j-1}, t_j] \neq \emptyset$, so that $D_i \cap E_j \neq \emptyset$, and suppose (for an eventual contradiction) that $f_j \neq g_j$ on $D_i \cap E_j$.

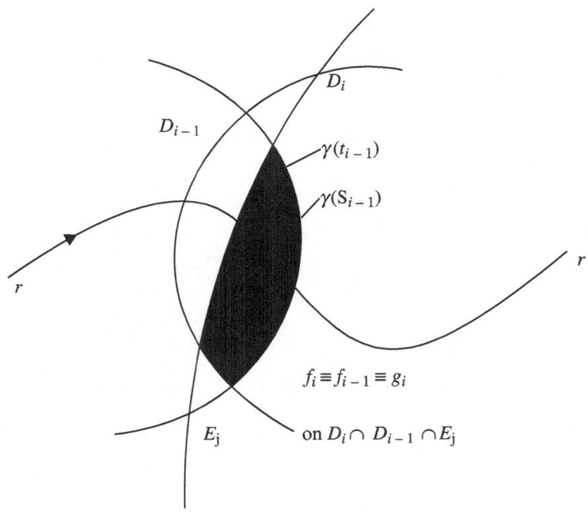

Fig. 16.6: The region $D_i \cap E_j$

We may assume that i and j are chosen so that $i+j$ is minimal with respect to this property and that $t_{j-1} \leq s_{i-1}$ (interchanging the two meromorphic continuations if necessary).

Since $(D_1, f_1) \sim (E_1, g_1)$ it follows that $i > 1$. Now $[s_{i-1}, s_i] \cap [t_{j-1}, t_j] \neq \emptyset$ implies $s_{i-1} \in [t_{j-1}, t_j]$, so $\gamma(s_{i-1}) \in E_j$. Clearly $\gamma(s_{i-1}) \in D_i \cap D_{i-1}$, so $D_i \cap D_{i-1} \cap E_j \neq \emptyset$. Now $s_{i-1} \in [s_{i-2}, s_{i-1}] \cap [t_{j-1}, t_j]$, so by the minimality of $i + j$ we have $f_{i-1} \equiv g_j$ on $D_{i-1} \cap E_j$. Since $(D_{i-1}, f_{i-1}) \sim (D_i, f_i)$ we have $f_{i-1} \equiv f_i$ on $D_{i-1} \cap D_i$, and hence $f_i \equiv g_j$ on $D_i \cap D_{i-1} \cap E_j$ which is a nonempty open subset of the region $D_i \cap E_j$. Thus $f_i \equiv g_j$ on $D_i \cap E_j$ by Lemma 16.3, contradicting our choice of i and j. Thus $(D_i, f_i) \sim (E_j, g_j)$

whenever $[s_{i-1}, s_i] \cap [t_{j-1}, t_j] \neq \emptyset$, and taking $i = m$ and $j = n$ (so that $1 \in [s_{i-1}, s_i] \cap [t_{j-1}, t_j]$) we have $(D_m, f_m) \sim (E_n, g_n)$, and hence $f_m(b) = g_n(b)$ since $b \in D_m \cap D_n$. From this it also follows that $[f_m]_b = [g_n]_b$ as $E_n \cap D_m \neq \emptyset$. $\square$

Example 16.11. *As an example of analytic continuation along paths, we consider the many-valued functions* $z^{\frac{1}{q}}$, *where* $q \in \mathbb{Z}$ *and* $q \geq 2$.

For any $z \in \mathbb{C} \setminus \{0\}$ there are q values of $z^{\frac{1}{q}}$, namely the solutions w of $w^q = z$; we can express these as $w = exp(\frac{1}{q} \log z)$, different values of $z^{\frac{1}{q}}$ corresponding to different values of $\log(z)$. More precisely, if (D, f) is a function element representing a branch of $\log(z)$, so that $\exp(f(z)) \equiv z$ on D, then the function $g(z) = \exp(\frac{1}{q} f(z))$ is analytic and satisfies $g(z)^q \equiv z$ on D, so the function element (D, g) represents a branch of $z^{\frac{1}{q}}$ on D. If γ is any path in $\mathbb{C} \setminus \{0\}$ then we can find the analytic continuation of $z^{\frac{1}{q}}$ along γ by considering the continuation of $\log(z)$. For example, if $(D, f) = L_A = (D_A, f_A)$ as above, giving the branch $g_A = \exp(\frac{1}{q} f_A)$ of $z^{\frac{1}{q}}$ on D_A satisfying $g_A(1) = \exp(0) = 1$, then by continuing from 1 to 1 along the path $\gamma^n(s) = e^{2\pi nis}$ we get

$$g_{(A+2n\pi)} \equiv \exp\left(\frac{1}{q} f_{(A+2n\pi)}\right) \equiv \exp\left(\frac{1}{q}(f_A + 2n\pi i)\right) \equiv \epsilon^n g_A,$$

where $\epsilon = \exp\left(\frac{2\pi i}{q}\right)$; this gives a branch of $z^{\frac{1}{q}}$ taking the value ϵ^n at 1. Since ϵ has order q under multiplication we have $g_{(A+2n\pi)} \equiv g_{(A+2m\pi)}$ if and only if $n \equiv m \mod (q)$; in particular, $g_{(A+2q\pi)} \equiv g_A$. Thus the q branches of $z^{\frac{1}{q}}$ on D_A are

$$g_A, g_{(A+2\pi)} \equiv \epsilon g_A, \cdots, g_{(A+2(q-1)\pi)} \equiv \epsilon^{q-1} g_A,$$

taking the values $1, \epsilon, \cdots, \epsilon^{q-1}$ at 1. Similarly, if we start with a particular value w of $z^{\frac{1}{q}}$ at any point $z \in \mathbb{C} \setminus \{0\}$, then by continuing around 0 repeatedly we obtain the values $w, \epsilon w, \cdots, \epsilon^{q-1} w$ of $z^{\frac{1}{q}}$ in succession. This shows that we cannot find an analytic (or even meromorphic)continuation of $z^{\frac{1}{q}}$ at 0; if we had a function element (D, g) representing a branch of $z^{\frac{1}{q}}$ on a region D containing 0, then g would be single-valued on D; however by continuing g analytically along a closed path which winds once around 0 within D, we replace g by $\epsilon g \neq g$, giving a contradiction. A similar argument shows that there is no continuation at ∞, so 0 and ∞ are singular points for $z^{\frac{1}{q}}$.

16.5 The Monodromy theorem

If a function element (D, f) can be continued meromorphically along a path γ from a to b, then by Theorem 16.6 the value $f_\gamma(b)$ of the resulting continuation at b is independent of the method of continuation along γ. However, different paths γ_0 and γ_1 from a to b may give different values at b, as we saw for $\log(z)$; we now consider sufficient conditions for $f_{\gamma_0}(b) = f_{\gamma_1}(b)$.

Example 16.12. *If γ_0 and γ_1 are the paths from 1 to -1 in $\mathbb{C}$ defined by $\gamma_0(s) = e^{\pi i s}$ and $\gamma_1(s) = e^{-\pi i s}$ then $\Gamma(s,t) = e^{\pi i s} - 2ti\sin(\pi s)$ gives a path homotopy from γ_0 to γ_1 in $\mathbb{C}$. However, it is intuitively clear that γ_0 and γ_1 are not homotopic in $\mathbb{C} \setminus \{0\}$: in this case some intermediate path γ_t would have to pass through 0 or ∞ which is impossible since $\Gamma(I \times I) \subseteq \mathbb{C} \setminus \{0\}$.*

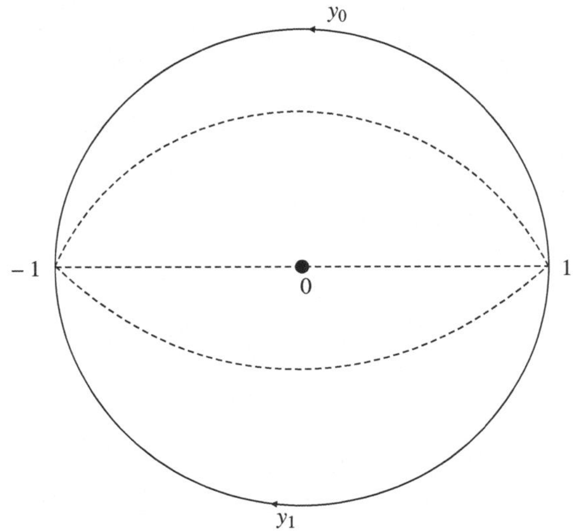

Fig. 16.7: Path homotopy from γ_0 to γ_1 in $\mathbb{C}$

Example 16.13. *Each closed path α from 1 to 1 in $\mathbb{C} \setminus \{0\}$ is homotopic to the path $\gamma^n(s) = e^{2n\pi i s}$ for some unique $n \in \mathbb{Z}$, called the winding number of α since it represents the number of times α winds around 0.*

The following result shows that if two paths γ_0 and γ_1 are path homotopic then they give rise to the same meromorphic continuation.

Theorem 16.7. *Let $X \subseteq \mathbb{C}_\infty$, let $a, b \in X$, and let Γ be a path homotopy in X between two paths γ_0 and γ_1 from a to b in X; if (D, f) is a function element which can be continued meromorphically along each path $\gamma_t : s \mapsto \Gamma(s,t), s, t \in I$, then the meromorphic functions at b, resulting from the continuations along γ_0 and γ_1, are identically equal in some neighborhood of b. That is, $[f_{\gamma_0}]_b = [f_{\gamma_1}]_b$ and in particular, $f_{\gamma_0}(b) = f_{\gamma_1}(b)$.*

Proof. For all $t, t' \in I$, we will write $t \sim t'$ if the meromorphic continuations of (D, f) along γ_t and $\gamma_{t'}$ produce functions which are identically equal in some neighborhood of b. By Theorem 16.6, this is independent of the particular continuations along these paths. Clearly '$\sim$' is an equivalence relation on

I, and we have to prove $0 \sim 1$. To do this, it is sufficient to prove that each equivalence class is open in I, for then each class is also closed (its complement, being a union of equivalence classes, is open), so the connectedness of I implies that there is just one equivalence class and hence $0 \sim 1$ as required.

To show that each equivalence class is open, we must show that for each $t \in I$

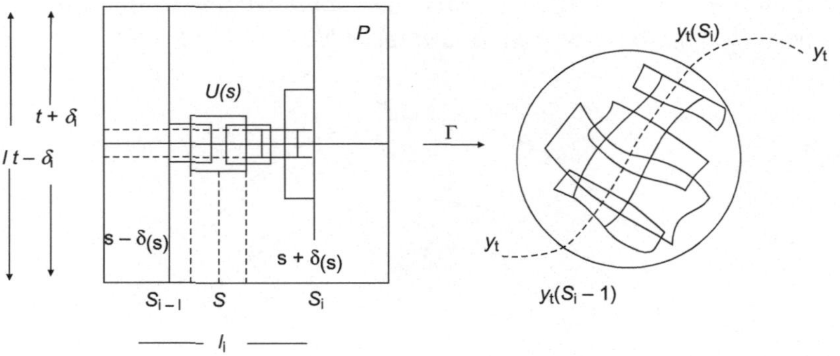

Fig. 16.8: The path homotopy Γ

there exists some $\delta > 0$ such that $t' \sim t$ for all $t' \in I$ satisfying $|t' - t| < \delta$. By hypothesis, there is a sequence of direct meromorphic continuations

$$(D, f) \sim (D_1, f_1) \sim \cdots \sim (D_m, f_m)$$

and a subdivision $0 = s_0 < s_1 < \cdots < s_m = 1$ of I such that $\gamma_t(I_i) \subseteq D_i$ for $1 \leq i \leq m$, where I_i denotes the closed subintervals $[s_{i-1}, s_i]$ of I. Since Γ is continuous and each D_i is open, the set $\Gamma^{-1}(D_i)$ is open in $I_i \times I$. For every $s \in I_i$ there exists an open neighborhood $U(s) = \{(s', t') \in I_i \times I : |s' - s| < \delta(s), |t' - t| < \delta(s)\}$ of $(s, t) \in I_i \times I$, with $\delta(s) > 0$, such that $(s', t') \in U(s)$ implies $\Gamma(s', t') \in D_i$, that is $\gamma_{t'}(s') \in D_i$.

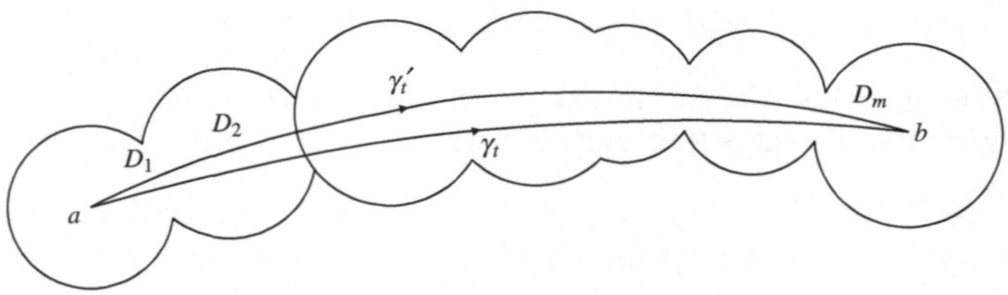

Fig. 16.9: Meromorphic continuation of (D, f) along $\gamma_{t'}$

Being compact, I_i is covered by finitely many such intervals $(s - \delta(s), s + \delta(s))$, so let δ_i be the least of the finitely many corresponding values of $\delta(s)$. Then we have $\delta_i > 0$, and if $t' \in I$ then $|t' - t| < \delta_i$ implies $\gamma_{t'}(I_i) \subseteq D_i$. Defining δ to be the least δ_i for $1 \leq i \leq m$, we have $\delta > 0$, and if $t' \in I$ then $|t' - t| < \delta$ implies $\gamma_{t'}(I_i) \subseteq D_i$ for all $i = 1, 2, \cdots, m$. For such values of t' we therefore have a meromorphic continuation of (D, f) along $\gamma_{t'}$ (see Fig. 16.9) given by $(D, f) \sim (D_1, f_1) \sim \cdots \sim (D_m, f_m)$, with the same subdivision $0 = s_0 < s_1 < \cdots < s_m = 1$ and the same function elements (D_i, f_i) as were used for continuations along γ_t.

Thus the continuations along $\gamma_t, \gamma_{t'}$ both give rise to the meromorphic function f_m near b, so $t' \sim t$, as required. □

Example 16.14. *We can now prove our earlier claim that the paths $\gamma_0(s) = e^{\pi i s}$ and $\gamma_1(s) = e^{-\pi i s}$ from 1 to -1 cannot be path homotopic in $\mathbb{C} \setminus \{0\}$.*

If we take (D, f) to be the function element L_A, the branch of $\log(z)$ defined before, then it is not hard to show that (D, f) can be continued analytically along any path from 1 to -1 in $\mathbb{C} \setminus \{0\}$ (using suitable function elements L_J, for example); since $f_{\gamma_0}(-1) = \pi i$ and $f_{\gamma_1}(-1) = -\pi i$, it follows from Theorem 16.7 that γ_0 and γ_1 are not path homotopic in $\mathbb{C} \setminus \{0\}$. Similarly, if γ^n is the closed path $\gamma^n(s) = e^{2\pi n i s}$ $(n \in \mathbb{Z})$, then by continuing $\log(z)$ analytically around γ^n we see that γ^m and γ^n are path homotopic in $\mathbb{C} \setminus \{0\}$ if and only if $m = n$.

In order to apply Theorem 16.7, we need a condition on X which guarantees that any two paths from a to b in X are path homotopic.

Definition 16.1. *For each $a \in X$, let $\gamma_{(a)}$ denote the constant path $\gamma_{(a)}(s) = a$ for all $s \in I$. A closed path γ from a to a in X is said to be null-homotopic if it is path homotopic in X to $\gamma_{(a)}$. A topological space X is said to be simply connected if it is path-connected and all closed paths in X are null-homotopic.*

Example 16.15. *The complex plane $\mathbb{C}$ is simply connected since if γ is any closed path from a to a in $\mathbb{C}$ then there is a path homotopy Γ between γ and $\gamma_{(a)}$ given by $\Gamma(s, t) = ta + (1 - t)\gamma(s)$.*

Example 16.16. *The space $\mathbb{C} \setminus \{0\}$ is not simply connected since the unit circle $\gamma(s) = e^{2\pi i s}$ is not null-homotopic in $\mathbb{C} \setminus \{0\}$.*

Theorem 16.8. *A topological space X is simply connected if and only if for each pair of points $a, b \in X$ there is a single homotopy class of paths from a to b in X.*

Proof. Suppose that for each $a, b \in X$ there is a single homotopy class of paths from a to b in X. Then X is path connected, and taking $a = b$ we see that any closed path from a to a in X is homotopic to $\gamma_{(a)}$, that is, X is simply connected.

Conversely, suppose that X is simply connected, so by definition, X is path-connected. Let γ_0 and γ_1 be paths from a to b in X, and let δ be the closed path from a to a given by

$$\delta(s) = \begin{cases} \gamma_0(2s) & \text{if } 0 \leq s \leq \frac{1}{2} \\ \gamma_1(2(1-s)) & \text{if } \frac{1}{2} \leq s \leq 1. \end{cases}$$

Since X is simply connected there is a homotopy Λ from δ to $\gamma_{(a)}$. Define $\Gamma : I \times I \to X$ as follows.

$$\Gamma(s,t) = \begin{cases} \gamma_0(\frac{3s}{3-4t}) & \text{if } 0 \le t \le \frac{1}{2}, 0 \le s \le 1 - \frac{4t}{3}; \\ \gamma_1(4-3s-4t) & \text{if } 0 \le t \le \frac{1}{2}, 1 - \frac{4t}{3} \le s \le 1 - \frac{2t}{3}; \\ \gamma_1(3s-2) & \text{if } 0 \le t \le \frac{1}{2}, 1 - \frac{2t}{3} \le s \le 1; \\ \Lambda(\frac{3s}{4-4t}, 2t-1) & \text{if } \frac{1}{2} \le t \le 1, 0 \le s \le \frac{4}{3}(1-t); \\ \gamma_1(\frac{4-3s-4t}{1-4t}), & \text{if } \frac{1}{2} \le t \le 1, \frac{4}{3}(1-t) \le s \le 1. \end{cases}$$

It is easy to check that $\Gamma(s,0) = \gamma_0(s), \Gamma(s,1) = \gamma_1(s), \Gamma(0,t) = a$ and $\Gamma(1,t) = b$ for all $s,t \in I$. By Pasting lemma (Theorem 2.54), the map Γ is continuous. Thus Γ is a path homotopy between γ_0 and γ_1 in X. So X has a single homotopy class of paths from a to b. □

We see that continuation of a function element onto a simply connected region E is independent of the path of continuation, so we have a single-valued meromorphic function on E.

Theorem 16.9. *(The monodromy theorem): Let E be a simply connected region in $\mathbb{C}_\infty$, and let (D, f) be a function element with $D \subseteq E$. If (D, f) can be continued meromorphically along all paths in E starting at some point $a \in D$, then there is a direct meromorphic continuation $(E, g) \sim (D, f)$.*

Proof. Since E is simply connected, E is path connected, so for each $b \in E$ there is a path γ in E from a to b. By hypothesis, we may continue (D, f) along γ; we denote the value of this continuation at b by $f_\gamma(b)$. Since E is simply connected by Theorem 16.8, any two paths in E from a to b are path homotopic. So by Theorem 16.7, they induce the same value of $f_\gamma(b)$. Thus $f_\gamma(b)$ is independent of the path γ from a to b. So we have a single-valued function $g : E \to \mathbb{C}$ given by $g(b) = f_\gamma(b)$. Now suppose that the value $g(b)$ is produced by a meromorphic continuation $(D, f) \sim (D_1, f_1) \sim \cdots \sim (D_m, f_m)$, along γ, so that $g(b) = f_m(b)$; there is no loss of generality in assuming that the disk D_m is contained in E.

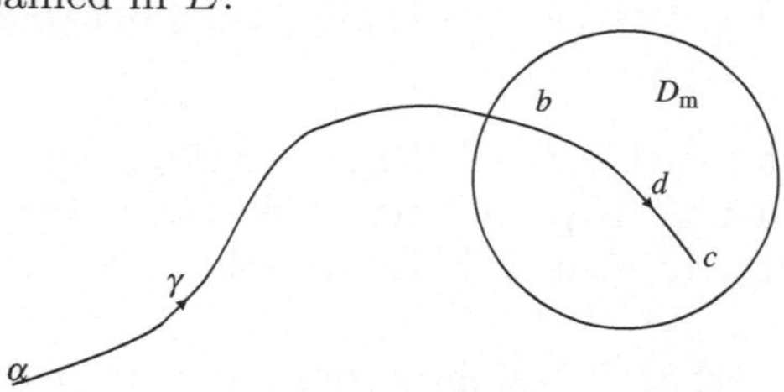

Fig. 16.10: The path δ in D_m from b to c

For each $c \in D_m$ there is a path δ in D_m from b to c, and hence there is a path α in E from a to c, following γ and then δ, given by

$$\alpha(s) = \begin{cases} \gamma(2s) & \text{if } 0 \le s \le \frac{1}{2} \\ \delta(2s-1) & \text{if } \frac{1}{2} \le s \le 1. \end{cases}$$

Since $\delta \subseteq D_m$, the sequence $(D, f) \sim (D_1, f_1) \sim \cdots \sim (D_m, f_m)$ is a meromorphic continuation along α, so $g(c) = f_\alpha(c) = f_m(c)$. Thus $g \equiv f_m$ on D_m, so g is meromorphic at b, since f_m is, and hence (E, g) is a function element. Since meromorphic continuation of (D, f) along paths within D must give $g(b) = f(b)$ for $b \in D$, we have $(E, g) \sim (D, f)$ as required. □

Thus we can continue a branch of $\log(z)$ from a neighborhood D of 1 onto a region $E = D_J$; where J is an open interval $(\alpha, \beta) \subset \mathbb{R}$ with $\beta - \alpha \leq 2\pi$, since such a region is simply connected. Taking $J = ((2n-1)\pi, (2n+1)\pi)$, with $n \in \mathbb{Z}$, we 'cut' $\mathbb{C}_\infty = \Sigma$ along the negative real line from 0 to ∞, as in Fig. 16.11, and the remaining region $E = D_j = \mathbb{C} \setminus \{z \in \mathbb{R} : z \leq 0\}$ is simply connected, so we have a branch f_J of $\log(z)$, defined on E. This branch, which

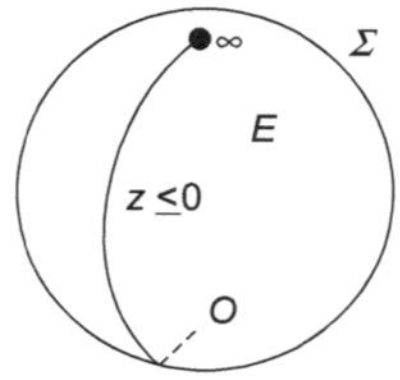

Fig. 16.11: The region $E = D_j = \mathbb{C} \setminus \{z \in \mathbb{R} : z \leq 0\}$

we shall denote by f_n, is determined by the value $f_n(1) = 2n\pi i$; the cut from 0 to ∞ prevents any path in E from winding around 0 (or equivalently, around ∞), so analytic continuation along paths in E is single-valued.

When we consider other many-valued functions we shall use a similar technique, cutting $\mathbb{C}_\infty$ along suitable lines to produce a simply connected region on which we can define a single-valued meromorphic branch of the function.

16.6 Exercises

1. Let f be bounded and holomorphic on the upper half plane $\text{Im}z \geq 0$ and real on the real axis. Prove that f is a constant.

2. If f is holomorphic on $\overline{B(0,1)}$ and real on the boundary then prove that f is a constant.

3. Let $U = \{z \in \mathbb{C} : \text{Im}z > 0\}$ and $f \in H(U)$. Assume that f is continuous on U along with the line segment $(a, b) \subset \mathbb{R}$. If $f(x) = x$ on (a, b), then $f(z) = z$ on U.

4. Let f be holomorphic on the half disk $\{|z| \leq 1\} \cap \{\text{Im}z > 0\}$ and real on the boundary. Define

$$g(z) = \begin{cases} f(z) & \text{if } \ |z| \leq 1 \text{ and } \text{Im}z > 0; \\ \overline{f(\frac{1}{\bar{z}})} & \text{if } \ |z| > 1 \text{ and } \text{Im}z > 0. \end{cases}$$

   Then g is holomorphic on the upper half plane.

5. If $f(z)$ is analytic in $|z| \leq 1$ and satisfies $|f| = 1$ on $|z| = 1$, show that $f(z)$ is rational.

6. Let (G, f) be a function element with $c \in \partial G$. If c is regular for (G, f) then show that $f(z)$ approaches some limit in $\mathbb{C}_\infty$ as $z \to c$ with $z \in G$.

7. Show that the series $f(z) = \sum_{n=1}^{\infty} z^{n!}$ has the unit circle $\mathbb{T} = \{z \in \mathbb{C} : |z| = 1\}$ as its natural boundary.

8. If a power series $f(z) = \sum_{n=0}^{\infty} a_n(z-a)^n$ has radius of convergence $\rho \neq 0, \infty$, then show that there is at least one point c on the circle $|z - a| = \rho$ at which direct analytic continuation is impossible.

9. Prove that the series $\sum_{n=0}^{\infty} z^{2^n}$ has the unit circle as its natural boundary.

10. Prove that the point $z = 1$ is a singular point for the power series $\sum_{n=1}^{\infty} \frac{z^n}{n^2}$.

11. The function $f(z) = \sum_{n=0}^{\infty} z^n$ is expanded into a Taylor series in the neighbourhood of the point $z = a(|a| < 1)$. For what values of a does this expansion permit analytic continuations of the function $f(z)$?

12. The sum of the power series $f(z) = \sum_{n=1}^{\infty} \frac{z^n}{n}$ is expanded into a Taylor series in the neighborhood of the point $z = -\frac{1}{2}$. What is the domain into which the function $f(z)$ will be thus continued?

13. The power series $f_1(z) = \sum_{n=1}^{\infty} \frac{z^n}{n}$ and $f_2(z) = i\pi + \sum_{n=1}^{\infty} (-1)^n \frac{(z-2)^n}{n}$ have no common domain of convergence. Prove, nevertheless, that the functions $f_1(z)$ and $f_2(z)$ are analytic continuations of one another.

14. Prove that the function represented by the following power series cannot be continued beyond the unit circle:

    (a) $f(z) = \sum_{n=0}^{\infty} z^{n!}$.

    (b) $f(z) = \sum_{n=0}^{\infty} \frac{z^{n!}}{n^2}$.

15. Prove that if α is an irrational number, then the series $\sum_{n=1}^{\infty} \frac{1}{2^n(z-e^{2in\pi\alpha})}$ represents, in the domains $|z| < 1$ and $|z| > 1$, analytic functions, for each of which the circle $|z| = 1$ is the natural boundary.

Chapter 17

Representation of Meromorphic Functions

17.1 Rational functions

A function $R(z)$ which can be written as $R(z) = \frac{P(z)}{Q(z)}$ where P and Q are polynomials is called a rational function. Suppose $P(z)$ and $Q(z)$ have no common factors. The function $R(z)$ takes the value ∞ at the zeros of $Q(z)$ and therefore be considered as a function with values in the extended complex plane, and as such it is continuous. Define $R(\infty)$ as the limit of $R(z)$ as $z \to \infty$. Thus $R(\infty) = R_1(0)$ where $R_1(z) = R\left(\frac{1}{z}\right)$. If $R_1(0) = 0$ or ∞, the order of the zero or pole at ∞ is defined as the order of the zero or pole of $R_1(z)$ at the origin. Let $R(z) = \frac{a_0+a_1z+\cdots+a_nz^n}{b_0+b_1z+\cdots+b_mz^m}$. Then $R_1(z) = z^{m-n}\frac{a_0z^n+a_1z^{n-1}+\cdots+a_n}{b_0z^m+b_1z^{m-1}+\cdots+b_m}$. If $m > n$, the rational function $R(z)$ has a zero of order $m-n$ at ∞, if $m < n$, the rational function $R(z)$ has a pole of order $n-m$ at ∞, and if $m = n$, $R(\infty) = \frac{a_n}{b_m} \neq 0, \infty$.

We can now count the total number of zeros and poles in the extended plane. The count shows that the number of zeros, including those at ∞, is equal to the greater of the numbers m and n. The number of poles is the same. This common number of zeros and poles is called the order of the rational function. Further, if a is any constant, the rational function $R(z) - a$ has the same poles as $R(z)$ and consequently they have the same order. Note that the zeros of $R(z) - a$ are roots of the equation $R(z) = a$, and if the roots are counted as many times as the order of the zero indicates, we have the following:

A rational function $R(z)$ of order p has p zeros and p poles, and every equation $R(z) = a$ has exactly p roots.

We will now show that every rational function has a representation by partial fractions. Let $R(z) = \frac{S(z)}{Q(z)}$ be a rational function where S and Q are two polynomials and $\deg Q < \deg S$. We carry out the division of $S(z)$ by $Q(z)$ until the degree of the remainder $P(z)$ is less than that of the denominator

$Q(z)$. Then we get

$$R(z) = G(z) + H(z)$$

where $H(z) = \frac{P(z)}{Q(z)}$. Without loss of generality, we assume that

$$Q(z) = (z - z_1)^{k_1} \cdots (z - z_n)^{k_n}.$$

Thus $H(z) = \frac{P(z)}{Q(z)} = \frac{P(z)}{(z-z_1)^{k_1} \cdots (z-z_n)^{k_n}}$. Since H has a pole of order at most k_1 at z_1,

$$H(z) = P_1 \left(\frac{1}{z - z_1} \right) + A_1(z),$$

where $P_1 \left(\frac{1}{z-z_1} \right)$ is the principal part of H around z_1 and A_1 is the analytic part. Furthermore, $A_1(z) = H(z) - P_1 \left(\frac{1}{z-z_1} \right)$ has a removable singularity at z_1 and the same principal parts as H at $z_2, \cdots, z_n$.

Thus if we take $P_2 \left(\frac{1}{z-z_2} \right)$ to be the principal part of H around z_2, and proceed inductively, we find that

$$A_n = H(z) - \left[P_1 \left(\frac{1}{z - z_1} \right) + \cdots + P_n \left(\frac{1}{z - z_n} \right) \right]$$

is an entire function (once it is correctly defined at $z_k, 1 \leq k \leq n$). Furthermore, A_n is bounded since H and all its principal parts approach zero as $z \to \infty$. Hence A_n is a constant, by Liouville's theorem. Indeed, $A_n \equiv 0$. Hence

$$H(z) = P_1 \left(\frac{1}{z - z_1} \right) + \cdots + P_n \left(\frac{1}{z - z_n} \right).$$

Definition 17.1. *A meromorphic function is rational meromorphic if $z = \infty$ is a regular point or a pole and transcendental meromorphic if $z = \infty$ is an essential singularity.*

Theorem 17.1. *A rational meromorphic function must be a rational function (i.e., quotient of two polynomials).*

Proof. Let f be a rational meromorphic function. Let the points $z_0, z_1, \cdots, z_n$ (one of which may be the point at ∞) include all the poles. The number of poles is finite, otherwise, they would possess a point of accumulation in the whole plane and the function would possess an essential singularity.

Let $P_0 \left(\frac{1}{z-z_0} \right), \cdots, P_n \left(\frac{1}{z-z_n} \right)$ denote the principal parts of the Laurent expansion of $f(z)$ about $z_0, \cdots, z_n$, respectively. Then $f(z) - \sum_{i=0}^{n} P_i \left(\frac{1}{z-z_i} \right)$ is analytic in the whole complex plane including the point at ∞. Thus it is uniformly bounded (since $\mathbb{C}_\infty$ is compact) and by Liouville's theorem it may be shown to be constant. Hence $f(z) = \sum_{i=0}^{n} P_i \left(\frac{1}{z-z_i} \right) + C$. However, the $P_i's$

are polynomials in the respective arguments, therefore, $f(z)$ is the quotient of two polynomials, i.e.,

$$f(z) = \frac{P(z)}{Q(z)};$$

and the equation gives the decomposition of $f(z)$ into its partial fractions where the z_r are the zeros of $Q(z)$. □

Lemma 17.1. *Suppose f is analytic in the entire (finite) complex plane $\mathbb{C}$ except for poles at $z_1, z_2, \cdots, z_k$. Let $P_r\left(\frac{1}{z-z_r}\right)$ be the principal part of $f(z)$ at $z = z_r, r = 1, \cdots, k$. Then there exists a function $\phi(z)$, analytic in the entire finite complex plane such that*

$$f(z) = \sum_{r=1}^{k} P_r\left(\frac{1}{z-z_r}\right) + \phi(z).$$

Moreover, $\phi(z)$ has the same principal part at $z = \infty$ as $f(z)$.

Proof. Consider $g(z) = \sum_{r=1}^{k} P_r\left(\frac{1}{z-z_r}\right)$ where $P_r\left(\frac{1}{z-z_r}\right) = \sum_{j=1}^{m_r} \frac{b_{jr}}{(z-z_r)^j}$ and m_r is the order of the pole of f at $z = z_r$. Clearly g is analytic in the entire finite complex plane except at points $z_1, \cdots, z_k$, where it has poles. Since for $s \neq r$, $P_s\left(\frac{1}{z-z_s}\right)$ is analytic at $z = z_r$, the principal part of $g(z)$ at $z = z_r$ is $P_r\left(\frac{1}{z-z_r}\right)$. Hence g has the same poles and principal parts at $z_1, \cdots, z_k$ as $f(z)$ and is analytic everywhere else. Consider $\phi(z) = f(z) - g(z),\ z \neq z_r,\ r = 1, 2, \cdots$. This function is analytic in the entire complex plane except for possible poles at $z_1, z_2, \cdots, z_k$. Since $g(z)$ and $f(z)$ have the same poles and principal parts, the difference $\phi(z)$ is such that the principal part at each pole is zero. Thus for

$$\begin{aligned} f(z) &= \tfrac{b_{-n}}{(z-a)^n} + \cdots + \tfrac{b_{-1}}{z-a} + b_0 + b_1(z-a) + \cdots, \\ g(z) &= \tfrac{b_{-n}}{(z-a)^n} + \cdots + \tfrac{b_{-1}}{z-a}, \\ \phi(a) &= b_0, \end{aligned}$$

and $\phi(z)$ has removable singularities at $z_1, \cdots, z_k$. Thus $\phi(z)$ is analytic. Since $g(z) \to 0, z \to \infty, g(z)$ has a removable singularity at $z = \infty$. Thus $\phi(z)$ has the same principal part at $z = \infty$ as $f(z)$. Writing $f(z) = g(z) + \phi(z)$, the lemma is proved. □

A rational meromorphic function is determined to within an additive constant by the behavior at its poles. We want to find how completely an arbitrary meromorphic function is determined by the character of its poles. We now prove the Mittag-Leffler theorem which represents a meromorphic function by an infinite decomposition into partial fractions.

Since a meromorphic function may be characterized by the nature of its poles only to within an added entire function and since the question has been answered for meromorphic functions with a finite number of poles, the only other possibility to answer is when ∞ is an accumulation point of poles. Clearly no finite point is an accumulation point of poles for in that case $\frac{1}{f} \equiv 0$.

Theorem 17.2. ***(Mittag-Leffler theorem):*** *Let $z_0, z_1, \cdots, z_n, \cdots$ be any sequence of distinct points tending to ∞. Suppose to each z_n there is associated a polynomial $P_n\left(\frac{1}{z-z_n}\right)$ in the variable $\frac{1}{z-z_n}$. It is possible to find a meromorphic function $f(z)$ having poles at the points z_n but no other points, and with corresponding principal parts $P_n\left(\frac{1}{z-z_n}\right)$. Then $f(z)$ may be represented in the form*

$$f(z) = \phi(z) + \sum_{\nu=0}^{\infty} \left\{ P_\nu\left(\frac{1}{z - z_\nu}\right) - q_\nu(z) \right\}$$

where $q_\nu(z)$ are polynomials and $\phi(z)$ is an entire function of z.

Proof. We must first assume that the given series converges. Suppose the sequence $\{z_n\}$ is ordered. Thus $|z_0| \leq |z_1| \leq \cdots$ since the only point of accumulation is infinity; possibly $z_0 = 0$, but all other points differ from zero. We suppose that $z_0 \neq 0$ and the function $P_\nu\left(\frac{1}{z-z_\nu}\right)$ being analytic everywhere except at z_ν, must itself be analytic at the origin. Thus P_ν has a Taylor series expansion at the origin given by $P_\nu\left(\frac{1}{z-z_\nu}\right) = a_0^{(\nu)} + a_1^{(\nu)} z + a_2^{(\nu)} z^2 + \cdots$. The radius of convergence is clearly $|z_\nu|$. The series converges uniformly in the disk $D_\nu = \{z \in \mathbb{C} : |z| \leq \frac{|z_\nu|}{2}\}$. Thus in D_ν, $P_\nu\left(\frac{1}{z-z_\nu}\right)$ can be approximated by finite sum as closely as we please. In particular, for $q_\nu(z) = a_0^{(\nu)} + a_1^{(\nu)} z + \cdots + a_{k_\nu}^{(\nu)} z^{k_\nu}$,

$$\left| P_\nu\left(\frac{1}{z - z_\nu}\right) - q_\nu(z) \right| < \frac{1}{2^\nu}$$

throughout D_ν. The series $\sum_{\nu=0}^{\infty} \left\{ P_\nu\left(\frac{1}{z-z_\nu}\right) - q_\nu(z) \right\}$ converges to the desired meromorphic function in every disk about the origin since any such disk can be contained within one of the $D_\nu' s$.

In D_ν, $\sum_{n=0}^{\nu-1} \left\{ P_n\left(\frac{1}{z-z_n}\right) - q_n(z) \right\}$ is well-behaved. It is an analytic function with no singularities but the prescribed poles. Thus

$$\sum_{n=\nu}^{\infty} \left\{ P_n\left(\frac{1}{z - z_n}\right) - q_n(z) \right\}$$

is an analytic function in D_ν and dominated by $\sum_{n=\nu}^{\infty} \frac{1}{2^n}$. The series $\sum_{n=\nu}^{\infty} \left\{ P_n\left(\frac{1}{z-z_n}\right) - q_n(z) \right\}$ then converges uniformly in D_ν. Thus by Weierstrass theorem (Theorem 7.10), since a uniformly convergent series of

analytic functions converges to an analytic function, the second part of the series
introduces no new singularities into D_ν. If $z_0 = 0$, we add on $P_0\left(\frac{1}{z}\right)$. The theorem is proved. □

Example 17.1. *Consider the function* $f(z) = \frac{\pi^2}{\sin^2 \pi z}$. *The function* f *has double poles at the points* $z = n, n \in \mathbb{Z}$. *The singular part at the origin is* $\frac{1}{z^2}$ *and at* $z = n$, *the singular part is* $\frac{1}{(z-n)^2}$. *This is so, as* $\sin^2 \pi(z-n) = \sin^2 \pi z$. *The series* $\sum_{n=-\infty}^{\infty} \frac{1}{(z-n)^2}$ *is convergent for* $z \neq n$, *as seen by comparing with the series* $\sum_{n=1}^{\infty} \frac{1}{n^2}$. *The series* $\sum_{n=-\infty}^{\infty} \frac{1}{(z-n)^2}$ *is uniformly convergent on any compact set after omission of the terms which become infinite on the set. So we can write*

$$\frac{\pi^2}{\sin^2 \pi z} = \sum_{n=-\infty}^{\infty} \frac{1}{(z-n)^2} + g(z)$$

where g *is an entire function. We shall show that* g *is identically zero. The entire function* g *is periodic with period* 1 *as* $\frac{\pi^2}{\sin^2 \pi z}$ *and the series* $\sum_{n=-\infty}^{\infty} \frac{1}{(z-n)^2}$ *are both periodic with the period* 1. *For* $z = x + iy$, *we have*

$$|\sin \pi z|^2 = \cosh^2 \pi y - \cos^2 \pi x$$

and hence $\frac{\pi^2}{\sin^2 \pi z}$ *tends uniformly to* 0 *as* $|y| \to \infty$. *The series* $\sum_{n=-\infty}^{\infty} \frac{1}{(z-n)^2}$ *also tend to zero as* $|y| \to \infty$. *Indeed, the convergence is uniform for* $|y| \geq 1$ *and the limit for* $|y| \to \infty$ *can thus be obtained by taking the limit in each term. We conclude that* $g(z)$ *tends uniformly to* 0 *for* $|y| \to \infty$. *This is sufficient to infer that* $|g(z)|$ *is bounded in a period strip* $0 \leq x \leq 1$, *and because of the periodicity* $|g(z)|$ *will be bounded in the whole plane. By Liouville's theorem* $g(z)$ *must reduce to a constant, and since the limit is* 0 *the constant must vanish. We have thus proved the identity*

$$\frac{\pi^2}{\sin^2 \pi z} = \sum_{n=-\infty}^{\infty} \frac{1}{(z-n)^2}. \tag{17.1}$$

If we now integrate the left-hand side of (17.1), we obtain $-\pi \cot \pi z$ *and the terms on the right are derivatives of* $-\frac{1}{z-n}$. *The series* $\sum_{n=-\infty}^{\infty} \frac{1}{z-n}$ *diverges and a partial sum of the Taylor series must be subtracted from all the terms with* $n \neq 0$. *In this case it is sufficient to subtract the constant terms, since*

$$\sum_{n \neq 0} \left(\frac{1}{z-n} + \frac{1}{n} \right) = \sum_{n \neq 0} \frac{z}{n(z-n)}$$

is comparable with $\sum_{n=1}^{\infty} \frac{1}{n^2}$ *and hence convergent. The convergence is uniform on every compact set, provided that we omit the terms which become infinite.*

Thus termwise differentiation is permissible, and we obtain

$$\pi \cot \pi z = \frac{1}{z} + \sum_{n \neq 0} \left(\frac{1}{z-n} + \frac{1}{n} \right) \tag{17.2}$$

except for an additive constant. If the terms corresponding to n and −n are bracketed together, (17.2) can be written as

$$\pi \cot \pi z = \lim_{m \to \infty} \sum_{n=-m}^{m} \frac{1}{z-n} = \frac{1}{z} + \sum_{n=1}^{\infty} \frac{2z}{z^2 - n^2}. \tag{17.3}$$

It becomes evident now that both members of the equation (17.2) are odd functions of z, and for this reason that integration constant must vanish. The equations (17.2) and (17.3) are therefore stated correctly.

17.2 Infinite products

Consider an infinite product of complex numbers $p_1 p_2 \cdots p_n \cdots = \prod_{n=1}^{\infty} p_n$. The product $P_m = p_1 p_2 \cdots p_m$ is called the mth partial product of the infinite product and p_n is called the nth factor of the infinite product.

Definition 17.2. *The infinite product $\prod_{n=1}^{\infty} p_n$ is said to converge if and only if at most a finite number of the factors are zero, and if the partial product formed by the nonvanishing factors tend to a finite limit P which is different from zero and we write $P = \lim\limits_{m \to \infty} P_m$.*

Note that there are good reasons for excluding the value zero. If the value $P = 0$ were allowed, any infinite product with one factor 0 would converge, and the convergence would not depend on the whole sequence of factors. Further the limit P can be zero without any of the terms p_n being zero (e.g., $p_n = \frac{1}{n}$). Convergence ought to depend on all of the factors (or at least all but finitely many). Thus for convergence of the product if we only require that the limit $\lim\limits_{n \to \infty} \prod\limits_{n=1}^{N} p_n = P$ exist then it will be inadequate for the above reasons.

Notice that in a convergent product the general factor p_n tends to 1. This follows from the expression that $p_n = \frac{P_n}{P_{n-1}}$, the zero factors being omitted. Thus we shall write all infinite products in the form $\prod_{n=1}^{\infty}(1 + a_n)$ so that $a_n \to 0$ is a necessary condition for convergence. The following holds.

Theorem 17.3. *The infinite product $\prod_{n=1}^{\infty}(1 + a_n)$ with $1 + a_n \neq 0$ converges if and only if the series*

$$\sum_{n=1}^{\infty} \log(1 + a_n) \tag{17.4}$$

whose terms represent the values of the principal branch of the logarithm converges.

Proof. Let $S_n = \sum_{k=1}^{n} \log(1+a_k)$. Then $P_n = e^{S_n}$. Thus if $S_n \to S$ it follows that P_n tends to $P = e^S \neq 0$. That is, the infinite product $\prod_{n=1}^{\infty}(1+a_n)$ converges. Now suppose that $P_n \to P \neq 0$. We shall denote the principal value of the logarithm by Log and its imaginary part by Arg. Because $\frac{P_n}{P} \to 1$, hence $Log\left(\frac{P_n}{P}\right) \to 0$ as $n \to \infty$. Further, there exists an integer h_n such that $Log\left(\frac{P_n}{P}\right) = S_n - Log P + h_n \cdot 2\pi i$. It follows therefore that

$$(h_{n+1} - h_n)\, 2\pi i = Log\left(\frac{P_{n+1}}{P}\right) - Log\left(\frac{P_n}{P}\right) - Log(1+a_n).$$

Hence $(h_{n+1} - h_n)\, 2\pi = Arg\left(\frac{P_{n+1}}{P}\right) - Arg\left(\frac{P_n}{P}\right) - Arg\,(1+a_n)$. Observe that $|Arg(1+a_n)| \leq \pi$, and we know that $Arg\left(\frac{P_{n+1}}{P}\right) - Arg\left(\frac{P_n}{P}\right) \to 0$ as $n \to \infty$.

Thus for large n, this is incompatible with the previous equation unless $h_{n+1} = h_n$. Hence h_n is equal to a fixed integer h ultimately and it follows from $Log\left(\frac{P_n}{P}\right) = S_n - Log P + h \cdot 2\pi i$ that $S_n \to Log P - h \cdot 2\pi i$. The theorem is proved. □

It thus follows from the Cauchy criterion for series that the infinite product $\prod_{n=1}^{\infty}(1+a_n)$ is convergent if for each $\epsilon > 0$ there is an integer N such that

$$\left| \prod_{n=n_1}^{n_2} (1+a_n) - 1 \right| < \epsilon$$

for $n_2 \geq n_1 \geq N$. Further, it also follows that the infinite product $\prod_{n=1}^{\infty}(1+a_n)$ is convergent if, and only if, $\alpha^{(m)} = \lim_{N\to\infty} \prod_{n=m}^{N} (1+a_n)$ exists and is nonzero for at least one value of m.

An infinite product $\prod_{n=1}^{\infty}(1+a_n)$ is said to be absolutely convergent if and only if the series (17.4) converges absolutely.

Theorem 17.4. *The infinite product $\prod_{n=1}^{\infty}(1+a_n)$ converges absolutely if and only if the series $\sum_{n=1}^{\infty} |a_n|$ converges.*

Proof. Observe that

$$\lim_{z\to 0} \frac{\log(1+z)}{z} = 1. \tag{17.5}$$

If either the series (17.4) or $\sum_{n=1}^{\infty} |a_n|$ converges, we have $a_n \to 0$, and from (17.5) it follows that given $\epsilon > 0$, there exists n_0 such that

$$(1-\epsilon)\,|a_n| < |\log(1+a_n)| < (1+\epsilon)\,|a_n|$$

for all $n \geq n_0$. Thus the two series are simultaneously absolutely convergent. □

Notice that if $\alpha = \prod_{n=1}^{\infty}(1 + a_n)$ is absolutely convergent, then for any rearrangement, $1 + \beta_m = 1 + a_{n_m}$, we have $\alpha = \prod_{m=1}^{\infty}(1 + \beta_m)$.

Products of functions

In the products to be considered the numbers $(1 + a_n)$ will be dependent analytically on the variable z. Usually $1 + a_n(z)$ will be entire for each n, and it will be assumed that the series $\sum_{n=1}^{\infty} |a_n(z)|$ converges uniformly on every disk of finite radius. For fixed z it follows from Theorem 17.4 that the product

$$P(z) = \prod_{n=1}^{\infty} (1 + a_n(z)) \tag{17.6}$$

is absolutely convergent. Letting z vary it will develop that $P(z)$ is itself an entire function. Our assumption on the sequence $\{a_n(z)\}$ is sufficient to guarantee for every $R, 0 \le R < \infty$, that

$$M_R = \sup_{|z| \le R} \sum_{n=1}^{\infty} |a_n(z)| \tag{17.7}$$

be finite. For each N, define

$$P_N(z) = \prod_{n=1}^{N} (1 + a_n(z)) . \tag{17.8}$$

Since $|1 + a_n(z)| \le 1 + |a_n(z)| \le e^{|a_n(z)|}$ for all n and z, it follows from (17.7) that if $|z| \le R$, then

$$\max_{|z| \le R} |P_N(z)| \le \max_{|z| \le R} e^{\sum_{n=1}^{N} |a_n(z)|} = e^{M_R}.$$

The family $\{P_N\}$ is therefore locally bounded, and for each z we have $\lim_{N \to \infty} P_N(z) = P(z)$. Thus from Vitali's theorem, it follows that the sequence $\{P_N(z)\}$ converges locally uniformly to an entire function $P(z)$. Moreover the series $\sum_{n=1}^{\infty} |a_n(z)|$ converges uniformly for $|z| \le R$, if, and only if, for some m, the sum

$$\sum_{n=m}^{\infty} |\log (1 + a_n(z))|$$

converges uniformly on the same set.

If we differentiate both sides of (17.8), we obtain letting $\alpha_n(z) = 1 + a_n(z)$, the relation

$$P'_N(z) = \alpha'_1(z) \prod_{n=2}^{N} \alpha_n(z) + \alpha'_2(z) \prod_{\substack{n=1 \\ n \ne 2}}^{N} \alpha_n(z) + \cdots + \alpha'_n(z) \prod_{n=1}^{N-1} \alpha_n(z)$$

which, if $P_N(z) \neq 0$, is equivalent to

$$\begin{aligned} P_N'(z) &= \alpha_1'(z) \cdot \frac{P_N(z)}{\alpha_1(z)} + \alpha_2'(z) \cdot \frac{P_N(z)}{\alpha_2(z)} + \cdots + \alpha_N'(z) \cdot \frac{P_N(z)}{\alpha_N(z)} \\ &= P_N(z) \left\{ \frac{\alpha_1'(z)}{\alpha_1(z)} + \cdots + \frac{\alpha_N'(z)}{\alpha_N(z)} \right\}. \end{aligned} \tag{17.9}$$

For the logarithmic derivative we obtain the expression (for $P_N(z) \neq 0$)

$$\frac{d}{dz} \log P_N(z) = \frac{P_N'(z)}{P_N(z)} = \frac{\alpha_1'(z)}{\alpha_1(z)} + \cdots + \frac{\alpha_N'(z)}{\alpha_N(z)}.$$

Since $\{P_N\}$ converges locally uniformly to P, the sequence $\{P_N'\}$ of derivatives converges locally uniformly to P'. Therefore, by (17.9) we have for $P(z) \neq 0$, $P'(z) = \lim_{N\to\infty} P_N'(z)$

$$= \lim_{N\to\infty} P_N(z) \left\{ \frac{\alpha_1'(z)}{\alpha_1(z)} + \cdots + \frac{\alpha_N'(z)}{\alpha_N(z)} \right\} = P(z) \sum_{n=1}^{\infty} \frac{\alpha_n'(z)}{\alpha_n(z)}.$$

Similarly, $\frac{d}{dz} \log P(z)$ is given by

$$\frac{P'(z)}{P(z)} = \sum_{n=1}^{\infty} \frac{\alpha_n'(z)}{\alpha_n(z)}$$

if $P(z) \neq 0$.

17.3 Infinite product representation of entire functions

In this section we shall discuss about the infinite product representation of an entire function. The concept of "genus" and "order" will be dealt with in more detail. This leads to a further object for studying entire functions, namely "type". We start with a well-known and very illuminating theorem of Weierstrass on factorization of entire functions.

The Weierstrass Factorization theorem

Since an entire function is an analytic function with no singularities except at ∞, consider an entire function $f(z)$ which has m zeros at the origin (m may be zero), and denote the other zeros by $a_1, a_2, \cdots, a_N$, multiple zeros being repeated. The function $f(z)$ can then be factored as

$$f(z) = z^m e^{g(z)} \prod_{n=1}^{N} \left(1 - \frac{z}{a_n}\right)$$

where g is an entire function.

Thus given any finite set of points $\{a_1, a_2, \cdots, a_n\}$, there always exists an entire function with zeros at $a_1, a_2, \cdots, a_n$. Consider now an infinite sequence of points $a_1, a_2, \cdots, a_n, \cdots$ such that

$$0 < |a_1| \leq |a_2| \leq \cdots \leq |a_n| \leq \cdots$$

and $\lim_{n\to\infty} |a_n| = \infty$. If we try to construct an entire function which vanishes at these points and nowhere else then we think of a similar representation by means of an infinite product. The obvious generalization would be

$$f(z) = z^m e^{g(z)} \prod_{n=1}^{\infty} \left(1 - \frac{z}{a_n}\right). \tag{17.10}$$

But the above product may diverge. This representation is valid only if the infinite product converges uniformly on every compact set. This is accomplished by Weierstrass factorization theorem, which demonstrates the construction and existence of such a function.

Note that the product in (17.10) converges absolutely if and only if $\sum_{n=1}^{\infty} \frac{1}{|a_n|} < \infty$, and in this case the convergence is uniform in every closed disk $|z| \leq R$. It is only under this special condition that we can obtain a representation of the form (17.10). In the general case convergence producing factor must be introduced. This is the content of the Weierstrass factorization theorem.

Theorem 17.5. ***(Weierstrass factorization theorem)***: *Let $\{a_n\}_{n=1}^{\infty}$ be an infinite sequence (this is not a sequence of distinct points, but no point is repeated an infinite number of times) of complex numbers different from zero such that $\lim_{n\to\infty} |a_n| = \infty$. Suppose m is a nonnegative integer. Then there exists an entire function with zeros at $a_n, n = 1, 2, 3, \cdots$ and a zero of multiplicity m at the origin and no other zeros. Every entire function with these and no other zeros can be written in the form*

$$f(z) = z^m e^{g(z)} \prod_{n=1}^{\infty} \left(1 - \frac{z}{a_n}\right) e^{\frac{z}{a_n} + \frac{1}{2}\left(\frac{z}{a_n}\right)^2 + \cdots + \frac{1}{m_n}\left(\frac{z}{a_n}\right)^{m_n}} \tag{17.11}$$

where g is an entire function and m_n are certain integers such that

$$\sum_{n=1}^{\infty} \left(\frac{r}{|a_n|}\right)^{m_n+1} < \infty$$

for all $r > 0$ and the infinite product

$$\prod_{n=1}^{\infty} \left(1 - \frac{z}{a_n}\right) e^{\frac{z}{a_n} + \frac{1}{2}\left(\frac{z}{a_n}\right)^2 + \cdots + \frac{1}{m_n}\left(\frac{z}{a_n}\right)^{m_n}}$$

converges uniformly on every compact subset of $\mathbb{C}$. *Further, if* $m_n = n-1$, *then (17.11) will be satisfied.*

Proof. Notice that the infinite product

$$\prod_{n=1}^{\infty}\left(1-\frac{z}{a_n}\right)e^{p_n(z)}$$

$$=\prod_{n=1}^{\infty}\left(1-\frac{z}{a_n}\right)e^{\frac{z}{a_n}+\frac{1}{2}\left(\frac{z}{a_n}\right)^2+\cdots+\frac{1}{m_n}\left(\frac{z}{a_n}\right)^{m_n}} \tag{17.12}$$

converges simultaneously with the series $\sum_{n=1}^{\infty} r_n(z)$ where

$$r_n(z)=\log\left(1-\frac{z}{a_n}\right)+p_n(z)$$

and the branch of the logarithm shall be chosen so that $-\pi < \mathrm{Im} r_n(z) \le \pi$. For a given R we consider only the terms with $|a_n| > R$. In the disk $|z| \le R$, the function can have only finitely many zeros. Expand the principal branch of $\log\left(1-\frac{z}{a_n}\right)$ as a Taylor series

$$\log\left(1-\frac{z}{a_n}\right)=-\frac{z}{a_n}-\frac{1}{2}\left(\frac{z}{a_n}\right)^2-\frac{1}{3}\left(\frac{z}{a_n}\right)^3-\cdots.$$

Let

$$p_n(z)=\frac{z}{a_n}+\frac{1}{2}\left(\frac{z}{a_n}\right)^2+\cdots+\frac{1}{m_n}\left(\frac{z}{a_n}\right)^{m_n}.$$

Then

$$r_n(z)=-\frac{1}{m_n+1}\left(\frac{z}{a_n}\right)^{m_n+1}-\frac{1}{m_n+2}\left(\frac{z}{a_n}\right)^{m_n+2}-\cdots$$

and

$$|r_n(z)| \le \frac{1}{m_n+1}\left(\frac{R}{|a_n|}\right)^{m_n+1}\frac{1}{1-\frac{R}{|a_n|}}.$$

If the series

$$\sum_{n=1}^{\infty}\frac{1}{m_n+1}\left(\frac{R}{|a_n|}\right)^{m_n+1} \tag{17.13}$$

converges then by comparison the series $\sum_{n=1}^{\infty} r_n(z)$ converges uniformly and absolutely for $|z| \le R$. Note that $r_n(z) \to 0$ and if n is sufficiently large, then $-\pi < \mathrm{Im} r_n(z) \le \pi$. Thus the infinite product (17.12) represents an analytic function in $|z| < R$. We have only considered those a_n such that $|a_n| > R$ and we have excluded the a_n's with $|a_n| \le R$. But the uniform convergence of (17.12) is not affected when the corresponding factors are again taken into account. If we take $m_n = n-1$, the series (17.13) converges for all R. This is so as, in this case (17.13) has a majorant geometric series with ratio less than 1 for any fixed value of R. □

The representation (17.11) becomes considerably more interesting if it is possible to choose all the m_n equal to each other. The preceding proof has shown that the product

$$\prod_{n=1}^{\infty}\left(1-\frac{z}{a_n}\right)e^{\frac{z}{a_n}+\frac{1}{2}\left(\frac{z}{a_n}\right)^2+\cdots+\frac{1}{h}\left(\frac{z}{a_n}\right)^h}$$

converges and represents an entire function provided that the series

$$\sum_{n=1}^{\infty}\frac{\left(\frac{R}{|a_n|}\right)^{h+1}}{h+1}$$

converges for all R, that is to say provided that $\sum_{n=1}^{\infty}\frac{1}{|a_n|^{h+1}}<\infty$. Assume that h is the smallest integer for which this series converges; the product above is then called the **canonical product** associated with the sequence $\{a_n\}$, and h is the **genus** of the canonical product. Whenever possible we use the canonical product in the representation (17.11), which is thereby uniquely determined. If in this representation $g(z)$ reduces to a polynomial, the function $f(z)$ is said to be of finite genus, and the **genus** of $f(z)$ is by definition equal to the degree of this polynomial or to the genus of the canonical product, whichever is the larger. If in the representation (17.11), $g(z)$ is not a polynomial, then the genus of f is infinite.

Example 17.2. *As an example of Weierstrass factorization theorem, we shall express* $\sin\pi z$ *as an infinite product. The zeros are* $z=\pm n$ *and all are of order one (simple zeros). Arrange the zeros in a sequence* $0,+1,-1,+2,-2,\cdots$. *In this case* $a_n=\pm n$. *Since* $\sum_{n=1}^{\infty}\frac{1}{n}=\infty$ *and* $\sum_{n=1}^{\infty}\frac{1}{n^2}<\infty$, *we have* $h=1$. *Further,* $m_j=1$ *(as each* a_n *is a simple zero of* $\sin\pi z$*). So we consider*

$$z\prod_{n=1}^{\infty}\left(1-\frac{z}{n}\right)e^{\frac{z}{n}}\left(1+\frac{z}{n}\right)e^{-\frac{z}{n}}=z\prod_{n=1}^{\infty}\left(1-\frac{z^2}{n^2}\right).$$

Thus

$$\sin\pi z=e^{g(z)}\cdot z\prod_{n=1}^{\infty}\left(1-\frac{z^2}{n^2}\right).$$

Taking the logarithmic derivative

$$\frac{\pi\cos\pi z}{\sin\pi z}=g'(z)+\frac{1}{z}+\sum_{n=1}^{\infty}\frac{2z}{z^2-n^2}.$$

If we use the fact that $\pi \cot \pi z = \frac{1}{z} + \sum_{n=1}^{\infty} \frac{2z}{z^2-n^2}$, *then* $g'(z) = 0$ *and* $g(z) =$ *constant. Since* $\frac{\sin \pi z}{\pi z} \to 1$ *as* $z \to 0$, *hence*

$$e^{g(z)} = \pi$$

and

$$\sin \pi z = \pi z \prod_{n=1}^{\infty} \left(1 - \frac{z^2}{n^2}\right).$$

17.4 Jensen's formula

If f is analytic in an open set $G \supset \overline{B(0,\rho)}$ and f does not vanish in $\overline{B(0,\rho)}$ then $\log|f|$ is harmonic in an open set in G containing $\overline{B(0,\rho)}$. Hence it satisfies the mean value property (MVP). That is,

$$\log|f(0)| = \frac{1}{2\pi} \int_0^{2\pi} \log|f(\rho e^{i\theta})| d\theta \tag{17.14}$$

and $\log|f(z)|$ can be expressed by Poisson's formula. Suppose f has exactly one zero at the point $a = \rho e^{i\theta_0}$ on the circle $|z| = \rho$. Consider the function $g(z) = \frac{f(z)}{z-a}$. Then g has no zero in $\overline{B(0,\rho)}$. Hence (17.14) can be applied to g to obtain

$$\log|g(0)| = \frac{1}{2\pi} \int_0^{2\pi} \left[\log\left|f\left(\rho e^{i\theta}\right)\right| - \log\left|\rho e^{i\theta} - \rho e^{i\theta_0}\right|\right] d\theta.$$

Now $\log|g(0)| = \log|f(0)| - \log\rho$ and

$$\begin{aligned}
\tfrac{1}{2\pi} \int_0^{2\pi} \log\left|\rho e^{i\theta} - \rho e^{i\theta_0}\right| d\theta &= \tfrac{1}{2\pi} \int_0^{2\pi} \left[\log\rho + \log\left|e^{i\theta} - e^{i\theta_0}\right|\right] d\theta \\
&= \log\rho + \tfrac{1}{2\pi} \int_0^{2\pi} \log\left|e^{i\theta} - e^{i\theta_0}\right| d\theta \\
&= \log\rho + \tfrac{1}{2\pi} \int_0^{2\pi} \log\left|1 - e^{i(\theta-\theta_0)}\right| d\theta \\
&= \log\rho + \tfrac{1}{2\pi} \int_0^{2\pi} \log\left|1 - e^{i\psi}\right| d\psi
\end{aligned}$$

$$\begin{aligned}
&= \log\rho + \tfrac{1}{2\pi} \int_0^{2\pi} \log\left(2\sin\frac{\psi}{2}\right) d\psi \\
&= \log\rho + \log 2 + \tfrac{1}{2\pi} \int_0^{2\pi} \log\sin\frac{\psi}{2} d\psi \\
&= \log\rho + \log 2 + \tfrac{2}{2\pi} \int_0^{\pi} \log\sin\phi d\phi \\
&= log\rho + \log 2 + \tfrac{1}{\pi}(-\pi\log 2) = \log\rho.
\end{aligned}$$

Thus we get

$$\log|f(0)| - \log\rho = \frac{1}{2\pi}\int_0^{2\pi} \log\left|f\left(\rho e^{i\theta}\right)\right| d\theta - \log\rho.$$

Hence

$$\log|f(0)| = \frac{1}{2\pi}\int_0^{2\pi} \log\left|f\left(\rho e^{i\theta}\right)\right| d\theta$$

and (17.14) remains valid if f has a single zero on $|z| = \rho$. In fact, (17.14) is valid as long as f has no zeros in $B(0,\rho)$.

We shall now examine what happens if f has zeros inside $B(0,\rho)$. In this case, $\log|f(z)|$ is no longer harmonic.

Theorem 17.6. *(Jensen's formula) Let f be analytic in a region containing $\overline{B(0,\rho)}$ and suppose that $a_1, \cdots, a_n$ are the zeros of f in $B(0,\rho)$ repeated according to multiplicity. If $f(0) \neq 0$ then*

$$\log|f(0)| = -\sum_{i=1}^{n} \log\left(\frac{\rho}{|a_i|}\right) + \frac{1}{2\pi}\int_0^{2\pi} \log\left|f\left(\rho e^{i\theta}\right)\right| d\theta. \tag{17.15}$$

Proof. Consider the function

$$F(z) = f(z)\prod_{i=1}^{n} \frac{\rho^2 - \bar{a}_i z}{\rho(z - a_i)}.$$

The function F has no zeros in $B(0,\rho)$, and $|F(z)| = |f(z)|$ on $|z| = \rho$. This is because if $|z| = \rho$, then

$$\left|\frac{\rho^2 - \bar{a}_i z}{\rho(z-a_i)}\right| = \left|\frac{z(\bar{z} - \bar{a}_i)}{\rho(z-a_i)}\right| = \frac{|z|}{\rho}\frac{|\bar{z} - \bar{a}_i|}{|z - a_i|} = 1$$

for all $i = 1, \cdots, n$. Further F is analytic in an open set containing $\overline{B(0,\rho)}$. So (17.14) applies to F to give

$$\log|F(0)| = \frac{1}{2\pi}\int_0^{2\pi} \log\left|F\left(\rho e^{i\theta}\right)\right| d\theta = \frac{1}{2\pi}\int_0^{2\pi} \log\left|f\left(\rho e^{i\theta}\right)\right| d\theta. \tag{17.16}$$

However

$$F(0) = f(0)\prod_{i=1}^{n}\left(-\frac{\rho}{a_i}\right).$$

Substituting the value of $F(0)$ in (17.16) we obtain the Jensen's formula. If $f(0) = 0$, the formula is somewhat more complicated. Writing $f(z) = mz^h + \cdots$ apply the Jensen's formula (17.15) to $f(z)(\frac{\rho}{z})^h$ and find that the left hand member of (17.15) must be replaced by $\log|m| + h\log\rho$. □

The importance of Jensen's formula lies in the fact that it relates the modulus $|f(z)|$ on a circle to the moduli of the zeros. There is a similar generalization of Poisson's formula. All that is needed is to apply the Poisson formula to $\log |F(z)|$.

Theorem 17.7. *(Poisson-Jensen formula): Let f be analytic in a region $G \supset \overline{B(0,r)}$ and let $a_1, \cdots, a_n$ be the zeros of f in $B(0,r)$ repeated according to multiplicity. If $|z| < r$ and $f(z) \neq 0$ then*

$$\log |f(z)| = -\sum_{i=1}^{n} \log \left| \frac{\rho^2 - \bar{a}_i z}{\rho(z - a_i)} \right| + \frac{1}{2\pi} \int_0^{2\pi} Re \left(\frac{\rho e^{i\theta} + z}{\rho e^{i\theta} - z} \right) \log \left| f\left(\rho e^{i\theta}\right) \right| d\theta.$$

17.5 Maximum modulus of an entire function

Consider a polynomial $P(z) = a_0 + a_1 z + \cdots + a_n z^n, a_n \neq 0, n \geq 1$. We have already seen in Example 7.25 that for arbitrary $\epsilon > 0$ and sufficiently large r,

$$|a_n| r^n (1 - \epsilon) \leq |P(z)| \leq |a_n| r^n (1 + \epsilon). \tag{17.17}$$

The inequality is true at a point in the disk $|z| \leq r$ at which $|P(z)|$ attains its maximum $M(r)$, i.e.,

$$M(r) \leq |a_n| r^n (1 + \epsilon), r > r_0(\epsilon).$$

Also since the value of $|P(z)|$ at a point z_0 on $|z| = r$ cannot exceed $M(r)$, thus $M(r) \geq |a_n| r^n (1 - \epsilon)$ and

$$1 - \epsilon \leq \frac{M(r)}{|a_n| r^n} \leq 1 + \epsilon.$$

Since ϵ is arbitrarily small, we have

$$\lim_{r \to \infty} \frac{M(r)}{|a_n| r^n} = 1,$$

i.e., the maximum modulus of a polynomial of degree n is asymptotically equal to the modulus of the highest degree term of the polynomial.

Let f be an entire function and let $M(r, f) = \max_{|z| \leq r} |f(z)|$. We shall now calculate the maximum modulus of $\cos z, \sin z$ and e^z.

Example 17.3. *Let $f(z) = \cos z$. Now $M(r; \cos z) = \max_{|z| \leq r} |\cos z|$. Since $\cos z = 1 - \frac{z^2}{2!} + \cdots, |z| \leq r < \infty$, then*

$$|\cos z| \leq 1 + \frac{|z|^2}{2!} + \frac{|z|^4}{4!} + \cdots.$$

Hence

$$|\cos z| \leq 1 + \frac{r^2}{2!} + \frac{r^4}{4!} + \cdots = \cosh r$$

and

$$|\cos z| \leq \frac{e^r + e^{-r}}{2}, |z| \leq r.$$

However, at $z = ir$ on the circle,

$$\cos z = \cos ir = \frac{e^{i(ir)} + e^{-i(ir)}}{2} = \frac{e^{-r} + e^r}{2}$$

and $\frac{e^r+e^{-r}}{2}$ coincides with the maximum modulus of $\cos z$ in $|z| \leq r$. Thus $M(r; \cos z) = \frac{e^r+e^{-r}}{2}$. Similarly, $M(r; \sin z) = \frac{e^r-e^{-r}}{2}$ and $M(r; e^z) = e^r$.

Theorem 17.8. *If an entire function is not a polynomial, its maximum modulus increases faster than the maximum modulus of any polynomial, that is, $\lim_{r\to\infty} \dfrac{M(r;P)}{M(r;f)} = 0$, where $P(z)$ is the polynomial and $f(z)$ is the entire function which is not a polynomial.*

Proof. Write $f(z) = a_0 + a_1 z + \cdots + a_n z^n + \cdots$, and let $P(z)$ be of degree m. Denote the coefficient of the highest degree term of the polynomial by $b_m \neq 0$ and for $\epsilon_0 > 0$,

$$M(r; P) \leq b_m r^m (1 + \epsilon_0), r > r_0.$$

Choose a term of degree $p \geq m + 1$ ($a_p \neq 0$) of the function $f(z)$. From Cauchy's estimate, $|a_p| \leq \frac{M(r;f)}{r^p}$ and we have

$$M(r; f) \geq r^p |a_p| \geq |a_p| r^{m+1}.$$

Choose $r > 1$, then for $r > r_0$,

$$\frac{M(r;P)}{M(r;f)} \leq \frac{|b_m| r^m (1 + \epsilon_0)}{|a_p| r^{m+1}} = \frac{|b_m| (1 + \epsilon_0)}{|a_p| r}.$$

Thus $\lim_{r\to\infty} \dfrac{M(r;P)}{M(r;f)} = 0$. □

We will now discuss about one of the methods by which we characterize an entire function, that is, the "order," which compares the rate of growth of the maximum modulus with the growth of the modulus of a simple entire function, namely e^z.

Since (i) $M(r; e^z) = e^r$, (ii) $M(r; exp\ z^k) = exp\ r^k$, (iii) $M(r; exp\ e^z) = exp\ e^r$, then

$$\log M(r; e^z) = r, \quad \log\log M(r; e^z) = \log r,$$

$$\log M(r; exp\ z^k) = r^k, \quad \log\log M(r; exp\ z^k) = k \log r,$$

$$\log M(r; exp\ e^z) = e^r, \quad \log\log M(r; exp\ e^z) = r.$$

These observations prompt us to formulate the following definition of the growth of an entire function.

Order and genus of an entire function

Definition 17.3. *The order of an entire function $f(z)$ not identically zero, is defined by*

$$\lambda = \overline{\lim_{r\to\infty}} \frac{\log\log M(r; f)}{\log\log M(r; e^z)} = \overline{\lim_{r\to\infty}} \frac{\log\log M(r; f)}{\log r},$$

that is, λ is equal to the limit supremum of the ratio of $\log\log M(r; f)$ *to* $\log\log M(r; e^z)$ *as $r \to \infty$.*

By definition of lim sup

$$\lambda = \overline{\lim_{r\to\infty}} \frac{\log\log M(r; f)}{\log r}$$

means given $\epsilon > 0$, there exists r_0 such that

$$M(r; f) = M(r) < exp\ r^{\lambda+\epsilon} \text{ for } |z| = r > r_0$$

and

$$M(r) > expr^{\lambda-\epsilon} \text{ for an infinite number of } r \to \infty.$$

That is,

$$\frac{\log\log M(r)}{\log r} < \lambda + \epsilon \text{ for } r > r_0$$

and

$$\frac{\log\log M(r)}{\log r} > \lambda - \epsilon$$

for an infinite number of $r \to +\infty$. Thus the entire function $f(z)$ is of finite order, if there exist constant $A > 0$ such that as $|z| = r \to \infty$,

$$|f(z)| = O\left(exp\ r^A\right).$$

Hence $f(z)$ is of order λ if

$$|f(z)| = O\left(exp\ r^{\lambda+\epsilon}\right)$$

for every $\epsilon > 0$ but not for any negative ϵ.

Example 17.4. *If* $f(z) = \cos z$,

$$M(r; \cos z) = \frac{e^r + e^{-r}}{2} = e^r \cdot \frac{1 + e^{-2r}}{2}$$

and

$$\frac{\log\log M(r;\cos z)}{\log r} = \frac{\log\left|r\left[1+r^{-1}(\log\frac{1+e^{-2r}}{2})\right]\right|}{\log r},$$

therefore

$$\lambda = \overline{\lim_{r\to\infty}}\left[1+\frac{\log\left[1+r^{-1}\left(\log\frac{1+e^{-2r}}{2}\right)\right]}{\log r}\right] = 1.$$

For arbitrary entire functions, order is somewhat difficult to calculate exactly. However, more often than not an estimate of the magnitude of this expression is good enough. In what follows, several definitions of order all of which are equivalent, will be presented as they are required.

Definition 17.4. *An entire function which is not a polynomial is called a transcendental entire function.*

Theorem 17.9. *If $f(z)$ is a transcendental entire function and $P(z)$ and $Q(z)$ are polynomials of degree m, n, respectively and if $P(z) \neq 0$, then the order λ_1 of $P(z)f(z) + Q(z)$ coincides with the order λ of $f(z)$, i.e., $\lambda_1 = \lambda$.*

Proof. Let $M(r) = \max_{|z|\le r}|f(z)|$ and $M_1(r) = \max_{|z|\le r}|P(z)f(z)+Q(z)|$. By maximum modulus principle we can only confine ourselves to values of the functions on the circle $|z| = r$. That is, $M(r) = \max_{|z|=r}|f(z)|$ and $M_1(r) = \max_{|z|=r}|P(z)f(z)+Q(z)|$. Let z_0 be a point on $|z| = r$ at which $|f(z)|$ attains its maximum $M(r)$. Let z_1 be a point on $|z| = r$ at which $|P(z)f(z) + Q(z)|$ attains its maximum $M_1(r)$.

Let $a_m z^m, b_n z^n$ denote, respectively, the leading terms in the polynomials $P(z)$ and $Q(z)$. In the inequalities (17.17) for polynomials, taking $\epsilon = \frac{1}{2}$ we obtain for $|z| = r > r_0$, that

$$\frac{1}{2}|a_m|r^m \le |P(z)| \le \frac{3}{2}|a_m|r^m$$

and

$$\frac{1}{2}|b_n|r^n \le |Q(z)| \le \frac{3}{2}|b_n|r^n.$$

Then

$$\begin{aligned} M_1(r) &\ge |P(z_0)f(z_0) + Q(z_0)| \ge |P(z_0)|\ |f(z_0)| - |Q(z_0)| \\ &\ge \frac{1}{2}|a_m|r^m M(r) - \frac{3}{2}|b_n|r^n. \end{aligned}$$

On the other hand,

$$\begin{aligned} M_1(r) &= |P(z_1)f(z_1) + Q(z_1)| \\ &\le \frac{3}{2}|a_m|r^m M(r) + \frac{3}{2}|b_n|r^n. \end{aligned}$$

Thus

$$\frac{1}{2}|a_m|r^m M(r)\left\{1-\frac{3|b_n|r^n}{|a_m|r^m M(r)}\right\}$$
$$\leq M_1(r) \leq \frac{3}{2}|a_m|r^m M(r)\left\{1+\frac{|b_n|r^n}{|a_m|r^m M(r)}\right\}.$$

Since f is a transcendental entire function it has been proved that $M(r)$ increases faster than the maximum modulus of any polynomial and hence faster than any power of r. Therefore each $\{\cdots\} \to 1$ as $r \to \infty$ and hence for r sufficiently large, the right-hand $\{\cdots\} < 2$ and the left hand $\{\cdots\} > \frac{2}{3}$, i.e.,

$$\frac{1}{3}|a_m|r^m M(r) \leq M_1(r) \leq 3|a_m|r^m M(r).$$

This implies

$$\log\left(\frac{1}{3}|a_m|r^m\right) + \log M(r) \leq \log M_1(r) \leq \log(3|a_m|r^m) + \log M(r).$$

Hence

$$\log M(r)\left\{1+\frac{\log\left(\frac{1}{3}|a_m|r^m\right)}{\log M(r)}\right\} \leq \log M_1(r) \leq \log M(r)\left\{1+\frac{\log(3|a_m|r^m)}{\log M(r)}\right\}$$

and each $\{\cdots\} \to 1$ as $r \to \infty$. Hence the right-hand $\{\cdots\} < 2$ and the left hand $\{\cdots\} > \frac{1}{2}$, and

$$\frac{1}{2}\log M(r) \leq \log M_1(r) \leq 2\log M(r).$$

Taking logarithms again, we obtain

$$\overline{\lim_{r\to\infty}}\frac{\log\log M(r)}{\log r} \leq \overline{\lim_{r\to\infty}}\frac{\log\log M_1(r)}{\log r} \leq \overline{\lim_{r\to\infty}}\frac{\log\log M(r)}{\log r}.$$

Thus if $P(z) \neq 0$ then the orders of $f(z)$ and $P(z)f(z)+Q(z)$ are equal. □

Theorem 17.10. *The order of an entire function $f(z) = P(z)e^{g(z)} + Q(z)$, where $P(z), Q(z)$, and $g(z)$ are polynomials and $P(z) \neq 0$, is equal to the degree of $g(z)$.*

Proof. From Theorem 17.9, it follows that the order of $f(z)$ is equal to the order of $\phi(z) = e^{g(z)}$. Suppose $g(z) = c_0 + c_1 z + \cdots + c_n z^n, c_n \neq 0$ (that is, degree of $g(z) = n$) and $M_1(r) = \max_{|z|=r}|\phi(z)|$. Further, suppose $c_k = \rho_k(\cos\alpha_k + i\sin\alpha_k)$ and $z = r(\cos\theta + i\sin\theta)$. Since $\rho_n = |c_n| \neq 0$, hence

$$g(z) = \sum_{k=0}^{n}\rho_k r^k(\cos\alpha_k + i\sin\alpha_k)(\cos k\theta + i\sin k\theta)$$

$$= \sum_{k=0}^{n} \rho_k r^k \{\cos(\alpha_k + k\theta) + i \sin(\alpha_k + k\theta)\}$$

and $|\phi(z)| = \exp \sum_{k=0}^{n} \rho_k r^k \cos(\alpha_k + k\theta)$. For fixed r and $0 \le \theta \le 2\pi$,

$$\log|\phi(z)| \le \sum_{k=0}^{n} \rho_k r^k$$

$$= \rho_n r^n \left(1 + \frac{\rho_{n-1}}{\rho_n}\frac{1}{r} + \cdots + \frac{\rho_0}{\rho_n}\frac{1}{r^n}\right),$$

and therefore, for ϵ in $(0,1)$ and $r > r(\epsilon)$,

$$\log|\phi(z)| < \rho_n r^n (1+\epsilon)$$

and

$$\log M_1(r) < \rho_n r^n (1+\epsilon).$$

Let z_0 be a point on $|z| = r$ such that $\cos(\alpha_n + n\theta) = 1$ (actually there are n such points). Then

$$\begin{aligned}\log M_1(r) \ge \log|\phi(z_0)| &= \rho_n r^n + \textstyle\sum_{k=0}^{n-1} \rho_k r^k \cos(\alpha_k + k\theta) \\ &\ge \rho_n r^n - \textstyle\sum_{k=0}^{n-1} \rho_k r^k \\ &= \rho_n r^n \left\{1 - \left(\frac{\rho_{n-1}}{\rho_n}\frac{1}{r} + \cdots + \frac{\rho_0}{\rho_n}\frac{1}{r^n}\right)\right\} \\ &> \rho_n r^n (1-\epsilon)\end{aligned}$$

for $r > r(\epsilon)$.

Thus

$$\rho_n r^n (1-\epsilon) < \log M_1(r) < \rho_n r^n (1+\epsilon)$$

and thus

$$\overline{\lim_{r\to\infty}} \frac{\log\log M_1(r)}{\log r} = n.$$

□

Lemma 17.2. *If $U(r,\theta)$ is the real part of an entire function $f(z)$, $\mu(r) = \max_{0\le\theta\le 2\pi} U(r,\theta)$ and*

$$U(r,\theta) \le \mu(r) \le Cr^\delta, \delta > 0$$

for all $r > N$, then $f(z)$ is a polynomial of degree not exceeding $n = [\delta]$, the integer part of δ.

Proof. Consider the entire function $f(z) = a_0 + a_1 z + \cdots + a_n z^n + \cdots$. Let $a_n = \alpha_n + i\beta_n$ and $z = r(\cos\theta + i\sin\theta), r \ge 0$. Then

$$f(z) = \sum_{n=0}^{\infty} (\alpha_n + i\beta_n) r^n (\cos\theta + i\sin\theta)^n.$$

If we write $f(z) = U(r, \theta) + iV(r, \theta)$, then

$$U(r, \theta) = \alpha_0 + \sum_{n=1}^{\infty} (\alpha_n \cos n\theta - \beta_n \sin n\theta)\, r^n. \tag{17.18}$$

Fixing r and integrating with respect to θ from 0 to 2π, we have

$$\alpha_0 = \frac{1}{2\pi} \int_0^{2\pi} U(r, \theta) d\theta.$$

To compute $\alpha_p, p \geq 1$ multiply (17.18) by $\cos p\theta$ and integrate with respect to θ from 0 to 2π. Thus

$$\int_0^{2\pi} U(r, \theta) \cos p\theta d\theta = a_p r^p \int_0^{2\pi} \cos^2 p\theta d\theta,$$

from which

$$\alpha_p = \frac{1}{\pi r^p} \int_0^{2\pi} U(r, \theta) \cos p\theta d\theta, p = 1, 2, \cdots$$

and similarly

$$\beta_p = \frac{1}{\pi r^p} \int_0^{2\pi} U(r, \theta) \sin p\theta d\theta, p = 1, 2, \cdots .$$

Also,

$$2\alpha_0 \pm \alpha_p r^p = \frac{1}{\pi} \int_0^{2\pi} U(r, \theta)(1 \pm \cos p\theta) d\theta$$

and

$$2\alpha_0 \pm \beta_p r^p = \frac{1}{\pi} \int_0^{2\pi} U(r, \theta)(1 \pm \sin p\theta) d\theta.$$

Notice that the factors multiplying $U(r, \theta)$ are now nonnegative. Since $\mu(r) = \max_{0 \leq \theta \leq 2\pi} U(r, \theta)$ on the circle of radius r, hence

$$2\alpha_0 \pm \alpha_p r^p \leq \frac{\mu(r)}{\pi} \int_0^{2\pi} (1 \pm \cos p\theta)\, d\theta = 2\mu(r).$$

Similarly $2\alpha_0 \pm \beta_p r^p \leq 2\mu(r)$. Hence $|\alpha_p| \leq \frac{2\{\mu(r) - \alpha_0\}}{r^p}$ and $|\beta_p| \leq \frac{2\{\mu(r) - \alpha_0\}}{r^p}$. Thus

$$|\alpha_p| \leq \frac{2\{Cr^\delta - \alpha_0\}}{r^p}$$

and

$$|\beta_p| \leq \frac{2\{Cr^\delta - \alpha_0\}}{r^p}.$$

If $p > [\delta]$, since p is an integer, $p \geq [\delta] + 1 > \delta$ hence $\alpha_p, \beta_p \to 0$, as $r \to \infty$, and $\alpha_p + i\beta_p = a_p = 0$ for $p > [\delta] = n$. □

Lemma 17.3. *If $g(z)$ is an entire function and if the order of the function $f(z) = e^{g(z)}$ is finite, then $g(z)$ is a polynomial and hence the order of $f(z)$ is an integer.*

Proof. If $z = r(\cos\theta + i\sin\theta)$ and $U(r,\theta) = \text{Re}g(z)$, then $\log|f(z)| = U(r,\theta)$. Let $M(r) = \max_{|z|=r}|f(z)|$ and $\mu(r) = \max_{0\le\theta\le 2\pi} U(r,\theta)$. Then

$$\mu(r) = \log M(r).$$

Suppose λ is the order of $f(z)$, that is, $\lambda = \overline{\lim_{r\to\infty}} \frac{\log\log M(r)}{\log r}$. Thus given $\epsilon > 0$, there exists $r(\epsilon) > 1$ such that for $r > r(\epsilon)$,

$$\frac{\log\log M(r)}{\log r} < \lambda + \epsilon.$$

Hence $\log\log M(r) < \log(r^{\lambda+\epsilon})$ or

$$\log M(r) < r^{\lambda+\epsilon}.$$

Thus it follows that $\mu(r) < r^{\lambda+\epsilon}$ for $r > r(\epsilon)$.

It follows from Lemma 17.2 that $g(z)$ is a polynomial of degree n and $n \le [\lambda+\epsilon]$ where ϵ is arbitrarily small. Thus $n \le [\lambda]$ and by Theorem 17.10, the order of $f(z)$ is equal to n. That is, $\lambda = n$. □

Theorem 17.11. *Let f be a transcendental entire function with order λ. Suppose λ is finite but is not an integer. If $P(z)$ is a nonzero polynomial, the equation $f(z) = AP(z)$ has infinitely many roots for every complex number A (no exceptions).*

Proof. Suppose there exist $A = A_0$ for which $f(z) = A_0P(z)$ has only finitely many roots if any at all. Then the entire function $f(z) - A_0P(z)$ has only finitely many zeros. Therefore we may write

$$f(z) - A_0P(z) = Q(z)e^{g(z)}$$

where $Q(z)$ is a nonzero polynomial (equal to 1 if $f - A_0P$ has no zeros). Further $g(z)$ is an entire function, and we have

$$f(z) = A_0P(z) + Q(z)e^{g(z)}.$$

Thus the order λ of $f(z)$ coincides with the order of $e^{g(z)}$ which, being finite, is an integer. The contradiction proves the theorem. □

Theorem 17.12. *If f is a transcendental entire function of finite integral order n and if $P(z)$ is a nonzero polynomial, then $f(z) = AP(z)$ has infinitely many roots for every complex number A, with the possible exception of one value.*

Proof. Suppose there exists complex numbers $a, b, a \neq b$ such that $f(z)-aP(z)$ and $f(z) - bP(z)$ has only finitely many zeros. Then

$$f(z) - aP(z) = Q_1(z)e^{g_1(z)} \tag{17.19}$$

and

$$f(z) - bP(z) = Q_2(z)e^{g_2(z)} \tag{17.20}$$

where $Q_1(z)$ and $Q_2(z)$ are nonzero polynomials and $g_1(z)$ and $g_2(z)$ are entire functions. Hence by Theorem 17.10, n = order of $f(z)$ = order of $e^{g_1(z)}$ = order of $e^{g_2(z)}$. Thus $g_1(z)$ and $g_2(z)$ are polynomials of degree n. Notice that $n \geq 1$; since if $n = 0$ then $g_1(z)$ and $g_2(z)$ are constants and therefore $f(z)$ is a polynomial and it is not a transcendental entire function.

Subtracting (17.19) from (17.20),

$$Q_1(z)e^{g_1(z)} - Q_2(z)e^{g_2(z)} = (b-a)P(z) = R(z), \tag{17.21}$$

where $R(z)$ is a nonzero polynomial since $b \neq a$ and $P(z) \neq 0$ by hypothesis. We need to show that this equation is not possible for $Q_1(z), Q_2(z)$ and $R(z)$ nonzero polynomials and $g_1(z), g_2(z)$ polynomials of degree greater than or equal to 1. Differentiating (17.21), we obtain

$$(Q_1' + Q_1g_1')e^{g_1} - (Q_2' + Q_2g_2')e^{g_2} = R' \tag{17.22}$$

Regarding (17.21) and (17.22) as a system of two equations in the unknowns e^{g_1} and e^{g_2}, the determinant of the system $\Delta(z)$ is given by

$$\Delta(z) = Q_1(Q_2' + Q_2g_2') - Q_2(Q_1' + Q_1g_1') = -Q_2Q_1' + Q_1Q_2' - Q_1Q_2(g_1' - g_2').$$

We shall now show that $\Delta(z) \neq 0$. Suppose to the contrary $\Delta(z) \equiv 0$, divide by $-Q_1Q_2$ to obtain

$$\frac{Q_1'}{Q_1} - \frac{Q_2'}{Q_2} + (g_1 - g_2)' = 0.$$

Integrating the above equation we obtain

$$\log\frac{Q_1}{Q_2} + g_1 - g_2 = constant = C_1.$$

Thus $\frac{Q_1}{Q_2}e^{g_1-g_2} = e^{C_1} = C \neq 0$. From equations (17.19) and (17.20), we obtain

$$C = \frac{Q_1}{Q_2}e^{g_1-g_2} = \frac{f-aP}{f-bP}.$$

Since $a \neq b$, we have $C \neq 1$. Now as $f(z)(1-C) = P(z)(a-bC)$, we obtain $f(z) = \frac{a-bC}{1-C}P(z)$ which contradicts the hypothesis that $f(z)$ is a transcendental entire function and $P(z)$ is a polynomial. Thus $\Delta(z) \neq 0$. Solving for e^{g_1} and e^{g_2}, using Cramer's rule we obtain

$$e^{g_1} = \frac{-R(Q_2' + Q_2 g_2') + R'Q_2}{\Delta(z)}$$

and

$$e^{g_2} = \frac{-R(Q_1' + Q_1 g_1') + R'Q_1}{\Delta(z)}.$$

However, these equations contain a contradiction since the left-hand side is a transcendental entire function and the function on the right-hand side is a rational function, and thus a polynomial since it is entire. This completes the proof. □

From the above theorem it follows (by taking $P(z) = 1$) that $f(z)$ takes on all values an infinite number of times with one possible exception.

Theorem 17.13. *Let h be a transcendental meromorphic function. For every complex number A (finite or infinite) with two possible exceptions, $h(z) = A$ has infinitely many roots.*

Proof. If h is an entire function, it does not become infinite at any z. Thus $A = \infty$ is an exceptional value for every entire function. By Theorem 17.12 the entire function f has at most one finite exceptional value. If f is a transcendental entire function and $P(z)$ is a nonzero polynomial , the meromorphic function $h(z) = \frac{f(z)}{P(z)}$ becomes infinite only at a finite number of points, namely the zeros of $P(z)$. Thus ∞ is an exceptional value, and further, one other finite exceptional value is possible. By Theorem 17.12, $\frac{f(z)}{P(z)} = A$ or $f(z) = AP(z)$ has infinitely many roots for every A except possibly a single finite value of A. Let $f(z)$ and $g(z)$ denote two transcendental entire functions whose ratio is not rational. Hence $\frac{f(z)}{g(z)}$ is meromorphic. Thus $\frac{f(z)}{g(z)} = A$ has infinitely many roots with two possible exceptions. □

We examine a theorem similar to Cauchy's estimate but involving the upper bound of $\text{Re} f(z)$ on $|z| = r$ rather than $M(r)$. Let $A(r)$ be the upper bound of the real part of $f(z)$ on $|z| = r$.

Theorem 17.14. *If $f(z) = \sum_{n=0}^{\infty} a_n z^n$, then*

$$|a_n| r^n \leq \max\{4A(r), 0\} - 2Re\{f(0)\}$$

for all $n > 0$ and r.

Proof. Let $z = re^{i\theta}$ and $f(z) = \sum_{n=0}^{\infty} a_n z^n = U(r, \theta) + iV(r, \theta)$ and let $a_n = \alpha_n + i\beta_n$. Then

$$U(r, \theta) = \sum_{n=0}^{\infty} (\alpha_n \cos n\theta - \beta_n \sin n\theta) r^n.$$

The series converges uniformly with respect to θ. Thus we may multiply by $\cos n\theta$ or $\sin n\theta$ and integrate term by term. Thus

$$\int_0^{2\pi} U(r,\theta)\cos n\theta d\theta = \int_0^{2\pi} \alpha_n r^n \cos^2 n\theta d\theta = \pi\alpha_n r^n.$$

Similarly

$$\int_0^{2\pi} U(r,\theta)\sin n\theta d\theta = -\pi\beta_n r^n, n > 0;$$

also,

$$\int_0^{2\pi} U(r,\theta) d\theta = 2\pi\alpha_0.$$

Hence

$$a_n r^n = (\alpha_n + i\beta_n) r^n = \frac{1}{\pi}\int_0^{2\pi} U(r,\theta) e^{-in\theta} d\theta, n > 0$$

and

$$|a_n| r^n \leq \frac{1}{\pi}\int_0^{2\pi} |U(r,\theta)| d\theta.$$

Thus

$$|a_n| r^n + 2\alpha_0 \leq \frac{1}{\pi}\int_0^{2\pi} \{|U(r,\theta)| + U(r,\theta)\} d\theta. \tag{17.23}$$

Note, $|u| + u = 0$ for $u < 0$. Hence if $A(r) < 0$, the right-hand side of (17.23) becomes zero and if $A(r) \geq 0$, the right-hand side does not exceed

$$\frac{1}{\pi}\int_0^{2\pi} 2A(r) d\theta = 4A(r).$$

Thus

$$|a_n| r^n \leq \max\{4A(r), 0\} - 2\text{Re}\{f(0)\}.$$

□

This may be improved as follows. Write

$$a_n = \alpha_n + i\beta_n = |a_n| e^{i\theta_n},$$

then

$$\begin{aligned} |a_n| r^n &= (\alpha_n + i\beta_n) r^n e^{-i\theta_n} \\ &= \tfrac{1}{\pi}\int_0^{2\pi} U(r,\theta) e^{-in\theta} e^{-i\theta_n} d\theta \\ &= \tfrac{1}{\pi}\int_0^{2\pi} U(r,\theta)\cos(n\theta + \theta_n) d\theta \quad (|a_n| r^n \text{real}). \end{aligned}$$

Thus

$$\begin{aligned}
|a_n|r^n + 2\alpha_0 &= \tfrac{1}{\pi}\int_0^{2\pi} U(r,\theta)\{1+\cos(n\theta+\theta_n)\}d\theta \\
&\le \tfrac{1}{\pi}\int_0^{2\pi} A(r)\{1+\cos(n\theta+\theta_n)\}d\theta \\
&= \tfrac{A(r)}{\pi}\int_0^{2\pi} \{1+\cos(n\theta+\theta_n)\}d\theta = 2A(r) \\
\text{and}\ \ |a_n|r^n &\le 2[A(r) - \mathrm{Re}\{f(0)\}].
\end{aligned}$$

The following result enables us to deduce an upper bound for the modulus of a function on $|z| = r$, from bounds for its real or imaginary parts on a larger concentric circle $|z| = R$.

Theorem 17.15. *Let $f(z)$ be analytic for $|z| \le R$ and let $M(r)$ and $A(r)$ denote the* $\max|f(z)|$ *and* $\max Re\{f(z)\}$ *on $|z| = r$. Then for $0 < r < R$,*

$$M(r) \le \frac{2r}{R-r}A(R) + \frac{R+r}{R-r}|f(0)| < \frac{R+r}{R-r}\{A(R) + |f(0)|\}.$$

Proof. The result is clearly true for $f(z) = \textit{constant}$. For $f(z)$ nonconstant suppose $f(0) = 0$. Then $A(R) > A(0) = 0$. Since by the Poisson-Jensen formula

$$U(0) = \frac{1}{2\pi}\int_0^{2\pi} U(R,\phi)d\phi,$$

and if we let $A(R) = \max_\phi U(R,\phi)$, then $U(R,\phi) \le A(R)$, for all ϕ and

$$U(0) = \frac{1}{2\pi}\int_0^{2\pi} U(R,\phi)d\phi \le \frac{1}{2\pi}\int_0^{2\pi} A(R)d\phi = A(R).$$

Thus $U(0) = A(R)$ implies $U(R,\phi) = A(R)$, for all ϕ and $A(0) \le U(0) \le A(R)$ unless the function is a constant. Let

$$\phi(z) = \frac{f(z)}{2A(R) - f(z)}.$$

Then $\phi(z)$ is analytic for $|z| \le R$ since the real part of the denominator does not vanish. Now $\phi(0) = 0$ and if $f(z) = u + iv$,

$$|\phi(z)|^2 = \frac{u^2+v^2}{\{2A(r)-u\}^2+v^2} \le 1$$

since $-2A(R) + u \le u \le 2A(R) - u$.

Also Schwarz's lemma gives

$$|\phi(z)| \le \frac{r}{R}$$

since $|\phi(z)|$ is bounded by 1. Thus

$$|f(z)| = \frac{|2A(R)\phi(z)|}{|1+\phi(z)|} \le \frac{2A(R)\cdot r}{R-r}$$

(use $|R+R\phi| \geq R-R|\phi|$). Thus we obtain the result for $f(0)=0$. If $f(0) \neq 0$, apply the result to $f(z)-f(0)$ and

$$|f(z)-f(0)| \leq \frac{2r}{R-r} \cdot \max_{|z|=R}\{\mathrm{Re}(f(z)-f(0))\} \leq \frac{2r}{R-r}\{A(R)+|f(0)|\}$$

since $f(0)$ may be negative. Adding $|f(0)|$ to both sides, we have that

$$|f(z)| \leq \frac{2r}{R-r}A(R) + \frac{R+r}{R-r}|f(0)|.$$

□

We also show that $\max\limits_{|z|=r}|f^{(n)}(z)| \leq \dfrac{2^{n+2}n!R}{(R-r)^{n+1}}\{A(R)+|f(0)|\}$ since

$$f^{(n)}(z) = \frac{n!}{2\pi i}\int_C \frac{f(w)-f(0)}{(w-z)^{n+1}}dw \tag{17.24}$$

where C is the circle $|w-z| = \delta = \frac{1}{2}(R-r)$ and

$$|w| \leq r + \frac{1}{2}(R-r) = \frac{1}{2}(R+r)$$

which ensures C is inside $|z|=R$ (see Fig. 17.1).

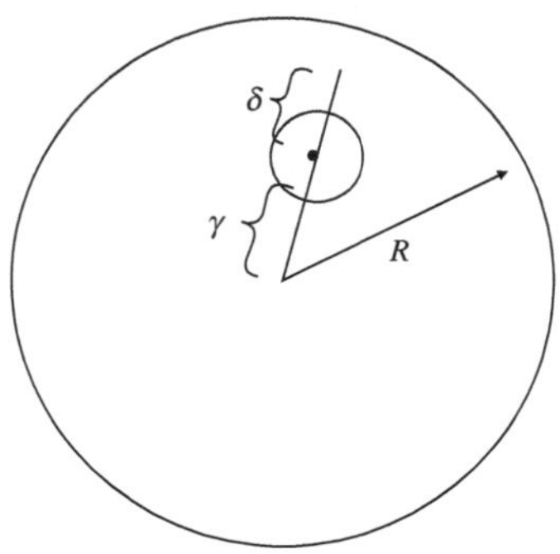

Fig. 17.1: Circle C inside $|z|=R$

The previous theorem gives

$$\max|f(z)-f(0)| \leq \frac{R+\frac{1}{2}(R+r)}{R-\frac{1}{2}(R+r)}\{A(R)+|f(0)|\}$$

and since $r<R$,

$$\max|f(z)-f(0)| < \frac{4R}{R-r}\{A(R)+|f(0)|\}.$$

Thus from equation (17.24) it follows that

$$|f^{(n)}(z)| \leq \frac{n!4R}{\delta^n(R-r)}\{A(R)+|f(0)|\} = \frac{2^{n+2}n!R}{(R-r)^{n+1}}\{A(R)+|f(0)|\}.$$

Theorem 17.16. *If f_1 and f_2 are entire functions of order λ_1 and λ_2, respectively, and if $\lambda_1 < \lambda_2$, then the order λ of $f(z) = f_1(z) + f_2(z)$ is equal to λ_2.*

Proof. Suppose λ_2 is finite. Since

$$\begin{aligned} M(r; f_1 + f_2) &\leq M(r; f_1) + M(r; f_2) \\ &\leq exp\ r^{\lambda_1+\epsilon} + exp\ r^{\lambda_2+\epsilon} \\ &< 2exp\ r^{\lambda_2+\epsilon} \text{ for } r > r_0(\epsilon). \end{aligned}$$

We have that $\lambda \leq \lambda_2 + \epsilon$ and hence $\lambda \leq \lambda_2$. On the other hand there exist a sequence of numbers $r_n \to \infty$ such that $M(r_n; f_2) > expr_n^{\lambda_2-\epsilon}$. Thus

$$\begin{aligned} M(r_n; f_1 + f_2) &\geq exp\ r_n^{\lambda_2-\epsilon} - exp\ r_n^{\lambda_1+\epsilon} \\ &= exp\ r_n^{\lambda_2-\epsilon}\{1 - exp\ (r_n^{\lambda_1+\epsilon} - r_n^{\lambda_2-\epsilon})\} \\ &> \tfrac{1}{2}exp\ r_n^{\lambda_2-\epsilon} \end{aligned}$$

provided ϵ is so small that $\lambda_1 + \epsilon < \lambda_2 - \epsilon$ and n is sufficiently large. Thus $\lambda \geq \lambda_2$ and the order of the sum, $f_1 + f_2$, equals λ_2. The proof also applies without significant changes when $\lambda_2 = \infty$. □

It may be noted that the theorem is sometimes false when $\lambda_1 = \lambda_2$.

Example 17.5. *Let $f_1(z) = e^z$ and $f_2(z) = -e^z$. Then $\lambda_1 = \lambda_2 = 1$ and $\lambda = 0$.*

Thus we can say that if f_1 and f_2 are entire functions of order λ_1 and λ_2, respectively, and if $\lambda_1 \leq \lambda_2$, then the order λ of $f(z) = f_1(z) + f_2(z)$ is less than or equal to λ_2.

Theorem 17.17. *If f_1 and f_2 are entire functions of order λ_1 and λ_2, respectively, where $\lambda_1 \leq \lambda_2$, then the order λ of $f(z) = f_1(z)f_2(z)$ is such that $\lambda \leq \lambda_2$.*

Proof. Given $\epsilon > 0$ and r sufficiently large, we have

$$\begin{aligned} M(r; f_1 f_2) &\leq M(r; f_1)M(r; f_2) \\ &\leq exp\ r^{\lambda_1+\epsilon} exp\ r^{\lambda_2+\epsilon} \\ &\leq exp\ 2r^{\lambda_2+\epsilon}. \end{aligned}$$

Thus $\lambda \leq \lambda_2 + \epsilon$ and hence $\lambda \leq \lambda_2$. □

Theorem 17.18. *If f is an entire function of order λ and $P(z)$ is a nonzero polynomial, then the product $f(z)P(z)$ is of order λ. If the quotient $\frac{f(z)}{P(z)}$ is an entire function, then it is also of order λ.*

Proof. Notice that the order of $f(z)P(z)$ does not exceed λ. Since $|P(z)| > 1$ for $|z|$ sufficiently large, for these values of z, $|f(z)P(z)| \geq |f(z)|$. Thus the order of $f(z)P(z)$ is not less than λ, which proves the first part of the theorem. Notice also that if $\frac{f(z)}{P(z)}$ is entire, then its order is the same as $P(z)\frac{f(z)}{P(z)} = f(z)$ and the second part of the theorem is proved. □

Let $f(z)$ be analytic for $|z| < R$. Suppose $f(0) \neq 0$ and let $r_1, r_2, \cdots, r_n, \cdots$ be the moduli of the zeros of $f(z)$ in $|z| < R$ arranged in a nondecreasing sequence and where a zero of order p is counted p times. Let $n(x)$ denote the number of zeros of $f(z)$ for $|z| \leq x$. Then if $r_n \leq r \leq r_{n+1}$,

$$\begin{aligned}\log \tfrac{r^n}{r_1 r_2 \cdots r_n} &= n \log r - \sum_{m=1}^{n} \log r_m \\ &= \sum_{m=1}^{n-1} m(\log r_{m+1} - \log r_m) + n(\log r - \log r_n) \\ &= \sum_{m=1}^{n-1} m \int_{r_m}^{r_{m+1}} \frac{dx}{x} + n \int_{r_n}^{r} \frac{dx}{x}.\end{aligned}$$

Notice that for $r_m \leq x < r_{m+1}$ we have $n(x) = m$ and for $r_n \leq x < r$, we have $n(x) = n$. Thus

$$\begin{aligned}\log \tfrac{r^n}{r_1 r_2 \cdots r_n} &= \int_0^r \frac{n(x)}{x} dx \\ &= \sum_{m=1}^{n-1} m \int_{r_m}^{r_{m+1}} \frac{dx}{x} + n \int_{r_n}^{r} \frac{dx}{x}.\end{aligned}$$

This is so because

$$\begin{aligned}\int_0^r \frac{n(x)}{x} dx &= \int_0^{r_1} 0 \cdot \frac{dx}{x} + \int_{r_1}^{r_2} 1 \cdot \frac{dx}{x} + \cdots + \int_{r_n}^{r} n \cdot \frac{dx}{x} \\ &= 1 \cdot \int_{r_1}^{r_2} \frac{dx}{x} + \cdot + (n-1) \int_{r_{n-1}}^{r_n} \frac{dx}{x} + n \int_{r_n}^{r} \frac{dx}{x}.\end{aligned}$$

By Jensen's formula if $r_n \leq r \leq r_{n+1}$ then

$$\log \frac{r^n |f(0)|}{r_1 r_2 \cdots r_n} = \frac{1}{2\pi} \int_0^{2\pi} \log |f(re^{i\theta})| d\theta.$$

It follows therefore that

$$\int_0^r \frac{n(x)}{x} dx = \frac{1}{2\pi} \int_0^{2\pi} \log |f(re^{i\theta})| d\theta - \log |f(0)|.$$

Note that both sides of the formula are equal for $r = 0$. If $f(z)$ has no zeros on $|z| = r$,

$$\begin{aligned} n(r) &= \tfrac{1}{2\pi i}\int_{|z|=r} \frac{f'(z)}{f(z)}dz \\ &= \tfrac{1}{2\pi}\int_0^{2\pi} \frac{f'(re^{i\theta})}{f(re^{i\theta})} re^{i\theta} d\theta. \end{aligned}$$

Theorem 17.19. *Suppose f is an entire function with infinite number of zeros, and of order $\lambda < \infty$. If $f(0) \neq 0$, then given $\epsilon > 0$, there exists R_0 such that for $R \geq R_0$,*

$$n\left(\frac{R}{3}\right) \leq \frac{1}{\log 2} \log\left\{\frac{expR^{\lambda+\epsilon}}{|f(0)|}\right\},$$

where $n(R)$ denotes the number of zeros of $f(z)$ in $|z| \leq R$.

Proof. Let f be analytic in $|z| \leq R$ and $a_1, a_2, \cdots, a_n$ are zeros of f in $|z| \leq \frac{R}{3}$. Then $|a_k| \leq \frac{R}{3}, k = 1, \cdots, n$. Define

$$g(z) = \frac{f(z)}{\left(1 - \frac{z}{a_1}\right) \cdots \left(1 - \frac{z}{a_n}\right)}.$$

Suppose $|f(z)| \leq M$ for $|z| = R$. Thus for $|z| = R$, $|\frac{z}{a_k}| \geq 3$ and $|1 - \frac{z}{a_k}| \geq |\frac{z}{a_k}| - 1 \geq 2$. Hence, by the maximum modulus principle, $|g(z)| \leq \frac{M}{2^n}$ in $|z| \leq R$. Thus $|g(0)| \leq \frac{M}{2^n}$. Therefore $|f(0)| \leq \frac{M}{2^n}$ and

$$n = n\left(\frac{R}{3}\right) \leq \frac{1}{\log 2} \log\left\{\frac{M}{|f(0)|}\right\}.$$

Since f has order λ, hence for $r > r_0$, $M(r) < \exp r^{\lambda+\epsilon}$ and the result follows. □

Remark 17.1. 1. One can in fact show that $n(R) = O\left(R^{\lambda+\epsilon}\right)$ .The argument is as follows: from Theorem 17.19, it follows that

$$n\left(\frac{R}{3}\right) \leq \frac{1}{\log 2} \log\left\{\frac{\exp R^{\lambda+\epsilon}}{|f(0)|}\right\} = \frac{1}{\log 2}\left\{R^{\lambda+\epsilon} - \log|f(0)|\right\}$$

and $-\log|f(0)| < R_1^{\lambda+\epsilon}$, say, then for $R \geq R_1$, $n\left(\frac{R}{3}\right) < \frac{2}{\log 2} R^{\lambda+\epsilon}$ and thus $n(R) = O\left(R^{\lambda+\epsilon}\right)$.

2. Since $n(R)$ denotes the number of zeros for which $|a_n| \leq R$, then $n(R)$ is a nondecreasing function of R which is constant in intervals. It is zero for $R < |a_1|$ if $f(0)$ is not zero. By virtue of Jensen's formula,

$$\int_0^R \frac{n(x)}{x} dx = \frac{1}{2\pi}\int_0^{2\pi} \log\left|f(Re^{i\theta})\right| d\theta - \log|f(0)|.$$

Since $f(z)$ is of order λ, that is, $|f(Re^{i\theta})| < k_1 \exp R^{\lambda+\epsilon}, \epsilon > 0$, then

$$\log \left| f(Re^{i\theta}) \right| < kR^{\lambda+\epsilon}.$$

Thus

$$\int_0^{2R} \frac{n(x)}{x} dx < k_2 R^{\lambda+\epsilon}.$$

Since $n(R)$ is nondecreasing,

$$\int_0^{2R} \frac{n(x)}{x} dx \geq n(R) \int_R^{2R} \frac{dx}{x} = n(R) \log 2$$

and $n(R) \leq \frac{1}{\log 2} \int_0^{2R} \frac{n(x)}{x} dx < k_3 R^{\lambda+\epsilon}$.

Thus the higher the order of a function the more zeros it may have in a given region.

Theorem 17.20. *If f is an entire function of order $\lambda < \infty$ and $r_1, r_2, \cdots$ are the moduli of its infinite number of zeros, then $\sum_{n=1}^{\infty} \frac{1}{r_n^\alpha} < \infty$ for $\alpha > \lambda$.*

Proof. Let β be a number between λ and α, that is, $\lambda < \beta < \alpha$. Then $n(r) < Ar^\beta$. Putting $r = r_n$, we have $n < Ar_n^\beta$ and $r_n^{-\alpha} < A_1 n^{-\frac{\alpha}{\beta}}$. Since $\frac{\alpha}{\beta} > 1$, $\sum_{n=1}^{\infty} \frac{1}{r_n^\alpha} < \infty$.

□

Definition 17.5. *The lower bound of the positive numbers α for which*

$$\sum_{n=1}^{\infty} \frac{1}{r_n^\alpha} < \infty$$

is called the exponent of convergence of the zeros and is denoted by λ_1. Formally, the empty set has $\lambda_1 = 0$ and if the series diverges for all positive α, then $\lambda_1 = \infty$. Thus

$$\sum_{n=1}^{\infty} \frac{1}{r_n^{\lambda_1+\epsilon}} < \infty, \epsilon > 0$$

and

$$\sum_{n=1}^{\infty} \frac{1}{r_n^{\lambda_1-\epsilon}} = \infty, \epsilon > 0.$$

We have thus proved that $\lambda_1 \leq \lambda$ since it may be possible to find numbers less than λ for which the series converges.

Remark 17.2. The series $\sum_{n=1}^{\infty} \frac{1}{r_n^{\lambda_1}}$ may either converge or diverge. For example, take $r_n = n$ or $r_n = n(\log n)^2$. If the zeros of $f(z)$ are finite or nil, $\lambda_1 = 0$. Thus $\lambda_1 > 0$ implies there exists infinitely many zeros. Further, we can have $\lambda_1 < \lambda$. For example, if $f(z) = e^z$ then order of $f(z) = \lambda = 1$ but since there are no zeros of e^z then $\lambda_1 = 0$.

Let f be an entire function of order $\lambda < \infty, f(0) \neq 0, f(z_n) = 0$, $n = 1, 2, \cdots$. Then there exists an integer $\mu + 1$ such that $\sum_{n=1}^{\infty} \frac{1}{|z_n|^{\mu+1}} < \infty$. By Theorem 17.20, any integer exceeding λ will serve as $\mu + 1$.

Definition 17.6. *The smallest integer p for which $\sum_{n=1}^{\infty} \frac{1}{|z_n|^{p+1}} < \infty$, is called the genus (rank) of the canonical product.*

Example 17.6. *If $z_n = n$, then $p = 1$; if $z_n = e^n$, then $p = 0$, and if $z_1 = \frac{1}{2}\log 2, z_n = \log n, n \geq 2$, there exists no finite p.*

Suppose for $\lambda_1 < \sigma$,

$$\sum_{n=1}^{\infty} \frac{1}{|z_n|^{\sigma}} < \infty, \sum_{n=1}^{\infty} \frac{1}{|z_n|^{p+1}} < \infty,$$

but

$$\sum_{n=1}^{\infty} \frac{1}{|z_n|^{p}} = \infty.$$

Thus p is the smallest integer for which the series $\sum_{n=1}^{\infty} \frac{1}{|z_n|^{p+1}} < \infty$. Hence if λ_1 is not an integer, $p = [\lambda_1]$.

Let f be an entire function, and denote its zeros by a_n; for the sake of simplicity we will assume that $f(0) \neq 0$. We recall that the genus of an entire function is the smallest integer h such that $f(z)$ can be represented in the form

$$f(z) = e^{g(z)} \prod_n \left(1 - \frac{z}{a_n}\right) e^{\frac{z}{a_n} + \frac{1}{2}(\frac{z}{a_n})^2 + \cdots + \frac{1}{h}(\frac{z}{a_n})^h}$$

where $g(z)$ is a polynomial of degree $\leq h$. If there is no such representation, the genus is infinite.

Theorem 17.21. *(Hadamard's theorem): Let f be an entire function of genus h then f is of finite order $\lambda \leq h + 1$. Further, if f is an entire function of finite order λ then f has finite genus $h \leq \lambda$.*

Proof. Suppose $\{a_1, a_2, \cdots\}$ are the zeros of f counted according to multiplicity and arranged such that $|a_1| \leq |a_2| \leq \cdots$. Without loss of generality we shall assume that $f(0) = 1$. Indeed if f has a zero at the origin of multiplicity m and $M(r) = \max\{|f(z)| : |z| = r\}$ then for any $\epsilon > 0$ and $|z| = r$,

$$\begin{aligned} \log \left| \frac{f(z)}{z^m} \right| &\leq \log \frac{M(r)}{r^m} \\ &= \log M(r) - \log r^m \leq r^{\lambda+\epsilon} - m \log r \\ &\leq r^{\lambda+2\epsilon} \end{aligned}$$

if r is sufficiently large. So $\frac{f(z)}{z^m}$ is an entire function of order λ with no zero at the origin. Since multiplication by a scalar does not affect the order and genus, the assumption that $f(0) = 1$ is justified.

For the first part let f be an entire function of genus h and suppose $h < \infty$. Then

$$f(z) = e^{g(z)} \prod_{n=1}^{\infty} E_h\left(\frac{z}{a_n}\right) \tag{17.25}$$

where g is a polynomial of degree $\leq h, E_h(u) = (1-u)\, e^{u+\frac{u^2}{2}+\cdots+\frac{u^h}{h}}$ and $E_0(u) = 1-u$. Note that the order of $(e^{g(z)}) \leq h$. We shall now show that order of the canonical product is less than or equal to $h+1$. This will imply by Theorem 17.17 that $\lambda = \text{order } (f(z)) \leq h+1$.

Since the canonical product $\prod_{n=1}^{\infty} E_h\left(\frac{z}{a_n}\right)$ converges, hence $\sum_{n=1}^{\infty} \frac{1}{|a_n|^{h+1}} < \infty$. Let $P(z) = \prod_{n=1}^{\infty} E_h\left(\frac{z}{a_n}\right)$. Notice that if $|z| < \frac{1}{2}$ then

$$\begin{aligned}
\log|E_h(z)| &= \operatorname{Re}\left\{\log(1-z) + z + \cdots + \frac{z^h}{h}\right\} \\
&= \operatorname{Re}\left\{-\frac{1}{h+1}z^{h+1} - \frac{1}{h+2}z^{h+2} - \cdots\right\} \\
&\leq |z|^{h+1}\left\{\frac{1}{h+1} + \frac{1}{h+2}|z| + \cdots\right\} \\
&\leq |z|^{h+1}\left\{1 + \frac{1}{2} + (\frac{1}{2})^2 + \cdots\right\} = 2|z|^{h+1}.
\end{aligned} \tag{17.26}$$

Also

$$|E_h(z)| \leq (1+|z|)\, e^{|z|+\cdots+\frac{|z|^h}{h}},$$

so that

$$\log|E_h(z)| \leq \log(1+|z|) + |z| + \cdots + \frac{|z|^h}{h}.$$

Hence,

$$\lim_{z\to\infty} \frac{\log|E_h(z)|}{|z|^{h+1}} = 0.$$

So if $A > 0$ then there is a number $R > 0$ such that

$$\log|E_h(z)| \leq A|z|^{h+1} \text{ whenever } |z| > R. \tag{17.27}$$

But on the set $\{z \in \mathbb{C} : \frac{1}{2} \leq |z| \leq R\}$ the function $\frac{\log|E_h(z)|}{|z|^{h+1}}$ is continuous except at $z = +1$, where it tends to $-\infty$. Hence there is a constant $B > 0$ such that

$$\log|E_h(z)| \leq B|z|^{h+1}, \frac{1}{2} \leq |z| \leq R. \tag{17.28}$$

Thus from (17.26), (17.27) and (17.28), it follows that

$$\log|E_h(z)| \leq M|z|^{h+1} \tag{17.29}$$

for all $z \in \mathbb{C}$, where $M = \max\{2, A, B\}$.

Since $\sum_{n=1}^{\infty} \frac{1}{|a_n|^{h+1}} < \infty$, given $\beta > 0$ there exists an integer N such that $\sum_{n=N+1}^{\infty} \frac{1}{|a_n|^{h+1}} < \frac{\beta}{4M}$. But from (17.29) it follows that

$$\sum_{n=N+1}^{\infty} \log |E_h(\frac{z}{a_n})| \leq M \sum_{n=N+1}^{\infty} \left|\frac{z}{a_n}\right|^{h+1} \leq \frac{\beta}{4}|z|^{h+1}. \tag{17.30}$$

Notice that in (17.27), A could be chosen as small as desired by taking R sufficiently large. So choose $r_1 > 0$ such that

$$\log |E_h(z)| \leq \frac{\beta}{4N}|z|^{h+1}, \text{ for } |z| > r_1.$$

If $r_2 = \max\{|a_1|r_1, |a_2|r_1, \cdots, |a_N|r_1\}$ then

$$\sum_{n=1}^{N} \log |E_h\left(\frac{z}{a_n}\right)| \leq \frac{\beta}{4}|z|^{h+1}, \text{ for } |z| > r_2.$$

Combining this with (17.30), we obtain

$$\log |P(z)| = \sum_{n=1}^{\infty} \log \left|E_h(\frac{z}{a_n})\right| < \frac{\beta}{2}|z|^{h+1}$$

for $|z| > r_2$. Thus order $(P(z)) \leq h+1$. Hence $\lambda =$ order $(f(z)) \leq h+1$.

For the second part assume that f is of finite order λ and let h be the largest integer $\leq \lambda$. Then $h+1 > \lambda$, and we have to show that $\sum_{n=1}^{\infty} \frac{1}{|a_n|^{h+1}} < \infty$.

Let $n(r)$ be the number of zeros of f in $B(0,r)$. It follows from Theorem 17.19 that $(\log 2)n(r) \leq \log M(r)$. Since f has order λ, $\log M(r) \leq r^{\lambda+\frac{\epsilon}{2}}$ for any $\epsilon > 0$ so that

$$\begin{aligned}\lim_{r\to\infty} \frac{n(r)}{r^{\lambda+\epsilon}} &\leq \lim_{r\to\infty} \frac{r^{\lambda+\frac{\epsilon}{2}}}{(\log 2)r^{\lambda+\epsilon}} \\ &= \lim_{r\to\infty} \frac{1}{\log 2}\frac{1}{r^{\frac{\epsilon}{2}}} = 0.\end{aligned}$$

Hence $n(r) \leq r^{\lambda+\epsilon}$ for sufficiently large r. Since $|a_1| \leq |a_2| \leq \cdots$, $k \leq n(|a_k|) \leq |a_k|^{\lambda+\epsilon}$ for all k larger than some integer k_0. Hence,

$$|a_k|^{-(h+1)} \leq k^{-(\frac{h+1}{\lambda+\epsilon})}$$

for $k > k_0$. We have noticed that $\lambda < h+1$.

We choose ϵ so that $\lambda+\epsilon < h+1$. Then $\sum_{k=1}^{\infty} \frac{1}{|a_k|^{h+1}} \leq \sum_{k=1}^{\infty} \frac{1}{k^{\left(\frac{h+1}{\lambda+\epsilon}\right)}} < \infty$. We have thus proved that $f(z)$ can be written in the form (17.25) where so far g is only known to be an entire function. It remains to prove that $g(z)$ is a polynomial of degree $\leq h$. According to the Poisson-Jensen formula

$$\log|f(z)| = -\sum_{k=1}^{n} \log\left|\frac{r^2-\overline{a}_k z}{r(z-a_k)}\right| + \frac{1}{2\pi}\int_0^{2\pi} \mathrm{Re}\left(\frac{re^{i\theta}+z}{re^{i\theta}-z}\right)\log|f(re^{i\theta})|d\theta$$

for $|z| < r$. If the operator $\left(\frac{\partial}{\partial x}\right) - i\left(\frac{\partial}{\partial y}\right)$ is applied to both sides of the identity and using Leibniz's rule for differentiation under an integral sign we obtain

$$\frac{f'(z)}{f(z)} = \sum_{k=1}^{n}\frac{1}{z-a_k} + \sum_{k=1}^{n}\frac{\overline{a}_k}{(r^2-\overline{a}_k z)} + \frac{1}{2\pi}\int_0^{2\pi}\frac{2re^{i\theta}}{(re^{i\theta}-z)^2}\log|f(re^{i\theta})|d\theta$$

for $|z| < r$ and $z \neq a_1, \cdots, a_n$. On differentiating h times with respect to z this yields

$$\frac{d^h}{dz^h}\left[\frac{f'(z)}{f(z)}\right] = -h!\sum_{k=1}^{n(r)}\frac{1}{(a_k-z)^{h+1}} + h!\sum_{k=1}^{n(r)}\overline{a}_k^{h+1}\frac{1}{(r^2-\overline{a}_k z)^{h+1}}$$

$$+(h+1)!\frac{1}{2\pi}\int_0^{2\pi}\frac{2re^{i\theta}}{(re^{i\theta}-z)^{h+2}}\log\left|f\left(re^{i\theta}\right)\right|d\theta. \tag{17.31}$$

Now as $r \to \infty$, $n(r) \to \infty$. We will show now that the last two summands in (17.31) tend to zero as $r \to \infty$.

To see that the second sum converges to zero, let $r > 2|z|$. If $|a_k| \le r$ then

$$|r^2 - \overline{a}_k z| \ge r^2 - |\overline{a}_k||z| \ge r^2 - \frac{r^2}{2} = \frac{r^2}{2}.$$

So

$$\left(\frac{|\overline{a}_k|}{|r^2-\overline{a}_k z|}\right)^{h+1} \le \left(\frac{r}{\frac{r^2}{2}}\right)^{h+1} = \left(\frac{2}{r}\right)^{h+1}.$$

Hence

$$h!\sum_{k=1}^{n(r)}\frac{(\overline{a}_k)^{h+1}}{(r^2-\overline{a}_k z)^{h+1}} \le h!n(r)\left(\frac{2}{r}\right)^{h+1}.$$

But by Theorem 17.19,

$$(\log 2)n(r) \le \log M(r).$$

Since f is of order λ, for any $\epsilon > 0$ and r is sufficiently large

$$(\log 2)n(r)r^{-(h+1)} \le \log M(r)r^{-(h+1)} \le r^{(\lambda+\epsilon)-(h+1)}.$$

But $\lambda < h+1$, so that ϵ may be chosen with $\lambda + \epsilon < h+1$ as we did before. Hence $h!n(r)(\frac{2}{r})^{h+1} \le \frac{h!2^{h+1}}{\log 2}r^{(\lambda+\epsilon)-(h+1)} \to 0$ as $r \to \infty$. Thus the second summand in (17.31) converges to zero.

To show that the integral in (17.31) converges to zero, notice that

$$\begin{aligned}\int_0^{2\pi} \frac{re^{i\theta}}{(re^{i\theta}-z)^{h+2}}d\theta &= \int_{|w|=r} \frac{w}{(w-z)^{h+2}}\frac{dw}{iw} \\ &= (-i)\int_{|w|=r} \frac{dw}{(w-z)^{h+2}} = 0\end{aligned}$$

since the integrand has a primitive. Thus for $|z| < \frac{r}{2}$,

$$\begin{aligned}&\left|\frac{(h+1)!}{2\pi}\int_0^{2\pi} \frac{2re^{i\theta}}{(re^{i\theta}-z)^{h+2}} \log|f(re^{i\theta})|d\theta\right| \\ =& \left|\frac{(h+1)!}{2\pi}\int_0^{2\pi} \frac{2re^{i\theta}}{(re^{i\theta}-z)^{h+2}} \left[\log|f(re^{i\theta})| - \log M(r)\right] d\theta\right| \\ \leq& (h+1)!\frac{2^{h+3}}{r^{h+1}}\frac{1}{2\pi}\int_0^{2\pi} \left[\log M(r) - \log|f(re^{i\theta})|\right] d\theta. \qquad (17.32)\end{aligned}$$

But according to Jensen's formula,

$$\frac{1}{2\pi}\int_0^{2\pi} \log|f(re^{i\theta})|d\theta \geq 0$$

since $f(0) = 1$.

Also $\log M(r) \leq r^{\lambda+\epsilon}$ for sufficiently large r so that (17.32) is dominated by $(h+1)!2^{h+3}r^{(\lambda+\epsilon)-(h+1)}$. As before, ϵ can be chosen so that $r^{(\lambda+\epsilon)-(h+1)} \to 0$ as $r \to \infty$. Thus

$$\frac{d^h}{dz^h}\left[\frac{f'(z)}{f(z)}\right] = -p!\sum_{k=1}^{\infty} \frac{1}{(a_k - z)^{h+1}}$$

for $z \neq a_1, a_2, \cdots$. Now $f(z) = e^{g(z)}P(z)$ where P is a canonical product defined before. Hence for $z \neq a_k$,

$$\frac{f'(z)}{f(z)} = g'(z) + \frac{P'(z)}{P(z)}$$

and

$$-h!\sum_{n=1}^{\infty} \frac{1}{(a_n - z)^{h+1}} = g^{(h+1)}(z) + \frac{d^h}{dz^h}\left[\frac{P'(z)}{P(z)}\right].$$

It is easy to show that

$$\frac{d^h}{dz^h}\left[\frac{P'(z)}{P(z)}\right] = -h!\sum_{n=1}^{\infty} \frac{1}{(a_n - z)^{h+1}}$$

for $z \neq a_1, a_2, \cdots$. Hence $g^{(h+1)} \equiv 0$ and g must be a polynomial of degree $\leq h$. So the genus of $f \leq h \leq \lambda$. □

As an application of Hadamard's theorem a special case of Picard's theorem can be proved.

Theorem 17.22. *Let f be an entire function of finite order, then f assumes each complex number with one possible exception.*

Proof. Suppose there are complex numbers a and b, $a \neq b$, such that $f(z) \neq a$ and $f(z) \neq b$ for all $z \in \mathbb{C}$. Then $h(z) = f(z) - a$ is an entire function and $h(z) \neq 0$ for all $z \in \mathbb{C}$. Hence there is an entire function g such that $f(z) - a = e^{g(z)}$. The function $f(z) - a$ has finite order, as f has finite order. So by Hadamard's theorem, g must be a polynomial. But $e^{g(z)}$ never assumes the value $b - a$ and this means that $g(z)$ never assumes the value $\log(b - a)$. But this is a contradiction to the fundamental theorem of algebra. □

Theorem 17.23. *Let f be an entire function of finite order λ where λ is not an integer. Then f has infinitely many zeros.*

Proof. Suppose f has only finite number of zeros $a_1, a_2, \cdots, a_n$ counted according to multiplicity. Then $f(z) = e^{g(z)}(z - a_1)\cdots(z - a_n)$ for an entire function g. Since f has finite order λ, by Hadamard's theorem g is a polynomial of degree $\leq \lambda$. But order(f) = order(e^g) = degree of g. Thus $\lambda \in \mathbb{Z}$. This is a contradiction. □

Corollary 17.1. *If f is an entire function of order λ and λ is not an integer then f assumes each complex value an infinite number of times.*

Proof. Let $\alpha \in \mathbb{C}$ and $g(z) = f(z) - \alpha$. Then g is an entire function and order (g) = order $(f) = \lambda \notin \mathbb{Z}$. By Theorem 17.23, g has infinitely many zeros. Thus f assumes the value α infinite number of times. □

Example 17.7. *Consider again Example 17.2. Since the order of $\sin \pi z$ is $\lambda = 1$, hence $g(z)$ is a polynomial of degree less than or equal to $\lambda = 1$ and hence $g(z) = a + bz$. Since*

$$e^{g(z)} = \frac{\sin \pi z}{z \prod_{n=1}^{\infty}(1 - \frac{z^2}{n^2})},$$

$e^{g(z)}$ is an even function and $e^{a+bz} = e^{a-bz}$, $e^{2bz} = 1$ or $b = 0$. Taking the limit as $z \longrightarrow 0$ we have $e^a = \pi$. Thus $e^{g(z)} = \pi$ and $\sin \pi z = \pi z \prod_{n=1}^{\infty}(1 - \frac{z^2}{n^2})$.

We shall now prove the following result using Hadamard's theorem.

Theorem 17.24. *If f is an entire function of order $\lambda < \infty$ and g is an entire function of order $\lambda' \leq \lambda$ and if the zeros of $g(z)$ are all zeros of $f(z)$, then $H(z) = \frac{f(z)}{g(z)}$ is of order λ at most.*

Proof. Let $P_1(z)$ and $P_2(z)$ be the canonical products of $f(z)$ and $g(z)$ respectively and suppose $Q_1(z)$ and $Q_2(z)$ are appropriate polynomials such that $f(z) = P_1(z)e^{Q_1(z)}$ and $g(z) = P_2(z)e^{Q_2(z)}$. Thus $H(z) = P(z)e^{Q_1(z)-Q_2(z)}$ where $P(z) = \frac{P_1(z)}{P_2(z)}$ is the canonical product formed from the zeros of $P_1(z)$ which are not zeros of $P_2(z)$. Since the exponent of convergence of a sequence is not increased by removing some of the terms, the exponent of convergence and hence the order of $P(z)$, does not exceed λ. Further, $Q_1(z) - Q_2(z)$ is a polynomial of degree not exceeding λ, thus the order of $H(z) = \frac{f(z)}{g(z)}$ is of order λ, at most. □

Theorem 17.25. *The order of a canonical product equals the exponent of convergence of its zeros.*

Proof. Since for any entire function $\lambda_1 \leq \lambda$, we require to prove that $\lambda \leq \lambda_1$ for a canonical product. Let the zeros be $z_1, z_2, \cdots$, and k be a constant greater than 1. Let $P(z)$ be the canonical product, and we have

$$\begin{aligned}\log|P(z)| &= \textstyle\sum_{|z_n|\leq kr} \log\left|\left(1-\frac{z}{z_n}\right)\exp\left(\frac{z}{z_n}+\cdots+\frac{1}{p}\left(\frac{z}{z_n}\right)^p\right)\right| \\ &+ \textstyle\sum_{|z_n|>kr} \log\left|\left(1-\frac{z}{z_n}\right)\exp\left(\frac{z}{z_n}+\cdots+\frac{1}{p}\left(\frac{z}{z_n}\right)^p\right)\right| \\ &= \textstyle\sum_1 + \sum_2 .\end{aligned}$$

For $\sum_2$, since $|z| = r$ and $|z_n| > kr$, $\left|\frac{z}{z_n}\right| < 1$ and

$$\log\left(1-\frac{z}{z_n}\right)\exp\left(\frac{z}{z_n}+\cdots+\frac{1}{p}\left(\frac{z}{z_n}\right)^p\right) = -\frac{1}{p+1}\left(\frac{z}{z_n}\right)^{p+1} - \cdots .$$

Thus

$$\begin{aligned}&\left|\log\left(1-\tfrac{z}{z_n}\right)\exp\left(\tfrac{z}{z_n}+\cdots+\tfrac{1}{p}\left(\tfrac{z}{z_n}\right)^p\right)\right| \\ &< \tfrac{1}{p+1}\left\{\left|\tfrac{z}{z_n}\right|^{p+1} + \left|\tfrac{z}{z_n}\right|^{p+2} + \cdots\right\} \\ &= \tfrac{1}{p+1}\left\{\frac{\left|\frac{z}{z_n}\right|^{p+1}}{1-\left|\frac{z}{z_n}\right|}\right\} \\ &< K\left|\tfrac{z}{z_n}\right|^{p+1} .\end{aligned}$$

Also, $\log|f| = \mathrm{Re}\{\log f\} \leq |\log f|$. Thus

$$\log\left|\left(1-\frac{z}{z_n}\right)\exp\left(\frac{z}{z_n}+\cdots+\frac{1}{p}\left(\frac{z}{z_n}\right)^p\right)\right| \leq A\left(\frac{z}{z_n}\right)^{p+1}$$

and

$$\begin{aligned}\textstyle\sum_{|z_n|>kr} \log|\cdots| &= \textstyle\sum_2 = O\left\{\sum_{|z_n|>kr} |\frac{z}{z_n}|^{p+1}\right\} \\ &= O\left\{|z|^{p+1} \textstyle\sum_{|z_n|>kr} \frac{1}{|z_n|^{p+1}}\right\} .\end{aligned}$$

If $p+1=\lambda_1$, $\sum_2 = O\left(|z|^{p+1}\right) = O\left(|z|^{\lambda_1}\right) = O\left\{r^{\lambda_1}\right\}$. Further recall that $p \le \lambda_1 \le p+1$. If $p+1 > \lambda_1+\epsilon$ (as $\sum_{n=1}^{\infty} \frac{1}{{r_n}^{\lambda_1+\epsilon}}$ converges), and ϵ is small enough, then

$$\begin{aligned} |z|^{p+1}\textstyle\sum_{|z_n|>kr} |z_n|^{-p-1} &= |z|^{p+1}\textstyle\sum_{|z_n|>kr} |z_n|^{\lambda_1+\epsilon-p-1}|z_n|^{-\lambda_1-\epsilon} \\ &< |z|^{p+1}(kr)^{\lambda_1+\epsilon-p-1}\textstyle\sum_{|z_n|>kr} |z_n|^{-\lambda_1-\epsilon} \\ &= O\left\{|z|^{\lambda_1+\epsilon}\right\}, \end{aligned}$$

since (i) $\frac{1}{|z_n|} < \frac{1}{kr}$, (ii)$r=|z|$ (iii)$\lambda_1+\epsilon-p-1<0$.

In $\sum_1$, $\left|\frac{z}{z_n}\right| \ge \frac{1}{k}$. Note that $\left|\frac{z}{z_n}\right|$ can be large but cannot be small. Since

$$\log\left|\left(1-\tfrac{z}{z_n}\right)\exp\left(\tfrac{z}{z_n}+\cdots+\tfrac{1}{p}\left(\tfrac{z}{z_n}\right)^p\right)\right|$$

$$\le \log\left(1+\left|\tfrac{z}{z_n}\right|\right)+\left|\tfrac{z}{z_n}\right|+\cdots+\tfrac{1}{p}\left|\tfrac{z}{z_n}\right|^p$$

and $\log\left(1+\left|\frac{z}{z_n}\right|\right) < \left|\frac{z}{z_n}\right|$ (as $1+|x|<e^{|x|}$), we have

$$\log\left|\left(1-\frac{z}{z_n}\right)\exp\left(\frac{z}{z_n}+\cdots+\frac{1}{p}\left(\frac{z}{z_n}\right)^p\right)\right| < K\left|\frac{z}{z_n}\right|^p$$

where K depends on k only. Thus,

$$\begin{aligned} \textstyle\sum_1 &= O\left\{\textstyle\sum_{|z_n|\le kr}\left|\frac{z}{z_n}\right|^p\right\} \\ &= O\left\{|z|^p\textstyle\sum_{|z_n|\le kr}|z_n|^{\lambda_1+\epsilon-p}|z_n|^{-\lambda_1-\epsilon}\right\} \\ &= O\left\{|z|^p(kr)^{\lambda_1+\epsilon-p}\textstyle\sum_{|z_n|\le kr}|z_n|^{-\lambda_1-\epsilon}\right\} \\ &= O\left\{|z|^{\lambda_1+\epsilon}\right\}, \end{aligned}$$

since (i) $|z_n|$ are bounded, (ii) $\sum_{|z_n|\le kr}|z_n|^{-\lambda_1-\epsilon}$ is a finite series, (iii) $p\le\lambda_1, p<\lambda_1+\epsilon$ (iv) $|z_n|\le k|z|=kr$.

Thus

$$\log|P(z)| = O\{|z|^{\lambda_1+\epsilon}\}$$

and

$$|P(z)| = O\{\exp r^{\lambda_1+\epsilon}\}$$

from which we conclude that the order of $P(z)=\lambda\le\lambda_1$. Since $\lambda_1\le\lambda$, we have $\lambda_1=\lambda$. □

A particularly useful result is the following lemma.

Lemma 17.4. *If $f(z)$ is an entire function of order $\lambda < \infty$ and if λ is not an integer then $\lambda_1 = \lambda$.*

Proof. Suppose $f(z) = e^{Q(z)}P(z)$ where $P(z)$ is the canonical product and $Q(z)$ is a polynomial. In any case $\lambda_1 \le \lambda$. Suppose $\lambda_1 < \lambda$. Then $P(z)$ is of order λ_1, that is, $P(z)$ is of order less than λ. If $Q(z)$ is of degree q, $e^{Q(z)}$ is of order $q \le \lambda$ but $q < \lambda$ since q is an integer and λ is not. Hence $f(z)$ is the product of two functions each of order less than λ. Thus $f(z)$ is of order less than λ which contradicts the hypothesis that $f(z)$ is an entire function of order λ. □

A consequence is that a function of nonintegral order must have an infinite number of zeros. Since if the number of zeros is finite, $\lambda_1 = 0 = \lambda$. Also, if the order is not an integer, the function is dominated by $P(z)$, whereas if the order is an integer, $P(z)$ may reduce to a polynomial or a constant and the order depends entirely on the factor $e^{Q(z)}$. In any case, since $P(z)$ is of order λ_1 and $e^{Q(z)}$ is of order q, hence $\lambda = \max\{q, \lambda_1\}$.

Definition 17.7. *Let f be an entire function and $f(z) = e^{Q(z)}P(z)$ where $Q(z)$ is a polynomial, deg$Q = q$, $P(z) = \prod_{n=1}^{\infty}(1 - \frac{z}{z_n})e^{\frac{z}{z_n}+\cdots+\frac{1}{p}(\frac{z}{z_n})^p}$ and p is the smallest integer for which $\sum_{n=1}^{\infty}\frac{1}{|z_n|^{p+1}} < \infty$. The genus of the entire function $f(z)$ is the greater of the two integers p and q and is therefore an integer. Since $p \le \lambda$ and $q \le \lambda$, the genus does not exceed the order.*

Example 17.8. *For the function*

$$\sin z = z\prod_{n=1}^{\infty}\left(1 - \frac{z^2}{n^2\pi^2}\right) = \prod_{n=1}^{\infty} z\left(1 \pm \frac{z}{n\pi}\right)e^{\pm\frac{z}{n}},$$

the order of $e^{Q(z)}$ is $q = 0$ and $p = 1$. The genus is $\max\{p, q\} = 1$. In this case $\lambda_1 = 1$ since the series $\sum_{n=1}^{\infty}\frac{1}{(n\pi)^{1+\epsilon}} < \infty, \epsilon > 0$, and the order $\lambda = \max\{q, \lambda_1\} = 1$. Hence the genus is 1 and the order is 1.

Example 17.9. *For the function*

$$f(z) = \prod_{n=2}^{\infty}\left(1 - \frac{z}{n(\log n)^2}\right),$$

the order of $e^{Q(z)} = q = 0$. The genus is $\max\{p, q\} = 0$ as $\sum_{n=2}^{\infty}\frac{1}{n(\log n)^2} < \infty$. The order is $\max\{0, 1\} = 1$ since in this case $\lambda_1 = 1$. This is so because from Gauss's test for infinite series it follows that $\sum_{n=2}^{\infty}\frac{1}{\{n(\log n)^2\}^r} < \infty$ for $r \ge 1$ only.

Theorem 17.26. *The derivative $f'(z)$, of an entire function $f(z)$ has the same order as $f(z)$.*

Proof. Let $M'(r) = \max\limits_{|z|=r} |f'(z)|$. Since $f(z) = \int_0^z f'(t)dt + f(0)$, by taking the integral along the straight line we have

$$M(r) \leq \int_0^{|z|} |f'(t)|dt + |f(0)| \leq rM'(r) + |f(0)|.$$

Thus we obtain

$$\frac{M(r) - |f(0)|}{r} \leq M'(r). \tag{17.33}$$

Further, for $|z| = r < R$, we have

$$f'(z) = \frac{1}{2\pi i} \int_\gamma \frac{f(w)}{(w-z)^2} dw$$

where γ is the circle $|w - z| = R - r$. By choosing z such that $|f'(z)| = M'(r)$, we get,

$$M'(r) \leq \frac{M(R)}{R-r}. \tag{17.34}$$

If we choose $R = 2r$ then we have the inequality

$$\frac{M(r) - |f(0)|}{r} \leq M'(r) \leq \frac{M(2r)}{r}.$$

Let

$$\lambda = \overline{\lim_{r\to\infty}} \frac{\log\log M(r)}{\log r}.$$

Notice that $\log M(r)$ is a convex function of $\log r$ and either $\frac{\log M(r)}{\log r} < A$ (constant) or $\frac{\log M(r)}{\log r}$ tends to infinity; therefore

$$\frac{\log\{\log M(2r) - \log r\}}{\log r} = \frac{\log\log M(2r)}{\log r} + \frac{\log\{1 - [\frac{\log r}{\log M(2r)}]\}}{\log r}.$$

If $\frac{\log M(r)}{\log r} \to \infty$, the second term tends to zero, and if $\frac{\log M(r)}{\log r} \to A > 1$, the second term tends to zero also. Similarly for the left hand side of the inequality. Thus

$$\lambda = \overline{\lim_{r\to\infty}} \frac{\log\log M'(r)}{\log r}.$$

□

Theorem 17.27. *If f is an entire function of order $\lambda < 2$, $f(\mathbb{R}) \subset \mathbb{R}$ and if f has only real zeros, then the zeros of f' are also real. Further, the zeros of $f'(z)$ are separated from each other by zeros of $f(z)$.*

Proof. The hypotheses give

$$f(z) = cz^k e^{az} \prod_{n=1}^{\infty} \left(1 - \frac{z}{z_n}\right) e^{\frac{z}{z_n}}$$

where k is zero or positive and $c, a, z_1, \cdots, z_n$ are all real. Taking logs and differentiating we obtain

$$\frac{f'(z)}{f(z)} = \frac{k}{z} + a + \sum_{n=1}^{\infty} \left(\frac{1}{z - z_n} + \frac{1}{z_n}\right).$$

If $z = x + iy$, then

$$Im\left\{\frac{f'(z)}{f(z)}\right\} = \begin{cases} -iy\left\{\frac{k}{x^2+y^2} + \sum_{n=1}^{\infty} \frac{1}{(x-z_n)^2+y^2}\right\} & \text{for} \quad y \neq 0; \\ 0 & \text{for} \quad y = 0 \text{ only.} \end{cases}$$

Thus $f'(z) = 0$ on the real axis only. Also

$$\frac{d}{dz}\left\{\frac{f'(z)}{f(z)}\right\} = -\frac{k}{z^2} - \sum_{n=1}^{\infty} \frac{1}{(z - z_n)^2}$$

is real and negative for real z. Thus $\frac{f'(z)}{f(z)}$ decreases steadily as z increases through real values from z_n to z_{n+1} and it cannot vanish more than once between z_n and z_{n+1} (there can be no inflections). Clearly it changes sign (since $f = 0$ at z_n, z_{n+1} and f does not change sign between z_n, z_{n+1} but f' does (see Fig. 17.2).)

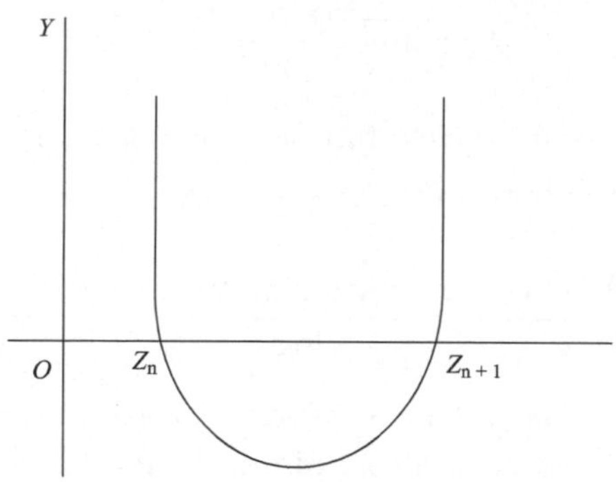

Fig. 17.2: No change sign of f between z_n, z_{n+1}

Consequently it vanishes exactly once in the interval and the theorem is proved. □

The above theorem is not true for functions of genus 2.

Example 17.10. *Let* $f(z) = ze^{z^2}$. *Then* $f'(z) = (2z^2 + 1)e^{z^2}$. *Here the zeros of* f *are real and the zeros of* f' *are complex.*

Example 17.11. *Let* $f(z) = (z^2-4)e^{\frac{z^2}{3}}$. *Then* $f'(z) = \frac{2}{3}z(z^2-1)e^{\frac{z^2}{3}}$. *The function* f *has order* 2, *the zeros of* f' *are real but not separated by the zeros of* f.

We shall now prove Lucas's theorem relating to the distribution of zeros of a polynomial and its derivative. For this we need to introduce the notion of convexity.

Recall that a set E is convex if whenever z_1 and z_2 are two points of E, then $\{z \in \mathbb{C} : z = z_1 + a(z_2 - z_1), 0 \le a \le 1\} \subset E$. Clearly a convex set is path connected, but need not be open (and therefore need not be a region).

Consider the interior of a triangle $\triangle$ with vertices at z_1, z_2, z_3. The interior is the point set,

$$\triangle^0 = \{z|z = a_1z_1 + a_2z_2 + a_3z_3; a_1 \ge 0, a_2 \ge 0, a_3 \ge 0; \sum_{r=1}^{3} a_r = 1\}.$$

Similarly, Γ is a convex polygon if Γ is the boundary of a convex domain. If a polygon Γ has vertices $z_1, z_2, \cdots, z_n$, the interior Γ^0 is such that

$$\Gamma^0 = \{z|z = a_1z_1 + \cdots + a_nz_n : a_r \ge 0, r = 1, \cdots, n; \sum_{r=1}^{n} a_r = 1\}.$$

Lemma 17.5. *If* $z = \sum_{r=1}^{n} a_rz_r$ *and* $a_r \ge 0, \sum_{r=1}^{n} a_r = 1$, *then* z *lies in the convex hull of the* z_r'*s.*

Proof. We shall use method of induction to prove the lemma. For the case $n = 2$, if z' is between z_1 and z_2 dividing the line joining z_1 and z_2 in the ratio $\lambda : 1-\lambda, 0 < \lambda < 1$ (see Fig. 17.3), then vectorially

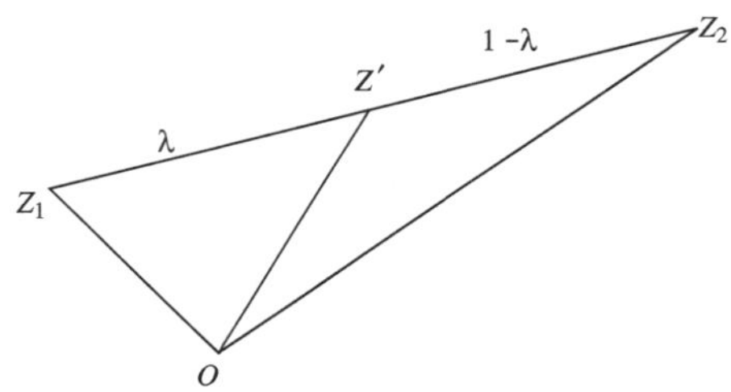

Fig. 17.3: Point z' between z_1 and z_2

$$\begin{aligned} \overline{oz_1} &= \overline{oz'} + \overline{z'z_1} \\ \overline{oz_2} &= \overline{oz'} + \overline{z'z_2} \\ \overline{oz'} &= \tfrac{1}{2}(\overline{oz_1} + \overline{oz_2} + \overline{z_2z'} + \overline{z_1z'}) \end{aligned}$$

$$\begin{aligned} \text{or} \quad z' &= \tfrac{1}{2}(z_1 + z_2 + (1-\lambda)(z_1 - z_2) + \lambda(z_2 - z_1)) \\ \text{and} \quad z' &= (1-\lambda)z_1 + \lambda z_2 : \textstyle\sum \text{coefficients} = 1. \end{aligned}$$

Conversely, if $z' = a_1z_1 + a_2z_2; a_1 + a_2 = 1; a_1, a_2 \ge 0$, then z is in the convex hull of z_1, z_2. Let $z = a_1z_1 + \cdots + a_nz_n; \sum_{r=1}^{n} a_r = 1; a_r \ge 0$. We assume that

$0 < a_n < 1$, and $z = (1 - a_n)\{\frac{a_1}{1-a_n} z_1 + \cdots + \frac{a_{n-1}}{1-a_n} z_{n-1}\} + a_n z_n$. Let

$$z'' = \frac{a_1}{1 - a_n} z_1 + \cdots + \frac{a_{n-1}}{1 - a_n} z_{n-1}.$$

Since

$$\frac{a_1}{1 - a_n} + \cdots + \frac{a_{n-1}}{1 - a_n} = 1,$$

then z'' is in the convex hull of $z_1, \cdots, z_{n-1}$. Now $z = (1 - a_n)z'' + a_n z_n$ and since $a_n, 1 - a_n \geq 0$ and $\{a_n + (1 - a_n)\} = 1$, then z is in the convex hull of z'' and z_n, i.e., $z_1, \cdots, z_n$. □

Theorem 17.28. *(Lucas's theorem): The zeros of the derivative $P'(z)$ of a polynomial $P(z)$, are contained within the convex hull of the zeros of $P(z)$.*

Proof. Let $P(z)$ have zeros $z_1, z_2, \cdots, z_n$. Let Γ be the least convex polygon containing these zeros. We show that $P'(z)$ cannot vanish anywhere in the exterior of Γ. Since $P(z) = (z - z_1)(z - z_2) \cdots (z - z_n)$, then

$$\frac{P'(z)}{P(z)} = \frac{d}{dz} \log P(z) = \sum_{k=1}^{n} \frac{1}{z - z_k}.$$

If $P'(z_0) = 0$, then $\sum_{k=1}^{n} \frac{1}{z_0 - z_k} = 0$ and $\sum_{k=1}^{n} \frac{1}{\bar{z}_0 - \bar{z}_k} = 0$. Thus

$$\sum_{k=1}^{n} \frac{(z_0 - z_k)}{|z_0 - z_k|^2} = 0$$

and

$$z_0 \sum_{k=1}^{n} \frac{1}{|z_0 - z_k|^2} = \sum_{k=1}^{n} \frac{z_k}{|z_0 - z_k|^2}.$$

Since $z_0 = \sum_{r=1}^{n} a_r z_r$ with $\sum_{r=1}^{n} a_r = 1, a_r \geq 0$, we deduce that z_0 lies within the convex hull of the $z_r's$. □

Lucas's theorem is not true for arbitrary entire functions, e.g., $f(z) = (z+1)e^{\frac{z^2}{2}}$ has only one zero $z = -1$, but the derivative $f'(z) = (1 + z + z^2)e^{\frac{z^2}{2}}$ has two zeros; $z_1 = e^{\frac{2\pi i}{3}}$ and $z_2 = e^{\frac{4\pi i}{3}}$ and Lucas's theorem clearly does not hold. The theorem does extend to functions whose order λ is such that $0 \leq \lambda < 1$. (For details see the paper of M. Marden, "On the zeros of the derivative of an entire function," Amer. Math. Monthly 25(8), 1968.)

Type

Entire functions may be further subdivided as follows. Given an entire function of order $\lambda < \infty$, suppose there exist a positive k such that $M(r) < \exp kr^{\lambda}$ for $r > R_0$ where $M(r) = \max_{|z|=r} |f(z)|$. Then $f(z)$ is said to be of finite type.

The greatest lower bound $\sigma = \inf k \geq 0$ of the $k's$ for which $M(r) < \exp kr^{\lambda}$ holds (starting from some sufficiently large r), is called the type of $f(z)$; for example, e^z has order $\lambda = 1$ and type $\sigma = 1$.

Definition 17.8. *The entire function f is called finite type if $0 < \sigma < \infty$ and f is called minimum type if $\sigma = 0$, and f is called maximum type or infinite type if $\sigma = \infty$ (that is, $M(r)$ exceeds $\exp kr^{\lambda}$ for arbitrary large r).*

As a consequence, we have the following result.

Theorem 17.29. *The type σ of an entire function of order $\lambda, 0 < \lambda < \infty$, is given by the formula*

$$\sigma = \overline{\lim_{r\to\infty}} \frac{\log M(r)}{r^{\lambda}}.$$

Proof. Since $\sigma = \inf k$, given $\epsilon > 0$, there exists $R(\epsilon) > 0$, such that

$$M(r) < \exp(\sigma + \epsilon)r^{\lambda}$$

for $r > R(\epsilon)$. Also, there is a sequence $\{r_n\}$ such that

$$0 < r_1 < \cdots < r_n < \cdots$$

and $M(r_n) > \exp(\sigma - \epsilon)r_n^{\lambda}$. Thus $\frac{\log M(r)}{r^{\lambda}} < \sigma + \epsilon$ for $r > R(\epsilon)$ while

$$\frac{\log M(r_n)}{r_n^{\lambda}} > \sigma - \epsilon$$

for arbitrary large r_n. This is precisely what is meant by $\sigma = \overline{\lim_{r\to\infty}} \frac{\log M(r)}{r^{\lambda}}$. □

Example 17.12.

$$\frac{e^r - 1}{2} \leq M(r) = \max_{|z|=r} |\sin z| \leq \frac{e^r + 1}{2}$$

and $\sin z$ *is of order* $\lambda = 1$ *and type* $\sigma = 1$.

Note that the maximum modulus $M(r)$ describes the growth of an entire function in the neighborhood of the point at infinity but gives no information about the behavior of $f(z)$ in various unbounded subdomains. Consider for example the function e^z which has order 1 and type 1. In this case, $|e^z| = e^x = e^{r\cos\theta}$. In the closed angle $-\frac{\pi}{2} + \epsilon \leq \theta \leq \frac{\pi}{2} - \epsilon, \epsilon > 0$,

$$e^{r\sin\epsilon} \leq |e^z| \leq e^r$$

and thus $e^z \to \infty$ as $r \to \infty$. However, in every closed angle

$$\left(\frac{\pi}{2}\right) + \epsilon \leq \theta \leq \left(\frac{3\pi}{2}\right) - \epsilon, \epsilon > 0,$$

$$|e^z| \leq e^{-r\sin\epsilon}$$

and $e^z \to 0$ as $r \to \infty$. Thus there exists two open angles each of π radians, in the right half plane Re$z > 0$ and the left half plane Re$z < 0$,such that $\lim\limits_{r\to\infty} e^z = \infty$ in every angle in Re$z > 0$ while $\lim_{r\to\infty} e^z = 0$ in every angle in Re$z < 0$.

If we have the power series representation of an entire function, we can calculate the order and type using the following two results.

Theorem 17.30. *A necessary and sufficient condition that* $f(z) = \sum_{n=0}^{\infty} a_n z^n$ *is an entire function of order* λ, *is that* $\underline{\lim}_{n\to\infty} \left(\dfrac{\log\left(\frac{1}{|a_n|}\right)}{n \log n} \right) = \dfrac{1}{\lambda}$.

Theorem 17.31. *If* $f(z)$ *is an entire function of finite order* λ $(0 < \lambda < \infty)$ *and type* σ, *then*

$$\sigma = \frac{1}{e\lambda} \overline{\lim}_{n\to\infty} n |a_n|^{\frac{\lambda}{n}}.$$

The proof of these theorems are left as exercises for the reader.

Example 17.13. *The function* $f(z) = \sum_{n=1}^{\infty} \left(\frac{e\lambda\sigma}{n}\right)^{\frac{n}{\lambda}} z^n$ *is of order* λ *and type* σ.

Example 17.14. *If* $f(z) = \sum_{n=0}^{\infty} a_n z^n$ *is an entire function such that*

$$\lim_{n\to\infty} \frac{\log n}{\log\left(\frac{1}{|a_n|^{\frac{1}{n}}}\right)} = 0$$

then f *is an entire function of order zero. From this it follows that any entire function* $f(z) = \sum_{n=0}^{\infty} a_n z^n$ *with coefficients* $|a_n| = \frac{1}{n^{\frac{n}{\epsilon_n}}}$ *where* $\{\epsilon_n\}$ *is a sequence of positive numbers converging to zero is of order zero. For example,* $f(z) = \sum_{n=1}^{\infty} \frac{z^n}{n^{n^{1+\delta}}}, \delta > 0$ *has order* $= \lambda = 0$ *(examine* $\log(\frac{1}{|a_n|^{\frac{1}{n}}})$*).*

Example 17.15. *Consider the condition*

$$\overline{\lim_{n\to\infty}} \frac{\log n}{\log\left(\frac{1}{|a_n|^{\frac{1}{n}}}\right)} = \infty$$

which is same as

$$\underline{\lim}_{n\to\infty} \frac{\log\left(\frac{1}{|a_n|}\right)}{n \log n} = 0.$$

This condition together with

$$\lim_{n\to\infty} |a_n|^{\frac{1}{n}} = 0$$

characterizes an entire function of infinite order; for example, consider $|a_n| = \frac{1}{n^{n\epsilon_n}}$, *where* $\{\epsilon_n\}$ *is a sequence of positive numbers converging to zero slowly enough that*

$$\lim_{n\to\infty} \epsilon_n \log n = \infty$$

(since we require $\lim_{n\to\infty} \dfrac{1}{|a_n|^{\frac{1}{n}}} = \infty$*). The sequence* $\epsilon_n = \frac{1}{(\log n)^{1-\delta}}$ $(n = 1, 2, \cdots)$ *meets these requirements if* $0 < \delta < 1$ *as* $\epsilon_n \to 0$ *but* $\lim_{n\to\infty} \epsilon_n \log n = \infty$*. Thus the series*

$$f(z) = \sum_{n=0}^{\infty} \frac{z^n}{\exp(n^{\delta} \log n)}, 0 < \delta < 1$$

represents an entire function of infinite order.

17.6 The Gamma function

Consider the canonical product

$$G(z) = \prod_{n=1}^{\infty} (1 + \frac{z}{n}) e^{-\frac{z}{n}}.$$

Let $z_n = -n$ for each integer $n \geq 1$. The exponent of convergence of the sequence $\{z_n\}$ is 1 and therefore G is an entire function of order 1. Now

$$G(1) = \lim_{N\to\infty} \prod_{n=1}^{N} (1 + \frac{1}{n}) e^{-\frac{1}{n}} = \lim_{N\to\infty} (N+1) e^{-(1+\frac{1}{2}+\cdots+\frac{1}{N})} = e^{-\gamma}$$

where $\gamma = \lim_{N\to\infty} (1 + \dfrac{1}{2} + \cdots + \dfrac{1}{N} - \log(N+1))$. The constant γ is called Euler's constant. The limit γ exists because $G(1)$ exists.

Define $H(z)$ by $H(z) = G(z-1)$. The entire function H has order 1, and has zeros at $z = 0$ and at the points $z_n = -n, n \in \mathbb{Z}, n \geq 1$. From the Weierstrass theorem it follows that the representation $G(z-1) = H(z) = ze^{az+b}G(z)$ for certain constants a and b. We can determine a and b by computing $H(1)$ and $H'(0)$ as follows. Notice that $G(0) = 1 = H(1) = e^{a+b}G(1) = e^{a+b-\gamma}$, and $a + b - \gamma = 2l\pi i$ for some $l \in \mathbb{Z}$. Further $H'(0) = \lim_{z\to 0} \dfrac{H(z) - H(0)}{z} = \lim_{z\to 0} \dfrac{H(z)}{z} = e^b$, and also $H'(0) = G'(-1)$. But

$$\begin{aligned}
G'(-1) &= \lim_{z\to -1} \frac{G(z)-G(-1)}{z+1} \\
&= \lim_{z\to -1} \frac{G(z)}{z+1} \\
&= \lim_{N\to\infty} \prod_{n=2}^{N} (1-\frac{1}{n})e^{\frac{1}{n}} \\
&= \lim_{N\to\infty} \frac{e^{\sum_{n=2}^{N}\frac{1}{n}}}{N} \\
&= e^{\gamma}.
\end{aligned}$$

Thus we can take $b=\gamma$, and $a = 2l\pi i$. Since $H(z)$ is real for real z, the factor $e^{2\pi liz}$ must be real for real z, and this implies $l=0$. We have obtained the formula $G(z-1) = ze^{\gamma}G(z)$.

Definition 17.9. *The gamma function, $\Gamma(z)$, is defined for $z \neq 0, -1, -2, \cdots$ by the equation*

$$\Gamma(z) = \frac{1}{ze^{\gamma z}G(z)}.$$

The function $\Gamma(z)$ is meromorphic with simple poles at the negative integers and zero. If $z \neq 0, -1, -2, \cdots$, there is the following reproducing formula for the function $\Gamma(z)$.

$$\begin{aligned}
\Gamma(z+1) &= \frac{1}{(z+1)e^{\gamma(z+1)}G(z+1)} \\
&= \frac{1}{(z+1)\frac{e^{\gamma(z+1)}G(z)}{(z+1)e^{\gamma}}} \\
&= \frac{1}{e^{\gamma z}G(z)} = z\Gamma(z).
\end{aligned}$$

Notice that $\Gamma(1) = \frac{1}{e^{\gamma}G(1)} = 1$ and for integers $n \geq 0, \Gamma(n+1) = n!$. The function $\Gamma(z)$ has residue 1 at $z=0$ since $\lim_{z\to 0} z\Gamma(z) = \lim_{z\to 0}\frac{1}{e^{\gamma z}G(z)} = 1$. Further at $z=-n, n\in\mathbb{N}$, the residue is obtained by the reproducing formula,

$$\begin{aligned}
\lim_{z\to -n}(z+n)\Gamma(z) &= \lim_{z\to -n}\Gamma(z)(z+n)\frac{z(z+1)\cdots(z+n-1)}{z(z+1)\cdots(z+n-1)} \\
&= \lim_{z\to -n}\frac{\Gamma(z+n+1)}{z(z+1)\cdots(z+n-1)} \\
&= \frac{(-1)^n}{n!}.
\end{aligned}$$

Notice that the function $\frac{1}{\Gamma(z)}$ is an entire function of order 1.

We shall establish some formulas concerning $\Gamma(z)$ which are very useful.

If z is not an integer, then $\Gamma(z)\Gamma(1-z) = \frac{\pi}{\sin \pi z}$. It is not so difficult to verify that

$$wG(w)G(-w) = \frac{\sin \pi w}{\pi}$$

(follows from the product representation of $\sin \pi w$). Let $w = z - 1$. Then we obtain

$$(z-1)G(z-1)G(1-z) = \frac{\sin \pi(z-1)}{\pi}.$$

Thus

$$\begin{aligned} G(1-z) &= \frac{\sin \pi(z-1)}{\pi} \frac{1}{(z-1)ze^{\gamma}G(z)} \\ &= \frac{\sin \pi z}{\pi} \frac{1}{(1-z)e^{\gamma}zG(z)}. \end{aligned}$$

Hence

$$\frac{\pi}{\sin \pi z} = \frac{1}{(1-z)G(1-z)} \frac{1}{ze^{\gamma}G(z)}.$$

On the other hand,

$$\begin{aligned} \Gamma(z)\Gamma(1-z) &= \frac{1}{ze^{\gamma z}G(z)} \frac{1}{(1-z)e^{\gamma(1-z)}G(1-z)} \\ &= \frac{1}{z(1-z)G(z)G(1-z)e^{\gamma}}. \end{aligned}$$

Thus

$$\Gamma(z)\Gamma(1-z) = \frac{\pi}{\sin \pi z}. \tag{17.35}$$

Putting $z = \frac{1}{2}$ in (17.35), we obtain $\left(\Gamma\left(\frac{1}{2}\right)\right)^2 = \pi$. Since $\Gamma\left(\frac{1}{2}\right) > 0$, hence $\Gamma(\frac{1}{2}) = \sqrt{\pi}$.

Theorem 17.32. *(Legendre Duplication formula): If $z \neq 0, -\frac{1}{2}, -1, -\frac{3}{2}, \cdots$ then*

$$\Gamma(z)\Gamma\left(z + \frac{1}{2}\right) = 2^{1-2z}\sqrt{\pi}\,\Gamma(2z).$$

Proof. Each of the functions $\frac{1}{\Gamma(2z)}$ and $\frac{1}{\Gamma(z)\Gamma\left(z+\frac{1}{2}\right)}$ is an entire function of order 1, and each of them vanishes at the points $z = 0, -\frac{1}{2}, -1, -\frac{3}{2}, \cdots$. The Weierstrass theorem implies $\frac{1}{\Gamma(z)\Gamma(z+\frac{1}{2})} = e^{az+b}\frac{1}{\Gamma(2z)}$. If $z = \frac{1}{2}$, then

$$\frac{1}{\Gamma\left(\frac{1}{2}\right)} = e^{\frac{a}{2}+b}.$$

If $z = w + \frac{1}{2}$, then

$$\frac{1}{\Gamma\left(w+\frac{1}{2}\right)}\frac{1}{\Gamma(w+1)} = \frac{e^{aw+b}e^{\frac{a}{2}}}{\Gamma(2w+1)},$$

or

$$\frac{1}{\Gamma(w)}\frac{1}{\Gamma\left(w+\frac{1}{2}\right)} = \frac{e^{aw+b}e^{\frac{a}{2}}}{2\Gamma(2w)}$$

since $\Gamma(z+1) = z\Gamma(z)$. Therefore, $e^{\frac{a}{2}} = 2$ and $e^b = \frac{1}{2\sqrt{\pi}}$. Hence $\dfrac{1}{\Gamma(z)\Gamma\left(z+\frac{1}{2}\right)} = \dfrac{2^{2z-1}}{\sqrt{\pi}\Gamma(2z)}$. □

Theorem 17.33. *If z is neither a negative integer nor zero then*

$$\Gamma(z) = \lim_{n\to\infty}\frac{n!n^z}{z(z+1)\cdots(z+n)}.$$

Proof. By definition

$$\begin{aligned}\frac{1}{\Gamma(z)} &= \lim_{N\to\infty} ze^{(1+\frac{1}{2}+\cdots+\frac{1}{N})z-(\log N)z}\prod_{n=1}^{N}(1+\frac{z}{n})e^{-\frac{z}{n}} \\ &= \lim_{N\to\infty} ze^{-(\log N)z}\prod_{n=1}^{N}(1+\frac{z}{n}) \\ &= \lim_{N\to\infty}\frac{z(z+1)\cdots(z+N)}{N^zN!}.\end{aligned}$$

Thus

$$\Gamma(z) = \lim_{N\to\infty}\frac{N^zN!}{z(z+1)\cdots(z+N)}$$

as claimed. □

Corollary 17.2.

$$\lim_{n\to\infty}\frac{\Gamma(z+n)}{n^z\Gamma(n)} = 1.$$

Proof. This follows by noting that

$$\begin{aligned}\frac{\Gamma(z+n)}{n^z\Gamma(n)} &= \frac{(z+n-1)\cdots(z+1)z\Gamma(z)}{n^z\Gamma(n)}\\ &= \frac{n}{z+n}\frac{(z+n)\cdots(z+1)z\Gamma(z)}{n^z n!}.\end{aligned}$$

By Theorem 17.33, the result follows. □

We shall now present the integral representation of $\Gamma(z)$. But first we prove the following lemma.

Lemma 17.6. *Let $\{\phi_n\}$ be a sequence of nonnegative Riemann integrable functions defined for $0 \le t < \infty$. If there exists a Riemann integrable function $\phi(t)$ such that*

(a) *$\phi_n(t) \le \phi(t)$ for all n, t, and*

(b) *For each $\delta, 0 < \delta < 1$, $\lim_{n\to\infty} \phi_n(t) = \phi(t)$ holds uniformly for $\delta \le t \le \frac{1}{\delta}$, then*

$$\lim_{n\to\infty} \int_0^\infty \phi_n(t)dt = \int_0^\infty \phi(t)dt.$$

Proof. Given $\epsilon > 0$, choose $\delta, 0 < \delta < 1$, such that

$$\int_\delta^{\frac{1}{\delta}} \phi(t)dt \ge \int_0^\infty \phi(t)dt - \frac{\epsilon}{2}.$$

Choose N large enough so that if $n \ge N$, then

$$\max_{\delta\le t\le\frac{1}{\delta}} |\phi_n(t) - \phi(t)| < \frac{\epsilon}{2}\frac{1}{\frac{1}{\delta} - \delta}.$$

Thus if $n \ge N$, we have

$$\begin{aligned}\int_0^\infty \phi_n(t)dt &\ge \int_\delta^{\frac{1}{\delta}} \phi_n(t)dt\\ &\ge \int_\delta^{\frac{1}{\delta}} \phi(t)dt - \frac{\epsilon}{2}\\ &\ge \int_0^\infty \phi(t)dt - \epsilon.\end{aligned}$$

Since also

$$\int_0^\infty \phi_n(t)dt \le \int_0^\infty \phi(t)dt,$$

it follows that if $n \geq n_0$, then

$$\left| \int_0^\infty \phi_n(t)dt - \int_0^\infty \phi(t)dt \right| < \epsilon,$$

and the lemma is proved. □

Theorem 17.34. *If Re$z > 0$, then $\Gamma(z) = \int_0^\infty t^{z-1}e^{-t}dt$.*

Proof. If Re$z > 0$, define

$$F(z) = \int_0^\infty t^{z-1}e^{-t}dt. \tag{17.36}$$

If $x =$ Rez, then $|t^{z-1}| = t^{x-1}$. If $x \geq 1$, it follows from (17.36) that $|F(z)| \leq F(x) < \infty$. This is so even for $0 < x < 1$. If $0 < x_1 < x < x_2 < \infty$, then $|F(z)| \leq F(x) = \int_0^1 t^{x-1}e^{-t}dt + \int_1^\infty t^{x-1}e^{-t}dt \leq \int_0^1 t^{x_1-1}e^{-t}dt + \int_1^\infty t^{x_2-1}e^{-t}dt \leq \int_0^\infty t^{x_1-1}e^{-t}dt + \int_0^\infty t^{x_2-1}e^{-t}dt = F(x_1) + F(x_2)$. Therefore, $F(z)$ is uniformly bounded in strips $x_1 \leq$ Re$z \leq x_2$. Also $F(z) = \lim_{n\to\infty} F_n(z)$, where $F_n(z)$ is the analytic function

$$F_n(z) = \int_{\frac{1}{n}}^n t^{z-1}e^{-t}dt.$$

Notice that $|F_n(z)| \leq F(x_1) + F(x_2)$. By Vitali's theorem $\{F_n\}$ converges locally uniformly to F, and so F is analytic. We shall now show that if $x \in \mathbb{R}$, and $x > 1$, then $F(x) = \Gamma(x)$. Fix x and for each n, let

$$\phi_n(t) = \begin{cases} \left(1 - \frac{t}{n}\right)^n t^{x-1} & \text{if } 0 \leq t \leq n; \\ 0 & \text{if } t > n. \end{cases}$$

If $0 < \delta \leq t \leq \frac{1}{\delta}$, then $\lim_{n\to\infty} \phi_n(t) = e^{-t}t^{x-1}$ holds uniformly. Furthermore, we claim $\phi_n(t) \leq e^{-t}t^{x-1}$ for all n, t. If $t \geq n$, this is clear. If $t < n$, then

$$\begin{aligned} \log \phi_n(t) &= n \log\left(1 - \tfrac{t}{n}\right) + \log t^{x-1} \\ &= -n \textstyle\sum_{m=1}^\infty \frac{\left(\frac{t}{n}\right)^m}{m} + \log t^{x-1} \\ &\leq -t + \log t^{x-1} \\ &= \log e^{-t}t^{x-1}. \end{aligned}$$

This implies $\phi_n(t) \leq e^{-t}t^{x-1}$ as claimed. By Lemma 17.6,

$$\lim_{n\to\infty} \int_0^\infty \phi_n(t)dt = \int_0^\infty e^{-t}t^{x-1}dt.$$

Now

$$
\begin{aligned}
\int_0^\infty \phi_n(t)dt &= \int_0^n \left(1-\frac{t}{n}\right)^n t^{x-1}dt \\
&= \left(1-\tfrac{t}{n}\right)^n \tfrac{t^x}{x}\Big|_0^n + \tfrac{1}{x}\int_0^n \left(1-\frac{t}{n}\right)^{n-1} t^x dt \\
&= \tfrac{1}{x}\int_0^n \left(1-\frac{t}{n}\right)^{n-1} t^x dt \\
&= \frac{1}{x}\frac{1}{x+1}\cdot\frac{n-1}{n}\int_0^n \left(1-\frac{t}{n}\right)^{n-2} t^{x+1}dt \\
&\;\;\vdots \\
&= \frac{1}{x}\frac{1}{x+1}\cdots\frac{1}{x+n-1}\cdot\frac{n-1}{n}\cdot\frac{n-2}{n}\cdots\frac{1}{n}\int_0^n t^{x+n-1}dt \\
&= \frac{1}{x}\frac{1}{x+1}\cdots\frac{1}{x+n}\frac{n!}{n^n}n^{x+n} \\
&= \frac{n^x n!}{x(x+1)\cdots(x+n)}.
\end{aligned}
$$

Hence

$$
\begin{aligned}
\lim_{n\to\infty}\int_0^\infty \phi_n(t)dt &= \lim_{n\to\infty}\frac{n^x n!}{x(x+1)\cdots(x+n)} \\
&= \Gamma(x)
\end{aligned}
$$

by Theorem 17.33. Since $\Gamma(z)$ and $F(z)$ are analytic for Re$z > 0$, and since $\Gamma(x) = F(x)$ for $x > 1$, the result follows from the identity theorem (Theorem 7.17). □

17.7 The Riemann Zeta function

The Riemann zeta function, $\zeta(s)$, is defined by the (Dirichlet) series

$$
\zeta(s) = \sum_{n=1}^\infty \frac{1}{n^s} \qquad (s = \sigma + it)
$$

for $\text{Re}s > 1$. Each term $\frac{1}{n^s}$ of the series is an analytic function. Also, if $\sigma = \text{Re}s \geq 1 + \delta$ (where δ is any fixed positive number) then,

$$\left|\frac{1}{n^s}\right| = \left|\frac{1}{e^{s\ln n}}\right| = \frac{1}{e^{x\ln n}} = \frac{1}{n^\sigma} \leq \frac{1}{n^{1+\delta}}.$$

Since $\sum \frac{1}{n^{1+\delta}}$ converges, we see by the Weierstrass M test that $\sum_{n=1}^{\infty} \frac{1}{n^s}$ converges uniformly for $\text{Re}s \geq 1 + \delta$. Hence by Definition 3.12, the partial sums $s_k(z) = \sum_{n=1}^{k} \frac{1}{n^s}$ is uniformly convergent for $\text{Re}s \geq 1 + \delta$. By Theorem 7.10, $\sum_{n=1}^{\infty} \frac{1}{n^s}$ is analytic in the region $\{s \in \mathbb{C} : \text{Re}s \geq 1 + \delta\}$ where δ is any fixed positive number. Thus the function $\zeta(s)$ is analytic in $\text{Re}s > 1$ with derivative $\zeta'(s) = -\sum_{n=1}^{\infty} \frac{\log n}{n^s}$. In this section we shall discuss about some of the elementary properties of $\zeta(s)$.

The Riemann zeta function plays an important role in research in number theory. In the study of number theory a sequence $f(1), f(2), \cdots$ of complex numbers is called a number-theoretic function. Just two of these functions will be of interest for our discussion. In what follows we shall discuss about some of the interplays between the Riemann zeta function and these number theoretic functions.

If $n \in \mathbb{Z}$ and $n > 1$, the fundamental theorem of arithmetic asserts that n can be decomposed into a product of nonzero powers of distinct prime numbers. That is,

$$n = p_1^{l_1} \cdots p_k^{l_k} \tag{17.37}$$

with $l_j > 0, 1 \leq j \leq k$. Furthermore, except for the ordering of $p_1, \cdots, p_k$ the representation is unique.

The first of our two number-theoretic functions is the Möbius function, $\mu(n)$. It is defined by

$$\mu(n) = \begin{cases} 1 & \text{if} \quad n = 1; \\ (-1)^k & \text{if} \quad n > 1 \text{ and } l_1 = \cdots = l_k = 1 \text{ in } (17.37); \\ 0 & \text{otherwise.} \end{cases}$$

The Möbius function is well defined because of the uniqueness in (17.37). Notice that $\mu(n) = 0$ just when n is divisible by a perfect square other than 1.

The symbol $d|n$, read "d divides n," means $\frac{n}{d}$ is an integer. Thus $\sum_{d|n}$ denotes a sum over only those integers $d, 1 \leq d \leq n$, which divide n.

Lemma 17.7. *If $n \in \mathbb{Z}, n > 1$, then*

$$\sum_{d|n} \mu(d) = \sum_{d|n} \mu\left(\frac{n}{d}\right) = 0. \tag{17.38}$$

Proof. To evaluate (17.38) it is sufficient to sum over those divisors of n for which $\mu(d) \neq 0$. If n is represented by (17.37), then unless $d = 1$, such a divisor must have the form

$$d = p_{i_1} \cdots p_{i_m}$$

for certain $i_1, \cdots, i_m$ with $1 \le i_1 < i_2 < \cdots < i_m \le k$. There are $\binom{k}{m}$ ways to choose $i_1, \cdots, i_m$, and for each choice $\mu(d) = (-1)^m$. Therefore,

$$\sum_{d|n} \mu(d) = 1 + \binom{k}{1}(-1) + \binom{k}{2}(-1)^2 + \cdots + \binom{k}{k}(-1)^k = (1 + (-1))^k = 0.$$

Since both d and $\frac{n}{d}$ are divisors of n if either is, the result follows. □

Let f be a number-theoretic function, and associate with f a second such function g, where g is defined by the formula

$$g(n) = \sum_{d|n} f(d). \tag{17.39}$$

The inverse transformation to (17.39) is given by the **Möbius inversion formula** which is proved below.

Theorem 17.35. *For each n,*

$$f(n) = \sum_{d|n} g(d)\mu\left(\frac{n}{d}\right)$$

if g and f are related as in (17.39).

Proof. Substituting the value of g from (17.39), we obtain

$$\sum_{d|n} g(d)\mu\left(\frac{n}{d}\right) = \sum_{d|n}\sum_{m|d} f(m)\mu\left(\frac{n}{d}\right).$$

If m is a fixed divisor of n, the integers $\frac{n}{d}$ which appear with m in the above expression are just the divisors of $\frac{n}{m}$. Unless $m = n$, the contribution of these terms to the sum is

$$f(m)\sum_{\frac{n}{d}|\frac{n}{m}} \mu\left(\frac{n}{d}\right) = 0$$

by Lemma 17.7. When $m = n$ the contribution is of course $f(n)$, and the theorem is proved. □

The second number-theoretic function we consider is $\Lambda(n)$, defined by

$$\Lambda(n) = \begin{cases} \log p & \text{if} \quad n = p^m, m > 0 \textit{ and } p \textit{ prime}; \\ 0 & \text{otherwise.} \end{cases}$$

Let $n \in \mathbb{Z}$ be such that $n > 1$ and write $n = p_1^{l_1} \cdots p_k^{l_k}$ as in (17.37). Since $\Lambda(d) = 0$ unless d is a nonzero power of a prime number, we have

$$\sum_{d|n} \Lambda(d) = \sum_{j=1}^{k} l_j \log p_j = \log n.$$

Therefore by Theorem 17.35 ,

$$\Lambda(n) = \sum_{d|n} \mu\left(\frac{n}{d}\right) \log d. \tag{17.40}$$

Define $\phi(s)$ by

$$\phi(s) = \sum_{n=1}^{\infty} \frac{\mu(n)}{n^s},$$

We compute the product $\zeta(s)\phi(s)$ by making use of the formula (17.38), getting

$$\zeta(s)\phi(s) = \sum_{k=1}^{\infty} \frac{1}{k^s} \sum_{l=1}^{\infty} \frac{\mu(l)}{l^s}$$

$$= \sum_{n=1}^{\infty} \sum_{k\cdot l=n} \frac{\mu(l)}{n^s} = \sum_{n=1}^{\infty} \frac{1}{n^s} \sum_{d|n} \mu(d) = 1. \tag{17.41}$$

It follows therefore that $\zeta(s) \neq 0$ for Re$s > 1$, and $\phi(s) = \frac{1}{\zeta(s)}$.

Lemma 17.8. *If Re$s > 1$, then*

$$\frac{\zeta'(s)}{\zeta(s)} = -\sum_{n=1}^{\infty} \frac{\Lambda(n)}{n^s}.$$

Proof. By (17.41), $\frac{\zeta'(s)}{\zeta(s)} = \zeta'(s)\phi(s)$. Using (17.40) we obtain

$$\zeta'(s)\phi(s) = -\sum_{n=1}^{\infty} \frac{1}{n^s} \sum_{d|n} \mu\left(\frac{n}{d}\right) \log d = -\sum_{n=1}^{\infty} \frac{\Lambda(n)}{n^s}.$$

The lemma is proved. □

The zeta function was known to Euler for certain values of s, and his name is attached to a well-known formula containing it.

Theorem 17.36. *(Euler product formula): If Re$s > 1, \zeta(s)$ can be expressed as a product,*

$$\zeta(s) = \prod_{n=1}^{\infty} (1 - p_n{}^{-s})^{-1}$$

where $\{p_n\}$ is the sequence of prime numbers.

Proof. For each n, $(1 - p_n^{-s})^{-1} = \sum_{m=0}^{\infty} p_n^{-ms}$. Thus substituting this expression in the finite product $\prod_{n=1}^{N}(1 - p_n^{-s})^{-1}$ we find

$$\prod_{n=1}^{N} (1 - p_n^{-s})^{-1} = \sum_{k=1}^{\infty} n_k^{-s} \tag{17.42}$$

the sum extending over those integers n_k whose prime decomposition contains only primes from the set $\{p_1, \cdots, p_N\}$. Because of unique factorization the coefficient of n_k^{-s} is 1 for each k. Now let $N \to \infty$ in (17.42). The right hand side approaches $\zeta(s)$ because every integer eventually appears in the sum and because the series $\sum_{n=1}^{\infty} \frac{1}{n^s}$ converges absolutely. Thus $\lim_{N\to\infty} \prod_{n=1}^{N} (1 - {p_n}^{-s})^{-1} = \zeta(s)$. Since $\zeta(s) \neq 0$, the infinite product is convergent. This completes the proof. □

Using the Euler product formula it is possible to prove $\sum_{n=1}^{\infty} \frac{1}{p_n} = \infty$.

Theorem 17.37. *The sum of the reciprocals of the prime numbers is infinite. That is,*

$$\sum_{n=1}^{\infty} \frac{1}{p_n} = \infty.$$

Proof. By Theorem 17.3, $\sum_{n=1}^{\infty} \frac{1}{p_n} = \infty$ if and only if $\sum_{n=1}^{\infty} \left|\log\left(1 - \frac{1}{p_n}\right)\right| = \infty$. Because $\log\left(1 - \frac{1}{p_n}\right) < 0$ for each n, hence it is sufficient to prove

$$-\sum_{n=1}^{\infty} \log\left(1 - \frac{1}{p_n}\right) = \infty.$$

Since the inequality $-\log\left(1 - \frac{1}{p_n}\right) \geq -\log\left(1 - \frac{1}{{p_n}^x}\right)$ is true for $1 < x < \infty$, using the Euler product formula we obtain

$$-\sum_{n=1}^{\infty} \log\left(1 - \frac{1}{p_n}\right) \geq -\sum_{n=1}^{\infty} \log\left(1 - \frac{1}{{p_n}^x}\right) = \log\zeta(x). \tag{17.43}$$

As $x \to 1, \zeta(x) \to \infty$ and $\log\zeta(x) \to \infty$.

The left side of (17.43), which is independent of x, must therefore be infinite and the theorem is proved. □

We now turn to the deeper analytic properties of $\zeta(s)$. It was proved by Riemann that $\zeta(s)$ can be analytically continued to a region much larger than the region $\{s \in \mathbb{C} : \text{Re}s > 1\}$. Riemann proved that there exists an entire function g such that $\zeta(s) = \frac{1}{s-1} + g(s)$. In other words $\zeta(s)$ can be defined to be analytic everywhere except $s = 1$ where there is a simple pole with residue 1. Here we will present Riemann's original argument of the above fact. But first we shall prove the following lemmas.

Lemma 17.9. *The integral*

$$f(s) = \int_0^{\infty} \frac{x^{s-1}}{e^x - 1} dx \tag{17.44}$$

defines an analytic function of s for Re$s > 1$. Here $x^{s-1} = e^{(s-1)\log x}$, $0 < x < \infty$.

Proof. If $-\infty < \sigma < \infty$, the function $\frac{x^{\sigma-1}}{e^x-1}$ is integrable on the interval $[1, \infty)$. If $\sigma > 1$, the function $\frac{x^{\sigma-1}}{e^x-1}$ is also integrable on $(0, 1)$ as $\frac{x^{\sigma-1}}{e^x-1} < x^{\sigma-2}$. Thus letting $\sigma = \text{Re}s$, (17.44) is convergent for Re$s > 1$. Now

$$f(s) = \lim_{\delta\to 0} \int_\delta^{\frac{1}{\delta}} \frac{x^{s-1}}{e^x - 1} dx = \lim_{\delta\to 0} f_\delta(s).$$

Thus f is the pointwise limit of entire functions $f_\delta, 0 < \delta < 1$. If $1 < \sigma_1 \le \text{Re}s \le \sigma_2 < \infty$, then trivially

$$\left|\frac{x^{s-1}}{e^x - 1}\right| \le \frac{x^{\sigma_1-1}}{e^x - 1} + \frac{x^{\sigma_2-1}}{e^x - 1}$$

on $(0, \infty)$. This implies $|f_\delta(s)| \le f(\sigma_1) + f(\sigma_2)$ and therefore $\{f_\delta\}$ is locally bounded in the half plane Re$s > 1$. By Vitali's theorem, the function f is analytic and the lemma is proved. □

Lemma 17.10. *If f is defined by (17.44), then $f(s) = \Gamma(s)\zeta(s)$. That is, for Re$s > 1$,*

$$\int_0^\infty \frac{x^{s-1}}{e^x - 1} dx = \Gamma(s)\zeta(s). \tag{17.45}$$

Proof. By the identity theorem it is enough to verify (17.45) for real values of s between 1 and ∞. Fixing such an s, we consider the function $\phi_n(x) = x^{s-1}\sum_{k=1}^n e^{-kx}$ for each $n \ge 1$. It is easy to see that $\lim_{n\to\infty} \phi_n(x) = \frac{x^{s-1}}{e^x-1}$ in such a way that the hypotheses of Lemma 17.6 are satisfied. Therefore

$$\begin{aligned} \int_0^\infty \frac{x^{s-1}}{e^x - 1} dx &= \lim_{n\to\infty} \sum_{k=1}^n \int_0^\infty x^{s-1} e^{-kx} dx \\ &= \lim_{n\to\infty} \sum_{k=1}^n \frac{1}{k^s} \int_0^\infty y^{s-1} e^{-y} dy \\ &= \Gamma(s)\zeta(s) \end{aligned}$$

which is (17.45). Applying the identity theorem, (17.45) holds for complex s with Re$s > 1$, and the lemma is proved. □

In order to define $\zeta(s)$ for Re$s \le 1$ it would be enough to define the left side of (17.45) for Re$s \le 1$. However $\frac{x^{s-1}}{e^x-1}$ is integrable near 0 only if Re$s > 1$. To avoid this difficulty at the origin 0 we will replace $(0, \infty)$ with a contour whose points stay away from the origin.

Let C_ρ be the contour in Fig. 17.4, and denote by γ_ρ the circular portion of C_ρ. Notice that C_ρ traverses the interval $[\rho, \infty)$ twice, once from ∞ to ρ and once from ρ to ∞.

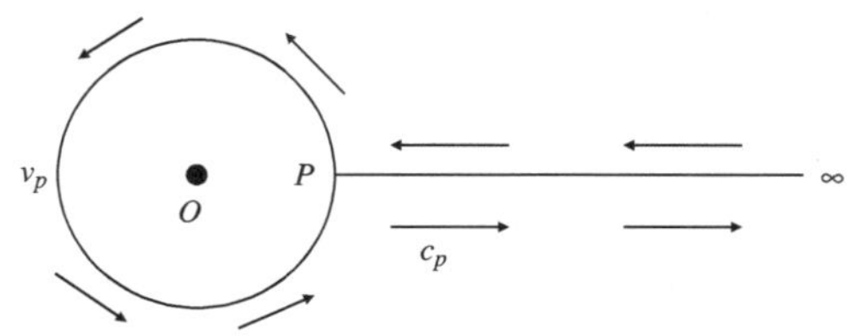

Fig. 17.4: The contour C_ρ with the circular portion γ_ρ

We define $\log z$ as follows on C_ρ. Let $\log z = ln z$ on the upper branch of $[\rho, \infty), \log z = ln\rho + i \arg z, 0 < \arg z < 2\pi$, on γ_ρ, and $\log z = ln z + 2\pi i$ on the lower branch of $[\rho, \infty)$. Then we let $z^{s-1} = e^{(s-1)\log z}$ and set up the function

$$I(s) = \int_{C_\rho} \frac{z^{s-1}}{e^z - 1} dz \tag{17.46}$$

where ρ is now taken to lie between 0 and 2π. The function $I(s)$, so defined is entire. The integral (17.46) can be written as a sum of three integrals, one for each component of C_ρ. Thus we obtain

$$I(s) = \int_{\gamma_\rho} \frac{z^{s-1}}{e^z - 1} dz + (e^{2\pi i s} - 1) \int_\rho^\infty \frac{x^{s-1}}{e^x - 1} dx, \tag{17.47}$$

where in the second integral $x^{s-1} = e^{(s-1)ln x}$. From residue theorem, it also follows that $I(1) = \int_{\gamma_\rho} \frac{dz}{e^z - 1} = 2\pi i$.

An important property of $I(s)$ is that it is independent of ρ, so long as $0 < \rho < 2\pi$.

Lemma 17.11. *The function $I(s)$ is independent of ρ, if $0 < \rho < 2\pi$.*

Proof. To prove this we fix ρ' and ρ with $0 < \rho' < \rho < 2\pi$, and let $C_\rho - C_{\rho'}$ be the contour of Fig. 17.5. It is easy to see that

$$\int_{C_\rho - C_{\rho'}} \frac{z^{s-1}}{e^z - 1} dz = \int_{C_\rho} \frac{z^{s-1}}{e^z - 1} dz - \int_{C_{\rho'}} \frac{z^{s-1}}{e^z - 1} dz.$$

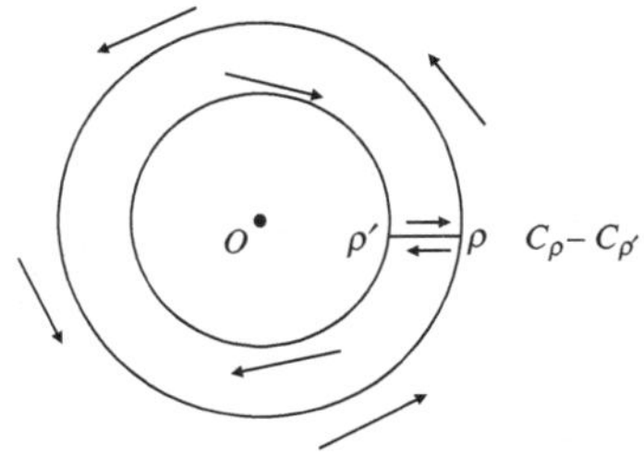

Fig. 17.5: The contour $C_\rho - C_{\rho'}$

Since the right side represents the difference between the definitions of $I(s)$ corresponding to ρ and ρ', our assertion will follow from the left side's being 0.

The contour $C_\rho - C_{\rho'}$ bounds a simply connected region (as in Fig. 17.5) in which $\frac{z^{s-1}}{e^z-1}$ is an analytic function of z. Therefore by Cauchy's theorem,

$$\int_{C_\rho - C_{\rho'}} \frac{z^{s-1}}{e^z - 1} dz = 0,$$

and $I(s)$ is independent of ρ. □

Lemma 17.12. *If Res > 1, then*

$$I(s) = (e^{2\pi i s} - 1)\Gamma(s)\zeta(s). \tag{17.48}$$

Proof. Fix $\sigma = \text{Re}s, \sigma > 1$. Since $e^z - 1 = zh(z)$ where h is continuous (in fact analytic) with $h(0) = 1$, there exists $\rho_0 > 0$ such that if $0 < |z| \le \rho_0$, $|\,|h(z)| - |h(0)|\,| \le |h(z) - h(0)| < \frac{1}{2}$. That is, $|h(z)| > \frac{1}{2}$ if $0 < |z| \le \rho_0$. Thus $\frac{1}{|e^z-1|} = \frac{1}{|zh(z)|} \le \frac{2}{|z|}$. From this inequality it follows for $0 < \rho \le \rho_0$ that

$$\left| \int_{\gamma_\rho} \frac{z^{s-1}}{e^z - 1} dz \right| \le 2 \int_{\gamma_\rho} \rho^{\sigma - 2} |dz| = 4\pi \rho^{\sigma-1}$$

which tends to 0 with ρ. Therefore as $\rho \to 0$ in (17.47), we conclude from Lemma 17.10 that $I(s) = (e^{2\pi i s} - 1)\Gamma(s)\zeta(s)$ which was to be proved.

□

Theorem 17.38. *There exists an entire function g such that $\zeta(s) = \frac{1}{s-1} + g(s)$.*

Proof. Notice that $(e^{2\pi i s} - 1)\Gamma(s)$ is an entire function of s as poles of the gamma function cancel with zeros of $(e^{2\pi i s} - 1)$. Further $I(s)$ is an entire function of s. Equation (17.48) is true for $\text{Re}s > 1$. Therefore $\zeta(s)$ can be defined by the equation $\zeta(s) = \frac{I(s)}{(e^{2\pi i s} - 1)\Gamma(s)}$ for all values of s such that $(e^{2\pi i s} - 1)\Gamma(s) \neq 0$.

Since $\Gamma(s)$ is never zero, and since $\Gamma(s)$ has simple poles at $s = 0, -1, -2, \cdots$, there are no zeros of $(e^{2\pi i s} - 1)\Gamma(s)$, except at $s = 1$, in the region $\{s \in \mathbb{C} : \text{Re}s \le 1\}$. We know from the definition that $\zeta(s)$ is analytic for $\text{Re}s > 1$, and therefore it follows that $\zeta(s)$ is analytic for all values of s but $s = 1$. At $s = 1$ there is a simple pole with residue

$$\lim_{s \to 1} (s-1) \frac{I(s)}{(e^{2\pi i s} - 1)\Gamma(s)} = \frac{I(1)}{2\pi i \Gamma(1)} = 1$$

as $I(1) = 2\pi i$ and $\Gamma(1) = 1$. The theorem follows. □

Much of the analytic interest in the zeta function comes from the equation

$$\zeta(s) = 2^s \pi^{s-1} \sin \frac{\pi s}{2} \Gamma(1-s)\zeta(1-s)$$

which is known as the **Riemann's functional equation**. It relates $\zeta(s)$ to $\zeta(1-s)$.

Theorem 17.39.

$$\zeta(s) = 2^s \pi^{s-1} \sin \frac{\pi s}{2} \Gamma(1-s)\zeta(1-s).$$

Proof. Let $C_n, n = 1, 2, \cdots$, be the contour arising from taking $\rho = (2n+1)\pi$ in Fig. 17.4. The function $\frac{z^{s-1}}{e^z-1}$ is finite on γ_n, the circular portion of C_n and the number

$$\sup_n \max_{z \in \gamma_n} \frac{1}{|e^z - 1|} = M \tag{17.49}$$

is finite. We assume, in what follows, that ρ is a fixed number between 0 and 2π. For each $n, C_n - C_\rho$ is a contour whose definition is similar to that of $C_\rho - C_{\rho'}$ defined earlier in this section. Further, unlike before, the function $\frac{z^{s-1}}{e^z-1}$ is not analytic for fixed s in the region bounded by the curve $C_n - C_\rho$. However, it is meromorphic in this region with simple poles at the points $z = \pm 2k\pi i, k = 1, \cdots, n$ and residues

$$(2k\pi i)^{s-1} = -i(2\pi)^{s-1} k^{s-1} e^{\frac{is\pi}{2}} \text{ at } 2k\pi i,$$

$$(-2k\pi i)^{s-1} = i(2\pi)^{s-1} k^{s-1} e^{\frac{is3\pi}{2}} \text{ at } -2k\pi i.$$

By the residue theorem,

$$\begin{aligned}
\int_{C_n - C_\rho} \frac{z^{s-1}}{e^z - 1} dz &= 2\pi i \times (\text{ sum of residues }) \\
&= (2\pi)^s e^{\frac{is\pi}{2}} \sum_{k=1}^{n} k^{s-1} - (2\pi)^s e^{\frac{is3\pi}{2}} \sum_{k=1}^{n} k^{s-1} \\
&= -2i(2\pi)^s e^{i\pi s} \sin \frac{\pi s}{2} \sum_{k=1}^{n} \frac{1}{k^{1-s}}. \qquad (17.50)
\end{aligned}$$

Suppose Re$s < 0$. Letting $n \to \infty$, the right side of (17.50) tends to $-2i(2\pi)^s e^{i\pi s} \sin \frac{\pi s}{2} \zeta(1-s)$, while the left we claim tends to

$$\int_{-C_\rho} \frac{z^{s-1}}{e^z - 1} dz = -I(s).$$

To prove our claim it is sufficient to note that by (17.49) the integral over γ_n in (17.50) is bounded by $2\pi M\{(2n+1)\pi\}^{\text{Re}s}$ which tends to 0 as $n \to \infty$. Thus we have proved

$$I(s) = 2i(2\pi)^s e^{i\pi s} \sin\frac{\pi s}{2}\zeta(1-s) \tag{17.51}$$

for Re$s < 0$. By the identity theorem, (17.51) holds for all values of s. Equating (17.51) and (17.48) we obtain

$$(e^{2\pi is}-1)\Gamma(s)\zeta(s) = 2i(2\pi)^s e^{i\pi s}\sin\frac{\pi s}{2}\zeta(1-s).$$

Since $e^{2\pi is}-1 = 2ie^{i\pi s}\sin\pi s$ and $\Gamma(s)\Gamma(1-s) = \frac{\pi}{\sin\pi s}$, the theorem follows. □

Consider the functional equation

$$\zeta(s) = 2^s\pi^{s-1}\sin\frac{\pi s}{2}\Gamma(1-s)\zeta(1-s). \tag{17.52}$$

We have already seen that the function $\zeta(s)$ is analytic everywhere except at $s = 1$. Thus it follows from the functional equation that every pole of $\Gamma(1-s)$, save $s = 1$, must be accompanied by a zero of either $2^s\pi^{s-1}\sin\frac{\pi s}{2}$ or $\zeta(1-s)$. The poles occur when $s = 2, 3, \cdots$, and only those at $s = 2, 4, 6, \cdots$ are cancelled by $2^s\pi^{s-1}\sin\frac{\pi s}{2}$. We are forced to conclude that $\zeta(1-s) = 0$ for $s = 3, 5, \cdots$. In other words, $\zeta(s) = 0$ when $s = -2, -4, \cdots$.

Since $\zeta(s) \neq 0$ for Re$s > 1$, another consequence of the functional equation is that every zero of $\zeta(1-s)$ with Re$s > 1$ (i.e., Re$(1-s) < 0$) must be accompanied by a pole of $2^s\pi^{s-1}\sin\frac{\pi s}{2}\Gamma(1-s)$. Therefore, it follows that the only zeros of $\zeta(s)$ with Re$s > 1$ or Re$s < 0$ are the points $s = -2, -4, \cdots$. These are called the trivial zeros.

In the context of $\zeta(s)$ the strip $\{s \in \mathbb{C} : 0 \leq \text{Re}s \leq 1\}$ is called the critical strip. It was conjectured by Riemann that the zeros of $\zeta(s)$ inside the critical strip have real parts equal to $\frac{1}{2}$. While a great deal is known of these zeros, this assertion, the Riemann hypothesis, remains an open question.

Thus we have seen as a consequence of the functional equation that those zeros of $\zeta(s)$ which lie outside of the critical strip are the trivial zeros $s = -2, -4, \cdots$. Inside the critical strip the situation is far more complicated, and the zeros are generally unknown. Indeed the Riemann hypothesis would give only the x-coordinates if it were true. There are however many existence and nonexistence theorems, and here we will discuss about one of them.Using the Weierstrass factorization theorem it will be proved that $\zeta(s)$ vanishes infinitely many times in the critical strip.

We introduce auxiliary functions

$$\xi(s) = \frac{1}{2}s(s-1)\pi^{-\frac{s}{2}}\Gamma\left(\frac{s}{2}\right)\zeta(s) \tag{17.53}$$

and

$$\Xi(s) = \xi\left(\frac{1}{2} + s\right). \tag{17.54}$$

Since each pole of $\Gamma\left(\frac{s}{2}\right)$ coincides with a trivial zero of $s\zeta(s)$, both $\xi(s)$ and $\Xi(s)$ are entire functions. Conversely, every trivial zero of $s\zeta(s)$ coincides with a pole of $\Gamma\left(\frac{s}{2}\right)$. It follows from (17.53) that a zero of $\xi(s)$ corresponds to a zero of $\zeta(s)$ which is not one of the trivial zeros, that is, to a zero in the critical strip. Thus, if $\xi(s)$ (or $\Xi(s)$) has an infinite number of zeros, $\zeta(s)$ has an infinite number of zeros in the critical strip.

Theorem 17.40. *The functions $\xi(s)$ and $\Xi(s)$ satisfy the relations*

$$\xi(s) = \xi(1-s) \tag{17.55}$$

and

$$\Xi(s) = \Xi(-s), \tag{17.56}$$

that is, Ξ is an even function.

Proof. First substitute in (17.53) the value of $\zeta(s)$ given by (17.52) to obtain

$$\xi(s) = \frac{1}{2}s(s-1)\pi^{\frac{s}{2}-1}2^s\Gamma\left(\frac{s}{2}\right)\sin\frac{\pi s}{2}\Gamma(1-s)\zeta(1-s).$$

Recall the formulas

$$\Gamma\left(\frac{s}{2}\right) = \frac{\pi}{\sin\frac{\pi s}{2}\Gamma(1-\frac{s}{2})} \tag{17.57}$$

and

$$\Gamma(s)\Gamma\left(s+\frac{1}{2}\right) = 2^{1-2s}\pi^{\frac{1}{2}}\Gamma(2s) \tag{17.58}$$

on gamma functions which were proved in the last section. Using (17.57) and (17.58) successively it follows that

$$\xi(s) = \frac{1}{2}s(s-1)\pi^{\frac{s}{2}}2^s\frac{\Gamma(1-s)}{\Gamma\left(1-\frac{s}{2}\right)}\zeta(1-s)$$

and

$$\xi(s) = \frac{1}{2}s(s-1)\pi^{\frac{s}{2}-\frac{1}{2}}\Gamma\left(\frac{1-s}{2}\right)\zeta(1-s) = \xi(1-s)$$

which was claimed in (17.55). Equation (17.56) is an obvious consequence of (17.55) and (17.54). □

Remark 17.3. Since $\frac{1}{2}s(s-1)$ is invariant under the substitution $s \to 1-s$, it is a consequence of Theorem 17.40 that $\pi^{-\frac{s}{2}}\Gamma\left(\frac{s}{2}\right)\zeta(s)$ is invariant under the same substitution. This is Riemann's original formulation of the functional equation.

Lemma 17.13. *As $\sigma \to \infty$ through real values, we have*

$$\limsup_{\sigma\to\infty} \frac{\log\log\Gamma(\sigma)}{\log\sigma} \leq 1 \tag{17.59}$$

or equivalently

$$\limsup_{\sigma\to\infty} \Gamma(\sigma)e^{-\sigma^{1+\epsilon}} = 0 \tag{17.60}$$

for every $\epsilon > 0$.

Proof. Equation (17.59) is a consequence of (17.60). Let $M = \max_{\frac{1}{2}\leq\sigma\leq\frac{3}{2}} \Gamma(\sigma)$. If $\sigma > \frac{3}{2}$, choose an integer n such that $\frac{1}{2} \leq \sigma - n < \frac{3}{2}$. Using the reproducing formula for the gamma function n times we have

$$\Gamma(\sigma) = (\sigma-1)(\sigma-2)\cdots(\sigma-n)\Gamma(\sigma-n) \leq (\sigma-1)^n M \leq \sigma^\sigma M.$$

Since $\sigma^\sigma = e^{\sigma\log\sigma}$, and since $\sigma\log\sigma - \sigma^{1+\epsilon} \to -\infty$ as $\sigma\to\infty$, (17.60) follows. The lemma is proved. □

Remark 17.4. It can be shown that

$$\lim_{\sigma\to\infty} \frac{\log\log\Gamma(\sigma)}{\log\sigma} = 1.$$

We shall now compute the order of $\xi(s) = \xi(1-s)$, which by definition is equal to

$$\lambda = \limsup_{\substack{|s|\to\infty \\ \mathrm{Re}s\geq\frac{1}{2}}} \frac{\log\log|\xi(s)|}{\log|s|}.$$

Let A be the set consisting of those points s such that $\mathrm{Re}\ s \geq \frac{1}{2}$, $|\mathrm{Im}\ s| \leq 1$, and $|s-1| \geq \frac{1}{2}$ (Fig. 17.6).

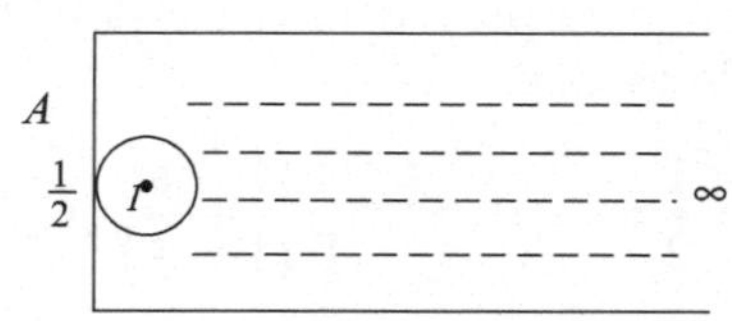

Fig. 17.6: The set $A = \{s \in \mathbb{C} : \mathrm{Re}\ s \geq \frac{1}{2}, |\mathrm{Im}\ s| \leq 1,\ \text{and}\ |s-1| \geq \frac{1}{2}\}$

Let B be the set consisting of those points s such that $\mathrm{Re}s \geq \frac{1}{2}$ and $|\mathrm{Im}s| \geq 1$ (Fig. 17.7).

We will prove the entire function $\xi(s)$ has order at most 1 by showing it has order at most 1 in the regions A and B. That is, we show

$$\limsup_{\substack{|s|\to\infty \\ s\in A}} \frac{\log\log|\xi(s)|}{\log|s|} \leq 1 \tag{17.61}$$

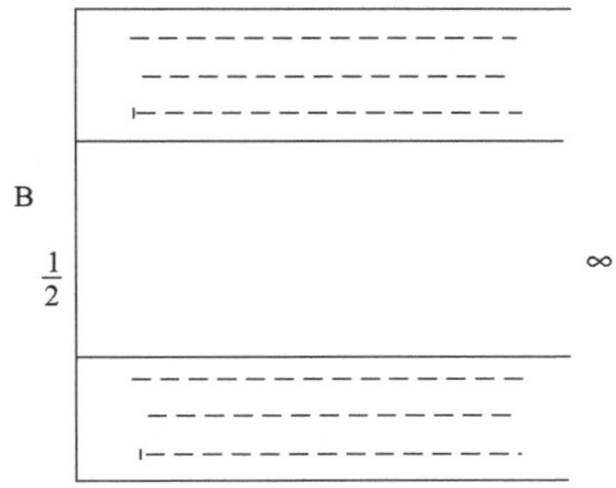

Fig. 17.7: The set $B = \{s \in \mathbb{C} : \text{Re}s \geq \frac{1}{2} \text{ and } |\text{Im}s| \geq 1\}$

and

$$\limsup_{\substack{|s|\to\infty \\ s\in B}} \frac{\log\log|\xi(s)|}{\log|s|} \leq 1. \tag{17.62}$$

To establish (17.61) we express $\xi(s)$ through its definition (17.53):

$$\xi(s) = \frac{1}{2}s(s-1)\pi^{-\frac{s}{2}}\Gamma(\frac{s}{2})\zeta(s).$$

In A both $\zeta(s)$ and $\pi^{-\frac{s}{2}}$ are bounded, while $\frac{1}{2}s(s-1)$ has order 0. Since $\left|\Gamma(\frac{s}{2})\right| \leq \Gamma(\frac{\sigma}{2})$ for $\sigma = \text{Re}s$, Lemma 17.13 implies $\Gamma(\frac{s}{2})$ has order at most 1 in A. A finite product of functions of order at most 1 has order at most 1, and therefore $\xi(s)$ has order at most 1 in A.

To prove (17.62) we will need the representation

$$\zeta(s) = \frac{I(s)}{(e^{2\pi is}-1)\Gamma(s)}$$

from Lemma 17.12, to express $\xi(s)$ as

$$\xi(s) = \frac{\frac{1}{2}s(s-1)\pi^{-\frac{s}{2}}\Gamma\left(\frac{s}{2}\right)I(s)}{\left(e^{2\pi is}-1\right)\Gamma(s)}.$$

Each of the functions $\frac{1}{2}s(s-1)$, $\pi^{-\frac{s}{2}}$, $\Gamma(\frac{s}{2})$, and $\frac{1}{\Gamma(s)}$ has order at most 1 in B. Furthermore $(e^{2\pi is}-1)^{-1}$ is bounded in B. Therefore we have only to prove $I(s)$ has order at most 1 in B.

Let ρ be a number between 0 and 2π. By (17.47)

$$I(s) = \int_{\gamma_\rho} \frac{z^{s-1}}{e^z-1}dz + (e^{2\pi is}-1)\int_\rho^\infty \frac{x^{s-1}}{e^x-1}dx.$$

If ρ has been chosen small enough, the first integral is bounded by $4\pi\rho^{\sigma-1}$. To estimate the second use the inequality

$$\frac{1}{e^x-1} \leq \frac{e^{-x}}{1-e^{-\rho}}$$

which is true for $\rho \le x < \infty$ to obtain

$$\left|\int_\rho^\infty \frac{x^{s-1}}{e^x - 1}dx\right| \le \int_\rho^\infty \frac{x^{\sigma-1}}{e^x-1}dx \le \frac{1}{1-e^{-\rho}}\int_0^\infty x^{\sigma-1}e^{-x}dx = \frac{\Gamma(\sigma)}{1-e^{-\rho}}.$$

Thus for $s = \sigma + it$,

$$|I(s)| \le 4\pi\rho^{\sigma-1} + \frac{|e^{2\pi is}-1|\,\Gamma(\sigma)}{1-e^{-\rho}}.$$

Since $e^{-|s|^{1+\epsilon}} \le e^{-\sigma^{1+\epsilon}}$, it follows that

$$\limsup_{\substack{|s|\to\infty \\ s\in B}} |I(s)|e^{-|s|^{1+\epsilon}} = 0$$

for $\epsilon > 0$, and (17.62) follows. We have thus proved that $\xi(s)$ has order at most 1. We will now prove it has order at least 1 and from that conclude the following.

Theorem 17.41. *The entire function $\xi(s)$ has order* 1.

Proof. We have to prove $\xi(s)$ has order at least 1, for which it suffices to observe

$$\lim_{n\to\infty} \frac{\log\xi(2n)}{2n} = \infty. \tag{17.63}$$

By definition (17.53) we have $\xi(2n) \ge \pi^{-n}(n-1)!$ for $n > 0$. Expressing $(n-1)!$ as

$$(n-1)! = n^{n-1}\prod_{k=1}^{n-1}\left(1-\frac{k}{n}\right)$$

and taking logarithms, we find

$$\frac{\log\xi(2n)}{2n} \ge -\frac{\log\pi}{2} + \frac{n-1}{2n}\log n + \frac{1}{2n}\sum_{k=1}^{n-1}\log\left(1-\frac{k}{n}\right).$$

The last sum on the right converges, by definition of the integral, to $\frac{1}{2}\int_0^1 \log(1-x)dx = -\frac{1}{2}$ as $n \to \infty$. Therefore since $\lim\limits_{n\to\infty} \frac{(n-1)}{2n}\log n = \infty$, (17.63) follows, and the theorem is proved. □

Theorem 17.42. *The entire function $\Xi(s)$ has infinitely many zeros.*

Proof. The function $\Xi(s)$ has order 1 because $\xi(s)$ does. The function $\Xi(s)$ is even, and therefore its power series expansion about 0 contains only even powers of s. It follows that $\Xi(s^{\frac{1}{2}})$ is entire and of order $\frac{1}{2}$. By the Weierstrass factorization theorem an entire function of order $\frac{1}{2}$ cannot have a finite canonical product. In other words $\Xi(\sqrt{s})$ and a fortiori $\Xi(s)$ have infinitely many zeros (see Theorem 17.23). The theorem is proved. □

Since every zero of $\Xi(s)$ corresponds to a zero of $\zeta(s)$ in the critical strip, we have proved the following:

Theorem 17.43. *The Riemann-zeta function $\zeta(s)$ has infinitely many zeros in the critical strip.*

We shall now prove the theorem of Hadamard and de la Vallée Poussin which asserts that $\zeta(s) \neq 0$ for $\mathrm{Re} s = 1$.

If f is analytic for $0 < |s - s_0| < r$, and if at s_0, the function f has either a pole of order $-k, k < 0$, or a zero of order $k, k \geq 0$, then

$$\lim_{s \to s_0} (s - s_0)\frac{f'(s)}{f(s)} = k,$$

that is, $\frac{f'}{f}$ has a simple pole (see Remark 11.4) at s_0 with residue k. Since k is real, it is also true that

$$\lim_{s \to s_0} \mathrm{Re}(s - s_0)\frac{f'(s)}{f(s)} = k.$$

Finally, if $s_0 = \sigma_0 + it_0$, and if $s \to s_0$ through values $s = \sigma + it_0$ of s with the same imaginary part, then $(s - s_0) = (\sigma - \sigma_0)$ is real, and $\lim_{\sigma \to \sigma_0} (\sigma - \sigma_0)\mathrm{Re}\frac{f'(s)}{f(s)} = k$. We shall use this fact in proving the following theorem.

Theorem 17.44. *$\zeta(s)$ has no zeros for $\mathrm{Re} s = 1$.*

Proof. From Lemma 17.8, it follows that if $\mathrm{Re} s > 1$, then

$$\frac{\zeta'(s)}{\zeta(s)} = -\sum_{n=1}^{\infty} \frac{\Lambda(n)}{n^s}.$$

Thus for $\mathrm{Re} s > 1$,

$$\mathrm{Re}\frac{\zeta'(s)}{\zeta(s)} = -\sum_{n=1}^{\infty} \mathrm{Re}\frac{\Lambda(n)}{n^s}. \tag{17.64}$$

If $s = \sigma + it$, a typical summand in (17.64) is

$$-\mathrm{Re}\frac{\log p}{p^{ms}} = -\frac{\log p}{p^{m\sigma}} \cos(mt \log p).$$

Therefore, it follows from the formula

$$3 + 4\cos(mt \log p) + \cos(2mt \log p) = 2(1 + \cos(mt \log p))^2 > 0$$

that

$$3\frac{\zeta'(\sigma)}{\zeta(\sigma)} + 4\mathrm{Re}\frac{\zeta'(\sigma + it)}{\zeta(\sigma + it)} + \mathrm{Re}\frac{\zeta'(\sigma + 2it)}{\zeta(\sigma + 2it)} < 0. \tag{17.65}$$

Let t_0 be a nonzero real number. Because $\zeta(s)$ is analytic at both $1+it_0$ and $1+2it_0$, the discussion prior to the theorem implies that the numbers

$$k = \lim_{\sigma\to 1}(\sigma-1)\mathrm{Re}\frac{\zeta'(\sigma+it_0)}{\zeta(\sigma+it_0)}$$

and

$$l = \lim_{\sigma\to 1}(\sigma-1)\mathrm{Re}\frac{\zeta'(\sigma+2it_0)}{\zeta(\sigma+2it_0)}$$

are nonnegative integers. Also because of the simple pole at $s=1$, we have

$$-1 = \lim_{\sigma\to 1}(\sigma-1)\frac{\zeta'(\sigma)}{\zeta(\sigma)}.$$

Combining the last three equations, we find

$$\lim_{\sigma\to 1}(\sigma-1)\left\{3\frac{\zeta'(\sigma)}{\zeta(\sigma)} + 4\mathrm{Re}\frac{\zeta'(\sigma+it_0)}{\zeta(\sigma+it_0)} + \mathrm{Re}\frac{\zeta'(\sigma+2it_0)}{\zeta(\sigma+2it_0)}\right\}$$

$$= -3+4k+l \geq -3+4k.$$

Since $\sigma-1>0$ for $\sigma>1$, this limit is ≤ 0 by (17.65). Therefore $4k\leq 0$ or $k=0$. Thus $\zeta(1+it_0)\neq 0$, and because t_0 is arbitrary, the theorem is proved. □

Remark 17.5. From the functional equation and from Theorem 17.44, it follows that $\zeta(s)$ also has no zeros for $\mathrm{Re}s=0$.

17.8 Exercises

1. Find the partial fraction expansion of $\frac{\pi}{\sin \pi z}$.
2. Show that $\prod_{n=2}^{\infty}\left(1-\frac{1}{n^2}\right)=\frac{1}{2}$.
3. Prove that $\prod_{n=1}^{\infty}\left(1-\frac{1}{n}\right)e^{\frac{1}{n}}$ converges absolutely.
4. For each sequence below, determine the convergence or divergence of $\prod_{n=1}^{\infty}(1-a_n)$.
   (i) $a_n=\frac{1}{n+1}$
   (ii) $a_n=\frac{1}{n^2+1}$
   (iii) $a_n=1-\cos\pi n$
   (iv) $a_n=1-\cos\frac{\pi}{n}$
   (v) $a_n=1-e^{2\pi i n}$
5. Show that the product $(1-\frac{1}{2})(1+\frac{1}{2})(1-\frac{1}{3})(1+\frac{1}{3})(1-\frac{1}{4})\cdots$ converges, but not absolutely.

6. Discuss the range of $\alpha' s, 0 < \alpha < \infty$, for which the following products are divergent, convergent, and absolutely convergent:

$$\prod_{n=1}^{\infty}\left(1+\frac{(-1)^{n+1}}{n^{\alpha}}\right).$$

7. Given complex numbers $a_1, \cdots, a_n$ show that

$$|1-(1+a_1)(1+a_2)\cdots(1+a_n)| \le e^{|a_1|+\cdots+|a_n|}-1.$$

8. Prove that for $|z|<1$, $(1+z)(1+z^2)(1+z^4)(1+z^8)\cdots = \frac{1}{1-z}$.

9. For which of the following sequences $\{a_n(z)\}$ is the product

$$\prod_{n=1}^{\infty}(1+a_n(z))$$

convergent to an entire function $P(z)$?

(a) $a_n(z) = q^n z, |q| < 1$.

(b) $a_n(z) = \frac{z}{\log(n+1)}$.

(c) $a_n(z) = \frac{z}{(n+1)\log^2(n+1)}$.

10. What is the genus and order of $\cos\sqrt{z}$?

11. Find the order of each of the following functions: (i) $\cosh\sqrt{z}$ (ii) $\sin z$.

12. Show that if $f(z)$ is of genus 0 or 1 with real zeros, and if $f(z)$ is real for real z, then all zeros of $f'(z)$ are real.

13. Find the order of the following entire functions:

(i) $\prod_{n=0}^{\infty}\{1-\left(\frac{2z}{(2n+1)\pi}\right)^2\}$

(ii) $f(z)=\prod_{n=0}^{\infty}\left(1-\frac{z}{n^{\alpha}}\right)e^{\frac{z}{n^{\alpha}}}, \frac{1}{2}<\alpha\le 1$

(iii) $f(z)=\prod_{n=2}^{\infty}\left(1-\frac{z}{n^2\log n}\right)$

(iv) $f(z)=\prod_{n=0}^{\infty}(1-q^n z), |q|<1$.

14. The beta function $B(p,q)$ is defined by

$$B(p,q)=\int_0^1 x^{p-1}(1-x)^{q-1}dx.$$

Show that $B(p,q)=B(q,p)$ and further prove that $B(p,q)=\frac{\Gamma(p)\Gamma(q)}{\Gamma(p+q)}$.

15. What are the residues of $\Gamma(z)$ at the poles $z=-n$?

16. Show that $\Gamma\left(\frac{1}{6}\right) = 2^{-\frac{1}{3}}\left(\frac{3}{\pi}\right)^{\frac{1}{2}}\Gamma\left(\frac{1}{3}\right)^2$.

17. Prove that $\frac{\zeta(s-1)}{\zeta(s)} = \sum_{n=1}^{\infty}\frac{\phi(n)}{n^s}$.

18. Use the product formula for $\zeta(s)$ to prove $\frac{1}{\zeta(s)} = \sum_{n=1}^{\infty}\frac{\mu(n)}{n^s}$.

19. Prove $\Xi(s)$ assumes every complex value an infinite number of times.

Chapter 18

Elliptic Functions

18.1 Periodic functions

Definition 18.1. *Let f be a function defined on the complex plane $\mathbb{C}$. Then a complex number ω is called a period of f if $f(z+\omega) = f(z)$ for all $z \in \mathbb{C}$, and f is called periodic if it has a period $\omega \neq 0$.*

Example 18.1. *The functions $\sin z$, $\cos z$ have period $2\pi, e^z$ has period $2\pi i$, and $\sin\left(\frac{2\pi z}{\omega}\right)$ has period ω for any complex number $\omega \neq 0$. Every function has period $\omega = 0$.*

The simplest function with period ω is the exponential function $e^{\frac{2\pi i z}{\omega}}$. Further, any function with period ω can be expressed in terms of the function $e^{\frac{2\pi i z}{\omega}}$. Two periods ω_1 and ω_2 of a function f are said to be independent if $\frac{\omega_2}{\omega_1} \notin \mathbb{R}$.

Let Ω_f be the set of periods of a function f. The set Ω_f has two important properties: one algebraic, valid for all f, and one topological, valid for nonconstant meromorphic functions f. These properties are given below.

Theorem 18.1. *Let Ω_f be the set of periods of a function f defined on $\mathbb{C}$; then Ω_f is a subgroup of the additive group $\mathbb{C}$.*

Proof. Let $\alpha, \beta \in \Omega_f$. Then $f(z+(\alpha+\beta)) = f((z+\alpha)+\beta) = f(z+\alpha) = f(z)$, and so $\alpha+\beta \in \Omega_f$. Moreover, $f(z-\alpha) = f((z-\alpha)+\alpha) = f(z)$, and so $-\alpha \in \Omega_f$. Finally, $f(z+0) = f(z)$, and so $0 \in \Omega_f$. Thus Ω_f is a subgroup of $\mathbb{C}$. □

Definition 18.2. *A subset $\triangle$ of a topological space is called discrete if every $x \in \triangle$ has a neighbourhood U such that $U \cap \triangle = \{x\}$.*

Example 18.2. *(i) The integers $\mathbb{Z}$ form a discrete subset of $\mathbb{R}$;*

(ii) Any finite subset of $\mathbb{R}^n$ is discrete;

(iii) The set $\{\frac{1}{n} : n \in \mathbb{Z}, n \neq 0\}$ is a discrete subset of $\mathbb{R}$.

However, $\{\frac{1}{n} : n \in \mathbb{Z}, n \neq 0\} \cup \{0\}$ is not discrete, since every neighborhood of 0 contains real numbers of form $\frac{1}{n}, n \in \mathbb{Z}$.

Theorem 18.2. *Let Ω_f be the set of periods of a non-constant meromorphic function f defined on $\mathbb{C}$; then Ω_f is a discrete subset of $\mathbb{C}$.*

Proof. If Ω_f is not discrete then there exists $\omega \in \Omega_f$ such that every neighborhood U of ω contains points of $\Omega_f \setminus \{\omega\}$. By taking these neighborhoods to be discs, centered at ω, with radii approaching 0, we obtain a sequence of periods $\omega_n \neq \omega$ with $\lim_{n\to\infty} \omega_n = \omega$. Since $\omega, \omega_n \in \Omega_f$ we have $f(\omega) = f(0) = f(\omega_n)$. If $f(0) = \infty$ then f has a convergent sequence of poles, which is impossible by Theorem 7.16 and Example 7.16. If $f(0) \neq \infty$ then the meromorphic function $g(z) = f(z) - f(0)$ has a convergent sequence of zeros, implying that g is identically zero (by Theorem 7.16), and so f is constant. In either case, we have a contradiction. □

Thus the set of periods of a non-constant meromorphic function is a discrete subgroup of $\mathbb{C}$. We shall show below that there are three types of discrete additive subgroups of $\mathbb{C}$, isomorphic to $\{0\}$, $\mathbb{Z}$ and $\mathbb{Z} \times \mathbb{Z}$ respectively.

Theorem 18.3. *Let Ω be a discrete subgroup of $\mathbb{C}$. Then one of the following holds:*

(i) $\Omega = \{0\}$;

(ii) $\Omega = \{n\omega_1 : n \in \mathbb{Z}\}$ *for some fixed* $\omega_1 \in \mathbb{C} \setminus \{0\}$, *and so* Ω *is isomorphic to* $\mathbb{Z}$;

(iii) $\Omega = \{m\omega_1 + n\omega_2 : m, n \in \mathbb{Z}\}$ *for some fixed* $\omega_1, \omega_2 \in \mathbb{C}$, *where* ω_1 *and* ω_2 *are linearly independent over* $\mathbb{R}$, *that is,* $\omega_1 \neq 0$, $\omega_2 \neq 0$ *and* $\frac{\omega_1}{\omega_2} \notin \mathbb{R}$; *in this case,* Ω *is isomorphic to* $\mathbb{Z} \times \mathbb{Z}$.

Proof. Suppose that $\Omega \neq \{0\}$. We first show that there exists $\omega_1 \in \Omega \setminus \{0\}$ with least value of $|\omega_1|$. Since Ω is discrete, there is some $\epsilon > 0$ such that the disc $B(0, \epsilon)$ contains no element of $\Omega \setminus \{0\}$. It follows that for any $\omega \in \Omega$, the disk $B(\omega, \epsilon)$ contains no element of $\Omega \setminus \{\omega\}$: for if $z \in \Omega$ satisfies $0 < |z - \omega| < \epsilon$, then $z - \omega$ is an element of Ω (since Ω is a group) within distance ϵ of 0, which is impossible. Thus each $\omega \in \Omega$ is the center of a disk of radius ϵ (independent of ω) containing no other elements of Ω. Therefore, disks of radius $\frac{\epsilon}{2}$, centered on the elements of Ω, are disjoint from each other. Now choose a disk $B(0, r)$ sufficiently large to contain at least one element of $\Omega \setminus \{0\}$ together with its surrounding disk of radius $\frac{\epsilon}{2}$; this disk has area πr^2, so, it can contain only finitely many disjoint disks of radius $\frac{\epsilon}{2}$ (at most $4r^2\epsilon^{-2}$, in fact); hence there are only finitely many elements of $\Omega \setminus \{0\}$ within distance r to 0, and out of this finite set we may choose ω_1 to have least modulus. (Of course ω_1 is not unique: $-\omega_1$ will do equally well.) Now let $L = \{\lambda\omega_1 : \lambda \in \mathbb{R}\}$ be the line through 0 and ω_1 in $\mathbb{C}$; then Ω contains $\{n\omega_1 : n \in \mathbb{Z}\}$, and this subgroup lies in L. First suppose that $\Omega \subseteq L$; we then claim that $\Omega = \{n\omega_1 : n \in \mathbb{Z}\}$, so that Ω is of type (ii). For if Ω contains some $\omega \neq n\omega_1$, then since $\Omega \subseteq L$ we have $\omega = \lambda\omega_1$ for some $\lambda \in \mathbb{R} \setminus \mathbb{Z}$, so $n < \lambda < n + 1$ for some $n \in \mathbb{Z}$.

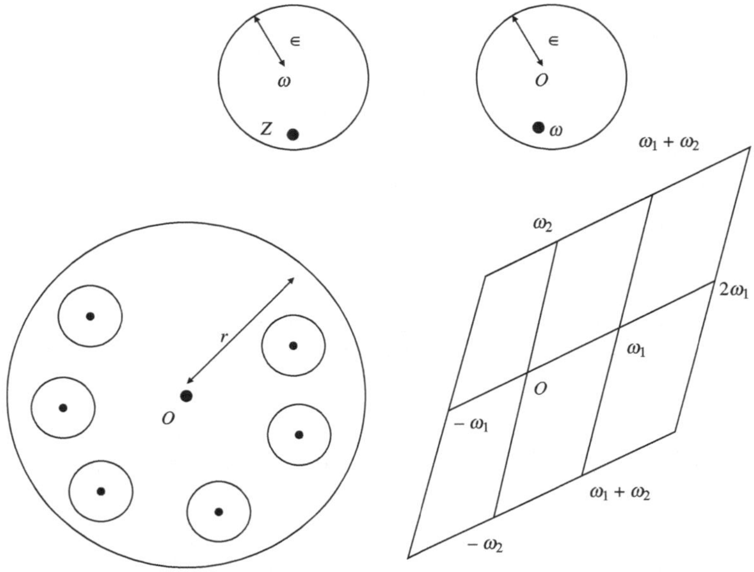

Fig. 18.1: The set of periods of a nonconstant meromorphic function

Since Ω is a group containing ω and $n\omega_1$, it contains $\omega - n\omega_1 = (\lambda - n)\omega_1$. However, $0 \neq |(\lambda - n)\omega_1| < |\omega_1|$ which contradicts the minimality of $|\omega_1|$. Hence $\Omega = \{n\omega_1 : n \in \mathbb{Z}\}$, and we are in case (ii).

Now suppose that $\Omega \not\subseteq L$. By an argument similar to that used for ω_1, we can show that $\Omega \setminus L$ has an element ω_2 with least modulus. Then Ω contains the subgroup $\Lambda = \{m\omega_1 + n\omega_2 : m, n \in \mathbb{Z}\}$, and since $\omega_2 \notin L$, ω_1 and ω_2 are linearly independent over $\mathbb{R}$, so Λ consists of the vertices of a tessellation of $\mathbb{C}$ by congruent parallelograms.

We now show that $\Omega = \Lambda$. First of all, because $\frac{\omega_1}{\omega_2}$ is nonreal any complex number ω can be written in the form $\lambda\omega_1 + \mu\omega_2$ with real λ and μ. To see this we need only solve the equations

$$\omega = \lambda\omega_1 + \mu\omega_2;$$

$$\bar{\omega} = \lambda\bar{\omega}_1 + \mu\bar{\omega}_2.$$

Since the determinant $\omega_1\bar{\omega}_2 - \omega_2\bar{\omega}_1$ is $\neq 0$ the system has a unique solution (λ, μ); but $(\bar{\lambda}, \bar{\mu})$ is also a solution, and we conclude that λ and μ are real. If $\Omega \neq \Lambda$, then there exists $\omega \in \Omega$ with $\omega \neq m\omega_1 + n\omega_2$ for any $m, n \in \mathbb{Z}$. Let $\omega = \lambda\omega_1 + \mu\omega_2$ with $\lambda, \mu \in \mathbb{R}$; then by adding or subtracting suitable multiples of ω_1 and ω_2 we may assume that $|\lambda| \leq \frac{1}{2}, |\mu| \leq \frac{1}{2}$. If $\mu = 0$ then $\omega = \lambda\omega_1 \in L$, with $|\omega| = |\lambda\omega_1| < |\omega_1|$; by minimality of $|\omega_1|$ we have $\omega = 0$ and hence $\omega \in \Lambda$, against our assumption. If $\lambda = 0$ then $\omega = \mu\omega_2$, and again $\omega = 0$, this time by minimality of $|\omega_2|$. Hence $\lambda\omega_1$ and $\mu\omega_2$ are non-zero and therefore linearly independent over $\mathbb{R}$, giving

$$\begin{aligned}|\omega| &< |\lambda\omega_1| + |\mu\omega_2| \\ &\leq \tfrac{1}{2}|\omega_1| + \tfrac{1}{2}|\omega_2| \\ &\leq \tfrac{1}{2}|\omega_2| + \tfrac{1}{2}|\omega_2| \\ &= |\omega_2|,\end{aligned}$$

the first inequality being strict because $\lambda\omega_1$ and $\mu\omega_2$ are linearly independent. Now $\omega \in \Omega \setminus L$ (since $\mu \neq 0$), so by minimality of $|\omega_2|$ we have $\omega = 0$, contradicting the fact that $\lambda, \mu \neq 0$. Thus $\omega \in \Lambda$, so $\Omega = \Lambda$ and we have proved case (iii). □

If a function f has its set Ω_f of periods of type (ii), then f is simply periodic; if Ω_f is of type (iii), then f is doubly periodic. Groups Ω of type (iii) are called lattices, and any pair ω_1, ω_2 such that $\Omega = \{m\omega_1 + n\omega_2 : m, n \in \mathbb{Z}\}$ is called a basis for the lattice. The above results show that periodic functions are either simply or doubly periodic (apart from constant functions, where $\Omega = \mathbb{C}$, and non-meromorphic functions, where Ω can be any subgroup of $\mathbb{C}$, not necessarily discrete). Thus any function with three or more independent periods is a constant.

18.2 Doubly periodic functions

Definition 18.3. *A meromorphic function $f : \mathbb{C} \to \mathbb{C}_\infty$ is said to be an elliptic function if f is doubly periodic.*

The connection between the doubly periodic functions and the ellipse is tenuous, but historically interesting. From the second half of the seventeenth century onwards, much attention was devoted to certain integrals, which, it seemed, could not be evaluated using 'elementary' functions. These were known as elliptic integrals, since they include the integral

$$\int \sqrt{\frac{a^2 - ex^2}{a^2 - x^2}}\,dx,$$

giving the circumference of the ellipse $\frac{x^2}{a^2} + \frac{y^2}{b^2} = 1$, where $e = 1 - \frac{b^2}{a^2}$. In the late 1790's, Gauss noticed that, just as the inverse functions of the integrals

$$\int \frac{dx}{1+x^2} \quad \text{and} \quad \int \frac{dx}{\sqrt{1-x^2}},$$

give the simply periodic trigonometric functions $\tan x$ and $\sin x$, the inverse functions of certain elliptic integrals, such as

$$\int \frac{dx}{\sqrt{1-x^4}},$$

give the doubly periodic functions. These ideas were unpublished, and were rediscovered in the 1820's by Abel and Jacobi, who gave the name 'elliptic' to these doubly periodic functions.

Let f be meromorphic in the whole plane. Let Ω_f be the set of all periods of f. If $\omega \in \Omega_f$ then $n\omega \in \Omega_f$ for all $n \in \mathbb{Z}$(the set of integers) and if $\omega_1, \omega_2 \in \Omega_f$ then $n_1\omega_1 + n_2\omega_2 \in \Omega_f$ for all $n_1, n_2 \in \mathbb{Z}$. Thus Ω_f is a module over $\mathbb{Z}$ and Ω_f is called the period module of f. A module with isolated points is said to be discrete. In this case Ω_f is a discrete module as points of Ω_f are isolated.

Let f be a doubly periodic function and Ω_f be the period module of f. The pair (ω_1, ω_2) $(\omega_1, \omega_2 \in M)$ has the property that any $\omega \in \Omega_f$ has a unique representation of the form $\omega = n_1\omega_1 + n_2\omega_2, n_1, n_2 \in \mathbb{Z}, \frac{\omega_2}{\omega_1} \notin \mathbb{R}$. Any pair with this property will be called a basis of Ω_f. We investigate the relation between two bases (ω_1, ω_2) and (ω_1', ω_2'). Because (ω_1, ω_2) is a basis there exist $a, b, c, d \in \mathbb{Z}$ such that

$$\begin{aligned} \omega_2' &= a\omega_2 + b\omega_1 \\ \omega_1' &= c\omega_2 + d\omega_1. \end{aligned} \tag{18.1}$$

Thus

$$\begin{aligned} \overline{\omega_2'} &= a\overline{\omega_2} + b\overline{\omega_1} \\ \overline{\omega_1'} &= c\overline{\omega_2} + d\overline{\omega_1}. \end{aligned} \tag{18.2}$$

If we write this in matrix notation, we get

$$\begin{pmatrix} \omega_2' & \overline{\omega_2'} \\ \omega_1' & \overline{\omega_1'} \end{pmatrix} = \begin{pmatrix} a & b \\ c & d \end{pmatrix} \begin{pmatrix} \omega_2 & \overline{\omega_2} \\ \omega_1 & \overline{\omega_1} \end{pmatrix}. \tag{18.3}$$

Now since (ω_1', ω_2') is a basis, we have similarly

$$\begin{pmatrix} \omega_2 & \overline{\omega_2} \\ \omega_1 & \overline{\omega_1} \end{pmatrix} = \begin{pmatrix} a' & b' \\ c' & d' \end{pmatrix} \begin{pmatrix} \omega_2' & \overline{\omega_2'} \\ \omega_1' & \overline{\omega_1'} \end{pmatrix}, \tag{18.4}$$

where $a', b', c', d' \in \mathbb{Z}$.

Hence from (18.3) and (18.4) we obtain

$$\begin{pmatrix} \omega_2 & \overline{\omega_2} \\ \omega_1 & \overline{\omega_1} \end{pmatrix} = \begin{pmatrix} a' & b' \\ c' & d' \end{pmatrix} \begin{pmatrix} a & b \\ c & d \end{pmatrix} \begin{pmatrix} \omega_2 & \overline{\omega_2} \\ \omega_1 & \overline{\omega_1} \end{pmatrix}. \tag{18.5}$$

Since $\frac{\omega_2}{\omega_1} \notin \mathbb{R}$, hence $\frac{\omega_2}{\omega_1} \neq \frac{\overline{\omega_2}}{\overline{\omega_1}}$. Thus $\omega_2\overline{\omega_1} - \overline{\omega_2}\omega_1 \neq 0$. Therefore the matrix $\begin{pmatrix} \omega_2 & \overline{\omega_2} \\ \omega_1 & \overline{\omega_1} \end{pmatrix}$ is invertible and if we multiply (18.5) by the inverse of $\begin{pmatrix} \omega_2 & \overline{\omega_2} \\ \omega_1 & \overline{\omega_1} \end{pmatrix}$ we obtain

$$\begin{pmatrix} a' & b' \\ c' & d' \end{pmatrix} \begin{pmatrix} a & b \\ c & d \end{pmatrix} = \begin{pmatrix} 1 & 0 \\ 0 & 1 \end{pmatrix}.$$

Thus the matrices

$$\begin{pmatrix} a & b \\ c & d \end{pmatrix} \text{ and } \begin{pmatrix} a' & b' \\ c' & d' \end{pmatrix}$$

are inverse to each other and

$$\begin{vmatrix} a' & b' \\ c' & d' \end{vmatrix} \cdot \begin{vmatrix} a & b \\ c & d \end{vmatrix} = 1.$$

Thus

$$\begin{vmatrix} a & b \\ c & d \end{vmatrix} = \begin{vmatrix} a' & b' \\ c' & d' \end{vmatrix} = \pm 1.$$

Linear transformation of the form (18.1) with integer coefficients and determinant ± 1 are said to be unimodular. Thus we proved that any two bases of the same module are connected by a unimodular transformation. In this chapter we consider the elementary properties of elliptic functions which arise naturally as the doubly periodic meromorphic functions on $\mathbb{C}$. It is not easy to construct nonconstant elliptic functions. For this purpose one should get acquainted with the concept of uniform convergence on compact sets and infinite products of functions. These techniques actually helps in constructing the Weierstrass functions $\mathcal{P}(z), \zeta(z)$ and $\sigma(z)$ and one can use these functions to deduce further properties of elliptic functions.

We shall now show that for the period module Ω_f of an elliptic function f there exists at least one basis. In fact, among all possible bases of Ω_f it is possible to single out one, almost uniquely which is called the canonical basis.

Theorem 18.4. *Let f be an elliptic function and let Ω_f be the period module of f. Then there exists a basis (ω_1, ω_2) of the module Ω_f such that the ratio $\tau = \frac{\omega_2}{\omega_1}$ satisfies the following conditions: (i) $Im\tau > 0, (ii)\ |\tau| \geq 1 (iii) -\frac{1}{2} < Re\tau \leq \frac{1}{2} (iv) Re\tau \geq 0$ if $|\tau| = 1$. The ratio τ is uniquely determined by these conditions, and there is a choice of two, four, or six corresponding bases.*

Proof. Let ω_1 and ω_2 be obtained as in Theorem 18.3. Then $|\omega_1| \leq |\omega_2|$. That is, $|\tau| \geq 1$. Now since $|\omega_2| \leq |\omega_1 + \omega_2|$ and $|\omega_2| \leq |\omega_1 - \omega_2|$, we have $|\tau| \leq |1+\tau|$ and $|\tau| \leq |1-\tau|$. That is, $\text{Re}\tau \geq -\frac{1}{2}$ and $\text{Re}\tau \leq \frac{1}{2}$. Thus $|\text{Re}\tau| \leq \frac{1}{2}$. If $\text{Im}\tau < 0$, we replace (ω_1, ω_2) by $(-\omega_1, \omega_2)$; this makes $\text{Im}\tau > 0$ without changing the condition on $\text{Re}\tau$. If $\text{Re}\tau = -\frac{1}{2}$, we replace the basis by $(\omega_1, \omega_1 + \omega_2)$, and if $|\tau| = 1, \text{Re}\tau < 0$, we replace it by $(-\omega_2, \omega_1)$. With these minor changes, all the conditions stated in the theorem are satisfied. The condition (i)-(iv) mean that the point τ lies in the part of the complex plane shown in Fig. 18.2.

It is bounded by the circle $|\tau| = 1$ and the vertical line $\text{Re}\tau = \pm\frac{1}{2}$, but only part of the boundary is included. This set is referred to as the 'fundamental region' of the unimodular group although the set is not open. We have seen that any two bases of the same module are connected by a unimodular transformation.

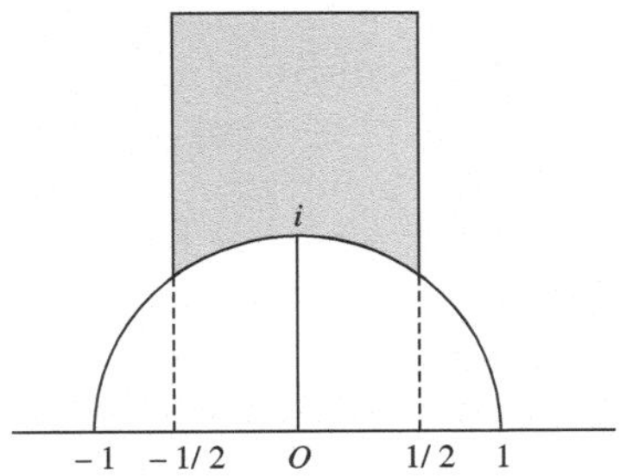

Fig. 18.2: The fundamental region

Let (ω_1', ω_2') be another basis of Ω_f and $\tau' = \frac{\omega_2'}{\omega_1'}$. Let $\omega_2' = a\omega_2 + b\omega_1$ and $\omega_1' = c\omega_2 + d\omega_1$. Then

$$\tau' = \frac{a\omega_2 + b\omega_1}{c\omega_2 + d\omega_1} = \frac{a\tau + b}{c\tau + d} \tag{18.6}$$

where $ad - bc = \pm 1$. Thus

$$\text{Im}\tau' = \frac{\tau' - \bar{\tau}'}{2i} = \frac{\frac{a\tau+b}{c\tau+d} - \overline{\frac{a\tau+b}{c\tau+d}}}{2i} = \frac{(ad-bc)\text{Im}\tau}{|c\tau+d|^2} = \frac{\pm\text{Im}\tau}{|c\tau+d|^2}$$

where the sign is the same as that of $ad - bc$. Suppose that both τ and τ' lie in the fundamental region. We shall show that $\tau = \tau'$. Since $\text{Im}\tau' > 0$ and $\text{Im}\tau > 0$, hence $ad - bc = 1$. Now either $\text{Im}\tau' \geq \text{Im}\tau$ or $\text{Im}\tau \geq \text{Im}\tau'$. We shall assume $\text{Im}\tau' \geq \text{Im}\tau$. (The other case is similar.) Thus it follows that $|c\tau + d| \leq 1$. Because c and d are integers, there are very few possibilities for this inequality to hold.

Suppose $c = 0, d = \pm 1$. Since $ad - bc = 1$, hence we have $ad = 1$. Because $a, d \in \mathbb{Z}$, either $a = d = 1$ or $a = d = -1$. Thus $\tau' = \tau \pm b$ and therefore $|b| = |\text{Re}\tau' - \text{Re}\tau| < 1$ by condition (iii). Since $b \in \mathbb{Z}$, we have $b = 0$ and $\tau' = \tau$.

Suppose now that $c \neq 0$. Then $|c\tau + d| \leq 1$ implies $|\tau + \frac{d}{c}| \leq \frac{1}{|c|}$. If $|c| \geq 2$, then $|\tau - \left(-\frac{d}{c}\right)| \leq \frac{1}{2}$; that is, τ would be at a distance less than or equal to $\frac{1}{2}$ from the real axis. This is not true as $|\tau| \geq 1$. Notice that the nearest point to the real axis in the fundamental region is at a distance $\frac{\sqrt{3}}{2}$. Thus $|c| = 1$ as $c \neq 0$. It follows therefore that $|\tau \pm d| \leq 1$. This can occur only if $d = 0$ or $d = \pm 1$. The inequality $|\tau + 1| \leq 1$ is never fulfilled, for the point $e^{\frac{2\pi i}{3}}$ is not in the fundamental region and $|\tau - 1| \leq 1$ only when $\tau = e^{\frac{\pi i}{3}}$. In the later case $|c\tau + d| = 1$. Thus $\text{Im}\tau' = \text{Im}\tau$. Since $\tau = e^{\frac{\pi i}{3}} = \cos\frac{\pi}{3} + i\sin\frac{\pi}{3}$, by the shape of the fundamental region, $\tau' = \tau$. In case $d = 0, |c| = 1$, we have $|\tau| = |c\tau + d| \leq 1$. Thus by (ii) we obtain $|\tau| = 1$. Now $1 = ad - bc = -bc$. Therefore $\frac{b}{c} = -1$ as b and c are integers and $\tau' = \pm a - \frac{1}{\tau} = \pm a - \bar{\tau}$. Hence $\text{Re}\,(\tau + \tau') = \pm a$. since $|\tau| = 1$ we have by (iv), $\text{Re}\tau \geq 0$. Further by (iii),

$0 \le \mathrm{Re}\tau \le \frac{1}{2}$. Since $\tau' = \pm a - \frac{1}{\tau}$ we have $\mathrm{Im}\tau' = \mathrm{Im}\tau$. If $\mathrm{Re}\tau' = \mathrm{Re}\tau$, then $\tau = \tau'$. So suppose $\mathrm{Re}\tau \neq \mathrm{Re}\tau'$ and $\mathrm{Re}(\tau + \tau') = \pm a$. Then this implies $a = 0$. In this case $\tau' = -\frac{1}{\tau}$. Thus $\tau = \tau' = i$ as $|\tau| = 1$ implies $\mathrm{Re}\tau \ge 0$.

We have thus proved that τ is unique. The canonical basis (ω_1, ω_2) can be replaced by $(-\omega_1, -\omega_2)$. There are other bases with the same τ if and only if τ is a fixed point of a unimodular transformation (18.6). This happens only for $\tau = i$ and $\tau = e^{\frac{\pi i}{3}}$; the former is a fixed point of $-\frac{1}{\tau}$, the later of $-\left(\frac{\tau+1}{\tau}\right)$ and of $-\frac{1}{\tau+1}$. These are the multiple choices referred to in the theorem. □

It is natural to consider geometrically the parallelogram spanned by a basis (ω_1, ω_2) in its relation to the lattice formed by all numbers in the module. Fig. 18.3 shows two bases of the same module. Observe that the parallelograms have equal area. Note that the unimodular matrices, or the corresponding linear transformations, form a group, the modular group.

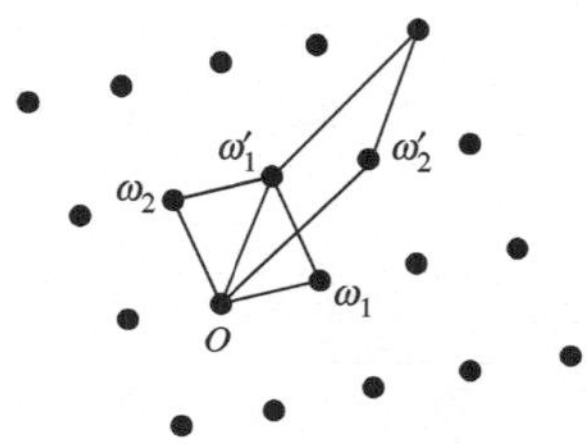

Fig. 18.3: Two bases of the same module

18.3 Geometric properties of elliptic functions

So far, the only elliptic functions we have met are the constant functions, and it is a substantial problem to construct non-constant elliptic functions. Before doing this, we will examine some of the elementary properties which elliptic functions must possess.

Let f be a meromorphic function and Ω_f be the period module of f. Let (ω_1, ω_2) be a basis for Ω_f (not necessarily the canonical basis). We define $z_1, z_2 \in \mathbb{C}$ to be congruent mod Ω_f written $z_1 \equiv z_2(\mathrm{mod}\Omega_f)$ if $z_1 - z_2 \in \Omega_f$, i.e., $z_1 = z_2 + n_1\omega_1 + n_2\omega_2, n_1, n_2 \in \mathbb{Z}$. The function f takes the same value at congruent points, and may thus be regarded as a function on the congruence classes.

Consider the parallelogram $\mathcal{P}_a$ with vertices $a, a+\omega_1, a+\omega_2, a+\omega_1+\omega_2$. By including part of the boundary we may represent each congruence class by exactly one point in $\mathcal{P}_a$, and then the behavior of f on $\mathbb{C}$ is completely determined by its values on $\mathcal{P}_a$. Hence we can regard f as a function defined on $\mathcal{P}_a$, and since f takes the same values at congruent boundary points we may as well identify them. Notice that in this case the choice of a is irrelevant but we shall chose a in such a way that no zeros or poles of f will lie on $\partial\mathcal{P}_a$.

Theorem 18.5. *An elliptic function without poles must be constant.*

Proof. Since $\overline{\mathcal{P}_a}$ is a bounded, closed subset of $\mathbb{C}$, hence it is a compact subset of $\mathbb{C}$. If f is an elliptic function without poles then f is analytic and therefore bounded in $\overline{\mathcal{P}_a}$. Since f is doubly periodic, the function f is an entire function and is bounded in $\mathbb{C}$. By Liouville's theorem, f is a constant function. □

Remark 18.1. Notice that the elliptic function f can have at most a finite number of poles in $\mathcal{P}_a$. This is so because poles have no accumulation point.

When we speak of the poles of an elliptic function f we mean a full set of mutually incongruent poles. Multiplicities are counted in the usual manner.

Theorem 18.6. *The sum of the residues of an elliptic function within $\mathcal{P}_a$ is zero.*

Proof. Let f be an elliptic function with independent periods ω_1 and ω_2 (i.e., $\frac{\omega_2}{\omega_1} \notin \mathbb{R}$). We shall chose a so that none of the poles of f fall on $\partial\mathcal{P}_a$, the boundary of $\mathcal{P}_a$. If $\partial\mathcal{P}_a$ is traced in the positive sense, then by Residue theorem, the sum of the residues at the poles in $\mathcal{P}_a$ equals $\frac{1}{2\pi i}\int_{\partial\mathcal{P}_a} f(z)dz$. But

$$\begin{aligned}
\frac{1}{2\pi i}\int_{\partial\mathcal{P}_a} f(z)dz &= \frac{1}{2\pi i}\int_a^{a+\omega_1} f(z)dz + \frac{1}{2\pi i}\int_{a+\omega_1}^{a+\omega_1+\omega_2} f(z)dz \\
&+ \frac{1}{2\pi i}\int_{a+\omega_1+\omega_2}^{a+\omega_2} f(z)dz + \frac{1}{2\pi i}\int_{a+\omega_2}^{a} f(z)dz \\
&= \frac{1}{2\pi i}\int_a^{a+\omega_1} f(z)dz + \frac{1}{2\pi i}\int_a^{a+\omega_2} f(z+\omega_1)dz \\
&+ \frac{1}{2\pi i}\int_{a+\omega_1}^{a} f(z+\omega_2)dz - \frac{1}{2\pi i}\int_a^{a+\omega_2} f(z)dz
\end{aligned}$$

$$\begin{aligned}
&= \frac{1}{2\pi i}\int_a^{a+\omega_1} f(z)dz + \frac{1}{2\pi i}\int_a^{a+\omega_2} f(z)dz \\
&- \frac{1}{2\pi i}\int_a^{a+\omega_1} f(z)dz - \frac{1}{2\pi i}\int_a^{a+\omega_2} f(z)dz = 0.
\end{aligned}$$

□

Remark 18.2. There does not exist an elliptic function with a single simple pole. For let f be an elliptic function with a single simple pole at $z = a \in \mathcal{P}_a$ and no other poles. Let

$$\frac{b_{-1}}{z-a} + b_0 + b_1(z-a) + b_2(z-a)^2 + \cdots, b_{-1} \neq 0$$

be the Laurent series expansion of f about a.

Then the sum of the residues of f within $\mathcal{P}_a$ is equal to b_{-1} which is nonzero. This is not possible by Theorem 18.6 as f is an elliptic function.

Theorem 18.7. *A nonconstant elliptic function has equally many poles as it has zeros.*

Proof. The poles and zeros of f are simple poles of $\frac{f'}{f}$ and the multiplicities are the residues of $\frac{f'}{f}$, counted positive for zeros and negative for poles (see Remark 11.4). Further since f is an elliptic function, hence $\frac{f'}{f}$ is also an elliptic function and it follows from argument principle that

$$\begin{aligned} 0 &= \text{sum of residues of } \tfrac{f'}{f} \text{ within } \mathcal{P}_a \\ &= \tfrac{1}{2\pi i}\int_{\partial\mathcal{P}_a}\frac{f'(z)}{f(z)}dz \\ &= \text{Number of zeros of } f \text{ in } \mathcal{P}_a - \text{ Number of poles of } f \text{ in } \mathcal{P}_a. \end{aligned}$$

Thus f has equal number of zeros and poles in $\mathcal{P}_a$. □

Remark 18.3. Let f be an elliptic function and c is any constant. Then $f(z)-c$ is also an elliptic function and $f(z)-c$ has the same poles as $f(z)$. By Theorem 18.7,

$$\begin{aligned} &\text{Number of zeros of } f(z)-c \text{ in } \mathcal{P}_a \\ &= \text{Number of poles of } f(z)-c \text{ in } \mathcal{P}_a \\ &= \text{Number of poles of } f \text{ in } \mathcal{P}_a \\ &= \text{Number of zeros of } f \text{ in } \mathcal{P}_a. \end{aligned}$$

Thus f assumes all values equally many times.

Remark 18.4. Suppose f is elliptic with period module $\Omega_f, c \in \mathbb{C}_\infty$ and f is not identically equal to c. Then the solution of $f(z) = c$ are isolated, and each solution has finite multiplicity, congruent solutions having the same multiplicity. Since the solutions are isolated, any parallelogram $\mathcal{P}_a$ contains only finitely many solutions (since $\overline{\mathcal{P}_a}$ is compact), and by replacing $\mathcal{P}_a$ by $\mathcal{P}_a + t$ $(t \in \mathbb{C})$ if necessary, we may assume that there are no solutions on the boundary $\partial\mathcal{P}_a$ of $\mathcal{P}_a$. Let the solutions within $\mathcal{P}_a$ be $z = z_1, \cdots, z_r$ with multiplicities $k_1, \cdots, k_r$, and let $N = k_1 + \cdots + k_r$. Then we say that there are N solutions of $f(z) = c$. Note that $z_1, \cdots, z_r$ are representatives of the congruence classes of solutions of $f(z) = c$, for $z \in \mathbb{C}$ in $\mathcal{P}_a$.

Definition 18.4. *The number of incongruent roots of the equation $f(z) = c$ is called the order of the elliptic function f.*

Theorem 18.8. *The zeros $a_1, \cdots, a_n$ and poles $b_1, \cdots, b_n$ of an elliptic function f satisfy*

$$a_1 + \cdots + a_n \equiv b_1 + \cdots + b_n (\ mod\ \Omega_f).$$

Proof. Notice that $\frac{f'}{f}$ is an elliptic function. Suppose none of the zeros and poles of f lie on the boundary of $\mathcal{P}_a$. Consider the integral

$$\frac{1}{2\pi i}\int_{\partial\mathcal{P}_a} z\,\frac{f'(z)}{f(z)}dz.$$

By generalized argument principle (Theorem 11.13),

$$(a_1+\cdots+a_n)-(b_1+\cdots+b_n)=\frac{1}{2\pi i}\int_{\partial\mathcal{P}_a} z\,\frac{f'(z)}{f(z)}dz,$$

provided that we choose the representative zeros and poles inside $\mathcal{P}_a$. But

$$\begin{aligned}
&\tfrac{1}{2\pi i}\int_{\partial\mathcal{P}_a} z\,\frac{f'(z)}{f(z)}dz\\
&= \tfrac{1}{2\pi i}\int_a^{a+\omega_1} z\,\frac{f'(z)}{f(z)}dz+\frac{1}{2\pi i}\int_{a+\omega_1}^{a+\omega_1+\omega_2} z\,\frac{f'(z)}{f(z)}dz\\
&\quad+\tfrac{1}{2\pi i}\int_{a+\omega_1+\omega_2}^{a+\omega_2} z\,\frac{f'(z)}{f(z)}dz+\frac{1}{2\pi i}\int_{a+\omega_2}^{a} z\,\frac{f'(z)}{f(z)}dz\\
&= \tfrac{1}{2\pi i}\left[\int_a^{a+\omega_1} z\,\frac{f'(z)}{f(z)}dz+\int_a^{a+\omega_2}(z+\omega_1)\,\frac{f'(z+\omega_1)}{f(z+\omega_1)}dz\right]\\
&\quad+\tfrac{1}{2\pi i}\left[\int_{a+\omega_1}^{a}(z+\omega_2)\,\frac{f'(z+\omega_2)}{f(z+\omega_2)}dz+\int_{a+\omega_2}^{a} z\,\frac{f'(z)}{f(z)}dz\right]
\end{aligned}$$

$$\begin{aligned}
&= \tfrac{1}{2\pi i}\left[\int_a^{a+\omega_1} z\,\frac{f'(z)}{f(z)}dz+\int_a^{a+\omega_2} z\,\frac{f'(z)}{f(z)}dz+\omega_1\int_a^{a+\omega_2}\frac{f'(z)}{f(z)}dz\right.\\
&\quad\left.-\int_a^{a+\omega_1} z\,\frac{f'(z)}{f(z)}dz-\omega_2\int_a^{a+\omega_1}\frac{f'(z)}{f(z)}dz-\int_a^{a+\omega_2} z\,\frac{f'(z)}{f(z)}dz\right]\\
&= \tfrac{\omega_1}{2\pi i}\int_a^{a+\omega_2}\frac{f'(z)}{f(z)}dz-\frac{\omega_2}{2\pi i}\int_a^{a+\omega_1}\frac{f'(z)}{f(z)}dz.
\end{aligned}$$

The integral $\frac{1}{2\pi i}\int_a^{a+\omega_1}\frac{f'(z)}{f(z)}dz$ represents the winding number around the origin of the closed curve described by $f(z)$ when z varies from a to $a+\omega_1$. It is therefore an integer. Similarly $\frac{1}{2\pi i}\int_a^{a+\omega_2}\frac{f'(z)}{f(z)}dz$ is also an integer. Thus

$$(a_1+\cdots+a_n)-(b_1+\cdots+b_n)=n_1\omega_1+n_2\omega_2$$

where $n_1,n_2\in\mathbb{Z}$ and the theorem is proved. □

Example 18.3. *Let f be an entire function such that $f(z+1) = f(z)$ and $f(z+i) = f(z)$ for all $z \in \mathbb{C}$. Then f is a constant function.*

Solution: Let $w_1 = 1$ and $w_2 = i$. Let $\mathcal{P}_a$ be the parallelogram with vertices $a, a+w_1, a+w_2, a+w_1+w_2$. The set $\overline{\mathcal{P}_a}$ is compact as it is closed and bounded. The function f is continuous and hence f is bounded in $\overline{\mathcal{P}_a}$. Since f is doubly periodic, hence f is bounded in the whole complex plane $\mathbb{C}$. By Liouville's theorem, f is a constant.

Example 18.4. *If f is entire, f is periodic then f has a fixed point.*

Solution: Suppose f is entire, f is periodic with period $w \neq 0$ but f has no fixed point. It therefore follows that $f(z) \neq z$ for all $z \in \mathbb{C}$. Let $g(z) = z - f(z), z \in \mathbb{C}$. Then $g(z) \neq 0$ for all $z \in \mathbb{C}$ and g is an entire function. Now $0 \neq g(z+w) = z+w-f(z+w) = z+w-f(z) = g(z)+w$ for all $z \in \mathbb{C}$. Thus $g(z) \neq -w$ for all $z \in \mathbb{C}$. Thus 0 and $-w$ are two lacunary values of the entire function g. Hence by Picard's theorem g is a constant function. Let $g(z) = c$ for all $z \in \mathbb{C}$. Then it follows that $z - f(z) = c$. Thus $f(z) = z - c$ for all $z \in \mathbb{C}$ and $f(z) = f(z+w) = z+w-c = f(z)+w$. Hence $w = 0$ which is a contradiction.

18.4 The Weierstrass $\mathcal{P}$ function

We need to construct nonconstant functions f which are elliptic with period module Ω_f. By Theorem 18.5, we know that such a function f cannot be analytic and so it must have poles in $\mathcal{P}_a$. By Theorem 18.6, we know that f cannot have just one simple pole in $\mathcal{P}_a$, so the simplest nonconstant elliptic function has order 2, with either two simple poles (having opposite residues) or else a single pole of order 2 in $\mathcal{P}_a$ with zero residue. In this section we shall introduce the Weierstrass function $\mathcal{P}(z)$ which is elliptic of order 2 with respect to the period module Ω_f having a basis $\{\omega_1, \omega_2\}$ and has a single pole of order 2 in $\mathcal{P}_a$. This will be our basic elliptic function in the sense that every function which is elliptic with respect to Ω_f is a rational function of $\mathcal{P}$ and its derivative $\mathcal{P}'$.

We shall construct an elliptic function of order 2, and it may be noted that all other elliptic function may be expressed in terms of $\mathcal{P}$, just as all simply periodic functions may be expressed in terms of $e^{2\pi i z}$. We have already seen that there does not exist an elliptic function with a single simple pole. So the simplest elliptic function are of order 2, and such functions have either a double pole with residue zero, or two simple poles with opposite residues. Suppose the function f has only a pole of order 2 at $z = 0$ and the singular part (of the Laurent expansion at $z = 0$) is $\frac{1}{z^2}$. If f is elliptic with two periods ω_1 and ω_2, $\frac{\omega_2}{\omega_1} \notin \mathbb{R}$ and f has only this singularity at the origin and its congruent points, it is not so difficult to check that f is an even function. Notice that $f(z) - f(-z)$ has the same periods ω_1 and ω_2 and no singularity. By Theorem 18.5,

$f(z)-f(-z)$ is a constant function. That is, $f(z)-f(-z)\equiv c$(a constant). Set $z=\frac{\omega_1}{2}$. Then

$$\begin{aligned} c &\equiv f\left(\tfrac{\omega_1}{2}\right)-f\left(-\tfrac{\omega_1}{2}\right) \\ &= f\left(\tfrac{\omega_1}{2}\right)-f\left(-\tfrac{\omega_1}{2}+\omega_1\right) \\ &= f\left(\tfrac{\omega_1}{2}\right)-f\left(\tfrac{\omega_1}{2}\right)=0. \end{aligned}$$

Thus we conclude that $f(z)\equiv f(-z)$, that is, f is an even function. Suppose the constant term in the Laurent development about the origin is zero and with this normalization, $f(z)$ is uniquely determined, and it is denoted by a symbol $\mathcal{P}(z)$. The Laurent development has the form

$$\mathcal{P}(z)=\frac{1}{z^2}+a_1z^2+a_2z^4+\cdots.$$

We shall show that

$$\mathcal{P}(z)=\frac{1}{z^2}+\sum_{\omega\neq 0}\left(\frac{1}{(z-\omega)^2}-\frac{1}{\omega^2}\right) \tag{18.7}$$

where the sum ranges over all $\omega = n_1\omega_1+n_2\omega_2$ except 0. Here $\frac{1}{(z-\omega)^2}$ is the singular part at ω, and that we have subtracted $\frac{1}{\omega^2}$ in order to produce convergence. First we verify that the series converges. If $|\omega|>2|z|$, then

$$\begin{aligned} \left|\tfrac{1}{(z-\omega)^2}-\tfrac{1}{\omega^2}\right| &= \left|\tfrac{z(2\omega-z)}{\omega^2(z-w)^2}\right| \\ &\leq \tfrac{|z|}{|\omega|^2}\left(\tfrac{2|\omega|+|z|}{(|\omega|-|z|)^2}\right) \\ &< \tfrac{|z|}{|\omega|^2}\left[\tfrac{2|\omega|+\frac{|\omega|}{2}}{\left(|\omega|-\frac{|\omega|}{2}\right)^2}\right] \\ &= \tfrac{|z|}{|\omega|^2}\tfrac{5|\omega|}{2.\frac{|\omega|^2}{4}} \\ &= \tfrac{10|z|}{|\omega|^3}. \end{aligned}$$

Therefore the series (18.7) converges, uniformly on every compact set if $\sum_{\omega\neq 0}\frac{1}{|\omega|^3}<\infty$. Since $\frac{\omega_2}{\omega_1}$ is nonreal there exists a $k>0$ such that

$$|n_1\omega_1+n_2\omega_2|\geq k(|n_1|+|n_2|) \tag{18.8}$$

for all real pairs (n_1,n_2). This is so because if both n_1,n_2 are real and positive, then the point $\dfrac{n_1}{n_1+n_2}\omega_1+\dfrac{n_2}{n_1+n_2}\omega_2$ will lie on the segment joining ω_1 and ω_2. Since $\frac{\omega_2}{\omega_1}$ is nonreal the points ω_1, ω_2 and origin does not lie on the same line. Further $\omega_1\neq 0$ and $\omega_2\neq 0$. Therefore there exists a $k_1>0$ such that $|\frac{n_1}{n_1+n_2}\omega_1+\frac{n_2}{n_1+n_2}\omega_2|\geq k_1$. Similarly if both n_1,n_2 are real and negative, then there exists a $k_2 > 0$ such that $|\frac{n_1}{n_1+n_2}\omega_1 + \frac{n_2}{n_1+n_2}\omega_2| \geq k_2$. If n_1 is real and negative and n_2 is real and positive then the point $\frac{-n_1}{|n_1|+|n_2|}(-\omega_1) + \frac{n_2}{|n_1|+|n_2|}\omega_2$ will lie on the segment joining $-\omega_1$ and ω_2. Since $-\frac{\omega_2}{\omega_1}$ is nonreal the points $-\omega_1$, ω_2 and origin does not lie on the same line. Therefore there exists a $k_3 > 0$ such that $|\frac{-n_1}{|n_1|+|n_2|}(-\omega_1) + \frac{n_2}{|n_1|+|n_2|}\omega_2| \geq k_3$. Similarly if n_2 is real and negative and n_1 is real and positive then there exists a $k_4 > 0$ such that $|\frac{n_1}{|n_1|+|n_2|}\omega_1 + \frac{-n_2}{|n_1|+|n_2|}(-\omega_2)| \geq k_4$. Let $k = \min\{k_1, k_2, k_3, k_4\}$. This constant k satisfies (18.8). If we consider only integers there are $4n$ pairs (n_1, n_2) with $|n_1| + |n_2| = n$. Thus $\sum_{\omega\neq 0} \frac{1}{|\omega|^3} \leq \sum_{n=1}^{\infty} \frac{4n}{k^3 n^3} = 4k^{-3} \sum_{n=1}^{\infty} \frac{1}{n^2} < \infty$. Thus the series (18.7) converges uniformly on every compact subsets of $\mathbb{C}$. We shall now show that the right hand side of (18.7) has periods ω_1 and ω_2. Since the series

$$\mathcal{P}(z) = \frac{1}{z^2} + \sum_{\omega\neq 0} \left(\frac{1}{(z-\omega)^2} - \frac{1}{\omega^2} \right)$$

defining $\mathcal{P}(z)$ is uniformly convergent on compact subsets of $\mathbb{C}$, we can differentiate term by term, and obtain $\mathcal{P}'(z) = \frac{-2}{z^3} - \sum_{\omega\neq 0} \frac{2}{(z-\omega)^3} = -2\sum_{\omega} \frac{1}{(z-\omega)^3}$. The function $\mathcal{P}'(z)$ is doubly periodic as $\sum_{\omega} \frac{1}{(z-\omega)^3}$ is doubly periodic. Hence it follows that $\mathcal{P}'(z+\omega_1) - \mathcal{P}'(z)$ and $\mathcal{P}'(z+\omega_2) - \mathcal{P}'(z)$ are identically zero. Therefore, $\mathcal{P}(z+\omega_1) - \mathcal{P}(z)$ and $\mathcal{P}(z+\omega_2) - \mathcal{P}(z)$ are constants. Let

$$\mathcal{P}(z+\omega_1) - \mathcal{P}(z) = c_1 \tag{18.9}$$

and

$$\mathcal{P}(z+\omega_2) - \mathcal{P}(z) = c_2 \tag{18.10}$$

for all $z \in \mathbb{C}$. From (18.7), it follows that the function $\mathcal{P}$ is even. Putting $z = -\frac{\omega_1}{2}$ in (18.9), we get

$$\begin{aligned} 0 &= \mathcal{P}\left(\tfrac{\omega_1}{2}\right) - \mathcal{P}\left(-\tfrac{\omega_1}{2}\right) \\ &= \mathcal{P}\left(-\tfrac{\omega_1}{2} + \omega_1\right) - \mathcal{P}\left(-\tfrac{\omega_1}{2}\right) \\ &= c_1. \end{aligned}$$

Putting $z = -\frac{\omega_2}{2}$ in (18.10), we obtain

$$\begin{aligned} 0 &= \mathcal{P}\left(\tfrac{\omega_2}{2}\right) - \mathcal{P}\left(-\tfrac{\omega_2}{2}\right) \\ &= \mathcal{P}\left(-\tfrac{\omega_2}{2} + \omega_2\right) - \mathcal{P}\left(-\tfrac{\omega_2}{2}\right) \\ &= c_2. \end{aligned}$$

Thus $\mathcal{P}$ is doubly periodic with periods ω_1 and ω_2. We thus proved the existence of $\mathcal{P}(z)$ and that it can be represented by the series (18.7). Further $\mathcal{P}'(z) = -2\sum_{\omega} \frac{1}{(z-\omega)^3}$. The function $\mathcal{P}$ is called the Weierstrass Pe function (or Weierstrass $\mathcal{P}$ function). Let Ω be the period module of $\mathcal{P}$. Then $\mathcal{P}$ is analytic in $\mathbb{C} \setminus \Omega$ and has poles of order 2 at each $\omega \in \Omega$.

18.5 The functions $\zeta(\mathbf{z})$ and $\sigma(\mathbf{z})$

The Weierstrass $\mathcal{P}$ function is an elliptic function and has only double pole at the origin and its congruent points. The function $\mathcal{P}(z)$ has zero residues and it is the derivative of a single-valued function. Denote the antiderivative of $\mathcal{P}(z)$by $-\zeta(z)$, and we shall normalize it so that it will be an odd function. Let

$$\zeta(z) = \frac{1}{z} + \sum_{\omega \neq 0} \left(\frac{1}{z-\omega} + \frac{1}{\omega} + \frac{z}{\omega^2} \right). \tag{18.11}$$

Let Ω be the period module of $\mathcal{P}$. Let $m = \min\{|\omega| : \omega \in \Omega \setminus \{0\}\}$, and let $D = \{z \in \mathbb{C} : |z| < m\}$, the largest open disk centered at 0 and containing no other points of Ω. Since $\frac{1}{z-\omega} + \frac{1}{\omega} + \frac{z}{\omega^2} = \frac{z^2}{\omega^2(z-\omega)}$, we see, by comparison with $\sum_{\omega\neq 0} \frac{1}{|\omega|^3}$, that $\sum_{\omega \neq 0} \left(\frac{1}{z-\omega} + \frac{1}{\omega} + \frac{z}{\omega^2}\right)$ is absolutely convergent for each $z \in \mathbb{C} \setminus \Omega$. (Notice that the series is obtained from (18.7) by integration from 0 to z along any path that does not pass through the poles. Thus the series (18.11) converges.)

Since $\zeta'(z) = -\mathcal{P}(z)$, we have $\zeta'(z + \omega_j) = \zeta'(z)$ for $j = 1, 2$, so that integration with respect to z gives $\zeta(z+\omega_j) = \zeta(z) + \eta_j$ $(j = 1, 2)$ where η_1, η_2 are constants, independent of z. Thus the function $\zeta(z)$ (known as Weierstrass zeta function) satisfies the conditions $\zeta(z + \omega_1) = \zeta(z) + \eta_1$ and $\zeta(z + \omega_2) = \zeta(z) + \eta_2$, where η_1 and η_2 are constants. By Residue theorem,

$$1 = \tfrac{1}{2\pi i} \int_{\partial \mathcal{P}_a} \zeta(z)dz$$

$$= \tfrac{1}{2\pi i} \left[\int_a^{a+\omega_1} \zeta(z)dz + \int_{a+\omega_1}^{a+\omega_1+\omega_2} \zeta(z)dz + \int_{a+\omega_1+\omega_2}^{a+\omega_2} \zeta(z)dz + \int_{a+\omega_2}^{a} \zeta(z)dz \right]$$

$$= \tfrac{1}{2\pi i} \left[\int_a^{a+\omega_1} \zeta(z)dz - \int_a^{a+\omega_2} \zeta(z)dz \right.$$

$$\left. + \int_a^{a+\omega_2} \zeta(z+\omega_1)dz - \int_a^{a+\omega_1} \zeta(z+\omega_2)dz \right]$$

$$= \tfrac{1}{2\pi i} \left[\int_a^{a+\omega_1} \zeta(z)dz - \int_a^{a+\omega_2} \zeta(z)dz \right]$$

$$+\frac{1}{2\pi i}\left[\int_a^{a+\omega_2}(\zeta(z)+\eta_1)dz-\int_a^{a+\omega_1}(\zeta(z)+\eta_2)dz\right]$$

$$=\frac{1}{2\pi i}\left[\int_a^{a+\omega_1}\zeta(z)dz-\int_a^{a+\omega_2}\zeta(z)dz\right]$$

$$+\frac{1}{2\pi i}\left[\int_a^{a+\omega_2}\zeta(z)dz-\int_a^{a+\omega_1}\zeta(z)dz+\eta_1\int_a^{a+\omega_2}dz-\eta_2\int_a^{a+\omega_1}dz\right].$$

Thus $\eta_1\omega_2-\eta_2\omega_1=2\pi i$ which is referred to as the Legendre's relation. It implies that at least one of η_1,η_2 is nonzero, so $\zeta(z)$ is not elliptic. It is not difficult to verify that the product

$$\sigma(z)=z\prod_{\omega\neq 0}\left(1-\frac{z}{\omega}\right)e^{\frac{z}{\omega}+\frac{1}{2}\left(\frac{z}{\omega}\right)^2}$$

converges and represents an entire function known as Weierstrass sigma function. Further $\frac{\sigma'(z)}{\sigma(z)}=\zeta(z)$.

We shall see how $\sigma(z)$ behaves under translation. Since $\zeta(z+\omega_1)=\zeta(z)+\eta_1$ and $\zeta(z+\omega_2)=\zeta(z)+\eta_2$,it follows that

$$\frac{\sigma'(z+\omega_1)}{\sigma(z+\omega_1)} = \frac{\sigma'(z)}{\sigma(z)}+\eta_1,$$

$$\frac{\sigma'(z+\omega_2)}{\sigma(z+\omega_2)} = \frac{\sigma'(z)}{\sigma(z)}+\eta_2.$$

Thus

$$\sigma(z+\omega_1)=c_1\sigma(z)e^{\eta_1 z}$$

and

$$\sigma(z+\omega_2)=c_2\sigma(z)e^{\eta_2 z}. \tag{18.12}$$

Note that $\sigma(z)$ is an odd function. On setting $z=-\frac{\omega_1}{2}$ and $z=-\frac{\omega_2}{2}$, the values of c_1 and c_2 can be determined. From (18.12), putting $z=-\frac{\omega_1}{2}$, we get

$$\sigma\left(\tfrac{\omega_1}{2}\right) = c_1\sigma\left(-\tfrac{\omega_1}{2}\right)e^{\eta_1\left(-\frac{\omega_1}{2}\right)}$$

$$= -c_1\sigma\left(\tfrac{\omega_1}{2}\right)e^{-\frac{\eta_1\omega_1}{2}}.$$

Thus $c_1=-e^{\frac{\eta_1\omega_1}{2}}$. Similarly $c_2=-e^{\frac{\eta_2\omega_2}{2}}$. Hence

$$\sigma(z+\omega_1)=-\sigma(z)e^{\eta_1\left(z+\frac{\omega_1}{2}\right)}$$

and

$$\sigma(z+\omega_2)=-\sigma(z)e^{\eta_2\left(z+\frac{\omega_2}{2}\right)}.$$

The functions $\zeta(z)$ and $\sigma(z)$ introduced here are not elliptic, for, as we have seen, they are not invariant under the translation $z\mapsto z+\omega_1$, $z\mapsto z+\omega_2$. However, an examination of their behavior under translations will enable us to construct elliptic functions.

18.6 Exercises

1. Prove that if an analytic function other than a constant, has distinct periods 2ω and $2\lambda\omega$, where λ is real, then λ is necessarily rational.

2. Show that any three periods $\omega_1, \omega_2, \omega_3$ of a multiply periodic analytic function are connected by a relation $l\omega_1 + m\omega_2 + n\omega_3 = 0$, where l, m and n are integers.

3. Show that if $f(z)$ is an elliptic function, so also is $f'(z)$.

4. Show that
$$\mathcal{P}(2z) = \frac{1}{4}\left\{\frac{\mathcal{P}''(z)}{\mathcal{P}'(z)}\right\}^2 - 2\mathcal{P}(z).$$
(This formula is called the duplication formula for the Weierstrass-$\mathcal{P}$ function $\mathcal{P}(z)$.)

5. Prove that $\mathcal{P}'(z) = -\frac{\sigma(2z)}{\sigma^4(z)}$.

6. Show that, if 3α is a period of $\mathcal{P}(z)$, then
$$\{\mathcal{P}(z) - \mathcal{P}(\alpha)\}\{\mathcal{P}(z+\alpha) - \mathcal{P}(\alpha)\}\{\mathcal{P}(z+2\alpha) - \mathcal{P}(\alpha)\} = -\mathcal{P}'^{\,2}(\alpha).$$

7. Show that a nonconstant rational function cannot be periodic.

Answers to Selected Exercises

Chapter 1

2. (i) For $n \in \mathbb{N}$,

$$i^n = \begin{cases} 1 & \text{if } n \equiv 0(\text{mod}4) \\ i & \text{if } n \equiv 1(\text{mod}4) \\ -1 & \text{if } n \equiv 2(\text{mod}4) \\ -i & \text{if } n \equiv 3(\text{mod}4). \end{cases}$$

   (ii) For $n \in \mathbb{Z}$,

$$\left(\frac{1-i}{1+i}\right)^n = \begin{cases} 1 & \text{if } n \equiv 0(\text{mod}4) \\ -i & \text{if } n \equiv 1(\text{mod}4) \\ -1 & \text{if } n \equiv 2(\text{mod}4) \\ i & \text{if } n \equiv 3(\text{mod}4). \end{cases}$$

   (iii) $(1+i)^n + (1-i)^n = 2^{\frac{n}{2}+1} \cos \frac{n\pi}{4}, n \in \mathbb{Z}$.

   (iv) $2+i$ (v)$-3-2i$

3. (a) (i) $\frac{2(1-|z|^2)}{1+|z|^2-2|z|\cos\theta}$ if $z = |z|(\cos\theta + i\sin\theta)$; (ii) $(|z| + \frac{1}{|z|})\cos\theta$ if $z = |z|(\cos\theta + i\sin\theta)$;

   (b) $x = 1, y = -2$; (c)(i)12; (ii) $\frac{6\sqrt{3}+4}{7}$; $(iii) - 7 + 3\sqrt{3} + i\sqrt{3}$.

4. (a) $\frac{1}{2}(1 \pm i\sqrt{3}), \frac{1}{2}(-1 \pm i\sqrt{3})$

5. $|z+1|^2 > |z-1|^2$ if and only if $(z+1)(\bar{z}+1) > (z-1)(\bar{z}-1)$. This is possible if and only if $2(\bar{z}+z) > 0$ or $\text{Re}z > 0$.

11. (b) (i) circle (ii) ellipse (iii) hyperbola (iv) circle (v) hyperbola

15. Suppose $P(w) = a_0 + a_1 w + \cdots a_n w^n, a_i \in \mathbb{R}$ for $i = 1, 2, \cdots, n$.
    If $P(z) = a_0 + a_1 z + \cdots + a_n z^n = 0$ then $0 = \bar{a}_0 + \bar{a}_1 \bar{z} + \cdots + \bar{a}_n \bar{z}^n = a_0 + a_1 \bar{z} + \cdots + a_n \bar{z}^n = P(\bar{z})$.

25. (a) (i) $\pm\frac{-1+i}{\sqrt{2}}$

    (ii) $2^{\frac{1}{6}}\{\cos(\frac{\frac{3\pi}{4}+2k\pi}{3}) + i\sin(\frac{\frac{3\pi}{4}+2k\pi}{3})\}, k = 0, 1, 2.$

    (iii) $4^{\frac{1}{4}}\{\cos(\frac{\frac{7\pi}{6}+2k\pi}{4}) + i\sin(\frac{\frac{7\pi}{6}+2k\pi}{4})\}, k = 0, 1, 2, 3.$

    (b) (i) $\cos 60° + i\sin 60°, \cos 180° + i\sin 180°, \cos 300° + i\sin 300°$

    (ii) $2(\cos 67.5° + i\sin 67.5°), 2(\cos 157.5° + i\sin 157.5°),$
    $2(\cos 247.5° + i\sin 247.5°), 2(\cos 337.5° + i\sin 337.5°)$

    (iii) $\sqrt{2}(\cos 27° + i\sin 27°), \sqrt{2}(\cos 99° + i\sin 99°),$

$\sqrt{2}(\cos 171^\circ + i\sin 171^\circ)$, $\sqrt{2}(\cos 243^\circ + i\sin 243^\circ)$,
$\sqrt{2}(\cos 315^\circ + i\sin 315^\circ)$

27. Notice that $w^n = 1$ and $w + w^2 + \cdots + w^n = 1 + w + w^2 + \cdots + w^{n-1} = \frac{1-w^n}{1-w} = 0$ since $w \neq 1$.

30. (a) Verify that $(w-1)(1 + 2w + 3w^2 + \cdots + nw^{n-1}) = n$ since $w^n = 1$.
    (b) We need to show that $(w-1)^2(1 + 4w + 9w^2 + \cdots + n^2w^{n-1}) = n^2(w-1) - 2n$. The left hand side is equal to $1 + 2w + 2w^2 + 2w^3 + \cdots + w^{n-1}[(n-2)^2 + n^2 - 2(n-1)^2] + w^n[(n-1)^2 - 2n^2] + n^2w^{n+1} = 1 + 2w(1 + w + w^2 + \cdots + w^{n-2}) + w^n(-n^2 + 1 - 2n) + n^2w = 1 + 2w(-w^{n-1}) + w^n(-n^2 + 1 - 2n) + n^2w = n^2(w-1) - 2n$ as $w^n = 1$.

37. $|z+1| = \sqrt{5}$

Chapter 2

1. (a) open (b)not open (c) not open (d) open (e) open (f) not open (points with Re $z = 0$ are not all interior points) (g) not open but closed (h) open (i) not open, not closed (j) not open but closed (k) not open, not closed (l) not open but closed (m)not open, not closed (n) not open but closed (p) not open, not closed.

2. $\mathbb{Z}$

5. By Theorem 2.12, $A = C \cap Y$ where C is closed in X. The space Y is closed in X by hypothesis and finite intersection of closed sets is closed. Hence A is closed in X.

6. Notice that $U - A = U \cap A^c$ and $A - U = A \cap U^c$.

7. Since $A \subset \overline{A}, B \subset \overline{B}$, hence $A \cup B \subset \overline{A} \cup \overline{B}$. Notice that $\overline{A} \cup \overline{B}$ is closed in X containing $A \cup B$ and $\overline{A \cup B}$ is the smallest closed set in X containing $A \cup B$. Hence $\overline{A \cup B} \subseteq \overline{A} \cup \overline{B}$.

   Now let $x \in \overline{A} \cup \overline{B}$. Then $x \in \overline{A}$ or $x \in \overline{B}$ and therefore every neighborhood U of x intersects A or B. Hence every neighborhood of x intersects $A \cup B$. This implies $x \in \overline{A \cup B}$. Therefore $\overline{A} \cup \overline{B} \subset \overline{A \cup B}$. Similarly show that $\bigcup \overline{A_\alpha} \subset \overline{\bigcup A_\alpha}$. Let $A_n = \{\frac{1}{n}\}, n \in \mathbb{N}$. The sets A_n are closed for all $n \in \mathbb{N}$ and therefore $\overline{\{\frac{1}{n}\}} = \{\frac{1}{n}\}$. Now $\overline{\bigcup A_n} = \overline{\{1, \frac{1}{2}, \frac{1}{3}, \cdots\}} = \{0, 1, \frac{1}{2}, \frac{1}{3}, \cdots\}$ and $\bigcup \overline{A_n} = \{1, \frac{1}{2}, \frac{1}{3}, \cdots\}$. Hence $\overline{\bigcup_{n=1}^{\infty} A_n} \neq \bigcup_{n=1}^{\infty} \overline{A_n}$.

8. $\bigcap_{n=1}^{\infty} A_n = \{z \in \mathbb{C} : |z| \leq 1\}$ which is closed in $\mathbb{C}$.

9. This follows from Theorem 2.15. Suppose A is a subset of $\mathbb{R}$ or $\mathbb{C}$ which is uncountable and discrete. Then by Theorem 2.15, A contains a limit point of itself. Hence A is not discrete. This is a contradiction.

12. Verify that $\rho(x,y) \le d(x,y) \le \sqrt{n}\rho(x,y)$ for all $x, y \in \mathbb{R}^n$.

15. Notice that $(X \times Y) - (A \times B) = [(X - A) \times Y] \cup [X \times (Y - B)]$.

16. Since $A \subset \overline{A}, B \subset \overline{B}$, hence $A \times B \subset \overline{A} \times \overline{B}$. By Exercise 15 of Chapter 2, $\overline{A} \times \overline{B}$ is a closed set in $X \times Y$ containing $A \times B$ and $\overline{A \times B}$ is the smallest closed set in $X \times Y$ containing $A \times B$. Hence $\overline{A \times B} \subset \overline{A} \times \overline{B}$. Now let $x \times y \in \overline{A} \times \overline{B}$. Then $x \in \overline{A}$ and $y \in \overline{B}$. We have to show $x \times y \in \overline{A \times B}$. Let W be a neighborhood of $x \times y$. Then $x \times y \in W$ and there exist sets U and V open in X and Y respectively such that $x \times y \in U \times V \subset W$. But $x \in \overline{A}$ implies $U \cap A \neq \emptyset$ and $y \in \overline{B}$ implies $V \cap B \neq \emptyset$. Hence $(U \times V) \bigcap (A \times B) = (U \cap A) \times (V \cap B) \neq \emptyset$. Thus $W \cap (A \times B) \neq \emptyset$.

17. The sets $A = (-\infty, \sqrt{2}) \cap Q$ and $B = (\sqrt{2}, \infty) \cap Q$ form a separation for Q.

21. (i) The open interval $(0,1)$ is not complete. Since $x_n = 1 - \frac{1}{n}, n \in \mathbb{N}$ is a Cauchy sequence in $(0,1)$ and it converges to $1 \notin (0,1)$.

    (ii) Q is not complete since $x_n = (1 + \frac{1}{n})^n, n \in \mathbb{N}$ is a Cauchy sequence in Q but it converges to $e \notin Q$.

    (iii) Consider $C[0,1]$, the space of all continuous complex valued functions on $[0,1]$. Define

$$d(f,g) = \left(\int_0^1 |f(x) - g(x)|^2 dx \right)^{\frac{1}{2}}.$$

Define

$$f_n(x) = \begin{cases} 0 & \text{if } 0 \le x \le \frac{1}{2} - \frac{1}{n}; \\ n(x - \frac{1}{2} + \frac{1}{n}) & \text{if } \frac{1}{2} - \frac{1}{n} \le x \le \frac{1}{2}; \\ 1 & \text{if } \frac{1}{2} \le x \le 1. \end{cases}$$

Now $f_n \in C[0,1]$ for all $n \in \mathbb{N}$ and (f_n) is a Cauchy sequence and

$$\int_0^1 |f_n(x) - f(x)|^2 dx \to 0$$

as $n \to \infty$ where

$$f(x) = \begin{cases} 0 & \text{if } 0 \le x < \frac{1}{2}; \\ 1 & \text{if } \frac{1}{2} \le x \le 1. \end{cases}$$

But the function f is not continuous on $[0,1]$.

30. (i) Let X be an infinite set. Define $d : X \times X \to \mathbb{R}$ as $d(x, y) = 1$ if $x \neq y$ and $d(x, y) = 0$ if $x = y$. It is not difficult to check that d is a metric on X. The metric d is called the discrete metric on X. With this metric verify that X is bounded. In fact if $a \in X$, is fixed, then $d(x, a) \leq 1$ for all $x \in X$. Further the set X is closed. But X is not compact. This can be verified as follows. Notice that if $x \in X$, then $B_d(x, 1) = \{x\}$. Hence $X \subset \bigcup_{x \in X} B_d(x, 1)$. Thus $\{B_d(x, 1)\}_{x \in X}$ is an open cover for X. But this open cover has no finite subcover as X is an infinite set.

    (ii) The set $[-\sqrt{2}, \sqrt{2}] \cap Q$ is a closed and bounded subset in Q, but not compact.

34. Let $h(x) = g(x) - f(x)$. Then h is continuous on X. Notice that $\{x : f(x) \leq g(x)\} = \{x \in X : h(x) \geq 0\} = h^{-1}([0, \infty))$. The set $[0, \infty)$ is closed in $\mathbb{R}$ as its complement $\mathbb{R} - [0, \infty) = (-\infty, 0)$ is open in $\mathbb{R}$. Since continuous function pull closed sets to closed sets (see Proposition 2.2) hence $h^{-1}([0, \infty)) = \{x : f(x) \leq g(x)\}$ is closed in X.

    (b) Use Pasting lemma (Theorem 2.54).

37. Suppose $n > 1$ and there exists a homeomorphism f between $\mathbb{R}^n$ and $\mathbb{R}$. Let $x \in \mathbb{R}^n$ and $f(x) = y \in \mathbb{R}$. Notice that $\mathbb{R}^n - \{x\}$ is connected and f is continuous on $\mathbb{R}^n - \{x\}$ and $f(\mathbb{R}^n - \{x\}) = \mathbb{R} - \{y\}$. Since continuous image of a connected set is connected hence it follows that $\mathbb{R} - \{y\}$ is connected. But $\mathbb{R} - \{y\}$ is not connected.

38. The function $f(x) = (\frac{x-a}{b-a})(d - c) + c$ is the required bijective map.

39. The function $g : (-1, 1) \to \mathbb{R}$ defined by $g(x) = \frac{x}{1-x^2}$ is the required homeomorphism.

40. See answer to Exercise 4, Chapter 10.

42. Define $\pi_1 : X \times Y \to X$ such that $\pi_1(x \times y) = x$ and $\pi_2 : X \times Y \to Y$ such that $\pi_2(x \times y) = y$. Show that π_1 and π_2 are continuous. Notice that $f_1 = \pi_1 \circ f$ and $f_2 = \pi_2 \circ f$. Hence f_1 and f_2 are continuous if f is continuous. Conversely, suppose that f_1 and f_2 are continuous. We show that for each basis element $U \times V$ for the topology of $X \times Y$, its inverse image $f^{-1}(U \times V)$ is open. A point $a \in f^{-1}(U \times V)$ if and only if $f(a) \in U \times V$, that is, if and only if $f_1(a) \in U$ and $f_2(a) \in V$. Therefore, $f^{-1}(U \times V) = f_1^{-1}(U) \cap f_2^{-1}(V)$. Since both of the sets $f_1^{-1}(U)$ and $f_2^{-1}(V)$ are open, so is their intersection.

Chapter 3

1. (i) converges to 0 (ii) diverges (iii) diverges (iv) converges to i (v) converges to 0 (vi) diverges (vii) diverges (viii) diverges (ix)converges to 0 (x) converges to 1 (xi) converges to $e^4 + 7i$ (xii) converges to $3i$ (xiii) converges to 0.

2. Since $||z| - |w|| \leq |z - w|$, the function $f(z) = |z|$ is a continuous function from $\mathbb{C}$ into $\mathbb{R}$. Use Theorem 2.49 to show that if $\lim_{n\to\infty} z_n = z$ then $\lim_{n\to\infty} |z_n| = |z|$. The converse is not true. For example, let $z_n = i^n$. Then $\lim_{n\to\infty} |z_n| = \lim_{n\to\infty} |i^n| = \lim_{n\to\infty} 1 = 1$. But the sequence (z_n) does not converge.

3. Suppose $\lim_{n\to\infty} z_n = 0$. Since modulus is a continuous function, hence $\lim_{n\to\infty} |z_n| = 0$. Suppose now $\lim_{n\to\infty} |z_n| = 0$. This implies given $\epsilon > 0$, there exists N such that $|z_n| < \epsilon$ for all $n \geq N$. Hence $\lim_{n\to\infty} z_n = 0$.

9. Use Theorem 2.49.

10. Let (X, d) be a metric space and (x_n) be a Cauchy sequence in X. Then given $\epsilon > 0$, there exists N such that $d(x_n, x_m) < \epsilon$ for all $n, m \geq N$. Let $\widetilde{M} = \sup\limits_{1\leq i,j\leq N} d(x_i, x_j)$ and $M = \max\{\widetilde{M}, \epsilon\}$. Then for any $n \in \mathbb{N}$, $d(x_n, x_1) \leq d(x_n, x_N) + d(x_N, x_1) \leq M + \widetilde{M} \leq 2M$. Thus $x_n \in B(x_1, 2M)$ for all $n \in \mathbb{N}$. Hence the sequence (x_n) is bounded.

17. (i) $R = \infty$ (ii) $R = 0$ (iii) $\mathbb{R}^k$ (iv) R if $|z_0| \leq 1$, $\frac{R}{|z_0|}$ if $|z_0| > 1$.

20. From the geometric series, we notice that the region of convergence of $\sum_{n=0}^{\infty} \left(\frac{z}{1+z}\right)^n$ is given by $|\frac{z}{1+z}| < 1$, that is, if Re $z > -\frac{1}{2}$. The region of uniform convergence of the series is obtained from $|z| \leq r$ for $0 < r < 1$, which is given by $|z - \frac{r^2}{z-r^2}| < \frac{r}{1-r^2}$.

    For $\sum_{n=0}^{\infty} \frac{z^n}{1+z^{2n}}$, use Theorem 3.7.

21. (i) $R = \infty$ if $p < -1; R = 0$ if $p \geq -1$.

    (ii) $R = 2$ (iii) $R = 1$ (iv) $R = 0$ (v) $R = 1$ (vi) $R = 1$ (vii) $R = 1$ (viii) $R = 1$ (ix) $R = 1$ (x) $R = e$ (xi) $R = 1$ (xii) $R = 1$ (xiii) $R = 3$ (xiv) $R = 1$ (xv) $R = \frac{1}{4}$ (xvi) $R = 0$ (xvii)$R = \frac{1}{4}$ (xviii) $R = \frac{1}{e}$ (xix)$R = 1$ if $|a| \leq 1$; $\frac{1}{|a|}$ if $|a| > 1$. (xx) $R = 0$ (xxi) $R = 1$ (xxii) $R = \infty$.

23. (i) $\frac{z}{(1-z)^2}$ (ii) $-\ln(1 - z)$ (iii) $\frac{1}{2} \ln \frac{1+z}{1-z}$ (iv) $\ln(1 + z)$.

24. $|e^z| = e^x < 1$ if and only if $x < 1$. Therefore, the set is open.

25. (a) $g(z) = \sum_{n=0}^{\infty} a_n z^n + \sum_{n=0}^{\infty} a_n(-z)^n = \sum_{n=0}^{\infty} a_n(1+(-1)^n)z^n$. Therefore

$$g^{(k)}(0) = \begin{cases} 2(k!)a_k & \text{if } k \text{ is even;} \\ 0 & \text{if } k \text{ is odd.} \end{cases}$$

(b) Set $g(z) = f(z) + f(-z) = 0$ since f is odd. So $g^{(k)}(0) = 0$ for all $k \geq 0$. By (a), $a_k = 0$ for all even k.

29. None

Chapter 4

2. $-\sin a$

3. $|\sin \pi z|^2 = |\sin(\pi x + i\pi y)|^2$ and $|\cos z|^2 = |\cos(x+iy)|^2$. Use the formulas (4) and (5) of Theorem 4.2.

4. $\sin i = \frac{(e^2-1)i}{2e}$; $\cos i = \frac{1+e^2}{2e}$; $\tan(1+i) = (-i)\frac{e^{i-1}-e^{-i+1}}{e^{i-1}+e^{-i+1}}$.

5. $e^{i(\ln 2 + 2k\pi)} = e^{2k\pi i}(\cos \ln 2 + i \sin \ln 2), k \in \mathbb{Z}$.

6. $z = i(2k + \frac{1}{2})\pi, k \in \mathbb{Z}; z = \ln \sqrt{5} + i(tan^{-1}2 + 2k\pi), k \in \mathbb{Z}$.

7. $1, \omega, \omega^2$ where $\omega = \frac{-1+\sqrt{3}i}{2}$.

9. $z^{1-i} = 4$ if and only if $e^{(1-i)\log z} = 4 = e^{\ln 4 + 2k\pi i}, k = 0, \pm 1, \pm 2, \cdots$. Thus $(1-i)\log z = 2\ln 2 + 2k\pi i$. Hence $\log z = \frac{2}{1-i}[\ln 2 + k\pi i] = (1+i)[\ln 2 + k\pi i] = [\ln 2 - k\pi] + i[\ln 2 + k\pi] = c$ say . Therefore, $z = e^{c+2m\pi i}, m = 0, \pm 1, \pm 2, \cdots$. Hence $z = 2e^{-k\pi}e^{i[\ln 2 + (k+2m)\pi]}, k, m \in \mathbb{Z}$.

10. Write the series of $\sin z$ and $\sinh z$. Notice that the series are uniformly convergent on any bounded subset of $\mathbb{C}$ and find the limit. One can also use the fact that $\lim_{z \to 0} \frac{f(z)-f(0)}{z} = f'(0)$ as in Chapter 5.

11. The kth term does not tend to 0 as $k \to \infty$.

13. See the proof of Theorem 18.3 and Exercise 28 of Chapter 7.

14. $\cos^2 z = \frac{1}{2}(\cos 2z + 1) = 1 - z^2 + \frac{z^4}{3} - \frac{2}{45}z^6 + \cdots$.

17. $z = \frac{1}{2}\log(a^2 + b^2) + iy$ where $\sin y = \frac{b}{\sqrt{a^2+b^2}}$ and $\cos y = \frac{a}{\sqrt{a^2+b^2}}$.

21. $z = \log(2 \pm \sqrt{3}) + (2n+1)\pi i, n \in \mathbb{Z}$ where $\log x$ is the real logarithm.

Chapter 5

2. $f'(0) = 0$

5. (a) The functions in (i), (ii), (iii) are differentiable nowhere.
   (b) (i) $\bar{z} = x - iy$ so $u_x = 1$ and $v_y = -1$. Hence the Cauchy-Riemann equations never hold, and $\bar{z}$ is holomorphic nowhere.
   (ii) On the given domain, $\arg z = \tan^{-1}\frac{y}{x}$. So $u = \tan^{-1}\frac{y}{x}$ and $v = 0$. Further, $u_y = +\frac{x}{(x^2+y^2)}$ and $v_x = 0$, so the Cauchy-Riemann equations hold nowhere on $\{\text{Re } z > 0\}$.

12. By the Cauchy-Riemann equation $u_x = v_y$ and $u_y = -v_x$. Thus, $au+bv = c$ implies $au_x + bv_x = 0 = au_y + bv_y$ and, therefore, $au_x - bu_y = 0 = au_y + bu_x$. Thus $\begin{pmatrix} a & -b \\ b & a \end{pmatrix}\begin{pmatrix} u_x \\ u_y \end{pmatrix} = \begin{pmatrix} 0 \\ 0 \end{pmatrix}$.

    Since the matrix $\begin{pmatrix} a & -b \\ b & a \end{pmatrix}$ has nonzero determinant $a^2 + b^2$, the homogeneous system has only the zero solution. Hence $u_x = u_y = 0$. By the Cauchy-Riemann equations, $v_x = v_y = 0$. Since G is connected, f is constant.

13. Suppose f is a holomorphic function with constant absolute value. By Theorem 5.8, f is constant. This contradicts the hypothesis that f is constant. Hence no such function exist.

14. z^3; No, there cannot be a holomorphic function whose real part is x^2.

16. $\arg f$ is constant, hence f is constant.

17. Suppose $f(z) = u(z) + iv(z)$ is holomorphic in $\mathbb{C}$. Then u and v have continuous first order partial derivatives and they satisfy the Cauchy-Riemann equations. That is, $u_x = v_y$ and $u_y = -v_x$. Suppose $g(z) = \overline{f(\bar{z})}$ $= u(\bar{z}) - iv(\bar{z}) = u(x,-y) - iv(x,-y) = u(x,\widetilde{y}) + iw(x,\widetilde{y})$. Since f is holomorphic, $u(x,\widetilde{y}) = u(x,-y)$ and $w(x,\widetilde{y}) = -v(x,-y)$ have continuous first order partial derivatives and $u_x = v_y = -v_{-y}$ and $-v_x = u_y = -u_{-y}$. That is, $u_x = w_{\widetilde{y}}$ and $u_{\widetilde{y}} = -w_x$. Thus g is holomorphic.

18. Suppose f and $\bar{f}$ are holomorphic in U. Then $\frac{\partial f}{\partial \bar{z}} = 0 = \frac{\partial \bar{f}}{\partial \bar{z}}$. Let $f = u+iv$. Then

$$0 = \frac{\partial f}{\partial \bar{z}} = \frac{\partial u}{\partial x}\frac{\partial x}{\partial \bar{z}} + \frac{\partial u}{\partial y}\frac{\partial y}{\partial \bar{z}} + i\left[\frac{\partial v}{\partial x}\frac{\partial x}{\partial \bar{z}} + \frac{\partial v}{\partial y}\frac{\partial y}{\partial \bar{z}}\right].$$

Further,

$$0 = \frac{\partial \bar{f}}{\partial \bar{z}} = \frac{\partial u}{\partial x}\frac{\partial x}{\partial \bar{z}} + \frac{\partial u}{\partial y}\frac{\partial y}{\partial \bar{z}} - i\left[\frac{\partial v}{\partial x}\frac{\partial x}{\partial \bar{z}} + \frac{\partial v}{\partial y}\frac{\partial y}{\partial \bar{z}}\right].$$

Hence

$$\frac{\partial u}{\partial x}\frac{1}{2} + \frac{\partial u}{\partial y}\left(-\frac{1}{2i}\right) + i\left[\frac{\partial v}{\partial x}\frac{1}{2} + \frac{\partial v}{\partial y}\left(-\frac{1}{2i}\right)\right] = 0 \tag{1}$$

and

$$\frac{\partial u}{\partial x}\frac{1}{2} + \frac{\partial u}{\partial y}\left(-\frac{1}{2i}\right) - i\left[\frac{\partial v}{\partial x}\frac{1}{2} + \frac{\partial v}{\partial y}\left(-\frac{1}{2i}\right)\right] = 0. \tag{2}$$

Thus $2\left[\frac{\partial u}{\partial x}\frac{1}{2} + \frac{\partial u}{\partial y}\left(-\frac{1}{2i}\right)\right] = 0$. Hence $\frac{\partial u}{\partial x} = 0$ and $\frac{\partial u}{\partial y} = 0$. That is, u is constant. Similarly, subtracting (2) from (1), one can show that v is constant. Therefore, f is constant.

19. In this case either $\arg f$ is a constant or $|f(z)-b| = c$ for some constants b and c. If $\arg f$ is constant then by Theorem 5.8, f is constant. Otherwise, let $g(z) = f(z) - b$. Then g is holomorphic and $|g(z)| = c$. Therefore by Theorem 5.8, g is constant and hence f is constant.

20. Proceed as in Example 5.20.

21. Notice that $x = \frac{z+\bar{z}}{2}$ and $y = \frac{z-\bar{z}}{2i}$. Let $u = u(x, y)$. Then

$$\frac{\partial u}{\partial z} = \frac{\partial u}{\partial x}\frac{\partial x}{\partial z} + \frac{\partial u}{\partial y}\frac{\partial y}{\partial z} = \frac{\partial u}{\partial x}\frac{1}{2} + \frac{\partial u}{\partial y}\frac{1}{2i}.$$

Hence

$$\frac{\partial}{\partial \bar{z}}\left(\frac{\partial u}{\partial z}\right) = \frac{1}{2}\left[\frac{\partial}{\partial x}\left(\frac{\partial u}{\partial x}\right)\cdot\frac{\partial x}{\partial z} + \frac{\partial}{\partial y}\left(\frac{\partial u}{\partial x}\right)\cdot\frac{\partial y}{\partial z}\right]$$

$$+\frac{1}{2i}\left[\frac{\partial}{\partial x}\left(\frac{\partial u}{\partial y}\right)\cdot\frac{\partial x}{\partial \bar{z}} + \frac{\partial}{\partial y}\left(\frac{\partial u}{\partial y}\right)\cdot\frac{\partial y}{\partial \bar{z}}\right] = \frac{1}{4}\left(\frac{\partial^2 u}{\partial x^2} + \frac{\partial^2 u}{\partial y^2}\right)$$

since second order mixed partial derivatives are equal.

22. $z^3 + 1$

Chapter 6

3. (i) 0 (ii)$-\frac{i}{\pi}\sinh\pi$ (iii)$1-i$

4. (i) $\frac{20(1+i)}{3}$ (ii) πi (iii) $\frac{-(1+i)}{2}$ (iv) $-\pi i$.

6. (i) $2\pi i$ (ii) $1-i$ (iii) $1-i$

7. $|\int_\gamma \frac{e^z}{z-1}dz| \leq L(\gamma)\sup_{z\in\gamma}|\frac{e^z}{z-1}|$. Now $L(\gamma) = 3\pi$, since γ is a semicircle of radius 3, and $|\frac{e^z}{z-1}| \leq \frac{e^3}{2}$, since $|e^z| = e^x$ and $|z-1| \geq 2$ on γ.

12. (i)$4\pi i$ (ii) 2π (iii) $\sin a$ (iv) $\frac{2\pi}{\sqrt{3}}$ (v)$-\frac{1}{2}(1+i)$.

13. (i) 8 (ii) $\frac{2\pi\rho}{\rho^2-|a|^2}$ (iii) $\pi r^2 i$ (iv) 0 (v) $\frac{2\pi r}{r^4-1}$.

Chapter 7

1. By Cauchy's integral formula, $\frac{1}{2\pi i}\int_{|z|=r}(f(z))^2\frac{dz}{z} = (f(0))^2$ for $r > 1$. That is, $\frac{1}{2\pi i}\int_0^{2\pi}(f(re^{i\theta}))^2 re^{i\theta}\frac{id\theta}{re^{i\theta}} = (f(0))^2$. Simplifying and taking real parts we obtain $\int_0^{2\pi}[(u(re^{i\theta}))^2 - (v(re^{i\theta}))^2]d\theta = 2\pi i(u(0)^2 - v(0)^2) = 0$.

2. The argument is same as in Example 7.20.

3. Suppose $f(z) = c$ (a constant) if $z \in \{x + iy : 0 \le x \le 1, y = 0\} = A$. Define for $z \in \mathbb{C}$, $g(z) = f(z) - c$. Then g is an entire function and $g(z) = 0$ if $z \in A$. Thus g has uncountable number of zeros. Then by Theorem 2.15 and Theorem 7.16, $g \equiv 0$ on $\mathbb{C}$. That is, $f \equiv c$ on $\mathbb{C}$.

4. If there exists such a function f then $f(z) = \frac{z^2}{z^2+1}$ on $\mathbb{C} \setminus \{i, -i\}$. Since f and the function $\frac{z^2}{z^2+1}$ agree on a uncountable set therefore $f(z) = \frac{z^2}{z^2+1}$ on $\mathbb{C}$. But $\frac{z^2}{z^2+1}$ is not an entire function. Thus there exists no entire function f such that $f(\frac{1}{n}) = \frac{1}{1+n^2}$ for all $n = 1, 2, 3, \cdots$.

5. It does not contradict the uniqueness theorem since the point $z = 1$ does not belong to the domain of analyticity of the function.

6. Let $h(z) = f(z) - kg(z)$. Then h is entire and $\text{Re}h(z) \le 0$. We then have $|e^{h(z)}| \le 1$ for all $z \in \mathbb{C}$. Therefore, by Liouville's theorem, e^h is constant, and so is h.

7. The functions f and g are analytic on a region G and $\bar{f}g$ is analytic. Suppose $f = u + iv$ and $g = \widetilde{u} + i\widetilde{v}$. Then $u_x = v_y$, $u_y = -v_x$ and $\widetilde{u}_x = \widetilde{v}_y$, $\widetilde{u}_y = -\widetilde{v}_x$. Further $\bar{f} = u - iv$ and $\bar{f}g = (u - iv)(\widetilde{u} + i\widetilde{v}) = (u\widetilde{u} + v\widetilde{v}) + i(u\widetilde{v} - v\widetilde{u})$.

   Since $\bar{f}g$ is analytic, $(u_x\widetilde{u}+\widetilde{u}_x u)+(v_x\widetilde{v}+v\widetilde{v}_x) = (u_y\widetilde{v}+u\widetilde{v}_y)-(v_y\widetilde{u}+v\widetilde{u}_y)$ and $(u_y\widetilde{u} + u\widetilde{u}_y) + (v_y\widetilde{v} + v\widetilde{v}_y) = -[(u_x\widetilde{v} + u\widetilde{v}_x) - (v_x\widetilde{u} + v\widetilde{u}_x)]$.

   Thus $2u_x\widetilde{u} = 2u_y\widetilde{v}$ and $2u_y\widetilde{u} = -2u_x\widetilde{v}$.

   That is, $u_x\widetilde{u} = u_y\widetilde{v}$ and $u_y\widetilde{u} = -u_x\widetilde{v}$. Hence $-u_x^2\widetilde{u}\widetilde{v} = u_y^2\widetilde{u}\widetilde{v}$. This implies $(u_x^2 + u_y^2)\widetilde{u}\widetilde{v} = 0$. If either $\widetilde{u}$ or $\widetilde{v}$ is 0 then $g \equiv 0$. If $u_x^2 + u_y^2 = 0$, then $|f'(z)|^2 = 0$. Hence f is a constant function.

8. (i) $\frac{81}{20}i\pi e^{-3}$ (ii) 0 (iii) $\frac{i\pi}{3}$ (iv) 0 (v) $2\pi i(-i + e^{-1})$, use Cauchy integral formula

11. Let $f(z) = \sum_{n=0}^{\infty} a_n(z - z_0)^n, z \in B(z_0, R)$. Let $s_n(z) = \sum_{k=0}^{n} a_k(z - z_0)^k$. Observe that $\int_0^{2\pi} |s_n(z_0+re^{it})|^2 dt = 2\pi \sum_{k=0}^{n} |a_k|^2 r^{2k}$. Recall also the fact that if f_n and g_n are complex-valued functions and converge uniformly to f and g on a region $G \subset \mathbb{C}$ and if f and g are bounded on X, then $f_n g_n$ converges uniformly on G to fg. Using this, we see that $\int_0^{2\pi} s_n(z_0 + re^{it})\overline{s_n(z_0 + re^{it})}dt$ converges to $\int_0^{2\pi} f(z_0 + re^{it})\overline{f(z_0 + re^{it})}dt$.

    But $\lim_{n\to\infty} 2\pi \sum_{k=0}^{n} |a_k|^2 r^{2k} = 2\pi \sum_{k=0}^{\infty} |a_k|^2 r^{2k}$. In particular, this infinite series is convergent and its sum is $\int_0^{2\pi} f(z_0 + re^{it})\overline{f(z_0 + re^{it})}dt$.

15. See Exercise 25 of Chapter 3.

16. $\max_{|z|\le 1} |f(z)| = \max_{|z|\le 1} \frac{|z|^2}{|z+2|} \le \max_{|z|\le 1} \frac{|z|^2}{2-|z|} = 1$ as $|z + 2| = |2 - (-z)| \ge 2 - |z|$ and $f(-1) = 1$. Thus $\max_{|z|\le 1} |f(z)| = 1$. The maximum is attained on the unit circle.

17. Here $f(z) = e^{z^2}$ and $\max_{|z|\leq 1} |f(z)| = \max_{|z|=1} |e^{z^2}| = e^{\max_{|z|=1} |z|^2} = e$ and $\min_{|z|\leq 1} |f(z)| = \min_{|z|=1} |e^{z^2}| = e^{i^2} = e^{-1} = \frac{1}{e}$.

18. Let $g(z) = f(z)f(-z)f(iz)f(-iz)$. The function g is holomorphic in U, continuous on $\overline{U}$ and $\partial U = \{z = x + iy : x = 1 \text{ and } -1 \leq y \leq 1\} \bigcup \{z = x + iy : -1 \leq x \leq 1 \text{ and } y = 1\} \bigcup \{z = x + iy : x = -1 \text{ and } -1 \leq y \leq 1\} \bigcup \{z = x + iy : -1 \leq x \leq 1 \text{ and } y = -1\} = A \bigcup B \bigcup C \bigcup D$ (let). Then $f(z) = 0$ if $z \in A$; $f(-z) = 0$ if $z \in C$; $f(iz) = 0$ if $z \in D$; $f(-iz) = 0$ if $z \in B$. Thus $g(z) = 0$ if $z \in \partial U$ (the boundary of U). By maximum modulus principle, $g \equiv 0$ on U. By Example 7.24 and since $f \in C(\overline{U})$, $f \equiv 0$ on $\overline{U}$.

19. Let $g(z) = f(z)f(-z)$. Then g is holomorphic in $B(0,1)$ and continuous in $\overline{B(0,1)}$. Notice that $g(z) = 0$ if $|z| = 1$. This is so as $f(z) = 0$ if $|z| = 1$ and $\text{Im}z \geq 0$; $f(-z) = 0$ if $\text{Im}z \leq 0$ and $|z| = 1$. By Corollary 7.12, either g is constant or $|g|$ attains its maximum on the boundary of $\overline{B(0,1)}$ which is the unit circle. Thus $|g(z)| = 0$ on $\overline{B(0,1)}$. That is, $g \equiv 0$ on $\overline{B(0,1)}$. By Example 7.24, $f \equiv 0$ on $\overline{B(0,1)}$.

21. Given that $f \in H(U) \cap C(\overline{U})$ and $\overline{U}$ is closed and bounded (hence compact). Therefore $|f|$ attains its minimum and maximum on $\overline{U}$. Suppose $\min_{z\in\overline{U}} |f(z)| = |f(a)|$ for some $a \in \partial U$. On the other hand by Corollary 7.12, since f is nonconstant, $|f|$ attains its maximum on the boundary of U, that is, on ∂U. Hence $|f(z)| = 1$ for all $z \in U$. But this implies, f is a constant function (by Theorem 5.8). This contradicts our hypothesis that f is nonconstant. Therefore, $\min_{z\in\overline{U}} |f(z)| = |f(a)|$ for some $a \in U$. Then by Theorem 7.19 (minimum modulus principle), $f(a) = 0$.

22. Proceed as in Exercise 21 of Chapter 7.

23. Let $h = f - g$. Then $h = 0$ on ∂U and $h \in H(U) \cap C(\overline{U})$. From Corollary 7.13 and Exercise 29 of Chapter 7, $h \equiv 0$ on $\overline{U}$. Hence $f = g$ on U.

24. The result follows from Theorem 2.35 and Theorem 7.16.

27. (a) Let $e^z = \sum_{n=0}^{\infty} b_n z^n$. By the product rule for power series, $e^z f(z) = \sum_{n=0}^{\infty} c_n z^n$ with $c_n = a_0 b_n + a_1 b_{n-1} + \cdots + a_n b_0 = \sum_{k=0}^{n} a_k b_{n-k}$. But $b_{n-k} = \frac{1}{(n-k)!}$, hence the result. Notice that both power series have infinite radius of convergence, so multiplication is valid everywhere.

    (b) $f'(z) = (e^{e^z})' = e^z e^{e^z} = e^z f(z)$. We proved earlier that we can differentiate power series term-by-term, so that $f'(z) = \sum_{n=0}^{\infty} n a_n z^{n-1} = \sum_{n=0}^{\infty} c_n z^n$ in (a) above. Equate coefficients of z^{n-1} (using the uniqueness property): $n a_n = \sum_{k=0}^{n-1} \frac{a_k}{(n-1-k)!}$ as required.

28. Let $\gamma : [a,b] \to \mathbb{C}$ be the contour which goes once counterclockwise round the square with corners $0, 1, 1+i, i$, and put

$$M = \sup\{|f(\gamma(t))| : t \in [a,b]\}.$$

Now $f \circ \gamma$ is a continuous function on a closed bounded interval $[a, b]$ and is therefore bounded, i.e., $|f(\gamma(t))| \leq M$ for $a \leq t \leq b$. By maximum modulus principle, $|f(z)| \leq M$ for all z inside γ. By hypothesis $f(z + m + ni) = f(z)$ for all $z \in \mathbb{C}, m, n \in \mathbb{Z}$. Now, let $z \in \mathbb{C}$; then we can choose $m, n \in \mathbb{Z}$ such that $z - m - ni$ lies inside the square with vertices $0, 1, i, 1 + i$. Then $|f(z)| = |f(z - m - ni)|$ and so $|f(z)| \leq M$. Thus f is a bounded entire function. Hence by Liouville's theorem, f is a constant function.

30. (i) $2\pi i$ (ii) $-4\pi i$ (iii) $-i\pi$

31. (i) No, since $|z|^2$ is not analytic;

    (ii) No, since $\frac{z^2}{z-2}$ is not analytic at $z = 2$, inside γ_1;

    (iii) No, since γ_2 is not a closed contour;

    (iv) Yes, since $\gamma_2 + \gamma_3$ is closed and $\sin(\frac{1}{z-1})$ is analytic everywhere inside it. The point $z = 1$ is not inside it.

32. (i) $2\pi i$ (ii) 0 (iii) $\frac{i\pi}{2}$

Chapter 8

5. (i) removable singularity at $z = \frac{\pi}{2}$

   (ii) pole of order 1

   (iii) pole of order 1

   (iv) $\frac{1-\cos z}{z} = \frac{z}{2!} - \frac{z^3}{4!} + \frac{z^5}{6!} - \cdots$, removable singularity

   (v) removable singularity at $z = 0$

   (vi) removable singularity at $z = 0$

   (vii) pole of order 2

   (viii) $\frac{e^{z^2}-1}{z^3} = \frac{1}{z} + \frac{z}{2!} + \frac{z^3}{3!} + \frac{z^5}{4!} + \frac{z^7}{5!} + \cdots$, pole of order 1 at $z = 0$

   (ix) pole of order 2

   (x) $\frac{e^z-1}{z^2} = \frac{1}{z^2}(1 + z + \frac{z^2}{2!} + \frac{z^3}{3!} + \cdots - 1) = \frac{1}{z} + \frac{1}{2!} + \frac{z}{3!} + \cdots$, pole of order 1 at $z = 0$.

6. (i) removable singularity at $z = 0$

   (ii) removable singularity at $z = 0$

   (iii) a pole at $z = 0$ of order 2

   (iv) $z = (2k + 1)\frac{\pi}{2}, k = 0, \pm 1, \pm 2, \cdots$ are poles of order 1; ($z = \infty$ is a limit point of poles, hence an essential singularity; see Chapter 13)

   (v) pole of order 1 at $z = 0$ (and $z = \infty$ is an essential singularity; see Chapter 13)

(vi) pole of order 1 at $z = 0$ (and $z = \infty$ is an essential singularity; see Chapter 13)

(vii) pole of order 1 at $z = 0$

(viii) pole of order 1 at $z = \infty$, see Chapter 13

(ix) $z = 0$ is an essential singularity

(x) simple poles at $z = 0$ and $z = 1$

(xi) $z = 0$ is an essential singularity; pole of order $n - 1$ at $z = \infty$ (see Chapter 13).

8. (i) $\frac{1}{z} + 2 + z$ for all $z \neq 0$

(ii) $\frac{1}{z(1-z)} = \frac{1}{z^2(\frac{1}{z}-1)} = -\frac{1}{z^2(1-\frac{1}{z}} = -\frac{1}{z^2}\sum_{n=0}^{\infty}(\frac{1}{z})^n$ in $|z| > 1$.

(iii) $\frac{1}{a-b}\sum_{n=0}^{\infty}(\frac{z^n}{b^{n+1}} + \frac{a^n}{z^{n+1}})$ for $|a| < |z| < |b|$.

(iv) $\frac{1}{z+1} + \frac{1}{z-3} = \frac{1}{z(1+\frac{1}{z})} + \frac{1}{3(\frac{z}{3}-1)} = \frac{1}{z}\sum_{n=0}^{\infty}\frac{(-1)^n}{z^n} - \frac{1}{3}\sum_{n=0}^{\infty}(\frac{z}{3})^n$

(v) $\frac{1}{(z-1)(z-2)} = \frac{-1}{2(1-z)(\frac{z}{2}-1)} = \frac{1}{2}(\sum_{n=0}^{\infty} z^n)(\sum_{n=0}^{\infty}(\frac{z}{2})^n)$ in $0 < |z| < 1$; expand the series and find the coefficients of $z^n, n \in \mathbb{Z}$

(vi) $\sum_{n=0}^{\infty}\frac{1}{n!}(\frac{1}{z})^n$

(vii) $2\sum_{n=1}^{\infty}\frac{(-1)^n}{z^{2n}} - \sum_{n=0}^{\infty}\frac{z^n}{2^{n+1}}$.

(viii) $e^{z+\frac{1}{z}} = \sum_{n=0}^{\infty} c_n z^n + \sum_{n=1}^{\infty} c_{-n} z^{-n}$ where $c_n = c_{-n} = \sum_{k=0}^{\infty}\frac{1}{k!(n+k)!}, n = 0, 1, 2, \cdots$ in the domain $0 < |z| < \infty$

(ix) $\cdots - \frac{1}{2z^4} + \frac{1}{2z^3} - \frac{1}{2z^2} + \frac{1}{2z} - \frac{1}{6} + \frac{z}{18} - \frac{z^2}{54} + \frac{z^3}{162} - \cdots$

(x) $\frac{1}{2(z+1)} - \frac{1}{4} + \frac{1}{8}(z+1) - \frac{1}{16}(z+1)^2 + \cdots$ in $0 < |z+1| < 2$

12. (i) simple poles at all integers n and no other singularities

(ii) $z = 1$ is a pole of order 4; $z = -1$ is a pole of order 6; $\omega, \widetilde{\omega}$ are poles of order 2 each where $\omega = \frac{1+\sqrt{3}i}{2}$ and $\widetilde{\omega} = \frac{1-\sqrt{3}i}{2}$.

(iii) $z = i, -i$ are poles of order 1

(iv) essential singularity at $z = 0$

(v) simple poles at $z = n, n \in \mathbb{Z}, n \neq 0$ (if $a \neq 0$); simple poles at $z = n, n \in \mathbb{Z}$ (if $a = 0$)

(vi) simple poles at $z = 2\pi i k, k \in \mathbb{Z}$

(vii) simple poles at $z = -(k + \frac{1}{4})i\pi, k \in \mathbb{Z}$

(viii) At $z = 0$, the function has an essential singularity since $z^{-1}\cosh z^{-1} = \frac{1}{z} - \frac{1}{2!z^3} + \frac{1}{4!z^5} - \cdots$

(ix) $e^{\frac{1}{z}} = 1 + \frac{1}{z} + \frac{1}{2!z^2} + \frac{1}{3!z^3} + \cdots$; the point $z = 0$ is an essential singularity.

(x) simple poles at $e^{(\frac{2k+1}{4})\pi i}, k = 0, 1, 2, 3$.

(xi) $z = -1$ is a pole of order 1 and $i, -i$ are poles of order 3 each.

(xii) simple poles at $e^{(\frac{2k+1}{4})\pi i}, k = 0, 1, 2, 3.$

(xiii) essential singularity at $z = 0$

(xiv) $z = -1$ is a pole of order 1

(xv) simple poles at $e^{(\frac{2k+1}{3})\pi i}, k = 0, 1, 2.$

(xvi) poles at $z = k\pi i, k \in \mathbb{Z}$; ($z = \infty$ is the limit point of poles)

(xvii) If $f(z) = \frac{\sin z}{z}$, then $z = 0$ is a removable singularity, since $f(0)$ is not defined, but $\lim_{z\to 0} \frac{\sin z}{z} = 1$. We define $f(0) = 1$. Notice that $\frac{\sin z}{z} = 1 - \frac{z^2}{3!} + \frac{z^4}{5!} - \frac{z^6}{7!} + \cdots$

(xviii) $z = 0$ is an essential singularity

(xix) poles at $\pi n, n \in \mathbb{Z}$ (∞ is an essential singularity, see Chapter 13)

(xx) simple poles at $z = e^{2k\pi i}, k \in \mathbb{Z}$

(xxi) $z = 3$ is an essential singularity

(xxii) simple poles at $e^{\frac{2\pi k i}{5}}, k = 1, 2, 3, 4$

(xxiii) $z = k\pi, k \in \mathbb{Z}$ are each poles of order 2, $z = \infty$ is a limit point of poles, hence an essential singularity, see Chapter 13.

13. Let $f(z) = \sum_{n=0}^{\infty} a_n z^n$ be the Taylor series expansion of f around origin. Since $f(\mathbb{R}) \subset \mathbb{R}$, it is not difficult to show that $a_n \in \mathbb{R}$ for all $n \in \mathbb{N} \cup \{0\}$ as $a_n = \frac{f^{(n)}(0)}{n!}$. Notice that $a_0 = f(0) \in \mathbb{R}$ and $a_1 = \frac{f'(0)}{1!} = \lim_{z\to 0} \frac{f(z)-f(0)}{z-0}$. Letting z approaches 0 through real values we obtain $a_1 \in \mathbb{R}$. Similarly show that $a_n \in \mathbb{R}$ for all $n \in \mathbb{N}$. Suppose w is a nonreal lacunary value of f. Then $\sum_{n=0}^{\infty} a_n z^n \neq w$ for all $z \in \mathbb{C}$. This implies $\sum_{n=0}^{\infty} \bar{a}_n \bar{z}^n \neq \bar{w}$ for all $z \in \mathbb{C}$. That is, $\sum_{n=0}^{\infty} a_n \bar{z}^n \neq \bar{w}$ for all $z \in \mathbb{C}$. Hence $\sum_{n=0}^{\infty} a_n z^n \neq \bar{w}$ for all $z \in \mathbb{C}$. Thus $\bar{w}$ is a lacunary value of f and $w \neq \bar{w}$. By Theorem 8.6, f is a constant. But f is nonconstant by hypothesis.

Chapter 9

2. (i) 3 (ii) 0

3. If z_0 is outside γ then z_0 is outside σ. Hence in that case $n(\gamma, z_0) = n(\sigma, z_0) = 0$. If z_0 is inside γ then z_0 is inside σ and $n(\gamma, z_0) = n(\sigma, z_0)$ as $|\sigma(t) - \gamma(t)| < \epsilon$ for all $t \in [a, b]$.

Chapter 10

1. The map $F(x, t) = (1 - t)f(x) + tg(x)$ is a homotopy between f and g. It is called a straight-line homotopy, because it moves the point $f(x)$ to the point $g(x)$ along the straight line segment joining them.

2. The map $F(s,t) = (1-t)f(s) + tg(s)$ is a path homotopy between f and g. But the straight-line homotopy between f and the path h is not acceptable, for its image does not lie in the space $X = \mathbb{R}^2 - \{0\}$. Indeed, there exists no path homotopy in X between paths f and h. It is intuitively clear that one cannot "deform f past the hole at 0" without introducing discontinuity. But it takes some work to prove. For details see [5].

3. Let $I = [0,1]$ and $F : X \times I \to Y$ be a homotopy between h and $\widetilde{h}$. Then F is continuous, $F(x,0) = h(x)$ and $F(x,1) = \widetilde{h}(x)$. Let $G : Y \times I \to Z$ be a homotopy between k and $\widetilde{k}$. Then G is continuous, $G(x,0) = k(x)$ and $G(x,1) = \widetilde{k}(x)$. Define $H : X \times I \to Z$ as $H(x,s) = G(F(x,s),s)$. Verify that H is continuous, $H(x,0) = (k \circ h)(x)$ and $H(x,1) = (\widetilde{k} \circ \widetilde{h})(x)$.

4. Notice that deleting a point from $\mathbb{R}^n, n > 2$, leaves a simply connected space, while deleting a point from $\mathbb{R}^2$ does not. Also use the fact that homeomorphic image of a simply connected space is simply connected. Recall that a topological space X is simply connected if it is path connected and if every closed path in X is homotopic to a constant path in X.

Chapter 11

3. $\arg f$ is constant, hence f is constant.

5. The sequence $\{f_n'\}$ converges locally uniformly to f' by Theorem 7.10. By Theorem 11.6, either $f' = 0$ or else $f'(z) \neq 0$ for all $z \in \Omega$. In the former case f is constant, and the result follows.

7. We first show that f one-to-one. Suppose $f(z_1) = f(z_2)$ for certain $z_1, z_2 \in \Omega$. We claim $z_1 = z_2$. By Theorem 11.5 if $n \geq \max\{n_1(z_1), n_2(z_2)\}$, there exist points $z_1^n, z_2^n \in \Omega$ such that (11.4) and (11.5) hold for z_1 and z_2. That is, $f_n(z_1^{(n)}) = f(z_1) = f(z_2) = f_n(z_2^{(n)})$ and $\lim_{n\to\infty} z_i^{(n)} = z_i, i = 1,2$. Since f_n is one-to-one, it must be that $z_1^{(n)} = z_2^{(n)}$ for each n, and therefore $z_1 = \lim_{n\to\infty} z_1^{(n)} = \lim_{n\to\infty} z_2^{(n)} = z_2$. Now by hypothesis f is not constant. Since f is one-to-one, f^{-1} is analytic on the region $f(\Omega)$. If $z \in \Omega$, and if $n \geq n_1(z)$, there exists exactly one point $z_n \in \Omega$ with $f_n(z_n) = f(z)$. Of course, it must be that $\lim_{n\to\infty} z_n = z$. Thus if $w \in f(\Omega)$, then for large n, $w \in f_n(\Omega)$ and $\lim_{n\to\infty} f_n^{-1}(w) = f^{-1}(w)$. Notice that f_n^{-1} is defined on D for each n and the result follows.

8. 100 (all simple zeros)

9. 3 zeros

10. f has simple poles at $z = \pm i, \operatorname{Res}(f, \pm i) = \pm\frac{1}{2i}$. Hence $\int_\gamma f = 0$

11. $\frac{2\pi}{\sqrt{a^2-1}}$

13. (i) poles at $\frac{2k+1}{2}\pi, k = 0, \pm 1, \pm 2, \cdots$; $\text{Res}(\tan z, \frac{2k+1}{2}\pi) = -1, k = 0, \pm 1, \pm 2, \cdots$

(iii) If $f(z) = e^{-\frac{1}{z}}$ then $z = 0$ is an essential singularity and $\text{Res}(f, 0) = -1$.

(v) $\text{Res}(\frac{1}{\sin z}, k\pi) = (-1)^k$ $(k = 0, \pm 1, \pm 2, \cdots)$; simple poles at $k\pi, k \in \mathbb{Z}$.

(vii) If $f(z) = \frac{z}{(z-1)(z+1)^2}$, $z = 1$ is a pole of order 1, $z = -1$ is a pole of order 2. $\text{Res}(f, 1) = \frac{1}{4}$ and $\text{Res}(f, -1) = -\frac{1}{4}$.

(ix) $z = -1$ is a double pole and $z = \pm 2i$ are simple poles; $\text{Res}(f, -1) = -\frac{14}{25}$, $\text{Res}(f, 2i) = \frac{7+i}{25}$, $\text{Res}(f, -2i) = \frac{7-i}{25}$.

(x) $f(z) = e^z \csc^2 z$. The function f has double poles at $z = 0, \pm\pi, \pm 2\pi, \cdots$ i.e., $z = m\pi$ where $m \in \mathbb{Z}$ and $\text{Res}(f, m\pi) = e^{m\pi}$.

14. (i) -2 (ii) $\frac{-i}{4e^2}$ (iii) $\frac{1}{2}$ (iv) $\frac{1}{4}$ (v) 1 (vi) -1 (vii) $\frac{-7}{45}$

15. (i) $-4\pi i$ (ii) $\frac{\pi}{2e^2}$ (iii) $\frac{i\pi}{2}$ (iv) $6\pi i$

Chapter 12

1. (a) (i) $\frac{2\pi}{\sqrt{a^2-1}}$ (ii) $-12\pi + \frac{17}{\sqrt{2}}\pi$ (iii) For $n = 1, 2, \cdots$,

$$\int_{-\pi}^{\pi} \frac{\sin n\theta}{\sin\theta} d\theta = \begin{cases} 0 & \text{if } n \text{ is even;} \\ 2\pi & \text{if } n \text{ is odd.} \end{cases}$$

(iv) $\frac{\pi a}{a^2-1}$ (v) $\frac{\pi}{2\sqrt{a(a+1)}}$ (vi) $\frac{2\pi}{\sqrt{3}}$

(b) (i) $\frac{\pi}{4a^3}$ (ii) $\frac{\pi}{6}$ (iii) $\frac{\pi}{\sqrt{2}}$ (iv) $\frac{\pi}{6}$ (v) $\frac{\pi}{18}$ (vi) $\frac{\pi}{2e}$ (vii) $\frac{\pi(a+1)e^{-a}}{4}$ (viii) $\frac{\pi}{e}$ (ix) $\frac{\pi}{b^2-a^2}\left(\frac{e^{-a}}{a} - \frac{e^{-b}}{b}\right)$ (x) $\frac{\pi}{\sqrt{2}}$

(c) (i) $\frac{\pi}{2e}$ (ii) $\frac{\pi}{4}$; consider $f(z) = \frac{z+ie^{iz}-i}{z^3}$ (iii) $\frac{\pi e^{-\frac{n}{\sqrt{2}}}}{\sqrt{2}}(\cos\frac{n}{\sqrt{2}} + \sin\frac{n}{\sqrt{2}})$

(d) (i) $\frac{\pi|a|}{2}$ (ii) $\frac{\pi(2n)!}{2^{2n}(n!)^2}, n \in \mathbb{N}$ (iii) 0 (iv) $-\frac{\pi}{4}$ (v) $\frac{\pi}{2\sin(\frac{2\pi}{3})}$

2. (iii)

$$\sum_{n=-\infty_{n\neq 0}}^{\infty} \frac{1}{n^4} = -\text{Res}(\frac{1}{z^4}\pi\cot\pi z, 0);$$

this is a pole of order 5, and for $0 < |h| < 1$,

$$\frac{\pi\cot\pi h}{h^4} = \frac{\pi\cos\pi h}{h^4\sin\pi h} = \frac{1}{h^5}\left[\frac{1 - \frac{\pi^2h^2}{2!} + \frac{\pi^4h^4}{4!} - \cdots}{1 - \frac{\pi^2h^2}{3!} + \frac{\pi^4h^4}{5!} - \cdots}\right] \qquad (*)$$

By Taylor's theorem we can write the term in brackets as a power series $b_0 + b_1h + b_2h^2 + \cdots$ in some disk about 0. Hence $(1 - \frac{\pi^2h^2}{2!} + \frac{\pi^4h^4}{4!} - \cdots) = (1 - \frac{\pi^2h^2}{3!} + \frac{\pi^4h^4}{5!} - \cdots)(b_0 + b_1h + b_2h^2 + \cdots)$.
We may equate coefficients, solving successively for $b_0, b_1, \cdots$, we get $1 = b_0, 0 = b_1, -\dfrac{\pi^2}{2} = b_2 - \dfrac{\pi^2}{6}b_0, 0 = b_3 - \dfrac{\pi^2}{6}b_1, \dfrac{\pi^4}{24} = b_4 - \dfrac{\pi^2}{6}b_2 + \dfrac{\pi^4}{120}b_0$ etc. Hence $b_0 = 1, b_1 = 0, b_2 = -\frac{\pi^2}{3}, b_3 = 0, b_4 = -\frac{\pi^4}{45}$, etc. The residue required is $b_4 = -\frac{\pi^4}{45}$ since this gives the coefficient of $\frac{1}{h}$ in (*). Hence $\sum_{n=1}^{\infty} \frac{1}{n^4} = \frac{1}{2}\sum_{n=-\infty_{n\neq 0}}^{\infty} \frac{1}{n^4} = \frac{1}{2}(\frac{\pi^4}{45}) = \frac{\pi^4}{90}$.

(vii) Apply the summation formula with $f(z) = \frac{z^2}{z^4+5z^2+4}$; simple poles at $\pm i, \pm 2i$, so

$$\sum_{n=-\infty}^{\infty} \frac{n^2}{n^4 + 5n^2 + 4}$$

$$= -\sum_{j=1}^{4} \text{Res}\left(\frac{z^2\pi\cot\pi z}{z^4 + 5z^2 + 4}, p_j\right) \text{ (where } p_1 = i, p_2 = -i, p_3 = 2i, p_4 = -2i)$$

$$= -\frac{i^2\pi\cot\pi i}{4i^3 + 10i} - \frac{(-i)^2\pi\cot(-\pi i)}{4(-i)^3 + 10(-i)}$$

$$-\frac{(+2i)^2\pi\cot(+2\pi i)}{4(2i)^3 + 10(2i)} - \frac{(-2i)^2\pi\cot(-2\pi i)}{4(-2i)^3 + 10(-2i)}$$

$$= \frac{\pi}{3}(2\coth(2\pi) - \coth\pi).$$

$$\text{So } \sum_{n=1}^{\infty} \frac{n^2}{n^4 + 5n^2 + 4} = \frac{\pi}{6}(2\coth 2\pi - \coth\pi).$$

Chapter 13

2. $\dfrac{z+i}{-3z+i}$

3. $\dfrac{-iz-1}{z-1}$

4. $\dfrac{i-z}{i+z}$

5. Let f be an entire function with negative real part. The function $g(z) = \frac{1+z}{1-z}$ is a conformal map from the left half plane onto the open unit disk

$\mathbb{D}$. Hence $g \circ f : \mathbb{C} \to \mathbb{D}$ is entire. By Liouville's theorem $g \circ f$ is a constant. Suppose $g \circ f \equiv c$. Then $\frac{1+f(z)}{1-f(z)} = c$ for all $z \in \mathbb{C}$. That is, $(1+c)f(z) = c-1$. Thus $f(z) = \frac{c-1}{c+1}$, a constant for all $z \in \mathbb{C}$.

6. (i) $e^{\frac{1}{z}}$ have essential singularity at 0.
   (ii) $\sin z$ has zeros at $n\pi, n \in \mathbb{Z}$. Thus ∞ is a limit point of zeros.
   (iii) $\cos z$ has zeros at $(n+\frac{1}{2})\pi, n \in \mathbb{Z}$. Hence ∞ is a limit point of zeros.
7. Same as Exercise 23 of Chapter 13.
8. See Proposition 13.5, Exercise 23 of Chapter 13.
9. (i) $z = \infty$ is a regular point (zero of third order).
   (ii) $z = \infty$ is a pole of third order.
   (iii) $z = \infty$ is an essential singularity.
11. $\text{Res}(f, \infty) = -\text{Res}(f, -1) = \cos 1$.
18. (i) a pole at ∞ of order 1 (ii) ∞ is a regular point (iii) ∞ is a regular point (iv) removable singularity at ∞.
19. $\frac{1}{\sin \pi z}$, poles at $z = n \in \mathbb{Z}$. No, it cannot be extended to a meromorphic function from $\mathbb{C}_\infty$ into $\mathbb{C}_\infty$ by Theorem 13.10.
20. See Exercise 28 of Chapter 7, Exercise 13 of Chapter 4 and Theorem 18.3 of Chapter 18.
21. $w = \frac{i-z}{i+z}$
22. upper half plane.
23. See Theorem 13.9 and Theorem 13.10. A proof can also be worked out in the following way:

    Without loss of generality we may assume that all singularities of f are poles. Let $F = \{a_1, a_2, \cdots, a_n\}$. Since $f : \mathbb{C}_\infty \setminus F \to \mathbb{C}$ is analytic; each $a_i, i = 1, 2, \cdots, n$ is a pole of f. Without loss of generality we may assume that $z = \infty$ is an element of F. Let $a_1 = \infty$. Let m_i be the order of the pole $z = a_i, i = 1, 2, 3, \cdots, n$. For $i = 2, 3, \cdots, n$, the point a_i is a pole of f of order m_i. Hence there exists function g analytic in $\mathbb{C}$ such that

$$f(z) = \frac{g(z)}{(z-a_2)^{m_2}(z-a_3)^{m_3}\cdots(z-a_n)^{m_n}}. \qquad (*)$$

    Since $a_1 = \infty$ is a pole of f of order m_1, so it is also a pole of g of order m_1. Hence $z = 0$ is a pole of $g(\frac{1}{z})$ of order m_1. Let $g(z) = \sum_{k=0}^{\infty} a_k z^k$ be the power series expansion of $g(z)$. This is possible as g is analytic in $\mathbb{C}$. Then $g(\frac{1}{z}) = \sum_{k=0}^{\infty} a_k z^{-k} = \sum_{k=-\infty}^{0} a_{-k} z^k$. Since $z = 0$ is a pole of $g(\frac{1}{z})$ of order m_1, we get $g(\frac{1}{z}) = \sum_{k=-m_1}^{0} a_{-k} z^k$. This implies $g(z) = \sum_{k=0}^{m_1} a_k z^k$.

Thus $g(z)$ is a polynomial of degree m_1. Hence from (*) we conclude that $f(z)$ is a rational function. This completes the proof.

Chapter 14

3. The set U is bounded. Therefore $\overline{U}$ is bounded. The functions $f = u+iv$ and $g = \widetilde{u}+i\widetilde{v}$ are continuous on $\overline{U}$ and holomorphic in U. Therefore, Ref and Reg are harmonic on U and so is $\text{Re}(f-g) = \text{Re}f - \text{Re}g = u - \widetilde{u}$ and the function $v - \widetilde{v}$ is the harmonic conjugate of $u - \widetilde{u}$. By hypothesis $\text{Re}(f-g) = 0$ on ∂U. Hence by Corollary 14.2, the function $u-\widetilde{u} = \text{Re}(f-g) \equiv 0$ on U. This implies $\frac{\partial(v-\widetilde{v})}{\partial x} = 0$ and $\frac{\partial(v-\widetilde{v})}{\partial y} = 0$. Therefore, $v - \widetilde{v} \equiv c$, a constant. Hence $f - g = (u+iv) - (\widetilde{u}+i\widetilde{v}) = (u-\widetilde{u}) + i(v-\widetilde{v}) \equiv ic$ on U.

4. Let $f = u + iv$. The functions f and f^2 are harmonic in Ω. Therefore $\frac{\partial^2 f}{\partial\bar{z}\partial z} = 0$ and $\frac{\partial^2 g}{\partial\bar{z}\partial z} = 0$ where $g = f^2$. Notice that $\frac{\partial g}{\partial z} = \frac{\partial g}{\partial f}\frac{\partial f}{\partial z} = 2f\frac{\partial f}{\partial z}$ and $\frac{\partial}{\partial\bar{z}}(\frac{\partial g}{\partial z}) = 2[\frac{\partial f}{\partial\bar{z}}\cdot\frac{\partial f}{\partial z} + f\frac{\partial^2 f}{\partial\bar{z}\partial z}] = 2[\frac{\partial f}{\partial\bar{z}}\cdot\frac{\partial f}{\partial z}]$.

   Thus $2[\frac{\partial f}{\partial\bar{z}}\cdot\frac{\partial f}{\partial z}] = 0$. That is, either $\frac{\partial f}{\partial\bar{z}} = 0$ or $\frac{\partial f}{\partial z} = 0$. If $\frac{\partial f}{\partial\bar{z}} = 0$ then f is holomorphic. Observe that $\dfrac{\partial f}{\partial z} = \dfrac{1}{2}\left(\dfrac{\partial f}{\partial x} - i\dfrac{\partial f}{\partial y}\right)$. Hence $\dfrac{\partial f}{\partial z} = 0$ if and only if $\frac{\partial f}{\partial x} = i\frac{\partial f}{\partial y}$. That is, if and only if $\frac{\partial u}{\partial x} + i\frac{\partial v}{\partial x} = i(\frac{\partial u}{\partial y} + i\frac{\partial v}{\partial y})$. Hence $\frac{\partial f}{\partial z} = 0$ if and only if $u_x = -v_y$ and $u_y = v_x$. That is, $\frac{\partial f}{\partial z} = 0$ if and only if $\bar{f}$ is holomorphic in Ω.

5. A linear fractional transformation $T(z) = \frac{az+b}{cz+d}$ maps the upper half plane onto itself if and only if a, b, c and d are real and $\begin{vmatrix} a & b \\ c & d \end{vmatrix} = 1.$

6. The function f is holomorphic and nonconstant on $B(z_0, R)$ and $z_0 \in B(z_0, R), 0 < r < R$. The circle $S(z_0, r) \subset B(z_0, R)$ and $\overline{B(z_0, r)} \subset B(z_0, R)$. Let $u = \text{Re}f$. Then u is harmonic in $B(z_0, R)$. Since $f(z_0) = 0$, we have $u(z_0) = 0$. Let $m = \min_{z\in\overline{B(z_0,r)}} u(z), M = \max_{z\in\overline{B(z_0,r)}} u(z)$. The values m, M are both attained by the function u on the boundary $S(z_0, r)$ and it cannot be attained by any point in the interior of $B(z_0, r)$ by Theorem 14.9 and Corollary 14.3. Thus $m < 0$ and $M > 0$. Hence $u = \text{Re}f$ assumes both positive and negative values.

7. Follows from the sub mean value property.

8. Follows from the definition of subharmonic functions.

12. (i) subharmonic (see Example 14.8).

    (ii) subharmonic (see Example 14.7).

(iii) superharmonic (verify using the same procedure as in Example 14.10).

(iv) subharmonic (see Example 14.10).

(v) subharmonic (see Example 14.10).

(vi) subharmonic (see Example 14.9).

Chapter 15

1. Let $f_n(z) = e^{nz}, n \in \mathbb{N}$. We claim $\mathcal{F} = \{e^{nz}\}_{n\in\mathbb{N}}$ is not normal on $\mathbb{C}$. Consider the compact set $\{z \in \mathbb{C} : |z| \leq 1\} = K$ in $\mathbb{C}$. The functions in the family $\mathcal{F}$ are analytic. Suppose $\mathcal{F}$ is a normal family. Let $\{e^{n_k z}\}$ is a subsequence of $\{e^{nz}\}$ which converges uniformly to f on K. By Definition 15.4, the limit function f either takes everywhere finite values or else it is identically infinite. But this does not happen for the sequence in K. For if $z = 1, e^{n_k z} \to \infty$ and $z = -1, e^{n_k z} \to 0$.

3. Let $f_n(z) = z^n, n = 0, 1, 2, \cdots$. Suppose $|z| < 1$. The family $\mathcal{F} = \{f_n\}$ contains analytic functions. Let K be a compact subset of $\{z \in \mathbb{C} : |z| < 1\}$. Then there exist $r < 1$ such that $K \subset \{z \in \mathbb{C} : |z| \leq r\}$. Now given $0 < \epsilon < 1$, $|z|^n \leq r^n < \epsilon$ if $n > \frac{|\log \epsilon|}{|\log r|} = N$. Notice that N depends only on ϵ and not on z. Hence $z^n \to 0$ uniformly on K. Further the sequence $\{z^n\}$ converges uniformly to ∞ on every compact subset of $\{|z| > 1\}$. Hence $\mathcal{F}$ is normal on $|z| > 1$. Suppose E is a subset of $\mathbb{C}$ such that $E \cap \{z \in \mathbb{C} : |z| = 1\} \neq \emptyset$. Let $w \in E$ and $|w| = 1$. Then every neighborhood U of w intersects the sets $\{z \in \mathbb{C} : |z| < 1\}$ and $\{z \in \mathbb{C} : |z| > 1\}$. Let C be a compact subset of E containing a neighborhood U of w. The family $\mathcal{F} = \{f_n(z)\} = \{z^n\}$ is a family of analytic functions. Now if $\mathcal{F}$ is normal on E then the sequence $\{z^n\}$ must have a subsequence $\{z^{n_k}\}$ which converges uniformly on C and the limit function is either everywhere finite or identically infinite. But it is impossible to find such a subsequence $\{z^{n_k}\}$ of $\{z^n\}$. For if $z \in U, |z| < 1, z^{n_k} \to 0$ and if $z \in U, |z| > 1, z^{n_k} \to \infty$. Thus by Definition 15.4, $\mathcal{F}$ is not normal.

7. Suppose f is analytic on an open set $G \supset \overline{B(a, R)}$. Let $0 < r < R$. By Cauchy's integral formula,

$$\begin{aligned} f(a) &= \tfrac{1}{2\pi i} \int_{|z-a|=r} \tfrac{f(z)}{z-a} dz \\ &= \tfrac{1}{2\pi} \int_0^{2\pi} f(a + re^{i\theta}) d\theta. \end{aligned}$$

Hence for $z = x + iy$,

$$\begin{aligned}\frac{1}{\pi R^2}\int\int_{B(a,R)} f(z)dxdy &= \frac{1}{\pi R^2}\int_0^R\int_0^{2\pi} f(a+re^{i\theta})rd\theta dr \\ &= \frac{1}{\pi R^2}\int_0^R r\left(\int_0^{2\pi} f(a+re^{i\theta})d\theta\right)dr \\ &= \frac{1}{\pi R^2}\int_0^R r2\pi f(a)dr \\ &= \frac{2}{R^2}f(a)\int_0^R rdr = f(a).\end{aligned}$$

Thus by Fubini's theorem, $f(a) = \frac{1}{\pi R^2}\int_0^{2\pi}\int_0^R f(a+re^{i\theta})rdrd\theta$. Applying Holder's inequality,

$$\begin{aligned}|f(a)| &= \left|\frac{1}{\pi R^2}\int_0^{2\pi}\int_0^R f(a+re^{i\theta})rdrd\theta\right| \\ &\le \frac{1}{\pi R^2}\left(\int_0^{2\pi}\int_0^R |f(a+re^{i\theta})|^2 rdrd\theta\right)^{\frac{1}{2}}\left(\int_0^{2\pi}\int_0^R rdrd\theta\right)^{\frac{1}{2}} \\ &= \frac{1}{R\sqrt{\pi}}\left(\int_0^{2\pi}\int_0^R |f(a+re^{i\theta})|^2 rdrd\theta\right)^{\frac{1}{2}}.\end{aligned}$$

Hence $|f(a)|^2 \le \frac{1}{\pi R^2}\int_0^{2\pi}\int_0^R |f(a+re^{i\theta})|^2 rdrd\theta$.

8. Let K be a compact subset of G. Let $a \in K$ be arbitrary. Suppose $R > 0$ is such that $\overline{B(a,R)} \subset G$. By Exercise 7 of Chapter 15 (see the solution)

$$\begin{aligned}|f(a)|^2 &\le \frac{1}{\pi R^2}\int\int_{B(a,R)} |f(z)|^2 dxdy \\ &\le \frac{M}{\pi R^2}.\end{aligned}$$

Thus $|f(a)| \le \frac{M}{R\sqrt{\pi}}$. Let $0 < \rho < \text{dist}(B(a,R), \partial G)$. By the first part, since $\overline{B(a,\rho)} \subset G$, hence $|f(z)| \le \frac{M}{\rho\sqrt{\pi}}$ for $|z-a| \le \rho$. Thus f is bounded in every closed disk contained in G. Given K is compact. We just now see that for each $a \in K$, there exist $\rho_a > 0$ such that $\overline{B(a,\rho_a)} \subset G$ and $|f(z)| \le \frac{M}{\rho_a\sqrt{\pi}}$ for all $z \in B(a,\rho_a)$. Now $K \subset \bigcup_{a\in K} B(a,\rho_a)$. Thus there exist $a_1, a_2, \cdots, a_n \in K$ such that $K \subset \bigcup_{i=1}^n B(a_i,\rho_{a_i})$ and $|f(z)| \le \frac{M}{\rho_{a_i}\sqrt{\pi}}$ for all $z \in B(a_i,\rho_{a_i}), i = 1, 2, \cdots, n$. Let $\widetilde{\rho} = \min\{\rho_{a_1}, \rho_{a_2}, \cdots, \rho_{a_n}\}$. Then for $z \in K$, $|f(z)| \le \frac{M}{\widetilde{\rho}\sqrt{\pi}}$. Thus the functions in $\mathcal{F}$ are holomorphic and are uniformly bounded on every compact set $K \subset G$. By Theorem 15.6 (Montel's theorem), $\mathcal{F}$ is normal.

Chapter 16

1. Follows from Theorem 16.1 and Liouville's theorem.

2. Notice that $\overline{B(0,1)}$ is compact and f is continuous. This implies $f(\overline{B(0,1)})$ is a compact subset of $\mathbb{C}$ and hence bounded. Therefore, f is bounded on $\overline{B(0,1)}$. Use Theorem 16.1 and Liouville's theorem.

3. Follows from Lemma 16.3.

4. Follows from Theorem 16.1.

5. Follows from Theorem 16.2 and Theorem 13.10. One can also make the following arguments:

   Suppose $f(z) \neq 0$ for all $z \in B(0,1)$. Then $|f|$ attains its maximum and minimum on the unit circle. Since $|f| = 1$ on $|z| = 1$, hence $f \equiv c$ (a constant) on $B(0,1)$. Otherwise notice that f can have only finitely many zeros in $\overline{B(0,1)}$. Let $a_1, a_2, \cdots, a_n$ be the zeros of f (counting multiplicities) in $B(0,1)$. Then $f(z) = (z-a_1)\cdots(z-a_n)h(z)$ where h is analytic and has no zeros in $B(0,1)$. We know if $|z| = 1$ then $|\frac{z-a_i}{1-\mathbf{\bar{a}_i} z}| = 1$ for all $i = 1, \cdots, n$.

   Since $|f| = 1$ on $|z| = 1$ we obtain

$$f(z) = (z-a_1)\cdots(z-a_n)\frac{g(z)}{(1-\mathbf{\bar{a}_1}z)\cdots(1-\mathbf{\bar{a}_n}z)}$$

   where g is analytic, it has no zeros in $B(0,1)$ and $|f(z)| = |g(z)| = 1$ on $|z| = 1$. By the first part of the proof $g \equiv d$ (a constant). Hence $f(z) = d\prod_{i=1}^{n} \frac{z-a_i}{1-\mathbf{\bar{a}_i}z}$.

9. The radius of convergence $\sum_{n=0}^{\infty} z^{2^n}$ of the power series is 1 and the function $f(z) = \sum_{n=0}^{\infty} z^{2^n}$ is analytic in $|z| < 1$. If $|z| \geq 1$, then

$$\lim_{n\to\infty} |z^{2^n}| \neq 0$$

   and so the series diverges for $|z| \geq 1$. Let $\rho = e^{\frac{2\pi i m}{2^n}}, m = 0, 1, \cdots, 2^n - 1$, $n \in \mathbb{N}$, be the 2^n-th root of unity. If $z = re^{\frac{2\pi i m}{2^n}}$, then

$$f(z) = \sum_{k=0}^{n-1} z^{2^k} + \sum_{k=n}^{\infty} r^{2^k}$$

   and so for $z = re^{\frac{2\pi i m}{2^n}} = r\rho$ and $r \to 1^-, |z| < 1$, we have

$$|f(\rho r)| \geq \sum_{k=n}^{\infty} r^{2^k} - |\sum_{k=0}^{n-1} z^{2^k}| \geq \sum_{k=n}^{\infty} r^{2^k} - \sum_{k=0}^{n-1} |z^{2^k}| \geq \sum_{k=n}^{\infty} r^{2^k} - n,$$

   and hence for every 2^nth root of unity ρ, we have $\lim_{r\to 1^-} |f(\rho r)| = \infty$. Therefore if G is a domain containing points of the open unit disk $\mathbb{D}$ and of its complement then G contains the points $\rho = e^{\frac{2\pi i m}{2^n}}$ and so any function F in the domain G which coincides with the function f in $G \cap \mathbb{D}$ cannot be continued analytically through $\rho^{2^n} = 1$ for any $n \in \mathbb{N}$. That is, any root of the equation $z^2 = 1, z^4 = 1, \cdots, z^{2^n} = 1$, for $n \in \mathbb{N}$ is a singular point of f and hence any arc, however small it may be, of the unit circle, contains an infinite number of singularities. Thus f on $\mathbb{D}$ cannot be continued analytically across the boundary $\partial\mathbb{D} = \{z \in \mathbb{C} : |z| = 1\}$ of $\mathbb{D}$. Hence the unit circle is the natural boundary for the power series $\sum_{n=0}^{\infty} z^{2^n}$.

11. $f(z) = \sum_{n=0}^{\infty} \frac{(z-a)^n}{(1-a)^{n+1}}$; by this expansion $f(z)$ is analytically continued inside the circle $|z-a| < |1-a|$, which does not lie entirely inside the circle $|z| < 1$ if a does not belong to the interval $[0, 1)$.

12. $f(z) = \ln\frac{2}{3} + \sum_{n=1}^{\infty}(\frac{2}{3})^n \frac{(z+\frac{1}{2})^n}{n}$; the circle of convergence of this series is $|z+\frac{1}{2}| < \frac{3}{2}$.

15. Hints: Observe that the sum of the series increases without bound along the radius vector as $z \to e^{2im\alpha\pi}$.

Chapter 17

1. $-\frac{1}{z} + \sum_{n=0}^{\infty} \frac{(-1)^n 2z}{z^2-n^2}$.

4. (i) divergent (ii) convergent (iii) divergent (iv) convergent (v) divergent.

6. Converges if $\alpha > \frac{1}{2}$, converges absolutely if $p > 1$, diverges if $p \leq \frac{1}{2}$.

9. (a) convergent (b) divergent (c) convergent.

10. genus $= 0$, order $= \frac{1}{2}$.

11. (i) $\frac{1}{2}$ (ii) 1.

12. see Theorem 17.27.

13. (i)1 (ii) $\frac{1}{\alpha}$ (iii) $\frac{1}{2}$ (iv)0.

15. $\frac{(-1)^n}{n!}, n \in \mathbb{N}$.

Chapter 18

1. It follows from Theorem 18.3 and Theorem 18.5 that there exists $\widetilde{\omega} \neq 0$ such that $2\omega = n\widetilde{\omega}$ and $2\lambda\omega = m\widetilde{\omega}, n, m \in \mathbb{Z}$. Hence $\lambda = \frac{m}{n} \in Q$.

2. It follows from Theorem 18.3. Since if ω_1 and ω_2 are two independent periods then $\omega_3 = n_1\omega_1 + n_2\omega_2, n_1, n_2 \in \mathbb{Z}$. If no two of these periods $\omega_1, \omega_2, \omega_3$ are independent then

$$\frac{\omega_2}{\omega_1} = \lambda = \frac{m}{n} \in Q$$

and

$$\frac{\omega_2}{\omega_3} = \lambda' = \frac{m'}{n'} \in Q,$$

$m', n', m, n \in \mathbb{Z}$. This implies $n\omega_2 = m\omega_1$ and $n'\omega_2 = m'\omega_3$. Thus we have $n\omega_2 - m\omega_1 = 0$ and $n'\omega_2 - m'\omega_3 = 0$. Therefore, $-m\omega_1 + (n+n')\omega_2 - m'\omega_3 = 0$.

References

[1] Ahlfors, L.V., *Complex Analysis*, McGraw-Hill Book Co., 1979.

[2] Conway, J.B., *Functions of one Complex variable*, Narosa publishing house, 1992.

[3] Holland, A.S.B., *Introduction to the theory of entire functions*, Academic Press, New York, 1973.

[4] Kumaresan, S., *Lecture Notes on Complex analysis*, University of Bombay, 1997.

[5] Munkres, J.R., *Topology, a first course*, Prentice Hall of India, New Delhi, 1988.

[6] Partington, J.R., *Complex analysis lecture notes*, University of Leeds, U.K., 1990.

[7] Ponnusamy, S., *Foundations of Complex analysis*, Narosa publishing house, 1995.

[8] Sarason, D., *Notes on Complex function theory*, Trim texts and readings in Mathematics 5, Hindustan book agency, 1994.

[9] Veech, W.A., *A second course in Complex analysis*, W.A. Benjamin, Inc., 1967.

Index